Bread making

Related titles from Woodhead's food science, technology and nutrition list:

Baking problems solved (ISBN 185573 564 4)
Written by two leading authorities on baking, this practical manual is designed to help busy baking professionals solve production and product quality problems quickly. It includes over 200 typical problems and questions together with their solutions. Individual chapters cover the key raw materials and types of bakery product.

Cereals processing technology (ISBN 185573 561 X)
Written by a distinguished team of contributors, this collection reviews the range of cereal products and the technologies used to produce them. The book includes chapters on cereals growing, milling, bread making and the processing of rice, pasta, noodles and breakfast cereals.

Cereal biotechnology (ISBN 185573 498 2)
This important collection reviews the key technologies in the genetic transformation of cereals and their applications in cereals breeding, milling and baking, malting, brewing and distilling.

Details of these books and a complete list of Woodhead's food science, technology and nutrition titles can be obtained by:

- visiting our web site at www.woodhead-publishing.com
- contacting Customer Services (e-mail: sales@woodhead-publishing.com; fax: +44 (0) 1223 893694; tel.: +44 (0) 1223 891358 ext. 30; address: Woodhead Publishing Limited, Abington Hall, Abington, Cambridge CB1 6AH, England)

Selected food science and technology titles are also available in electronic form. Visit our web site (www.woodhead-publishing.com) to find out more.

If you would like to receive information on forthcoming titles in this area, please send your address details to: Francis Dodds (address, telephone and fax as above; e-mail: francisd@woodhead-publishing.com). Please confirm which subject areas you are interested in.

Bread making

Improving quality

Edited by
Stanley P. Cauvain

CRC Press
Boca Raton Boston New York Washington, DC

WOODHEAD PUBLISHING LIMITED
Cambridge England

Published by Woodhead Publishing Limited
Abington Hall, Abington
Cambridge CB1 6AH
England
www.woodhead-publishing.com

Published in North America by CRC Press LLC
2000 Corporate Blvd, NW
Boca Raton FL 33431
USA

First published 2003, Woodhead Publishing Limited and CRC Press LLC

British Library Cataloguing in Publication Data
A catalogue record for this book is available from the British Library.

Library of Congress Cataloging in Publication Data
A catalog record for this book is available from the Library of Congress.

Woodhead Publishing Limited ISBN 1 85573 553 9 (book); 1 85573 712 4 (e-book)
CRC Press ISBN 0-8493-1762-2
CRC Press order number: WP1762

Cover design by The ColourStudio
Project managed by Macfarlane Production Services, Markyate, Hertfordshire
(e-mail: macfarl@aol.com)
Typeset by MHL Typesetting Limited, Coventry, Warwickshire
Printed by TJ International Limited, Padstow, Cornwall, England

Contents

Contributor contact details

Chapters 1 and 2

S. Cauvain
Campden and Chorelywood Food Research Association
Chipping Campden
GL55 6LD
UK

Tel: +44(0)1386 842135
Fax: +44(0)1386 842150
E-mail: s.cauvain@campden.co.uk

Chapter 3

Prof H. Cornell
Dept of Applied Chemistry
RMIT University
GPO Box 2476 V
Melbourne, Victoria 3001
Australia

Fax: (613) 9639 1321
E-mail: hugh.cornell@rmit.edu.au

Chapter 4

Dr C. Wrigley
Food Science Australia
PO Box 52
North Ryde
NSW 1670
Australia

Tel: 612-9490-8401
Fax: 612-9490-8499
E-mail: Colin.Wrigley@csiro.au

Chapter 5

Dr A. M. Gil
Department of Chemistry
University of Aveiro
3810-193 Aveiro
Portugal

Tel: +351-234-370707
Fax: +351-234-370084
E-mail: agil@dq.ua.pt

Chapter 6

Dr E. N. C. Mills
Institute of Food Research
Norwich Research Park
Colney
Norwich
NR4 7UA
UK

Tel: +44(0)1603 255000
Fax: +44(0)1603 507723
E-mail: clare.mills@bbsrc.ac.uk

Chapter 7

Prof. A.-C. Eliasson
Department of Food Technology, Engineering and Nutrition
Center for Chemistry and Chemical Engineering
Lund University
PO Box 124
S-221 00 Lund
Sweden

Tel: +46-46-222-9674
Fax: +46-46-222-9517
E-mail: ann-charlotte.eliasson@livsteki.lth.se

Chapter 8

Prof. P. R. Shewry
Rothamsted Research
Harpenden
AL5 2JQ
UK

Tel: +44(0)1582 763133
Fax: +44(0)1582 763010
E-mail: peter.shewry@bbsrc.ac.uk

Chapter 9

M. Hajšelová
Pump Cottage
Blacksmiths Lane
Beckford
Tewksbury
GL20 7AH
UK

Tel: +44(0)1386 882105
E-mail: anton.mirka@hron.fsnet.co.uk

A. J. Alldrick
Campden and Chorleywood Food Research Association
Chipping Campden
GL55 6LD
UK

Tel: +44(0)1386 842127
Fax: +44(0)1386 842150
E-mail: a.alldrick@campden.co.uk

Chapter 10

Prof. Colin Webb
Satake Centre for Grain Process Engineering
UMIST
PO Box 88
Manchester
M60 1QD
UK

Tel: +44 (0)161-200-4379
Fax +44 (0)161-200-4399
E-mail: colin.webb@umist.ac.uk

Chapter 11

Prof. R. Appels
Department of Agriculture
Murdoch University
Locked Bag 4
Bentley Delivery Centre
WA 6983
Australia

Tel: 61-08-9368-3544
Fax: 61-08-9474-2840
E-mail:
rappels@central.murdoch.edu.au
rappels@agric.wa.gov.au

Chapter 12

Dr C. M. Rosell
Cereal Laboratory
Instituto de Agroquimica y Tecnologia de Alimentos (CSIC)
PO Box 73
Burjasot 46100
Valencia
Spain

Tel: 34-96-3900022
Fax: 34-96-3636301
E-mail: crosell@iata.csic.es

Chapter 13

Prof. P. S. Belton
School of Chemical Sciences and Pharmacy
University of East Anglia
Norwich
NR4 7TJ
UK

Tel: +44-1603-593984
Fax: +44-1603-592003
E-mail: P.belton@uea.ac.uk

Chapter 14

Prof. Yrjo H. Roos
University College Cork,
Department of Food and Nutritional Sciences
Cork
Ireland

Tel: +353-21-4902386
Fax: +353-21-4270213
E-mail: yrjo.roos@ucc.ie

Chapter 15

Prof. A. Schiraldi
DISTAM, Universita di Milano
Via Celoria 2
20133 Milano
Italy

Tel: +39-02-5031-6634
Fax: +39-02-5031-6632
E-mail: alberto.schiraldi@unimi.it

Chapter 16

Dr Peter Wilde
Food Materials Science Division
Institute of Food Research
Norwich Research Park
Norwich
NR4 7UA
UK

Tel: +44 (0)1603-255258
Fax: +44 (0)1603-507723
E-mail: peter.wilde@bbsrc.ac.uk

Chapter 17

Dr Grant Campbell
Satake Centre for Grain Process Engineering

Department of Chemical Engineering
UMIST
PO Box 88
Manchester
M60 1QD
UK

Tel: +44 (0)161-200-4472
Fax: +44 (0)161-200-4399
E-mail: grant.campbell@umist.ac.uk

Chapter 18

Dr Bogdan Dobraszczyk
School of Food Biosciences
The University of Reading
PO Box 226
Whiteknights
Reading
RG6 6AP
UK

Tel: +44(0)1189 318714
Fax: +44(0)1189 310080
E-mail: b.dobraszczyk@reading.ac.uk

Chapter 19

Dr S. Millar
Baking Technology Manager
Department of Baking and Cereals Processing
Campden and Chorleywood Food Research Association
Chipping Campden
Gloucestershire
GL55 6LD
UK

Tel: +44 (0)1386-842157
Fax: +44 (0)1386-842150
E-mail: s.millar@campden.co.uk

Chapter 20

Dr Herbert Wieser
Deutsche Forschungsanstalt für Lebensmittelchemie
Lichtenbergstrasse 4
DE-85748 Garching
Germany

Fax: +49 89 289 14183
E-mail: h.wieser@Lrz.tum.de

Chapter 21

L. S. Young
Cereals & Cereal Processing Division
Campden and Chorleywood Food Research Association
Chipping Campden
Gloucestershire
GL55 6LD
UK

Tel +44 (0)1386 842134
Fax: +44 (0)1386 842150
E-mail: lys@baketran.demon.co.uk

Chapter 22

Dr R. L. Wirtz
664 Chimney Hill Circle
Evans
GA 30809
USA

E-mail: walther@knology.net

Chapter 23

Dr K. Katina
VTT Biotechnology
Tietotie 2
PO Box 1500
02044 VTT
Finland

E-mail: kati.katina@vtt.fi

Chapter 24

Prof. N. Magan
Biotechnology Centre
Cranfield University
Silsoe
Bedford MK45 4DT
UK

Tel: +44 (0)1525-863539 Ext 3786
Fax: +44 (0)1525-863540
E-mail: n.magan@cranfield.ac.uk

Chapter 25

C. Waalwijk
Plant Research International
Droevendaalsesteeg 1
6704 PB Wageningen
The Netherlands

Tel: +31 317 476 286
Fax: +31 317 418 094
E-mail: cees.waalwijk@wur.nl

Chapter 26

Dr O. K. Chung, Dr M. Tilley and Dr G. L. Lookhart
Agricultural Research Service
USDA
Grain Marketing and Production Research Center
1515 College Ave.
Manhattan, KS66502
USA

Tel: 785-776-2703
Fax: 785-537-5534
E-mail: okchung@gmprc.ksu.edu

Dr S.-H. Park
Kansas State University
USA

Tel: 785-776-2708
Fax: 785-537-5534
E-mail: seokho@gmprc.ksu.edu

Chapter 27

Prof P. Chinachoti
Department of Food Science
University of Massachusetts
Amherst
MA 01003
USA

Tel: (413)-545-1025
Fax: (413)-545-1262
E-mail: pavinee@foodsci.umass.edu

1

Introduction

S. Cauvain, Campden and Chorleywood Food Research Association, UK

Mention the word 'quality' and on a good day my eyes normally glaze over as pictures of detailed and often meaningless 'procedures' apparently designed to ensure the quality of a process or a product spring to mind. On a bad day I might run screaming from thc room. The big problem with most 'quality systems' is that they fail to take sufficient account of the 'real' purpose of the procedure or product concerned. This is especially the case in breadmaking where the perception of product quality is very personal. We all have our selection of criteria which classify bread as 'good' or 'bad', 'correct' or 'wrong' and while there are some common characteristics on which many of us would agree (for example, the bread should not be mouldy) the final judgement will always be highly personal.

In the final analysis if we define quality as being 'fitness for purpose' then for the each of us bread quality may be described as being the sum of those quality traits which confer the sensory pleasures associated with smell, taste and texture. In summary, when we taste and eat the product we experience pleasure. Our individual perception will then depend on the combination of those parameters which characterise the mix of product appearance, texture, smell and flavour which best suits each of us. We must also recognise that our perception of quality will change with the passage of time and personal circumstances, not least our health.

While there are many factors which characterise the quality of bread products that can be easily defined and measured, e.g. loaf height and volume, there are many others which are ephemeral in nature and therefore more difficult to define, e.g. smell and flavour. In defining quality, the term 'consistency' is often used as frequently as fitness for end use. Consistency is indeed a desirable quality trait with most products but for the highly individual perception of bread quality even this 'simple' quality trait is difficult to apply with absolute certainty. If you are a sandwich maker then you certainly want consistency of volume, shape and cell structure but in a craft bakery context customers may still want to choose between light and dark crust coloured products. The variability in bread qualities sought by consumers demands that bakers be able to meet all of the quality requirements concerned. All in all, breadmaking requires a deep understanding of the many complex raw material and process interactions that all play a part in determining final product quality.

1.1 Wheat and its special properties

Almost every discussion of bread and its quality will start with a reference to the special nature of wheat, given that wheat flour is the largest ingredient in dough and bread formulations it is hardly surprising. In the contributions which follow much will be made of the special properties of wheat flour proteins to form gluten after hydration and during mixing. Dough mixing is the process which starts dough aeration as the gluten forms a network which traps and retains bubbles of air for inflation by carbon dioxide gas from yeast fermentation.

It is because of the special properties of wheat proteins that much research has been devoted to them. Chapter 3 provides a comprehensive insight to the many different aspects of wheat chemistry and biochemistry and this theme is continued in Chapters 6 and 26. Improvements in wheat quality to make it better suited to its end uses are considered in Chapters 8 and 11, along with opportunities for future development. The key role of the gluten forming proteins rightly receives considerable attention in several chapters but to address the balance the role of starch is considered in Chapter 7.

Fundamental to achieving a given bread quality is the ability to measure wheat quality and predict the likely breadmaking performance of any given wheat sample. This has been the 'holy grail' for farmers, millers and bakers for centuries. Cereal science has developed many techniques and tools to help but the complexity of those little wheat grains and the impact of farming practices on grain quality means that the goal still remains largely unattained. Much progress has been made and is considered in several of the chapters which follow. However, if you read on expecting to encounter a single test to make the perfect prediction you will be disappointed. It remains unlikely that there will be a single test which will predict bread quality with certainty and that is hardly surprising given that there are many different types of bread and breadmaking processes, each requiring a degree of variation in the 'quality' predictors. For the moment we will have to content ourselves with using a range of analytical techniques but I hope that the chapters, for example 4, 5 and 9, which address this general aspect will enable you to decide which combination best suit your particular needs.

Cereals and cereal-based products are amongst the safest of foods that we have available. However, growing and processing grain is not without its hazards. Some of the most deadly toxins are those which occur naturally. In cereals such unwanted materials may come in the form of mycotoxins. With the widespread transport and longer term-storage of cereal products issues of food safety have become increasingly important and for these reasons the detection and control of mycotoxins materials are the subject of Chapter 25.

1.2 Converting wheat to flour

In order to make an aerated bread structure it is necessary to process the wheat grains into another, more suitable form. Over the centuries techniques have evolved which permit the separation of the white endosperm from the darker coloured bran skins and germ. The progression towards greater availability of white flour has spawned the large number of bread products that we see in bakeries today. There are a number of factors which have driven the move to white flours, not least the greater gluten forming potential of the wheat proteins in the starchy endosperm. As discussed in Chapter 10 the flour milling process has evolved into a sophisticated process but its efficiency and economic viability still depend on the quality of the raw material entering the flour mill. Because of this flour millers remain acutely aware of the need to assess wheat quality and link its qualities with the final product for which it is destined. Flour millers therefore continue to use a range of tests to assess both the reliability and consistency of their own operations, as discussed in Chapter 9.

Since the nutritional properties of the wheat grain are not homogeneously distributed throughout the wheat grain the separation of the endosperm from the other components to yield white flour is not without its penalties. In those circumstances where nutrition is at a premium the fortification of flour- and wheat-based products has become a political and humanitarian issue. The role of fortification and the means by which it may be achieved are discussed in Chapter 12 and they provide an insight to another important aspect of bread quality.

The nutritional-quality link is also covered in Chapter 23 where the role for increasing fibre levels in bread are discussed. It is somewhat ironic that the Western world approach to breadmaking which has become dominated by white flour-based products should be advocating a return to higher extraction flours. The conversion of all of the wheat grain to flour has always been with us but the quality attributes of wholemeal or wholewheat bread have not had sufficient appeal for the mass market. In the UK when the technology for making wholemeal and similar breads as large and similar in softness to white bread were developed sales rose from 2 to 20% of breads sold. This increase in sales was truly quality driven since the fibre hypothesis had been around for quite some years with little impact on the dietary habits of the average UK consumer.

1.3 Making bread

Breadmaking is a centuries old traditional craft, practised in any country capable of growing or importing wheat. This has meant the evolution of a diverse range of breadmaking processes designed to achieve a wide range of bread products. There are a number of central themes which are common to all bread products and breadmaking processes. They are, the mixing of wheat flour, water, yeast and other functional ingredients and the expansion of the dough mass through the generation of carbon dioxide gas.

In the *Technology of Breadmaking* (Cauvain and Young, 1998, 2002) it has been stated that for no-time doughmaking processes that 'About 90% of final bread quality is decided by what bakers chose to do in the mixer'. This aspect of quality embraces the choice of raw materials and formulation as well as decisions on how to mix and develop the gluten structure in the dough. The relationship between mixing and dough development is still not fully understood. This is a theme which is visited in many of the chapters in this book.

It is well known that simply blending the bread recipe ingredients is not enough to initiate the development of the gluten structure. The technological aspects associated with gluten development and their place in the different breadmaking processes are discussed in Chapters 2, 13 and 14. If you want convincing of the relative importance of mixing and dough development, try mixing your own bread dough by hand. The harder you work the dough (that is the more energy you put into the mixing/development process) the greater will be the gas retention in the dough, the larger the loaf and the softer its crumb. However, the mixing times concerned may last for 30 minutes so be prepared for some hard work!

In-depth considerations of the molecular changes during dough mixing will be encountered in several chapters. The molecular interactions involved depend very significantly on the key quality traits of the proteins in the wheats and as discussed in Chapters 13 and 14 we can see how the genetic puzzle that is wheat protein is slowly being solved.

The role of water in gluten development is commonly taken for granted with water being seen simply as an ingredient which varies in level of addition with flour properties. It is true that the level of water addition to bread flour is critical in providing a dough rheology which is suitable for subsequent processing but as discussed in Chapters 15, 16 and 19 it is also part of the underpinning essential molecular changes which occur during the mixing/development process. Experiments with powdered ice or pre-hydrated flours reveal the complex relationship between hydration and development.

Water plays a key role all through the breadmaking process starting with mixing and ending with contributions to end product eating and keeping quality. These pivotal roles are described in Chapter 21. Critical reference to dough rheology, its control and contribution to final product quality is made in many chapters. Techniques for assessing dough rheology have changed and the latest developments are described in Chapter 18.

There was a time when many bakers considered water to be a 'free' ingredient or at least a cheap one. Those days have gone and the only truly cheap ingredient left for the baker is air. It is ironic that a plentiful, cheap ingredient plays such a key role in the breadmaking processes. The role of air assumes equal importance with that of wheat flour, water and yeast. It has been said that gases are the neglected ingredients in breadmaking but after reading Chapter 17 you would be forgiven for thinking that this is far from true. The study of the contribution of gases goes back over 60 years and the latest research is providing a fascinating insight into the role of gas bubbles in bread doughs and how they change during breadmaking. Gases may well have been taken for granted and overlooked by the baker but that is certainly not the case for the cereal scientist.

The concept of bread flavour is perhaps the most contentious of all the quality issues associated with bread. The formation of bread flavour arises in part from fermentation processes and in part from the complex interactions between the heat of the oven and the recipe ingredients. Factors which influence bread flavour are discussed in Chapter 22 and readers are left to form their own opinions on this highly individual subject.

1.4 Functional ingredients

An alternative term for functional ingredients in common usage is 'improvers'. Every ingredient used in breadmaking has a function and in the alternative sense 'improves' bread quality. An 'improving' effect could be put forward for the use of yeast added to flour and water to produce leavened rather than unleavened bread (although as discussed in Chapter 17 leavened bread in some contexts is considered impure). On the one hand an 'improving' effect can be claimed for salt additions (sodium chloride) which contribute to control of gluten rheology and yeast fermentation, but a counter argument would certainly be put by the advocates for lower salt levels in bread for medical reasons, though their case is not fully proven.

The foregoing comments show the ambiguous position that functional ingredients (improvers) hold in the psychology of breadmaking. While purists may argue for no 'artificial' additives (whatever that means) the demands of the modern consumer for consistent quality and safe food continue to make a strong case for additions of functional ingredients. Loss of bread quality is commonly associated with microbial spoilage as discussed in Chapter 24. Consumer shopping patterns in many parts of the world have changed such that the daily trip to the bakery is not the norm and such consumers expect that bread products remain spoilage free for significant periods since they have no desire to throw away uneaten product. Given these constraints it is not surprising that bakers have turned to the use of preservatives to limit microbial growth. 'Natural' anti-microbial agents may become more readily available as scientific studies continue.

From the moment that bread leaves the oven its qualities begin to change. This is the 'staling' process. Bread staling is most commonly associated with the progressive firming of the crumb, with or without moisture loss. This is the process discussed in Chapter 27. Again consumer expectations are high and they see firming of the crumb as unacceptable in quality terms. Again bakers have evolved strategies to limit the changes through the additions of functional ingredients such as emulsifiers and enzymes. Again the purists may see this an unacceptable practice.

Modern no-time dough making processes of the type described in Chapter 2 make use of mechanical development of gluten structures in the presence of oxidising agents. As concerns over food safely increase (rightly so) the list of permitted oxidants continues to shrink. The key roles for oxidising agents are discussed in Chapter 20. While one may argue that such ingredients are unacceptable, the baker is faced with the demand for consistent quality and so ingredients like ascorbic acid continue to find a place in modern breadmaking.

1.5 Bread in the future

Bread has a long history and undoubtedly has a long future. It is hard to imagine a world evolving in which bread does not play a part. However, this does not mean that bread products themselves will not change. The history of breadmaking is one of continuous product development and process innovation. For many years the pace of development was slow but in the 20th century improved communication and wider travel opportunities have exposed the majority of consumers to a wider range of food experiences than those of the previous centuries would have enjoyed.

A wider variety of breads will continue to be available for the 'average' consumer. Whether this will arise because of consumer demands or marketer's push would be an interesting debate. Bread 'price wars' are all too common and they devalue the standing of this noble product but we must recognise that they also provide the drive for product and process innovation. Consumers may want different products for their table but consistent quality remains a pre-requisite. This is not to argue that all breads must be based on modern processes. Quite the reverse, 'traditional' products will remain in demand because they add to the range of available bread varieties.

Today even the word 'traditional' when applied to bread has begun to take on a new meaning, especially in the UK. When I joined the UK cereals industry the Chorleywood Bread Process (CBP) was relatively new and facing the hostility of 'traditional' craft bakers. Ironically the CBP was originally developed for the craft baker but the economic and consistent quality arguments for its use were quickly identified by the larger plant bakeries. It is with plant-produced bread that CBP has mostly been associated. Like them or not, the innovations in breadmaking which occurred during the 1960s were a watershed for the cereals industry, the legacy of which we continue to employ today. It is over 40 years since the launch of the CBP and now the UK Food Standards Agency has come to the startling (and perhaps unpalatable for some) conclusion that the term 'traditional' may be applied to bread made by the CBP. Surely that is only fair given that we now have at least a whole generation of bakers who have grown up knowing little other than no-time dough making processes like the CBP.

So where will breadmaking go in the future? Answers to that question will come from two sources. The first is from the authors of the different chapters in this book who have identified their own views of the past, present and future prospects for their own specialist area. I wish to personally thank each of them for their time and efforts. Writing a book chapter is seldom easy and trying to second guess where the cereals industry will go in the future needs a very special type of crystal ball.

The second source of inspiration for future developments and innovation in breadmaking is from you, the reader. Hopefully each chapter of this book will provide you with inspiration and encouragement to make your own personal contribution to making bread of the 'right' quality for consumers of the future. I hope that you enjoy reading the book.

1.6 References

CAUVAIN, S.P. and YOUNG, L.S. (1998) *Technology of Breadmaking*, Blackie Academic and Professional, London, UK.

CAUVAIN, S.P. and YOUNG, L.S. (2002) *Fabricacion de Pan*, Editorial Acribia, S.A., Zaragoza, Spain.

2

Breadmaking: an overview

S. Cauvain, Campden and Chorleywood Food Research Association, UK

2.1 Introduction

Bread is a staple foodstuff, which is made and eaten in most countries around the world. Bread products have evolved to take many forms, each based on quite different and distinctive characteristics. Over the centuries craft bakers have developed our traditional bread varieties using their accumulated knowledge as to how to make best use of their available raw materials to achieve the desired bread quality. In some countries the nature of breadmaking has retained its traditional form while in others it has changed dramatically. The proliferation of bread varieties derives from the unique properties of wheat proteins to form gluten and from the bakers' ingenuity in manipulating the gluten structures formed within the dough. The rubbery mass of gluten with its ability to deform, stretch, recover shape and trap gases is very important in the production of bread and all fermented products. Of all the cereals, wheat is almost unique in this respect.

The term 'bread' is used to describe such a wide range of products with different shapes, sizes, textures, crusts, colours, softness, eating qualities and flavours that the terms 'good' or 'bad' quality tend to have no real meaning, except to the individual making the assessment. A baguette is not a baguette without a crisp crust, while the same crust formation would be unacceptable on north American pan bread and the fine cell structure of sandwich bread in the UK has no relevance to the flat breads of the Middle East.

The character of bread and other fermented products depends heavily on the formation of a gluten network which traps gas from yeast fermentation and makes a direct contribution to the formation of a cellular crumb structure which, after baking, confers texture and eating qualities quite different from other baked products. Look closely at the crumb structures of most baked breads and you will see that the common linking theme is that they are formed of holes of differing shapes, sizes and distributions. Each hole is embraced by a network of connected strands, coagulated gluten, in which starch granules and bran particles are firmly embedded. When this crumb is subjected to pressure with the fingers it deforms, and when the force is removed it springs back to assume its original shape, at least when the product is fresh. This combination of a cellular crumb with the ability to recover after being compressed largely distinguishes breads from other baked products: these are the very characteristics that bakers seek to achieve in most bread products.

While there are many different breadmaking processes, they have the common aim of converting wheat flour and other ingredients into a light, aerated and palatable food. The move to improve the digestibility of the wild grass seed forerunners of early wheat types through fermentation and baking represents a major step in the development of human food production. The unique properties of the proteins in wheat with their ability to form a cohesive mass of dough once the flour has been hydrated and subjected to the energy of mixing, even by hand, provides the basis of the transition from flour to bread. This cohesive mass is the one bakers call 'gluten' and once it has formed into a dough it has the ability to trap gases during fermentation, proof and baking which allows the mass to expand to become a softer, lighter and more palatable as a food after baking. The discovery that dough left for long periods of time would increase in volume without being subjected to the high temperatures of baking identified the basis of fermentation. The combined effect of these rheological changes is for the baked mass to increase in volume and give a product with an even softer, more digestible character and different flavour.

There are a few basic steps that form the basis of all breadmaking. They can be listed as follows:

- The mixing of wheat flour and water, together with yeast and salt, and other specified ingredients in appropriate ratios.
- The development of a gluten structure in the dough through the application of energy during mixing.
- The incorporation of air bubbles within the dough during mixing.
- The continued 'development' of the gluten structure created in order to modify the rheological properties of the dough and to improve its ability to expand when gas pressures increase during fermentation.
- The creation and modification of particular flavour compounds in the dough.
- The subdivision of the dough mass into unit pieces.
- A preliminary modification of the shape of the divided piece.
- A short delay in processing to further modify physical and rheological properties of the dough pieces.
- The shaping of the dough pieces to their required shape.
- The fermentation and expansion of the shaped dough pieces during proof.
- Further expansion of the dough pieces and fixation of the final bread structure during baking.
- Cooling and storage of the final product before consumption.

Loss of product freshness is as much about what we expect a product character to be as it is about its age since original manufacture. Whatever the criteria we use to judge bread staleness it becomes clear that the single most common requirement of fermented products is that it should ideally retain all of the attributes that it had when it left the oven; above all else we expect our bread to be 'fresh'. When we collect our bread from the baker and it is still warm to the touch we have no doubt as to its freshness, but when we purchase it cold from the store shelf we need convincing as to its freshness.

Raw materials and the processes used change and are time and temperature sensitive. To be able to make our particular bread type we must have an understanding of the complex interactions between our raw materials and the methods we will use in the conversion processes from ingredients to baked product.

2.2 Bread dough development

Dough development is a relatively undefined term that covers a number of complex changes that begin when the ingredients first become mixed. These changes are associated with the formation of gluten, which requires both the hydration of the proteins in the flour and the application of energy through the process of kneading. The role of energy in the formation of gluten is not always fully appreciated but it is a significant contributor to the breadmaking process.

There is more to dough development than a simple kneading process. The process of developing bread dough brings about changes in the physical properties of the dough and in particular improvement in its ability to retain the carbon dioxide gas which will later be generated by yeast fermentation. This improvement in gas retention ability is particularly important when the dough pieces reach the oven. In the early stages of baking before the dough has set, yeast activity is at its greatest and large quantities of carbon dioxide gas are being generated and released from solution in the aqueous phase of the dough. If the dough pieces are to continue to expand at this time then the dough must be able to retain a large quantity of that gas being generated and it can do this only if we have created a gluten structure with the appropriate physical properties.

It is important to distinguish between gas production and gas retention in fermented dough. Gas production refers to the generation of carbon dioxide gas as a natural consequence of yeast fermentation. Provided the yeast cells in the dough remain viable and there is sufficient substrate, then gas production will continue, but expansion of the dough can occur only if that carbon dioxide gas is retained in the dough. Not all of the gas generated during the breadmaking process will be retained within the dough before it finally sets in the oven. The proportion that will be retained depends on the development of a suitable gluten matrix within which the expanding gas can be held. Gas retention in dough is therefore closely linked with the degree of dough development. The most commonly considered factors are those related to the protein component of wheat flour; however, dough development will be affected by a large number of ingredients and processing parameters, many of which are not necessarily independent of one another.

2.3 Breadmaking processes

The development of no-time (i.e. no resting time in bulk before dividing) dough-making processes changed traditional (pre-1960) breadmaking. Foremost in these changes was the invention of the Chorleywood Bread Process (CBP) in which the development of optimum dough qualities is achieved in the mixer by measuring a defined energy expenditure rather than through the effects of fermentation (Cauvain, 1998). The result of the introduction of the CBP was to eliminate the need for bulk fermentation periods with considerable raw material and time savings, as well as to initiate changes in ingredient and processing technologies. The principles of the CBP were adopted in many countries around the world (Gould, 1998). Even in those bakeries that did not adopt the CBP there has been a similar trend away from long periods of bulk fermentation to shorter processing times and the use of more functional ingredients to achieve more consistent bread quality (Cauvain, 1998).

2.3.1 Bulk fermentation

The key controlling factor in optimising consistent bread quality with bulk fermentation was the skill of the baker who judged when sufficient change had taken place in dough rheology to yield the desired characters in the final bread. The process was commonly referred to as 'ripening'. Key issues with such fermentation processes are as follows:

- The quantity of protein in the flour, with lower protein flours requiring shorter fermentation times to achieve maturity (i.e. be ready for further processing) and vice versa.
- The rheological character of the gluten network first formed and its gradual modification during ripening. In this case the traditional baker assessed the changes by stretching a portion of the dough between the fingers. This assessment method is still practised today, even when bulk fermentation processes are not being used.
- The modification of bread flavour because of the fermentation processes.

2.3.2 The Chorleywood Bread Process

The basic principles involved in the production of bread and fermented goods by the CBP remain the same as those first published by the Chorleywood team in 1961, though the practices have changed with changes in ingredients and mixing equipment. The essential features of the CBP are as follows:

- Mixing and dough development in a single operation lasting between 2 and 5 minutes to a fixed energy expenditure during mixing.
- The addition of an oxidising improver above that added in the flour mill.
- The inclusion of a high melting point fat, emulsifier or fat and emulsifier combination.
- The addition of extra water to adjust dough consistency to be comparable with those from bulk fermentation.
- The addition of extra yeast to maintain final proof times comparable with that seen with bulk fermentation.
- The control of mixer headspace atmosphere to achieve given bread cell structures.

As the level of energy per kg dough in the mixer increases bread volume increases and with the increase in bread volume comes a reduction in cell size, increased cell uniformity and improved crumb softness. The role of energy during CBP mixing has yet to be fully explained but may be likened to the effects of natural or chemical reduction and, as such, will increase the available sites for oxidation. Chamberlain (1985) considered that only about 5% of the available energy was required to break the disulfide bonds, with the rest being consumed by mixing of the ingredients and the breaking of weaker bonds. The input of energy during mixing causes a considerable temperature rise to occur and typically final dough temperatures fall in the region of 27 to 32 °C.

The cell structure in the final bread does not become finer (smaller average cell size) as the result of processing CBP dough. In the case of CBP dough final bread crumb cell structure is almost exclusively based on an expanded version of that created during the initial mixing process (Cauvain *et al.*, 1999). The creation of bubble structures in CBP dough and indeed for many other no-time processes, depends on the occlusion and subdivision of air during mixing. The numbers, sizes and regularity of the gas bubbles depend in part on the mixing action, energy inputs and control of mixer headspace atmospheric conditions. Collins (1983) illustrated how bread cell structure improved (in the sense of becoming finer and more uniform) with increasing energy input up to an optimum level with subsequent deterioration beyond that optimum. He also showed how different mechanical mixing actions yielded breads with varying degrees of crumb cell size.

The requirement to add extra water to provide softer, more machinable dough is particularly true when the dough is mixed under partial vacuum in the CBP (Cauvain, 1998). The lower the pressure during mixing, the 'drier' the dough feels and the more water that needs to be added to achieve the same dough consistency as dough at the end of a bulk fermentation period. This increased dryness with CBP doughs comes in part from the lower volume of gas occluded in the dough at the end of mixing. In practice the reduction of oxygen available for ascorbic acid conversion and the need for some air to be occluded to provide gas bubble nuclei (Baker and Mize, 1941) places a lower limit of about 0.3 bar in the mixer.

2.3.3 Sponge and dough

The sponge and dough process is most common in the USA (Cauvain, 1998). Elements of the processes are similar to those for bulk fermentation in that a prolonged period of fermentation is required to effect physical and chemical changes in the dough. Only part of the ingredients are fermented – the sponge. Sponge fermentation times may vary considerably, as may their compositions. The key features of sponge and dough processes are as follows:

- A two-stage process in which part of the total quantity of flour, water and other ingredients from the formulation are mixed to form a homogeneous soft dough – the sponge.
- The resting of the sponge so formed, in bulk for a prescribed time, mainly depending on flavour requirements.
- Mixing of the sponge with the remainder of the ingredients to form an homogeneous dough.
- Immediate processing of the final dough after mixing.

The sponge contributes to flavour modification and the development of the final dough. The process of flavour development in the sponge, though complex, is observed as an increase in the acidic flavour notes arising from the fermentation by the added yeast and other microorganisms naturally present in the flour. To maintain the right flavour profile in the finished product the sponge fermentation conditions are closely controlled and care is taken to avoid a build-up of unwanted flavours. In many cases the addition of the sponge changes the rheological character of the final dough sufficiently to warrant further bulk resting time unnecessary so that dividing and moulding can proceed without further delay.

Improver additions are commonly made in the dough rather than the sponge. Flours used in typical sponge and dough production will be at least as strong as those used in bulk fermented dough with protein contents not less than 12% and high Falling Numbers (typically >250 seconds). High alpha-amylase activity could be a problem in the sponge because of excessive softening but is less likely to be a problem in the dough.

2.3.4 Spiral mixing and no-time dough processes

In many of the smaller bakeries around the world a no-time dough-making process has evolved based on mixers running with a single, vertical S-shaped mixing tool. Typically the bowl will rotate and often there is a single, vertical, fixed bar to enable greater energy input to the dough. Mixing speeds are somewhat lower than those seen with the CBP or sponge and dough breadmaking methods and energy inputs are also somewhat lower. After mixing the dough may be given a short period of fermentation in bulk before dividing. In the case of the Dutch green dough process (Cauvain, 1998) more than one rounding stage may be employed with extended resting periods in between (of about 40 minutes).

Spiral mixers are more effective at incorporating air during mixing than the CBP-compatible type (Marsh, 1998) and therefore have a greater potential to make more effective use of ascorbic acid. However, the bread cell structure which comes from this type of mixing is more open (i.e. larger average cell size) than would be obtained with CBP-compatible or horizontal bar mixers and the most common applications are those in which fine, uniform cell structures are not required, e.g. baguette. Improver additions are common with this type of mixer.

2.4 What determines bread quality?

While there are as many opinions on what makes 'good' bread as there are bakers and consumers, it is true to say that certain quality characteristics are required for individual bread varieties to be acceptable to the widest cross-section of consumers. For example, baguettes are characterised by a hard and crisp crust and without it we would reject the product, often describing a baguette with a soft crust as 'stale'. On the other hand, sliced pan breads in the USA, the UK and elsewhere are characterised by a thin but soft crust and if the crust were thick and hard it would be rejected by consumers, ironically, also being described as 'stale'.

Bread quality is determined by the complex interactions of the raw materials, their qualities and quantities used in the recipe and the dough processing method. It is therefore not possible to point to a single aspect of breadmaking and identify with clarity the factor which will predict bread quality. Some of the essential inputs to bread quality are as follows.

2.4.1 Flour

Since the formation of gluten is an essential component of breadmaking processes and wheat is the contributor of the proteins necessary for its formation, it follows that a significant factor that determines final bread quality comes from the wheat via the flour from the mill. The level and quality of the gluten-forming proteins depend heavily on the wheat variety, agricultural practices and environmental effects.

The protein content of wheat flour varies according to the wheats that are used by the millers and adjustments they may make in the mill. In general, the higher the protein content in the wheat, the higher the protein content of the flours produced from it. The higher the protein content of a flour, the better is its ability to trap and retain carbon dioxide gas and the larger can be the bread volume. Many North American and Australian wheat varieties have higher protein contents than most European wheats and this has led to the common view that you will get better bread from such wheats. However, with the changes that have occurred in dough-making processes this view is out of date. European wheats are well suited to modern breadmaking and large quantities of

North American wheats are required in European milling grists only where the product or breadmaking process demands their special qualities, e.g. wholemeal flour.

Protein quality also influences final product quality. It is most often judged by some form of dough rheological test though the prediction of final product quality is less certain because most dough rheological testing methods are carried out using conditions that have a limited relationship to the breadmaking process in which the flour will be used. Protein quality testing relies heavily on the interpretation of the rheological data by experts.

The grade colour figure (GCF), ash or Branscan values of flour are measures of the amount of bran that is present in a white flour. The higher the GCF, ash or Branscan value, the lower will be bread volume, in part because of the dilution effect on the functional protein content.

During the growing cycle for the wheat plant there are a large number of enzymes at work. Of interest to us are the ones known collectively as amylases, and especially alpha-amylase. The term alpha-amylase is used to describe a range of enzymes capable of breaking down damaged starch granules into dextrins and, in combination with beta-amylase, they will produce maltose. Alpha-amylase is produced during the growing cycle and can achieve quite high levels if the period around harvesting is wet.

Large numbers of the starch granules are damaged during milling. These damaged starch granules absorb more water than the undamaged granules, so that the larger the proportion of damaged starch the higher the water absorption of the flour (Stauffer, 1998).

2.4.2 Yeast

Bakers' yeast *(Saccharomyces cerevisiae)* comes in a number of different forms (Williams and Pullen, 1998). Compressed yeast comprises around 28–30% dry matter. The other main form is cream or pumpable yeast. The yeast produces carbon dioxide gas to expand the dough at its various processing stages, particularly during proof and the early stages of baking.

2.4.3 Salt

A basic function of salt in bread dough is to contribute flavour but it also has an inhibiting effect on the formation of gluten during mixing. In high-speed mixing systems the effect is quite small but it increases as the mixer speed becomes lower.

2.4.4 Sugar (sucrose)

In the UK and many other countries, little or no sugar is used in basic breads, while around 6% flour weight may be present in the sponge and dough breads of the USA. Rolls and other small fermented products may have up to 15%

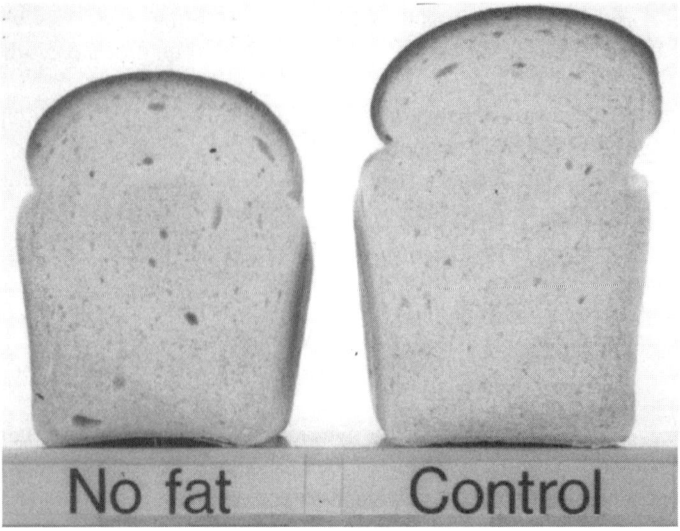

Fig. 2.1 Effect of fat addition on bread quality.

sugar. High levels of sugar inhibit yeast activity even though it is fermentable. In modern breadmaking sugars contribute to product sweetness and crust colour.

2.4.5 Fat

Compound bakery fats (mixtures of oil and solid fat at a given temperature) are used to improve the gas retention of dough and thereby increase volume and softness; see Fig. 2.1. The level used will vary according to the type of flour, with wholemeal flours requiring higher levels of fat addition than white, often two or three times more (Williams and Pullen, 1998). A proportion of the fat should remain solid in bread dough at the end of final proof, i.e. at 45 °C.

2.4.6 Water

The properties of the dough will vary according to the level of added water. Too little and the dough will be firm and difficult to mould (Cauvain and Young, 2001), producing breads that have small volume and poor external appearance. Too much and the dough will be soft and also difficult to mould; it will flow in the prover and give poor-quality bread. The 'optimum' level of water is really the maximum quantity we can get into the dough and still be able to mould the pieces and give bread of acceptable quality. It depends on the flour properties discussed above.

2.4.7 Improvers

This term covers any ingredient added to 'improve' the breadmaking potential of a given flour. Different breadmaking processes use different flours and different improver formulations. The functional ingredients used in improvers vary but typically contain one or more of the following ingredients:

- Oxidising agents to improve the gas retention abilities of the dough. The functions of the oxidant are complex and at the protein molecule level are currently thought to be mostly related to 'cross-linking' of proteins. By improving dough development we will get larger product volume and improved crumb softness.
- Reducing agents such as L-cysteine may be added to 'weaken' the dough structure. It will be used only at low levels in improvers but by reducing dough resistance to deformation it helps in moulding and shape forming without structural damage.
- Emulsifiers may be added to bread to improve its quality, each one acting slightly differently and having its own special effects. There are four commonly used emulsifiers: DATA (diacetyl tartaric acid esters of mono- and di-glycerides) esters, sodium stearoyl lactylate, distilled monoglycerides and lecithins (Kamel and Ponte, 1993; Williams and Pullen, 1998).
- Enzyme-active materials have become important to many sectors of the baking industry following the limitations placed on the use of oxidants. Those most commonly used are the alpha-amylases (fungal and cereal) and the hemicellulases. Proteolytic enzymes may be used in the USA (Kulp, 1993).
- Full-fat, enzyme-active soya flour may be used as a functional dough ingredient. It has two principal beneficial functions, both arising from its lypoxygenase enzyme system. They are to bleach the flour and assist in dough oxidation.

2.5 Dough mixing and processing

In essence, mixing is the homogenisation of the ingredients, whereas kneading is the development of the dough (gluten) structure by 'work done' after the initial mixing. However, with several breadmaking methods both processes take place within the same mixing machine and so can be considered as one rather than two processes. This is especially true of no-time dough processes since around 90% of the final bread is determined by the mechanics of mixing and the reactions between the ingredients which take place in the mixer.

The subprocesses taking place during mixing can be summarised as:

1. The uniform dispersion of the recipe ingredients.
2. Dissolution and hydration of those ingredients, in particular the flour proteins and the damaged starch.
3. The development of a gluten (hydrated flour protein) structure in the dough arising from the input of mechanical energy by the mixing action.

4 The incorporation of air bubbles within the dough to provide the gas bubble nuclei for the carbon dioxide which will be generated by yeast fermentation and oxygen for oxidation and yeast activity.
5 The formation of a dough with suitable rheological properties for subsequent processing.

Mixing machines vary widely from those that virtually mimic a hand mixing action to high-speed machines able to intensively work the mix to the required dough condition within a few minutes. Many mixing machines still work the dough as originally done by hand through a series of compressing and stretching operations (kneading) while others use a high-speed and intensive mechanical shearing action to impart the necessary work to the dough. In both the CBP and sponge and dough mixing processes the velocity of the dough being moved around within the mixing chamber is used to incorporate the full volume of ingredients into the mix and impart energy to the dough from the mixing tool. The essential features of the CBP have been described above. In the UK energy levels of around 11 Wh/kg of dough in the mixer are common, while in other parts of the world or with products, such as breads in the USA, this may rise to as much as 20 Wh/kg of dough (Tweedy of Burnley Ltd, 1982; Gould, 1998).

In many CBP-compatible mixers control of the headspace atmosphere is incorporated into the mixing arrangements. In its 'classic' form this consisted of a vacuum pump capable of reducing the headspace pressure to 0.5 bar. With the loss of potassium bromate as a permitted oxidising agent in UK breadmaking the relationship between headspace atmosphere and ascorbic acid became more critical. In response to deficiencies in product quality in some breads a CBP-compatible mixer was developed in which mixer headspace pressures could be varied sequentially above and below atmospheric (APV Corporation Ltd, 1992). In another possible variation of mixer headspace control it is possible with some CBP-compatible mixers to replace the atmospheric headspace gas with different gas mixtures. Most successful has been the application of a mixture of 60% oxygen and 40% nitrogen based on the principles of providing improved ascorbic acid oxidation (Chamberlain, 1979).

Horizontal bar mixers are usually capable of mixing large quantities of dough in one batch. Mixing speeds typically range up to a maximum speed of 150 rpm. The horizontal mixer is most often used with the sponge and dough process (Stear, 1990). The mixing action of the horizontal bar mixer depends on the design of the beater arms in the chamber. The two main variations are based on roller bars and elliptical-shaped beaters. In both cases the mixing action is strongly influenced by the relatively small size of the gap between the outer edge of the beaters and the sides of the bowl. The main action tends to be one of stretching and folding of the dough. The dough is picked up by the mixer blades and thrown against the outer side of the bowl but because of the slower speed, less energy is transferred to the dough than with CBP-compatible types. Gravity also plays a role in that the bulk of the dough will fall to the base of the mixer where it is partly picked up for further mixing and partly stretched as the mixing tool moves through the dough.

The lower mixing speed in horizontal bar mixers means that a longer mixing time is required than with the CBP-type in order to develop the gluten structure of the stronger flours which tend to be used with sponge and dough processes. The slightly longer mixing time also allows for longer contact times with the mixing bowl and so cooling jackets can therefore be more effective at removing the heat generated from dough mixing.

2.6 Cell creation during mixing

The production of a defined cellular structure in the baked bread depends entirely on the creation of gas bubbles in the dough during mixing and their retention during subsequent processing. After mixing has been completed the only 'new' gas available is the carbon dioxide generated by the yeast fermentation. Carbon dioxide gas has high solubility relative to other gases and in bread dough cannot form gas bubbles (Baker and Mize, 1941). As the yeast produces carbon dioxide, the gas goes into solution in the aqueous phase within the dough until saturation is achieved. Thereafter, continued fermentation causes dough expansion as the gas is retained within its structure. The two other gases present in the dough after mixing are oxygen and nitrogen. The residence time for oxygen is relatively short since it is quickly used up by the yeast cells within the dough (Chamberlain, 1979). Indeed, so successful is yeast at scavenging oxygen that none remains in the dough by the end of the mixing cycle, or shortly after. With the removal of oxygen the only gas that remains entrapped is nitrogen and this plays a major role by providing bubble nuclei into which the carbon dioxide gas can diffuse as the latter comes out of solution. The numbers and sizes of gas bubbles in the dough at the end of mixing are strongly influenced by the mechanism of dough formation, mixer design and mixing conditions in a particular machine. Recent work to measure bubble distributions in CBP bread doughs (Cauvain *et al.*, 1999) has confirmed that different mixing machines do yield different bubble sizes, numbers and distributions, see Fig. 2.2.

The modification of bubble populations through the control of mixer headspace atmospheric conditions has been known for many years, commonly through the application of partial vacuum to CBP-compatible mixers (Pickles, 1968). This control was useful in the creation of the fine and uniform cell structures typically required for UK sandwich breads but was unsuited to the production of open cell structure breads. In more recently developed CBP-compatible pressure-vacuum mixers which are able to work sequentially at pressures above and below atmospheric it has become possible to obtain a wider range of cell structures in the baked product. When the dough is mixed under pressure, larger quantities of air are occluded, which give improved ascorbic acid oxidation but more open cell structures. In contrast, dough bubble size becomes smaller as the pressure in the mixer headspace reduces and ascorbic

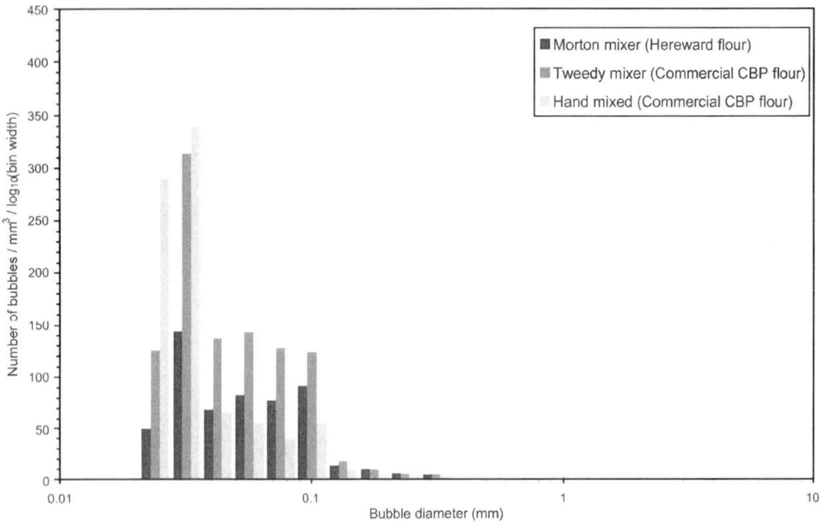

Fig. 2.2 Comparison of gas bubble size distributions in dough prepared in different mixers.

acid oxidation decreases as the pressure decreases. The greater control of dough bubble populations realised in these mixers allows a wide range of bubble structures to be created in the dough (Cauvain *et al.*, 1999). In addition to the fine and uniform structure created from the application of partial vacuum open cell structure for baguette and similar products can take place in the mixing bowl by mixing at above atmospheric pressure (Cauvain, 1994, 1995). Examples of product structures that can be achieved are illustrated in Fig. 2.3.

Similar considerations to those discussed above apply to the horizontal bar mixers which are typically used with sponge and dough processes. Air is incorporated in the sponge during the mixing stage and oxygen is lost because of the yeast activity, leaving only nitrogen gas bubble nuclei. When the sponge is mixed with the other ingredients at the dough-mixing stage quantities of these gas bubbles may be lost as the dough matrix ruptures. However, at the same time fresh air bubbles are incorporated and the process of oxygen depletion by yeast action again takes place. At the end of mixing the gas bubble population will be dominated by nitrogen, though carbon dioxide will be present in larger quantities that in CBP doughs. Nevertheless the same principle applies after mixing, namely that the gas bubble structure created during mixing will largely be the one expanded during proving and baking.

2.7 Dough processing

The stages required to process the bulk dough into bread may be described as follows.

Fig. 2.3 Examples of bread product structures obtained using a pressure-vacuum mixer.

2.7.1 Dividing

The bulk dough is divided to generate the shape and size of product required. Dough is generally divided volumetrically with portions of a given size cut either by filling a chamber with dough and cutting off the excess (piston dividing) or by pushing the dough through an orifice at a fixed rate and cutting billets from the end at regular intervals (extrusion dividing). In either case the accuracy of the system depends on the homogeneity of the dough. Different dividers need to be matched to different dough types in order to give optimum dividing accuracy with minimal compression damage. For example, 'strong' North American bread doughs can withstand high compression loads, whereas more delicate French baguette doughs are readily damaged.

2.7.2 Rounding and first moulding

Modification of the shape of the divided dough piece is common. Mechanical moulding subjects the dough to stresses and strains and may lead to damage to the existing gas bubble structure present. Some breadmaking processes require the rounder to have a degassing effect; however, if the dough comes from a breadmaking process which leaves little gas in the dough at the end of mixing (e.g. the CBP) then this requirement is unnecessary.

2.7.3 Intermediate or first proving

A period of rest between the work carried out by dividing and rounding and before final shaping may be used. The length of time chosen for this process is related to the dough rheology required for final moulding. Changes occur in dough rheology as it rests, the longer that it rests the greater will be the changes. In no-time dough-making processes (e.g. the CBP), the changes in dough which occur in first proof can have a considerable effect on final bread quality. The elimination of first proof can lead to a reduction in loaf volume (see Fig. 2.4) and poorer crumb cell structure because of damage to the bubble structure in the dough.

2.7.4 Final moulding

The basic functions of this stage are to shape the dough to fit the product concept and to re-orientate the cell structure. This usually requires:

- Passage of the round dough piece through sets of parallel rolls moving at high speed to reduce its thickness.
- Curling of the ellipse which has been created by sheeting by trapping the leading edge underneath a static chain which creates a 'Swiss roll' of dough.
- Compression and shaping of the Swiss roll to give a uniform cylinder of dough. This is achieved by compressing the dough piece underneath a pressure board while it is still being moved along the length of the moulder by the action of a moving belt.

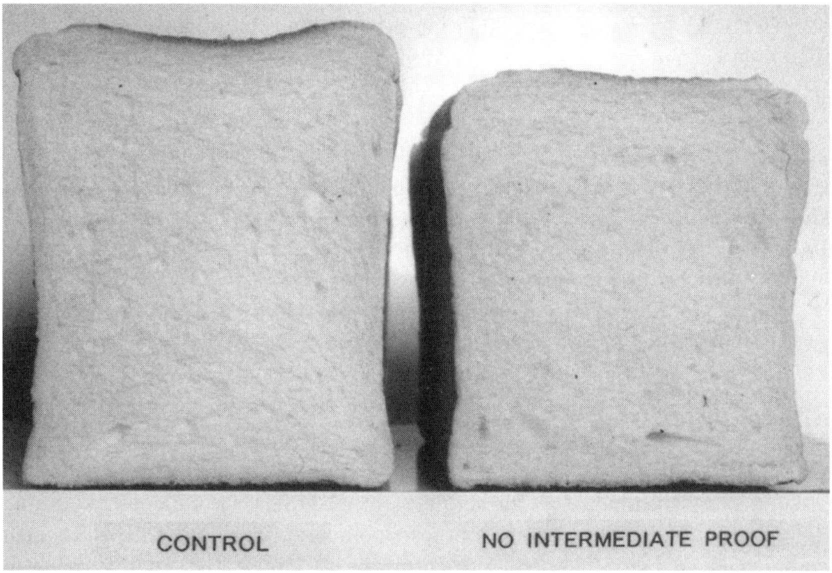

Fig. 2.4 Effect of first proof on bread quality.

- In some cases the dough pieces may be re-oriented to further modify the final bread cell structure; for example, by cutting the cylinder coming from the pressure board into four equal pieces and turning these through 90° before placing them in the pan (Cauvain and Young, 2001).

Final moulding clearly requires the dough to have the appropriate rheology in order to achieve the required change in shape. This is particularly true when starting from a rounded dough ball. The dough pieces should have low resistance to deformation and minimal elasticity, otherwise the high pressures required to change dough shape can cause loss of product quality (Cauvain and Young, 2000).

2.8 Gas bubble control during dough processing

A key feature of no-time doughs is that major degassing of the dough is not required after mixing. Little change occurs to gas bubble populations during dividing and first moulding operations. During intermediate proof the size of the gas bubbles increases as the carbon dioxide gas diffuses into the gas bubbles present (Whitworth and Alava, 1999).

It is in the final moulding stages that one of two significant changes may occur in the gas bubble populations. They are the potential elongation of gas bubbles and a slight, though potentially important degassing, during sheeting. As the round dough piece passes through the sheeting rolls, some elongation of gas bubbles in the direction of sheeting is likely to occur and this orientation is likely to be retained during subsequent curling. Elongation is most likely to occur with the larger gas bubbles located nearer to the surface of the dough during sheeting. It is unlikely that the pressures applied during sheeting will affect the smaller gas bubbles located in the centre of the dough. Nevertheless the elongation of gas bubbles does affect bread quality because when baked into the bread they tend to be shallower than other surrounding gas bubbles and, since they cast less shadow in the cut bread surface, they will make the crumb appear whiter. Elongation also contributes to the physical strength of the breadcrumb during slicing and its eating qualities.

The degree to which a dough may be degassed during the sheeting stages of final moulding depends on its rheology and interactions with the equipment. Whitworth and Alava (1999) have shown that the de-gassing of no-time doughs is small, but examination of computer tomography X-ray scans of CBP doughs shows that it does occur. In such scans the sheeted dough surfaces are visible as white lines because the dough is denser at this point and therefore there is greater X-ray absorbance. A further problem which may be encountered during dough sheeting is the rupture of gas-stabilising films and the subsequent coalescence of two gas bubbles to form one of larger size. Such damage to dough bubble in structures is thought to be a major factor in the formation of large, unwanted holes in breadcrumb (Cauvain and Young, 2000).

2.9 Proving and baking

Proving is the name given to the dough resting period, after the moulded pieces have been put into tins or placed in trays, during which fermentation continues in a controlled atmosphere, typically 40–45 °C and 85% relative humidity. Bakers' yeast is at its most active at 35 to 40 °C and so running the prover around 40 °C minimises the time required for proof. During proof the starch from the flour is progressively converted into dextrins and sugars by enzyme action. Yeast feeds on the sugars to produce carbon dioxide and alcohol. The carbon dioxide diffuses into the gas bubbles in the dough, causing them to grow and the dough to expand. Progressively the size of the gas bubbles increases (Whitworth and Alava, 1999). An example of the changes in gas bubble structure seen with X-ray tomography is shown in Fig. 2.5.

After proof the dough must be heat-set, that is baked. The process is one of conversion of a foam to a sponge. Baking temperatures will vary from oven to oven and with product. Typically oven temperatures lie in the region of 220–250 °C. A key parameter of loaf quality is to achieve a core temperature of about 92–96 °C by the end of baking to ensure that the product structure is fully set.

For the centre of the dough piece, the move from prover to the oven has little impact because it is so well insulated by surrounding dough. This means that the centre of the dough gets additional proof. The driving force for heat transfer is the temperature gradient from regions near the crusts, where the temperature is limited to the boiling point of water, to the centre. The heat transfer mechanism

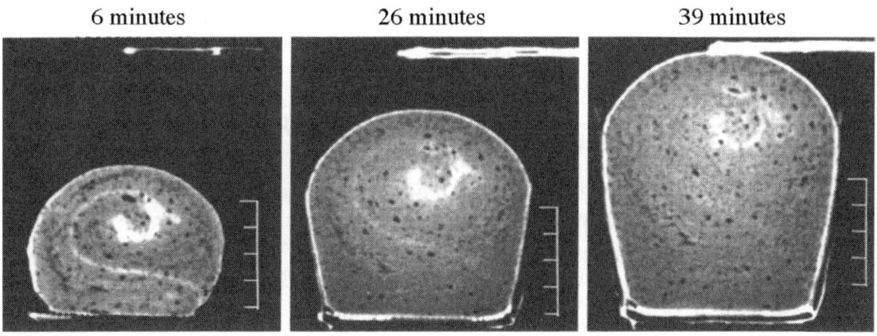

Fig. 2.5 CT X-ray images of dough during proof (note the image resolution of this technique only reveals bubbles of about 1 mm or larger).

is conduction along the cell walls and the centre temperature will rise independently of the oven temperature and approach boiling point asymptotically. There is no significant movement of moisture and the moisture content will be the same at the end of baking as at the beginning.

As dough warms up it goes through a complex progression of physical, chemical and biochemical changes. Yeast activity decreases from 43 °C and ceases by 55 °C. Structural stability is maintained by the expansion of the trapped gases. Gelatinization of the starch starts at about 60 °C and initially the starch granules absorb any free water in the dough. Alpha-amylase activity converts the starch into dextrins and then sugars and reaches its maximum activity between 60 and 70 °C.

The formation of a crust provides much of the strength of the finished loaf and the greater part of the flavour. Condensation on the surface of the loaf at the start of baking is essential for the formation of gloss, but quite soon the temperature of the surface rises above the local dew point temperature and evaporation starts. Soon after that the surface reaches the boiling point of the free liquid and the rate of moisture loss accelerates. The heat transfer mechanisms at the evaporation front are complex. There is conduction within the cell walls and water evaporates at the hot end of the cell. Some is lost to the outside but the rest moves across the cell towards the centre and condenses at the cold end of the cell. In doing so it transfers its latent heat before diffusing along the cell wall to evaporate again at the hot end. The evaporation front will develop at different rates depending on the bread types. The crust is outside the evaporation front and here the temperature rises towards the air temperature in the oven. As water is driven off and the crust acquires its characteristic crispness and colour, flavour and aroma develop from the Maillard reactions, which start at temperatures above 115 °C.

The other contributor to crust formation is the continuing expansion of the inside of the dough piece from the final burst of carbon dioxide production from yeast fermentation and the thermal expansion of the gases trapped in the cellular structure of the dough. If the dough is contained in a pan then it can only expand upwards. This effect is most obvious at the top edges of the loaf, where the displacement is greatest and where a split develops as the top crust lifts, exposing a band of elongated inner crust cells, called 'oven spring' or 'shred'. The degree of oven spring shown is an indicator of the 'strength' of the dough system, with weak flours showing little or no oven spring.

Some breads are characterised by the crispness of their crust, e.g. baguette. The first few moments in the oven are vital for the formation of a glossy crust. To obtain gloss, it is essential that vapour condenses on the surface to form a starch paste that will gelatinise, form dextrins and eventually caramelise to give both colour and shine. If there is excess water, *paste-type gelation* takes place, while with insufficient water *crumb-type gelation* occurs. To deliver the necessary water, steam is introduced into the oven.

2.10 Future trends

Many factors affect the quality of bread and fermented products and there is still significant opportunity for future developments. The nature of such developments depends on the continued evolution of the understanding of the complex interactions between ingredients, formulation and processing. Given that the formation of gluten structures is essential to breadmaking it is inevitable that we should anticipate further modification of wheat and flour functionality. It is important that functionality changes take full account of the intended purpose of the flour product. In the past moves to produce wheat varieties with increasingly higher proteins and stronger glutens have led to problems in those breadmaking processes where such strength is not required, with subsequent rejection of the wheat in the commercial world.

Bakery equipment will also continue to develop, often without consideration of the requirements for either the dough or the final product. All too often one encounters equipment that requires the dough to be modified so that it can be successfully processed. This cannot be the best way to improve bread quality.

Breadmaking as a skill, craft, technology and science has been around for many thousands of years and many of the key ingredient and process technologies have been established through much trial and error. Tradition in different parts of the world has evolved a wide range of bread products with many different attributes. The modernisation of breadmaking is really the product of the past 50 years. In more recent years there has been a trend to using fewer 'chemical' additives to deliver the necessary product qualities. This trend has placed a greater emphasis on the understanding of the interactions of ingredients and processing methods. Improvements in this understanding will undoubtedly lead to further changes in the production of what is probably the original processed food.

2.11 Sources of further information and advice

There are many reference books on breadmaking. Among the most useful are:

- *Technology of Breadmaking*, editors S.P. Cauvain and L.S. Young (1998) Blackie Academic & Professional, London, UK.
- Spanish version, *Fabricacion de pan*, editors S.P. Cauvain and L.S. Young (2002), Acribia, SA, Zaragoza, Spain.
- *Advances in Baking Technology*, editors B.S. Kamel and C.E. Stauffer (1993) Blackie Academic & Professional, London, UK.
- *Master Bakers Book of Breadmaking* (1996), 3rd Edition, National Association of Master Bakers, Ware, UK.

Useful sources of advice include:

- Campden & Chorleywood Food Research Assocation; www.campden.co.uk
- American Institute of Baking; www.aibonline.org
- BRI Australia Ltd; www.bri.co.au

2.12 References

APV CORPORATION LTD (1992) *Dough Mixing*, UK Patent GB 2,264,623A, HMSO, London, UK.

BAKER, J.C. and MIZE, M.D. (1941) The origin of the gas cell in bread dough, *Cereal Chem.*, **18**, Jan., 19–34.

CAUVAIN, S.P. (1994) New mixer for variety bread production, *Europ. Food Drink Rev.*, Autumn, **51**, 53.

CAUVAIN, S.P. (1995) Creating the structure: the key to quality, *South African Food Review*, **22**, April/May, 33, 35, 37.

CAUVAIN, S.P. (1998) Breadmaking processes, in (Eds. S.P. Cauvain and L.S. Young), *Technology of Breadmaking*, Blackie Academic & Professional, London, UK, pp. 18–44.

CAUVAIN, S.P. and YOUNG, L.S. (2000) *Bakery Food Manufacture and Quality: Water Control and Effects*, Blackwell Science, Oxford, UK.

CAUVAIN, S.P. and YOUNG, L.S. (2001) *Baking Problems Solved*, Woodhead Publishing, Cambridge, UK.

CAUVAIN, S.P., WHITWORTH, M.B. and ALAVA, J.M. (1999) The evolution of bubble structure in bread doughs and its effects on bread cell structure in (Eds. G.M. Campbell, C. Webb, S.S. Pandiella and K. Niranjan), *Bubbles in Food*, Eagen Press, St. Paul, USA, pp. 85–8.

CHAMBERLAIN, N. (1979) Gases – the neglected bread ingredients, *Proceedings of the 49th Conference of the British Society of Baking*, Stratford-on-Avon, UK, pp. 12–17.

CHAMBERLAIN, N. (1985) Dough formation and development, in (Ed. J. Brown), *The Master Bakers Book of Breadmaking*, 2nd Edition, Turret-Wheatland Ltd, Rickmansworth, UK, pp. 47–57.

COLLINS, T.H. (1983) The Creation and Control of Bread Crumb Cell Structure, FMBRA Report No. 104, July, CCFRA, Chipping Campden, UK.

GOULD, J.T. (1998) Breadmaking around the world in (Eds. S.P. Cauvain and L.S. Young),

Technology of Breadmaking, Blackie Academic & Professional, London, UK, pp. 197–213.

KAMEL, B.S. and PONTE, J.G. (1993) Emulsifiers in baking, in (Eds. B.S. Kamel and C.E. Stauffer) *Advances in Baking Technology*, Blackie Academic & Professional, London, UK, pp. 179–222.

KULP, K. (1993) Enzymes as dough improvers, in (Eds. B.S. Kamel and C.E. Stauffer), *Advances in Baking Technology*, Blackie Academic & Professional, London, UK, pp. 152–78.

PICKLES, K. (1968) *Tweedy (Chipping) Ltd. Improvements in or relating to dough production*, UK Patent No. 1,133,472, HMSO, London, UK.

MARSH, D. (1998) Mixing and dough processing, in (Eds. S.P. Cauvain and L.S. Young), *Technology of Breadmaking*, Blackie Academic & Professional, London, UK, pp. 81–119.

STAUFFER, C.E. (1998) Principles of dough formation, in (Eds. S.P. Cauvain and L.S. Young), *Technology of Breadmaking*, Blackie Academic & Professional, London, UK, pp. 262–95.

STEAR, C.A. (1990) *Handbook of Breadmaking Technology*, Elsevier Applied Science, London, UK.

TWEEDY OF BURNLEY LTD (1982) *Dough mixing for farinaceous foodstuffs*, UK Patent GB 2,030,883B, HMSO, London, UK.

WHITWORTH, M.B. and ALAVA, J.M. (1999) The imaging and measurement of bubbles in bread doughs in (Eds. G.M. Campbell, C. Webb, S.S. Pandiella and K. Niranjan), *Bubbles in Food*, Eagen Press, St. Paul, USA, pp. 221– 31.

WILLIAMS, A. and PULLEN, G. (1998) Functional ingredients, in (Eds. S.P. Cauvain and L.S. Young), *Technology of Breadmaking*, Blackie Academic & Professional, London, UK, pp. 45–80.

Part I

Wheat and flour quality

3

The chemistry and biochemistry of wheat

H. Cornell, RMIT University, Australia

3.1 Introduction: the structure of the wheat kernel

The main themes of this chapter revolve around the component parts of the wheat kernel and their chemical composition. Like all products of living organisms, wheat is an extremely complex material in terms of the number of components present and the chemical structures of these components. Needless to say, the myriad chemical reactions that occur in the grain during its development and those that occur in the doughmaking and baking processes are still not well understood, but progress is being made steadily.

The nutritional value of wheat is extremely important and a good deal of emphasis has also been placed on this aspect. This section covers the types of nutritional factors present in the kernel and their value to human health. Hence, carbohydrates and proteins, together with the minor components, such as lipids, vitamins and minerals, are also discussed in this context. The understanding of the types of components present and their relative amounts is provided by the inclusion of purification and analytical techniques. Modern technology has allowed us great insight into the factors affecting the quality of wheat products and this, of course, is the key theme of the book.

Common wheat is a member of the wild grasses (Gramineae family) native to parts of Western Asia. It belongs to one of the groups of the genus *Triticum* known as *Triticum aestivum*. Remarkably, it has been cultivated for about 10 000 years. Although flat or unleavened bread is commonly consumed in the Levant and Western Asia, the presence of a unique elastic protein complex in wheat, known as the gluten complex, ensures that a matrix is provided for the gases to form an even and open texture that is preferred for leavened bread.

3.1.1 Components of the kernel and their composition

The structure of a kernel of wheat is quite complex, consisting of many readily discernible entities. The caryopsis, as it is known botanically, is ovoid in shape with a longitudinal crease. The length of the kernel varies with the type of wheat and conditions of growth, but is generally about 4–8 mm. The outer layer, or epidermis, is a sheath for other layers of cells, which constitute a pericarp about 50 μm thick. Beneath a layer of nucellar tissue is the aleurone layer, before we encounter the starch-rich endosperm from which white flour is made. The endosperm containing the stored food for the plant, represents over 80% of the weight of the kernel. Apart from carbohydrates, the endosperm contains the albumins, globulins and the major proteins of the gluten complex – glutenins and gliadins, discussed in Section 3.3. The germ is situated towards the lower end of the kernel and on the opposite side to the crease. It consists of a plumule, to which is attached the scutellum – the larger part of the germ which acts as an absorbing organ for food – and a stem attached to the coleoptile, which serves as a protective sheath. The plumule forms the shoot when the seed germinates (see review of Evers and Bechtel, 1988).

The wheat germ represents only 2–3% by weight of the kernel, but it is rich in protein (≅25%) and lipid (8–13%). Wheat germ is available as a separate entity because it is an important source of Vitamin E (Fulcher *et al*., 1972). Wheat germ has only one half the glutamine and proline of flour, but the levels of alanine, arginine, asparagine, glycine, lysine and threonine are double.

The bran acts as a barrier to protect the grain and makes up over 8% of the weight of the kernel. The aleurone layer encases the endosperm (Section 3.1.2) and part of the embryo. It is of similar mass to the bran and is considered by millers as being part of the bran.

In order to protect the grain and endosperm material against the elements, the bran comprises water-insoluble fibre. Its chemical composition is complex, but it contains, essentially, cellulose – a polymer based on glucose – and pentosans, polymers based on xylose and arabinose, which are tightly bound to proteins. These substances are typical of polymers present in the cell walls of wheat and layers of cells such as the aleurone layer. There are large differences between the levels of certain amino acids in the aleurone layer and those in flour. Glutamine and proline levels are only about one half, while arginine is treble and alanine, asparagine, glycine, histidine and lysine are double those in wheat flour (Fulcher *et al*., 1972).

3.1.2 Endosperm materials

The endosperm represents the major part (80–85% by weight) of the kernel and consists of an intimate mixture of proteins and starch. The proteins are present as discrete particles and as interstitial material. Many different proteins are present in the endosperm, but the four main groups of proteins are the gliadins, glutenins, albumins and globulins. They represent the storage proteins of the wheat and usually make up about 10–14% of the weight of the kernel. The

Table 3.1 Amino acid (E signifies essential amino acid) levels in hard red wheat (% by dry weight) and recoveries in flour (% of wheat)

Amino acid	Content (%)	Recovery in flour (%)
Lysine (E)	0.43	69
Histidine (E)	0.36	85
Ammonia	0.55	106
Arginine (E)	0.76	72
Aspartic acid/asparagine	0.79	76
Theconine	0.47	86
Serine	0.83	95
Glutamic acid/glutamine	4.98 (30.2%)	103
Proline	1.62 (9.8%)	110
Glycine	0.65	80
Alanine	0.56	79
Cystine	0.31	94
Valine (E)	0.74	90
Methionine (E)	0.25	103
Isoleucine (E)	0.62	98
Leucine (E)	1.07 (6.5%)	97
Tyrosine	0.46	94
Phenylalanine (E)	0.77	101
Tryptophan (E)	0.27	83
Total	**16.49**	

Adapted from Teopfer *et al.* (1972).

endosperm proteins, as represented in flour, have a similar amino acid composition to that of the whole wheat (since the yield of white flour is about 75% of the wheat). The amino acid composition of a hard red wheat and the recovery of each amino acid on milling are given in Table 3.1.

The wheat and endosperm are very rich in glutamine (30%) and proline (10%), while leucine (6.5%) is the next important amino acid. All the other amino acids are at levels between 1% and 5% of the wheat. The recoveries of certain amino acids in the endosperm are lower than others because they are present at higher levels in other fractions, such as the bran and germ. The bran has only about two-thirds of the glutamine and proline in the endosperm, but has higher levels of glycine, lysine, arginine and asparagine/aspartic acid. Glutamine and proline are even lower in the germ, but lysine, arginine, asparagine/aspartic acid, threonine and alanine are all higher than their levels in the endosperm. Nevertheless, the protein content of white flour is similar to that of the wheat, while the lipid content is about 2%, this figure having been reduced by removal of the germ.

Albumins have lower glutamine/glutamic acid (23%) and proline (9%) contents than the glutenins and gliadins and higher cysteine (6%). Globulins tend to be higher in the basic amino acids – lysine and arginine – and even lower in glutamine/glutamic acid (15%) and proline (5%), while cysteine, at 5%, is higher than it is in glutenins and gliadins (Wieser *et al.*, 1994).

Flour from the endosperm contains about 82% starch, which is present as granules of various size and shape, and starch is the major carbohydrate present. Starch, pentosans and cellulosic materials are discussed in more detail in Section 3.2.

3.2 Wheat carbohydrates

3.2.1 Biosynthesis

The embryo of plants such as wheat uses starch as a source of energy until it can germinate and make its carbohydrates by photosynthesis. Starch, a mixture of polysacharides, is biosynthesised later than proteins, but becomes rapid as the plant matures. Photosynthesis converts about 10^{11} tonnes of carbon dioxide into organic polymers such as starch and cellulose annually. In this process, sunlight and water are utilised to convert CO_2 into carbohydrates by means of chlorophyll.

The equation required to allow for the production of the major carbohydrates, cellulose and starch, and for the formation of molecular oxygen from oxygen atoms ($O^{\cdot}$) in water and light, is:

$$6nCO_2 + 12nH_2O^{\cdot} \longrightarrow (C_6H_{10}O_5)_n + 6nO^{\cdot}_2 + 7nH_2O$$

Photosynthesis involves two series of light reactions and a dark reaction which are linked (Cornell and Hoveling, 1998).

3.2.2 Molecular structures

The chief wheat carbohydrate, starch, is a mixture of two polymers, amylose and amylopectin. Both are classified as D-glycans, or more specifically, D-glucans, as they are polymers of glucose. The amylose content of most natural starches is around 25%. However mutant varieties of corn can contain up to 85% amylose and by contrast, waxy starches contain only amylopectin (Whistler and Daniel, 1984). Amylose, in overall shape, is a fairly linear molecule, with atoms in a helical arrangement, because of the glucosidic bonds being mostly of the α-1,4′ type. There is a small degree of branching. Amylopectin, by contrast, is a highly branched molecule, with glucosidic bonds of the α-1,4′ type and the α-1,6′ type, the latter being responsible for branching. Simplified structures are shown in Fig. 3.1.

Structures such as Haworth representations are ideal for showing the types of glucosidic bonds and are seen in Fig. 3.2. They illustrate how branching occurs in amylopectin. However, the more realistic representation is where there is an attempt to show the rings in three dimensions and this is seen in Fig. 3.3. The rings shown are of the pyranose type (five carbon atoms, one oxygen atom) and are represented as chair forms. The oxygen attached to carbon 1 points downward in the case of α-1,4′ bonds and the group so formed is called an axial type, as is the group formed from the α-1,6′ bond.

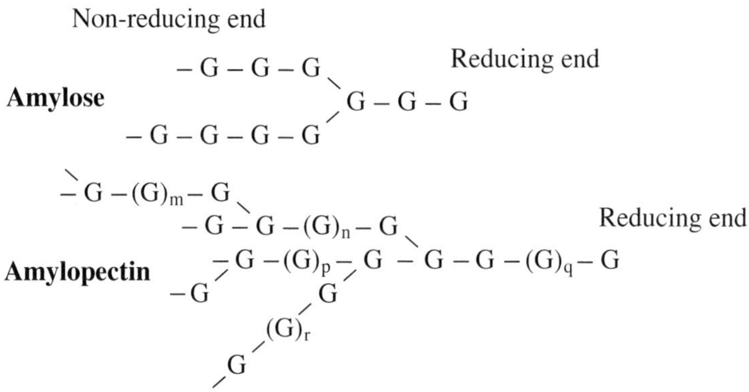

Fig. 3.1 Simplified structures of amylose (linear) and amylopectin (branched) based on anhydroglucose units (G). Branches occur, on average, about every 200–400 glucose units in the case of amylose and every 15–30 glucose units in the case of amylopectin.

Structural studies of amylose and amylopectin can be carried out in different solvents using viscometry as a measure of any changes. Banks and Greenwood (1975) showed that 0.5 mol/l KOH, dimethyl sulfoxide (DMSO) and formamide were good solvents for amylose. Information obtained from the use of the Mark–Houwink–Sakurada equation (Kitamura *et al.*, 1989) relates to the molecular size and to the molecular shape. The equation shows the relationship between

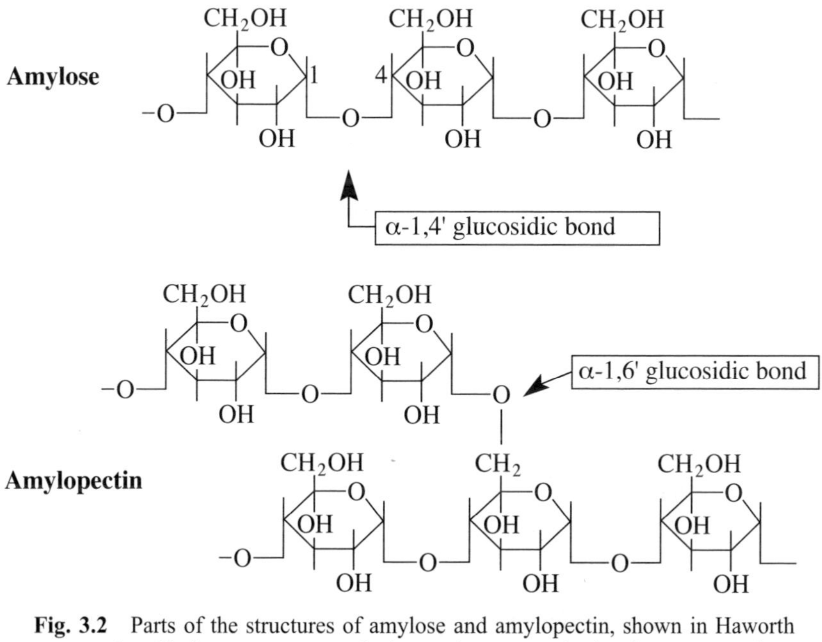

Fig. 3.2 Parts of the structures of amylose and amylopectin, shown in Haworth representation. Hydrogen atoms attached to carbon atoms in the rings are not shown.

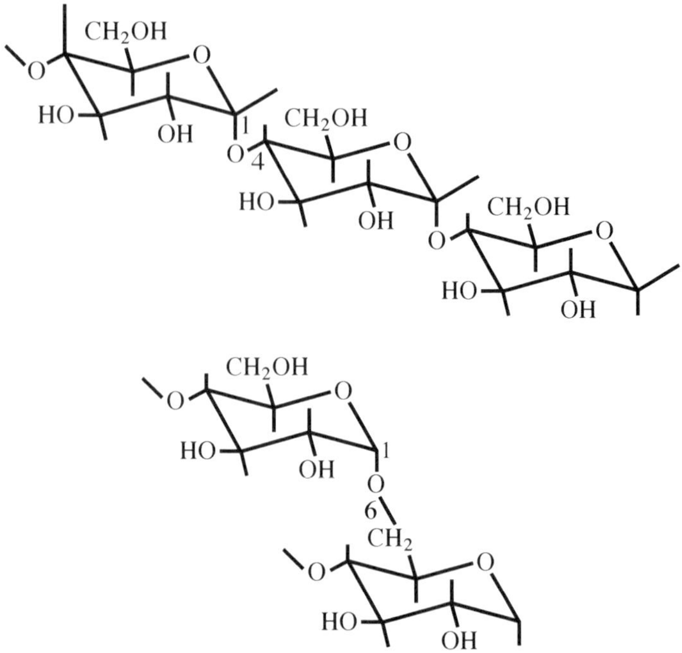

Fig. 3.3 Part of the structure of amylose shown in 3-dimensional representation (top) and (below) an α-1,6′ linkage as found more commonly in amylopectin. Branches occur regularly, with an average of about every 15–30 glucose units for most starches. Pyranose rings are shown in the chair form.

the molecular weight, limiting viscosity number and an exponent, α, in the following way:

$$[\eta] = KM^{\alpha}$$

where $[\eta]$ is the limiting viscosity number, obtained from a plot of specific viscosity against concentration, K is a constant, M is molecular weight (weight average) and α is an exponent which depends upon the shape of the polymer and the solvent used.

If molecular weight standards are available, the molecular weight of the unknown can be calculated by solving simultaneous equations for K and α and using these values to calculate M for the unknown sample. Values of the exponent α in the Mark–Houwink–Sakurada equation of 0.7–0.8 indicate that some type of coiled molecule is likely for amylose in solution. The work of Cheetham and Tao (1997) showed that the limiting viscosity of amylose was affected by addition of DMSO in DMSO–water mixtures and the ability of the amylose to form the iodine inclusion complex was lessened. This was attributed to the formation of random coils.

The molecular size distribution of amylose varies with the botanical source. For most starches, the range of weight-average molecular weights of amylose is from 150 000 to 400 000 (800 to 2200 glucose units). The figure obtained will depend to some degree on both the solvent used and the method used for the estimation. Commonly, viscometric methods using an Ubberlohde capillary flow viscometer, are employed. Light scattering, as well as size-exclusion chromatography, are also useful methods because they are based on different principles. Values for α of 0.70 and molecular weight 209 700 have been obtained for potato amylose (Banks and Greenwood, 1968).

Estimations of the molecular size of amylopectin are more difficult and it is a much larger molecule than amylose. There appears to be a wide variation in the distribution of molecular size (i.e. heterodispersity) in amylopectin, as with amylose. The molecules are less coiled than those of amylose and the more highly branched structure, where there are more –OH groups on the surface, accounts for the greater solubility in water compared with amylose. In this way, amylopectin is different from amylose, which forms metastable solutions in aqueous solvents. Cornell *et al.* (2002a) have estimated the average molecular weight of samples of fractionated corn amylopectin and waxy corn to be about 12 and 14 million respectively. These estimates were based on viscometric studies in a mixture of DMSO-2 mol/l KSCN (25 : 75 v/v) where the KSCN has limited the intramolecular hydrogen bonding between –OH groups on amylopectin and also bonding between amylopectin and water. The structural properties of amylose and amylopectin can also be studied using the well-known reaction with iodine in which it is believed that the iodine, as I_5^- ion, fits into the helical structure of the amylose to form a blue-coloured complex. This complex is most likely to be formed when the I_5^- ion (present in solutions of iodine in aqueous KI) fits inside the intermediate-length helix (McGrane *et al.*, 1998). The blue colour does not form in the solutions of amylose in DMSO or readily in mixtures of water and DMSO containing high amounts of the latter, indicating that DMSO alters the structure of the amylose. Studies of the structure of amylopectin can also be made in this way.

Other significant differences in the properties of amylose and amylopectin result from these molecular differences. The most notable of these properties is gel formation. In DMSO and DMSO–water mixtures up to 40% (v/v) water, potato amylose does not form gels, but at 60% (v/v) water content, gels form on standing (Cornell *et al.*, 2002a). Amylopectin, by comparison, does not form gels, whether in DMSO, water or mixtures thereof.

There is also the property of retrogradation, which is much more evident with amylose than with amylopectin in aqueous solutions. Retrogradation follows the formation of hydrogen bonds between amylose molecules and water molecules. Figure 3.4 is a simplified equation showing the release of water on storage from H-bonded amylose gels. This process is called syneresis.

Wheat starch is obtained by wet milling of white flour, preferably that from softer grade wheats. The 'Martin Process' and the 'Batter Process' are two such processes (Cornell and Hoveling, 1998). Wheat starch is present at about 63–66% of the weight of the wheat kernel, figures being higher for the soft wheats than for the hard wheats (Toepfer *et al.*, 1972).

Starches from different botanical sources have different size distribution shown in the different sized granules, with the largest being about 50 μm and the

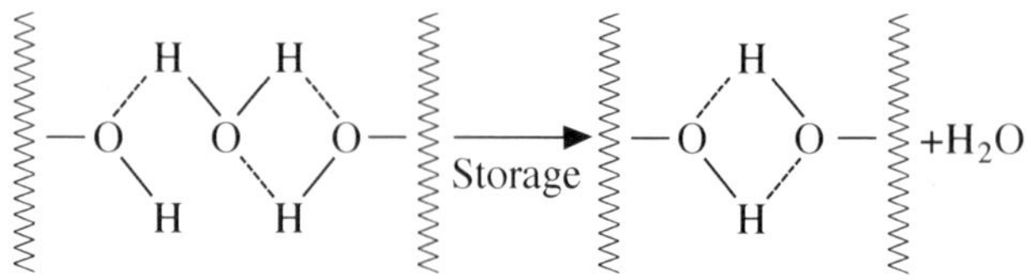

Fig. 3.4 Syneresis of starch gel, showing release of hydrogen-bonded water from amylose gel.

smallest 2 μm. Scanning electron microscopy of wheat starch granules shows concentric shells, exposed after treatment with enzymes (French, 1984). The lipid contents of wheat starch fractions of different size are roughly proportional to the specific surface areas of the fractions, which strongly suggests that the lipid is concentrated near the surface of the granules (Whattham and Cornell, 1991).

Several methods are available for particle size analysis of wheat starch and other fine powders. Sedimentation is one of these and various types of centrifuges still form the basis of modern wet-milling processes for recovery of industrial starches (Cornell and Hoveling, 1998).

Modern methods usually rely on laser beam diffraction methods, especially as they are more rapid, use less sample and are applicable to samples with a wide range of particle size. Low-angle laser light scattering (LALLS) relies on a gas laser passing through a stirred suspension of the material being tested, the focused rays passing to a photosensitive detector for measurement of intensity. Figure 3.5 shows typical patterns for wheat starch, in comparison with the coarser potato starch and the finer rice starch (black areas). Particle size may also be determined by gas adsorption methods and acoustic spectroscopy (Cornell and Hoveling, 1998).

3.2.3 Purification and separation of components of starch

The minor constituents of the wheat starch, as commonly manufactured by wet milling of wheat flour, are proteins, sugars, complex carbohydrates, lipids, traces of other organic compounds and minerals. Proteins such as albumins and globulins and some carbohydrates and minerals are readily removed by washing the starch with water and dilute salt solutions. Removal of the glutenins and gliadins require dilute alkali at pH 12 (Cornell *et al.*, 1994a).

Other types of protein, e.g. friabilins, may have been extracted by this process. These proteins are strongly associated with the polar lipids, which explains why permeation of water appears to be controlled by the surface layer of the starch granules (Greenblatt *et al.*, 1995). Levels of protein are often reported to be less than 0.25% in good quality wheat starch, but this figure is misleading owing to nitrogen in complex lipids. Insoluble pentosans remain in the tailing starch. The total pentosans in the prime starch may only be about 1% by weight but the tailing starch may contain up to 4% pentosans (Cornell and Hoveling, 1998).

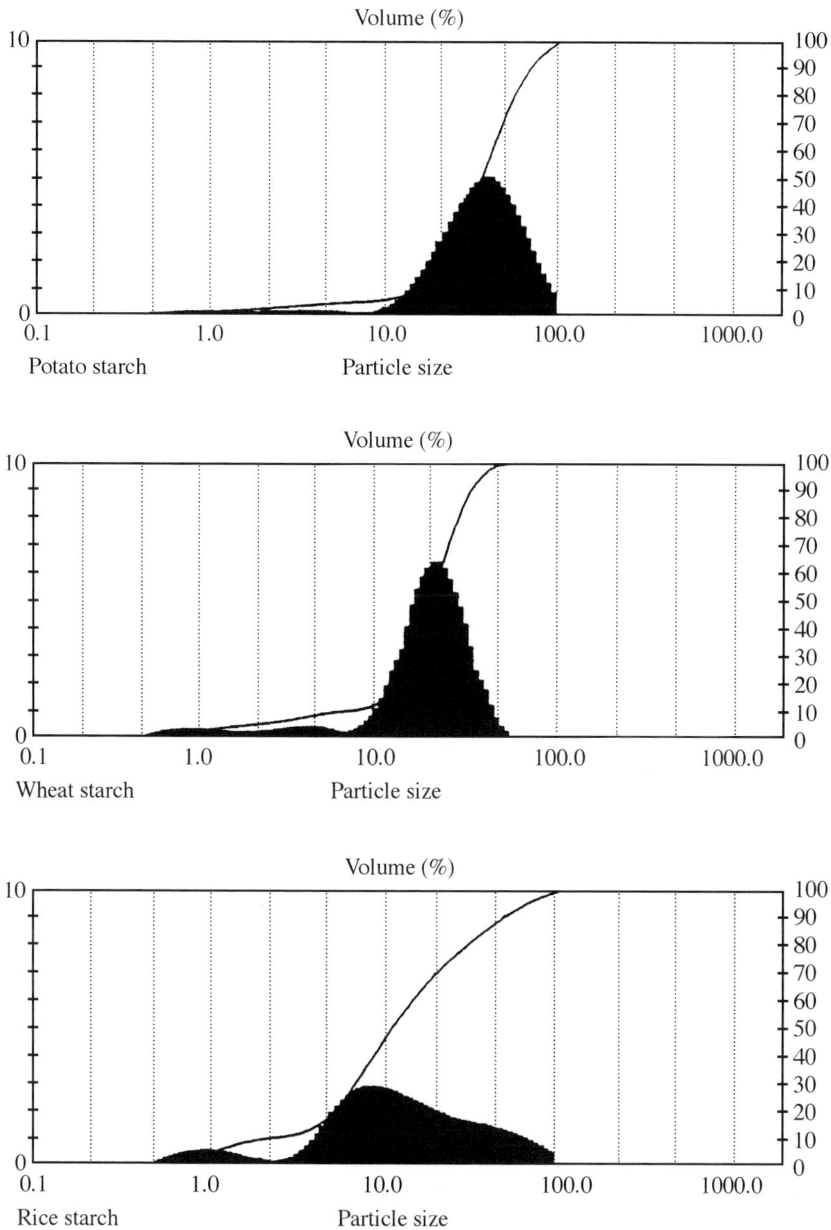

Fig. 3.5 Particle size distributions of potato starch (top), wheat starch (centre) and rice starch (bottom) as determined by the use of the Malvern Mastersizer X. The lines on the charts show the cumulative results. (Reprinted from Cornell HJ *et al.*, *Starch/Staerke*, **46**, 203–207, 1994, with kind permission of Wiley-VCH, Weinheim, Germany.)

Lipids in wheat starch are reported as being mainly lysophosphatidyl choline (Morrison, 1978). They are strongly associated with the starch as amylose inclusion complexes, and are not completely extracted even by hot polar solvents. Separation of the linear amylose from the branched amylopectin of starch without degradation of either component has been difficult. Added to that is the need for speed and efficiency of recovery. The basis of many methods has been the selective precipitation of amylose from starch sols by 1-butanol (Schoch, 1942). The hot sol containing 1-butanol is allowed to cool slowly and the amylose-rich complex is separated by centrifugation and then heated to yield amylose. Amylopectin remains in solution and can be recovered by precipitation with an excess of ethanol. More recently, the 1-butanol precipitation method has been carried out in the presence of potassium thiocyanate, KSCN, resulting in good yields of products from wheat starch (Cornell *et al.*, 1999).

3.2.4 Physical and chemical properties of wheat starch

Intact starch granules are almost insoluble in water, but the granules do swell because of the disruption of surface membranes. Starch in grains contains about 12–18% water. About half of this is bound chemically, forming a 'sphero-crystal', a type of crystal lattice. One of the most important physical properties of starch is its ability to form pastes on heating in water. The temperature at which the granules swell and burst to form these pastes depends upon the botanical source of the starch. These changes to the granules are detected readily by microscopy and differential scanning calorimetry (DSC) and are affected by the concentration of the starch, the rate of heating and the presence of sugars, fats and other food components.

DSC has been used to determine the change in enthalpy during gelatinisation of starch. Experiments with wheat starch suspensions of different water/starch ratios were also carried out using light microscopy to examine loss of birefringence (Ghiasi *et al.*, 1982). At a water/starch ratio of 2 : 1, birefringence was lost over 57–64 °C but as the water/starch ratio decreased to 0.5 : 1, the birefringence was lost over a wider range of temperature (57–87 °C). The thermograms are shown in Fig. 3.6. Another method, relevant to these changes, is to constantly monitor the viscosity while stirring. It is eminently suitable for high water/starch ratios. An increase in viscosity occurs as gelatinisation begins and this increase continues until the process is complete, after which time some decrease occurs due to breakdown of the structure of the paste. This breakdown is pH dependent and is very evident at low pH values (due to hydrolysis) and less so at high pH values (see Fig. 3.7). Temperatures at which the viscosity suddenly increases are generally a little higher than the changes observed by DSC or microscopy. This is seen with the Brabender Amylograph, an instrument used for evaluating the strength of starch pastes, their stability and their 'setback' on cooling. Starch pasting characteristics can also be measured using a Rapid Visco Analyser (Newport Scientific, Sydney, Australia) using as little as 2.5 g of starch sample.

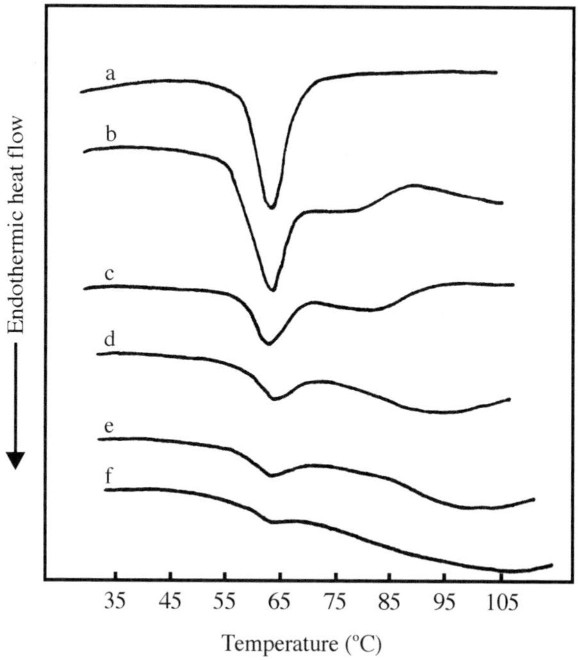

Fig. 3.6 Differential scanning calorimeter thermograms of heated wheat starch suspensions with water-to-starch ratios of 2.0 (a), 1.0 (b), 0.75 (c), 0.5 (d), 0.44 (e) and 0.35 (f). (Reprinted from Ghiasi K *et al.*, *Cereal Chem*, **59**, 258–262, 1982, with permission of American Association of Cereal Chemists, St. Paul, MN, USA.)

Wheat starch pastes at any given concentration have lower viscosities than pastes from potato or corn starch. When these pastes are allowed to cool to room temperature, they form gels due to the amylose. Amylose molecules tend to 'zip' together as hydrogen bonds are formed between amylose molecules and water molecules. Solutions of starch in DMSO or 0.5 mol/l KOH are clearer and do not 'set back' to form gels. The properties of aqueous starch gels are very important in the food industry where starch is used as a thickener. The relationship between the viscosity of starch pastes and the concentration of starch in the mixture is approximated by the relationship

$$\log_{10} V_T = kC + d$$

where V_T = the viscosity at a particular temperature,
C = percentage concentration of starch (w/w),
and k and d are constants (Cornell and Hoveling, 1998).

Starch gels display thixotropic properties – they do not display Newtonian behaviour, as do, say, mineral oils. Under shear, the viscosity decreases, which is thought to be due to progressive orientation of molecules in the direction of flow and the rupture of hydrogen bonds in the paste structure. When the shear force is removed, viscosity values return to those before the shear force was

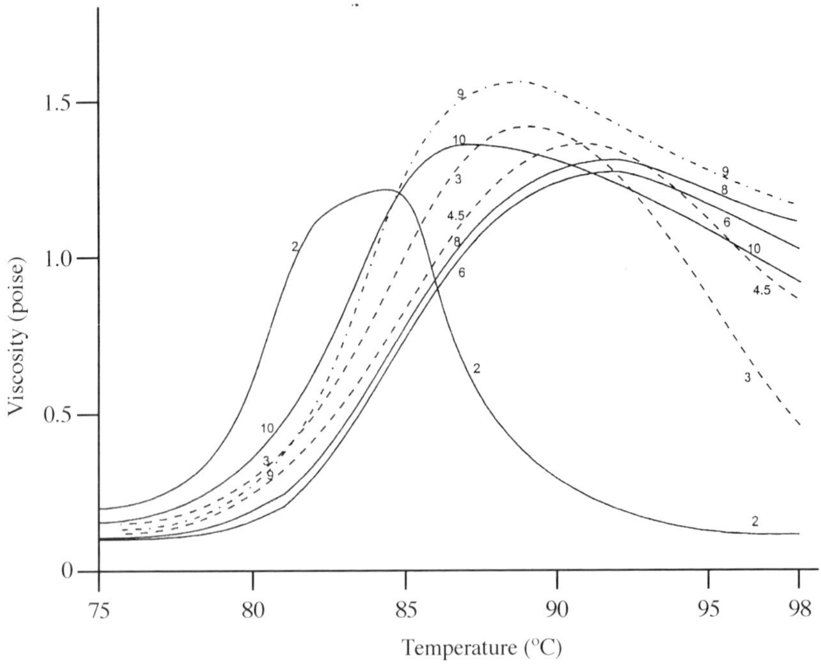

Fig. 3.7 Viscosity of wheat starch slurries (6% w/w) on cooking at different pH values. (From Cornell HJ and Hoveling AW, *Wheat: Chemistry* and *Utilization*, p. 156, 1998, with kind permission of CRC Press, Cleveland, Ohio, USA.)

applied. Rotational spindle viscometers are used to study rheological properties, but modern instruments such as the Rheometrics Fluids Spectrometer RFS II (Rheometrics, New Jersey, USA) are preferred. This type of instrument is equipped with parallel plates, allowing the shear stress to be monitored accurately.

One of the most important chemical properties of starch is its hydrolysis, using acids or enzymes. When starch pastes are treated with acids, amylose and amylopectin are hydrolysed to lower molecular weight carbohydrates known as dextrins and finally to oligosaccharides and simple sugars. This breakdown occurs because of attack on the glycosidic bonds by H^+ ions. The complete breakdown to glucose can be represented, stoichiometrically, as follows:

$$(C_6H_{10}O_5)_n + nH_2O \longrightarrow nC_6H_{12}O_6 \text{ (glucose)}$$

Starch pastes are also acted upon by amylases, resulting in the production of smaller molecules. In the case of α-amylase, the products are mainly dextrins, but with β-amylase, the product is maltose. With the enzyme β-glucosidase, dextrins can be further broken down to glucose.

Rain at harvest time increases the risk that crops such as wheat and barley will be damaged by the development of excess α-amylase in the grain, causing a reduction in viscosity of starch pastes produced from such grains. The test commonly used for this effect is the Hagberg Falling Number.

3.2.5 Commercial uses of wheat starch

Starch is used in a wide number of industries and in the home. In the food industry its pasting properties are utilised for gravies, soups, custards and desserts of various types. Wheat, corn and potato starches are widely used and their derivatives which have special properties, such as starch ethers and esters (e.g. phosphates) employed for better clarity and stability. Starch is a common addition to various types of baked goods, in particular biscuits and cakes. The addition of starch to low-protein flours improves the lightness of texture required in baked goods such as sponges and pastry.

A large proportion of starch is converted into syrups for use in the confectionery and brewing industries. The adhesive industry still uses starch because of its low cost and bonding properties to make cardboard boxes. Starch and modified starches are also used for the sizing of paper and fabrics.

3.2.6 Cellulose and pentosans

Cellulose occurs in wheat to the extent of about 3% (Toepfer *et al.*, 1972). It is a β-1,4′ glycan, where the oxygen attached to carbon 1 is more in the plane of the ring and is referred to as an equatorial group. This makes a considerable difference to the physical properties of the polymer as well as the chemical properties. Cellulose is more fibrous than starch and is not susceptible to attack by α-amylase, unlike starch. White flour has normally less than 1% of cellulose, but other non-starch polysaccharides are also present to the extent of about 3%. Bran contains much more of these non-starch materials, typically about 9% cellulose and 30% of other non-starch polysaccharides. Obviously, wholemeal flour is a better source of these polysaccharides than is white flour.

Pentosans are typical of these non-starch polysaccharides, which are the main constituents of cell walls of the lignified bran layer. They are classified as dietary fibre, i.e. they are not digested by endogenous secretions of the human digestive tract. Pentosans are mainly mixed arabinoxylans, protective polymers formed by chains of anhydro-D-xylopyranoxyl residues linked by β-1,4′ glycosidic bonds and single α-L-arabinofuranosyl residues to the 2- and 3-positions of xylosyl residues (Lineback and Rasper, 1988). Other mixed polymers are present.

Flour pentosans can be fractionated into water-soluble types, by extraction with cold and hot water, and water-insoluble types, which can be recovered by centrifugation of the tailing starch. Treatment with α-amylase converts these types of pentosans into water-soluble products. Water-soluble pentosans are also rich in arabinoxylans. Some pentosans contribute to water absorption of flour and viscosity of doughs and batters. They increase loaf volume and can improve crumb and crust characteristics, especially if treated with pentosanase enzymes (Higgins, 2002).

3.3 Wheat proteins

Proteins are a diverse group of compounds ranging from very high molecular weight, as in the case of structural proteins, through to low molecular weight hormones such as insulin. The reason for this huge diversity is, of course, their differences in chemical structure. All proteins contain particular sequences of amino acids, referred to as the 'building blocks' of proteins. The 20 common amino acids are shown in Table 3.2. The specific arrangements of amino acids connected by peptide bonds is referred to as the primary structure of the protein. Where only a small number of amino acids are connected together in this way, the term peptide is used. As the size of the peptide is increased, the molecule begins to assume a helical conformation because of the formation of hydrogen bonds (H-bonds) between groups such as the C=O and the N–H and are designated by interrupted lines.

Many of these H-bonds stabilise the structure in an α-helical conformation where the helix is coiled in a clockwise or right-handed manner with transoid (planar) peptide bonds and the hydrogen bonds are roughly parallel to the axis of the helix with the side-chains pointing outwards. This structure is referred to as 'secondary structure' (Fig. 3.8). Pleated sheet (β-sheet) structures can also be formed by H-bonds formed by overlap of several sections of the protein chain.

Other structures are formed by interactions between the side-chains, producing folding of the structures, which again may be numerous and result in a stable tertiary structure being formed. Acidic and basic side-chains are capable of causing these interactions and forming salt linkages (electrovalent bonds), aromatic rings and other non-polar side chains causing hydrophobic interactions, dipole–dipole interactions, e.g. between two –OH groups and most importantly, sulfhydryl groups of cysteine forming strong disulfide bonds (covalent bonds).

The structural organisation of protein molecules governs their properties. Protein biosynthesis begins early in the life of the plant and is directed by DNA. Thus proteins are molecules encoded by genes. The gene is copied, a process called 'transcription', and the copy messenger, mRNA, directs the assembly of amino acids in the required sequence for a particular protein. This process, called 'translation', occurs on particles called 'ribosomes' and is achieved by the

Table 3.2 Classification and one-letter symbols for the common amino acids

Neutral, aliphatic	Glutamine (Q), glycine (G), leucine (L), isoleucine (I), serine (S), alanine (A), asparagine (N), threonine (T), valine (V), proline (P) – pyrrole ring
Neutral, aromatic	Tryptophan (W), tyrosine (Y), phenylalanine (F)
Acidic, aliphatic	Aspartic acid (D), glutamic acid (E)
Basic, aliphatic	Arginine (R), lysine (K), histidine (H) – has imidazole ring
Sulfur-containing	Methionine (M), cysteine (C) – both aliphatic

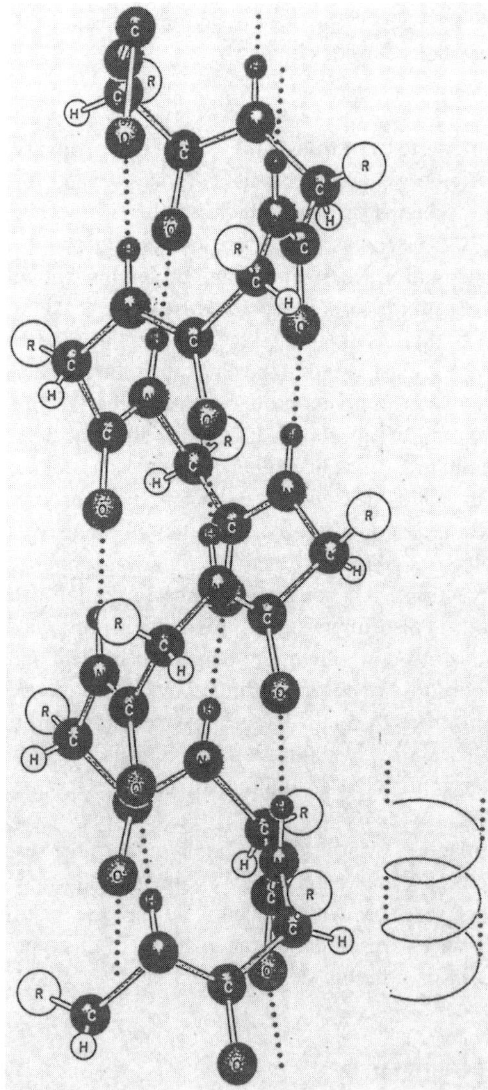

Fig. 3.8 A representation of the α-helical structure of a polypeptide. Hydrogen bonds are denoted by dotted lines. (Reprinted from Solomons TWG, *Organic Chemistry*, p. 975, 1976, with permission of John Wiley & Sons, New York, USA.)

involvement of enzymes. The four bases – uracil (U), cytosine (C), adenine (A) and guanine (G) arranged in sets of three 'codons' – determine which amino acid is to be selected. The codon $^{5}UAA^{3}$, which codes for methionine, is selected to start the biosynthesis, while others are selected to continue the process and still others for the termination. Transfer RNA (tRNA) molecules are the carriers for the amino acids selected for the synthesis. Hence there are more than 20 different tRNAs, at least one for each of the amino acids. ATP is also required to supply the necessary energy for these reactions. The book by Callewaert and Genyea (1980) provides further details.

3.3.1 The wheat gluten complex and functional properties of proteins

Wheat is an extremely heterogeneous commodity and this is exemplified particularly in the proteins present, which number in the hundreds. The wheat gluten complex is a viscoelastic mixture containing about equal amounts of glutelins (glutenins in the case of wheat) and prolamins, known as gliadins in the case of wheat. This mixture, as it separates from the washing of a wheat flour dough with water or dilute salt solution, contains about 33–35% solids, but it can be dried carefully, e.g. flash dried, to produce the gluten of commerce which can be reconstituted in water to give a rubbery mass of texture similar to the undried material, and is termed 'vital' gluten. It contains about 80% protein, 10% starch, 5% lipids, plus minerals, fibre and other impurities.

Wheat proteins have ideal functional properties for use in Western-style bread. The glutenin function is able to form an extensive three-dimensional network of molecules through disulfide bonding, hydrogen bonding and hydrophobic interactions. All contribute to the formation of a cohesive elastic dough. The gliadin is also important in this network of reactions.

Proteins, such as those in gluten, are denatured by heat and extreme changes in pH because both these conditions bring about conformational changes in the protein. Hence, in order to retain its strong viscoelastic properties, wet gluten should be dried carefully. Denaturation is often associated with a decreased solubility in water and decreased biological activity. Organic solvents can also cause denaturation.

The Maillard reaction (Maillard, 1912) is typical of the reactive properties of proteins. Flour proteins and sugars such as glucose (formed during doughing) participate in this reaction which occurs during the baking of bread. The browning that we see as crust colour is an example of this type of reaction and, if carefully controlled, is a highly desirable feature.

3.3.2 Uses of wheat gluten

Vital dry gluten, as well as the wet gluten, is used to fortify weaker flours for breadmaking and to increase the protein content of breakfast cereals and other nutritional products. The strong viscoelastic properties are the key to its usefulness. Gluten is added to bread doughs to improve their strength, enrich the protein and to produce better loaf volume and crumb texture. Wet gluten is preferable to dry vital gluten.

Fortification of wheat flour is still the main outlet for gluten, with the addition of some 3–5% of dry gluten to a weaker flour. Mixing tolerance and fermentation tolerance are improved and a higher yield of bread is obtained because of the addition of more water to the dough. Slimming loaves, which are also useful for people with diabetes, contain even higher levels of protein. The use of gluten as meat analogues (substitutes) caters for the needs of vegetarians. Where it is legal, gluten is added to sausage meat to boost protein and reduce the fat content.

3.3.3 Glutenins, composition and properties

The glutenins are responsible for the elasticity of the gluten complex. The types of proteins present are:

- Unassociated fractions of molecular weight 15 000–150 000.
- Associated fractions of molecular weight 150 000–3 000 000 (some more than this), many having extensive disulfide bonding.

Reduction of the intermolecular disulfide bonds yields subunits. The glutenin subunits have been shown to consist of high molecular weight (HMW) and low molecular weight (LMW) subunits (Shewry and Miflin, 1985). The degree of cross-linking by intermolecular disulfide bonds varies considerably giving proteins with a wide range of molecular weights. In wheat, glutenins make up about 30–45% of the total protein, about the same as corn, barley and rye, and about half that of rice (Mosse, 1968). The properties of glutenins and their key structural features are summarised in Table 3.3.

Both the gliadins and the glutenins are capable of combining with lipids to form lipoproteins and similar complexes. Such complexes could be expected to display some degree of solubility in 70% v/v ethanol. There are certainly some glutenin-type proteins which are soluble in this solvent. Differences in the amino acid composition of the (LMW and HMW subunits have been reported (Wieser *et al.*, 1990). The LMW ones are lower in glycine (5% compared with 18–20% for HMW) but higher in valine, isoleucine, leucine and phenylalanine, compared with the HMW subunits. Intermolecular disulfide bonds feature predominantly in glutenin and their presence is the main reason for the viscoelastic properties of the wheat gluten. The reaction is simply

Table 3.3 Properties of the glutenins

Property	Behaviour/characteristics
Solubility	Low solubility in water and neutral buffers. Products are viscoelastic materials. Small amount of material soluble in 70% (v/v) ethanol.
Molecular structure	Protein chains cross-linked by intermolecular disulfide bonds, and insoluble in 70% (v/v) ethanol. High molecular weight and low molecular weight subunits present. Regions of α-helix, β-structure and random coil structure.
Amount in wheat	30–45% of total protein.
Amino acid content	High in glutamine ($\cong$30%) and proline ($\cong$13%) Lower amounts of all the other amino acids. Cysteine content $\cong$ 2.4% (as ½ cystine).*

* Ewart (1967).

$$R{-}SH + R^{*}{-}SH + [O] \longrightarrow R{-}S{-}S{-}R^{*} + H_2O$$

Oxidation of sulfhydryl groups to disulfide groups thus occurs and is an important reaction during both doughing and baking. There are also intramolecular disulfide bonds, which are formed between cysteine side chains in the same molecule.

3.3.4 Gliadins: composition and properties

The different gliadins, termed α-, β, and ω-gliadins, in order of their electrophoretic mobility, are all essentially monomeric proteins. Many forms of these are present to the extent of 40–50% of the total protein content of wheat, much the same as corn and barley, and higher than rye and oats, rice being the lowest at 1–5% of the total protein, based on solubility in 70% v/v ethanol (Mosse, 1968). A summary of their properties is presented in Table 3.4.

Some of the different gliadin proteins from one variety of wheat have a similar sequence of amino acids, these showing greatest homology in the N-terminal domain. High contents of proline are present in many of the fractions of gliadin and this is believed to be responsible for the high numbers of β-bends (β-turns), where the helix changes direction. Gliadins have a higher content of proline residues than glutenin and yet do not display such strong viscoelastic behaviour, indicating that the disulfide bonds contribute more significantly than the β- bends. Very long molecules are likely to be formed during dough-making because the ends HMW subunits containing cysteine residues react to form disulfide bonds with other similar molecules. LMW subunits can also be integrated into these structures.

A large percentage of amide nitrogen is present due to the high glutamine content. Hydrophobic side-chains of valine, leucine, isoleucine, phenylalanine

Table 3.4 Properties of the gliadins

Property	Behaviour/characteristics
Solubility	Extremely low solubility in water and neutral buffers. Products are of sticky texture. Dry products are mostly soluble in 70% v/v ethanol.
Molecular structure	Single polypeptide chains capable of some intramolecular disulfide bonding. Considerable amount of α-helical and random coil structure. High incidence of β-turns.
Molecular weight	30 000–50 000
Amount in wheat	40–50% of total protein
Amino acid content	Very high in glutanine ($\cong$35%). Proline also high ($\cong$20%). Low levels of arginine, lysine and histidine. Low levels of aspartic acid and glutamic acid. Cysteine content 3% (as ½ cystine).*

* Ewart (1967).

and proline are collectively quite high in number and contribute to the lack of solubility in water and dilute salt solutions. Ewart (1967) found that glutenin had higher contents of lysine, glycine, tryptophan and somewhat higher contents of arginine, tyrosine, threonine, aspartic acid/asparagine, serine and alanine, compared with gliadin. The relatively small difference in cysteine between glutenin and gliadin is puzzling, but can be explained by the fact that the disulfide bonds from cysteine are mainly intermolecular in the case of glutenin compared with intramolecular types in the case of gliadin (Table 3.4).

The amino acid composition of the gliadins shows higher amounts of glutamine, proline, phenylalanine and isoleucine than glutenins (Ewart, 1967). However, the composition varies in the different electrophoretic fractions of the gliadins. This is particularly so for the γ- and ω-gliadins compared with α- and β-gliadins. The γ-gliadins have low levels of tyrosine, but their amino acid analysis and their molecular weights are similar (Wieser *et al.*, 1991).

The ω-gliadins are low in the basic amino acids (lysine, histidine and arginine) as well as asparagine, glycine, alanine, cysteine, valine, methionine, isoleucine, leucine and tryptophan. On the other hand they are higher in glutamine, proline and phenylaline compared with α- and β-gliadins. Wieser *et al.* (1991) having two glutamine/glutamic acid contents of 43% and 54% and proline 29% and 20% respectively, while the phenylalanine contents were each about 9%. Furthermore, the molecular weights of these ω-gliadins were about 50% higher than the α- and γ-gliadins. Gliadins, more than the other wheat proteins, have been shown to be toxic to individuals with coeliac disease (see Section 3.3.10).

3.3.5 Isolation and purification of proteins

Isolation of proteins from any part of the wheat kernel requires appropriate methods of extraction based on preferential solubility of certain components over others, together with more refined methods which depend upon physical and/or chemical properties.

Extractions of flour are generally made in order to obtain: (a) prolamins (gliadins) and (b) glutenins. Initial removal of albumins and globulins is usually advised. This can be done with water and buffers containing salt. Extraction of lipids should also be carried out with solvents such as chloroform/methanol (70 : 30 v/v) or water-saturated butanol. Agitation of the flour for about eight hours at room temperature in 70% ethanol is considered a good procedure for removal of the prolamins. Some glutenins are also extracted, the ethanol-soluble glutenins (ESGs).

A scheme for separation of the main groups of proteins, based on the work of Jones *et al.* (1959), involves washing of the gluten in water and sodium chloride, dispersal in 70% (v/v) ethanol containing 0.05 mol/l acetic acid and precipitation of the glutenin by neutralisation of the mixture. After centrifugation, gliadin is recovered from the supernatant liquor by precipitation with absolute ethanol (Cornell and Hoveling, 1998).

Glutenins are much more difficult to dissolve than gliadins because of their high molecular weight brought about by extensive disulfide bonding. Hence, reduction procedures on dispersions of the glutenins in 8 mol/l urea (or 6 mol/l guanidine hydrochloride) can be carried out with 2-mercaptoethanol or dithiothreitol. Borate buffers (pH 9.95) containing sodium dodecylsulfate (SDS) (0.1%) and 2-mercaptoethanol (1–3%) can also be used to extract glutenins. If further separations are carried out without reducing agent, alkylation (e.g. with 4-vinyl pyridine or iodoacetic acid), needs to be performed. Singh *et al.* (1990a) showed that sonication could be used to achieve rapid and complete extraction of proteins from flour in a 0.5 mol/l phosphate buffer (pH 6.9) containing 2% SDS. Addition of urea (3–4 mol/l) aids solubility, but denaturation occurs.

3.3.6 Fractionation of wheat proteins

Many methods are available for fractionation of wheat proteins, based on ionic charge, polarity, molecular size of other differences. A brief description of these methods follows.

Ion exchange (IE) chromatography

Wheat proteins can be chromatographed directly on ion exchange columns. Glutenin causes difficulties because of its high molecular weight and requires reduction prior to chromatography. Carboxymethyl cellulose (Cornell, 1990) has been used on gliadin with success, providing fractions for further tests.

Reversed phase (RP) chromatography

In reversed phase high-performance liquid chromatography, alkyl groups such as C_8 and C_{18} are chemically bonded to pressure-stable, porous silica and after bonding of the sample has taken place, the various components in the mixture are eluted using mixtures of a buffer with a water-miscible organic solvent such as methanol or acetonitrile. Acids such as phosphoric acid or trifluoro acetic acid favour ion-pair formation with basic groups of the peptide with concomitant suppression of carboxyl groups. By changing the pH by addition of triethylamine, optimisation of the separation is achievable (Hancock *et al.*, 1978). Elution is best carried out by using a linear gradient, e.g. from 10 to 40% v/v acetonitrile concentration in buffer. In RPLC the mobile phase is always more hydrophilic than the stationary phase. Detection of peaks is usually by absorbance between 200 and 230 nm.

A typical example applied to wheat proteins is shown in Fig. 3.9. The profiles form the basis for reliable identification of wheat varieties and for detecting genes relating to quality. RP-HPLC is complementary to capillary electrophoresis as it depends upon different characteristics, i.e. surface hydrophobicity rather than molecular charge, or size, as is the case for SDS-PAGE (polyacrylamide gel electrophoresis).

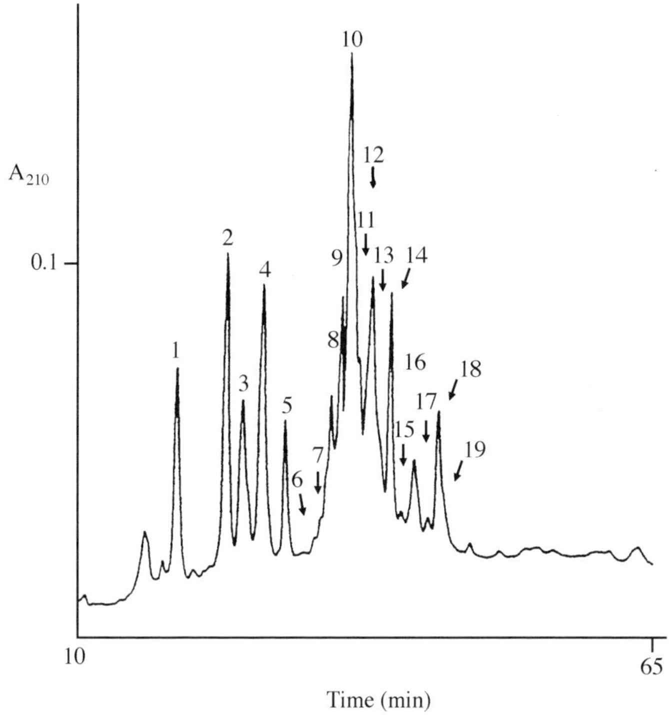

Fig. 3.9 RP-HPLC separation of pyridylethylated glutenin subunits from the wheat variety Chinese Spring. Peaks 1–4 correspond to the HMW subunits from this cultivar; later eluting polypeptides are LMW subunits. (From Bietz JA and Simpson DG, *J Chromatogr* **624**, 66, 1992, with kind permission of Elsevier Science, Amsterdam, Netherlands.)

Size exclusion (SE) chromatography

HPLC based on size exclusion (HP-SEC) of wheat proteins was used for prediction of the baking quality of the flour (Singh *et al.*, 1990b). However, there is some controversy over the relative importance of the high molecular weight and low molecular weight subunits (Gupta *et al.*, 1991). HP-SEC was used by Seilmeier *et al.* (1987) to study the HMW and LMW subunits of reduced glutenin. It was able to show significant differences in amino acid composition between the subfractions. In preparative procedures, the purification can be monitored by assays of biological activity, e.g. enzymes, while in other cases the number of peaks, bands or spots obtained is the criterion of purity, with final confirmation by amino acid sequence, hence the need also for analytical techniques.

3.3.7 Analytical techniques

Electrophoretic techniques and mass spectrometry are worthy of mention because of their analytical power in assessing the purification of proteins.

Polyacrylamide gel electrophoresis (PAGE)

Apart from standard PAGE, which has been used to characterise wheat proteins, there is also gradient PAGE, which makes use of slabs of gel in which the polyacrylamide concentration increases with the distance of migration. These gels, which are available commercially, sharpen the bands by reducing the mobility as the proteins migrate.

SDS-PAGE

SDS-PAGE has been eminently suitable for characterisation of glutenins. The SDS denatures the proteins forming random coils which carry a negative charge and allows separations on the basis of size and thus the estimation of molecular weight (Weber and Osborn, 1969). Bietz and Wall (1972) resolved reduced-gliadin into components of different molecular weight, the largest of these being ω-gliadins (60 000–80 000), and the others being α-, β- and γ-gliadins (30 000–40 000). Glutenins, after reduction, contained subunits of molecular weight 100 000–140 000 and 30 000–50 000. Comparisons with the native proteins indicated that 'polymers' of smaller proteins joined by disulfide bonds were present.

Isoelectric focusing (IEF)

IEF separates proteins on the basis of their isoelectric points, and molecular size is irrelevant. Frequently, resolution is better in IEF than PAGE, especially where narrow range ampholytes are used. IEF is therefore best used together with SDS-PAGE for two-dimensional electrophoresis.

Capillary electrophoresis (CE)

Separation of components by CE is accomplished by their differential migration in an electrical field in a capillary containing a solution of the analyte in buffer. Capillaries are typically 10–100 μm internal diameter and 50–100 cm in length and made of fused silica, borosilicate (Pyrex) glass or Teflon (Thibault and Dovichi, 1998).

Potential differences of between 100 and 600 V/cm are applied across the capillary. For peptides and proteins, the preferred buffers are acidic phosphate types. Detection of components is usually by UV absorbance or diode array, but fluorescence and mass spectrometry can be utilised. The advantages of CE over conventional electrophoresis are that it is completely automated, has high efficiency of resolution in rapid time and requires only nano-litre amounts of sample. Quantitative assessment of components based on peak area is also readily achieved.

Bietz and Schmalzried (1995) have demonstrated the effectiveness of CE in separating proteins from flour extracted with 30% (v/v) ethanol (see Fig. 3.10). The early peaks are primarily albumins and globulins while the later peaks comprise α-, β, γ and ω-gliadins.

CE is a valuable technique for complementing RP-HPLC and the combination, in particular, is useful for identifying varieties of wheat and for predicting baking quality.

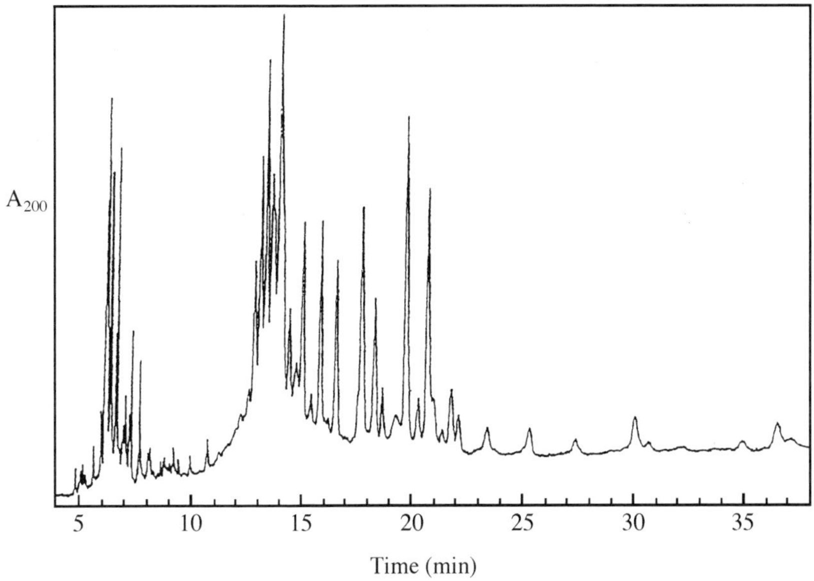

Fig. 3.10 Capillary electrophoresis separation of proteins extracted from the wheat variety Centurk with 300 ml/l ethanol. (Reprinted from Bietz JA and Schmalzried E, *Lebensm.-Wiss. u. Technol.* **28**, 174–184, 1995, with kind permission of Academic Press Ltd.)

Mass spectrometry
Matrix-assisted laser desorption/ionisation mass spectrometry (MALDI-MS) is a method in which ions are formed directly from the solid state by the impact of photons generated by a laser. It enables high molecular weight analytes to be detected intact at sub-picomole concentrations (Kaufmann, 1995). For this purpose, time of flight (TOF)-MS can be used in the linear mode in which ions are accelerated in a field-free region of the instrument and strike a detector at the opposite end of the flight tube. A reflectron can be added to a TOF-MS to improve resolution. This device uses an electrostatic field to reflect ions through a small angle towards a second detector. An example of MALDI-TOF-MS applied to analysis of wheat proteins is the work done by Cornell *et al.* (2002b) on cereal prolamin extracts. Prolamins from wheat, rye, barley and oats were extracted in 70% v/v ethanol–water and purified by precipitation in 93% v/v ethanol–water. Wheat, rye and barley showed similar patterns with peaks between 30 000 and 32 000 mass/charge (*m/z*) whereas oats showed peaks in a lower range 28 000–30 000 *m/z* (Fig. 3.11). The high resolution and sensitivity of MALDI-TOF-MS have allowed the elucidation of protonated molecular masses of most of the gliadins, hordeins, secalins and avenins of the above cereals, thus making it an ideal method for detecting their presence in 70% ethanol extracts of food products (Mendez *et al.*, 1995).

Another valuable application of MALDI-TOF-MS is in the amino acid sequencing of peptides. It requires the interpretation of post-source decay (PSD)

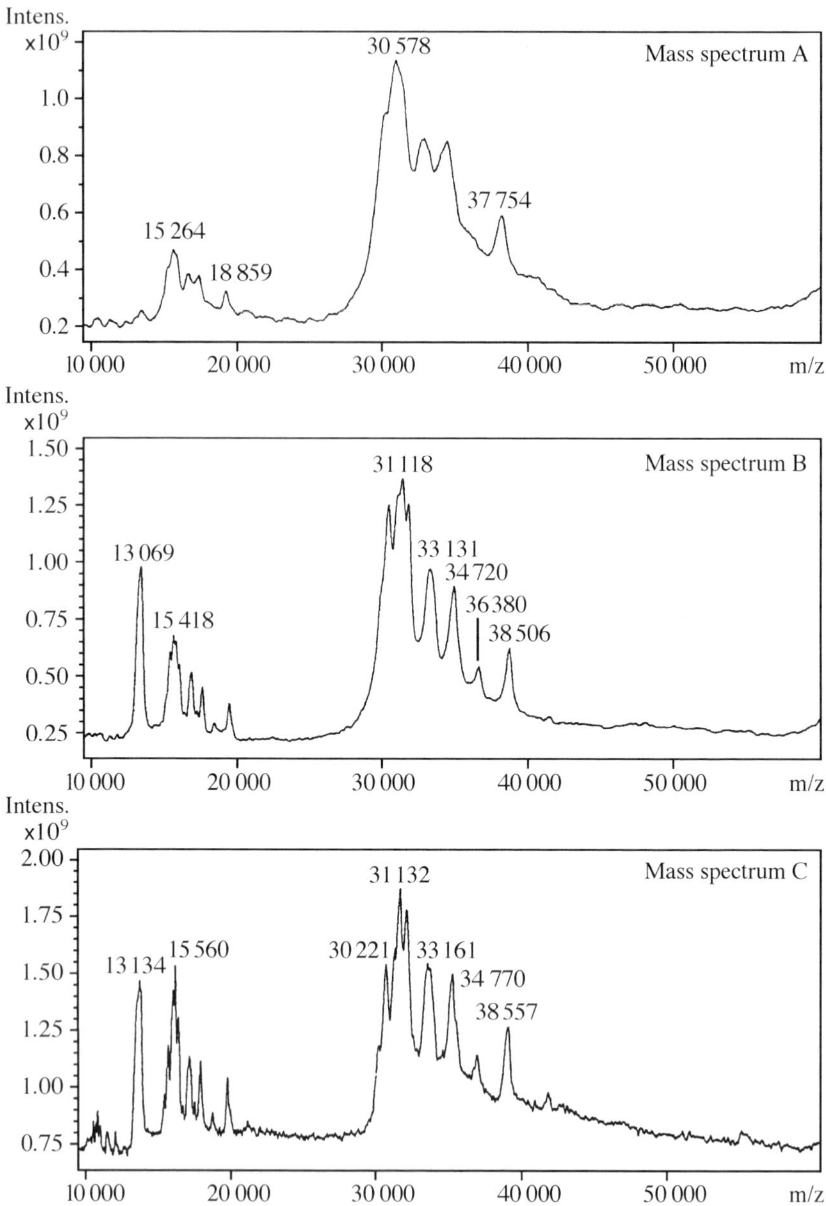

Fig. 3.11 MALDI-TOF mass spectra of wheat prolamin extracts: (a) crude extract, (b) ethanol-precipitated extract and (c) purified prolamin extract. (From Cornell HJ *et al.*, *J. Biochem. Mol. Biol. Biophys.*, **6**, 151–158, 2002, with kind permission of Taylor & Francis, Abingdon, UK.)

data. It relies on fragment ions being formed by metastable decay after they have left the source. The reflectron is able to separate parent ions from fragment ions because of their different residence times. Amino acid sequences can be checked against the standard Edman degradation method, where one amino acid at a time is removed from the peptide and detected as its phenylthiohydantoin derivative.

3.3.8 Purothionins

Purothionins are proteins rich in cystine and lysine and also, unlike the other wheat proteins, are low in glutamine and proline. The cystine is the result of intramolecular disulfide bonds, purothionins being single chain proteins. Jones and Mark (1977) have sequenced α-purothionins of hexaploid wheat and found all three to have very similar sequences with molecular weights in the region of 6000 (45 amino acids in length). Purothionins are present as lipid complexes (lipoproteins) in endosperm.

3.3.9 Glycoproteins

Glycoproteins are present in glutenin and complexes which contain glycosylated proteins are extracted with aqueous solvents. They appear to be associated with the high molecular weight subunits of the glutenin (Bollecker *et al.*, 1998).

3.3.10 Bioactive wheat proteins

Wheat allergy is the result of abnormal immunological reactions to certain wheat proteins. It has a totally different mechanism from that in coeliac disease and the proteins involved are not gliadins but albumins and globulins. These proteins cause Type I hypersensitivity reactions which are mediated by allergen-specific immunoglobulin E (IgE). Bakers' asthma is a typical condition in which water-soluble flour proteins bond to serum IgE as a result of inhalation of flour particles (Baldo and Wrigley, 1984). Coeliac disease (CD), or gluten-sensitive enteropathy, is a condition that results in damage to the small intestine, resulting in malabsorption. The symptoms are commonly poor growth, diarrhoea and abdominal pain and are brought about by the presence of wheat, rye or barley in the diet (Dicke, 1950). Oats can also be a causative agent, but it is certainly much less toxic than the above-mentioned cereals.

CD is largely confined to people of European descent and in most countries of Europe the incidence is around 1 in 300 of the population. The causative agents in wheat were first shown to be due to the gluten proteins (Anderson *et al.*, 1952) and this finding was soon confirmed by van de Kamer *et al.* (1953), who showed that rye and barley, were also toxic.

Frazer *et al.* (1959) demonstrated that peptic-tryptic digests of the whole gluten complex were also toxic and that this toxicity was abolished by pre-incubation of the digests with hog intestinal mucosa. This led to the hypothesis that individuals with coeliac disease lacked an enzyme necessary for complete digestion of the gluten proteins (the 'enzymopathic' hypothesis). Cornell and Townley (1973) added to this evidence by showing that only one fraction of a peptic-tryptic-pancreatic digest of wheat gliadin, obtained by chromatography (Fraction 9), was incompletely digested *in vitro* by homogenates of duodenal mucosa from patients with CD in remission. Fraction 9 and all the other fractions were completely digested by homogenates of mucosa from normal individuals. They went on to show that Fraction 9 was toxic *in vivo*, whereas the other fractions, when pooled, were non-toxic.

The other major hypothesis of the cause of CD is the immunological hypothesis. It is based on the idea that gluten peptides bind to the intestinal mucosa, triggering damaging immunological reactions (Strober, 1976). Cornell *et al.* (1994b) showed that peripheral blood lymphocytes from people with coeliac diseaese and some people without were stimulated to produce γ-interferon. Hence it is clear that immunological reactions can be triggered by abnormally high concentrations of gluten peptides (resulting from defective digestion) but they may not be the primary cause (aetiology) of the disease. Many different types of antibodies are produced in the sera and at the sites of damage in patients with active CD but antibody levels return to normal after the patient has been placed on a strict gluten-free diet.

More precise knowledge of the causative agents in coeliac disease has been obtained since the determination of the structure of an α-gliadin, called A-gliadin, by Kasarda *et al.* (1984). Figure 3.12 shows that it contains 266 amino acid residues, with glutamine (Q) and proline (P) being the most prominent. Certain short sequences of amino acids, termed 'motifs' appear at certain residues and have been associated with toxicity, e.g. the PSQQ and QQQP motifs (De Ritis *et al.*, 1988) and the QQPY motif (Cornell and Mothes, 1993).

```
1                10                 20                 30
V-R-V-P-V-P-Q-L-Q-P-Q-N-P-S-Q-Q-Q-P-Q-E-Q-V-P-L-V-Q-Q-Q-Q-F-
                 40                 50                 60
L-G-Q-Q-Q-P-F-P-P-Q-Q-P-Y-P-Q-P-Q-P-F-P-S-Q-Q-P-Y-L-Q-L-Q-P-
                 70                 80                 90
F-P-Q-P-Q-L-P-Y-S-Q-P-Q-P-F-R-P-Q-Q-P-Y-P-Q-P-Q-P-Q-Y-S-Q-P-
                100                110                120
Q-Q-P-I-S-Q-Q-Q-Q-Q-Q-Q-Q-Q-Q-Q-Q-Q-Q-Q-Q-Q-Q-I-L-Q-Q-I-L-Q-
                130                140                150
Q-Q-L-I-F-C-M-D-V-V-L-Q-Q-H-N-I-A-H-G-R-S-Q-V-L-Q-Q-S-T-Y-Q-
                160                170                180
L-L-Q-E-L-C-C-Q-H-L-W-Q-I-P-E-Q-S-Q-C-Q-A-I-H-N-V-V-H-A-I-I-
                190                200                210
L-H-Q-Q-Q-K-Q-Q-Q-Q-P-S-S-Q-V-S-F-Q-Q-P-L-Q-Q-Y-P-L-G-Q-G-S-
                220                230                240
F-R-P-S-Q-Q-N-P-Q-A-Q-G-S-V-Q-P-Q-Q-L-P-Q-F-E-E-I-R-N-L-A-L-
                250                260          266
Q-T-L-P-A-M-C-N-V-Y-I-A-P-Y-C-T-I-A-P-F-G-I-F-G-T-N-
```

Fig. 3.12 The amino acid sequence of A-gliadin according to Kasarda *et al.* (1984).

Kocna *et al.* (1991) demonstrated activity in peptide 8–19, containing the PSQQ and QQQP motifs (overlapping). Sturgess *et al.* (1994) found the peptide 31–49 to be toxic *in vivo*. This peptide contains all three motifs associated with toxicity.

Mucosal digestion studies have shown that all the active peptides within the sequences 8–19 and 75–86 can be digested down to smaller non-active peptides with normal mucosa. However, mucosa from coeliac patients in remission is unable to abolish the toxicity of these peptides (Cornell, 1998) and this is thought to be due to the lack of an enzyme such as prolyl oligopeptidase (EC 3.4.21.26) which cleaves peptide bonds on the C-terminal side of proline residues.

Recent work (Cornell and Wills-Johnson, 2001) shows that the toxicity of their serine-containing peptides (containing the PSQQ and QQQP motifs) appears to be mainly due to direct toxic action, while the toxicity of the tyrosine-containing peptides (QQPY motif) is chiefly immunologically mediated. Information from protein databases shows that these motifs and some extensions of these motifs, such as QQPYP, are also present in rye and barley (McLachlan *et al.*, 2002).

Schizophrenia is another condition in which the effects of gluten have been studied. The term covers a number of diseases in which the functioning of the brain is disturbed, giving rise to the psychiatric symptoms observed. Dohan (1988) has suggested that wheat may be damaging to people with schizophrenia and has pointed to the higher incidence of CD in people with schizophrenia. It appears that, in some people, sufficient amounts of incompletely digested peptides may cross the damaged mucosa and also the blood–brain barrier, causing abnormal reactions in the central nervous system. The use of a gluten-free diet in schizophrenia was shown to be beneficial by Singh and Kay (1976). It is well known that many children with untreated CD have psychiatric symptoms somewhat similar to schizophrenia, thus strengthening the peptide hypothesis.

3.4 Wheat lipids

Lipids are distributed throughout the wheat kernel and represent components of membranes, organelles and spherosomes (membrane-bound droplets) in tissues such as the aleurone layers, scutellum and embryo. In the endosperm, lipids are associated with proteins mostly as fully acylated glycerides, while those associated with the starch are monoacyl lipids (Morrison, 1978). Outside the starch they can exist as triglycerides, free fatty acids and other free forms. Although the lipids are widely distributed in the kernel, they are minor constituents in all but the germ. Lipid bilayers form the basis of biomembranes, in which polar lipids, particularly phospholipids, are prominent. In wheat, bilayers are able to associate with proteins and these structures are in a dynamic state. Such membranes regulate the passage of substances in and out of the cell.

Lipids are present to only a small extent in cereals but they have a significant effect on the quality and texture of foods because of their ability to associate with proteins due to their amphipathic nature (hydrophilic and hydrophobic groups present) and with starch, forming inclusion complexes.

In wheat, the maturing seed synthesises fatty acids at different rates and by the dormant period, the fatty acid composition is the same as that of the parent seed. The biosynthesis of lipids depends upon acetyl coenzyme A (acetyl CoA). This important compound is involved in synthesis of the acyl lipids such as glycerides, phospholipids, glycolipids, waxes, sphingosine lipids as well as the isoprenoid series, e.g. sitosterols and carotenoids. Malonyl-CoA is also utilised, together with NADPH, and further dehydration and condensation reactions occur to produce palmitic acid ($C_{16:0}$), which can then be extended to stearic acid by another reaction. Synthesis of linoleic acid occurs in higher plants by two separate pathways in the presence of microsomal enzymes (Conn and Stumpf, 1976). Isopentenyl pyrophosphate is formed from mevalonic acid which in turn is synthesised from acetyl CoA. It is able to form β-squalene and, from this compound, the characteristic steroid structures, of which cholesterol is a member, are produced, although only in very small amounts.

Lipids need to be isolated in order to study their nutritional significance and role in breadmaking. They can be extracted from milled grain using organic solvents. Soxhlet extraction with solvents of low polarity such as petroleum spirit or dichloromethane removes up to 1.5% lipid (free lipids) from flour but leaves a significant amount of lipid behind, bound to the flour. More polar solvents such as 1-butanol saturated with water (WSB), remove more lipid (bound lipid) but some non-lipid material is also extracted (e.g. free sugars).

McKillican *et al.* (1968) used low-temperature solvent extraction procedures to bring about less physical and chemical changes to the lipids. They used extraction with hexane followed by WSB at 1–5 °C under nitrogen. For hard red spring wheat, soft white spring wheat and amber durum wheats, the percentages of free lipid in the total lipid from each flour were 50.3%, 57.6% and 61.8% respectively.

The germ has the highest amount of lipids ($\cong$11%), but significant amounts are also associated with the bran and the starch and proteins of the endosperm. Complex polar lipids extracted by WSB account for about half the total lipids in the endosperm compared with about 23% in the bran and 17% in the germ, but the latter two contain more triglycerides. The bound lipids are mostly phosphatides such as phosphatidyl choline, phosphatidyl ethanolamine and phosphatidyl serine, as well as lysophosphatidyl derivatives, where there is one free hydroxyl group on the glycerol moiety.

Thin layer chromatography has been employed to separate the various classes of lipid using different solvent mixtures. MacMurray and Morrison (1970) identified 23 classes of lipids in wheat flour using WSB as the extractant. Steryl esters (7.5%), triglycerides (20.8%), diglycerides (12.2%) and free fatty acids (7.0%) were the major free lipids. Digalactosyl diglyceride and monogalactosyl diglyceride (4.9%) were the major glycolipids, while lysophosphatidyl choline

(7.1%), phosphatidyl choline (5.8%) and *N*-acyl phosphatidyl ethanolamine (2.9%) were the major phospholipids. Overall, free lipids amounted to 50.9%, glycolipids 26.4% and phospholipids 22.7% of the total lipids obtained. The principal sterols were identified as β-sitosterol, campesterol and C_{28} and C_{29} saturated sterols. Sphingolipids present are mostly ceramide hexosides, containing glucose and mannose.

The well-known chloroform/methanol (2 : 1 v/v) mixture of Folch *et al.* (1957) first used to extract lipids from brain and other tissue, can also be used for removal of similar lipids from wheat products. Hargin and Morrison (1980) concluded that triglycerides are the main storage lipids in spherosomes of the mature scutellum and aleurone layer. Glycolipids are probably membrane lipids present in the starchy endosperm while the acylphospholipids are widely distributed in all membranes.

Fatty acids in lipids can be determined quantitatively by two main methods. Morrison and Smith (1964) heated the lipids with boron trifluoride in methanol (140 g BF_3/l) and extracted the methyl esters (FAMEs) with pentane after addition of water. The pentane extracts were then analysed on gas–liquid chromatography (GLC) columns. The other method uses a transesterification method involving tetramethylammonium hydroxide in methanol (Misir *et al.*, 1985). The FAMEs are then extracted into petroleum spirit (b.p. 30–60 °C) after acidification and the extracts analysed as above. Numerous studies of the fatty acid composition of the lipids in wheat bran, germ and endosperm have been carried out showing a high level of linoleate ($C_{18:2}$) in both the total lipid and the triglycerides from the three fractions with lower amounts of palmitate ($C_{16:0}$) and oleate ($C_{18:1}$).

The results of baking tests carried out on flours from which certain types of lipids have been removed and replaced with other types of lipids have given variable results. Lipids extracted with WSB have not been able to restore loaf volume diminished by their extraction. The free lipids extracted with less polar solvents have been fractionated into polar and non-polar types. Bound polar lipids are not able to match the good results obtained with free polar lipids. Graybosch *et al.* (1993) have confirmed that free polar lipids are beneficial, especially to crumb texture. The role of lipid oxidation in bread doughs has also been studied. The peroxidation of unsaturated lipids by lipoxidase is a type of reaction which can lead to oxidation of sulfhydryl groups with an effect on baking quality (van Oort *et al.*, 1995).

Lipoproteins are the result of bonding of lipid to protein to form complexes. Bonding can occur by association of hydrophobic and/or ionic groups of protein side-chains with ionic groups such as those in phospholipids with hydrophobic fatty acid chains. Practically all the phospholipids in gluten are present as lipoprotein complexes. Frazier *et al.* (1981) reported the isolation of a lipoprotein, ligolin, from an acetic acid (0.05 mol/l) extract of lyophilised dough, with a lipid to protein ratio of about 1 : 1, molar.

3.5 Wheat enzymes

The enzymes of any plant are vital to the synthesis of food for the plant and its growth. In the case of wheat, the mature plant provides nourishing food for humans and animals and is a renewable source of energy. The enzymes in wheat are of some importance to the performance of the flour in breadmaking, especially the amylases.

3.5.1 Amylases

The amylase are examples of hydrolases as they catalyse the hydrolysis of the polysaccharides in the starch. One of the main enzymes of this type is α-amylase (EC 3.2.1.1). It is known as an endo-hydrolase. During breadmaking it reacts slowly with damaged starch granules, but rapidly with gelatinised starch to produce a mixture of dextrins, other oligosaccharides and small amounts of maltose. Calcium ions act as activators of the enzyme, while phytic acid, present in bran, acts as an inhibitor due to its reaction with calcium ions.

The other major amylase in wheat is β-amylase (EC 3.2.1.2) which acts on α-1,4′ glycosidic bonds near the non-reducing ends of amylose and amylopectin molecules to produce maltose. However, with amylopectin, the enzyme can achieve only about 60% conversion due to the α-1,6′ linkages, the material remaining being referred to as 'limit' dextrin. The amount of maltose in the flour is indicative of the extent of the reaction of the β-amylase with damaged starch.

In high-quality wheat, the content of α- and β-amylases is low, but if rain comes before harvesting, the inactive (bound) form of β-amylase is converted to the active (free) form and α-amylase also increases.

Apart from α- and β-amylases, α-1, 4′-α-1, 6′-D-glucosidases (pullulanases), exoglycosidases capable of hydrolysing α-1,4′- and α-1, 6′-glycosidic bonds, are present. Other amylases and pentosanases are also present at low levels of activity.

3.5.2 Proteases

Proteases in wheat reduce the consistency of doughs and batters after mixing and resting. The endosperm contains very little proteolytic activity, with the aleurone layer, the pericarp and the embryo being the main sources. Levels of activity increase rapidly on germination. Small amounts of fungal proteases may be added to strong flours in order to reduce mixing times and improve dough extensibility. The proteases are very important in malting, where their role is more clearly defined than it is in sound grain.

3.5.3 Other enzymes

Cytases are enzymes which act on various glucans, pentosans and poly-saccharide–protein complexes in cell walls and starch granules and belong to the family of endo- and exo-β-glucanases and pentosanases. They are important in malting processes because, by reacting in cell walls, they allow the entry of amylases and proteases for starch and protein hydrolysis.

Lipases are present in significant amounts in wheat germ. They are examples of esterases which hydrolyse ester groups, particularly those in glycerides which are common in lipids.

Phosphatases are other esterases, belonging to the family of hydrolases which hydrolyse phosphate esters. In wheat, an important example is phytase, which acts upon phytic acid (inositol hexaphosphoric acid), present in high amounts in bran. Phytase is thus able to lower the levels of phytic acid, thus counteracting the complexation of valuable minerals such as calcium, zinc and iron by phytic acid. The ubiquitous acid phosphatase is also present.

Oxidases in wheat are represented by lipoxygenase, present at high levels in the germ, and other enzymes of this type capable of catalysing reactions in which molecular oxygen is utilised to oxidise specific substrates. It has been shown to improve the rheological properties of the dough, probably because of its activity with sulfhydryl groups. Other oxidases such as phenol oxidases are present, particularly in the bran, where the substrates are polymeric substances.

3.6 Pigments

Wholewheat contains significant amounts of flavone pigments such as tricin and xanthophylls). Wheats of different origin and variety and grown under different conditions show variable amounts of the pigments, but can be expected to contain about 1.5 ppm of xanthophyll and its esters and over 2 ppm of flavones. The xanthophylls are at their highest concentration in the embryo ($\cong$11 ppm). Xanthophyll is a dihydroxy α-carotene.

Carotenoids are members of a class of hydrocarbons (carotenes) and their oxygenated derivatives (xanthophylls) consist of eight isoprenoid units. Lutein is a xanthophyll pigment which is abundant in wheat, and present also as its esters.

Flavones are almost insoluble in water but soluble in 85% (v/v) ethanol and WSB. Extraction with WSB and measurement of absorbance at 435.9 nm is used for estimation of these materials as a measure of flour colour. Alternatively, solvents containing *n*-hexane can be used with measurement of absorbance at 447 nm, which is more characteristic of xanthophylls. They are present mainly in the bran and are yellow to brown in colour. Carotene (provitamin A) is easily oxidised by the bleaching agents used in the milling industry. However, flour is not a significant source of vitamin A.

3.7 Future trends in wheat utilisation

The emphasis in recent years on wheat utilisation has been on value-added products, and has been largely the result of a drive from the highly competitive food industry. International trade in processed foods has been enhanced by globalisation of food companies, increased demand for a wide variety of value-added goods offered and, connected with the latter, the desire to try food of other cultures. In the future, the range of end-uses for wheat will need to be much greater to meet the needs of international food markets.

3.7.1 New developments in plant breeding, improved crop yields, protein quality

Yields of wheat are always dependent upon soil quality, the use of legumes in crop rotation, sowing time and climatic conditions. In the last decade or so, yields of wheat grown in particular areas have increased as the result of this local knowledge. The quality of the storage proteins is determined by the genotype of the plant and reflected in alleles of the particular cultivar. Research in this area is able to distinguish between the alleles of good baking quality wheats and those of poor baking quality wheats. The isolation of cDNAs and genes for most gluten proteins is still far from routine. This work has been hampered by the large number of repetitive amino acid sequences in wheat and the complex structures present in the glutenins, in particular.

3.7.2 Role of biotechnology

Biotechnology is the application of living organisms for the development of new products. Genetic engineering of plants allows the transfer of genetic information in a controlled way. Unlike traditional plant breeding, which involves the crossing of perhaps thousands of genes to a hybrid plant, genetic engineering involves one or a few desirable genes, so that the offspring does not have the undesirable trait or traits. Transgenic cereals thus contain foreign DNA from one organism which has been moved into another. Agronomic traits commonly sought are resistance to disease and insect pests and tolerance to environmental stress factors and herbicides.

Genetically modified (GM) foods allow food producers to alter certain characteristics of a food crop by introducing genes from another source. The gene introduced may be from another species of plant. Gene technology accelerates the modification process by identifying those genes that produce these special traits and their introduction into the crop.

Wheat and barley crops are now of sufficient magnitude to encourage research and development of transgenic techniques and their commercialisation, as has happened with corn. Endosperms of transgenic wheat plants have been used to demonstrate that a gene encoding a desirable HMW-glutenin subunit could be stably incorporated into the wheat genome by genetic transformation (Blechl and Anderson, 1995).

3.7.3 Functional cereal foods – nutritional research for prevention of disease

It seems as though it will be possible to genetically modify wheat to increase its yields of antioxidants, capable of reducing the amounts of free-radicals in the body. Whole grains contain polyphenolic compounds which are strong antioxidants due to their ability to scavenge free radicals. Increased consumption of phenolics has been correlated with a reduced risk for certain types of cancer (Bravo, 1998).

Wheat samples with the ability to kill human colon carcinoma cells in culture, presumably due to their high content of polyphenolic acids such as ferulic acid and caffeic acid, were able to prevent tumour formation in Min mice (Drankhan *et al.*, 2002). Madison, a soft white winter cultivar, reduced tumour formation by about 60% after eight weeks. As all the wheats tested had equal fibre content, the differences were attributed to the high levels of polyphenolic acids in this cultivar. The same principles should also be able to be applied to nutrients and other desirable factors in the wheat, such as vitamins (Table 3.5).

3.7.4 Dietary foods

There is a strong need for the public, especially the young, to be more informed about the food they eat and how this affects their health. It should include knowledge of medical problems such as obesity, diabetes, heart disease, osteoporosis and cancer. Otherwise, we will be driven by market forces to eat food lacking in nutritional value, commonly referred to as 'junk food'. Added to this are the various 'crash diets' designed to help one lose weight quickly.

Dietary foods, on the other hand, are those that cater for specialised health needs. Low-fat dairy foods are of great assistance in helping to reduce obesity and blood cholesterol levels. We need to look at how cereals can be modified in a similar way to dairy products in order to bring about health benefits. For those who suffer from coeliac disease, the range of dietary food labelled as 'gluten-free' is expected to widen in the future.

3.7.5 Improved starch quality

Starch quality may not be so important as protein quality for conventional bread making, but it is of great importance for the manufacture of Asian noodles (Miskelly and Moss, 1985). Asian noodles require firm texture after cooling, which was found to be highly correlated with protein content (Baik *et al.*, 1994). Italian-style pasta is now so popular that durum wheat will demand more detailed quality investigations. Ratios of amylose to amylopectin in the various cultivars and transgenic varieties will become increasingly important in future developments in these areas. 'Waxy' wheat, which gives pastes of soft texture, is one such example of a recent development.

Food biotechnology must become an important subject to take at university for all food scientists and agricultural scientists. The public will demand

Table 3.5 Important vitamins, recommended daily intake (RDI) and levels in hard red wheat

Vitamin	Function	RDI (mg)	Content in hard red wheat * (mg/100 g)	Deficiency symptoms
A (A_1 is called retinol)	Vision, skin, growth immunity	0.8	Very low	Poor night vision, skin problems, frequent infections
B_1 (thiamine)	Brain function, metabolism of carbohydrates	1.4	0.6	Poor concentration, irritability
B_2 (riboflavin)	Metabolises fats, proteins and carbohydrates	1.6	0.1	Skin problems
B_3 (niacin)	Metabolises carbohydrates, fats, proteins	10	7.4	Poor growth (especially children)
B_5 (pantothenic acid)	Part of coenzyme A required for acylation	5	0.7	Neuromuscular problem
B_6 (pyridoxine)	Antidepressant, hormone balance	2.0	0.4	Depression, irritability
B_{12} (cyano cobalamin)	Haematopoiesis, maintenance of myelin	0.002	Very low	anaemia, debility
C (ascorbic acid)	Immunity, healthy skin, bones, teeth	60	Very low	Frequent infections bleeding gums
D (D_3 is cholecalciferol)	Regulates Ca, Mg, P for healthy bones, heart and nervous system	0.005	Very low	Joint pains, muscle cramps
E (α-tocopherol)	Protects cell membranes, helps healing and promotes fertility	10	5.8 (total tocopherols and tocotrienols)	Lack of muscle tone, infertility
K	Blood clotting	0.07–0.14	Very low	Prolonged bleeding of wounds
Folic acid	Co-factor with vitamin B_{12} for production of blood cells	0.2	0.04	Neural tube defects in women during pregnancy

Adapted from Toepfer *et al.* (1972).

knowledge of genetically modified foods, so scientists will need to play a vital role in public education in this field.

Marketing of wheat in the future will need to be much more comprehensive, taking genetics, microbiology, DNA-based assays and nutritional information into account to ensure that what we eat is the best product that can be produced.

3.8 Sources of further information and advice

In these times of the information explosion, it is not difficult to obtain information on wheat and its processing. Wheat features prominently on web sites of agricultural and scientific institutions. Various bibliographic databases are available on the WWW, e.g. 'Uncover' (University of Colorado), 'Chemical Abstracts' and 'Current Contents' and so on.

Collections of mathematical data, mass spectrometry or infra-red data are available on the non-bibliographic databases. Home pages of various institutions are available through Netscape, Internet Explorer and others. A search engine such as AltaVista provides the necessary links. Key word combinations such as 'wheat' and 'proteins', together with another key word will provide the researcher with a number of 'hits' on the desired topic.

Many institutions around the world are renowned for their traditions in cereal chemistry through the publications of their experts. Their contribution to our knowledge about wheat and the breadmaking in particular demands a close look at their publications and reports. They include:

- Campden and Chorleywood Food Research Association, Chipping Campden, UK
- Bread Research Institute, North Ryde, Sydney, Australia
- CSIRO (Australia) – Division of Food Science – Wheat Research Unit
- Kansas State University, USA, Dept of Agriculture
- Ohio State University, USA, Dept of Agriculture
- University of Manchester Institute for Science and Technology (UMIST), UK
- USA Departments of Agriculture, e.g. at Peoria (Illinois) and Albany (California)
- VTT Biotechnology, Espoo, Finland

Journals worth perusing for carbohydrate and protein research are:

- *Carbohydrate Research*
- *Journal of Chromatography*
- *Lipid Research*
- *Makromolekular Chemie*
- *Methods in Carbohydrate Chemistry*
- *Protein Science*
- *Starch/Staerke*

Examples of journals meeting the needs of technical personnel with an interest in the science of cereals, their applications and processing, and nutrition are:

- *Baker's Digest*
- *British Journal of Nutrition*
- *Cereal Chemistry*
- *Cereal Foods World*
- *Food Technology*
- *International Milling*
- *Journal of Cereal Science*
- *Journal of the Science of Food and Agriculture*

Some of the following books are for the technical minded and scientists who are keen to link the chemistry of wheat and other cereals with the way in which they can be utilised by modern technology:

- *Food Chemistry*, 2nd edn, ed. Fennema O, 1985 (Marcel Dekker Inc., New York)
- *Gluten Proteins*, eds Bushuk W and Tkachuk R, 1990 (Amer. Ass. Cereal Chemists)
- *Handbook of Cereal Science and Technology*, 2nd edn, eds Lorenz KJ and Kulp IG, 2000 (Marcel Dekker Inc., New York)
- *Starch: Chemistry and Technology*, 2nd edn, Whistler RL, Miller JN and Paschall EF, 1984 (Academic Press, New York)
- *Wheat – Chemistry and Technology*, Vols 1 & 2, ed. Pomeranz V, 1988 (Amer. Ass. Cereal Chemists, St. Paul, MN)
- *Wheat – Chemistry and Utilization*, Cornell HJ and Hoveling AW, 1998 (CRC Press, Cleveland, OH)

3.9 References

ANDERSON CM, FRAZER AC, FRENCH JM, GERRARD JW, SAMMONS HG and SMELLIE JM (1952) Coeliac disease: gastrointestinal studies and the effect of dietary wheat flour, *Lancet* **i**, 836–842.

BAIK BK, CZUCHAJOWSKA Z and POMERANZ Y (1994) Role and contribution of starch and protein contents and quality to texture profile analysis of oriental noodles, *Cereal Chem*, **71**, 315–320.

BALDO BA and WRIGLEY CW (1984) Allergies to cereal in, *Advances in Cereal Science and Technology*, Vol. 6, Pomeranz Y ed, St. Paul, American Association of Cereal Chemists, 289–356.

BANKS W and GREENWOOD CT (1968) The hydrodynamic behaviour of native amylose in good solvents, *Carbohydrate Res*, **7**, 414–420.

BANKS W and GREENWOOD CT. (1975) *Starch and its components*, Edinburgh, University Press, 141–152.

BIETZ JA and SCHMALZRIED E (1995) Capillary electrophoresis of wheat gliadin, *Lebensm-*

Wiss Technol., **28**, 174–184.

BIETZ JA and WALL JS (1972) Wheat gluten subunits: molecular weight determined by sodium dodecyl sulphate polyacrylamide gel electrophoresis, *Cereal Chem*, **49**, 416–430.

BLECHL AE and ANDERSON OD (1995) Wheat structure, biochemistry and functionality, in *Wheat Structure, Biochemistry and Functionality*, ed Schofield JD, Cambridge, Royal Soc. Chem., 206–210.

BOLLECKER SSJ, KAISER K-P, KOEHLER P, WIESER H and SCHOFIELD JD (1998) Re-examination of the glycosylation of high M_r subunits of wheat glutenin, *J Agric Food Chem*, **46,** 481–482.

BRAVO L (1998) Polyphenols: chemistry, dietary sources, metabolism and nutritional significance, *Nutr Res.*, **56**, 317–333.

CALLEWAERT DM and GENYEA J (1980) *Fundamentals of Organic and Biological Chemistry*, New York, Worth Publishers Inc.

CHEETHAM NWH and TAO L. (1997) Amylose conformational transitions in binary DMSO/water mixtures, *Starch/Staerke*, **49**, 407–415.

CONN EE and STUMPF PK (1976) *Outlines of Biochemistry*, 4th edn, New York, John Wiley and Sons Inc.

CORNELL HJ (1990) Mucosal digestion studies of whole gliadin fractions in coeliac disease, *Ann Clin Biochem*, **27**, 44–49.

CORNELL HJ (1998) Partial *in vitro* digestion of active gliadin-related peptides in coeliac disease, *J Prot Chem*, **17**, 739–744.

CORNELL HJ and HOVELING AW (1998) *Wheat – Chemistry and Utilization*, Cleveland, CRC Press.

CORNELL HJ and MOTHES T (1993) The activity of wheat gliadin peptides in *in vitro* assays for coeliac disease, *Biochem Biophys Acta*, **1181**, 169–173.

CORNELL HJ and TOWNLEY RRW (1973) Investigation of possible peptidase deficiency in coeliac disease, *Clin Chim Acta*, **43**, 113–125.

CORNELL HJ and WILLS-JOHNSON G (2001) Structure-activity relationships in coeliac-toxic peptides, *Amino Acids*, **21**, 243–253.

CORNELL HJ, HOVELING AW, CHRYSS A and ROGERS M (1994a) Particle size distribution in wheat starch and its importance in processing, *Starch/Staerke,* **46**, 203–207.

CORNELL HJ, SKERRITT JH, PUY R and JAVADPOUR M (1994b) Studies of *in vitro* γ-interferon production in coeliac disease as a response to gliadin peptides, *Biochem Biophys Acta*, **1226**, 126–130.

CORNELL HJ, MCGRANE SJ and RIX CJ (1999) A novel and rapid method for the partial fractionation of starch using 1-butanol in the presence of thiocyanate, *Starch/Staerke*, **51**, 335–341.

CORNELL HJ, RIX CJ and MCGRANE SJ (2002a) Viscometric properties of solutions of amylose and amylopectin in aqueous potassium thiocyanate, *Starch/Staerke*, **54**, 517–526.

CORNELL HJ, MCLACHLAN A and CULLIS PG (2002b) Extraction of cereal prolamins and their toxicity in coeliac disease, *J Biochem Mol Biol Biophys*, **6**, 151–158.

DE RITIS G, AURICCHIO S, JONES HW, LEW EJ-L, BERNARDIN JE and KASARDA DD (1988) *In vitro* (organ culture) studies of the toxicity of specific A-gliadin peptides in coeliac disease, *Gastroenterology*, **94**, 41–49.

DICKE WK (1950) Coeliakie, MD Thesis, Utrecht, Netherlands.

DOHAN FC (1988) Genetic hypothesis of idiopathetic schizophrenia: its exorphin connection *Schizophr Bull.*, **14**, 489–494.

DRANKHAN K, CARTER J, MADL R, KLOPFENSTEIN C, PADULA F, LU Y, WARREN T, SCHMITZ N and TAKEMOTO DJ (2002) *Anti-cancer Activity of Wheats with High Polyphenolic Content*, Publication No 02-461-J, Kansas Agricultural Experiment Station.

EGGUM BO (1977) Nutritional aspects of cereal proteins, in *Genetic Diversity in Plants*, Muhammed A, Aksel R and von Borstel eds, New York, Plenum Press, 349–369.

EVERS AD and BECHTEL BD (1988) Microscopic structure of the wheat grain, in *Wheat: Chemistry and Technology* 3rd edn, Vol 1, Pomeranz Y ed., St Paul, MN, AACC, 17–95.

EWART JAD (1967) Amino acid analysis of glutenins and gliadins, *J Sci Fd Agric*, **18** 111–116.

FOLCH J, LEES M and SLOANE-STANLEY GH (1957) A simple method for the isolation and purification of total lipids from animal tissue, *J Biol Chem*, **226,** 497–509.

FRAZER AC, FLETCHER RF, ROSS CAC, SHAW B, SAMMONS HG and SCHNEIDER R (1959) Gluten-induced enteropathy, *Lancet*, **2**, 252–261.

FRAZIER PJ, DANIELS NWR and RUSSEL-EGGIT PW (1981) Lipid-protein interactions during dough development, *J Sci Fd Agric,* **32**, 877–897.

FRENCH D (1984) Organization of starch granules, in *Starch: Chemistry and Technology* 2nd edn, Whistler RL, Be Miller JN and Paschall EF eds, New York, Academic Press, 183–247.

FULCHER RG, O'BRIEN TP and SIMMONDS DH (1972) Localization of arginine-rich proteins in mature seeds of some members of the Gramineae, *Aust J Biol Sci*, **25**, 487–497.

GHIASI K, HOSENEY RC and VARRIANO-MARSTON E (1982) Gelatinization of wheat starch III. Comparison by differential scanning calorimetry and light microscopy, *Cereal Chem*, **59**, 258–262.

GRAYBOSCH R, PETERSON CJ. MOORE KJ, STEARNS M and GRANT DL (1993) Comparative effects of wheat flour protein, lipid and pentosan composition in relation to baking and milling quality, *Cereal Chem,* **70**, 95–101.

GREENBLATT GA, BETTGE AD and MORRIS CF (1995) Relationship between endosperm texture and the occurrence of friabilin and bound polar lipids in wheat starch, *Cereal Chem*, **72,** 172–176.

GUPTA RB, BEKES F and WRIGLEY CW (1991) Prediction of physical dough properties from glutenin subunit composition in bread wheats: correlation studies, *Cereal Chem.,* **68,** 328–333.

HANCOCK WS, BISHOP CA, PRESTIDGE RL, HARDING DRK and HEARN MTW (1978) High pressure liquid chromatography of peptides II, *J Chromatogr,* **151**, 391–398.

HARGIN KD and MORRISON WR (1980) The distribution of acyl lipids in the germ, aleurone starch and non-starch endosperm of four wheat varieties, *J Sci Fd Agric*, **31**, 877–888.

HIGGINS JM (2002) The role of carbohydrate hydrolases in breadbaking, PhD Thesis, Victoria University, Australia.

JONES BL and MARK AS (1977) Amino acid sequence of the α-purothionins of hexaploid wheat, *Cereal Chem*, **54**, 511–523.

JONES RW, TAYLOR NW and SENTI FR (1959) Electrophoresis and fractionation of wheat gluten, *Arch Biochem Biophys,* **84**, 363–376.

KASARDA DD, OKITA TW, BERNARDIN JE, BAECKER PA, NIMMO CC, LEW EJL, DIETLER MD and GREENE FC (1984) Nucleic acid (cDNA) and amino acid sequences of α-type gliadins from wheat (*Triticum aestivum*), *Proc Natl Acad Sci* USA, **81**, 4712–4716.

KAUFMANN R (1995) Matrix-assisted laser desorption ionization (MALDI) mass spectrometry: a novel analytical tool in molecular biology, *J. Biotechnology*, **41**,

155–175.

KITAMURA S, KOBAYASHI K, TANAHASHI H, OZAKI T and KUGA T (1989) On the Mark–Houwink–Sakurada equation for amylose in aqueous solvents, *Denpur Kagaku*, **36**, 257–264.

KOCNA P, MOTHES T, KRCHNAK V and FRIC P (1991) Relationship between gliadin peptide structure and their effects on the fetal chick duodenum, *Z Lebenson Unters Forsch*, **192**, 116–119.

LINEBACK DR and RASPER VF (1988) Wheat carbohydrates, in *Wheat: Chemistry and Technology*, Pomeranz Y ed., St Paul, MN, Amer Assoc Cereal Chemists, 377–372.

MCGRANE SJ, CORNELL HJ and RIX CJ (1998) A simple and rapid colorimetric method for the determination of amylose in starch products, *Starch/Staerke*, **50**, 158–163.

MCKILLICAN ME, SIMS RPA, JOHNSTON FB and MES JC (1968) Low-temperature anaerobic extraction of free and bound lipid from wheat flour, *Cereal Chem*, **45**, 512–519.

MCLACHLAN A, GULLIS P and CORNELL HJ (2002) The use of extended amino acid motifs for focussing on toxic peptides in coeliac disease, *J Biochem Mol Biol Biophys*, **6**, 319–324.

MACMURRAY TA and MORRISON WR (1970) Composition of wheat-flour lipids, *J Sci Fd Agric*, 21 520–528.

MAILLARD LC (1912) Action des acides amines sur les sucres; formation des melanoidines par voie methodique, *Compt Rend*, **154**, 66–68.

MENDEZ E, CAMAFECTA E, SAN SEBASTIAN J, VALLE I, SOLIS J, MAYER-POSNER FJ, SUCKAU D, MARFISI C and SORIANO F (1995) Direct identification of wheat gliadins and related cereal prolamins by MALDI-TOF mass spectrometry, *J Mass Spect Rapid Comm Mass Spect*, Supplement S123–S128.

MISIR R, LAARVELD B and BLAIR R (1985) Evaluation of a rapid method for preparation of fatty acid methyl esters for analysis by gas-liquid chromatography, *J Chromatogr*, **331**, 141–148.

MISKELLY DM and MOSS HJ (1985) Flour quality requirements for Chinese noodle manufacture, *J Cereal Sci*, **3**, 379–387.

MORRISON WR (1978) Wheat-lipid composition, *Cereal Chem*, 55 548–558.

MORRISON WR and SMITH LM (1964) Preparation of fatty acid methyl esters and dimethylacetals from lipids with boron fluoride-methanol, *J Lipid Res*, **5**, 600–608.

MOSSE J (1968) Les proteines des cereales, *Actual Sci Ind*, **1305**, 47–81.

SCHOCH TJ (1942) Fractionation of starch by selective precipitation with butanol, *J Amer Chem Soc*, **64**, 2957–2960.

SEILMEIER W, WIESER H and BELITZ H-D (1987) High performance liquid chromatography of reduced glutenin: amino acid composition of fractions and components, *Z-Lebensm Unters Forsch*, **185**, 487–489.

SHEWRY PR and MIFLIN BJ (1985) Seed storage proteins of economically important cereals, in *Advances in Cereal Science and Technology*, Vol 7, Pomeranz Y ed, St Paul MN, Am Assn Cereal Chem, 1–78.

SINGH MM and KAY SR (1976) Wheat gluten as a pathogenic factor in schizophrenia, *Science,* **191**, 401–402.

SINGH NK, DONOVAN GR, BATEY IL and MACRITCHIE F (1990a) Use of sonication and size exclusion high-performance liquid chromatography in the study of wheat flour proteins. I., *Cereal Chem.*, **67**, 150–161.

SINGH NK, DONOVAN GR and MACRITCHIE F (1990b) Use of sonication and size exclusion high-performance liquid chromatography in the study of wheat and flour proteins

II, *Cereal Chem.*, **67**, 161–170.

STROBER W (1976) Gluten-sensitive enteropathy, *Clin Gastroenterol*, **5**, 429–452.

STURGESS R, DAY P, ELLIS HJ, LUNDIN KEA, GJERTSEN HA, KONTAKOV M and CICLITIRA PJ (1994) Wheat peptide challenge in coeliac disease, *Lancet*, **343**, 758–761.

THIBAULT P and DOVICHI NN (1998) General instrumentation and detection systems including mass spectrometric interfaces, in *Capillary Electrophoresis* 2nd edn, Camilleri P ed, Baton Rouge, CRC Press, 23–89.

TOEPFER EW, POLANSKY MM, EHEART JF, GLOVER HT, MORRIS ER, HEPBURN FN and QUACKENBUSH FW (1972) Nutrient composition of selected wheats and wheat products. XI Summary, *Cereal Chem.*, **49**, 173–186.

VAN DE KAMER JH, WEIJERS HA and DICKE WK (1953) An investigation into the injurious constituents of wheat in connection with their action on patients with coeliac disease, *Acta Paediatr*, **42**, 223–231.

VAN OORT M, HENNINK H, SCHENKELS P and LAANE C (1995) Peroxidases in breadmaking, in *Wheat Structure, Biochemistry and Functionality*, ed Schofield JD, Cambridge, Royal Soc Chem Food Chem Group, 350–360.

WEBER K and OSBORN M (1969) The reliability of molecular weight determinations by dodecyl sulfate-polyacrylamide gel electrophoresis, *J Biol Chem*, **244**, 4406–4412.

WHATTAM J and CORNELL HJ (1991) Distribution and composition of the lipids in starch fractions from wheat flour, *Starch/Staerke*, **43**, 152–156.

WHISTLER RL and DANIEL JR (1984) Molecular structure of starch, in *Starch: Chemistry and Technology*, 2nd edn, Whistler RL, Be Miller JN and Paschall EF, eds, New York, Academic Press, 153–182.

WIESER H, SEILMEIER W and BELITZ H-D (1990) Characterisation of ethanol-extractable reduced sub-units of glutenin separated by reversed phase high performance liquid chromatography, *J Cereal Sci*, **12**, 63–71.

WIESER H, SEILMEIER W and BELITZ H-D (1991) Klassifizierung der Protein Komponenten des Weizenklebers, *Getreide, Mehl, Brot,* **45**, 35–38.

WIESER H, SEILMEIER W and BELITZ H-D (1994) Quantitative determination of gliadin subgroups from different wheat cultivars, *J Cereal Sci*, **19**, 149–153.

4

Assessing grain quality

C. Wrigley and I. Batey, Wheat CRC and Food Science Australia

4.1 Introduction: the interaction of genotype with the environment

Basic to the success of breadmaking is the suitability of the raw materials, primarily, the wheat that is milled into flour for baking. In this context, 'grain quality' thus means that the grain used must be suitable for the efficient production of flour with the attributes that are appropriate to the specific baking process. A list of essential considerations is provided in Fig. 4.1. Many of these aspects of grain quality can be summarised by consideration of the variety involved. This is because the wheat breeder has already considered the needs of the miller and baker by 'building in' the necessary qualities; or more correctly, by building into the variety (the 'genotype') the genetic potential for the appropriate qualities.

However, that is where the breeder's influence may appear to cease. Beyond the breeding process, there is the considerable contribution of the wheat grower's management and the conditions of growth environment, plus the further environmental contributions of the conditions of harvesting, transport and storage, even before the miller contributes by turning the grain into flour. Grain quality at the mill is thus the result of the interaction of genotype with environment (G $\times$ E), namely, all the environmental conditions from sowing to delivery to the mill.

The interaction of genotype with environment is potentially different for each aspect of grain quality. Accordingly, the placement of each quality attribute in Fig. 4.1 is intended to provide a graphical indication of the relative contributions of these two major factors, namely:

- Genotype – the 'responsibility of the breeder'.
- Environment – the 'responsibility of the grower'.

Influence on each grain-quality attribute

Genotype .... Environment

Variety
Starch properties
Milling quality
Grain hardness
Protein content
Protein quality
Dough characteristics
Pre-harvest sprouting
Defects
Test weight
Moisture content
Pesticide residues
Contaminants

Fig. 4.1 Grain-quality attributes, listed in position from left to right according to the relative influences of genotype and growth environment on each attribute.

For example, the presence of weed seeds or stones in the harvested grain and the possible use of grain-protectant chemicals are the sole responsibility of the grower, with no influence from the variety of wheat sown; thus 'contaminants' and 'pesticide residues' appear at the extreme right of Fig. 4.1. On the other hand, 'grain hardness' (towards the left of Fig. 4.1) is primarily determined by the variety.

However, it is not correct to suggest that the breeder has no influence over growth environment. More and more, it is proving possible for breeding to succeed in 'building in' the further attributes of tolerance to the damaging effects of environmental circumstances that would otherwise downgrade the grain. A good example is tolerance to pre-harvest sprouting, which for some varieties means that rain at harvest does not cause a rise in the alpha-amylase activity of the grain.

4.1.1 Protein content and protein quality as examples of the G × E interaction

Grain-protein content and protein quality are further important aspects of grain quality that are the result of G × E interactions. Grain-protein content is often critical in determining the market value of grain. The protein level is partly determined by the nitrogen nutrition of the plant, but in addition, varieties differ in their ability to translocate amino acids to the grain for deposition as storage protein. These differences translate, in turn, to different abilities for producing gluten of appropriate quality for breadmaking.

An example of this scenario is provided in Fig. 4.2, which contrasts two genotypes that differ in their genetic potential for protein content and 'protein

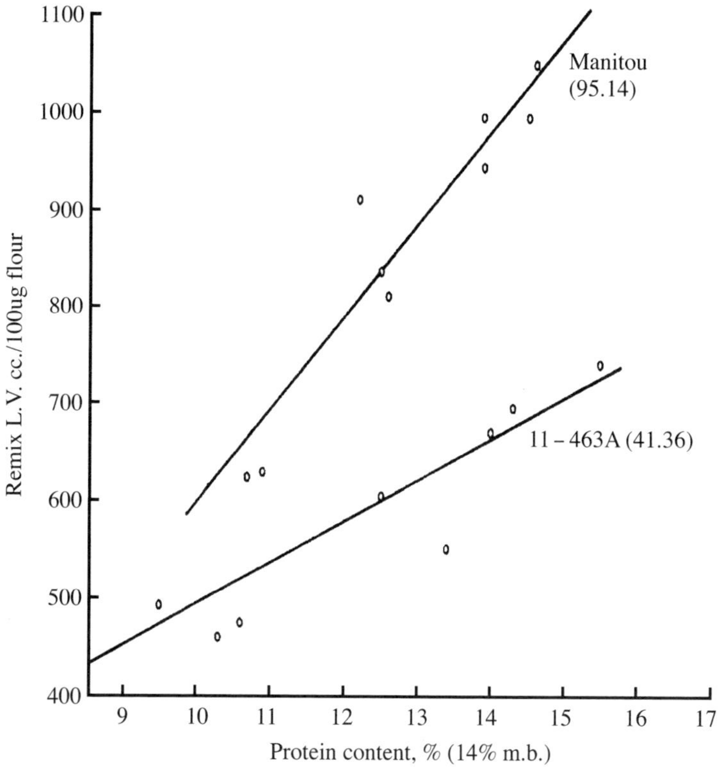

Fig. 4.2 Remix loaf volume results for flour samples of two Canadian genotypes (Manitou and the unnamed line 11-463A), with a range of protein contents. Baking quality is indicated as loaf volume (L.V.) and protein content is expressed on the basis of 14% moisture basis (m.b.). Reproduced with permission from Bushuk *et al.* (1969).

quality'. Results are shown for several samples of the two varieties. As plant nutrition has improved, so grain protein content has increased, providing a range of samples of both genotypes covering protein contents from about 9 to 16%. When samples of equivalent protein content are compared, the genotype Manitou has consistently produced better bread (based on loaf volume), indicating the influence of genetic potential, interacting with the effect of environment, which has largely determined protein content.

4.2 The importance of variety

The choice of variety is of great importance in determining the final quality of the grain at harvest. This is primarily because of the influence of genetic potential on the several aspects of grain quality that appear on the left half of Fig. 4.1. In addition, the choice of variety is vital with respect to many agronomic factors, which in turn affect grain quality.

For example, the maturity of the variety must be considered in relation to the choice of sowing date; an early-maturing variety might be sown later than a late-maturing one, so as to avoid the damaging effects of spring frosts on the flowering process. In some wheat-growing regions, the risk of heat stress in late spring must also be avoided by the selection of sowing date in relation to the expected maturity of the variety. Computer programs have been developed to assist the grower in making such decisions.

Other agronomic considerations include biotic stresses, especially root diseases and leaf pathogens. The avoidance of such stresses is critical to the production of grain with a high test weight of appropriate protein content. The selection of pathogen resistance (or at least tolerance) in a variety is a basis for avoiding such stresses, to which must be added appropriate farm-management practice to minimise the effects of diseases on grain quality.

The interaction of genotype with environment is the crux of these considerations. For the breeder, this interaction must be tested in the late stages of the process of selecting the best lines for commercial release. This process involves growing these lines at a range of sites, and evaluating the results. Inevitably, there will be differences between sites as to which lines perform best for yield and quality, because of the $G \times E$ interactions. The task of interpreting these interactions is difficult for the breeder, and various statistical strategies have been developed for this purpose, e.g. Lukow (1991), Kang and Gauch (1996) and Basford and Cooper (1998). These same considerations continue into the process of deciding what varieties to sow in a specific locality, according to known soil conditions and climate expectations.

Finally, it is important that the seed for sowing is pure, containing only grain of the declared variety, without foreign admixture, and without contaminating seeds that would later cause the harvested grain to be downgraded.

4.2.1 Methods for the identification of variety

Given the importance of variety in wheat growing, there must be certainty about the varietal identity of seed for sowing and of grain that is delivered after harvest. Methods are thus needed for the identification (or at least for the verification) of variety. At the time of registration of a new variety, there is the requirement in many wheat-growing countries for the provision of evidence that the variety is distinct from all others, that it is uniform and that its genotype is stable from one generation to the next (Cooke, 1995a). This process requires the nomination of identification methods whereby these claims are substantiated. These may range from visual inspection to complex laboratory analysis.

Traditionally, varietal identification has involved visual inspection of plants and grain, often with the help of a reference manual listing systematic descriptions of national sets of varieties, e.g. Ferns *et al.* (1975), Jarman (1995) and Anon (2000). Although visual examination offers the great advantages of requiring no equipment, it is subjective, requiring great experience, and the potential for discrimination is poor. For these reasons, varietal identification of

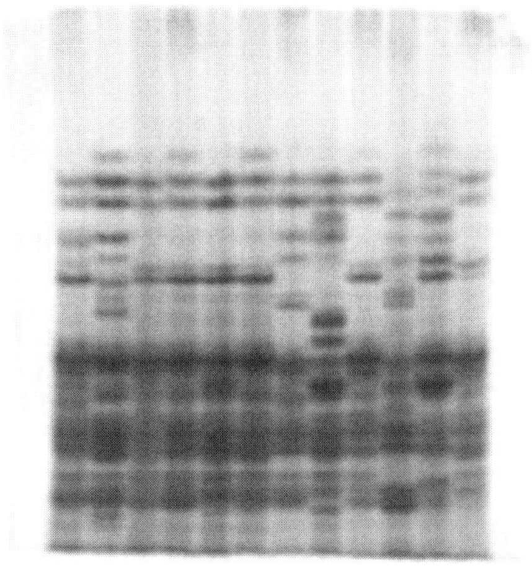

Fig. 4.3 Identification of wheat varieties by gel electrophoresis of the gliadin proteins, extracted from the grain.

grain samples generally requires laboratory analysis, most frequently by the determination of protein composition. Gel electrophoresis of the grain proteins provides a pattern of distinct bands (parallel lines), each one representing a different protein (Fig. 4.3). The arrangement of these bands is specific for the variety, and is largely unaffected by differences in growth conditions. Several variations of the gel electrophoresis method have been devised depending on what class of grain proteins are analysed (Cooke, 1992; Lookhart and Wrigley, 1995). However, the gel electrophoresis methods are labour intensive, and closely related varieties may not be distinguishable.

In recent years, several different methods of protein fractionation have been adapted to varietal identification. These include versions of reversed phase high-performance liquid chromatography and capillary electrophoresis (Wrigley and Bekes, 2002). The equipment for these methods is much more expensive than that needed for gel electrophoresis, but these methods offer the advantages that analysis times are much shorter (overall about an hour), samples may be loaded automatically, and the immediate acquisition of fractionation data facilitates the automatic interpretation of results.

The added difficulties of checking the varietal identity include sampling and statistical evaluation of the results. Clearly, the identification of only a few grains from a large consignment might provide little information about the composition of the whole consignment, unless it could be assumed that the consignment is uniform. It is thus important first to take a subsample that is representative of the whole, using appropriate sampling methods. Secondly, analysis of a ground sample is recommended if protein composition is to be used for identification, thereby causing the range of varieties present to be averaged, if the consignment is a mixture of varieties. Finally, it may be necessary to conduct the identification process on many grains one at a time, followed by statistical analysis to determine the significance of the results (Wrigley and Batey, 1995).

4.3 Environmental factors affecting grain quality

4.3.1 Soil nutrition

A wide range of factors in the growth environment may cause changes in the quality of the harvested grain. Soil nutrition obviously determines the extent to which nutrients are available to the growing plant, and thus to the developing grain. This factor is complemented by any treatment with fertiliser, possibly at sowing, or during the growth of the wheat plants. Of critical importance for grain quality is the availability of nitrogen as a factor that determines the combination of grain yield and grain protein content. If grain-protein content is low (e.g. below 8%), gluten content will also be low, leading to the likelihood that the grain will be unsuitable for breadmaking due to lack of dough strength. On the other hand, grain with a protein content in the lower range may be suitable for the manufacture of biscuits (cookies), if the variety is soft-grained, or for grocery flour.

In addition, sulfur-based fertilisers may be needed, especially in recent decades when pollution controls have reduced the adventitious deposition of sulfur from the atmosphere in the farmlands of industrialised countries. If the level of sulfur in harvested grain is lower than a ratio of nitrogen-to-sulfur of 17 : 1, there is the indication that sulfur is limiting, and it is likely that the properties of the dough made from this grain will be deficient, especially in extensibility (Randall and Wrigley, 1986; Byers *et al.*, 1987; Withers and Sinclair, 1994). This situation also indicates that sulfur-containing fertilisers are needed before wheat is grown again at this site. Furthermore, surveys of grain grown in Britain during the 1980s and 1990s have shown that there has been a gradual fall in the sulfur status of the harvested grain and thus of the soils (Zhao *et al.*, 1995). A range of micronutrients, e.g. phosphorus, potassium and copper, may also be needed to maximise yield, but there is no general agreement that these elements are critical to grain quality.

4.3.2 Drought and water-logging

Lack of water is an over-riding environmental problem for some wheat-growing regions, while in other localities excess water can affect grain yield and quality due to water-logging. Drought is likely to reduce grain yield, while increasing the protein content of the resulting grain. There have been no indications that drought alters the dough properties, apart from the changes that would be expected from the higher protein content (Plaut *et al.*, 1999).

4.3.3 Disease

Plant diseases, such as root nematodes and stem or leaf rust, restrict the ability of the plant to contribute fully to the filling of the developing grains in the head. As a result, smaller-than-expected grains are harvested, resulting in a low test weight. In such cases, this aspect of grain quality will probably fall below the levels specified for milling grades, so that it will be used for lower-value uses such as feed grain.

4.3.4 Frost

Low temperatures at the stages of flowering and soon afterwards can seriously affect the ability of the wheat florets to set seed, or the immature grain may be prevented from developing correctly. As a result, these grains may be small and pinched when harvested, thereby reducing the test weight of the grain and causing it to be rejected from milling grades. However, in such cases, there may also be a large proportion of properly filled, plump grains, so that separation of the pinched grain from the plump grains may be economically justified. In a recent study, this well-filled grain was separated from the shrivelled grain, and it was shown to perform as well as normal grain in milling and baking (Allen *et al.*, 2001). However, the positive outcome of this report may not apply in cases where the severity of the frost is greater (Dexter *et al.*, 1985, 1994). A complementary study demonstrated that the shrivelled grain, separated from the plump grain, had adequate feed value for ruminants (Richardson *et al.*, 2001).

4.3.5 Variations in temperature and in CO_2 levels during grain filling

Several studies have demonstrated that there is a progressive strengthening of dough made from wheat grown at a range of temperatures, as growth conditions warm from 15 °C to 30 °C (Schipper *et al.*, 1986; Randall and Moss, 1990; Uhlen *et al.*, 1998). This may help to explain variations in dough strength through different parts of Europe, for example, where growth temperatures may vary through this range.

At the extreme upper end of the temperature spectrum, there is the risk of high temperatures during grain filling in many wheat-growing regions, namely, a few days with maximum temperatures of over 35 °C. These heat-shock conditions have been shown to cause weakening of the dough in the resulting mature grain (Blumenthal *et al.*, 1993; Ciaffi *et al.*, 1996; Corbellini *et al.*, 1998). The heat-affected grain is generally higher in protein content. Instead of causing greater dough strength, as would normally be expected, the resulting dough is weaker, as shown by the mixing curves in Fig. 4.4 by the shorter time to peak resistance, the lower peak and the rapid breakdown in dough properties following the peak. There are reports that this environmental factor can even cause the complete loss of breadmaking quality.

These dramatic results appear to be a consequence of reduced synthesis of large glutenin polymers, the part of the gluten complex that makes the greatest

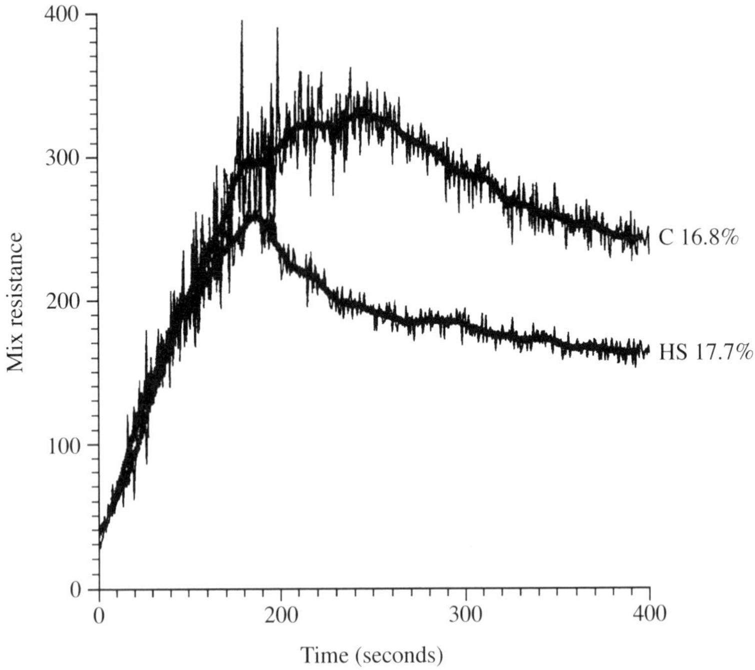

Fig. 4.4 Mixograph curves for dough from grain of the variety Ella that has experienced heat shock during grain filling (indicated by 'HS') and for dough from control grain ('C').

contribution to dough strength. However, some varieties have been identified that are tolerant to the dough-weakening effect of heat shock (Blumenthal *et al.*, 1995). This further example of G × E interaction offers a possible solution to this problem. Heat shock has the added effect of altering the natural proportions of small and large starch granules by reducing the proportion of the small (B-type) granules (Blumenthal *et al.*, 1995), but this alteration does not have significant consequences for breadmaking quality.

The importance of the heat-shock response in wheat has been accentuated by forecasts that the frequency of such heat-shock episodes is likely to increase with the progressive onset of global warming. These increases in temperature are linked to the increasing levels of atmospheric carbon dioxide. Considerable increases in grain yield (6–35%) have been obtained for wheat grown in an atmosphere enriched with CO_2 to double the present level. Of most concern was the reduction in grain-protein content, ranging down to levels (below 8%) at which normal processing would be difficult (Blumenthal *et al.*, 1996). Dough testing of the grain showed that dough properties were reduced (especially extensibility) but interpretation was difficult owing to the low protein content. A dramatic change in grain composition was the considerable increase in the proportion of large (A-type) starch granules.

4.3.6 Rain at harvest

Quality wheat is dry and sound. There is thus a great incentive for the grower to harvest the mature grain under dry conditions, partly to ensure that the moisture content specifications are met. It is also important to avoid the possibility that rain on the standing crop will cause the grain to start the germination process in the heads. In the case of a modestly moist harvest, the grower may choose to harvest the grain and attempt to dry it to the required moisture content by artificial means. Not only is this costly, but it introduces the risk that at least some of the grain will be over-heated while it is still moist, thereby causing damage to the functional properties of the gluten-forming proteins.

The evidence for such damage is difficult to find by visual examination, but some indication is provided by the presence of 'bin-burn' grains, those grains that have had excess heat such that they are actually darkened. On the other hand, this form of damage is readily apparent if the affected grain is milled to flour and tested for dough properties. Procedures have been established to ensure that the drying conditions are 'safe' with respect to the effect on baking quality, as modelled by Bruce (1992).

The extreme scenario of the second problem (sprout damage) is evident as roots and shoots hanging from the head, or evident in the harvested grain. However, there can be significant sprout damage without obvious visible signs that the germination process has commenced. In this case, the evidence is provided by analyses for the increased level of hydrolytic enzymes, particularly alpha-amylase, which is produced by the grain to break down its starch reserves, thus providing energy for the germination process. However, the amylase action causes unacceptably poor-quality bread and sticky crumb in the baking process, as well as causing the production of inferior products for other uses such as noodles and Arabic breads (Edwards *et al.*, 1989).

4.4 Storage and transport

4.4.1 Moisture and temperature

A great advantage of the edible grains is that they can be stored for considerable periods of time without significant loss of quality, as distinct from fresh fruit and vegetables. Nevertheless, to realise this advantage fully, adequate attention must be paid to the conditions of the grain during storage and transport (Desmarchelier and Ghaly, 1993). The grain must be kept dry, to ensure that the moisture level required at harvest is not exceeded. Grain that acquires a higher moisture is much more susceptible to attack by insects, mites and fungi, thereby introducing the added risk that mycotoxins may develop, making the grain unacceptable for human use or for animal feed.

These deleterious effects of moist storage are accentuated by an increase in the temperature of the grain, because the proliferation of many of the threatening organisms is greater as the temperature rises, say, through the range of 20–35 °C. In addition, storage of wheat at higher temperatures (over 30 °C) causes changes in the quality of the dough that is made from the stored grain (Haig *et al.*, 1997). Under these conditions, dough strength increased gradually and extensibility decreased. These changes were attributed to increases in the amount of large glutenin, due to a slow continuation of the process of disulfide-bond formation, which had occurred during the earlier stages of grain filling (Wrigley and Bekes, 1999).

4.4.2 Odour from materials stored with grain

Another hazard to grain quality can occur during storage and transport owing to the risk of a taint or odour being acquired by the grain from other materials nearby. One such source can be the packaging material used, in the case of grain that is stored and transported in bags. For example, hessian bags have been known to cause this problem, but in the cases where this has occurred, it was difficult at first for the source of the taint to be identified.

4.5 Critical quality attributes and their analysis

Wheat-growing countries have grain-receival standards that specify a range of quality attributes, according to which grain received is allocated to a specific quality grade. In other cases, where the grain is delivered directly to the flour mill, a similar set of specifications govern the price payable, or whether the grain will even be received.

As described above, many of these quality attributes are indicated by knowledge of the variety (or the variety mix) that is being delivered, assuming that the quality characteristics of relevant varieties are known. This information relates to those aspects of quality on the left of Fig. 4.1, but it is still necessary to check for the aspects to the right side of this figure.

In addition, it is valuable to know the growth site of the grain consignment, and thus to be aware of the climate that has prevailed during grain filling and harvest. This information about growth environment will assume even greater importance as our knowledge accumulates about how climate affects quality. An example of the value of this awareness is the knowledge that there has been rain at the time and place of harvest. This indicates the need to check for the possibility of incipient sprouting.

Various quality considerations are discussed in this section, although it may not be possible to assess all of them when the grain is delivered to the mill or storage point (grain elevator or silo).

4.5.1 Test weight

Plump, well-filled grain is a basic requirement of grain quality. It is generally measured as a bulk density, specified as kilograms per hectolitre, or pounds per bushel. Test weight is the most efficient means of measuring grain plumpness under the rushed conditions that often restrict the opportunities for thorough testing at the point of grain receival. Alternatively, the measurement of average grain mass is undertaken, generally expressed as thousand kernel weight, requiring the use of an automatic seed counter. These measurements are presumed to indicate likely yield of white flour in milling, on the basis that plump grains provide a maximum of endosperm (flour) and a minimum of non-flour material (Marshall *et al.*, 1986). On the other hand, the presence of shrunken kernels reduces both these estimates of grain quality, thereby providing less endosperm and a reduced yield of flour.

4.5.2 Grain moisture content

Moisture content is also a basic aspect of grain quality that is universally considered. Grain that is too moist, e.g. >15% moisture, is likely to deteriorate on storage and transport, because of the possibility of 'caking' and the increased risk of attack by insects and moulds. The maximum moisture for grain that may be received varies between wheat-growing regions, mainly in the range 12–14.5%. Moisture is generally determined by near infra-red (NIR) analysis. Currently NIR is applied to whole grain, but sometimes NIR equipment requires that grain is ground to wholemeal (Batten, 1998). Alternative instrumental methods of moisture determination are available, but they all require calibration against a set of samples whose moisture has been determined by oven drying.

Under dry and hot conditions of harvest, grain moisture may be as low as 8%. In contrast, in other climates, artificial drying may be required to meet the maximum moisture requirements. Further combinations of circumstances may dictate that grain is harvested as soon as possible during the ripening stage. It is even likely that some loss of yield is involved in delaying harvesting until a very low moisture content is reached. In addition, the grower is faced with the possibility that payment is based on actual mass of grain, not its dry weight, and this consideration warrants the delivery of grain at the upper limit of moisture. The range of factors that relate to these considerations are discussed by Desmarchelier and Ghaly (1993).

Grain that is harvested very moist, e.g. over 17% moisture, may suffer from the very serious condition of the production of mycotoxins, produced by moulds such as *Fusarium* and *Alternaria* (Christensen *et al.*, 1977). Mycotoxins are active at very low concentrations, even at a few parts per billion. They are not always produced in mouldy grain, but the risk is obviously best avoided by keeping grain dry and free from moulds. Immunoassay kits are now available for the determination of various mycotoxins in 'field' situations within minutes, providing the opportunity of checking for this risk 'on-the-spot', in contrast to the need some years ago of reliance on laboratory-based analysis taking some days for results.

4.5.3 Contaminants

As the miller's aim is to maximise the yield of white flour from the grain delivered, the presence of non-grain material ('dockage' or 'screenings') is undesirable because of its ability to reduce flour yield. Accordingly, contaminants of any type provide the buyer with grounds for reducing the price to be paid for a grain consignment. However, contaminants differ in how serious they are in their effects on milling and on further processing. The least serious is material that may be removed relatively easily, such as very small grains, stones, glumes (husks) and the unfilled tips of wheat heads.

More critical may be foreign grains if they interfere with milling or if removal is essential, involving an added processing cost. Oilseeds, for example, pose the threat of disrupting milling equipment with the build-up of oil residues, although contaminants such as canola are relatively easy to separate before milling.

On the other hand, there appears to be no great need to remove other cereal grains, such as barley and oats (up to about 5%), according to studies by Dexter *et al.* (1984). They showed that flour yield decreased by about 0.4% for each addition of 1% of barley added to the base wheat sample. Milling stocks fed well during milling, despite the barley contamination. Likewise, the effects of contamination with cultivated oats (or even wild oats) were similar to that of barley, except that at 5% addition, oats caused difficulty with feed rate near the end of the milling process. At 10% contamination with either two- or six-row barley, loaf volume was reduced by only 5%. On the other hand, oat-containing samples were inferior for loaf volume and appearance, the presence of 1% hulled oats reducing loaf volume by about 2%.

Weed seeds may present problems of greater severity, especially because some are actually toxic, and others are tainting seeds, whose scent may taint a whole consignment of grain. Furthermore, seeds with a dark seed coat are likely to cause dark specks in the flour. The seeds of noxious weeds are restricted for seed wheat, but they are destroyed in the milling process, thereby posing no threat to agriculture after milling.

The presence of insects is a serious case of contamination, whether they are dead or alive, since there is a nil tolerance for insects in many regions. The presence of insect infestation (past or current) is also indicated by the presence of grains that have evidence of being eaten by insects. The identification of the species of insect is also warranted, because they differ in the severity and consequences of infestation. NIR analysis has been used to detect the presence of insect infestation, probably because this spectroscopic method can detect the distinctive presence of insect protein and/or chitin (Ridgway and Chambers, 1996).

In any wheat-growing region, there is likely to be many types of contaminants, with different levels of contamination being permitted for each in quality specifications. It is thus critical for these to be correctly identified and for the level of contamination to be determined. The species of contaminating seeds obviously varies considerably from one region to another. Accordingly, manuals illustrating the range of contaminants are provided for staff training, to inform growers and for use when grain is delivered (Ferns *et al.*, 1978).

4.5.4 Grain defects

There are many defects of the grain that devalue it for market and processing, some being more serious than others. They include grains that are affected by ergot, bunt, black point, smut, bleaching, bin-burn, frost, immature harvesting, mottling, insect-damage and 'bug' damage. Infection by microorganisms detracts directly from grain quality by spoiling its appearance, by tainting it and even by making it toxic to humans and animals.

For example, ergot (*Claviceps purpurea*) infects the flowers of several cereal grains, producing an ergot body in place of the grain. Whole ergot bodies are readily recognisable in grain samples, but obviously not after milling. Ergot from wheat, and contaminating ergot bodies from other grasses, produce toxic alkaloids, which may cause injury in cattle when present at a level as low as 0.05%. On the other hand, some poultry species appear to be much more tolerant to ergot poisoning.

Discoloration of the grain at the embryo end is known as black point, fungal staining or fusarium-affected grain (Dexter and Matsuo, 1982). Black point has been commonly thought to be caused by infection with *Alternaria alternata*, but this has been questioned by Williamson (1997) who failed to find an association between infection and the symptoms.

More serious than black point is infection with bunt, also known as ball smut or stinking smut, caused by *Tilletia caries* or *T. foetida*. This defect involves the replacement of the endosperm of the grain by bunt spores, which have an unpleasant odour, resulting in the tainting of sound grain when the bunt-affected grain is admixed. Bunt contamination is also detectable by the presence of black bunt spores caught in the brush hairs of sound grains.

Bug damage may affect wheat grown in much of southern Europe and Russia (Paulian and Popov, 1980). If immature grain is attacked in the field by insect species of *Heteropterous*, there is damage to the mature grain. Affected grain is detectable visually by the presence of the puncture mark where the insect has penetrated the bran layer of the grain. The dough made from this grain is very weak, owing to the action of a protease presumably injected by the insect. Because of the severe results on dough formation and baking quality, bug-damaged grain is severely downgraded, generally warranting only feed grade prices.

Bug damage is also a problem for the wheat industry in New Zealand. There are differences in the genus and species of insect causing the damage in various countries where there is the bug-damage problem. In New Zealand, the causal insect is *Nysius huttoni* (Every, 1993), whereas in southern Europe, the insect is *Eurygaster* spp. and/or *Aelia* spp. Differences have been reported for the specificity of the proteases associated with the various species of insect, but the net effects of the damage are similar. Apart from visual inspection, there is no diagnostic test to identify bug-damaged grain at harvest. However, detection is important, so as to avoid the admixture of sound grain with damaged grain. Because of the involvement of enzymic action, the effects of mixing bug-damaged grain with sound grain are disproportionate to the proportions of samples mixed.

Other defects of wheat may not involve the action of other organisms. Dry-green grain, for example, is a defect that is detectable by the colour of the grain. The presence of such grains indicates the harvesting of a crop of mixed maturity, or the premature harvest of an immature crop, followed by artificial drying to achieve maximum moisture requirements. The processing problem likely to result from dry-green grains is a raised level of amylase activity.

The defects of mottling and bleaching of grain are visible as patches or overall opaqueness of the grain, probably due to the effects of rain at harvest on grain of hard varieties that would otherwise be uniformly vitreous. These defects detract from the appearance of the grain, and thus from its market value, but they would not generally be considered to be detrimental to processing quality.

Most of the above defects involve visual identification by grain-receival staff at the point of grain delivery, usually aided by manuals with illustrations of the defects. This system has operated for as long as there has been attention to quality issues. In an effort to remove the subjective nature of visual inspection, instrumental methods have been developed, based on image analysis. For example, a system of colour machine vision has been developed by Luo *et al.* (1999) to provide objective identification of a range of grain defects. Progressively, such systems may replace the need for operator inspection, especially in cases of dispute, but there will always be a place for visual inspection as the method for use on-the-spot.

4.5.5 Pre-harvest sprouting

Sprouting of grain due to rain at harvest is visually evident if it has progressed far enough for the germ to be swollen and split, but it is still possible for incipient sprouting to have progressed far enough for the grain to be inadmissible for milling grades and for the grain to cause difficulties in processing, but without visible signs of sprouting being apparent. The traditional method for assessment of sprout damage is the Falling Number method, which involves the heating of wholemeal sample with water in a long precision-bore tube, followed by measuring the time for the stirrer to fall through the mass of gelatinised starch. The fall rate depends on the extent to which the alpha-amylase of sprouting has hydrolysed the starch of the wheat sample (Anon, 1983).

Alternative equipment, better suited to the rough conditions of grain receival, is the Rapid ViscoAnalyser (RVA), which produces results correlated to Falling Number. The RVA analysis takes only a few minutes using the distinct principle of stirring the heated sample and measuring changes in the energy required for the stirring action (Wrigley *et al.*, 1996). In addition, the RVA provides

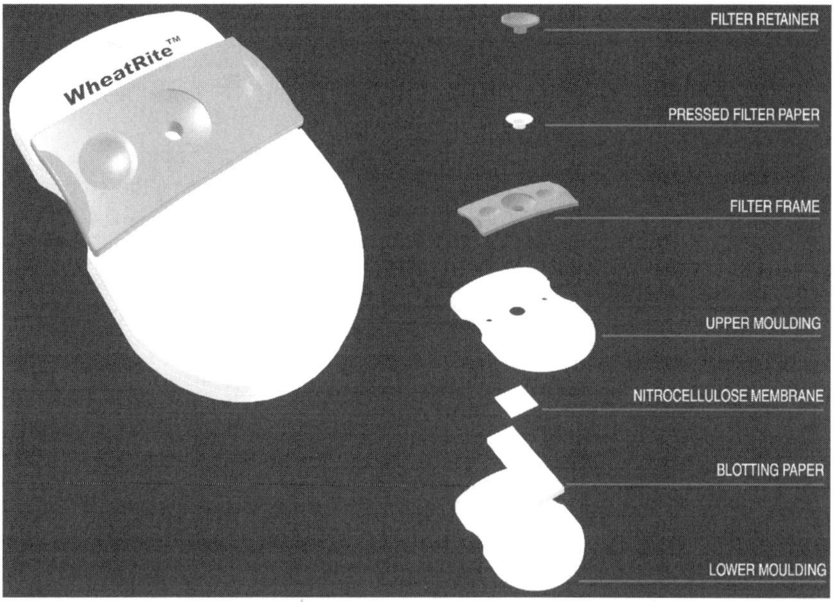

Fig. 4.5 Analysis of sprout damage using the WheatRite test kit. A small amount of grain extract is placed into the central reaction chamber, plus a small volume of reagent (gold-labelled antibody to alpha-amylase). Snapping the coloured bar down onto the body of the kit brings the mixture into contact with the nitrocellulose membrane, which is imprinted with the second antibody. Within a few minutes of starting the test, the result is displayed on the nitrocellulose membrane as a line, whose intensity indicates the degree of sprout damage. This can be read visually (with reference to a colour card) or with the small ReadRite scanner.

indications of the pasting properties of the starch in the grain sample, a property that may be indicative of the digestibility of the starch content of the grain.

Immunoassay is a further option, with the advantage that it requires no expensive equipment – only a test card and the means of roughly grinding the grain sample. The WheatRite test card (Fig. 4.5) provides rapid analysis of amylase, correlated to Falling Number using immunochromatography with an antibody to the amylase enzyme (Skerritt and Heywood, 2000). Within about five minutes of placing two drops of an extract of the sample onto the card, a test line appears, the intensity of which indicates the degree of sprout damage. This can be judged by eye, or using a small scanner.

4.5.6 Grain hardness

The hardness of the wheat grain is a basic quality attribute that distinguishes grades in international trade, owing to its importance in determining the suitability of the resulting flour for breadmaking (hard grain) *versus* the need for soft grain for a range of specific products, including biscuits (cookies), cakes, pastries and grocery flour. Grain hardness is a varietal characteristic (Fig. 4.1), so this attribute is established if varietal identity is known. Hardness is usually evident visually as a vitreous (horny) texture for the grain, in contrast to the opaque texture of the soft grain. This difference is evident in the whole grain, and it is more obvious in the surface of the cut grain.

Grain hardness is determined in the laboratory as the Particle Size Index (PSI), by grinding grain and sieving under controlled conditions, the proportion that passes through the sieve being the basis of the Index. The endosperm of soft wheats fragments readily, allowing a high proportion of the sample to pass through (and thus a high PSI), whereas hard wheats give a lower PSI. Routinely, NIR analysis is generally used to determine hardness, after calibration with a suitable set of standard samples.

The difference that makes grain hardness important is that, during milling, the starch particles of soft grain fall apart readily, leaving the granules intact. In contrast, when hard grains are milled, fractures are likely to pass right through the starch granules. As a result, 'starch damage' is much greater in flours milled from hard wheats than it is in flour from soft varieties. The attribute of starch damage relates in turn to baking quality, a high level being desirable for breadmaking, to provide ready access of amylase to the starch and also to promote water absorption.

4.5.7 Milling quality

The main aim of flour milling is to maximise the yield of white flour suited to the process for which the flour is required. The determination of milling quality requires tedious test milling in the laboratory, generally needing about one kilogram of grain. Recently, a micro mill has been developed capable of milling very small amounts of grain, but best suited to the testing of about five grams of grain (Salgo *et al.*, 2001). This mill is the result a collaboration between Australian and Hungarian cereal scientists. In theory, therefore, it could be used to assess milling quality rapidly, for example at grain receipt, but in practice its application has mainly been for quality selection in breeding and in research applications.

Significant factors affecting flour yield can be determined with the single-kernel characterisation system (SKCS), equipment designed to determine the mass, thickness, moisture content and hardness of individual grain kernels (Osborne *et al.*, 1997). Grains are picked up individually from a hopper and deposited one at a time onto a weighing arm, from which they drop onto a crescent arm for diameter measurement and crushing, the crush-force profile being recorded by a load cell. Measurement of the conductivity of the crushed grain provides an estimate of moisture content. Because the equipment is capable of analysing over 100 kernels per minute, it can provide a profile of attribute distribution for some hundreds of grains within a few minutes. This approach to quality testing has good potential for the prediction of milling quality (Osborne *et al.*, 1997).

4.5.8 Protein content

As Fig. 4.6 shows, protein content (with hardness) is a basic aspect of grain quality, determining the type of product to which the grain is suited. Traditionally, protein content is based on the total nitrogen content of the grain, this value being multiplied by the factor 5.7 to obtain an estimate of protein content. This assumes that protein is by far the main nitrogen-containing component of the grain.

Nitrogen content is determined by the Kjeldahl digestion method (by which nitrogen is estimated as ammonia) or by the Dumas method (in which nitrogen is determined as the gas). Routinely, protein content is determined by NIR analysis, based on its calibration using a set of relevant samples whose protein content has been determined by either the Kjeldahl or Dumas method.

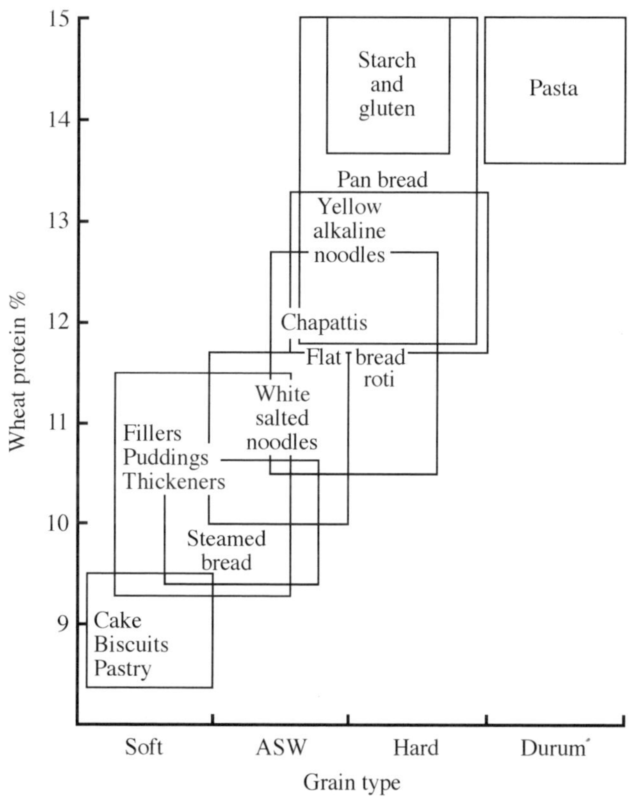

Fig. 4.6 Relationship between the protein content and grain hardness of wheats suited to the many products made from wheat.

4.5.9 Protein quality

Figure 4.2 illustrates the distinction between protein content and protein quality (with respect to baking properties). Both attributes involve contributions from genotype and environment. Basic to the genetic aspect of protein quality is the concept that some of the gluten proteins are more effective than others. In particular, a ranking has been established for the effectiveness of the polypeptides (subunits) that are disulfide cross-linked to form the very large glutenin polymers (Shewry *et al.*, 1992). Of these, it is the high-molecular-weight subunits 5 and 10 that make the greatest contribution to dough properties suited for breadmaking. In addition, other subunits also contribute to various extents. Gel electrophoresis (Fig. 4.7) resolves the subunits so that the composition of a particular sample can be analysed for its glutenin subunits (Ng *et al.*, 1989). The composition of these subunits has been published for about 700 varieties from 20 countries by Cooke (1995b).

4.5.10 Dough characteristics

Most breadmaking processes require a strong extensible dough to provide best bread quality. In contrast, a weaker but extensible dough is required for most types of biscuits (cookies). Provision of grain that will yield the most suitable dough properties is critical to the efficiency of the milling and baking process. Attention to the many aspects of quality that have been described above should ensure that this aim is achieved.

There are several types of equipment for analysing dough properties in the laboratory. One of these is the Mixograph, the mixing curves from which are shown in Fig. 4.4. These show the progressive rise in resistance to mixing as the dough structure forms, reaching a peak, after which there may be a slow or faster

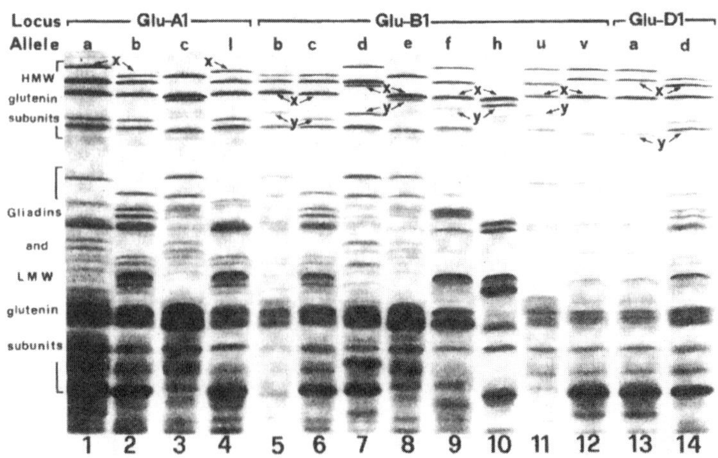

Fig. 4.7 SDS gel electrophoresis of the polypeptides of flour of several Canadian wheat varieties. The high-molecular-weight subunits of glutenin appear at the top of the patterns. Reproduced with permission from Ng *et al.* (1989).

decline in resistance to mixing, depending on whether the dough is stronger or weaker, respectively.

The Extensograph, another type of dough-testing equipment, shows the results of stretching dough, in terms of the resistance to this stretching (as the height of the curve) and the extensibility of the dough (as the length of the curve). In this case, the dough is mixed in a Farinograph (with inter-meshing Z-shaped arms), which provides a different mixing action from the pin-mixing mechanism of the Mixograph.

4.5.11 Starch properties

Starch is by far the major component of the wheat grain but until relatively recent times, it has received less attention than the other components. This may have been because early reports of reconstitution experiments suggested that starch did not play a role in dough and bread quality, despite recognition for a long time that wheat starch from different varieties did have varying properties such as viscosity and granule-size distribution (Rask and Alsberg, 1924; Moss, 1980). More recently, studies have shown that starch is an important, perhaps the most important, contributor to the quality of noodles made from wheat flour, particularly the soft Japanese Udon noodles. At the breeding stage, selection of wheats for this purpose is now done principally on the basis of starch properties (Crosbie and Lambe, 1993). At the receival stage in Australia, suitable cultivars are segregated because of their genetic potential for the correct starch properties.

Starch occurs naturally as discrete particles or granules, which in wheat have a bimodal size distribution. About 80% of the mass but only about 20% of the number of starch granules are larger than 10 μm. There is variation in the actual amount of large granules in different varieties but for most purposes, the size distribution does not matter (Stoddard, 1999). There is also a variation resulting from environmental effects. In the industrial separation of starch and gluten, size distribution is important because a high yield of large starch granules is required. Small granules require more energy to separate from other components and also require more water to wash residual protein from their surface. These factors result in significant amounts being lost in the effluent from the process. Wheat with a higher proportion of large granules would be preferable for this use. On the other hand, there is evidence that a larger proportion of small granules affects dough mixing and baking (P.W. Gras, personal communication). The greater surface area-to-mass ratio of the small granules allows the absorption of more water, which may be beneficial in the breadmaking process. With the genetic potential for wheats with differing proportions of large and small starch granules, there is the possibility that wheats with a more desirable starch-granule size distribution for specific end uses may become available.

At the molecular level of organisation, the starch granule comprises two types of polymers; one an essentially linear molecule of glucose subunits called amylose and the other a highly branched polymer of glucose called amylopectin. Amylose has a molecular weight of less than one million daltons while amylopectin has a molecular weight of many millions. Most wheat starches contain around 20–25% amylose (Rahman *et al.*, 2000). Amylose biosynthesis is controlled by an enzyme called granule-bound starch synthase (GBSS). In wheat, this enzyme is coded for on chromosomes 7A, 7D and 4A. The gene on chromosome 4A has arisen from a natural translocation from chromosome 7B in prehistoric times. There are cultivars and lines in which one or more of these three enzymes are missing. Cultivars with the 4A-GBSS missing have been shown to be particularly suited for the manufacture of Japanese Udon noodles (Zhao *et al.*, 1998). These have a slightly reduced amylose content compared with those lines with all three enzymes (Graybosch *et al.*, 1998). Similar reductions in amylose content are reported for lines with one of the other GBSS enzymes missing (Zhao and Sharp, 1998). When two forms of the enzyme are missing, a further small reduction in the amount of amylose is observed, but all three forms have to be absent to reduce the amylose content to be effectively zero. Lines of this type have been reported by a number of groups (Nakamura *et al.*, 1995; Zhao and Sharp, 1998). The gelatinisation temperature and granule-size distribution of these so-called waxy starches are similar to those of normal wheat starch containing amylose, but pasting viscosity in particular is quite different (Fig. 4.8).

The development of wheat starches with higher amylose content has yet to be reported but there is much work being done to understand the biosynthetic pathways that lead to high-amylose starches. There is considerable potential to improve product quality if starch properties can be manipulated. In products such as bread, for example, the recrystallisation of the amylose and amylopectin

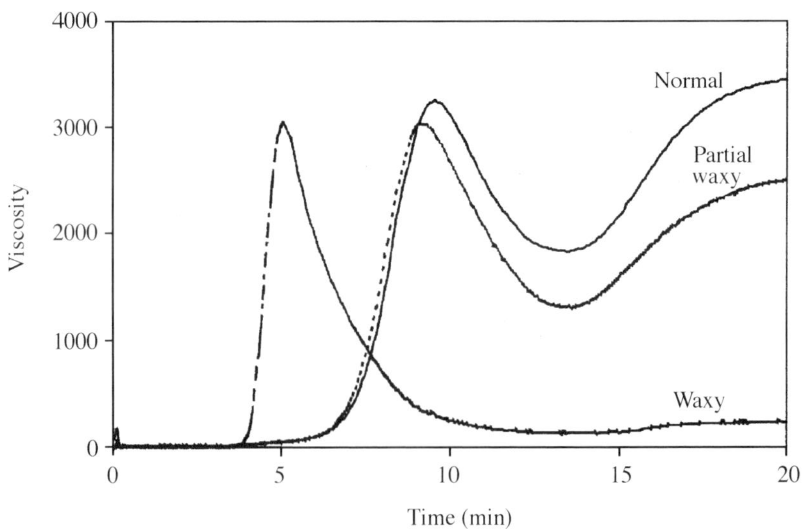

Fig. 4.8 Pasting curves for waxy, partial waxy and normal starches, determined in the Rapid ViscoAnalyser.

components occurs at different rates. This process, also known as retrogradation, affects the staling of bread. There have been reports that the inclusion of different starches from other botanical sources into bread doughs can improve its staling properties (Furcsik, 1992).

4.6 Grain quality bargaining

Inevitably, grain defects reduce the market value of grain, because of the assumption that the defects will cause difficulties during milling and baking. However, some aspects of defective grain may not necessarily reduce processing efficiency significantly. The economic effects of some defects can readily be assessed. For example, the presence of 2% screenings (by weight) at a 5% reduction in price would be worthwhile, provided the cost of cleaning is covered in the price difference. In another case, it may be possible to meet quality specifications by intelligently blending diverse grain samples.

Furthermore, the severity of black point has been questioned by Rees *et al.* (1984). They concluded that 'the use of black-pointed grain is likely to have little effect in breadmaking', based on experiments with 12 samples, having black-point contamination levels from 10% to over 50%. More recent studies by Allen *et al.* (2002) have reinforced this conclusion, demonstrating that the discoloration of the germ end of the grain does not necessarily lead to abnormal flour colour, nor to poorer bread-crumb colour. However, the bran resulting from milling black-point grain is obviously unsuitable for food uses with a high proportion of the bran layers, such as wholemeal products.

Whenever a commodity is sold, any variation in its appearance provides grounds for bargaining and for price reduction. In the sale of grain, appearance can thus be an over-riding consideration, but the actual consequences of such defects need to be considered objectively, taking into account how serious will be the effects of any defects for processing envisaged. Thus, there may be bargains to be obtained when grain is being sold, provided the buyer is convinced that the risk taken is justified.

4.6.1 The overall need for consistency of quality

A common complaint of bakers is that the quality of the flour provided varies in its dough-handling properties, and thus in its overall baking quality. Some bakers would even claim that they can cope with many types in flour quality, but that they want consistency in quality – 'no surprises'. It is thus common for the baker to blame the miller, and for the miller to blame the grain supplied. Many of the sources of variation in grain quality have been described above. To do so comprehensively is not possible, because there is still much to be elucidated. To perform all the quality testing described above is probably impractical, because of the time and cost involved. There is thus the subjective decision required of the grain buyers, in the context of commercial realities, as to how far to go in attempting to fulfil the needs of millers and bakers in providing consistent and acceptable grain quality.

4.7 Future trends

No doubt the future holds more information of great value to grain buyers to assist them in achieving these goals of consistent and acceptable grain quality. The buyers, in turn, must place pressure further 'upstream' to convince growers and breeders to use our increasing knowledge of the interactions of genotype, environment and farm management to achieve quality targets. For the breeder, new opportunities are already becoming available to achieve quality goals, by making better selections of parent lines and of progeny, through such innovations as the application of molecular markers for specific quality attributes. These markers take the form of both DNA sequences and proteins, for which efficient screening procedures are becoming available, with new insights into their application (Galili *et al.*, 2002).

4.8 Sources of further information and advice

Relevant web sites are listed below, as valuable sources of further information.

- American Association of Cereal Chemists — www.aaccnet.org
- AWB Ltd, Melbourne, Australia — www.awb.com.au
- New Zealand Institute of Crop & Food Research — www.crop.cri.nz
- Canadian Grain Commission, Winnipeg — www.cgc.ca
- Stored Grain Research Laboratory, Canberra, Australia — www.sgrl.csiro.au
- CSIRO Plant Industry, Australia — www.pi.csiro.au
- Graingenes — www.wheat.pw.usda.gov/index.shtml
- International Association for Cereal Science and Technology — www.icc.or.at
- Campden & Chorleywood Food Research Association — www.campden.co.uk
- United States Department of Agriculture — www.usda.gov
- BRI Australia Ltd, Sydney — www.bri.com.au

4.9 References

ALLEN HM PUMPA JK and BATTEN GD (2001), 'Effect of frost on the quality of samples of Janz wheat', *Aust J Exp Agric*, **41**, 641–647.

ALLEN HM PLEMING DK and PAN HY (2002), 'Blackpoint and product quality of wheat', in

Wootton M Batey IL and Wrigley CW, *Cereals 2001. Proc. 51st Aust. Cereal Chem. Conference*, Melbourne, Royal Aust Chem Inst, 29–32.

ANON. (1983), *Approved Methods of the American Association of Cereal Chemists*. Eighth Edition, St. Paul, MN, USA, American Association of Cereal Chemists Inc.

ANON. (2000, and regular issues) *Botanical Descriptions of Cereal Varieties*. Cambridge, England, National Institute of Agricultural Botany.

BASFORD KE and COOPER M (1998), 'Genotype X enviroment interactions and some considerations of their implications for wheat breeding in Australia', *Aust J Agric Res*, **49**, 153–174.

BATTEN GD (1998), Plant analysis using near infrared reflectance spectroscopy: the potential and the limitations', *Aust J Exp Agric*, **38**, 697–706.

BLUMENTHAL C BARLOW EWR and WRIGLEY CW (1993), 'Growth environment and wheat quality; the effect of heat stress on dough properties and gluten proteins', *J Cereal Sci*, **18**, 3–21.

BLUMENTHAL C BEKES F GRAS PW BARLOW EWR and WRIGLEY CW (1995), 'Identification of wheat genotypes tolerant to the effects of heat stress on grain quality', *Cereal Chem*, **72**, 539–544.

BLUMENTHAL C RAWSON HM MCKENZIE E GRAS PW BARLOW EWR and WRIGLEY CW (1996), 'Changes in the grain quality of field-grown wheat due to doubling the level of atmospheric carbon dioxide', *Cereal Chem*, **73**, 762–776.

BRUCE DM (1992), 'A model of the effect of heated-air drying on the bread baking quality of wheat', *J Agric Engng Res*, **52**, 53–76.

BUSHUK W BRIGGS KG and SHEBESKI LH (1969), 'Protein quantity and quality as factors in the evaluation of bread wheats', *Can J Plant Sci*, **49**, 113–122.

BYERS M FRANKLIN J and SMITH SJ (1987), 'The nitrogen and sulphur nutrition of wheat and its effect on the composition and baking quality of the grain', *Aspects Appl Biol*, **15**, 337–344.

CHRISTENSEN CM MIROCHA CJ and MERONUCK RA (1977), 'Molds, mycotoxins, and mycotoxicoses', *Cereal Foods World*, **22,** 513–520.

CIAFFI M TOZZI L BORGHI B CORBELLINI M and LAFIANDRA D (1996), 'Effect of heat shock during grain filling on the gluten protein composition of bread wheat', *J Cereal Sci*, **24**, 91–100.

COOKE RJ (1992), 'Electrophoresis handbook: Variety identification', in *Handbook of Variety Testing*, Zurich, Switzerland, International Seed Testing Association, 1–50.

COOKE RJ (1995a), 'Introduction: The reasons for variety identification', in Wrigley CW, *Identification of Food-Grain Varieties*. St. Paul, MN, USA, American Association of Cereal Chemists Inc., 1–17.

COOKE RJ (1995b), 'Allelic variability at the *Glu-1* loci in wheat varieties', *Plant Var Seeds*, **8**, 97–106.

CORBELLINI M MAZZA L CIAFFI M LAFIANDRA D and BORGHI B (1998), 'Effect of heat shock during grain filling on protein composition and technological quality of wheats', *Euphytica*, **100**, 147–154.

CROSBIE GB and LAMBE WJ (1993), 'The application of flour swelling volume test for potential noodle quality to wheat breeding lines affected by sprouting', *J Cereal Sci*, **18,** 267–276.

DESMARCHELIER JM and GHALY T (1993), 'Effects of raising the receival moisture content on the storability of Australian wheat', *Aust J Exp Agric*, **33**, 909–914.

DEXTER JE and MATSUO RR (1982), 'Effect of smudge, blackpoint, mildewed kernels, and ergot on durum wheat quality', *Cereal Chem*, **59**, 63–69.

DEXTER JE PRESTON KR and TIPPLES KH (1984), 'The effects of various levels of barley, wild oats and domestic oats on the milling and baking performance of hard red spring wheat', *Canad J Plant Sci*, **64**, 275–283.

DEXTER JE MARTIN DG PRESTON KR TIPPLES KH and MACGREGOR AW (1985), 'The effects of frost damage on the milling and baking quality of red spring wheat', *Cereal Chem*, **62**, 75–80.

DEXTER JE MARCHYLO BA and MELLISH VJ (1994), 'Effect of frost damage and immaturity on the quality of durum wheat', *Cereal Chem*, **71**, 494–501.

EDWARDS RA ROSS AS MARES DJ ELLISON FW and TOMLINSON JD (1989), 'Enzymes from rain-damaged and laboratory-germinated wheat', I. Effects on product quality. *J Cereal Sci*, **10**, 157–167.

EVERY D (1993), 'Purification and characterization of a glutenin hydrolysing proteinase from wheat damaged by the New Zealand wheat bug, *Nysius huttoni*', *J Cereal Sci*, **18**, 239–250.

FERNS GK FITZSIMMONS RW MARTIN RH SIMMONDS DH and WRIGLEY CW (1975), *Australian Wheat Varieties: Identification According to Growth, Head and Grain Characteristics*, Melbourne, CSIRO.

FERNS GK FITZSIMMONS RW MARTIN RH and WRIGLEY CW (1978), *Australian Wheat Varieties: Supplement No 1*, Melbourne, CSIRO.

FURCSIK S (1992), 'Process for improving the shelf life of baked goods', US Patent 5,147,665.

GALILI G GALILI S LEWINSOHN E and TADMOR Y (2002), 'Genetic, molecular and genomic approaches to improve the value of plant foods and feeds', *Critical Rev Plant Sci*, **21**(3), 167–204.

GRAYBOSCH RA PETERSON CJ HANSEN LE RAHMAN S HILL A and SKERRITT J (1998), 'Identification and characterisation of US wheats carrying null alleles at the *wx* loci', *Cereal Chem*, **75**, 162–165.

HAIG KJ SUTER DAI HIGGINS P O'RIORDAN B GRAS PW and BEKES F (1997), 'Effect of wheat storage conditions on bread quality', in Tarr AW Ross AS and Wrigley CW, '*Cereals '97*', Melbourne, Royal Aust Chem Inst, 47–51.

JARMAN RJ (1995), 'Visual identification of grain characteristics', in Wrigley CW, *Identification of Food-grain Varieties*, St. Paul, MN, USA, American Association of Cereal Chemists Inc., 35–54.

KANG MS and GAUCH HG (1996), *Genotype-by-environment Interaction*, Boca Raton, FL, USA, CRC Press.

LOOKHART GL and WRIGLEY CW (1995), 'Electrophoretic analysis', in Wrigley, CW, *Identification of Food-grain Varieties*, St. Paul, MN, USA, American Association of Cereal Chemists Inc., 55–71.

LUKOW OM (1991), 'Screening of bread wheats for milling and baking quality – a Canadian perspective', *Cereal Foods World*, **36**, 497–501.

LUO X JAYAS DS and SYMONS SJ (1999), 'Identification of damaged kernels in wheat using a colour machine vision system', *J Cereal Sci*, **30**, 49–59.

MARSHALL DR MARES DJ MOSS HJ and ELLISON FW (1986), 'Effects of grain shape and size on milling yields in wheat. II. Experimental studies', *Aust J Agric Res*, **37**, 331–342.

MOSS HJ (1980), 'The pasting properties of some wheat starches free from sprout damage', *Cereal Res Commun*, **8**, 297–302.

NAKAMURA T YAMAMORI M HIRANO H HIDAKA S and NAGAMINE T (1995), 'Production of waxy (amylose-free) wheats', *Molecular and General Genetics*, **37**, 639–649.

NG PKW POGNA NE MELLINI F and BUSHUK W (1989), '*Glu-1* allele composition of the wheat cultivars registered in Canada', *J Genet Breeding*, **43**, 53–59.

OSBORNE BG KOTWAL Z BLAKENEY AB O'BRIEN L SHAH S and FEARN T (1997), 'Application of the single-kernel characterization system to wheat receiving testing and quality prediction', *Cereal Chem,* **74**, 467–70.

PAULIAN F and POPOV C (1980), 'Sunpest or cereal bug', in Hafliger E, *Wheat*, Basel, Ciba-Geigy, 69–74.

PLAUT Z BLUMENTHAL C GRAS PW and WRIGLEY CW (1999), 'Drought stress and wheat quality', in O'Brien L Blakeney AB Ross AS and Wrigley CW, *Cereals '98. Proc 48th Australian Cereal Chem Conf*, Melbourne, Royal Aust Chem Instit, 63–66.

RAHMAN S LI Z BATEY IL COCHRANE MP APPELS R and MORELL MK (2000), 'Genetic alteration of starch functionality in wheat', *J Cereal Sci,* **31**, 91–110.

RANDALL PJ and MOSS HJ (1990), 'Some effects of temperature regime during grain filling in wheat quality', *Aust J Agric Res*, **41**, 603–617.

RANDALL PJ and WRIGLEY CW (1986), 'Effects of sulfur deficiency on the yield, composition and quality of grain from cereals, oil seeds and legumes', *Adv Cereal Sci Technol*, **8**, 171–206.

RASK OS and ALSBERG CL (1924), 'A viscometric study of wheat starches', *Cereal Chem*, **1**, 7–26.

REES RG MARTIN DJ and LAW DP (1984), 'Black point in bread wheat: effects on quality and germination, and fungal associations', *Aust J Exp Agric Anim Husb*, **24**, 601–605.

RICHARDSON EC KAISER AG and PILTZ JW (2001), 'The nutritive value of frosted wheat for ruminants', *Aust J Exp Agric*, **41**, 205–210.

RIDGWAY C and CHAMBERS J (1996), 'Detection of external and internal insect infestation in wheat by near-infrared reflection spectroscopy', *J Sci Food Agric*, **71**, 251–264.

SALGO A VARGA J TOMOSLOZI S GRAS P RATH C BEKES F NANASI J FODOR D and SOUTHAN M (2001), 'Novel lab micro mill – a tool for small scale testing', in Wootton M Batey IL and Wrigley CW, *Cereals 2000. Proc. 11th ICC Cereal and Bread Congress*, Melbourne, Royal Aust Chem Inst, 35–40.

SCHIPPER A JAHN-DEESBACH W and WEIPERT D (1986), 'Untersuchungen zum Klimateinfluss auf die Weizenqualitat', *Getreide, Mehl und Brot*, **40**, 99–103.

SHEWRY PR HALFORD NG and TATHAM AS (1992), 'High molecular weight subunits of wheat glutenin', *J Cereal Sci*, **15**, 105–120.

SKERRITT JH and HEYWOOD RH (2000), 'A five-minute field test for on-farm detection of pre-harvest sprouting in wheat', *Crop Sci*, **40**, 742–756.

STODDARD FL (1999) 'Survey of starch particle-size distribution in wheat and related species', *Cereal Chem*, **76**, 145–149.

UHLEN AK HAFSKJOLD R KALHOVD AH SAHLSTROM S LONGVA A and MAGNUS EM (1998), 'Effects of cultivar and temperature during grain filling on wheat protein content, composition, and dough mixing properties', *Cereal Chem*, **75**, 460–465.

WILLIAMSON PM (1997), 'Black point of wheat: *in vitro* production of symptoms, enzymes involved, and association with *Alternaria alternata*', *Aust J Agric Res*, **48**, 13–19.

WITHERS PJA and SINCLAIR AH (1994), *Sulphur Nutrition of Cereals in the UK: Effects on Yield and Grain Quality*. Research Review No. 30, London, Home-Grown Cereals Authority.

WRIGLEY CW and BATEY IL (1995), 'Efficient strategies for variety identification', in Wrigley CW *Identification of Food-grain Varieties*, St. Paul, MN, USA, American Association of Cereal Chemists Inc, 19–33.

WRIGLEY CW and BEKES F (1999), 'Glutenin-protein formation during the continuum from anthesis to processing', *Cereal Foods World*, **44**, 562–565.

WRIGLEY CW and BEKES F (2002), 'Grain-protein composition as a document of wheat-quality type; new approaches to varietal identification', in Ng PKW and Wrigley CW, *Wheat Quality Elucidation: The Bushuk Legacy*, St. Paul, MN, USA, American Association of Cereal Chemists Inc, 65–86.

WRIGLEY CW BOOTH RI BASON ML and WALKER CE (1996), 'Rapid Visco Analyser: progress from concept to adoption', *Cereal Foods World*, **41**, 6–11.

ZHAO F-J MCGRATH SP and CROSLAND AR (1995), 'Changes in the sulphur status of British wheat grain in the last decade, and its geographical distribution', *J Sci Food Agric*, **68**, 507–514.

ZHAO XC and SHARP PJ (1998), 'Production of all eight genotypes of null alleles at waxy loci in bread wheat, *Triticum aestivum* L.', *Plant Breeding*, **117**, 488–490.

ZHAO XC SHARP PJ CROSBIE G BARCLAY I WILSON R BATEY IL and APPELS R (1998), 'A single genetic locus associated with starch granule and noodle quality in wheat', *J Cereal Sci*, **27**, 7–13.

5

Techniques for analysing wheat proteins

A. M. Gil, University of Aveiro, Portugal

5.1 Introduction

Cereal proteins constitute about 10% of the grain dry weight and are an important source of protein in diet. In addition, they play a determinant role in the processing properties of cereal flours, namely the ability of wheat to be baked into leavened bread. The wheat storage proteins (gluten) are particularly important in the latter aspect and their chemistry and structure began being investigated as early as the 18th century. Gluten comprises a complex mixture of proteins (prolamins) differing in molecular size and structure. These may be classified into gliadins (monomeric, soluble in aqueous alcohols) and glutenins (high M_r, polymeric, only reduced forms are soluble in aqueous alcohols) and a classification in terms of genetic and amino acid sequence has also been proposed (Shewry *et al.*, 1994): (a) high-molecular-weight (HMW) prolamins, (b) sulfur-poor prolamins comprising ω-gliadins and (c) sulfur-rich prolamins comprising α-, β-, γ-gliadins and low-molecular-weight (LMW) glutenins. It is known that both gluten quantity and composition determine dough viscoelasticity and, hence, performance. However, the molecular origins of gluten and dough viscoelasticity are not fully understood.

The quest for knowledge about structure/function relationships in wheat gluten proteins has increasingly revealed the need for analytical methods with the ability to tackle the high complexity and insolubility of these systems. Moreover, an interest has developed in studying the proteins as close as possible to their functional environment, i.e. in the hydrated solid state. In this chapter, mention will be made of separation methods, methods to determine molecular size, shape and secondary structure and rheological methods. The focus will, however, be placed on the applications of spectroscopic methods (infrared spectroscopy, solid-state nuclear magnetic resonance (NMR) spectroscopy and electron spin resonance (ESR) spectroscopy), since these are useful probes for molecular properties of wheat proteins in their hydrated solid state.

5.2 Separation methods

The first step in the study of wheat proteins is usually the fractionation of flour into fractions, based on their solubility in different solvents. The initial Osborne classification (Osborne, 1924), into albumin, globulin, prolamin and glutelin, has been the basis of many modified and improved separation and extraction procedures (Byers *et al.*, 1983; Macritchie, 1985; Czuchajowska and Pomeranz, 1993). However, the high heterogeneity of the protein fractions obtained has called for more effective separation techniques. Electrophoretic studies have been carried out for wheat protein fractions since the 1960s and have been invaluable for identifying single proteins, with basis on their M_r values. Sodium dodecyl sulfate polyacrylamide gel electrophoresis (SDS-PAGE) has been one of the most extensively used methods for the separation and identification of wheat proteins, having, for instance, given M_r values of 44 000–73 000 for ω-gliadins, 30 000–45 000 for γ-gliadins, and forming the basis for the identification and differentiation of several glutenin HMW subunits ($M_r > 10^7$) (Kasarda *et al.*, 1983; Shewry *et al.*, 1994; Jood *et al.*, 2001).

Similarly, many chromatographic methods have enabled fractionation of wheat storage proteins to be achieved. Size-exclusion high-performance liquid chromatography (SE-HPLC) and reverse-phase HPLC (RP-HPLC), for instance, have proved to be excellent methods for fractionation of wheat proteins (Kruger and Bietz, 1994; Huebner and Bietz, 1995; Dachkevitch and Autran, 1989; Ueno *et al.*, 2002). SE-HPLC has been used to characterise quantity and size distribution of wheat proteins in gluten fractions differing in gliadin/glutenin ratio and in glutenin size distribution and the effects of these compositional differences on the rheological behaviour have been investigated (Cornec *et al.*, 1994). RP-HPLC chromatographic characterisation of wheat gliadins has been used to identify different wheat varieties (Huebner and Bietz, 1995) and the use of smaller columns has led to shorter analysis times, reduced sample quantities required (mg flour), reduced analysis cost.

Although HPLC methods have undoubtedly become a powerful tool for the analysis of cereal proteins, the resolution quality and information provided by electrophoretic methods have stimulated the steady improvement of such methods. In fact, the field of capillary electrophoresis (CE) (Lookhart and Bean, 1996) and, more recently, high-performance CE (HPCE) (Bean and Lookhart, 1999, 2001a,b) is one that has seen some of the most remarkable developments in recent years, aiming at achieving higher resolution, reproducibility and rapidity when applied to food protein separations, including wheat proteins. CE methods emerged in the 1980s with the separation of solutes in an electrical field using capillary tubes. The use of high voltage and efficient cooling permitted rapid, reproducible, high-resolution separations. The potential of HPCE was recognised, since it combines the electrophoresis high resolution with the automation and ease of HPLC; this led to significant development in the HPCE methods and their applications (Bean and Lookhart, 1998a, 2001c). The subject of HPCE of food proteins including cereal proteins has been reviewed (Bean and Lookhart, 1998b, 2001b), addressing in detail the uses of Free Zone CE (FZCE) and SDS-CE. Of these methods, FZCE has been the most frequently employed for wheat proteins, its experimental conditions having been improved so as to provide rapid high-resolution separation, at a lower cost and with lower solvent waste than HPLC methods (Bean and Lookhart, 2000). Also, different polymer/buffer systems have been evaluated for the use of SDS-CE as a means of separating wheat proteins (Bean and Lookhart, 2001b) and several SDS-CE studies of wheat HMW subunits have been carried out, including, for instance, studies of the protein changes occurring during maturation (Werner *et al.*, 1994; Bean and Lookhart, 1999; Scholz *et al.*, 2000). Figure 5.1 shows some examples of the separation of wheat glutenins achieved by SDS-CE. The coupling of FZCE and RP-HPLC may be used to produce 2D separation of wheat proteins (Bean and Lookhart, 1997, 1998b). This consists of collecting samples fractionated by HPLC and then analysing them by HPCE in order to obtain 2D maps of gliadins and glutenins.

The coupling of HPLC with mass spectrometry (MS) methods has taken advantage of the latest MS developments, becoming one of the most useful techniques for the characterisation of wheat proteins. Off-line and on-line HPLC-MS applications to food proteins, including cereal proteins, have been reviewed by Leonil *et al.* (2000). Examples of such studies include the development of LC-MS methods for fast and sensitive fingerprinting of gliadins and glutenins (Mamone *et al.*, 2000). The analysis of protein fractions from several durum wheat varieties enabled about 40 components in each fraction to be identified. It has been suggested that MS detection of specific gliadin types may enable those proteins to be used as markers for wheat traces in gluten-free foods for celiac patients. The importance of differentiation between wheat varieties has been the subject of other studies, for instance using matrix-assisted laser desorption/ionisation mass spectrometry (MALDI/MS) to characterise HPLC separated fractions of common and durum wheat varieties (Dworschak *et al.*, 1998). The complexity of the mass spectra of gliadins and LMW glutenins was found to preclude the identification of individual components, holding, however, potential for differentiation between wheat varieties. The mass spectral profiles of glutenins were much simpler, potentially enabling the identification of lines containing subunits associated with superior wheat quality.

5.3 Analysing molecular properties

Purified wheat protein fractions or single proteins may be evaluated in terms of molecular size and shape by methods such as intrinsic viscosity measurements,

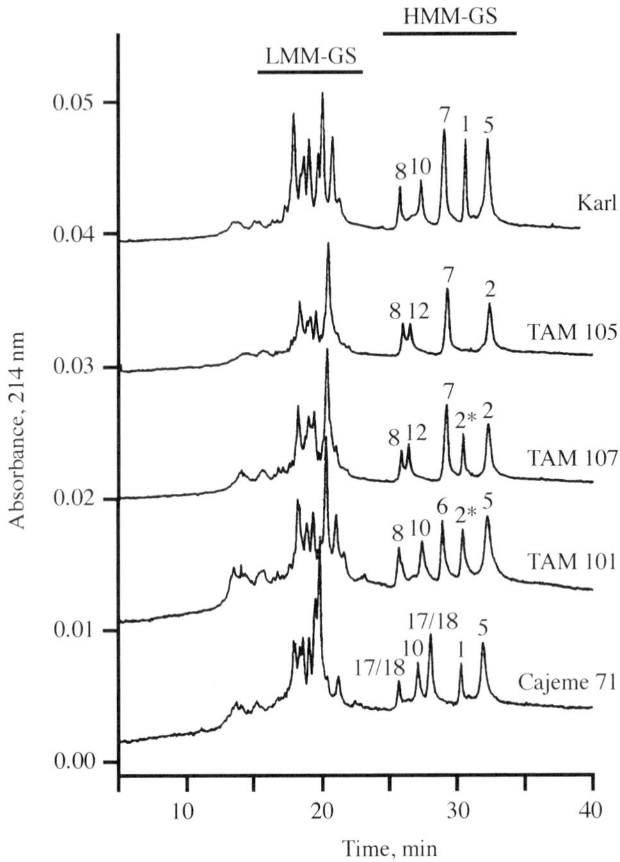

Fig. 5.1 SDS-CE separation of wheat glutenins. Numbers indicate specific HMW glutenin subunits. Reprinted with permission from Bean and Lookhart (1999). Copyright 1999 American Chemical Society.

small-angle X-ray scattering (SAXS) and scanning tunnelling microscopy (STM). STM has, since the 1980s, provided important information on the solid-state structure of wheat proteins, offering a number of advantages over light and electron microscopy, e.g. higher resolution, non-invasiveness (not requiring staining or coating of the sample) and the ability of studying the proteins in the hydrated state. This technique is based on the tunnelling of electrons through a gap between the sample surface and the tip of a needle attached to a piezoelectric ceramic piece, the effect being sensitive to the gap size, on an atomic scale. STM images of gliadin molecules deposited onto graphite from a trifluoroethanol solution have shown rod-shaped molecules with dimensions of about 10×3 nm in solution. In addition, a structure of about 16×3 nm has been observed for hydrated solid ω-gliadins, reflecting the importance of water content in the protein structure. STM studies in tandem with SAXS suggested asymmetric shapes for γ-gliadin molecules (Thomson *et al.*, 1992), whereas a set of studies have indicated that HMW subunits have extended conformations (molecular lengths of about 50 nm), either in excess solvent or in the hydrated solid state (Field *et al.*, 1987; Miles *et al.*, 1991; Matsushima *et al.*, 1992). In a more recent study, the application of both STM and AFM (atomic force microscopy) to wheat proteins is described, providing images of dry and hydrated gliadins and glutenins in order to study their aggregative behaviour (Tatham *et al.*, 1999).

Useful information on the secondary structure of purified wheat proteins has been obtained by circular dichroism (CD), a technique based on absorption differences for right- and left-polarised light caused by circular birefringence of the medium. Interpretation of experimental results obtained for wheat proteins in terms of protein secondary structure have been often compared and supported by predictive methods based on the known amino acid sequences. The CD spectra of ω-gliadins in dilute acetic acid have indicated a predominance of β-turns conformation and, in addition, γ-gliadins have been shown to form both β-turns and poly-L-proline II structures, with the C-terminal adopting α-helical conformation (Tatham and Shewry, 1985). Predictive studies of the secondary structures of glutenin HMW subunits and of synthetic model peptides have also suggested a central repetitive motif based on repetitive β-turns with α-helical structure occurring at the terminals (Tatham *et al.*, 1984; Field *et al.*, 1987; Tatham *et al.* 1990; Miles *et al.*, 1991; Shewry *et al.*, 1998).

Some conformational information may also be obtained by fluorescence measurements arising from tryptophan residues. For instance, γ-gliadins have been studied by fluorescence spectroscopy, indicating the absence of significant conformational changes with denaturation and suggesting the location of tryptophan residues at the molecule surface, probably involved in protein–protein interactions (Yeboah *et al.*, 1994). Time-resolved fluorescence provides information on molecular motion: it uses polarised excitation radiation and the occurrence of depolarisation reflects motion of the tryptophan residues in the time between absorption and emission (nanoseconds). Rotational correlation times of 3–6 ns were found for γ-gliadins in solution, reflecting a high degree of molecular mobility.

Spectroscopic methods that have the ability to probe molecular structure and/or dynamics and that can handle samples in the hydrated solid state naturally hold great potential for the study of gluten and its functionality. In fact, an increasing number of studies based on infrared (IR), nuclear magnetic resonance (NMR) and electron spin resonance (ESR) spectroscopies have emerged, as will be described in detail in Section 5.5.

5.4 Rheological measurements

Rheological methods measure the macroscopic viscoelastic properties of dough and its components, having been widely applied to the analysis of wheat flour. It was recognised long ago that gliadins and glutenins have different effects on the rheology of gluten and dough, respectively contributing for dough viscosity and viscoelasticity. The rheological behaviour of gluten fractions differing in glutenin/gliadin ratio and in glutenin size distribution has been determined by dynamic assay in shear (Cornec *et al.*, 1994). The results obtained for the fractions were similar to whole gluten rheological behaviour, although large differences were observed in the values of storage and loss moduli, G′ and G″. It became apparent, however, that gliadin-rich fractions exhibit low viscoelastic behaviour, consistent with the conclusion that gliadins contribute mostly to viscosity. On the other hand, viscoelasticity correlated strongly with the proportion of the largest glutenin polymers (as evaluated by SE-HPLC). The rheological effect of different gliadin/glutenin ratios was addressed in a parallel study of two wheat cultivars of good and poor breadmaking quality (Khatkar *et al.*, 1995), confirming the importance of both glutenin nature and gliadins/glutenin ratio towards gluten viscoelasticity (Fig. 5.2). In addition, lipids and non-prolamin proteins were seen not to contribute significantly towards gluten rheology (Hargreaves *et al.*, 1994a, 1995b). The viscoelasticity of glutens from genetic variants of bread wheat has been studied by rheometrics (Hargreaves *et al.*, 1996) and was shown to correlate closely to the composition of HMW glutenins.

Since a comprehensive coverage of the large number of applications of rheometrics to dough and gluten is beyond the scope of this text, the interested reader is recommended to consult more specific literature (Janssen *et al.*, 1996; Tsiami *et al.*, 1997; Edwards *et al.*, 2001; Khatkar *et al.*, 2002; Uthayakumaran

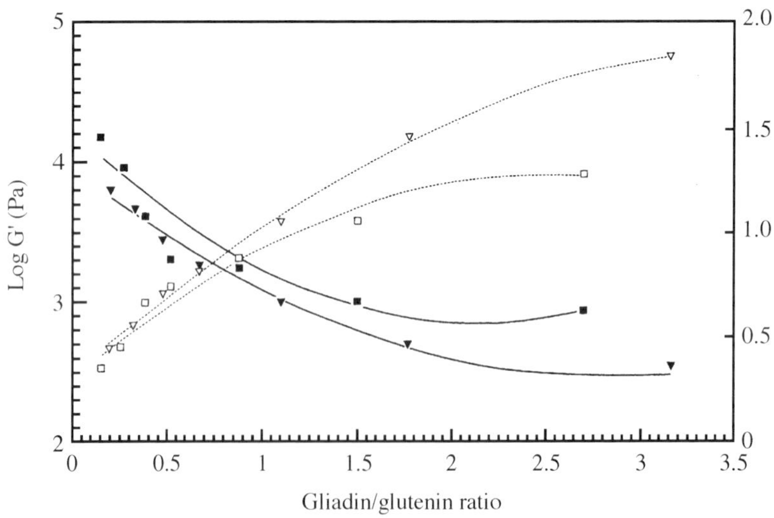

Fig. 5.2 The effect of varying gliadin/glutenin ratio on the G′ and tan δ values of cv. Hereward and cv. Riband glutens. Frequency 1 Hz, stress amplitude 203.7 Pa: ❑, cv. Hereward, ▽, cv. Riband; –, G′; . . ., tan δ. Reprinted from *J Cereal Sci*, **22**, Khatkar B S, Bell A E, Schofield J D, 'The dynamic rheological properties of glutens and gluten sub-fractions from wheats of good and poor breadmaking quality', 29–44, Copyright 1995, with permission from Elsevier Science.

et al., 2002). It should still be mentioned, however, that contrary to standard rheometrics which study the effects of shear stress, recent developments in dough rheometric methods have measured rheological parameters under biaxial extension. Inflation has been used to study the effects of biaxial extension and, indeed, bubble inflation is the basis of the Chopin Alveograph test for dough and gluten extensibility, and hence dough quality. Furthermore, biaxial extension is the relevant deformation taking place in the dough material surrounding the expanding gas bubbles during baking. A modification of the Alveograph has led to a new rheometer (Dobraszczyk and Roberts, 1994; Dobraszczyk, 1997), the Dobraszczyk–Roberts dough inflation system, which enables the rheological properties of dough and gluten to be measured under biaxial extension. The extensional properties measurable in this way seem to provide an effective baking quality control test, as well as rheological information on the dough/gluten system in a situation very close to the real baking situation. This development is an important step towards the understanding of the mechanisms of dough baking.

5.5 Infrared spectroscopy

Infrared (IR) spectroscopy is based on the analysis of molecular vibrational motions. A vibrating diatomic molecule may be modelled by a simple harmonic oscillator whose energy is quantised so that only discrete vibrations occur for: $\nu = 1/2\pi(k/\mu)^{1/2}$, and $E_\mathrm{v} = (\mathrm{v} + 1/2)h\nu$, where ν is the oscillation frequency, k is the force constant, μ is the reduced mass, v is the vibrational quantum number and E_v is the energy of the vibrational level. Absorption of electromagnetic radiation stimulates transitions between energy levels corresponding to set values of ΔE characteristic of the molecule. The transition energy values fall in the IR range of the electromagnetic spectrum, with frequencies in the 10^{14}–10^{12} Hz range (or 100–10^4 cm^{-1} in wavenumbers), although the regions most frequently used for IR spectroscopy are mid-IR (MIR, for 200–5000 cm^{-1}) and near-IR (NIR, for 5000–10 000 cm^{-1}). Not all transitions are allowed and the selection rule $\Delta\mathrm{v} = \pm 1$ applies. This condition alone is not enough to enable the observation of a transition since the vibration only interacts with the incoming radiation if the vibration is characterised by a change in dipole moment. In practice, the picture of an anharmonic oscillator provides a more realistic description of vibrating molecules and gives rise to a change in the selection rules so that $\Delta\mathrm{v} = \pm 1, \pm 2, \pm 3 \ldots$ The overtones ($\Delta\mathrm{v} > \pm 1$) are less probable and give rise to less intense bands, typically in the NIR region. The number of vibrations occurring in a molecule depends on its number of degrees of freedom ($3N$–6 for a non-linear molecule with N atoms) so that, for instance, the water molecule has three vibrational degrees of freedom and hence undergoes three independent types of vibration.

Fourier transform IR (FTIR) is now most commonly used, having replaced the use of dispersive elements by the Michelson interferometer, thus improving rapidity and quality of the spectra. Furthermore, a variety of sampling methods currently exist, allowing suitable analysis of samples in all physical states. The sampling method most currently used for the studies of paste-like materials such as hydrated gluten or dough is the Attenuated Total Reflectance (ATR) cell. The sample is directly placed on a crystal (e.g. of zinc selenide or diamond) through which internal single or multiple reflections of the infrared radiation occur. At the points of contact between the sample and the crystal, light is absorbed by the sample thus giving rise to an absorption spectrum. There is a variety of manuals specifically about vibrational spectroscopy (Brown, 1998; Diem, 1993; Nyquist, 2001).

The subject of NIR spectroscopy applied to the study and evaluation of wheat proteins will be only briefly mentioned here, since the subject is addressed in detail elsewhere in this book. NIR spectroscopy is one of the spectroscopic methods most extensively employed in the food industry owing to its sensitivity to food quality, ease of use, rapidity and low cost. In the subject of wheat and wheat protein analysis, NIR has for instance proved useful for determination of glutenin and gliadin contents in flour of different types (Delwiche *et al.*, 1998; Anjum and Walker, 2000).

MIR studies of proteins usually rely on the observation of the following typical protein bands: Amide I (1650–1660 cm^{-1}), a combination band due to C=O stretching and CNH bending in amide linkages and Amide II (1512–1550 cm^{-1}), a combination band due to NH bending and CN stretching in amide linkages. A third combination band characteristic of amide linkages, Amide III (1230–1260 cm^{-1}), occurs rather weakly in infrared. Some of the first MIR studies of gluten and gluten proteins have consisted of solution studies of some of the soluble wheat proteins. Since the shape of the amide I band has proved particularly informative with respect to protein conformation, FTIR studies of wheat gliadin solutions have relied on the detailed study of the Amide I, suggesting the predominance of β-turn conformation in ω-gliadins (Purcell *et al.*, 1988; Tatham *et al.*, 1989; Pézolet *et al.*, 1992). Interestingly, increasing protein concentrations led to an increase in the proportion of β-sheet conformation (both intra- and intermolecular) (Tatham *et al.*, 1989; Pézolet *et al.*, 1992).

The development of new sampling methods such as ATR cells have triggered the infrared study of solid and semi-solid materials. An interesting FTIR study of gluten and gluten fractions has compared solution state and dough-state conformations, using transmission and ATR cells respectively (Popineau *et al.*, 1994). This study showed that β-sheet content was higher in the dough-state than in solution and that β-sheet amount increased as the glutenin content increased. It was suggested that interactions between glutenin subunits may occur through segments in β-sheet conformation. Other studies have focused on the amide I and II changes upon gradual hydration of purified HMW subunits (Belton *et al.*, 1995) and, more recently, the effects of residual starch and lipids on gluten hydration has also been investigated by FTIR in tandem with NMR (Grant *et al.*, 1999). Figure 5.3 shows the changes occurring with hydration in the Amide I and Amide II bands for a wheat HMW subunit. The combination of

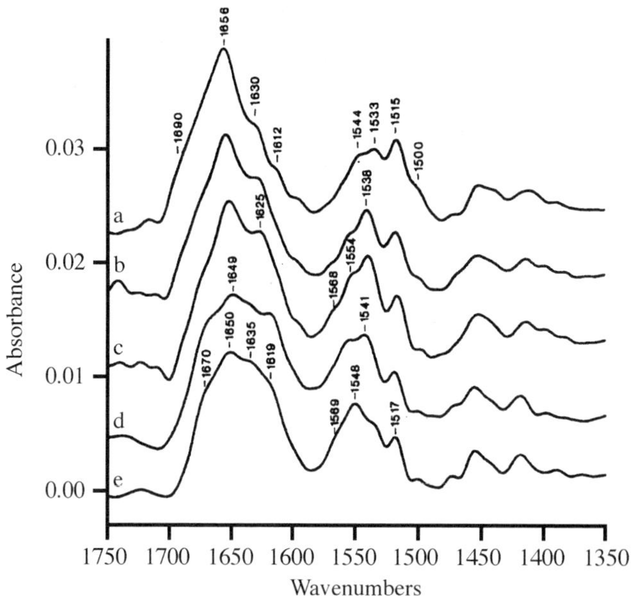

Fig. 5.3 Fourier-deconvoluted FTIR spectra of unalkylated HMW subunits: (a) dry protein film, (b) hydrated to 9% w/w H_2O, (c) hydrated to 37% w/w H_2O, (d) submerged in H_2O, the water spectrum subtracted, (e) in 0.1M acetic acid, acetic acid spectrum subtracted. Reprinted from *Int J Biol Macromol*, **17**(2), Belton P S, Colquhoun I J, Grant A, Wellner N, Field J M, Shewry P R, Tatham A S, 'FTIR and NMR studies on the hydration of a high-M_r subunit of glutenin', 74–80, Copyright 1995, with permission from Elsevier Science.

different vibrational spectroscopy techniques NIR, MIR and Raman, in the form of 2D spectroscopic techniques has proved an interesting approach. This has been applied to the evaluation of cereal quality, including wheat (Barton *et al.*, 1996) and rice (Barton *et al.*, 2002). The technique enables correlations between the different spectroscopic dimensions to be established, providing molecular level information on the changes of flour components occurring, for instance, during the cooking process.

5.6 NMR spectroscopy

The fundamental property of the atomic nucleus that forms the basis of NMR is the nuclear spin (I), which may take values of 0, ½, 1, 3/2, etc., the actual spin value depending on the mass and atomic numbers of the nucleus. A nuclear magnetic moment (μ) results from the occurrence of spin and is directly proportional to the spin quantum number I so that $\mu = \gamma Ih/2\pi$, where γ is the magnetogyric ratio of the nucleus. When an external magnetic field B_0 (from 4.7 to 21 T) is applied, the nuclear moments of nuclei I orient themselves with 2I + 1 possible orientations, given by the value of the magnetic quantum number m_I

($m_I = -I, -I+1 \ldots I-1, I$), i.e. for a 3/2 nucleus, m_I takes the values −3/2, −1/2, +1/2, +3/2. The energy of the interaction between μ and B_0 is given by $E = -(\gamma h/2\pi)m_I B_0$ and, since the selection rule for an NMR transition to occur is $\Delta m_I = \pm 1$, the transition energy ΔE is given by $\Delta E = \gamma h B_0/2\pi$. For detection of this transition energy, radiation with a frequency $\nu_0 = \gamma B_0/2\pi$, the Larmor frequency, must be applied (radiowave range: 10^6–10^8 Hz).

In a simple NMR experiment the first step is the irradiation of the sample with a radiofrequency pulse, followed by the recording of the sample response translating the loss of absorbed energy as a function of time, or Free Induction Decay (FID). Subsequently, this time domain signal is converted by a Fourier transformation into the energy domain, the spectrum. In the spectrum, peak positions (or chemical shifts) express the deviations in the value of ΔE_{abs} due to the interactions between the nucleus and the surrounding environment. This dependence of the NMR peak position on the molecular environment is the basis of the value of NMR as a probe for molecular structure. In addition, spin relaxation mechanisms also affect the spectrum, providing valuable information about molecular dynamics. Some examples of basic and medium level texts are Abraham *et al.* (1990); Harris (1986); Derome (1987) and Roberts (1993).

The study of wheat proteins calls for the use of NMR for the study of systems in both solution state (for soluble proteins) and solid state (or dough state). Although the basic principles of the NMR experiment remain unchanged, the instrumentation and techniques differ significantly from liquid to solid-state NMR. Indeed, some factors affecting the NMR response of solids do not arise, or average to zero in the liquid-state. These are: (1) dipole–dipole interactions, (2) chemical shift anisotropy (CSA) and (3) the generally long relaxation times of nuclei in solids. Dipole–dipole interactions and CSA result in broad and complex lineshapes and, although these effects do occur in solution they are generally averaged to zero as a result of the random motion of molecules. In addition, the typical long relaxation times in solids lead to lengthy experimental times. Therefore, if a resolved spectrum is to be obtained for a solid, techniques such as high power dipolar decoupling, magic angle spinning (MAS) and cross-polarisation (CP) have to be employed to improve resolution and signal-to-noise ratio (Mehring, 1983; Stejskal and Memory, 1994; Bovey and Mirau, 1996).

A large part of solid-state NMR studies do not, however, necessarily imply the analysis of the spectrum, involving instead the direct analysis of the time domain information. These studies are referred to as low-resolution NMR. Relaxation time measurements often fall into this category (although not necessarily) and, since they reflect the mobility of the nuclei in different frequency ranges, they are good indicators of molecular dynamics. In the solid state, however, some relaxation parameters are also influenced by spin diffusion between neighbouring nuclei, thus reflecting inter-nuclei proximity. Depending on the measuring pulse sequence chosen, several parameters may be measured, e.g. transverse relaxation time T_2 and longitudinal relaxation times T_1, $T_{1\rho}$ and T_{1D}, each having different sensitivities to molecular motions and spin diffusion. The measurement of relaxation times for dilute spins, e.g. ^{13}C, are more reliable indicators of mobility alone, since spin diffusion effects are no longer significant due to the nuclei low abundance. The following discussion about NMR application to wheat proteins will be organised in terms of low- and high-resolution applications, the latter including both liquid- and solid-state studies. Finally, a short mention of rheo-NMR of wheat proteins is included.

5.6.1 Solid-state low-resolution NMR studies of wheat proteins

The measurement of average T_{2H} relaxation times in dry gluten has enabled the identification of a protein solid-like environment making up 80–90% of the protons and a mobile sub-population of protons arising from lipids and some protein groups (Belton *et al.*, 1988a,b). In addition, T_{1H} measurements revealed two protein motional domains, the nature of which was, however, left unclear. The hydration of ω-gliadins was followed by low-resolution measurement of T_{1H}, $T_{1\rho H}$ and T_{2H} (Belton *et al.*, 1998) and methyl and amino group rotation, together with proline ring puckering, were identified as the motions mainly responsible for transverse relaxation. Interestingly, it was also found that T_{2H} was a particularly good indicator of the occurrence of glass transition, in this type of protein. ^{1}H and ^{2}H relaxation time measurements performed on a HMW subunit at different hydrations and temperatures have been compared with measurements reported for a synthetic peptide based on the tissue protein elastin (Belton *et al.*, 1994). Elastin has well-known elastic properties and the aim of its comparison with HMW subunits was to discover the origins of such properties in both systems. However, results indicated that the ^{2}H T_2 component associated with free D_2O decreased in proportion with increasing temperature, a different behaviour from that registered for elastin. Therefore, no similarity in elasticity mechanisms was found.

The use of relaxation times measurements has been particularly useful for the characterisation of the states of water in gluten and this has been pursued through ^{2}H and ^{17}O measurements. Water mobility may be monitored in this way and processes such as freezing and glass transition in gluten have been addressed (Cherian and Chinachoti, 1996, 1997; Grant *et al.*, 1999). Furthermore, the great sensitivity of $T_{1\rho H}$ to proton proximity has made it possible to monitor the extent of mixing between flour components, e.g. starch and gluten. This has been attempted for samples with different starch : gluten ratios and to samples subjected to heating treatment (Li *et al.*, 1996).

5.6.2 Liquid-state and solid-state high-resolution NMR studies of wheat proteins

The potential of NMR for the study of individual atomic sites in protein solutions was extended to soluble wheat proteins in the early 1980s, with the first ^{1}H and ^{13}C NMR spectra obtained for gliadin extracts in CD_3COOD/DCl

(Baianu, 1981; Baianu *et al.*, 1982). The difficulty in assigning the highly overlapped ^{1}H NMR spectra was noted and a preliminary assignment of the ^{13}C NMR spectra of gliadins was achieved. The need for the use of purified protein fractions and higher field spectrometers was emphasised by the authors. More recently, liquid-state NMR has been useful in characterising secondary structures in natural or synthetic peptides representative of specific wheat proteins. Examples of this approach are the NMR studies of synthetic cyclic and linear peptides which contain the consensus peptides PGQGQQ and GYYPTSPQQ found for HMW subunits, as well as the study of the repetitive HMW domains (VanDijk *et al.*, 1997a,b). Evidence was obtained for the existence of β-turns of both types I and II characterising the repetitive domain of the HMW subunits studied.

With the increasing interest in understanding gluten and its proteins in the hydrated solid state, solution-state NMR was gradually replaced by solid-state NMR. Most high-resolution solid-state NMR studies dedicated to wheat proteins have relied on ^{13}C observation through CP/MAS and single pulse excitation (SPE) experiments. The CP step at the beginning of the CP/MAS experiment involves the magnetisation transfer from ^{1}H to ^{13}C and relies on non-zero ^{1}H/^{13}C dipolar interactions. This condition is satisfied for rigid solids but for semi-solids such as hydrated gluten, the CP/MAS experiment selects the information on the rigid part(s) of the sample alone. Complementary information on the mobile part(s) may be obtained by direct SPE of the ^{13}C nuclei, as long as short waiting times are used between consecutive scans. If ^{1}H observation is intended, fast MAS is required to average out the ^{1}H/^{1}H dipolar interactions. This is achieved successfully if the sample has some inherent molecular mobility, as is the case of hydrated gluten, and if specific high-resolution MAS (HR-MAS) conditions are used.

One of the first solid-state NMR studies of wheat proteins was the ^{13}C CP/MAS characterisation of solid powder gluten, glutenin-enriched and gliadin-enriched samples (Schofield and Baianu, 1982). This was followed by other CP/MAS studies of whole gluten and subfractions (Belton *et al.*, 1985, 1988a,b; Moonen *et al.*, 1985), in some cases exploring the occurrence of dynamic heterogeneity (Belton *et al.*, 1985, 1988a,b). NMR studies of dry and hydrated gluten led to the identification of lipids in very mobile environments and of different gluten populations differing in molecular mobility (Belton *et al.*, 1987; Ablett *et al.*, 1988). The effects of heating at 80 °C on the ^{13}C CP/MAS spectra and T_{2H} relaxation times of gluten were investigated, no significant changes having been, however, observed (Ablett *et al.*, 1988). The effects of heating were again studied more recently, comparing conventional and microwave heating, and using solid-state ^{13}C NMR to study the molecular mobility of gluten. However, few differences in protein mobility could be found (Umbach *et al.*, 1998).

Accompanying the developments in preparative separation methods, solid-state NMR has been also applied to the investigation of the hydration process of purified wheat storage proteins, namely ω-gliadins (Gil *et al.*, 1997; Belton *et al.*, 1998) and HMW subunit glutenins (Alberti *et al.*, 2001, 2002a,b). The changes in the ^{13}C CP/MAS spectrum of ω-gliadins with 0–50% water (Fig. 5.4)

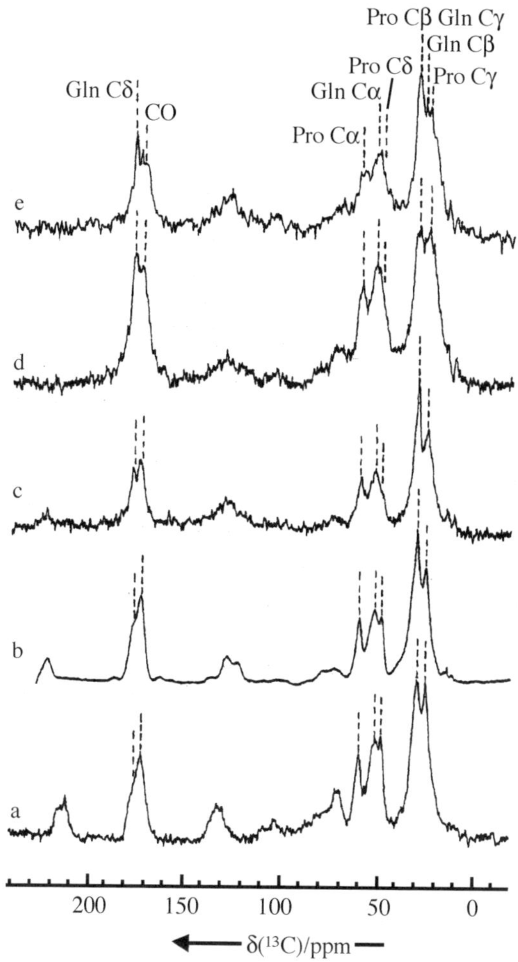

Fig. 5.4 ^{13}C CP/MAS spectra of omega-gliadins at different hydrations: (a) 0% water, 464 scans, (b) 7% water, 11 720 scans, (c) 19% water, 32 305 scans, (d) 35% water, 5500 scans, (e) 50% water, 17 732 scans. Reprinted with permission from *Magnetic Resonance in Chemistry*, **35**, Gil A M, Alberti E A, Tatham A S, Belton P S, Humpfer E, Manfred M (1997), 'Magic angle spinning NMR study of the hydration of the wheat seed storage protein omega-gliadins', S101–S111, © copyright 1997 (copyright owner as specified in the journal).

indicated a gradual mobility increase with hydration, and carbon T_1 relaxation times and CP dynamic parameters T_{CH} and $T_{1\rho H}$ provided additional dynamic information. In addition, the positions of certain peaks (e.g. glutamine and proline α carbons) particularly sensitive to protein conformation have indicated a predominance of β-sheet in the drier states and the conversion to a different conformation at high hydrations.

In spite of the general mobility increase with hydration, a subpopulation of glutamine side-chains was seen to remain relatively immobilised even at 50%

hydration. These groups had comparable mobility to backbone sites (Gil *et al.*, 1997; Belton *et al.*, 1998), suggesting that inter-glutamine hydrogen bonding occurs in hydrated ω-gliadins. ^{1}H MAS complements ^{13}C NMR since it provides information on the whole sample, including the water present and the glutamine side-chain NH_2 groups. ^{1}H HR-MAS of hydrated gliadins enabled 2D NMR methods to be employed (Gil *et al.*, 1997) and spectral assignment was achieved. Hydration of gliadins with D_2O, rather than H_2O, enabled the identification of solvent-protected glutamine NH_2s, consistently with T_2-weighted spectra which confirmed such groups as relatively hindered. In tandem with T_{1H}, $T_{1\rho H}$ and T_{2H} measurements (Belton *et al.*, 1998), these results have provided support for the hypothesis that hydrated ω-gliadins form mobile protein loops co-existing with some regions of strong protein–protein interaction.

Following the recognition that gluten viscoelasticity is seen to correlate to the HMW subunits, the hydration of the 1Dx5 subunit, believed to be associated with good bread quality, was investigated by ^{1}H and ^{13}C MAS (Alberti *et al.*, 2001, 2002a,b). Both ^{1}H and ^{13}C NMR results have provided evidence that a network, in some ways similar to that suggested for ω-gliadins, is formed by 1Dx5 upon hydration. In the case of 1Dx5, the network seems to comprise mobile segments rich in glutamine and glycine and hindered segments rich in hydrophobic residues and containing a few glutamine residues too. These latter glutamines were suggested to stimulate inter-segment hydrogen bonding, probably acting cooperatively with inter-segment hydrophobic interactions (Alberti *et al.*, 2001, 2002a). As to the role played by cross-linking on the hydration behaviour of 1Dx5 subunit, it has been shown by NMR that disulfide bonds promote easier protein plasticisation and the formation of a more mobile network, probably comprising larger and/or higher number of loops.

Comparison of whole proteins with model peptides that represent specific parts of the protein chain has proved a valuable approach in many studies of cereal proteins. This approach was followed for 1Dx5, in order to investigate the importance of the non-repetitive terminal domains and the length of the main chain (Alberti *et al.*, 2002b), again by ^{1}H and ^{13}C solid-state NMR. The non-repetitive terminal domains were found to induce water insolubility and a generally higher network hindrance. Shorter chain lengths were shown to increase plasticisation and water-solubility. One of the main conclusions was that the ability of HMW subunits to form hindered hydrogen-bonded segments must be strongly dependent on the peptide primary structure. This is, therefore, an important property which, besides determining the structural properties and mobility of the hydrated network, may well be also determinant for the functionality of these proteins.

5.6.3 Rheo-NMR studies of wheat proteins

Once the molecular level behaviour of wheat protein doughs is adequately understood and the macroscopic behaviour (e.g. measured by rheometrics or extensional tests) is characterised, the bridging of the two levels of information should follow. It often happens that, owing to different requirements of each analytical method (in terms of sample quantity and/or preparation, measurement time and timing), the actual samples characterised at the macroscopic level are not those characterised at the molecular level. In the case of flour and its components, this fact may have important consequences since it is known that flour functionality is exquisitely sensitive to aspects like the dough-making process, standing time, dough mixing time and process. This calls for the development of methods that can simultaneously measure quantities indicative of macroscopic behaviour and of molecular level properties. Efforts are currently being expressed along these lines by exploring the coupling of spectroscopic methods with some type of stress application mechanism.

Applications of NMR in rheological studies of complex fluids, or rheo-NMR, can be employed to obtain two types of information: flow profiles may be obtained in various shear geometries through NMR velocimetry providing direct information on flow curves and hence non-linear, shear-dependent viscosity (Xia and Callaghan, 1991) and, through NMR spectroscopy, insight is obtained regarding molecular order and dynamics under shear (Nakatani *et al.*, 1990) or extension (Callaghan and Gil, 1999; Gil *et al.*, 2000). Preliminary results of rheo-NMR applied to hydrated gluten (Callaghan and Gil, 1999; Gil *et al.*, 2000) have illustrated the potential of the method to probe the direct effects of shear and extensional deformations on the structure of gluten. The resulting changes in gluten ^{1}H NMR spectra (Fig. 5.5) suggest that inter-glutamine hydrogen bonds (reflected by the 7.5 ppm peak) are broken when stress is applied and re-formed after stress cessation. Such effects have been clearly shown for a sample of soft flour gluten, unlike the results obtained for a sample of hard flour gluten, suggesting a possible relationship between the different technological properties of flours and the role of hydrogen-bonded glutamine residues in the gluten network.

5.7 Electron spin resonance spectroscopy

Atoms or molecules that contain one or more electrons with unpaired spins are expected to show electron spin resonance (ESR) spectra. These substances may either arise naturally (e.g. NO, O_2, NO_2, Fe^{3+} and its complexes) or be produced artificially. Unstable paramagnetic materials, or free radicals, may form as intermediates in a chemical reaction or by irradiation of a stable molecule with UV or X-ray radiation, or with a beam of nuclear particles. When subjected to an external magnetic field B_0 (usually of about 0.3 T), an unpaired electron spin behaves in a similar way as the nuclear spin, i.e. the electron magnetic moment interacts with B_0 to give two energy levels of energies separated by ΔE: $\Delta E = h\nu = g\,\beta\, B_0$, where β is the Bohr magneton (9.273×10^{-24} J T^{-1}) and g is named as the g-value of the radical or complex. A resonance absorption will thus occur at a frequency $\nu = \Delta E/h$, of the order of 10^8–10^{10} Hz, i.e. in the microwave range, and its position is referred to in terms of its g-value ($g = \Delta E/\beta\, B = h\nu/\beta\, B$). For instance, a resonance observed at 8388.255 MHz

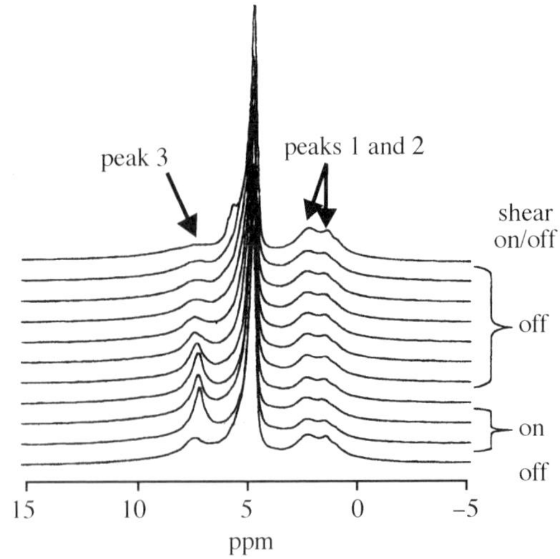

Fig. 5.5 ^{1}H NMR spectra of Amazonia (soft flour) gluten hydrated to 50% water. Spectra were registered successively from the bottom spectrum (prior to shear) to the top spectrum (after shear cessation). Note that in the first spectrum the effect of pre-shearing has resulted in a slight enhanced amplitude for the 7.5 ppm peak. Reprinted with permission from *Rheological Acta*, '^{1}H NMR spectroscopy of polymers under shear and extensional flow', Callaghan P T, Gil A M, **38**, 528–536, figure 8, 1999, Copyright 1999 Springer-Verlag GmbH&Co.

in a 0.30 T field would be recorded as a resonance at $g = 2.0023$, the g-value for a free electron. The g-values may, however, deviate considerably and may be highly anisotropic leading to broad and asymmetric ESR spectra. In the ESR spectrum, usually presented in the form of the first derivative of the absorption spectrum, peak intensity is proportional to the concentration of free radical or paramagnetic material in the sample and peak width reflects the electron spin relaxation process. The coupling between the electron spin and the nuclear spins of neighbouring nuclei gives rise to hyperfine structure in the spectra. In addition, the anisotropy and broadening of the ESR spectrum are strongly dependent on molecular dynamics and therefore the appearance of the spectrum may provide useful information on molecular mobility.

In the case of dough and cereal proteins, no intrinsic paramagnetism activity exists so that spin-probe or spin-labelling techniques need to be used. In the former case, paramagnetic compounds are non-covalently introduced in the system, their spectra reflecting the motion of the probe which depends on its size and polarity but, most importantly, on the characteristics of the environment. In the latter case, paramagnetic compounds are covalently attached to particular amino acid side-chains, thus directly reflecting the local dynamic environment of the polypeptide chain.

ESR spectroscopy has proved useful in providing information on the mobility of wheat protein molecules in solution or hydrated solid state materials. Several isolated HMW subunits have been studied by ESR in the solution state, after reduction and cysteine alkylation with a maleimide spin label (Moonen *et al.*, 1985). Spectra obtained in denaturing conditions (in 3 M urea) were identical for all HMW subunits and showed the sharp three-line pattern typical of isotropically reorienting systems. From the peak intensities and linewidth, the isotropic rotational correlation times τ_c were calculated, giving similar values of 0.22 ± 0.02 nanoseconds for all proteins. As expected, these values were found to increase for a less denaturing medium, 10 mM acetic acid. Differences in τ_c observed for different subunits were interpreted in terms of different cross-linking abilities and a correlation between ESR properties and breadmaking quality is suggested, although attention is drawn to the fact that spin labelling may significantly affect and change the original cross-linking ability of the systems. Several ESR applications to hydrated solid-state gluten samples have been carried out (Pearce *et al.*, 1988; Hargreaves *et al.*, 1994b), often in tandem with other techniques such as rheology or separation methods to further characterise the samples under study. This was the case of a study on a set of gluten subfractions differing in gliadin/glutenin ratio for which spin-probing with nitroxide radicals of different sizes was used, as well as specific spin-labelling of protein groups (either sulfhydryl groups or amino and hydroxyl groups), thus directly probing protein mobility (Hargreaves *et al.*, 1994b). It turned out that spin-probing and spin-labelling give useful complementary information in that, whereas spin-probes reflect different sized water pores (depending on the probe size), spin-labels enable selective observation of molecular variations in the gluten proteins. For instance, spin-labels at the cysteine residues showed an increase in rigidity as the amount of largest glutenins, and hence viscoelasticity, increases in the sample.

Extensions of these studies have investigated the effects of lipids and non-prolamin content (Hargreaves *et al.*, 1995b) and of heating (Hargreaves *et al.*, 1995a) on the properties and structure of gluten. Figure 5.6 shows typical ESR spectra of spin-labelled gluten as a function of heating. Results were consistent with lipids and non-prolamic proteins acting simply as fillers in the gluten network and, in addition, it was observed that the effect of heating on molecular mobility in gluten was reversible. Such reversibility was, however, not noted for glutenin-enriched fractions, raising the question of fractionation possibly altering the protein's behaviour compared with that in the original state, i.e. in complex gluten. The molecular flexibility properties as viewed by spin-labelling ESR were compared for glutens originating from genetic variants of bread wheat (Hargreaves *et al.*, 1996) and the results were correlated with their different viscoelastic behaviour.

5.8 Future trends

The many studies dedicated to characterising structure and behaviour of particular single gluten proteins have led to the suggestion that the primary structure of these proteins, namely the particular distribution of glutamines, glycines and hydrophobic residues, is determinant for functionality. This

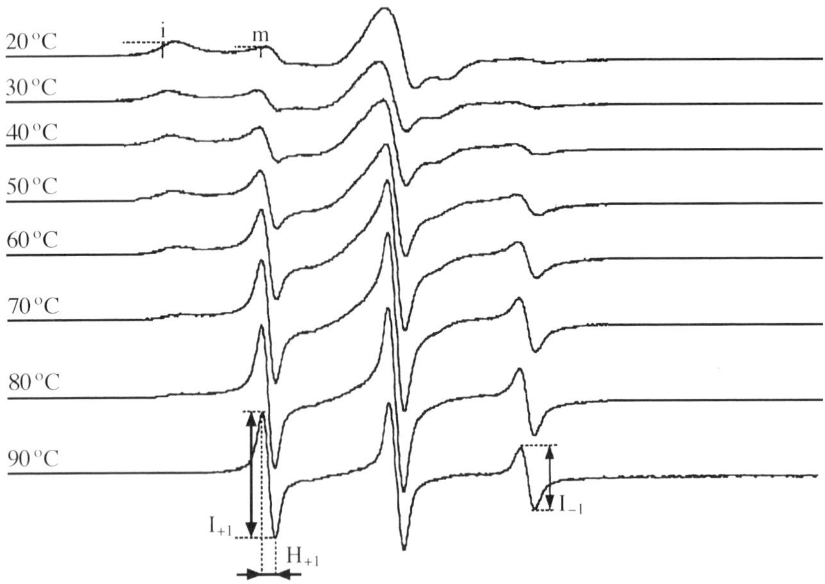

Fig. 5.6 Conventional ESR spectra of 4-maleimido-TEMPO spin-labelled gluten recorded at different temperatures. The spectral features used to calculate the *R*-value (parameter which reflects the proportion of slow-moving to fast-moving population of spin labels) and the rotational correlation time τ_c are indicated. Reprinted with permission from *FEBS Lett*, **43**, Hargreaves J, Popineau Y, Le Meste M, Hemminga M A, 'Molecular mobility in wheat gluten proteins submitted to heating', 1170–1176, Copyright 1995.

suggestion needs now to be verified, for instance, through systematic studies of model polypeptides with controlled primary structures. The potential identification of particular primary sequences as promoters of functionality should open important ways ahead in the control of gluten and dough quality.

However, the continuing work on dough and gluten fractionation, in the attempt to understand the system as a whole, continues to pose the question of how faithfully subfractions reflect the behaviour and functionality of the whole. This question has, so far, been left unanswered and it is important to find ways to address this.

Finally, regarding the analytical methods, it is clear that the way towards understanding a complex system such as wheat proteins and dough is through devising new methods with the ability to simultaneously deliver information on the functionality (macroscopic level) and on the molecular level behaviour. Examples of possible ways forward are the coupling of spectroscopic methods, microscopic methods and diffraction methods with the application of the types of stress involved in the dough-making and dough-baking processes, i.e. biaxial extension. Such methods would therefore investigate the changes taking place at the molecular level during mixing and baking, processes during which the functional quality of the system reveals itself.

5.9 References

ABLETT S, BARNES D J, DAVIES A P, INGMAN S J and PATIENT D W (1988), '^{13}C and pulse nuclear magnetic resonance spectroscopy of wheat proteins', *J Cereal Sci*, **7**, 11–20.

ABRAHAM R J, FISHER J and LOFTUS P (1990), *Introduction to NMR Spectroscopy*, Chichester, John Wiley & Sons.

ALBERTI E, HUMPFER E, SPRAUL M, GILBERT S M, TATHAM A S, SHEWRY P R and GIL A M (2001), 'A high resolution ^{1}H magic angle spinning NMR study of a high-M_r subunit of wheat glutenin', *Biopolymers*, **58**(1) 33–45.

ALBERTI E A, GILBERT S M, TATHAM A S, SHEWRY P R and GIL A M (2002a), 'Study of wheat high molecular weight 1Dx5 subunit by ^{13}C and ^{1}H solid state NMR spectroscopy: I. The role of covalent crosslinking', *Biopolymers, Biospectroscopy section*, **67**(6), 487–498.

ALBERTI E A, GILBERT S M, TATHAM A S, SHEWRY P R, NAITO A, OKUDA K, SAITÔ H and GIL A M (2002b), 'Study of wheat high molecular weight 1Dx5 subunit by ^{13}C and ^{1}H solid state NMR: II. The roles of non-repetitive terminal domains and length of repetitive domain', *Biopolymers, Biospectroscopy section*, **65**(2), 158–168.

ANJUM F M and WALKER C E (2000), 'Grain, flour and breadmaking properties of eight Pakistani hard white spring wheat cultivars grown at three different locations for 2 years', *Int J Food Sci Technol*, **35**(4), 407–416.

BAIANU I C (1981), 'Carbon-13 and proton nuclear magnetic resonance studies of wheat proteins. Spectral assignments for Flanders gliadins in solution', *J Sci Food Agric*, **32**, 309–313.

BAIANU I C, JOHNSON L F and WADDELL D K (1982), 'High-resolution proton, C-13 and N-15 nuclear magnetic-resonance studies of wheat proteins at high magnetic-fields – spectral assignments, changes with concentration and heating treatments of flinor gliadins in solution – comparison with gluten spectra', *J Sci Food Agric*, **33**(4), 373–383

BARTON F E II, HIMMELSBACH D S and ARCHIBALD D D (1996), 'Two-dimensional vibrational spectroscopy. V. Correlation of mid and near infrared of hard red winter and spring wheats', *J Near Infr Spectrosc*, **4**, 139–152.

BARTON F E II, HIMMELSBACH D S, MCCLUNG A M and CHAMPAGNE E L (2002), 'Two-dimensional vibrational spectroscopy of rice quality and cooking', *Cereal Chem*, **79**(1), 143–147.

BEAN S R and LOOKHART G L (1997), 'Separation of wheat proteins by two-dimensional reversed-phase high-performance liquid chromatography plus free zone capillary electrophoresis', *Cereal Chem*, **74**, 758–765.

BEAN S R and LOOKHART G L (1998a), 'Faster capillary electrophoresis separations of wheat proteins through modification to buffer composition and sample handling', *Electrophoresis*, **19**, 3190–3198.

BEAN S R and LOOKHART G L (1998b), 'High performance capillary electrophoresis of cereal proteins', *J Chromatogr. A*, **814**, 25–41.

BEAN S R and LOOKHART G L (1999), 'Sodium dodecyl sulfate capillary electrophoresis of wheat proteins. 1. Uncoated capillaries', *J Agr Food Chem*, **47**(10), 4246–4255.

BEAN S R and LOOKHART G L (2000), 'Ultrafast capillary electrophoretic analysis of cereal storage proteins and its applications to protein characterisation and cultivar differentiation', *J Agric Food Chem*, **48**, 344–353.

BEAN S R and LOOKHART G L (2001a), 'Optimising quantitative reproducibility in high-performance capillary electrophoresis (HPCE) separations of cereal proteins', *Cereal Chem*, **78**(5), 530–537.

BEAN S R and LOOKHART G L (2001b), 'High-performance capillary electrophoresis of meat dairy and cereal proteins', *Electrophoresis*, **22**(19), 4207–4215.

BEAN S R and LOOKHART G L (2001c), 'Recent developments in high-performance capillary electrophoresis of cereal proteins', *Electrophoresis*, **22**(8), 1503–1509.

BELTON P S, SHEWRY P R and TATHAM A S (1985), '^{13}C Solid state nuclear magnetic resonance study of wheat gluten', *J Cereal* Sci, **3**, 305–317.

BELTON P S, DUCE S L and TATHAM A S (1987), '13C solution state and solid state n.m.r. of wheat gluten', *Int J Biol Macromol*, **9**, 357–362.

BELTON P S, DUCE S L, COLQUHOUN I J and TATHAM A S (1988a), 'High-power ^{13}C and ^{1}H nuclear magnetic resonance in dry gluten', *Magn Reson Chem*, **26**, 245–251.

BELTON P S, DUCE S L and TATHAM A S (1988b), 'Proton nuclear magnetic resonance relaxation studies of dry gluten', *J Cereal Sci,* **7**, 113–122.

BELTON P S, COLQUHOUN I J, FIELD J M, GRANT A, SHEWRY P R and TATHAM A S (1994), '^{1}H and ^{2}H NMR relaxation studies of a high M_r subunit of wheat glutenin and comparison with elastin', *J Cereal* Sci, **19**, 115–121.

BELTON P S, COLQUHOUN I J, GRANT A, WELLNER N, FIELD J M, SHEWRY P R and TATHAM A S (1995), 'FTIR and NMR studies on the hydration of a high-M_r subunit of glutenin', *Int J Biol Macromol*, **17**(2), 74–80.

BELTON P S, GIL A M, GRANT A, ALBERTI E and TATHAM A S (1998), 'Proton and carbon NMR measurements of the effects of hydration on the wheat protein omega-gliadin', *Spectrochim Acta A*, **54**(7), 955–966.

BOVEY F A and MIRAU P A (1996), *NMR of Polymers*, San Diego, Academic Press.

BROWN J M (1998), *Molecular Spectroscopy*, New York, Oxford University Press.

BYERS M, MIFLIN B J and SMITH S J (1983), 'A quantitative comparison of the extraction of protein fractions from wheat grain by different solvents and of the polypeptide and amino acid composition of the alcohol-soluble proteins', *J Sci Food Agric,* **34**, 447–462.

CALLAGHAN P T and GIL A M (1999) '^{1}H NMR spectroscopy of polymers under shear and extensional flow', *Rheol Acta*, **38**, 528–536.

CHERIAN G and CHINACHOTI P (1996), 'H-2 and O-17 nuclear magnetic resonance study of water in gluten in the glassy and rubbery state', *Cereal Chem.*, **73**(5), 618–624.

CHERIAN G and CHINACHOTI P (1997), 'Action of oxidants on water sorption, H-2 nuclear magnetic resonance mobility and glass transition behaviour of gluten', *Cereal Chem*, **74**(3), 312–317.

CORNEC M, POPINEAU Y and LEFEBVRE J (1994), 'Characterisation of gluten subfractions by SE-HPLC and dynamic rheological analysis in shear', *J Cereal Sci*, **19**(2), 131–139.

CZUCHAJOWSKA Z and POMERANZ Y (1993), 'Protein-concentrates and prime starch from wheat flours', *Cereal Chem*, **70**(6), 701–706.

DACHKEVITCH T and AUTRAN J C (1989), 'Prediction of baking quality of bread wheats in breeding programs by size-exclusion high-performance liquid-chromatography', *Cereal Chem*, **66**, 448–456.

DELWICHE S R, GRAYBOSCH R A and PETERSON C J (1998), 'Predicting protein composition, biochemical properties and dough-handling properties of hard red winter wheat flour by near-infrared reflectance', *Cereal* Chem, **75**(4), 412–416.

DEROME A E (1987), *Modern NMR Techniques for Chemistry Research*, Oxford, Pergamon Press.

DIEM M (1993), *Introduction to Modern Vibrational Spectroscopy*, New York, John Wiley & Sons.

DOBRASZCZYK B J (1997), 'Development of a new dough inflation system to evaluate doughs', *Cereal Foods World*, **42**(7), 516–519.

DOBRASZCZYK B J and ROBERTS C A (1994), 'Strain-hardening and dough gas cell-wall failure in biaxial extension', *J Cereal Sci*, **20**(3), 265–274

DWORSCHAK R G, ENS W, STANDING K G, PRESTON K R, MARCHYLO B A, NIGHTINGALE M J, STEVENSON S G and HATCHER D W (1998), 'Analysis of wheat gluten proteins by matrix-assisted laser desorption/ionisation mass spectrometry', *J Mass Spectrometry*, **33**(5), 429–435.

EDWARDS N M, PERESSINI D, DEXTER J E and MULVANEY S J (2001), 'Viscoelastic properties of durum wheat and common wheat dough of different strengths', *Rheol Acta*, **40**(2), 142–153.

FIELD J M, TATHAM A S and SHEWRY P R (1987), 'The structure of high M_r subunit of durum wheat (*T. durum*) gluten', *Biochem J*, **247**, 215–221.

GIL A M, ALBERTI E A, TATHAM A S, BELTON P S, HUMPFER E and MANFRED M (1997), 'Magic angle spinning NMR study of the hydration of the wheat seed storage protein omega-gliadins', *Magn Reson Chem*, **35**, S101–S111.

GIL A M, ALBERTI E and CALLAGHAN P T (2000) 'The viscoelasticity of gluten by NMR', *Proceed. 2nd Int. Symp. Food Rheol. Struct.*, Eds. P Fischer, I Marti, E J Windhab, ETH, Zurich, 295–299.

GRANT A, BELTON P S, COLQUHOUN I J, PARKER M L, PLIJTER J J, SHEWRY P R and TATHAM A S, WELLNER N (1999), 'Effects of temperature on sorption of water by wheat gluten determined using deuterium nuclear magnetic resonance', *Cereal Chem*, **76**(2), 219–226.

HARGREAVES J, LE MESTE M and POPINEAU Y (1994a), 'ESR studies of gluten–lipid systems', *J Cereal Sci*, **19**, 107–113.

HARGREAVES J, LEMESTE M, CORNEC M and POPINEAU Y (1994b), 'Electron-spin-resonance studies of wheat protein fractions', *J Agr Food Chem*, **42**(12), 2698–2702.

HARGREAVES J, POPINEAU Y, LE MESTE M and HEMMINGA M A (1995a), 'Molecular mobility in wheat gluten proteins submitted to heating', *FEBS Lett*, **43**, 1170–1176.

HARGREAVES J, POPINEAU Y, MARION D, LEFEBVRE J and LE MESTE M (1995b), 'Gluten viscoelasticity is not lipid-mediated. A rheological and molecular flexibility study on lipid and non-prolamin protein depleted glutens', *J Agric Food Chem*, 43, 1170–1176.

HARGREAVES J, POPINEAU Y, CORNEC M and LEFEBVRE J (1996), 'Relations between aggregative, viscoelastic and molecular properties in gluten from genetic variants of bread wheat', **18**(1/2), 69–75.

HARRIS R K (1986), *Nuclear Magnetic Resonance Spectroscopy*, New York, John Wiley & Sons.

HUEBNER F R and BIETZ J A (1995), 'Rapid and sensitive wheat-protein fractionation and varietal identification by narrow-bore reversed-phase high-performance liquid-chromatography', *Cereal Chem*, **72**(5), 504–507.

JANSSEN A M, VAN VLIET T and VEREIJKEN J M (1996), 'Rheological behaviour of wheat glutens at small and large deformations. Comparison of two glutens differing in breadmaking potential', *J Cereal Sci*, **23**(1), 19–31.

JOOD S, SCHOFIELD J D, TSIAMI A A and BOLLECKER S (2001), 'Effect of glutenin subfractions on bread-making quality of wheat', *Int J Food Sci Technol*, **36**(5), 573–584.

KASARDA D D, AUTRAN J-C, LEW E J-L, NIMMO C C and SHEWRY P R (1983), 'N-Terminal amino acid sequences of the ω-gliadins and ω-secalins: implications for the evolution of prolamin genes', *Biochim Biophys Acta,* **747**, 138–150.

KHATKAR B S, BELL A E and SCHOFIELD J D (1995), 'The dynamic rheological properties of glutens and gluten sub-fractions from wheats of good and poor breadmaking quality', *J Cereal Sci*, **22**, 29–44.

KHATKAR B S, FIDO R J, TAHAM A S and SCHOFIELD J D (2002), 'Functional properties of wheat gliadins. II. Effects on dynamic rheological properties of wheat gluten', *J Cereal Sci*, **35**(3), 307–313.

KRUGER J E and BIETZ J A (Eds) (1994), *High Performance Liquid Chromatography of Cereal and Legume Proteins*, St. Paul MN, American Association of Cereal Chemists.

LEONIL J, GAGNAIRE V, MOLLE D, PEZENNEC S and BOUHALLAB S (2000), 'Application of chromatography and mass spectrometry to the characterisation of food proteins and derived peptide', *J Chromatogr A,* **881**(1/2), 1–21.

LI S, DICKINSON L C and CHINACHOTI P (1996), 'Proton relaxation of starch and gluten by solid-state nuclear magnetic resonance spectroscopy', *Cereal Chem*, **73**(6), 736–743.

LOOKHART G L and BEAN S R (1996), 'Improvements in cereal protein separations by capillary electrophoresis: resolution and reproducibility', *Cereal Chem*, **73**(1), 81–87.

MACRITCHIE F (1985), 'Studies of the methodology for fractionation and reconstitution of wheat flours', *J Cereal Sci,* **3**, 221–230.

MAMONE G, FERRANTI P, CHIANESE L, SCAFURI L and ADDEO F (2000), 'Qualitative and quantitative analysis of wheat gluten proteins by liquid chromatography and electronspray mass spectrometry', *Rapid Commun Mass Spectrom*, **14**(10), 897–904.

MATSUSHIMA N, DANNO G, SASAKI N and ZUMI Y (1992), 'Small and X-ray scattering study by synchrotron orbital radiation reveals that high molecular weight subunit of glutenin is a very anisotropic molecule', *Biochem Biophys Res Commun*, **186**, 1057–1064.

MEHRING M (1983), *Principles of High Resolution NMR in Solids*, 2nd edn, Berlin, Springer-Verlag.

MILES M J, CARR H J, MCMASTER T C, LANSON K J, BELTON P S, MORRIS V J, FIELD J M, SHEWRY P R and TATHAM A S (1991), 'Scanning tunnelling microscopy of a wheat seed storage proteins reveals details of an unusual supersecondary structure'. *Proceed. Nat. Acad. Sci.Uni.St.Am.*, **88**(1), 68–71.

MOONEN J H, HEMRNINGA M A and GRAVELAND A (1985), 'Magnetic resonance spectroscopy of wheat proteins: a magic angle spinning ^{13}C nuclear magnetic resonance and an electron spin resonance spin label study'. *J Cer. Sci.*, **3**, 319–327.

NAKATANI A I, POLLIKS M D and SAMULSKI E T (1990), 'NMR investigation of chain deformation in sheared polymer fluids', *Macromolecules*, **23**, 2686–2692.

NYQUIST R A (2001), *Interpreting Infrared, Raman and Nuclear Magnetic Resonance Spectra*, San Diego, Academic Press.

OSBORNE T B (1924), *The vegetable proteins*, 2nd ed, London, Longmans, Green and Co.

PEARCE L E, DAVIS E A, GORDON J and MILLER W G (1988), 'Electron spin resonance studies of isolated gluten systems', *Cereal Chem*, **65**(1), 55–58.

PÉZOLET M, BONENFANT S, DOUSSEAU F and POPINEAU Y (1992), 'Conformation of wheat gluten proteins. Comparison between functional and solution states as determined by infrared spectroscopy', *FEBS Letts*, **299**, 247–250.

POPINEAU Y, BONENFANT S, CORNEC M and PEZOLET M (1994), 'A study by infrared-spectroscopy of the conformations of gluten proteins differing in their gliadin and glutenin compositions'. *J. Cer. Sci.*, *20*, 15–22.

PURCELL J M, KASARDA D D and WU C-S C (1988); 'Secondary structures of wheat a- and w-gliadin protems:Fourier transform infrared spectroscopy', *J. Cer. Sci.*, **7**, 21–32.

ROBERTS G C K (ED) (1993), *NMR of Macromolecules – A Practical Approach*, New York, Oxford University Press.

SCHOFIELD D and BAIANU I C (1982), 'Solid-state cross-polarisation and magic angle spinning carbon-13 nuclear magnetic resonance and biochemical characterisation of wheat proteins', *Cereal Chem*, **59**(4), 240–245.

SCHOLZ E, GANZLER K and GERGELY S (2000), 'Separation of the unique proteins of wheat protein fractions by capillary electrophoresis', *Chromatographia*, **51**, S130–S134.

SHEWRY P R, MILES J M and TATHAM A S (1994), 'The prolamin storage proteins of wheat and related cereals', *Prog Biophys Molec* Biol, **61**, 37–59

SHEWRY P R, GREENFIELD J, BUONOCORE F, WELLBER N, BELTON P S, PARCHMENT O, OSGUTHORPE D and TATHAM A S (1998), 'Conformational studies of the repetitive sequences of the HMW subunits of wheat glutenin', in Gueguen J and Popineau Y, *Plant Proteins from European Crops,* Berlin, Springer-Verlag, 64–69.

STEJSKAL E O and MEMORY J D (1994), *High resolution MR in the solid state-fundamentals of CAP-XMS.* New York-Oxford, Oxford University Press.

TATHAM A S and SHEWRY P R (1985), 'The conformations of wheat gluten proteins. The secondary structures and thermal stabilities of the α β γ and ω-gliadins', *J Cereal* Sci, **3**, 103–113.

TATHAM A S, SHEWRY P R and MIFLIN B J (1984), 'Wheat gluten elasticity: a similar mechanism to elastin?', *FEBS Letts*, **177**, 205–208.

TATHAM A S, DRAKE A F and SHEWRY P R (1989), 'Conformational studies of a synthetic peptide corresponding to the repeat motif of C hordein, *Biochem* J, **259**, 471–476.

TATHAM A S, MASSON P and POPINEAU Y (1990), 'Conformational studies of peptides derived from the enzymatic hydrolysis of a gamma-type gliadin', *J Cereal* Sci, **11**, 1–13.

TATHAM A S, THOMSON N H, MCMASTER T J, HUMPHRIS A D L, MILES M J and SHEWRY P R (1999), 'Scanning probe microscopy studies of cereal seed storage protein structures'. *Scanning*, **21**(5), 293–298.

THOMSON N, MILES M J, TATHAM A S and SHEWRY P R (1992), 'Molecular images of cereal prolamins by STM, *Ultramicroscopy*, **42–44**, 1204–1213.

TSIAMI A A, BOT A and AGTEROF W G M (1997), 'Rheology of mixtures of glutenin subfractions', *J Cereal Sci*, **26**(3), 279–287.

UENO T, STEVENSON S G, PRESTON K R, NIGHTINGALE M J and MARCHYLO B M (2002), 'Simplified dilute acetic acid based extraction procedure for fractionation and analysis of wheat flour protein by size exclusion HPLC and flow field flow fractionation', *Cereal Chem*, **79**, 155–161.

UMBACH J L, DAVIS E A and GORDON J (1998), 'C-13 NMR spectroscopy of conventional and microwave heated vital wheat gluten'. **28**(3), 233–242.

UTHAYAKUMARAN S, NEWBERRY M, PHAN-THIEN N and TANNER R (2002), 'Small and large strain rheology of wheat gluten', *Rheol Acta*, **41**(1/2), 162–172.

VANDIJK A A, VANWIJK L L, VANVLIET A, HARIS P, VANSWIETEN E, TESSER G I and ROBILLARD G T (1997a), 'Structure characterization of the central repetitive domain of high molecular weight gluten proteins. 1. Model studies using cyclic and linear peptides', *Protein Sci*, **6**(3), 637–648.

VANDIJK A A, DEBOEF E, BEKKERS A, VANWIJK L L, VANSWIETEN E, HAMER RJ and ROBILLARD G T (1997b), 'Structure characterization of the central repetitive domain of high molecular weight gluten proteins. 2. Characterization in solution and in the dry state', *Protein Sci*, **6**(3), 649–656.

WERNER W E, WIKTOROWICZ J E and KASARDA D D (1994), 'Wheat varietal identification by capillary electrophoresis of gliadins and high-molecular-weight gluterin subunits', *Cereal Chem*, **71**, 397–402.

XIA Y, CALLAGHAN P T (1991), 'Study of Shear Thinning in High Polymer Solution using Dynamic NMR Microscopy', *Macromolecules*, **24**, 4777–4786.

YEBOAH N A, POPINEAU Y, FREEDMAN R B, SHEWRY P R and TATHAM A S (1994), 'Fluorescence studies of two γ-gliadin fractions from bread wheat', *J Cereal Sci,* **19**(2), 141–148.

6

Wheat proteins and bread quality

E. N. Clare Mills, N. Wellner, L. A. Salt, J. Robertson and J. A. Jenkins, Institute of Food Research, UK

6.1 Introduction: cereal protein classification

Cereal grain proteins were among the first proteins to be systematically studied, beginning with the historic report written in 1745 by Beccari when he described the isolation of gluten by washing wheat flour with dilute salt solutions. The next great landmark in cereal protein characterisation was the work of Osborne, who, 150 years later, brought together a number of other extraction procedures to give rise to what is now known as the 'Osborne fractionation' (Osborne, 1924). Based on the differential solubilities of proteins in a variety of solvents it exploits sequential extraction to separate proteins into a number of groups as follows:

- **Albumins** – water-soluble proteins.
- **Globulins** – proteins soluble in 0.5–1.0 M salt.
- **Prolamins** – proteins soluble in 60–70% aqueous ethanol.
- **Glutelins** – dilute acid or alkali extractable proteins.

Subsequently this procedure has been modified by, for example, replacing ethanol with other alcohols. Thus, 50% (v/v) propan-l-ol is now used to solubilise certain prolamin fractions, while reducing agents have also been included to improve the extraction of prolamin subunits present in polymers associated by inter-chain disulfide bonds. The use of detergents, such as SDS, and/or chaotropic agents like urea has also been adopted to solubilise glutelins instead of acids or alkalis, which can cause partial degradation of proteins (Miflin *et al.*, 1983). Detailed studies of fractionation indicate that the albumin and globulin fractions account for around 30% of wheat flour protein, with 50% of the remainder comprising the storage prolamins. The latter are characteristically rich in the amino acids proline and glutamine, from which the name 'prolamin' is derived. The prolamins consist of two main fractions: (1) the monomeric gliadins, which are characteristically soluble in 70% (v/v) ethanol or dilute acetic acid; and (2) the polymeric glutenins, which are soluble in 2% (w/v) sodium dodecyl sulfate (SDS) or 50% (v/v) propan-1-ol containing a reducing agent such as 2% (v/v) 2-mercaptoethanol.

Electrophoretic analysis has shown that each of these 'Osborne' prolamin fractions comprises a heterogeneous mixture of polypeptides. Thus, lactic acid polyacrylamide gel electrophoresis (PAGE) can be used to separate the gliadins into four mobility groups termed α-, β-, γ- and ω-gliadins, the ω-gliadins being further split into 'fast' and 'slow' components. In general the α-, β- and γ-gliadins have M_r values of 35–45 000 by SDS-PAGE, the ω-gliadins being much larger with M_r values ~ 60 000. Following reduction the polymeric glutenins can be resolved into high-molecular weight (HMW) subunits of glutenin, with M_r values ~ 100 000, and the somewhat smaller low-molecular weight (LMW) subunits of glutenin which have M_r values ~45–50 000. LMN subunits have also been split into B-, C- and D-types on the basis of their mobilities (Shewry and Tatham, 1999).

6.1.1 Molecular classification

In the succeeding years since Osborne's work, it has become apparent that many proteins have overlapping solubilities, and consequently the same proteins may be present to differing extents in several Osborne fractions (Miflin *et al.*, 1983). Thus, some proteins can be extracted, albeit to differing extents, in both water and dilute salt, a situation that is even more complex for the prolamins. It was only with the advent of gene cloning in the 1970s and 1980s, which led to an explosion in the availability of protein sequences, that proteins could begin to be classified on the basis of both their molecular structures and their solubility properties. This has been done to particular effect for the prolamin storage proteins by Shewry and co-workers who placed them into three groups (Shewry *et al.*, 1986). The largest of these is the sulfur-rich (S-rich) prolamins, which include the monomeric α-type gliadins, γ-gliadins and polymeric LMW B and C-type subunits of glutenin, while the monomeric ω-gliadins plus D-type LMW subunits, and the polymeric HMW subunits of glutenin, form the S-poor and HMW prolamin groups respectively.

6.1.2 Protein families

Unlike the prolamins, there has been no systematic classification of the soluble proteins found in the albumin and globulin fractions of wheat flour, largely because of their apparent heterogeneity. However, in recent years, stimulated by the avalanche of information from determining the three-dimensional structures of proteins, it has become evident that living organisms have used a number of three-dimensional scaffolds to generate proteins with a huge diversity of function. Many of these have been classified into families or superfamilies, some of which have been conserved throughout evolution, being found in bacteria, animals and plants, others having being found only in one type of organism.

In tandem with efforts to sequence the genomes of major organisms, new bioinformatic tools have been developed to predict the gene function, which capitalise on protein classification into protein families. These make use of either the full protein sequence or characteristic domains, such as calcium binding motifs or nucleotide binding domains. Databases such as Pfam (Bateman *et al.*, 2002) and SMART (Letunic *et al.*, 2002) have been created using approaches where 'seed' sequences are aligned, a process that includes an element of knowledge-based manipulation to incorporate structural information. These seed sequence alignments can then be used to search for homologues in the protein and nucleotide sequence databases using algorithms such as the hidden Markov models, to produce alignments of a complete family. Several types of this tool have been combined in the INTERPRO database (Apweiler *et al.*, 2001; http://www.ebi.ac.uk/interpro/) and they offer the means of classifying the proteins predicted in an entire genome. As more information about the structural and functional properties of proteins is published, deposited and analysed, it is likely that the predictive value of these methods will improve. However, they have been developed to classify well-ordered globular proteins and it may be more difficult to detect homology or predict function in disordered proteins or highly repetitive sequences.

The application of such an approach to classifying wheat endosperm proteins can be illustrated by the so-called prolamin superfamily first identified by Kreis *et al.* (1985), members of which are not solely confined to cereals. It includes a number of proteins, which have quite different solubilities and functions in the plant and brings together the seed storage proteins, α-amylase/trypsin inhibitors, puroindolines (PINs) and non-specific lipid transfer proteins (nsLTPs) (Fig. 6.1(a)). Despite the diversity of their functions these proteins all share a conserved pattern of cysteine residues containing characteristic Cys-Cys and Cys-X-Cys motifs, where X represents any other residue. Although the alignment shown in Fig. 6.1(a) is imperfect, with only the CC sequence on the second helix absolutely conserved, the conserved cysteine skeleton can be defined by the formula:

$$\text{Cys-(X}=7-13\text{)-Cys-(X}=8-26\text{)-Cys-Cys-(X}=8-30\text{)-Cys-X-Cys-}$$

$$\text{(X}=20-48\text{)-Cys}$$

A dendrogram of the wheat prolamin superfamily members shows it is divided into three deep branches corresponding to different arrangements of the disulfide bridges formed by the cysteine skeleton, the length of the branches indicating the degree of relatedness (Fig. 6.1(b)). The proteins shown in Fig. 6.1 were chosen to represent the most divergent sequences in the wheat prolamin superfamily in order to maximise the branch length, groups of closely related

```
tr|Q9FVJ5|GSP1                  --MKIFFLLAFLALVVNTAIAQYAEVPSPAAQAPTADGSGEWVAIAPSAS--GSENC---
tr|Q43659|15kDaGSP              --MKTFFLLAFLALVVSTTIAQYAEVPSPAMQAPTADVFGEWVAIAPSAS--GFEDC---
tr|Q43658|15kDaGSP              --MKTFFLLAFLALVVSTAIAQYAEVPSPAAQRPTADGFGEWVAIAPSAS--GSENC---
tr|Q43657|15kDaGSP(frag)        -----------VVLVVSTAIAQYAEVPSPAAQAPTADGFGEWVAIAPSAS--GSENC---
sp|Q10464|PINb                  --MKTLFLLALLALVASTTFAQYSEV--------------GGWYNEVGGG--GSQQC---
sp|P33432|PINa                  --MKALFLIGLLALVASTAFAQYSEV--------------VGSYD-VAGGG--GAQQC---
sp|P21292|gamma-gliadin         ---------------------------------------------AIQSF----------
sp|P08079|gamma-gliadin (frag)  ------------------------------------------------------------
sp|P04729|gamma-gliadin         ------------------------------------------------------------
sp|P04726|alpha/beta-gliadin    ------------------------------------------------------------
tr|Q41540|CM17                  MASKSNY-NLLFTALLVFIF------------------------AAVAAVG--NEDCTPW
sp|P16159|CM16                  MASKSNC-VLLLAAVLVSIF------------------------AAVAAIG--NEDCTPW
sp|P16851|CM2                   MASKSSITHLLLAAVLVSVF------------------------AAAAATG--PY-CYPG
sp|P17314|CM3                   MACKSSCSLLLLAAVLLSVL------------------------AAASASG--S--CVPG
sp|P01085|0.19 inhibitor        --------------------------------------------------SGPW-MCYPG
sp|P01084|0.53 inhibitor        --------------------------------------------------SGPW-MCYPG
tr|P93602|PUP88 protein         MASNHRR-FLLSGAVLLSVL------------------------AAVAALESVEDECQPG
sp|P01083|0.28 inhibitor        --------------------------------------------------SGPWSWCNPA
sp|P24296|9kDa ns LTP           ------AQVMLMAVALVLML-----------------------AAVPRAA--VAIDC---
tr|Q9S876|ns LTP                ----------------------------------------------------AVANC---
tr|P82900|7kDa ns LTP           ------------------------------------------------------------
tr|P93611|Cold acclimation prot ----------------------------------------------------STGQC---
sp|P10387|HMW su 1Dy10          --MAKRLVLFAAVVIALV------------------------ALTTAEGEASRQLQCE--
```

```
tr|Q9FVJ5|GSP1                  -EEEQPKLDSCSDYVMDRCVT-KDMPLSWV-----FPQTWGKRSCE-EVQNQCCQQLRQT
tr|Q43659|15kDaGSP              -EEEHPKLDSCSDYVMDRCVM-KDMPLSWI-----FPRTWGKRSCE-EVRNQCCQQLRQT
tr|Q43658|15kDaGSP              -EEEQPKVDSCSDYVMDRCVM-KDMPLSWF-----FPRTWGKRSCE-EVRNQCCKQLRQT
tr|Q43657|15kDaGSP(frag)        -EEEQPKVDSCSDYVMDRCVM-KDMPLSWF-----FPRTWGKRSCE-EVRNQCCKQLRQT
sp|Q10464|PINB                  -PQERPKLSSCKDYVMERCFTMKDFPVTWP-----T-K-WWKGGCEHEVREKCCKQLSQI
sp|P33432|PINA                  -PVE-TKLNSCRNYLLDRCSTMKDFPVTWR-----WWK-WWKGGCQ-ELLGECCSRLGQM
sp|P21292|gamma-gliadin         ---LQQQMNPCKNFLLQQCNHV-SLVSSLV------SIILPRSDCQ-VMQQQCCQQLAQI
sp|P08079|gamma-gliadin(frag)   -------VNPCKNFLLQQCKPV-SLVSSLW------SMIWPQSDCQ-VMRQQCCQQLAQI
sp|P04729|gamma-gliadin         -------LNPCKVFLQQQCSPV-AMPQRLA-----RSQMWQQSSCH-VMQQQCCQQLQQI
sp|P04726|alpha/beta-gliadin    -------LIPCRDVVLQQHNIA-H-----A-----RSQVLQQSTYQ-PLQQLCCQQLWQI
tr|Q41540|CM17                  TSTLITPLPSCRNYVEEQACRI-EMPGP----------------P-YLAKQECCEQLANI
sp|P16159|CM16                  MSTLITPLPSCRDYVEQQACRI-ETPGS----------------P-YLAKQQCCGELANI
sp|P16851|CM2                   MGLPSNPLEGCREYVAQQTCGV-GIVGSPVS----------TEPG-NTPRDRCCKELYDA
sp|P17314|CM3                   VAFRTNLLPHCRDYVLQQTCGT-FTPGSKLPEWMTSASIYSPGKP-YLAKLYCCQELAEI
tr|P93602|0.19 inhibitor        VAFPHNALATCHTYVIKRVCGR-GP-----------------SRP-MLVKERCCRELAVV
sp|Q43691|0.53 inhibitor        REITYESLNARREYAVRQTCGY-YL------------------SA-ERQKRRCCDELSKV
sp|P01085|PUP88 protein         QAFQVPALPACRPLLRLQCNGS-QVP--------------------EAVLRDCCQQLAHI
sp|P01083|0.28 inhibitor        TGYKVSALTGCRAMVKLQCVGS-QVP--------------------EAVLRDCCQQLADI
sp|P24296|9kDa ns LTP           -GHVDSLVRPCLSYVQG-G---------------------------PGPSGQCCDGVKNL
tr|Q9S876|ns LTP                -GQVVSYLAPCISYAMGRV---------------------------SAPGGGCCSGVRGL
tr|P82900|7kDa ns LTP           ---------ACQASQLAVCASA-------------------ILSG-AKPSGECCGNL-RA
tr|P93611|Cold acclimation prot -PIDTLKLLGCVDGLNGLVH---------------------AVIG-SSASDSCCPLLSGV
sp|P10387| HMW su 1Dy10         RELQESSLEACRQVVDQQLAGR--LPWST------------------GLQMRCCQQLRDV
```

```
tr|Q9FVJ5|GSP1                  T-PR----CRCKAIWTSIQGDLS-----GFKGLQQG------------------------
tr|Q43659|15kDaGSP              T-PR----CRCKAIWTSIQGDLS-----GFKGVQQG------------------------
tr|Q43658|15kDaGSP              T-SR----CRCKAIWTSIQGDLS-----GFKGLQQG------------------------
tr|Q43657|15kDaGSP(frag)        T-SR----CRCKAIWTSIQGDLS-----GFKGLQQG------------------------
sp|Q10464|PINB                  A-PQ----CRCDSIRRVIQGRLG-----GFLGIWRG------------------------
sp|P33432|PINA                  P-PQ----CRCNIIQGSIQGDLG-----GIFGFQRD------------------------
sp|P21292|gamma-gliadin         P-QQ----LQCAAIHSVAHSIIMQQEQQ------------QGVPILRPLFQLAQ------
sp|P08079|gamma-gliadin(frag)   P-QQ----LQCAAIHTIIHSIIMQQEQQ---------EQQQGMHILLPLYQQQQ------
sp|P04729|gamma-gliadin         P-EQ----SRYEAIRAIIYSIILQEQQQGFVQPQQQQPQQSGQGVSQSQQQSQQQLGQCS
sp|P04726|alpha/beta-gliadin    P-EQ----SRCQAIHNVVHAIILHQQQR---------QQQPSSQVSLQQPQQQYPSGQGF
tr|Q41540|CM17                  P-QQ----CRCQALRYFMG-----PKS---------RPDQSGL-----------------
sp|P16159|CM16                  P-QQ----CRCQALRYFMG-----PKS---------RPDQSGL-----------------
sp|P16851|CM2                   S-QH----CRCEAVRYFIG-----RTS---------DPNSGVL-----------------
sp|P17314|CM3                   S-QQ----CRCEALRYFIALPVPSQPVDPRSG----NVGESGL-----------------
tr|P93602|0.19 inhibitor        P-DY----CRCEALRVLMDGVRA------EEG----HVVEGRL-----------------
sp|Q43691|0.53 inhibitor        P-EL----CWCEVLRILMDRRVT------KEG----VVKDSLL-----------------
sp|P01085|PUP88 protein         S-EW----CRCGALYSMLDSMYKEHGAQ---------EGQAGT-----------------
sp|P01083|0.28 inhibitor        NNEW----CRCGDLSSMLRSVYQELGVR---------EGKE-------------------
sp|P24296|9kDa ns LTP           H-NQ----ARSQSDRQSACN--------CLKGIARG------------------------
tr|Q9S876|ns LTP                N-AA----AATPADRKTTCT--------CLKQQASG------------------------
tr|P82900|7kDa ns LTP           Q-QG----CFCQYAKDPTYG----------------------------------------
tr|P93611|Cold acclimation prot A-DLDAALCLCTTIKLKALN----------------------------------------
sp|P10387|HMW su 1Dy10          S-AK----CRSVAVSQVARQ----------------------------YEQTVVPPKGGS
```

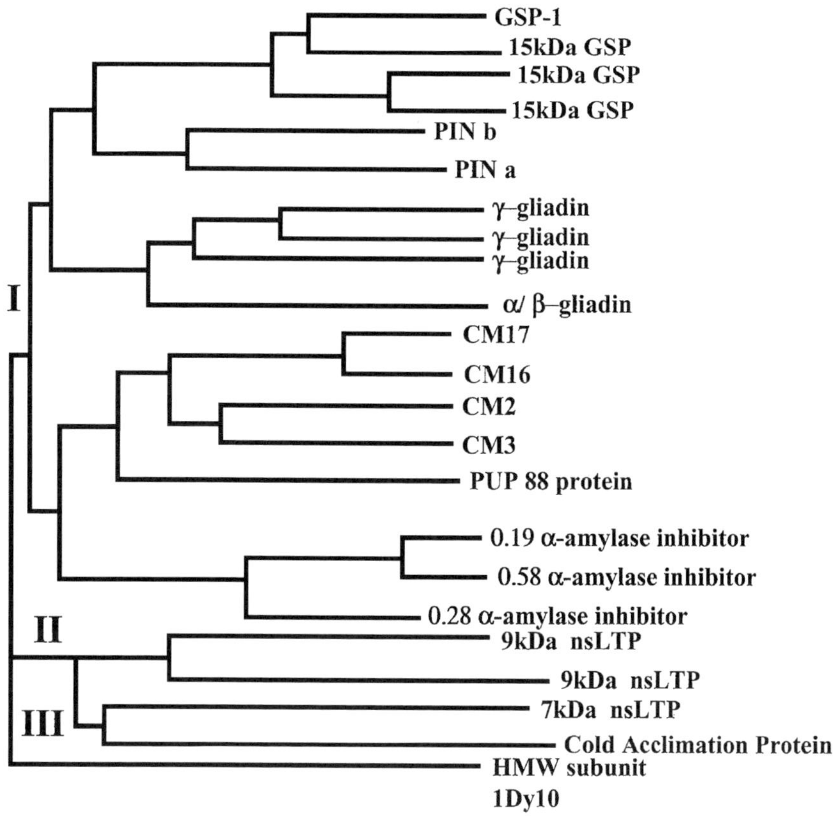

Fig. 6.1 A major wheat endosperm protein family – the prolamin superfamily. (a) (opposite): Aligned sequences of members of the prolamin superfamily in wheat showing the core conserved cysteine residues highlighted in bold. The alignment of the prolamin superfamily in wheat was calculated using T_Coffee Version 1.41 (Notredame *et al.*, 2000). The sequences are listed with their SwissProt or translated EMBL codes. All cysteines are shown in bold and highlighted. The following regions were removed from the prolamin storage protein and cold acclimation protein sequences prior to alignment: P21292, 1–158 residues; P08079, 1–152 residues; P04729, 1–130; P04726, 1–151; P10387, 146–648; P93611 227–315.
(b) (above): Dendrogram showing the relationships of members of the prolamin superfamily in wheat calculated using T_Coffee version 1.41 (Notredame *et al.*, 2000) and Treeview (Page, 1996). Only single representatives of closely related sequences have been retained.

proteins sequences being reduced to a single representative. Only the *C*-terminal portions of the gliadins, and the *N*-terminal region of the HMW subunit 1Dy10 were used in the alignment. Similarly only residues 1–226 of the cold acclimation protein were used, as this represents one of the two prolamin-like domains found in this protein. Branch I comprises the PINs and grain softness proteins, together with the S-rich prolamins, and the α-amylase/trypsin inhibitor family, branch II the ns LTPs, and branch III the HMW subunits of glutenin. For most members of the prolamin superfamily the conserved cysteine skeleton accounts for almost the whole protein. The exceptions are the prolamin storage proteins, which contain a repetitive domain inserted into the skeleton at either the *N*- or *C*-terminal end. The HMW subunits lie on branch III of the dendrogram, which is the most divergent part of the dendrogram and have only one of the cysteines in the CXC motif. The remainder of the cysteine skeleton was probably lost during evolution when the repetitive domain was inserted, making it impossible to construct a reliable model of the folding of the globular regions of these sequences, although successful models have been built for other members of the superfamily.

Although the overall degree of sequence identity between the conserved regions of various members of the prolamin superfamily is low, a comparison of known three-dimensional structures of the prolamin superfamily members demonstrates striking similarity at the structural level. Structures have been determined for certain members of branches I (α-amylase/trypsin inhibitors) and II (nsLTPs) and are illustrated by those of the structure of the 0.19 α-amylase inhibitor and 9 kDa nsLTP of wheat shown in Fig. 6.2. They share a related fold consisting of bundles of four α-helices stabilised by disulfide bonds, the positions of the α-helices being approximately conserved, although there are some significant shifts, particularly in relation to the disulfide connectivities. The CC pair is the best-conserved feature. It is located on helix 2 and forms conserved disulfides. The α-amylase/trypsin inhibitors have an additional disulfide bridge when compared with nsLTPs, illustrating that differences in the arrangements of the disulfide bridges underlie the divisions on the dendrogram between branches II and III. Thus, between the α-amylase inhibitors and the ns-LTPs, the CXC sequence is displaced by two turns of helix 3 and is connected to other cysteine residues in reverse order.

The wheat nsLTP branch is also subdivided into a branch of M_r 9000 nsLTPs and a branch with the M_r 7000 nsLTPs, together with a cold acclimation protein. The 7k nsLTP of rice has been shown to have the disulfide connectivity of soybean hydrophobic protein (Liu *et al.*, 2002; Samuel *et al.*, 2002; Baud *et al.*, 1993). Both of the two LTP-like domains of the cold acclimation protein also resemble soybean hydrophobic protein, having 40% sequence identity and a similar hydrophobic character. In the soybean hydrophobic protein the CXC sequence is positioned on the three-dimensional scaffold in the same position as is found in the nsLTPs, but forms disulfide links in the same way as the α-amylase inhibitors. Thus, it is likely that the divergence in the 9k and 7k nsLTP sub-branches of branch II resulted when an insertion of two turns in helix 3 occurred, which was followed by a reorganisation of the disulfide connectivity. Such alterations in cysteine connectivities and shifts in the basic three-dimensional scaffold across the prolamin superfamily show that it is a rare example of protein sequence being more highly conserved than protein structure.

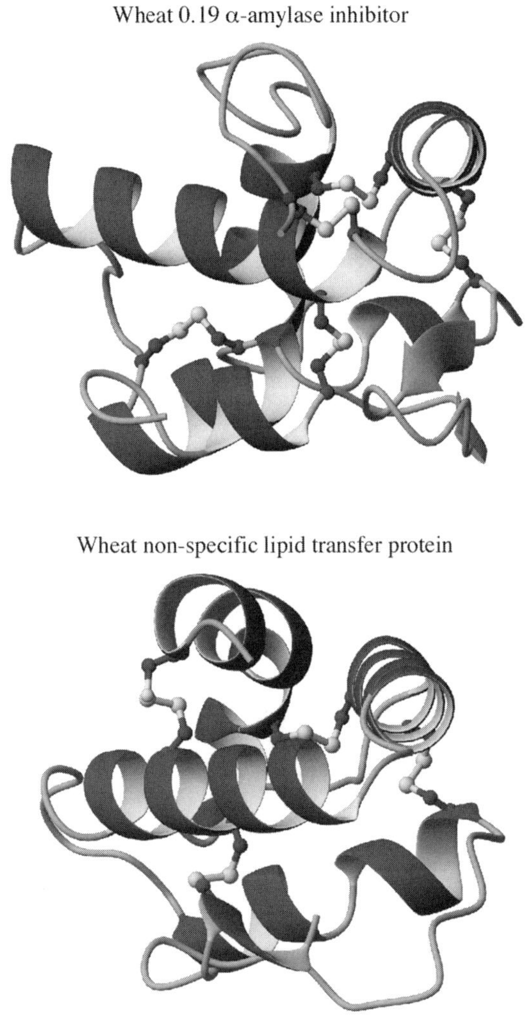

Fig. 6.2 Three-dimensional structures of selected members of the prolamin superfamily from wheat. Ribbon diagrams of the experimentally determined three-dimensional structures of two members of the wheat prolamin superfamily, in similar orientations with the disulfide bridges shown as 'ball and stick' drawings. The 0.19 α-amylase inhibitor is shown at the top (Oda *et al.* 1997). The wheat non-specific lipid transfer protein (Charvolin *et al.*, 1999) is shown at the bottom, using coordinates from the complex with two molecules of phospholipid, which are not shown. The helices move further apart in the complex to accommodate the phospholipid. The conserved CC sequence is on the second conserved α-helix to the top right in each model and the CXC sequence is on the third α-helix towards the viewer. The *C*-terminus is near the bottom of both drawings.

6.2 Cereal proteins and breadmaking quality

The viscoelastic properties of wheat doughs are a major determinant of baking quality, properties that are largely the result of the structures and interactions of the seed storage prolamins (Shewry and Tatham, 1999). A characteristic of the prolamins is their central domain comprising repeated sequences based on one or two short motifs rich in proline (P) and glutamine (Q). Cysteine residues are present in the non-repetitive *N*- and *C*-terminal domains, and it is these which are responsible for the inter-molecular disulfide bonds linking LMW and HMW subunits of glutenin into glutenin polymers. Over many years, it has become apparent that prolamins are not amenable to conventional methods of three-dimensional structure determination, as they cannot be crystallised, and are too large for high-resolution NMR studies, the high proportions of glutamine and proline also hampering spectral assignment. As a consequence we still lack a precise molecular mechanism for the viscoelastic properties of wheat doughs. A summary is given below regarding the role prolamin structure has on breadmaking functionality, a more detailed overview regarding the role of HMW subunits being given in Chapter 8.

6.2.1 HMW subunits of glutenin

All HMW subunits possess an extensive repetitive domain containing a number of characteristic repeat motifs, corresponding to Pro-Gly-Gln-Gly-Gln-Gln and Gly-Tyr-Tyr-Pro-Thr-Ser-Pro/Leu-Gln-Gln being found in x- and y-type subunits, x-type subunits also containing the tripeptide repeat Gly-Gln-Gln (Shewry *et al.*, 1992, 1994). The non-repetitive *C*-terminal region (81–104 residues) contains one cysteine residue, whereas the *N*-terminal region can have either three (x-type HMW subunits) or five (y-type HMW subunits), with certain subunits, such as 1Dx5, having an additional single cysteine residue in the repetitive domain. Certain of these cysteine residues can participate in inter-molecular disulfide bonds, allowing the HMW subunits to form part of the glutenin polymers (Shewry *et al.*, 1992; Shewry and Tatham, 1997).

Secondary structure prediction has indicated that the *N*- and *C*-terminal domains of HMW subunits are probably α-helical in nature, the repeat sequences having a high propensity to form β-turns. These predictions have been borne out by low-resolution spectroscopic studies using circular dichroism (CD) and Fourier transform infrared (FT-IR) which have shown that HMW subunits contain significant proportions of both α-helical and β-turn structures (Belton *et al.*, 1995; Field *et al.*, 1987). It is tempting to speculate that the *N*-terminal domain may adopt an α-helical structure reminiscent of the other members of the prolamin superfamily, although this may be highly disrupted, especially in HMW subunits of glutenin, given that the repetitive domain breaks up the conserved cysteine skeleton more extensively than is the case for other prolamin storage proteins (Fig. 6.1(a)).

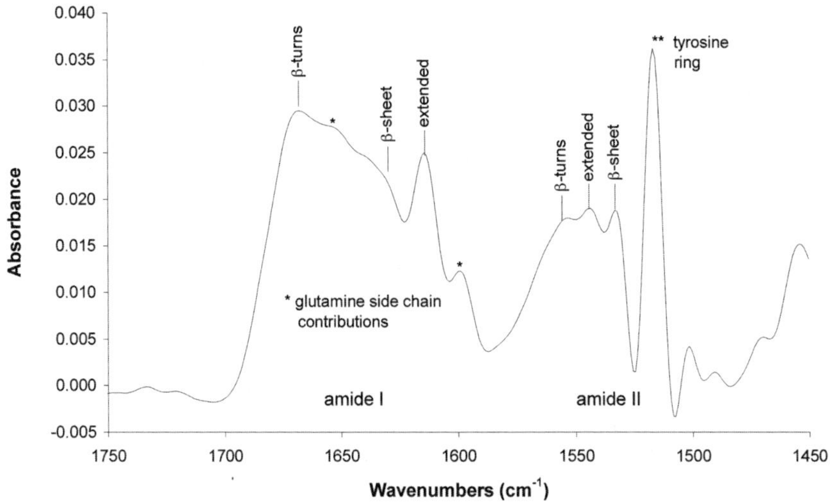

Fig. 6.3 Secondary structure of the repetitive domain of HMW subunits of glutenin. Deconvoluted water-subtracted FT-IR spectrum of a peptide (P45) with the sequence PGQGQQGYYPTSLQQPGQGQQGYYPTSLQQPGQGQQGYYPTSLQQ, corresponding to part of the y-type HMW subunit consensus repeat dissolved in water.

As β-turn type structures have been observed in both synthetic and recombinant peptides corresponding to various sections of the central repetitive domain, it is clear that this domain contributes most of the β-type structure observed in the intact protein (Gilbert *et al.*, 2000; van Djik *et al.*, 1997a,b). This is illustrated by Fig. 6.3, which shows the resolution enhanced amide region in the FT-IR spectrum of a 'perfect repeat' peptide corresponding to three copies of the consensus HMW repeat motif Pro-Gly-Gln-Gly-Gln-Gln-Gly-Tyr-Tyr-Pro-Thr-Ser-Leu-Gln-Gln. It shows the peak assigned to β-turn structures and contributions from β-sheet and extended hydrated structures. Solid-state NMR studies have also indicated that the repetitive domain is highly mobile (Belton *et al.*, 1995; Gilbert *et al.*, 2000) and probably exists as a dynamic ensemble of β-type conformations.

HMW subunits have been shown to be rod-shaped molecules both in solution with dimensions ranging from 49 nm × 1.8 nm in 50% propanol, to 62 nm × 1.5 nm in solvent, such as trifluroethanol (Field *et al.*, 1987), a structure also indicated by scanning probe microscopy studies (Miles *et al.*, 1991; Thomson *et al.*, 1992). On the basis of these data a molecular model has been proposed for the repetitive domain of the HMW subunits which comprises overlapping β-turns adopting a loose supersecondary spiral structure with about 13.5 residues per turn and a diameter of 1.7–1.8 nm (Tatham *et al.*, 1985; Parchment *et al.*, 2001).

6.3 Prolamin structure and bread quality

LMW subunits together with the α-, β- and γ-gliadins constitute about 80% of wheat endosperm storage protein. Their sequence similarity has led to them being grouped together as they all contain a short unique *N*-terminal domain, followed by a short repetitive domain, finishing with a *C*-terminal domain which comprises three homologous subdomains flanked by intermediate regions. The LMW subunits have been classified into B- (the major form), C- and D-type subunits and appear to be structurally very similar to gliadins, differing only in having additional cysteine residues, which allow them to participate in polymer formation. The B-type contain characteristic *N*-terminal sequences, and have been classified as being either LMW-s or LMW-m types depending on whether they have the *N*-terminal sequences Met-Glu-Thr-Arg-Cys-Ile-Pro or Ser-His-Ile-Pro (Lew *et al.*, 1992). The consensus repeat motifs differ from those found in HMW subunits and include:

- Pro-Gln-Gln-Gln-Pro-(Phe-Pro) and Pro-Gln-Gln-Pro-Tyr for the α-gliadins;
- Pro-Gln-Gln-Pro-Phe-Pro-Gln-(Pro) for γ-gliadins; and
- Pro-Gln-Gln-Pro-Pro-Phe-Ser and (Gln-)Gln-Gln-Gln-Gln-(Ile/Val)Leu for the LMW subunits.

CD and FT-IR spectroscopy of intact proteins and fragments has shown that, like HMW subunits, γ-gliadins contain a high proportion (30–35%) of α-helical structures which are mainly attributed to the *C*-terminal domains. The repetitive domains, which contain a high proportion of β-turn structures, may form either β-turn helices and/or poly-Pro II helices (Tatham and Shewry, 1985; Areas and Cassiano, 2001). The monomeric α-gliadins are rather more compact and have a higher content of α-helical structure than the γ-gliadins (Purcell *et al.*, 1988). A subfraction of the α-gliadins is known as A-gliadin because of its propensity to aggregate, forming fibrillar structures 300–400 nm in length (Purcell *et al.*, 1988; McMaster *et al.*, 2000). It has regions of polyglutamine in the *C*-terminal domain and it is probably these, which are responsible for the ability of A-gliadin to aggregate as they have the potential to form polar zippers (Belton *et al.*, 2000).

The secondary structures adopted by LMW subunits are very similar to those of α- and γ-gliadins, comprising about 35% α-helical structures, again thought to be located primarily in the *N*- and *C*-terminal domains, with a slightly lower proportion of β-turn structures originating from the repetitive domain. Unlike gliadins, which are essentially monomeric proteins, LMW subunits are able to form large disulfide-linked aggregates by virtue of two additional cysteines in the *N*- and *C*-terminal region of the polypeptide chain. These are probably unable to form intramolecular disulfide bonds because of their spatial separation, participating rather in the formation of intermolecular disulfide bonds with both other LMW subunits and HMW subunits of glutenin (Shewry *et al.*, 1994; Shewry and Tatham, 1997). Although they make up 50% of gluten protein, the role LMW subunits play in determining the breadmaking potential of wheat is still unclear, mainly because of their higher variability and polymorphism which has made quality correlations with individual subunits difficult.

6.3.1 S-poor prolamins

The S-poor prolamins correspond to the monomeric prolamin fraction known as ω-gliadins, and lack both cysteine and methionine residues. Much larger than the other gliadin species, ω-gliadins have M_r values of ~ 44–78 000. The recent publication of an ω-gliadin sequence showing the high level of homology between the S-poor prolamins of the Triticeae has justified the approach of using C hordein as a model for studying the structure of S-poor prolamins (Masoudi-Nejad *et al.*, 2002). This sequence confirms that ω-gliadin, like S-poor prolamins from other cereals, consists of a large repetitive domain made up almost entirely of penta- and octapeptide repeats with consensus sequences of Pro-Gln-Gln-Pro-Tyr and Pro-Gln-Gln-Pro-Phe-Pro-Gln-Gln flanked by very short unique *N*- and *C*-terminal regions (Tatham and Shewry, 1995).

It was from the sequence of C hordein that the first predictions were made that the repetitive domains of prolamins would contain several overlapping β-turns. CD and FT-IR spectroscopy have shown that the S-poor prolamins exist in a temperature-dependent equilibrium of β-turn and poly-Pro II structures with little α-helix, the poly-Pro II structures dominating at lower temperatures (Tatham *et al.*, 1989; Brett *et al.*, 2001). Like the repetitive domains of other prolamins, the structures adopted vary depending on the solvent, trifluoroethanol (TFE) promoting β-turn formation, although in water or aqueous buffers they exist as a concentration-dependent mixture of β-turns, unordered and β-sheet structures. ω-Gliadins appear to be rod-shaped molecules with dimensions of 36 $\times$ 1.7 nm to 26.5 $\times$ 2 nm, depending on solvent and temperature, as determined by viscometry (Field *et al.*, 1986), X-ray scattering studies of C hordein have shown it to adopt a stiff worm-like coil structure (I'Anson *et al.*, 1992).

6.3.2 Role of prolamin composition and structure in breadmaking quality

How does our understanding of prolamin structure and properties bring us closer to understanding the molecular mechanisms underlying the unique viscoelastic properties of wheat doughs and gluten? While there are numerous reports regarding the relationship between individual gliadin and/or LMW subunits or alleles and their effects on baking quality, the most clear-cut association has been made between HMW subunits, even though they constitute only around 8–10% of the gluten protein (Halford *et al.*, 1992). It has been proposed that the β-spiral structure of HMW subunits may contribute to the elastic mechanism of gluten, although it is probable that interactions (disulfide cross-linking and non-covalent hydrogen bonding) between individual subunits are also involved (Belton, 1999; Shewry *et al.*, 1992). Both links between HMW subunits, and HMW to LMW subunit links are important for formation of the disulfide-bonded polymers of several million daltons in size found in glutenin. The good correlation between the amount of disulfide-linked aggregated prolamins and dough strength (Field *et al.*, 1983b) shows the importance of these polymers in determining the viscoelastic properties of dough.

While prolamins probably exist as disulfide-bonded polymers within the wheat grain itself (Field *et al.*, 1983a), the ability of the disulfide bonds to rearrange during mixing is probably an important feature in the development of a viscoelastic gluten network. Thus, bread ingredients, such as ascorbate, are thought to mediate their positive effects on dough rheology by catalysing this rearrangement. An ω-gliadin-like D-type LMW subunit of glutenin with a single free cysteine residue has been identified, which may affect the formation of the disulfide-linked polymer network by acting as a chain terminator, thus weakening the dough (Masci *et al.*, 1993). The position and nature of the intermolecular disulfide bonds formed between prolamins may also be important. Thus, it has been proposed that the extra cysteine residue present in the repetitive domain of subunit 1Dx5 is responsible, in part at least, for the quality associated with subunits 1Dx5 + 1Dy10, when compared with subunits 1Dx2 + 1Dy12 (Shewry and Tatham, 1997).

In addition to covalent links, it is increasingly evident that the non-covalent protein–protein interactions between prolamins are also important in determining gluten viscoelasticity. Many of the structural studies on prolamins have been carried out with isolated soluble proteins, yet one of the characteristics of gluten is its insolubility in water and dilute salt solutions. In the past, it was suggested that prolamin aggregation was driven by hydrophobic interactions. However, prolamins are actually rather hydrophilic, and it is now apparent that the formation of intermolecular β-sheet structures is the dominant factor underlying prolamin aggregation and insolubility. Thus, solid-state prolamins contain a higher proportion of β-sheet structures than soluble protein (Pézolet *et al.*, 1992). Similar changes are seen when solid-state ω-gliadins and HMW subunits are hydrated, the proportion of protein aggregation (as indicated by intermolecular β-sheet), decreasing with hydration, this being accompanied by an increase in molecular mobility (Belton *et al.*, 1995; Wellner *et al.*, 1996).

Formed by hydrogen bonding of the repetitive domains via both backbone amides and glutamine side-chains, such intermolecular β-sheets provide a source of non-covalent cross-links in glutenin polymers, in addition to disulfide bonds. It has been proposed that these may also contribute to the elastic properties of dough (Belton, 1999). Polymeric glutenins contain more intermolecular β-sheet than monomeric gliadins (Popineau *et al.*, 1994). Incorporation experiments have shown that addition of both total gliadins and purified gliadin fractions to gluten increased the proportions of un-cross-linked material, which were greatest for α- and ω-gliadin fractions (Khatkar *et al.*, 2002b). These fractions can improve the mixing and baking properties of doughs, although this is least pronounced for the ω-gliadins (Khatkar *et al.*, 2002a). Such data suggest that the gliadins may be able to modify intermolecular β-sheet formation in gluten polymers, leading to the proposition that they may act as plasticisers.

6.4 Soluble proteins, xylanase inhibitors and bread quality

The soluble, non-gluten protein albumin and globulin fractions of wheat endosperm are a complex mixture of proteins. This is illustrated in Fig. 6.4 by a two-dimensional PAGE of salt extractable proteins from wheat flour which has been annotated using modern proteomic methods. This soluble protein fraction contains a number of metabolic proteins, in addition to those with an anti-pathogen role. Unlike gluten proteins, a clear-cut relationship between soluble protein components and baking quality has not been readily demonstrated, even though it is generally accepted that gluten proteins do not account for all the observed variations in baking quality. Thus, it has been found that β-amylase, an abundant non-gluten protein in flour, can form disulfide-linked oligomers in barley (Shewry *et al.*, 1988), and can become disulfide-linked into the glutenin polymers (Peruffo *et al.*, 1996). There is also some evidence that the amount of β-amylase is inversely correlated with the size of glutenin polymers, implying that it acts in some way to limit polymer formation (Curioni *et al.*, 1996). Other soluble proteins that may affect baking quality indirectly are the xylanase inhibitors, which are described in more detail below.

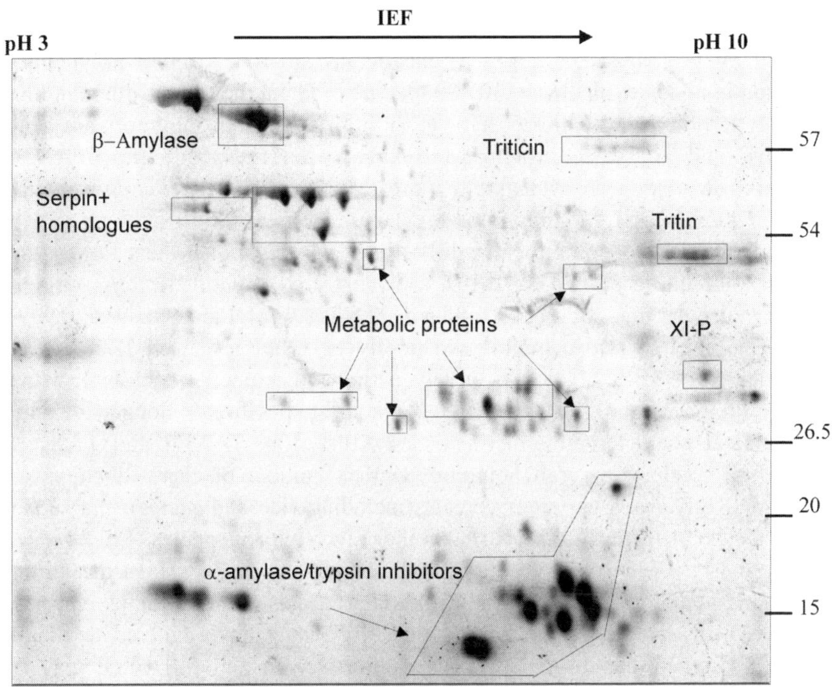

Fig. 6.4 Two-dimensional PAGE of salt-extractable proteins from wheat endosperm. Proteins were extracted from white flour (cv Hereward) in 0.5 M NaCl. First dimension was isoelectric focusing performed on a pH 3–10 gradient followed by a 15% SDS-PAGE separation for the second dimension. Annotation was performed using a combination of MALDI-TOF and Q-TOF mass spectrometry.

6.4.1 Xylanase inhibitors

There is an increasing use of endoxylanases (also known as pentosanases) as processing aids in the baking industry to improve the rheological properties of wheat-based doughs, oven spring and final loaf volume. They have also found a role in other end-uses such as gluten manufacture and particularly as additives in the animal feed industry to improve the nutritional quality of cereal-based feeds for farm animals (Courtin and Delcour, 2002). Such aids are thought to mediate their effect through the solubilisation and modification of the structure and properties of cell wall non-starch polysaccharides, known as arabinoxylans (AX) or pentosans. The endoxylanases used are generally of microbial origin, many coming from fungi such as *Trichoderma viride* or *Aspergillus niger*. However, it has been found that AX solubilisation is lower in wheat doughs than in model reconstituted flour systems, and subsequently the agents responsible for these effects have been identified as endoxylanase inhibitors (Rouau and Surget, 1998).

To date two main types of endoxylanase inhibitor have been characterised in wheat and other cereal species, particularly barley. One of the inhibitors from wheat has been named TAXI (*T. aestivum* L. endoxylanase inhibitor), an M_r ~ 40 000 protein which exists as two molecular forms, A and B, now termed TAXI I and II respectively (Debyser *et al.*, 1999). The TAXI II form is thought to be derived by proteolysis of TAXI I, existing as an M_r ~ 10,000 polypeptide disulfide-bonded to an M_r ~ 30,000 polypeptide in the seed, although this proteolytic processing pathway has yet to be proven. Both inhibitors are heat-sensitive and are basic proteins, but have different specificities, TAXI I inhibiting *A. niger* endoxylanase to a greater extent than the *Bacillus subtilis* enzyme, TAXI II only inhibiting endoxylanase from *B. subtilis*.

A second class of inhibitor which has been purified from wheat flour is the xylanase-inhibiting protein I (XIP-I), a glycosylated, monomeric basic protein with a molecular mass of 29 kDa, and a p*I* of 8.7–8.9 (McLauchlan *et al.*, 1999). XIP-I has been shown to inhibit two family-11 xylanases, from *Trichoderma viride* and *Aspergillus niger* respectively, acting as a competitive inhibitor with a strict preference for fungal xylanases. This strict specificity, along with other biochemical characteristics, distinguishes XIP-I from TAXI I and TAXI II. Indeed XIP-I shows overall homology with a number of class III chitinases (family 18 glycosidases) from cereals (including rice and maize), the rubber plant, *Hevea brasiliensis*, and various other diverse plant species (Elliott *et al.*, 2002). It is evident from the aligned sequences that XIP-I shares the chitin-binding domain of these proteins, although it has no detectable chitinase activity. Consequently, it is thought that XIP-I may have evolved to inhibit xylanases rather than to function as a chitinase, an activity which may be relevant to its physiological role in the wheat plant, since cell wall hydrolases and their inhibitors are thought to form part of the plant's defence response to pathogen attack.

6.5 Detergent-solubilised proteins and bread quality

In addition to salt and alcohol soluble proteins, there is a small proportion of wheat endosperm proteins, which require detergents for solubilisation, many of which are extrinsic or intrinsic membrane proteins. A recent proteomic study of such proteins extracted with Triton X114 identified around 276 polypeptides which were common to two wheat cultivars, together with a further 170 which were associated with one or another of these cultivars (Amiour *et al.*, 2002). Of these proteins the best characterised and most important with regards to its influence on breadmaking quality, are puroindolines (PINs) and the related grain softness proteins.

6.5.1 Puroindolines

PINs are $M_r \sim 13\,000$ proteins and are found as two major isoforms, PIN-a and PIN-b which exhibit around 55% sequence homology (Gautier *et al.*, 1993; Morris *et al.*, 1994). Both are found in the starchy endosperm, but only PIN-b is found in the aleurone layer. Basic proteins with p*I* values of around 10, PINs are members of the prolamin superfamily (Fig. 6.1). They also have a unique lipid-binding tryptophan-rich domain comprising five tryptophan residues in PINa (Trp-Arg-Trp-Trp-Lys-Trp-Trp-Lys) which is reduced to three in PINb (Trp-Pro-Thr-Lys-Trp-Trp-Lys) (Douliez *et al.*, 2000). Their association with breadmaking quality relates to their potential role in determining grain hardness as well as effects that are more direct, which may be mediated through their lipid-binding properties.

Grain hardness has long been a wheat character that breeders have selected for, as the milling properties of hard wheats enable more efficient industrial milling to obtain white flour with a high extraction rate. The greater starch damage obtained in hard milling wheats also improves the water-holding capacity of a flour, a factor of economic importance to the baking industry. Grain texture is a trait, which is largely controlled by the *Ha* locus on the short arm of chromosome 5D (Law *et al.*, 1978), and while the genetic basis of endosperm texture is well established, the molecular basis is not. One factor thought to be important in determining endosperm hardness is the adhesion between the starch granules and the protein matrix of the endosperm. An $M_r \sim 15\,000$ protein 'friabilin' was found to be associated with the starch granule surface in soft milling wheat cultivars by Greenwell and Schofield (1986). They proposed that the protein acted to produce a 'non-stick' surface on starch granules from soft milling wheats, reducing the adhesion with the endosperm protein matrix. It has subsequently been found that a major component of friabilin corresponds to proteins extracted from wheat flour using the detergent Triton X114, namely PINs. Other workers have identified a mixture of around four polypeptides termed grain softness protein (GSP), which also includes PINs, together with an additional protein termed GSP-1 which are all found at the *Ha* locus and are also members of the prolamin superfamily (Fig. 6.1) (Rahman *et al.*, 1994; Jolly *et al.*, 1996).

Transformation experiments have shown that while expression of PIN in plant such as rice results in a softer grain texture (Krishnamurthy and Giroux, 2001), there is no clear-cut relationship between levels of PIN-a and PIN-b and milling quality (Greenblatt *et al.*, 1995; Igrejas *et al.*, 2001). The binding of friabilin (and by association the constituent PINs) to starch granules is associated with higher levels of bound phospholipids and glycolipids, indicating that PINs may well bind via the lipids that constitute the remains of the amyloplast membrane (Greenblatt *et al.*, 1995). A number of PIN mutants have been identified, including a null form of PINa associated with harder milling texture, and a mutant which results in a substitution of ^{46}Gly adjacent to the tryptophan-rich region with a serine residue. This may affect the adherence of this protein to the starch granule surface, although this has not been confirmed (Turnbull *et al.*, 2000). Thus, the actual mechanisms whereby PINs influences grain texture may be much more complex than was originally envisaged by Greenwell and Schofield (1986).

In addition to influencing milling texture, PINs may also play a role in determining the crumb structure of bread. This is formed by an interconnected sponge-like network of gas-cells which expand during proving as they fill with CO_2, undergoing coalescence and disproportionation to form a fragile foam-like structure in the risen dough. The gas-cell walls are formed from the starch-gluten matrix of dough and are thought to be lined by a thin liquid film (MacRitchie and Gras, 1973; Gan and Schofield, 1995). During cell expansion discontinuities form in the starch-gluten walls, coalescence being prevented by the liquid film, which ensures gas is retained. The properties and composition of this liquid layer, including the soluble proteins, may be important in determining the crumb structure of bread. Of these, PINs are one of the potentially important candidates, having been shown to improve the foaming properties of lipid-damaged foams and affect the crumb structure of bread (Dubriel *et al.*, 1998). It may be that PINs function to bind wheat flour lipids (Wilde *et al.*, 1993), and hence prevent them from destabilising the fragile liquid films which are thought to line the gas-cells. In other systems, notably that of beer foam, the non-specific lipid transfer proteins (nsLTPs) have also been found to have foam-stabilising properties (Bech *et al.*, 1995), although a role for them in gas-cell stabilisation in bread has yet to be shown.

6.6 Genomics and the wheat grain proteome

The sequence of the model dicotyledonous plant species, *Arabidopsis thaliana*, was published in 2000 (Arabidopsis Genome Initiative), with around 25 498 predicted genes having been identified. This was followed in 2002 by that of the monocotyledonous cereal crop, rice, which has around 32–50 000 predicted genes (Goff *et al.*, 2002; Yu *et al.*, 2002). A large proportion of genes (~ 13 000) appear to be common to *Arabidopsis* and rice, of which 8000 are plant specific and are likely to be found in all plant species. The number of genes and gene families present in hexaploid wheat is inevitably larger than that of either *Arabidopisis* or rice, although the numbers and types of families are probably more similar to those found in rice.

On the basis of our current knowledge of protein structure and function, around 70% of the genes in *Arabidopsis* can be assigned a putative function, leaving the remaining 30% unclassified, indicating that a large proportion of highly conserved plant proteins remain to be characterised. Many of the proteins for which a putative function can be assigned are involved in metabolism, cell signalling, plant defence, cell communication and signal transduction, with a significant proportion (at least ~17% in *Arabidopsis*) involved in transcription (Fig. 6.5(a)). Given the commonalities between *Arabidopsis* and rice genomes, it is likely that many genes will code for proteins involved in core processes and are probably found in all plant species. Wheat endosperm contains the products of a mixture of active genes, including both metabolic and seed storage proteins, as shown by expression and proteomic analysis of developing grain. The endosperm architecture is set between 8 and 12 days post-anthesis (dpa), the grain then going on to synthesise the starch and proteins in the grain filling stages prior to the onset of senescence and desiccation. At this stage between 4500–8000 genes are active in wheat (Clark *et al.*, 2000), and analysis of around 1000 of these genes showed that around 60% could be identified and classified according to their predicted function (Fig. 6.5(b)).

Others have begun to characterise the proteome of developing and mature grain (Skylas *et al.*, 2000) and have estimated that at 17 dpa there are around 1298 detectable proteins, while at 28 dpa this had dropped slightly to around 1125. Of those that could be annotated around 60% of the identified proteins were prolamins of various types, the S-rich α-, β- and γ-gliadins/LMW subunits of glutenin being among the most abundant. The second largest group were those with some type of protective function, of which 38 were different forms of α-amylase/trypsin inhibitors.

Given the hexaploid nature of wheat it is unlikely that its genome will be sequenced. However, the synteny between cereal species is allowing the rice genome to be used as a model genome, since homologues of almost all the sequenced wheat genes can be found in the rice genome (Goff *et al.*, 2002). In addition large libraries of expressed sequence tags (ESTs) from wheat are being assembled, and at present there are 420 070 (wheat) or 433 046 (all *Triticum* species) ESTs deposited in the National Centre for Biological Information (NCBI) database. Over the next few years almost all the major expressed genes in wheat will be represented in this database, although as it is degenerate the fraction of the wheat genome it represents cannot be estimated. A proportion of the ESTs overlap sufficiently to represent complete protein sequences, and hence can be considered equivalent to a complete gene. An analysis of 94 076 ESTs in the NCBI database identified 14 594 overlapping 'clusters', together with approximately 50 000 less stringent 'clusters' and singletons, some 'clusters' being present in as many as 50 copies (Jenkins and Barker, unpublished observations). However, the numbers of such 'clusters' of ESTs

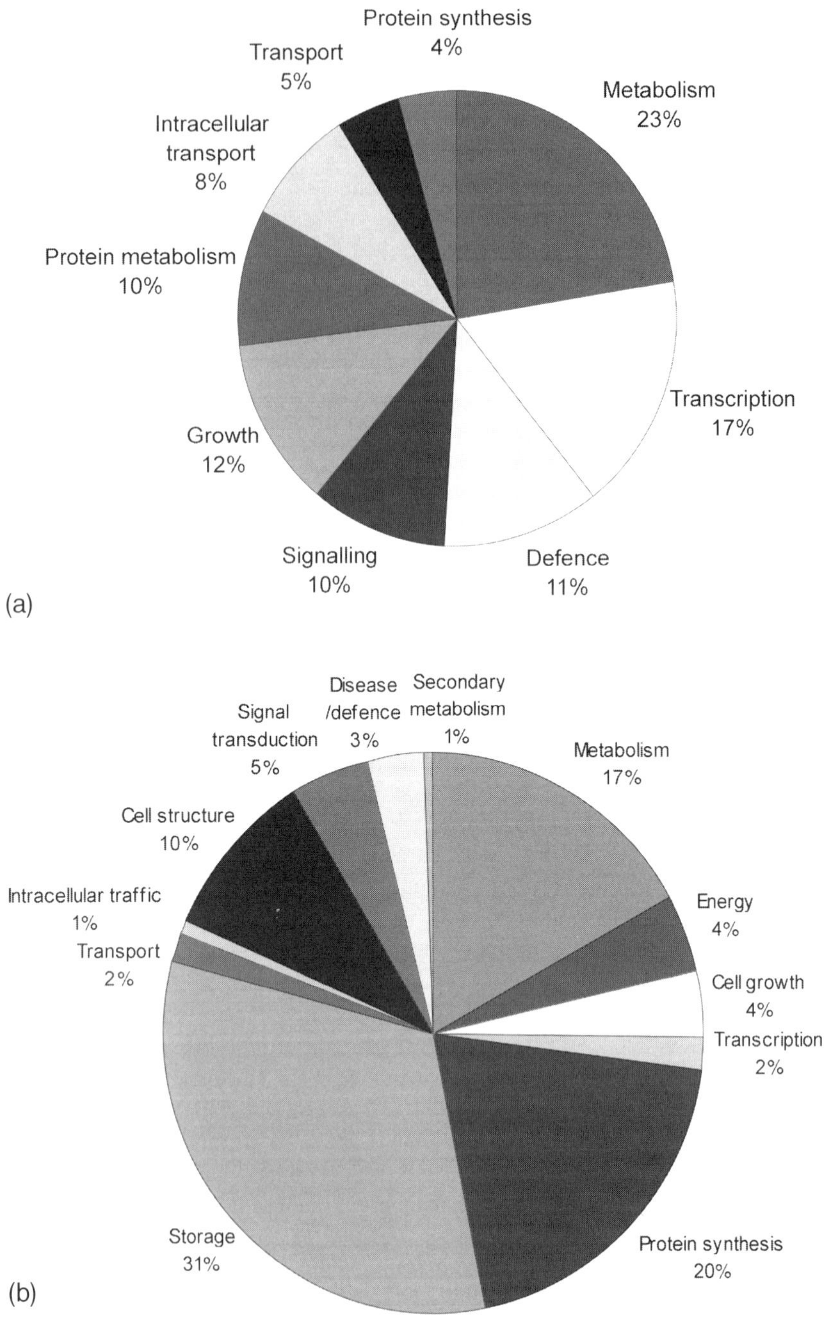

Fig. 6.5 The distribution and function of plant genes.
(a): The distribution of plant genes from the genome of *Arabidopsis thalinia* (after the Arabidopsis Genome Initiative, 2000). (b): The distribution of expressed genes in developing wheat endosperm 17 days post-anthesis (Clark *et al*., 2000).

are likely to be an underestimate as sequencing errors may reduce clustering. Furthermore, EST databases, while representing the bulk of expressed genes, are not representative of genes with low levels of expression. Nevertheless, given its rapid rate of expansion the wheat EST database will shortly represent most of the genes expressed in a given tissue, and certainly those that can be detected using current proteomic methods. The integration of such information from the EST database, together with proteomic profiling and genetic mapping data on the rice genome will, over the coming years, allow the identification of the majority of wheat genes and their functions, including their role in determining breadmaking quality.

6.7 Conclusion and future trends

The availability of sequenced plant genomes, particularly that of rice, will undoubtedly prove a powerful tool for investigating wheat protein function in the future. Such information is set to revolutionise our understanding of wheat endosperm proteins, their structural and evolutionary relationships and their biological function. It will undoubtedly give rise to new avenues for investigating and unravelling the complexities of how wheat endosperm proteins affect the technological properties of wheat doughs and hence the breadmaking quality of wheat. In particular it will allow us to investigate the way in which events during grain development affect the synthesis and deposition of endosperm components and the consequences of this for end-use quality. Such advances will allow the development of better predictive tests for quality, which can take account of both the genetic and environmental factors, as well as more effective knowledge-based strategies for its manipulation.

6.8 Acknowledgements

The work in this chapter was partly funded through a BBSRC CSG grant to IFR, NW was funded through BBSRC grant (218/D14544) and LS through a BBSRC CASE award with RHM Technology.

6.9 References

AMIOUR N MERLINO M LEROY P and BRANLARD G (2002) ‘Proteomic analysis of amphiphilic proteins in hexaploid wheat kernels’, *Proteomics*, **2**(6), 632–641.

APWEILER R ATTWOOD TK BAIROCH A BATEMAN A BIRNEY E BISWAS M BUCHER P CERUTTI L CORPET F CRONING MD DURBIN R FALQUET L FLEISCHMANN W GOUZY J HERMJAKOB H HULO N JONASSEN I KAHN D KANAPIN A KARAVIDOPOULOU Y LOPEZ R MARX B MULDER NJ OINN TM PAGNI M SERVANT F SIGRIST CJ and ZDOBNOV EM (2001) ‘The InterPro database, an integrated documentation resource for protein families, domains and functional sites', *Nucleic Acids Res*, **29**(1), 37–40.

ARABIDOPSIS GENOME INITIATIVE (2000) 'Analysis of the genome sequence of the flowering plant *Arabidopsis thaliana*', *Nature*, **408**(6814), 796–815.

AREAS EPG and CASSIANO MM (2001), 'Folding interpenetration in a gliadin model: the role of the characteristic octapeptide motif', *Biophys Chem,* **90**(2), 135–146.

BATEMAN A BIRNEY E DURBIN R EDDY SR HOWE KL and SONNHAMMER EL (2002) 'The Pfam protein families database', *Nucleic Acids Res,* **28**(1), 263–266.

BAUD F PEBAY-PEYROULA E COHEN-ADDAD C ODANI S and LEHMANN MS (1993) 'Crystal structure of hydrophobic protein from soybean; a member of a new cysteine-rich family', *J Mol Biol*, **231**(3), 877–887.

BECCARI DE FRUMENTO (1745) *De Bononiensi Scientiarum et Artium Instituto atque Academia Commentarii, II.* Part I, 122–127.

BECH LM VAAG P HEINEMANN B and BREDDAM K (1995) 'Throughout the brewing process barley lipid transfer protein 1 is transformed into a more foam-promoting form', in *Proc Congr Eur Brew Conv Lisbon*, 561–568.

BELTON PS (1999) 'On the elasticity of wheat gluten', *J Cereal Sci,* **29**(2), 103–107.

BELTON PS COLQUHOUN I J GRANT A WELLNER N FIELD JM SHEWRY PR and TATHAM AS (1995) 'FTIR and NMR studies on the hydration of a high-$M(_r)$ subunit of glutenin', *Int J Biol Macromol*, **17**(2), 74–80.

BELTON PS WELLNER N MILLS ENC GRANT A and JENKINS J (2000) 'Do high molecular weight subunits of glutenin form "polar zippers"?', in Shewry PR and Tatham AS *Wheat Gluten*, Royal Society Chemistry, Cambridge, 363–367.

BRETT GM MILLS ENC BACON J WELLNER N HUSAIN RD TATHAM AS SHEWRY PR and MORGAN MRA (2001) 'Temperature-dependent binding of monoclonal antibodies to C hordein', *Biochim Biophys Acta*, **1594**(1), 17–26.

CHARVOLIN D DOULIEZ JP MARION D COHEN-ADDAD C and PEBAY-PEYROULA E (1999) 'The crystal structure of a wheat nonspecific lipid transfer protein (ns-LTP1) complexed with two molecules of phospholipid at 2.1 Å resolution', *Eur J Biochem*, **264**(2), 562–568.

CLARK BC HOBBS M SKYLAS D and APPELS R (2000) 'Genes active in developing wheat endosperm', *Funct Integr Genomics*, **1**(1), 44–55.

COURTIN CM and DELCOUR JA (2002) 'Arabinoxylans and endoxylanases in wheat flour bread-making', *J Cereal Sci*, **35**(3), 225–243.

CURIONI A POGNA NE and PERUFFO ADB (1996) 'The quantity of bound β-amylases is related to the size of glutenin polymers', in Wrigley CW ed, *Gluten '96* RACI, Melbourne, Australia, 307–311.

DEBYSER W PEUMANS WJ VAN DAMME EJM and DELCOUR JA (1999) 'Triticum aestivum xylanase inhibitor (TAXI), a new class of enzyme inhibitor affecting breadmaking performance', *J Cereal Sci,* **30**(1), 39–43.

DOULIEZ J-P MICHON T ELMORJANI K and MARION D (2000) 'Structure, biological and technological function of lipid transfer proteins and puroindolines, the major lipid binding proteins from cereal kernels', *J Cereal Sci*, **32**(1) 1–20.

DUBRIEL L MÉLINOLE S CHIRON H COMPOINT JP QUILLEN L BRANLARD G and MARION D (1998) 'Effect of puroindolines on the bread making properties of wheat flour', *Cereal Chem*, **75**(2), 222–229.

ELLIOTT GO HUGHES RK JUGE N KROON PA and WILLIAMSON G (2002) 'Functional identification of the cDNA coding for a wheat endo-1,4-beta-D-xylanase inhibitor', *FEBS Lett*, **519**(1–3), 66–70.

FIELD JM SHEWRY PR BURGESS SR FORDE J PARMAR S and MIFLIN BJ (1983a) 'The presence of high molecular weight aggregates in the protein bodies of developing endosperms of wheat and other cereals', *J Cereal Sci*, **1**(1), 33–41.

FIELD J M SHEWRY PR and MIFLIN BJ (1983b) 'Solubilisation and characterisation of wheat gluten proteins: correlations between the amount of aggregated proteins and baking quality', *J Sci Food Agric*, **34**(4), 370–377.

FIELD JM TATHAM AS BAKER AM and SHEWRY PR (1986) 'The structure of C hordein', *FEBS Lett*, **200**(1), 76–80.

FIELD JM TATHAM AS and SHEWRY PR (1987) 'The structure of a high-M_r subunit of durum wheat (*Triticum durum*) gluten', *Biochem J*, **247**(1), 215–221.

GAN Z ELLIS PR and SCHOFIELD JD (1995) 'Mini-review: gas cell stabilisation and gas retention in wheat bread dough', *J Cereal Sci*, **21**(3), 215–230.

GAUTIER MF ALEMAN ME GUIRAO A MARION D and JOUDRIER P (1993) '*Triticum-aestivum* puroindolines, 2 basic cystine-rich seed proteins – cDNA sequence-analysis and developmental gene-expression', *Plant Mol Biol*, **25**(1), 43–57.

GILBERT SM WELLNER N BELTON PS GREENFIELD JA SILIGARDI G SHEWRY PR and TATHAM AS (2000) 'Expression and characterisation of a highly repetitive peptide derived from a wheat seed storage protein', *Biochim Biophys Acta*, **1479**(1–2), 135–146.

GOFF SA RICKE D LAN TH PRESTING G WANG RL DUNN M GLAZEBROOK J SESSIONS A OELLER P VARMA H HADLEY D HUTCHINSON D MARTIN C KATAGIRI F LANGE BM MOUGHAMER T XIA Y BUDWORTH P ZHONG JP MIGUEL T PASZKOWSKI U ZHANG SP COLBERT M SUN WL CHEN LL COOPER B PARK S WOOD TC MAO L QUAIL P WING R DEAN R YU YS ZHARKIKH A SHEN R SAHASRABUDHE S THOMAS A CANNINGS R GUTIN A PRUSS D REID J TAVTIGIAN S MITCHELL J ELDREDGE G SCHOLL T MILLER RM BHATNAGAR S ADEY N RUBANO T TUSNEEM N ROBINSON R FELDHAUS J MACALMA T OLIPHANT A and BRIGGS S (2002) 'A draft sequence of the rice genome (*Oryza sativa* L. ssp. *Japonica*)', *Science*, **296**(5565), 92–100.

GREENBLATT GA BETTIGE AD and MORRIS CF (1995) 'Relationship between endosperm texture and the occurrence of "friabilin" and bound polar lipids in wheat starch', *Cereal Chem*, **72**(2), 172–176.

GREENWELL P and SCHOFIELD JD (1986) 'A starch granule protein associated with endosperm softness in wheat', *Cereal Chem*, **63**(4), 378–380.

HALFORD NG FIELD JM BLAIR H URWIN P MOORE K ROBERT L THOMPSON R FLAVELL RB TATHAM AS and SHEWRY PR (1992) 'Analysis of HMW glutenin subunits encoded by chromosome-1A of bread wheat (*Triticum aestiuum* L) indicates quantitative effects on grain quality', *Theor Appl Genet*, **83**(3), 373–378.

I'ANSON KJ MORRIS VJ SHEWRY PR and TATHAM AS (1992) 'Small-angle X-ray-scattering studies of the C-hordeins of barley (*Hordeum vulgare*)', *Biochem J*, **287**(1), 183–185.

IGREJAS G GABORIT T OURY F-X CHIRON H MARION D and BRANLARD G (2001) 'Genetic and environmental effects on puroindoline-a and puroindoline-b content and their relationship to technological properties in french bread wheats', *J Cereal Sci*, **34**(1), 37–47.

JOLLY CJ GLEN GM and RAHMAN S (1996) '*GSP-1* genes are linked to he grain hardness locus (*Ha*) on wheat chromosome 5D', *Proc Nat Acad Sci (USA)*, **93**, 2408–2413.

KHATKAR BS FIDO RJ TATHAM AS and SCHOFIELD JD (2002a) 'Functional properties of wheat gliadins. I. Effects on mixing characteristics and bread making quality', *J Cereal Sci*, **35**(3), 299–306

KHATKAR BS FIDO RJ TATHAM AS and SCHOFIELD JD (2002b) 'Functional properties of wheat gliadins. II. Effects on dynamic rheological properties of wheat gluten', *J*

Cereal Sci, **35**(3), 307–313.

KREIS M SHEWRY PR FORDE BG FORDE J and MIFLIN BJ (1985) 'Structure and evolution of seed storage proteins and their genes with particular reference to those of wheat, barley and rye', *Oxford Surv Plant Mol Cell Biol*, **2**, 253–317.

KRISHNAMURTHY K and GIROUX MJ (2001) 'Expression of wheat puroindoline genes in transgenic rice enhances grain softness', *Nat Biotechnol*, **19**(2), 162–166.

LAW CN YOUNG CF BROWN JWS SNAPE JW and WORLAND AJ (1978) 'The study of grain protein control in wheat using whole chromosome substitution lines', in *Seed Improvement by Nuclear Techniques*, 483–502, Internat Atomic Energy Agency, Vienna.

LETUNIC I GOODSTADT L DICKENS NJ DOERKS T SCHULTZ J MOTT R CICCARELLI F COPLEY RR PONTING CP and BORK P (2002) 'Recent improvements to the SMART domain-based sequence annotation resource', *Nucleic Acids Res*, **30**(1), 242–244.

LEW EJL KUZMICKY DD and KASARDA DD (1992) 'Characterization of low-molecular-weight glutenin subunits by reversed-phase high-performance liquid-chromatography, sodium dodecyl sulfate-polyacrylamide gel-electrophoresis, and N-terminal amino-acid sequencing', *Cereal Chem*, **69**(5), 508–515.

LIU YJ SAMUEL D LIN CH and LYU PC (2002) 'Purification and characterization of a novel 7-kDa non-specific lipid transfer protein-2 from rice (*Oryza sativa*)', *Biochem Biophys Res Comm*, **294**(3), 535–540.

MACRITCHIE F and GRAS PW (1973) 'The role of lipids in baking', *Cereal Chem*, **50**, 292–302.

MCLAUCHLAN WR GARCIA-CONESA MT WILLIAMSON G ROZA M RAVESTEIN and MAAT PJ (1999) 'A novel class of protein from wheat which inhibits xylanases', *Biochem J*, **338**(2), 441–446.

MCMASTER TJ MILES MJ KASARDA DD SHEWRY PR and TATHAM AS (2000) 'Atomic force microscopy of A-gliadin fibrils and *in situ* degradation', *J Cereal Sci*, **31**(3), 281–286.

MASCI S LAFIANDRA D PORCEDDU E LEW E J-L TAO HP and KASARDA DD (1993) 'D-Glutenin subunits: N-terminal sequence and evidence for the presence of cysteine', *J Agric Food Chem*, **70**(5), 581–585.

MASOUDI-NEJAD A NASUDA S KAWABE A and ENDO TR (2002) 'Molecular cloning, sequencing and chromosome mapping of a 1A-encoded omega-type prolamin sequence from wheat', *Genome*, **45**(4), 661–669.

MIFLIN BJ FIELD JM and SHEWRY PR (1983) 'Cereal storage proteins and their effect on technological properties', in Daussant J Mossé J and Vaughan J, *Seed Proteins*, Academic Press, London, 255–319.

MILES M CARR H J MCMASTER T BELTON PS MORRIS VJ FIELD JM SHEWRY P R and TATHAM A S (1991) 'Scanning tunnelling microscopy of a wheat seed storage protein reveals details of an unusual supersecondary structure', *Proc Natl Acad Sci (USA)*, **88**(1), 68–71.

MORRIS CF GREENBLATT GA BETTGE AD and MALKAWI HI (1994) 'Isolation and characterization of multiple forms of friabilin', *J Cereal Sci*, **20**(2), 167–174.

NOTREDAME C HIGGINS DG and HERINGA J (2000) 'T-Coffee: a novel method for fast and accurate multiple sequence alignment', *J Mol Biol*, **302**(1), 205–217.

ODA Y MATSUNAGA T FUKUYAMA K MIYAZAKI T and MORIMOTO T (1997) 'Tertiary and quaternary structures of 0.19 alpha-amylase inhibitor from wheat kernel determined by X-ray analysis at 2.06 Å resolution', *Biochemistry*, **36**(44), 13503–13511.

OSBORNE TB (1924) *The Vegetable Proteins*, 2nd edn. Longmans Green and Co.: 1924 London.

PAGE RD (1996) 'TreeView: an application to display phylogenetic trees on personal computers', *Comput Appl Biosci*, **12**(4), 357–358.

PARCHMENT O SHEWRY PR TATHAM AS and OSGUTHORPE DJ (2001) 'Molecular modeling of unusual spiral structure in elastomeric wheat seed protein', *Cereal Chem*, **78**(6), 658–662.

PERUFFO ADB POGNA NE and CURIONIN A (1996) 'Evidence for the presence of disulphide bonds between β-amylases and low molecular weight glutenin subunits', in Wrigley CW ed, *Gluten '96* RACI Melbourne, Australia, 312–315.

PÉZOLET M BONENFANT S DOUSSEAU F and POPINEAU Y (1992) 'Conformation of wheat gluten proteins – comparison between functional and solution states as determined by infrared spectroscopy', *FEBS Lett*, **299**(3), 247–250.

POPINEAU Y BONENFANT S CORNEC M and PÉZOLET M (1994) 'A study by infrared spectroscopy of the conformations of gluten proteins differing in their gliadin and glutenin compositions', *J Cereal Sci*, **20**(1), 15–22.

PURCELL JM KASARDA DD and WU CSC (1988) 'Secondary structures of wheat α-gliadin and ω-gliadin proteins: Fourier transform infrared spectroscopy', *J Cereal Sci*, **7**(1), 21–32.

RAHMAN S JOLLY CJ SKERRIT JH and WALLOSHECK A (1994) 'Cloning of a wheat 15-kDa grain softness protein (GSP): GSP is a mixture of puroindoline-like polypeptides', *Eur J Biochem*, **223**(3), 917–925.

ROUAU X and SURGET A (1998) 'Evidence for the presence of a pentosanase inhibitor in wheat flours', *J Cereal Sci*, **28**(1), 63–70.

SAMUEL D LIU YJ CHENG CS and LYU PC (2002) 'Solution structure of plant nonspecific lipid transfer protein-2 from rice (*Oryza sativa*)', *J Biol Chem*, **277**(38), 35267–35273.

SHEWRY PR and TATHAM AS (1997) 'Disulphide bonds in wheat gluten proteins', *J Cereal Sci*, **25**(3), 207–227.

SHEWRY PR and TATHAM AS (1999) 'The characterisation, structures and evolutionary relationships of prolamins', in Shewry PR and Casey R, *Seed Proteins*, Kluwer Academic Publishers, Dordrecht, 11–33.

SHEWRY PR TATHAM AS FORDE J KREIS M and MIFLIN BJ (1986) 'The classification and nomenclature of wheat gluten proteins – a reassessment', *J Cereal Sci*, **4**(2), 97–106.

SHEWRY PR PARMAR S BUXTON B GALE MD LIU CJ HEJGAARD J and KREIS M (1988) 'Multiple molecular-forms of β-amylase in seeds and vegetative tissues of barley', *Planta*, **176**(1), 127–134.

SHEWRY PR HALFORD NG and TATHAM AS (1992) 'High molecular weight subunits of wheat glutenin', *J Cereal Sci*, **15**(5), 105–120.

SHEWRY PR MILES M J and TATHAM AS (1994) 'The prolamin storage proteins of wheat and related cereals', *Prog Biophys Molec Biol*, **61**(1), 37–59.

SKYLAS DJ MACKINTOSH JA CORDWELL SJ BASSEAL DJ WALSH BJ HARRY J BLUMENTHAL C COPELAND L WRIGLEY CW and RATHMELL W (2000) 'Proteome approach to the characterisation of protein composition in the developing and mature wheat-grain endosperm', *J Cereal Sci*, **32**(2), 169–188.

TATHAM AS and SHEWRY PR (1985) 'The conformation of wheat gluten proteins. The secondary structures and thermal stabilities of α-, β-, γ- and ω-gliadins', *J Cereal Sci*, **3**(2), 103–113.

TATHAM AS and SHEWRY PR (1995) 'The S-poor prolamins of wheat, barley and rye', *J*

Cereal Sci, **22**(1), 1–16.

TATHAM AS DRAKE AF and SHEWRY PR (1989) 'Conformational studies of a synthetic peptide corresponding to the repeat motif of C-hordein', *Biochem J*, **259**(2), 471–476.

TATHAM AS MIFLIN BJ and SHEWRY PR (1985) 'The β-turn conformation in wheat gluten proteins – relationship to gluten elasticity', *Cereal Chem*, **62**(5), 405–412

THOMSON NH MILES MJ TATHAM AS and SHEWRY PR (1992) 'Molecular images of cereal proteins by STM', *Ultramicro*, **42**(B), 1204–1213.

TURNBULL KM GABORIT T MARION D and RAHMAN S (2000) 'Variation in puroindoline polypeptides in Australian wheat cultivars in relation to grain hardness', *Aust J Plant Physiol*, **27**(2), 153–158.

VAN DIJK AA DEBOEF E BEKKERS A VANWIJK LL VAN SWIETEN E HAMER RJ and ROBILLARD GT (1997a), 'Structure characterization of the central repetitive domain of high molecular weight gluten proteins. 2. Characterization in solution and in the dry state', *Protein Sci*, **6**(3), 649–656.

VAN DIJK AA VAN WIJK LL VANVLIET A HARIS P VAN SWIETEN E TESSER GI and ROBILLARD GT (1997b), 'Structure characterization of the central repetitive domain of high molecular weight gluten proteins. 1. Model studies using cyclic and linear peptides', *Protein Sci*, **6**(3), 637–648.

WELLNER N, BELTON PS and TATHAM AS (1996) 'Fourier transform IR spectroscopic study of hydration-induced structure changes in the solid state of ω-gliadins', *Biochem J*, **319**(3), 741–747.

WILDE PJ CLARKE DC and MARION D (1993) 'Influence of competitive adsorption of a lysopalmitolphosphatidylcholine on the functional properties puroindoline, a lipid binding protein from wheat', *J Agr Food Chem*, **41**(10), 1570–1576.

YU J HU SN WANG J WONG GKS LI SG LIU B DENG YJ DAI L ZHOU Y ZHANG XQ CAO ML LIU J SUN JD TANG J, CHEN YJ HUANG XB LIN W YE C TONG W CONG LJ GENG JN HAN YJ LI L LI W HU GQ HUANG XG LI WJ LI J LIU ZW LI L LIU JP QI QH LIU JS LI L LI T WANG XG LU H WU TT ZHU M NI PX HAN H DONG W REN XY FENG XL CUI P LI XR WANG H XU X ZHAI WX XU Z ZHANG JS HE SJ ZHANG JG XU JC ZHANG KL ZHENG XW DONG JH ZENG WY TAO L YE J TAN J REN XD CHEN XW HE J LIU DF TIAN W TIAN CG XIA HG BAO QY LI G GAO H CAO T WANG J ZHAO WM LI P CHEN W WANG XD ZHANG Y HU JF WANG J LIU S YANG J ZHANG GY XIONG YQ LI ZJ MAO L ZHOU CS ZHU Z CHEN RS HAO BL ZHENG WM CHEN SY GUO W LI GJ LIU SQ TAO M WANG J ZHU LH YUAN LP and YANG HM (2002) 'A draft sequence of the rice genome (*Oryza sativa* L. ssp. *Indica*)' *Science*, **296**(5565), 79–92.

7

Starch structure and bread quality

A.-C. Eliasson, Lund University, Sweden

7.1 Introduction: the importance of starch structure to bread quality

Starch is the most abundant component in wheat flour, but is seldom discussed in relation to wheat quality and baking performance. It might therefore seem unnecessary to include a chapter on starch structure when discussing improving breadmaking quality. However, the role of starch is slowly being reconsidered, and new ways of changing starch structure using modern molecular biology techniques have increased interest in the relationships between starch structure and function. Banks and Greenwood stated in 1975 that 'in fact, it is probably no exaggeration to say that each granule in a population is unique, differing from its neighbours in terms of both fine structure and properties' (Banks and Greenwood, 1975). There is thus much to be learned about the structure–function relationships and thus also enormous possibilities to changing starch properties.

Although our understanding of the role of starch in the baking process is far from complete we know that any starch will not do. Reconstitution experiments with a common gluten and different starches have shown that the best results were obtained with wheat starch, and almost as good results with rye and barley starches (Hoseney *et al.*, 1971). Also, in baking of cookies and cakes, wheat, rye and barley starches perform the best (Sollars and Rubenthaler, 1971). Different types of wheat starches actually give differences in baking results (D'Appolonia and Gilles, 1971). With a better knowledge about starch structure and the chemical composition of amylose and amylopectin it might be possible to relate these results to the fine structure of starch.

Starch is involved in the staling of bread. Water redistribution (Breaden and Willhoft, 1971) starch–gluten interactions (Martin *et al.*, 1991) and gluten firming (Mita, 1990) have been suggested as explanations for staling, but the recrystallisation of amylopectin is the most important factor (Zobel and Kulp, 1996). The relation between firmness and recrystallisation of amylopectin has been shown in several studies (Russell, 1983; Krog *et al.*, 1989; Morgan *et al.*, 1997). The influence of amylose on the recrystallisation of amylopectin and the ability of amylopectin to form a three-dimensional network are factors that affect the staling of bread. A challenge would thus be to modify the wheat starch so that it behaves like ordinary starch during dough mixing, fermentation and baking, but then does not retrograde during the storage of bread.

With the recognition of starch in bread as a 'rapid' carbohydrate resulting in high glycaemic and insulin levels after the meal, food factors that moderate the glycaemic responses to starch have been looked for. Such food factors, related to the starch, could be retrogradation, increased amylose content or a reduced degree of gelatinisation (Björck, 1996). From a nutritional point of view, retrograded starch would thus be beneficial. Another challenge would thus be to make bread with a huge amount of retrograded starch (causing low glycaemic and insulin levels), but without having the negative sensory attributes of stale bread.

Relations between starch properties and end-product quality have been found for products other than bread. In cake-baking, chlorinated wheat flour is often used, and the improving effect of chlorination seems to be related to the starch (Stauffer, 1994), perhaps owing to an influence on the hydrophobicity of the starch granule surface (Seguchi, 1987). It is known that starch contributes to overall texture of cooked Japanese noodles, and amylose–reduced varieties have been shown to give the best noodle quality (Yasui *et al.*, 1996; Yoo and Jane, 2002; Abdel-Aal *et al.*, 2002). Australian wheat samples have been investigated for the relation to white salted noodle quality (Yun *et al.*, 1996). Durum is regarded as the most appropriate wheat for pasta-making, which could, at least to some extent, be related to starch properties (Vansteelandt and Delcour, 1999). It is thus time to investigate also the relation between bread-baking performance and starch properties.

7.2 Starch properties and baking performance

Starch could be expected to influence baking performance because of physicochemical properties such as crystallinity, granule size distribution, and gelatinisation and retrogradation behaviour. These properties depend on starch granule structure, and on the molecular composition of the starch polysaccharides amylose and amylopectin. The starch properties might thus depend on the genetic background, but also on wheat processing (cultivation, drying, milling, etc.). The result of a process might, in turn, depend on the wheat variety. For example, milling causes different levels of starch damage and different granulation depending on whether a hard or soft wheat variety is milled. Although growing conditions might influence hardness, this variety characteristic is mostly under genetic control (Svensson, 1981). Certain properties could be described as 'starch properties' and any starch would do (such as contributing carbohydrates for the yeast). Other properties might be 'wheat starch properties', meaning that any wheat starch would do (as in bread-baking today, as long as the Falling Number is acceptable). It remains to find out whether there are starch properties that in fact differ between wheat varieties, and if therefore certain wheat varieties are better for bread-baking purposes than others depending on these starch properties.

When studying starch structure–function relationships starch is often extracted from wheat flour or kernel before the physicochemical properties are studied. It can be questioned whether this is a good approach for finding relations between starch properties and end-product quality. In the study of noodle quality for Australian wheat samples, mentioned before, better correlations were obtained between starch properties and eating quality with flour samples than with starch samples (Yun *et al.*, 1996). The gelatinisation endotherm, observed in differential scanning calorimetry (DSC), for example, differs, depending on whether a wheat starch–water mixture or a wheat flour–water mixture is being studied (Eliasson, 1989). The retrogradation behaviour for extracted starch (measured as the melting enthalpy for recrystallised amylopectin in the DSC) seems not to be a good prediction for starch retrogradation in bread (Lee *et al.*, 2001). The rheological behaviour of a wheat flour suspension is not exactly the same as for a wheat starch suspension (Champenois *et al.*, 1998b). In a concentrated system, such as dough, it is observed that the frequency dependence in dynamic rheological measurements decreases when the starch content of the dough increases (Hibberd, 1970; Larsson *et al.*, 2000). Such differences between wheat flour and extracted starch are often attributed to starch–protein interactions. However, the nature of these interactions is seldom discussed, and it might be added that the flour does contain also other components, beside the protein, that might influence the behaviour of starch.

7.2.1 Flour quality

Falling Number and starch damage are two starch-related parameters often used as a measure of flour quality (AACC, 1983a,b). Too low a Falling Number and too high a level of starch damage are both indications that the starch properties are such that the flour is too poor to be used for bread-baking. A low Falling Number is indicative of starch damage due to excessive enzymatic activity, whereas starch damage relates to mechanical damage of starch, obtained, for example, during milling.

The Falling Number is an indirect measurement, based on viscosity, of α-amylase activity (AACC, 1983a). However, the viscosity development may be related to other factors than the enzymatic degradation of starch, as, for example, inherent differences in starch viscosity (Wong and Lelievre, 1981). For the new waxy wheat starches it has been observed that these starches give rise to poor viscosity development in the Falling Number analysis, although the enzyme activity is low (Abdel-Aal *et al.*, 2002). A low Falling Number is thus not necessarily indicative of a high α-amylase activity, and the Falling Number cannot be used to predict α-amylase activity for the wheat varieties.

The mechanical damage of starch granules is evident when the granules are placed in water. They then seem to swell and gelatinise at room temperature, i.e. they lose optical birefringence and crystallinity and form a translucent gel, and soluble glucans are leached into the water (Morrison *et al.*, 1994). Gelatinisation parameters (temperatures and enthalpies determined with DSC) were found to be shifted to lower values for damaged starch, with the changes being smaller for a soft wheat compared to a hard wheat (Morrison *et al.*, 1994; Yoo and Jane, 2002). Whereas amylopectin molecules were converted to low-molecular-weight fragments at the beginning of the milling, amylose molecules were affected only after severe milling (Morrison and Tester, 1994). Damaged starch is rapidly hydrolysed by amylases, and damaged starch granules are thus the substrates for amylases during fermentation. This enzymatic availability is also the basis for the enzymatic determination of starch damage (AACC, 1983b).

Heat damage can also affect starch properties. This type of damage might occur during hot air drying of wheat seeds (Zamponi *et al.*, 1990; Köksel *et al.*, 1993). Heat damage of wheat has an adverse effect on baking performance, and this adverse effect was also observed when starch was isolated from heat-damaged wheat and used in a baking test in a starch–gluten system (Lorenz *et al.*, 1993). However, there were no apparent changes in physicochemical properties (microscopic appearance, amylograph, gelatinisation, X-ray, swelling power, solubility, water hydration capacity), but the Falling Number increased for the heat damaged wheat. Thus, one consequence of heat damage might be that α-amylase activity is destroyed. Another effect of the application of heat is annealing of the starch, which occurs when starch is kept at a water content that is high enough for gelatinisation, but at a temperature that is below the gelatinisation onset temperature (Knutson, 1990; Larsson and Eliasson, 1991; Hoover and Vasanthan, 1994). The result of the annealing is usually an increase in the gelatinisation temperature measured in the DSC. This might as well occur during proper conditions in the field, i.e. an unusal hot and dry summer might result in wheat starch with increased gelatinisation temperature (Tester *et al.*, 1995).

7.2.2 Mixing

During mixing the rheological properties of the wheat flour will manifest themselves (Bloksma, 1990). Although differences in rheological properties between wheat varieties often are attributed to proteins the presence of starch granule has an influence – the rheological behaviour of dough and gluten is not exactly the same (Dreese *et al.*, 1988). The structure of the dough is more complicated than simply gluten with added particles. It has been suggested that starch and protein form two independent and bicontiuous aqueous phases in the dough (Eliasson and Larsson, 1993), and a partial segregation of starch from the protein phase was observed in light microscopy (Hug-Iten *et al.*, 1999).

The baking absorption, i.e. the amount of water needed to give the optimal mixing treatment, depends both on the composition of the wheat flour and on the starch properties. Starch influences baking absorption owing to the level of mechanical or enzymatic damage (Bushuk, 1966), and also to granule size. Small granules give a higher water absorption than large granules, and, moreover, the mixing time seems to be longer for the large A-granules and shorter for the small B-granules (D'Appolonia and Gilles, 1971; Petrofsky and Hoseney, 1995). Not only do non-wheat starches such as corn, oat, rye, rice or potato change the water absorption and the mixing behaviour (Petrofsky and Hoseney, 1995), but waxy wheat flours also influence the mixing process. Waxy wheat flours have been shown to decrease both mixing time and stability in the farinograph (Bhattacharya *et al.*, 2002).

The rheological properties have been studied for doughs made of starch isolated from different wheat cultivars and a common gluten (Petrofsky and Hoseney, 1995; Miller and Hoseney, 1999). One out of the four starches studied gave significantly different rheological parameters (lower storage modulus (G') and loss modulus (G''), higher phase shift ($\tan \delta$)), and it was concluded that this special wheat starch did not interact as strongly with gluten as the other starches (Miller and Hoseney, 1999). A difference in rheological properties was also observed between starches from soft and hard wheats, with lower values of G' and G'' for the hard wheat than for the soft wheat variety (Petrofsky and Hoseney, 1995).

7.2.3 Bread quality

The influence of starch properties on loaf volume, porosity and other bread characteristics has not been very much studied, especially when compared with how much the influence of proteins has been studied. It was concluded that starch granule composition, swelling and gelatinisation properties were not related to baking quality for ten Greek bread wheat varieties (Matsoukas and Morrison, 1991). On the other hand, in a study of nine different wheat varieties it was concluded that starch quality parameters in addition to protein quality are of importance for the quality of hearth bread (Sahlström *et al.*, 2003).

From baking experiments, where the wheat flour was diluted with wheat, barley or waxy barley starch, it was concluded that amylose is responsible for setting of the crumb structure (Ghiasi *et al.*, 1984). Loaf specific volume did not change, but loaves with waxy barley had a soft and sticky crumb. Also, the loaf with waxy barley did shrink excessively. It was also found that the waxy barley starch granules were more disrupted than normal wheat starch (Ghiasi *et al.*, 1984). The proportion of large granules has been discussed in relation to the gas cell stability, and crumb grain in the final bread, and it was suggested that a greater proportion of large starch granules causes gas cell coalescence and, thus, results in an open crumb grain (Hayman *et al.*, 1998). Based on light microscopy it was suggested that in the baked loaf amylose and amylopectin in the continuous starch phase are phase separated, with amylose rich zones outside the granules (Hug-Iten *et al.*, 1999). It was further observed that amylose in the centre of the granules was strongly birefringent, whereas amylopectin in the outer parts of the granules was less so. Ordering of the amylose could increase the rigidity of starch granules, and thus contribute both to the setting of the crumb, and to the development of firmness of the bread during storage.

The use of waxy wheat flour in a flour mixture will lead to a crumb that is more open and irregular than with normal wheat flour (Lee *et al.*, 2001). The gas production was found to increase when the proportion of waxy wheat increased in the flour blend, but the gas retention decreased. Also waxy durum flour has been mixed with ordinary wheat flour in baking, and up to 20% could be used in the mixture without adverse effects on loaf volume (Bhattacharya *et al.*, 2002).

7.2.4 Staling

Although other factors contribute, the recrystallisation of amylopectin is the most important parameter for the staling of bread (Zobel and Kulp, 1996). The addition of emulsifiers, and the addition of amylases, both additives that are known to reduce the staling of bread, affect the amylopectin recrystallisation (Dragsdorf and Varriano-Marston, 1980; Russell, 1983; Krog *et al.*, 1989; Morgan *et al.*, 1997). There are several studies indicating that complex formation occurs between amylopectin and monoacyl lipids, for example emulsifiers, and that such complexation will reduce the staling of bread (Batres and White, 1986; Evans, 1986; Krog *et al.*, 1989; Villwock *et al.*, 1999; Lundqvist *et al.*, 2002a). The explanation of the effect would be that the outer branches of the amylopectin molecule form the complex, thus preventing the formation of a three-dimensional network (Lundqvist *et al.*, 2002a,b). The same effect would be created using α-amylase, i.e. if the external amylopectin branch chains are degraded the creation of a continuous three-dimensional network is prevented (Dragsdorf and Varriano-Marston, 1980; Wursch and Gumy, 1994; Lundqvist *et al.*, 2002c).

It is known that the presence of amylose influences the recrystallisation of amylopectin, i.e. more amylopectin recrystallisation is measured in the presence of amylose than would be expected from the presence of only amylopectin (Gudmundsson and Eliasson, 1990). It might then be concluded that although staling of bread is not directly related to the recrystallisation of amylose (Zobel and Kulp, 1996); there might be an indirect effect of amylose due to its influence on amylopectin recrystallisaton. It has therefore been speculated that the incorporation of waxy wheat flours could reduce the problem with staling. Reduced staling was also observed when waxy durum wheat was mixed with ordinary wheat flour (Bhattacharya *et al.*, 2002). The firmness for all breads was the same on the day of baking, but the increase was less during storage for the breads with waxy wheat (up to 30% were used). The enthalpy values for melting of recrystallised amylopectin, measured by DSC, were in accordance with the firmness values. It was found that the increase in enthalpy over time was much lower in waxy wheat starch samples than in control samples, and it was concluded that waxy wheat starch was more resistant to retrogradation (Bhattacharya *et al.*, 2002). Similar results were obtained when breads were baked from starch (waxy wheat and normal wheat starches) and gluten blends, i.e. the firmness was lower for breads with the waxy wheat flour (Lee *et al.*, 2001). However, the enthalpy measured for the endotherm related to recrystallised amylopectin was highest for the bread with 100% waxy wheat. The relation between firmness and amylopectin recrystallisation is thus not straightforward, and the retrogradation of waxy wheat starch will be discussed later in this chapter.

7.3 Starch structure

Starch structure can mean the crystalline organisation of the starch granule as well as details in the chemical structure of amylopectin or amylose. In order to understand relations between starch structure and function it is thus necessary to define the level of structure that is referred to. Here, the relevance of granular structure, crystalline organisation and molecular structure will be discussed.

7.3.1 Granules: particles and surfaces

The presence of starch particles as such can be of relevance for dough and bread quality. The starch granule can affect rheological properties of dough by acting as a filler, and then size, shape and size distribution are all important particle properties (Soulaka and Morrison, 1985a; Sahlström *et al.*, 1998; van Vliet, 1988). However, glass beads do not work as replacement for starch granules, so there is more to starch than being a particle of certain shape, size and size distribution (Ghiasi *et al.*, 1984). The starch granule surface might influence the breadmaking performance due to interactions with the surrounding aqueous phases in the dough, and it has been suggested that the surface composition of starch granules is important for starch properties (Cauvain *et al.*, 1977).

The size distribution of wheat starch granules is bimodal (Soulaka and Morrison, 1985a), and the size distribution might influence the baking performance as mentioned above, i.e. because of the particle size and the differences in surface area depending on the starch granule size. However, the chemical composition of the different size classes might also differ, causing differences in properties that are of relevance for the baking performance. Information about the starch granule surface is obtained at different levels of resolution using microscopic techniques. In ordinary light microscope pores and channels into the interior can be detected (Fannon *et al.*, 1992), whereas at the atomic resolution of AFM (atomic force microscopy) 10–50 nm structures have been observed, interpreted as being amylopectin side chain clusters (Baldwin *et al.*, 1998). Non-contact AFM showed a rough surface of wheat starch granules with protruding surface structures of 200 nm in size or below (Juszczak *et al.*, 2003).

The presence of carbohydrates, lipids and proteins on the wheat starch granule surface has been verified with many techniques, including ESCA (electron spectroscopy for chemical analysis) (Russell *et al.*, 1987), dye binding (Seguchi, 1986), extraction of components (Seguchi and Kanenaga, 1997) and confocal laser scanning microscopy in combination with a protein-specific dye (Han and Hamaker, 2002). The proteins present on the starch granule surface are a mixture of storage proteins, starch biosynthetic enzymes, friabilin/puroindolines ($\approx$15 kDa), a 30 kDa protein, and a 60 kDa starch granule bound starch synthase (Baldwin, 2001). Using time-of-flight secondary ion mass spectrometry (TOF-SIMS) it was possible to identify not only the presence of lipids at the surface, but also their fatty acid chain composition (Baldwin *et al.*, 1997). It was further observed that certain peaks in the TOF-SIMS spectra were characteristic of the type of starch, i.e. the surface composition is unique to the starch.

The protein film present on the wheat starch granule has been suggested to be necessary for the tertiary structure of the starch granule, and it has even been suggested that if the protein film is removed the starch granule will gelatinise (Seguchi, 1986). Others have concluded that the proteins (namely granule-bound starch synthetase) is important for maintaining the integrity of the structures after gelatinisation ('ghost' structures) (Han and Hamaker, 2002). By changing the composition of the starch granule surface it is possible to influence the rheological properties of the dough (Larsson and Eliasson, 1997). The addition of protein-coated starch granules increased G' of the dough more than the same addition of uncoated starch granules together with the same amount of protein added to the dough.

The starch granules in wheat, rye and barley show a bimodal size distribution with, in case of wheat, about 24% small B-granules (Soulaka and Morrison, 1985a). The specific surface area is of course larger for the B-granules than for the A-granules, 0.788 m^2/g for B-granules and 0.265 m^2/g for A-granules (Soulaka and Morrison, 1985a). An optimum in baking performance has been found at 25–35% B-granules (Soulaka and Morrison, 1985b). Baking tests using blends of different starches with a single gluten preparation showed that the small granules have a lower baking potential than the regular ones (Kulp, 1973). There might also be a different optimum starch size fraction depending on the protein concentration in the dough (Lelievre *et al.*, 1987). Starch granule size distribution might explain bread weight (55%) and form ratio (48%); high weights and form ratios were promoted by small A-granules (size around 12 μm) (Sahlström *et al.*, 1998). Mixing speed and work input were also found to be related to starch granule size distribution. Moreover, starch might also influence the baking absorption owing to the starch granule size distribution, as more water is used to coat small granules because of their larger surface area (Sahlström *et al.*, 1998). Small granules resulted in increased water absorption and mixing time also for gluten–starch doughs compared with a mixture with the

Table 7.1 Chemical composition of large A-granules and small B-granules in wheat starch

Parameter	A-granules	B-granules
Amylose (%)[a]	29.89–35.54	24.55–27.37
Amylose (%)[b]	18.7–20.2	16.0 19.2
Amylose (%)[c]	28.6–31.0	25.1–31.5
Lysolecithin (%)[c]	0.70–1.0	1.06–1.40

[a] Peng *et al.* (1999)
[b] Meredith (1981)
[c] Soulaka and Morrison (1985b)

normal starch (Petrofsky and Hoseney, 1995). A higher rate of water absorption has also been reported for B-granules (Chiotelle and LeMeste, 2002).

A difference in the composition between small and large granules (Table 7.1) could be one explanation for the influence of starch granule size distribution on baking results. Differences in results could be related to the difficulties in complete separation of the different size classes before investigation of properties. The B-granules, which are more irregular, have a tendency to be highly agglomerated (Chiotelle and LeMeste, 2002). Usually, the different size classes are separated by gravity (Meredith *et al.*, 1977). Recently, different methods for fractionation were investigated, and it was found that micro-sieving could not completely separate A- and B-granules, whereas centrifugation in aqueous sugar solutions or Pecoll gave 100% A-granules and 100% B-granules (Peng *et al.*, 1999). The lipid content of B-granules is higher than of A-granules, and there seems to be a tendency that the amylose content is higher for A-granules.

Although the differences are small the gelatinisation peak temperature (T_P) seems to be higher for the B-granules, whereas the gelatinisation enthalpy seems to be lower (Table 7.2). The enthalpy of the transition of the amylose–lipids complex is larger for the B-granules, which reflects the higher lipid content in these granules (Tables 7.1 and 7.2). The non-amylopectin content of the B-granules might also explain their lower gelatinisation enthalpy, as this is expressed in J/g starch, and not J/g amylopectin. The DSC-endotherm was found to be narrow for A-granules and broad for the B-granules (Chiotelle and LeMeste, 2002). There are thus certain differences both in the composition and thermal properties between A- and B-granules.

7.3.2 The crystalline organisation of starch granules

Crystallinity of wheat starch has been determined to 35.5% by the X-ray diffraction technique, and the double-helix content to 46% by NMR (Morrison *et al.*, 1994). The large A-granules are more crystalline than B-granules (Chiotelle and LeMeste, 2002). The X-ray diffraction pattern is of the A-type, typical for cereal starches (Zobel, 1988). Small-angle X-ray scattering has shown that there is a constant structural periodicity in starch granules, independent of starch

Table 7.2 DSC characteristics of A- and B-granules in wheat starch at excess water conditions

Sample	T_p (°C)	ΔH (J/g starch)	T_{cx} (°C)	ΔH_{cx} (J/g starch)
A-granules[a]	61.2±0.5	14.0±0.17	102.7±0.7	1.63±0.17
B-granules[a]	62.4±0.2	13.2±0.12	102.3±1.6	3.23±0.04
A-granules[b]	58.0–62.2	10.0–12.2		
B-granules[b]	59.6–63.4	8.0–10.0		
A-granules[c]	60.15±0.73	11.06±0.20	107.33±0.63	0.99±0.09
B-granules[c]	64.61±0.51	8.12±0.08	107.25±0.51	1.70±0.04
A-granules[d]	59.2	9.81	99.0	1.68
B-granules[d]	61.2	9.46	98.3	2.44

[a] Eliasson and Karlsson (1983)
[b] Peng *et al.* (1999)
[c] Chiotelle and LeMeste (2002)
[d] Sahlström *et al.* (2003)

source, and with a repeat distance corresponding to the average size of an amylopectin cluster (Jenkins *et al.*, 1993). The repeat distance of 9 nm is a combination of crystalline and amorphous regions. The amylopectin clusters are organised in 'blocklets' with a diameter of 20 to 500 nm depending on the botanical source (Gallant *et al.*, 1997). There are also amorphous channels present. The size of the blocklet depends also on where in the granule the blocklet is located; the blocklets are larger in crystalline regions. Pores, 100 nm in diameter, have been observed along the equatorial groove of large granules of wheat starch (and of rye and barley starches, but not of oat starch) (Fannon *et al.*, 1992).

There is evidence for amylose–lipid inclusion complexes from 13C-CP/MAS-NMR (cross-polarisation magic angle spinning nuclear magnetic resonance) for a range of starches and lintner residues of starches (Morrison, 1995). The amylose could thus be present in the granule either as lipid-complexed amylose (LAM) or lipid-free amylose (FAM). Independent of whether the amylose is present as LAM or FAM it seems to be amorphous. Therefore, FAM, LAM and some amylopectin are located in the broad amorphous zones between well-organised rings (Morrison, 1995). The amylose is interspersed between amylopectin molecules, at least in potato and corn starch granules (Jane *et al.*, 1992).

The organisation of the starch granule in alternating amorphous and crystalline regions is of uttermost relevance for the properties of the starch, and thus for its relevance for baking performance. For example, the gelatinisation can be described as a process related to the plasticisation and mobility of the amorphous regions within the starch granule, and it can thus vary from an all swelling-driven process in excess water to an all melting process at very low levels of water (Perry and Donald, 2002). The starch gelatinisation in bread is usually complete, as judged from DSC, birefringence studies, etc.

However, the process to arrive to the complete gelatinisation might differ, depending on the starch and the water availability. Therefore the rheological properties of the starch phase could differ, which would be of relevance for the setting of the crumb and the crumb structure of the baked bread.

7.3.3 Amylose and amylopectin

The amylose–amylopectin ratio influences functional properties, but also the amylopectin fine structure plays a critical role in the characteristics of starch (Fredriksson *et al.*, 1998; McPherson and Jane, 1999). Wheat starches differing in amylose–amylopectin ratios are available and can give some insight about what is important for baking performance. The amylopectin fine structure has been studied for starches from different botanical sources, but no comparisons of different wheat starch amylopectins seem to have been made. The influence of molecular structure on functional properties has thus to be extrapolated from other starches than wheat.

Some properties of wheat amylose and amylopectin are given in Table 7.3. The chain length distribution of wheat amylopectin has been determined with the degree of polymerisation (dp): dp 6–9, 5.18%; dp 6–12, 19.0%; dp 13–24, 41.7%; dp 25–36, 16.2%; dp > 37, 13.0% (Jane *et al.*, 1999). The chain-length distribution seems to be characteristic of the botanical source of the starch. For example, cereal starches have been found to display very few short chains of dp 6 and a gradual increase in chains of dp 7–9 (Jane *et al.*, 1999).

Waxy wheat varieties are now available, and some basic studies have been performed on the properties of waxy wheat starch (Nakamura *et al.*, 1995; Hayakawa *et al.*, 1997; Fujita *et al.*, 1998; Grant *et al.*, 2001). There are no differences in the granule-size distribution and granule morphology between waxy wheat starch and normal wheat starch (Yoo and Jane, 2002; Abdel-Aal *et al.*, 2002). They all give the A-pattern in X-ray diffraction studies, but the waxy wheat starch did not give any indication of the presence of an amylose–lipid complex. The crystallinity was estimated to 21.2% for waxy wheat and to 15.7% for normal wheat (Abdel-Aal *et al.*, 2002). Amylopectin of the waxy starch had

Table 7.3 Characteristics of wheat starch amylose and amylopectin

	Amylopectin	Amylose
Average DPn[a]	13 000–18 000	1200–1500
Average DP of chains[b]	22.1	
Average DP of chains[c]	21–25	
Limiting viscosity number $[\eta]$, ml/g[a]	147–154	652–656
Number of branch linkage, %[a]		0.24–0.32

[a] Shibanuma *et al.* (1996)
[b] Jane *et al.* (1999)
[c] Hayakawa *et al.* (1997)

the largest molecular weight, whereas the molecular weight distribution of amylose was similar (Yoo and Jane, 2002). Also amylopectin branch chain-length distributions were similar to each other, although there were extra-long chains in the amylopectin from normal starches (Yoo and Jane, 2002).

Compared with the normal wheat starch, waxy wheat starch has a higher swelling power, a lower pasting temperature, a higher peak viscosity and a poorer consistency of the cold paste (Abdel-Aal *et al.*, 2002). However, the waxy wheat starch showed less syneresis after freezing and thawing. In DSC-measurements higher gelatinisation temperatures have been observed for waxy wheat starch, as well as higher gelatinisation enthalpies (Yasui *et al.*, 1996; Hayakawa *et al.*, 1997; Sasaki *et al.*, 2000; Abdel-Aal *et al.*, 2002). However, when recalculated on a common amylopectin basis the enthalpies are similar.

7.4 Starch structure and bread quality

During recent years the methods for analysing amylopectin and amylose fine structure have improved (Hizukuri, 1996), and it should thus be possible to investigate the relation between baking parameters and molecular properties such as chain-length distribution, branch structure, molecular weight and gyration radius (Yoo and Jane, 2002). Although these possibilities have increased there are still few investigations comparing starch from different wheat varieties, and relating molecular structure to baking. With the arrival of waxy wheats a broader range of amylose contents has been investigated (Sasaki *et al.*, 2000), but still relations between starch structure and functional properties have been mostly studied for starches from different botanical sources (Fredriksson *et al.*, 1998; Jane *et al.*, 1999). Another way to gain information about starch structure and functional properties is to study starches where molecular characteristics have been modified using enzymes (Wursch and Gumy, 1994; Lundqvist *et al.*, 2002c; Frigård *et al.*, 2003).

7.4.1 Crystallinity and gelatinisation

The gelatinisation of starch is important for the fixation of the crumb, and the onset of starch gelatinisation will be the end of the oven spring (Eliasson and Larsson, 1993). Therefore, gelatinisation at the proper temperature and time during baking is important (Eliasson *et al.*, 1995). The gelatinisation of starch might be affected due to, for example, changes in crystallinity, chemical composition of amylose/amylopectin, presence of lipids, and phosphorylation.

The crystalline organisation of the starch granule has been linked to the amylopectin branch chain distribution profile, which is typical for the source of starch (Jane *et al.*, 1999), and it has been suggested that short chains will give rise to the A-pattern in X-ray diffraction, and longer chains will give the B-pattern (Hizukuri, 1986). Independent of the amylose content (at least in the range 0–30% amylose), wheat starches show the A-pattern (Hayakawa *et al.*,

between starch granules and protein matrix (Edwards *et al*., 2002). It could thus be expected that the total volume occupied by granules, their size distribution and any interactions between them will influence the rheological behaviour.

Starches were isolated from wheat cultivars and combined with a gluten isolated from commercial flour to determine whether the starches affected the dough's rheological properties. One of the four starches studied gave significant different values, with lower G' and G'' and higher tan δ compared with the other starches (Miller and Hoseney, 1999). The authors explained the finding about the starch with that this starch seemed to interact more strongly with the gluten than the other starches. Waxy wheat flours formed intermediate strength doughs that required significantly less time and work to develop, as indicated by the mixograph and farinograph tests. The dough showed less stability and was sensitive to mechanical mixing. The waxy wheat flours absorbed more water than other flours (Abdel-Aal *et al*., 2002).

In small amplitude oscillatory rheological measurements on starch gels it was found that the starch with the highest proportion of small granules gave the highest G' and lowest tan δ (Chiotelle and LeMeste, 2002). During heating of the starch suspensions an increase in rigidity occurred at a lower temperature for B-granules than for A-granules. Dynamic viscoelastic measurements of starch gels at the concentration of 30 and 40% (w/w) showed that starch gels with waxy wheat starch gave lower G' values than the normal starches. These gels also had a stronger frequency dependence (Sasaki *et al*., 2002). However, the frequency dependence of the waxy starch decreased during storage, although the G' increased more during storage for waxy wheats, and approached the values for normal starches. In a concentrated system amylose leaching is suppressed, and it is the recrystallisation of the outer branches of amylopectin that governs the retrogradation of the starch gel. Viscoelastic properties of dispersions of wheat starch and gluten blends were studied, and it was found that when G' was measured as a function of temperature the profile depended on starch concentration (Champenois *et al*., 1998b). The increase in G' occurred at lower temperature with increasing concentration, and tan δ decreased strongly. Addition of gluten delayed the increase in G'. The effect of α-amylase was found to be less in the presence of gluten (or lipids) (Champenois *et al*., 1998a).

Pasting properties of starch or flour suspensions have been studied using the Brabender amylo/viscograph and the rapid visco-analyser (RVA) has been used to study differences in viscous properties between starches in more diluted suspensions (compared with the dough). Correlations between certain RVA parameters and noodle quality were found in a study of Australian wheats (Yun *et al*., 1996). For eating quality the set-back value was most important. It was also found that the correlation coefficient depended on the experimental profile in the RVA.

Studies have been performed in order to elucidate the effect of different molecular parameters such as amylose–amylopectin ratio and properties of individual amylose and amylopectin molecules. There is a relation between viscosity and concentration, and this relation depends on the wheat starch variety (Shibanuma *et al.*, 1996). High-viscosity starches had larger molecules of both amylose and amylopectin than those of low-viscosity starches. There was a good correlation between peak viscosity and average degree of polymerisation for both amylose and amylopectin. Also, the high-viscosity amylopectin seemed to contain less of the extra-long chains, whereas the low-viscosity starches seemed to contain more of this fraction. In pasting, a higher and earlier peak is obtained for some waxy wheat starches (Yoo and Jane, 2002; Abdel-Aal *et al.*, 2002), but lower or similar peak values have also been observed (Hayakawa *et al.*, 1997; Sasaki *et al.*, 2000). The set-back values are reported to be lower for the waxy wheat starches in all studies. This is a result of the lower amylose content of the starch paste. Pasting properties of starch are also affected by amylose and lipid contents. The amylose–lipid complexes caused an increase in pasting temperature and increased resistance to shear-thinning. Therefore, wheat starches have high pasting temperature and very low peak viscosity (Jane *et al.*, 1999).

7.5 Future trends

Wheat starch affects wheat flour quality due, for example, to starch content, grain hardness, granule size distribution and shape, the presence of endogenous lipids in the granule, amylopectin structure, and the ratio of amylose to amylopectin. Each of these factors might be possible to modify by molecular/genetic changes in genomic DNA (Rahman *et al.*, 2000). It is therefore likely that in the future research will be performed in order to modify starch properties with means of molecular biology. However, for these modifications to be useful we need better knowledge about the starch–structure function relationships. Because molecular biologists can change the starch structure in a different way, or to a degree that is not possible in ordinary wheat breeding, this might also help us to gain understanding about the structure–function relationships. It will thus be possible to obtain a wider range of starch structures, and thus also of starch properties.

Knowledge is increasing about the role of starch in the diet due to its influence on glucaemic and insulin levels, and to the formation of resistant starch. There will certainly be an increased interest in manipulating starch behaviour in a way so that it can be regarded as a functional food. The demands on starch properties and behaviour might also increase owing to an increased interest in wholemeal products. Higher content of dietary fibre is advantageous, but with increasing dietary fibre content the baking performance of the flour will be reduced. One reason for this is the simple dilution effect of gluten, and it can thus be expected that starch will play an important role in this kind of bread.

Retrogradation is still a big problem in the baking industry, and the role of amylopectin and amylose needs to be clarified. Relations between amylopectin branch chain lengths and retrogradation are found, but what will happen with the baking properties if the amylopectin is changed in a way that will reduce retrogradation and staling? Another aspect of the retrogradation is the use of starches in frozen products. It is already common to use amylases as well as emulsifiers in order to reduce the retrogradation of starch. In the future this will probably increase, owing to the availability of more and new enzymes. Moreover, the function of the enzymes might be better optimised in relation to time and temperature in the baking process. With the knowledge that amylopectin complexation is important for the retardation of retrogradation new emulsifiers might be found that are better complexing agents for amylopectin.

7.6 Sources of further information and advice

For basic knowledge about starch and starch behaviour:

WHISTLER, R. L., BEMILLER, J. N. and PASCHALL, E. F. (eds) (1984) *Starch Chemistry and Technology*, Vol. 2. Academic Press, Orlando.

ELIASSON, A.-C. (ed.) (1996) *Carbohydrates in Food*, Marcel Dekker, Inc., New York.

For those interested in the staling of bread the book edited by Hebeda and Zobel is a real gold mine:

HEBEDA, R. E. and ZOBEL, H. F. (eds), (1996) *Baked Goods Freshness. Technology, Evaluation, and Inhibition of Staling*. Marcel Dekker, Inc, New York.

Some recent review papers might also be useful, for different aspects of starch:

BALDWIN, P. M. (2001) Starch granule-associated proteins and polypeptides: a review, *Starch/Stärke,* **53,** 475–503.

BULEON, A., COLONNA, P., PLANCHOT, V. and BALL, S. (1998) Starch granules: structure and biosynthesis, *Int. J. Biol. Macromol.,* **23,** 85–112.

PARKER, R. and RING, S. G. (2001) Aspects of the physical chemistry of starch, *Journal of Cereal Science*, 34, 1–17.

RAHMAN, S., LI, Z., BATEY, I., COCHRANE, M. P., APPLES, R. and MORELL, M. (2000) Genetic alteration of starch functionality in wheat, *J. Cereal Sci.,* **31,** 91–110.

The Danish–Swedish joint research programme, Öresund Starch Profile (www.osp.kvl.dk) might in time give useful data.

7.7 References

AACC (1983a), *Approved Methods of the American Association of Cereal Chemists,* AACC method 56–81B, American Association of Cereal Chemists, St. Paul, MN.

AACC (1983b), *Approved Methods of the American Association of Cereal Chemists,* AACC method 76–30A, American Association of Cereal Chemists, St. Paul, MN.

ABDEL-AAL, E. S. M., HUD, P., CHIBBAR, R. N., HAN, H. L. and DEMEKE, T. (2002) Physicochemical and structural characteristics of flours and starches from waxy and nonwaxy wheats, *Cereal Chem.,* **79**, 458–464.

BALDWIN, P. M. (2001) Starch granule-associated proteins and polypeptides: a review, *Starch/Stärke,* **53**, 475–503.

BALDWIN, P. M., MELIA, C. D. and DAVIES, M. C. (1997) The surface chemistry of starch granules studied by time-of-flight secondary ion mass spectrometry, *J. Cereal Sci.,* **26**, 329–346.

BALDWIN, P. M., ADLER, J., DAVIES, M. C. and MELIA, C. D. (1998) High resolution imaging of starch granule surfaces by atomic force microscopy, *J. Cereal Sci.,* **27**, 255–265.

BANKS, W. and GREENWOOD, C. T. (1975) *Starch and its Components*, Edinburgh University Press, Edinburgh.

BATRES, L. R. and WHITE, P. J. (1986) Interaction of amylopectin with monoglycerides in model systems, *J.A.O.C.S.,* **63**, 1537–1540.

BHATTACHARYA, M., ERAZO-CASTREJÓN, S. V., DOEHLERT, D. C. and MCMULLEN, M. S. (2002) Staling of bread as affected by waxy wheat flour blends, *Cereal Chem.,* **79**, 178–182.

BJÖRCK, I. (1996) Starch: nutritional aspects. In *Carbohydrates in Food* (ed., Eliasson, A.-C.), Marcel Dekker, New York, pp. 505–553.

BLOKSMA, A. H. (1975) The effect of temperature on some rheological properties of wheat flour doughs, *J. Texture Stud.,* **6**, 343–361.

BLOKSMA, A. H. (1990) Rheology of the breadmaking process, *Cereal Foods World,* **35**, 228–236.

BREADEN, P. W. and WILLHOFT, E. M. A. (1971) Bread staling Part III – measurement of the re-distribution of moisture in bread by gravimetry, *J. Sci. Fd. Agric.,* **22**, 647–649.

BUSHUK, W. (1966) Distribution of water in dough and bread, *Bakers' Dig.,* **40(5)**, 38–40.

CAUVAIN, S. P., GOUGH, B. M. and WHITEHOUSE, M. E. (1977) The role of starch in baked goods. Part 2. The influence of purification procedure on the surface properties of the granule, *Starch/Stärke,* **29**, 91–95.

CHAMPENOIS, Y., RAO, M. A. and WALKER, L. P. (1998a) Influence of α-amylase on the viscoelastic properties of starch-gluten pastes and gels, *J. Sci. Food Agric.,* **78**, 127–133.

CHAMPENOIS, Y., RAO, M. A. and WALKER, L. P. (1998b) Influence of gluten on the viscoelastic properties of starch pastes and gels, *J. Sci. Food Agric.,* **78**, 119–126.

CHIOTELLE, E. and LEMESTE, M. (2002) Effect of small and large wheat starch granules on thermomechanical behaviour of starch, *Cereal Chem.,* **79**, 286–293.

D'APPOLONIA, B. L. and GILLES, K. A. (1971) Effect of various starches in baking, *Cereal Chem.,* **48**, 625–636.

DRAGSDORF, R. D. and VARRIANO-MARSTON, E. (1980) Bread staling: X-ray diffraction studies on bread supplemented with alfa-amylases from different sources, *Cereal Chem.,* **57**, 310–314.

DREESE, P. C., FAUBION, J. M. and HOSENEY, R. C. (1988) Dynamic rheological properties of flour, gluten, and gluten-starch doughs. I. Temperature-dependent changes during heating, *Cereal Chem.,* **65**, 348–353.

EDWARDS, N. M., DEXTER, J. E. and SCANLON, M. G. (2002) Starch participation in durum dough linear viscoelastic properties, *Cereal Chem.,* **79**, 850–856.

ELIASSON, A.-C. (1989) Some physico-chemical proerties of wheat starch. In *Wheat End-use Properties. Wheat and Flour Characterization for Specific End-use* (ed. Salovaara, H.) University of Helsinki, Helsinki, pp. 355–364.

ELIASSON, A.-C. and KARLSSON, R. (1983) Gelatinization properties of different size classes of wheat starch granules measured with differential scanning calorimetry, *Starch/Stärke,* **35**, 130–133.

ELIASSON, A.-C. and LARSSON, K. (1993), *Cereals in Breadmaking: A Molecular/Colloidal Approach,* Marcel Dekker, New York.

ELIASSON, A.-C., GUDMUNDSSON, M. and SVENSSON, G. (1995) Thermal behaviour of wheat starch in flour-relation to flour quality, *Lebensm.-Wiss. u.-Technol.,* **28**, 227–235.

EVANS, I. D. (1986) An investigation of starch/surfactant interactions using viscometry and differential scanning calorimetry, *Starch/Stärke,* **38**, 227–235.

FANNON, J. E., HAUBER, R. J. and BEMILLER, J. N. (1992) Surface pores of starch granules, *Cereal Chem.,* **69**, 284–288.

FREDRIKSSON, H., SILVERIO, J., ANDERSSON, R., ELIASSON, A.-C. and ÅMAN, P. (1998) The influence of amylose and amylopectin characteristics on gelatinization and retrogradation properties of different starches, *Carbohydr. Polym.,* **35**, 119–134.

FRIGÅRD, T., LUNDQVIST, H., ANDERSSON, R., ÅMAN, P. and ELIASSON, A.-C. (2003) Retrogradation properties of enzymatically modified amylopectin from potato and barley. Submitted for publication.

FUJITA, S., YAMAMOTO, H., SUGIMOTO, Y., MORITA, N. and YAMAMORI, M. (1998) Thermal and crystalline properties of waxy wheat (*Triticum aestivum* L.) starch, *J. Cereal Sci.,* **27**, 1–5.

GALLANT, D. J., BOUCHET, B. and BALDWIN, P. M. (1997) Microscopy of starch: evidence of a new level of granule organization, *Carbohydr. Polym.,* **32**, 177–191.

GHIASI, K., HOSENEY, R. C., ZELEZNAK, K. and ROGERS, D. E. (1984) Effect of waxy barley starch and reheating on firmness of bread crumb, *Cereal Chem.,* **61**, 281–285.

GRANT, L. A., VIGNAUX, N., DOEHLERT, D. C., MCMULLEN, M. S., ELIAS, E. M. and KIANIAN, S. (2001) Starch characteristics of waxy and nonwaxy tetraploid (*Triticum turgidum* L. var. durum) wheats, *Cereal Chem.,* **78**, 590–595.

GUDMUNDSSON, M. and ELIASSON, A.-C. (1989) Some physicochemical properties of oat starches extracted from varieties with different oil content, *Acta. Agric. Scand.,* **39**, 101–111.

GUDMUNDSSON, M. and ELIASSON, A.-C. (1990) Retrogradation of amylopectin and the effects of amylose and added surfactants/emulsifiers, *Carbohydr. Polym.,* **13**, 295–315.

HAN, X.-Z. and HAMAKER, B. R. (2002) Association of starch granule proteins with starch ghosts and remnants revealed by confocal laser scanning microscopy, *Cereal Chem.,* **79**, 892–896.

HAYAKAWA, K., TANAKA, K., NAKAMURA, T., ENDO, S. and HOSHINO, T. (1997) Quality characteristics of waxy hexaploid wheat (Triticum aestivum L.): properties of starch gelatinization and retrogradation, *Cereal Chem.,* **74**, 576–580.

HAYMAN, D., SIPES, K., HOSENEY, R. C. and FAUBION, J. M. (1998) Factors controlling gas cell failure in bread dough, *Cereal Chem.,* **75**, 585–589.

HIBBERD, G. E. (1970) Dynamic viscoelastic behaviour of wheat flour doughs. Part III The influence of the starch granules, *Rheol. Acta.,* **9**, 501–505.

HIZUKURI, S. (1986) Polymodal distribution of the chain lengths of amylopectins, and its significance, *Carbohydr. Res.,* **147**, 342–347.

HIZUKURI, S. (1996) Starch: analytical aspects. In *Carbohydrates in Food* (ed. Eliasson, A.-C.) Marcel Dekker, New York, pp. 347–429.

HOOVER, R. and VASANTHAN, T. (1992) Studies on isolation and characterization of starch from oat (*Avena nuda*) grains, *Carbohydr. Polym.,* **19**, 285–297.

HOOVER, R. and VASANTHAN, T. (1994) The effect of annealing on the physicochemical properties of wheat, oat, potato and lentil starches, *J. Food Biochem.,* **17**, 303–325.

HOSENEY, R. C., FINNEY, K. F., POMERANZ, Y. and SHOGREN, M. D. (1971) Functional (breadmaking) and biochemical properties of wheat flour components. VIII. Starch, *Cereal Chem.,* **48**, 191–201.

HUG-ITEN, S., HANDSCHIN, S., CONDE-PETIT, B. and ESCHER, F. (1999) Changes in starch microstructure on baking and staling of wheat bread, *Lebensm.-Wiss. u.-Technol.,* **32**, 255–260.

JANE, J., XU, A., RADOSAVLJEVIC, M. and SEIB, P. A. (1992) Location of amylose in normal starch granules. I. Susceptibility of amylose and amylopectin to cross-linking reagents, *Cereal Chem.,* **69**, 405–409.

JANE, J., CHEN, Y. Y., LEE, L. F., MCPHERSON, A. E., WONG, K. S., RADOSAVLJEVIC, M. and KASEMSUWAN, T. (1999) Effects of amylopectin branch chain length and amylose content on the gelatinization and pasting properties of starch, *Cereal Chem.,* **76**, 629–637.

JENKINS, P. J., CAMERON, R. E. and DONALD, A. M. (1993) A universal feature in the structure of starch granules from different botanical sources, *Starch/Stärke,* **45**, 417–420.

JUSZCZAK, L., FORTUNA, T. and KROK, F. (2003) Non-contact atomic force microscopy of starch granules surface. Part II. Selected cereal starches, *Starch/Stärke,* **55**, 8–18.

KNUTSON, C. A. (1990) Annealing of maize starches at elevated temperatures, *Cereal Chem.,* **67**, 376–384.

KÖKSEL, H., SAHBAZ, F. and ÖSBOY, Ö. (1993) Influence of wheat-drying temperatures on the birefringence and X-ray diffraction patterns of wet-harvested wheat starch, *Cereal Chem.,* **70**, 481–483.

KROG, N., OLESEN, S. K., TOERNAES, H. and JOENSSON, T. (1989) Retrogradation of the starch fraction in wheat bread, *Cereal Foods World,* **34**, 281–285.

KULP, K. (1973) Characteristics of small-granule starch of flour and wheat, *Cereal Chem.,* **50**, 666–679.

LARSSON, H. and ELIASSON, A.-C. (1997) Influence of the starch granule surface on the rheological behaviour of wheat flour dough, *J. Text. Stud.,* **28**, 487–501.

LARSSON, I. and ELIASSON, A.-C. (1991) Annealing of starch at an intermediate water content, *Starch/Stärke,* **43**, 227–231.

LARSSON, H., ELIASSON, A.-C., JOHANSSON, E. and SVENSSON, G. (2000) Influence of added starch on mixing of dough made with three wheat flours differing in high molecular weight subunit composition: rheological behaviour, *Cereal Chem.,* **77**, 633–639.

LEE, M.-R., SWANSON, B. G. and BAIK, B.-K. (2001) Influence of amylose content on properties of wheat starch and breadmaking quality of starch and gluten blends, *Cereal Chem.,* **78**, 701–706.

LELIEVRE, J., LORENZ, K., MEREDITH, P. and BARUCH, D. W. (1987) Effects of starch particle size and protein concentration on breadmaking performance, *Starch/Stärke,* **39**, 347–352.

LORENZ, K., KULP, K., EVERY, D. and LARSEN, N. (1993) Effect of heat damage on the baking quality of starch extracted from wheat, *Starch/Stärke,* **45**, 25–30.

LUNDQVIST, H., ELIASSON, A.-C. and OLOFSSON, G. (2002a) Binding of hexadecyltrimethylammonium bromide (CTAB) to starch polysaccharides. Part I. Surface tension measurements, *Carbohyd. Polym.,* **49**, 43–55.

LUNDQVIST, H., ELIASSON, A.-C. and OLOFSSON, G. (2002b) Binding of hexadecyltrimethylammonium bromide to starch polysaccharides. Part II. Calorimetric study,

Carbohyd. Polym., **49**, 109–120.

LUNDQVIST, H., NILSSON, G., ELIASSON, A.-C. and GORTON, L. (2002c) Changing the amylopectin-sodium dodecyl sulphate interaction by modifying the exterior chain length, *Starch/Stärke,* **54**, 100–107.

MARTIN, M. L., ZELEZNAK, K. J. and HOSENEY, R. C. (1991) A mechanism of bread firming I. Role of starch swelling, *Cereal Chem.,* **68**, 498–503.

MATSOUKAS, N. P. and MORRISON, W. (1991) Breadmaking quality of ten Greek breadwheats. II. Relationships of protein, lipid and starch components to baking quality, *J. Sci. Food Agric.,* **55**, 87–101.

MCPHERSON, A. E. and JANE, J. (1999) Comparison of waxy potato with other root and tuber starches, *Carbohydr. Polym.,* **40**, 57–70.

MEREDITH, P. (1981) Large and small starch granules in wheat – are they really different? *Starch/Stärke,* **33**, 40–44.

MEREDITH, P., BARUCH, D. W. and JENKINS, L. D. (1977) The developing starch granule. Part V. Techniques for description of size distribution of granules, *Starch/Stärke,* **29**, 217.

MILLER, K. A. and HOSENEY, R. C. (1999) Dynamic rheological properties of wheat starch-gluten doughs, *Cereal Chem.,* **76**, 105–109.

MITA, T. (1990) Effect of aging on the rheological properties of gluten gel, *Agric. Biol. Chem.,* **54**, 927–935.

MORGAN, K. R., GERRARD, J., EVERY, D. and ROSS, M. G., M. (1997) Staling in starch breads: the effect of antistaling α-amylase, *Starch/Stärke,* **49**, 54–59.

MORRISON, W. R. (1995) Starch lipids and how they relate to starch granule structure and functionality, *Cereal Foods World,* **40**, 437–446.

MORRISON, W. R. and TESTER, R. F. (1994) Properties of damaged starch granules. IV. Composition of ball-milled wheat starches and of fractions obtained on hydration, *J. Cereal Sci.,* **20**, 69–77.

MORRISON, W. R., TESTER, R. F. and GIDLEY, M. J. (1994) Properties of damaged starch granules. II. Crystallinity, molecular order and gelatinisation of ball-milled starch, *J. Cereal Sci.,* **19**, 209–217.

NAKAMURA, T., YAMAMORI, M., HIRANO, H., HIDAKA, S. and NAGAMINE, T. (1995) Production of waxy (amylose-free) wheats, *Mol. Gen. Genet.,* **248**, 253–259.

PENG, M., GAO, M., ABDEL-TAL, E. S. M., HUEL, P. and CHIBBAR, R. N. (1999) Separation and characterization of A- and B-type starch granules in wheat endosperm, *Cereal Chem.,* **76**, 375–379.

PERRY, P. A. and DONALD, A. M. (2002) The effect of sugars on the gelaltinisation of starch, *Carbohydr. Polym.,* **49**, 155–165.

PETROFSKY, K. E. and HOSENEY, R. C. (1995) Rheological properties of dough made with starch and gluten from several cereal sources, *Cereal Chem.,* **72**, 53–58.

RAHMAN, S., LI, Z., BATEY, I., COCHRANE, M. P., APPLES, R. and MORELL, M. (2000) Genetic alteration of starch functionality in wheat, *J. Cereal Sci.,* **31**, 91–110.

RASPER, V. F. and DEMAN, J. M. (1980) Effect of granule size of substituted starches on the rheological character of composite doughs, *Cereal Chem.,* **57**, 331–340.

ROULET, P., MACINNES, W. M., GUMY, D. and WURSCH, P. (1990) Retrogradation kinetics of eight starches, *Starch/Stärke,* **42**, 99–101.

RUSSELL, P., GOUGH, B. M., GREENWELL, P., FOWLER, A. and MUNRO, H. S. (1987) A study by ESCA of the surface of native and chlorine-treated wheat starch granules: the effects of various surface treatments, *J. Cereal Sci.,* **5**, 83–100.

RUSSELL, P. L. (1983) A kinetic study of bread staling by differential scanning calorimetry and compressibility measurements. The effect of added monoglycerides, *J. Cereal Sci.,* **1**, 297–303.

SAHLSTRÖM, S., BRÅTHEN, E., LEA, P. and AUTIO, K. (1998) Influence of starch granule size distribution on bread characteristics, *J. Cereal Sci.,* **28**, 157–164.

SAHLSTRÖM, S., BRÅTHEN, E., LEA, P. and AUTIO, K. (2003) Impact of starch properties on hearth bread characteristics. I. Starch in wheat flour, *J. Cereal Sci.*, **37**, 275–284.

SASAKI, T., YASUI, T. and MATSUKI, J. (2000) Effect of amylose content on gelatinisation, retrogradation, and pasting properties of starches from waxy and nonwaxy wheat and their F1 seeds, *Cereal Chem.,* **77**, 58–63.

SASAKI, T., YASUI, T., MAATSUKI, J. and SATAKE, T. (2002) Comparison of physical properties of wheat starch gels with different amylose content, *Cereal Chem.,* **79**, 861–866.

SEGUCHI, M. (1986) Dye binding to the surface of wheat starch granules, *Cereal Chem.,* **63**, 518–520.

SEGUCHI, M. (1987) Effect of chlorination on the hydrophobicity of wheat starch, *Cereal Chem.,* **64**, 281–282.

SEGUCHI, M. and KANENAGA, K. (1997) Study of three-dimensional structure of wheat starch granules stained with remazolbrilliant blue dye and extracted with aqueous sodium dedecyl sulfate and mercaptoethanol solution, *Cereal Chem.,* **74**, 548–552.

SHI, Y.-C. and SEIB, P. A. (1992) The structure of four waxy starches related to gelatinization and retrogradation, *Carbohydr. Res.,* **227**, 131–145.

SHIBANUMA, Y., TAKEDA, Y. and HIZUKURI, S. (1996) Molecular and pasting properties of some wheat starches, *Carbohydr. Polym.,* **29**, 253–261.

SOLLARS, W. F. and RUBENTHALER, G. L. (1971) Performance of wheat and other starches in reconstitued flours, *Cereal Chem.,* **48**, 397–410.

SOULAKA, A. B. and MORRISON, W. R. (1985a) The amylose and lipid contents, dimensions, and gelatinization characteristics of some wheat starches and their A-and B-granule fraction, *J. Sci. Food Agric.,* **36**, 709–718.

SOULAKA, A. B. and MORRISON, W. R. (1985b) The bread baking quality of six wheat starches differing in composition and physical properties, *J. Sci. Food Agric.,* **36**, 719–727.

STAUFFER, C. E. (1994) Redox systems in cookie and cracker dough. In *The Science of Cookie and Cracker Production* (ed. Faridi, H.) Chapman & Hall, New York, pp. 227–251.

SVENSSON, G. (1981) Varietal and environmental effects on wheat milling quality, *Agri Hortique Genetica,* **XXXIX**, 1–103.

TESTER, R. F., SOUTH, J. B., MORRISON, W. R. and ELLIS, R. P. (1991) The effects of ambient temperature during the grain-filling period on the composition and properties of starch from four barley genotypes, *J. Cereal Sci.,* **13**, 113–127.

TESTER, R. F., MORRISON, W. R., ELLIS, R. H., PIGGOTT, J. R., BATTS, G. R., WHEELER, T. R., MORISON, J. I. L., HADLEY, P. and LEDWARD, D. A. (1995) Effect of elevated growth temperature and carbon dioxide levels on some physicochemical properties of wheat starch, *J. Cereal Sci.,* **22**, 63–71.

VAN VLIET, T. (1988) Rheological properties of filled gels. Influence of filler matrix interaction, *Colloid Polym. Sci.,* **266**, 518–524.

VANSTEELANDT, J. and DELCOUR, J. A. (1999) Characterisation of starch from durum wheat (*Triticum durum*), *Starch/Stärke,* **51**, 73–80.

VILLWOCK, V. K., ELIASSON, A.-C., SILVERIO, J. and BEMILLER, J. N. (1999) Starch-lipid interactions in common, waxy, *ae du*, and, *ae su2* maize starches examined by differential scanning calorimetry, *Cereal Chem.,* **76**, 292–298.

WANG, L. Z. and WHITE, P. J. (1994a) Functional properties of oat starches and relationships among functional and structural characteristics, *Cereal Chem.,* **71**, 451–458.

WANG, L. Z. and WHITE, P. J. (1994b) Structure and physicochemical properties of starches from oats with different lipid contents, *Cereal Chem.,* **71**, 443–450.

WARD, K. E. J., HOSENEY, R. C. and SEIB, P. (1994) Retrogradation of amylopectin from maize and wheat starches, *Cereal Chem.,* **71**, 150–155.

WONG, R. B. K. and LELIEVRE, J. (1981) Viscoelastic behaviour of wheat starch pastes, *Rheol. Acta,* **20**, 299–307.

WURSCH, P. and GUMY, D. (1994) Inhibition of amylopectin retrogradation by partial beta-amylolysis, *Carbohydr. Res.,* **256**, 129–137.

YASUI, T., MAATSUKI, J., SASAKI, T. and YAMAMORI, M. (1996) Amylose and lipid contents, amylopectin structure and gelatinisation properties of waxy wheat (*Triticum aestivum*) starch, *J. Cereal Sci.,* **24**, 131–137.

YOO, S.-H. and JANE, J. (2002) Structural and physical characteristics of waxy and other wheat starches, *Carbohyd. Polym.,* **49**, 297–305.

YUN, S.-H., QUAIL, K. and MOSS, R. (1996) Physicochemical properties of Australian wheat flours for white salted noodles, *J. Cereal Sci.,* **23**, 181–189.

ZAMPONI, R. A., GINER, S. A., LUPANO, C. E. and ANON, M. C. (1990) Effect of heat on thermal and functional properties of wheat, *J. Cereal Sci.,* **12**, 279–287.

ZOBEL, H. F. (1988) Molecules to granules: a comprehensive starch review, *Starch/Stärke,* **40**, 44–50.

ZOBEL, H. F. and KULP, K. (1996) The staling mechanism. In *Baked Goods Freshness. Technology, Evaluation, and Inhibition of Staling* (eds, Hebeda, R. E. and Zobel, H. F.) Marcel Dekker, Inc, New York, pp. 1–64.

8

Improving wheat quality: the role of biotechnology

P. R. Shewry, Rothamsted Research, UK

8.1 Introduction

White flour, as used for breadmaking, is derived from the starchy endosperm cells of the mature grain. The sole function of these cells is to provide storage reserves to support germination and seedling growth with two major types of storage compound being present: starch and protein. It is well established that the flour proteins are the major determinants of breadmaking performance and they will therefore form the subject of this chapter.

The precise protein content of white flour can vary widely, depending on the amount of fertilizer nitrogen applied by the farmer and, to a lesser extent, the crop genotype. This variation facilitates the use of the flour for a range of purposes, with higher protein contents (generally above about 12% dry matter) being required for breadmaking and lower contents (as low as 8%) for other baked products and noodles. However, protein content is not sufficient to determine the suitability of wheat for a specific end use: protein quality is also important and this is determined by the properties of the gluten proteins.

8.2 Wheat gluten proteins

The gluten proteins have been reported to account for about 85% of the total protein in white flour. They correspond to the grain storage proteins, which are initially deposited in discrete protein bodies within the endosperm cells. However, these protein bodies coalesce during the later stages of grain development, and the gluten proteins form a continuous matrix in the mature endosperm cells and flour derived from them. This is illustrated in Fig. 8.1,

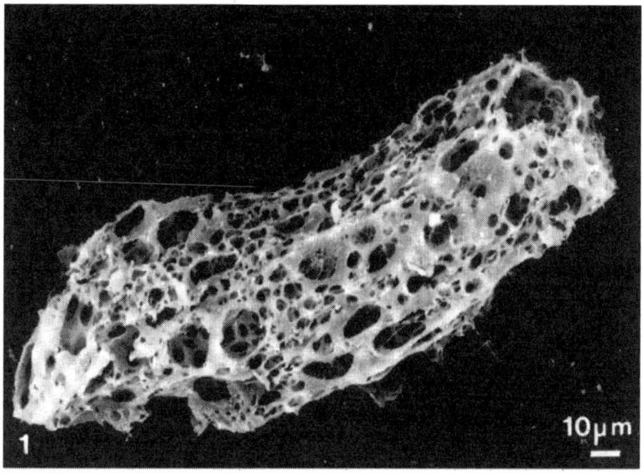

Fig. 8.1 Micrograph of a flour particle, after digestion to remove starch. Taken from Amend (1995) with permission.

which shows a flour particle after digestion to remove starch, leaving a three-dimensional proteinaceous network. When flour is hydrated and mixed to form dough, the individual gluten networks present in individual flour particles come together to form a continuous network of gluten in the dough.

Wheat gluten is most readily isolated by washing dough with water to remove the bulk of the starch. Left behind is a cohesive mass comprising about 80% protein, 10% starch and 10% other components (lipid, minerals, fibre, etc.). It is probable that some of these 'contaminants' (e.g. starch, fibre) are physically entrapped within the gluten network but others (notably lipid) may interact with the gluten proteins and influence their properties (see Belton *et al.*, 1987).

8.2.1 Gliadins and glutenins

Wheat gluten proteins are classically divided into two fractions termed gliadins and glutenins. Traditionally this separation was achieved by extracting gluten or flour sequentially with aqueous ethanol (70% v/v) followed by dilute (0.5 M) acetic acid. The first fraction corresponds essentially to the gliadins which are present as monomers while the second contains glutenins which are polymers stabilized by inter-chain disulfide bonds. However, this procedure is certainly not straightforward as the 'gliadin' fraction may contain some polymeric components, which can be removed by gel filtration chromatography and bulked together with glutenins. Similarly, the glutenins are not completely extracted by dilute acetic acid. Complete extraction requires the use of chaotropic agents and/or detergents, such as a combination of 3 M urea and cetyltrimethyl ammonium bromide (CTAB) (Meredith and Wren, 1966). Alternatively, the component subunits of glutenin can be extracted as reduced monomers by including a reducing agent (usually 2-mercaptoethanol) and dilute acetic acid (0.5 M) in 50%

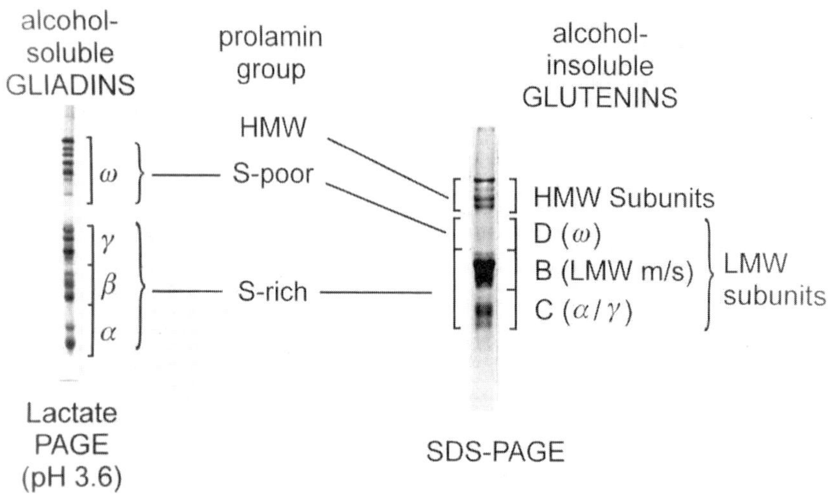

Fig. 8.2 The groups of gliadins and glutenin subunits separated by lactate-PAGE and SDS-PAGE, respectively. Taken from Shewry *et al.* (1999), with permission.

(v/v) propan-l-ol. There is not sufficient space to consider extraction protocols in detail here and the reader is referred to the voluminous primary literature on the subject, published, particularly, in the *Journal of Cereal Science* and *Cereal Chemistry*.

Once extracted it is usual to separate the gliadins and glutenins by electrophoresis, with two systems being widely used. The gliadins can be separated in their unreduced state by electrophoresis at low pH (pH 3.0–3.2) using acid polyacrylamide gel electrophoresis (PAGE) systems based on lactic acid and aluminium and/or sodium lactate. A typical separation, as shown in Fig. 8.2, resolves gliadins into four groups of components called α, β, γ and ω-gliadins in order of decreasing mobility. In contrast, the reduced glutenin subunits are usually separated by sodium dodecyl sulfate (SDS)-PAGE, which resolves groups of high molecular weight (HMW) and low molecular weight (LMW) subunits, the latter being subdivided into B, C and D-type subunits (Fig. 8.2). Detailed biochemical, molecular and genetic studies have elucidated the structural and evolutionary relationships of individual gliadins and glutenin subunits of all the above types, as discussed in recent review articles (Shewry *et al.*, 1999; Shewry, 2002). In brief, these studies have shown that all gluten proteins can be divided into three groups: the sulfur-rich (S-rich) prolamins which comprise the α-, β- and γ-gliadins and the B-type and C-type LMW subunits, the S-poor prolamins which comprise the ω-gliadins and D-type LMW subunits and the HMW prolamins which comprise the HMW subunits. Of these, the HMW subunits play the major role in determining the functional properties.

8.3 HMW subunits and bread quality

Strong (i.e. highly elastic) doughs are required for breadmaking and there is clear evidence that dough strength is related to the amount and properties of the high molecular mass glutenin polymers. This relationship was initially indicated by early work in which measures of quality were shown to be positively correlated with the amounts of insoluble glutenins, i.e. the fraction remaining after extraction with solvents such as 3 M urea (Pomeranz, 1965) or dilute acetic acid (Orth and Bushuk, 1972; Mecham *et al.*, 1972). However, the use of more efficient solvents such as AUC (0.01 M acetic acid, 6 M urea, 0.055 M cetyltrimethyl ammonium bromide) followed by gel filtration chromatography allowed direct positive correlations to be established between breadmaking quality and the amount of high molecular weight polymers while SDS-PAGE showed that these polymers are enriched in HMW subunits of glutenin (Huebner and Wall, 1976; Field *et al.*, 1983) (Fig. 8.3).

Complementary evidence for the importance of the HMW subunits came from genetical studies which were initiated by Payne and colleagues in the late 1970s (Payne *et al.*, 1979, 1981a) and have been confirmed in a number of other laboratories (e.g. Burnouf and Bouriquet 1980; Moonen *et al.*, 1982; Cressey *et al.*, 1987; Lawrence *et al.*, 1987). These studies demonstrated that allelic variation in HMW subunit composition was correlated with differences in dough strength and breadmaking quality, or with various indirect measurements of these functional properties. However, before reviewing these studies in detail it is necessary to discuss the genetics of wheat HMW subunits.

8.3.1 The genetics of HMW subunits

The HMW subunits are encoded by the *Glu-1* loci, which are present on the long arms of the group 1 chromosomes (chromosomes 1A, 1B and 1D of hexaploid bread wheat). Each locus consists of two tightly linked genes, encoding one x-type and one y-type subunit. Although these subunit types were initially defined based on their mobility on SDS-PAGE (y-type subunits being faster) (Payne *et al.*, 1981) they were subsequently shown to differ fundamentally in their amino acid sequences, as discussed below. The presence of three x-type and three y-type HMW subunit genes means that bread wheat could theoretically express six different HMW subunit proteins. However, this occurs only rarely with commercial cultivars containing only three, four or five subunits. This results from the silencing of specific genes, with 1Bx, 1Dx and 1Dy genes being expressed in all cultivars and 1Ax and/or 1By subunits being present in some cultivars only. 1Ay subunits are never present in cultivated hexaploid bread wheat but may be present in wild and cultivated diploid and tetraploid wheats containing the A genome (Waines and Payne, 1987; Levy *et al.*, 1988; Margiotta *et al.*, 1998).

In addition to varying in number, the HMW subunits also occur in allelic forms which differ in their mobility on SDS-PAGE (resulting from differences in molecular mass) (Fig. 8.4). These allelic forms were initially numbered in

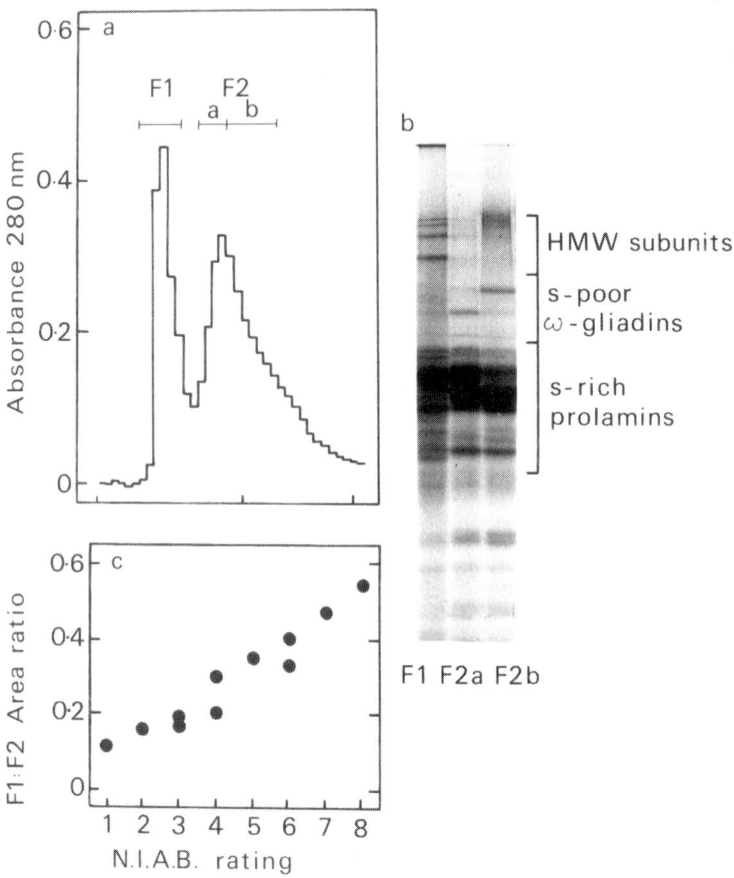

Fig. 8.3 Association of the amount of high molecular weight aggregated gluten proteins with baking quality in British wheats. (a) Fractionation of total gluten proteins of the cv. Copain on a column of controlled pore glass: fraction F1 consists of high molecular weight (probably over 1×10^6) aggregated proteins. (b) SDS-PAGE, after reduction, of fractions F1, F2a and F2b from (a): note that the HMW subunits of glutenin are only present in the F1 fraction. (c) The relationship of the ratio of the F1 and F2 (a + b) fractions and the baking quality (taken as the NIAB scores for the year of harvest) of 11 cultivars of winter and spring wheat. Taken from Field *et al.* (1983) with permission.

order of their mobility on SDS-PAGE (subunit 1Ax1 being slowest) (Payne and Lawrence, 1983) but the subsequent identification of additional subunits with faster, slower or intermediate mobilities has required the use of some additional symbols (e.g. subunits 1Dx2.2 and 1Dx2.2* are both slower allelic forms of subunit 1Dx2 while 1Ax2* has similar mobility to 1Dx2 on many SDS-PAGE systems).

The x-type and y-type HMW subunit genes are tightly linked and hence the allelic forms of x-type and y-type subunits are inherited as 'allelic pairs'. For example, subunits 1Dx5 and 1Dy10 are allelic to subunits 1Dx2 and 1Dy12.

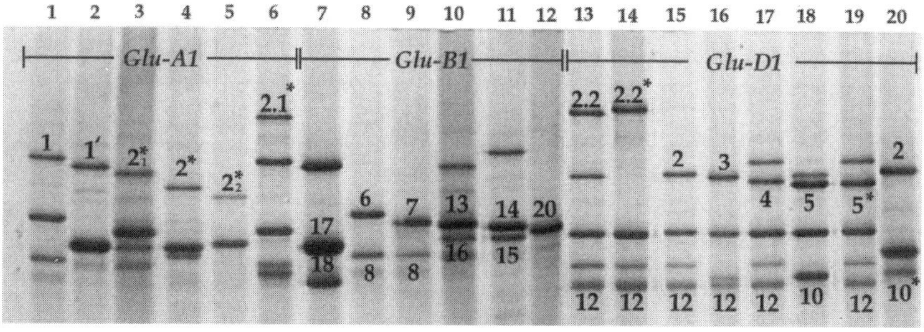

Fig. 8.4 SDS-PAGE of HMW subunits from a range of genotypes of wheat showing allelic variation in the mobilities of proteins encoded by the *Glu-A1*, *Glu-B1* and *Glu-D1* loci. The numbers are according to Payne and Lawrence (1983) with subsequent modifications by other workers. Taken from Shewry *et al.* (2003b) with permission.

8.3.2 Correlation between HMW subunit composition and breadmaking quality

The high level of polymorphism in HMW subunit composition has allowed the presence of specific allelic forms to be correlated with breadmaking performance or parameters that relate to this (e.g. SDS sedimentation, dough strength, mixing properties). The literature on this topic is vast, and the reader is referred to review articles for full details (Payne, 1987; Shewry *et al.*, 1992; Shewry, 2002). Three types of experimental material have been used for these studies, with broadly consistent results. These are collections of cultivars, the progeny of crosses between lines with different HMW subunit compositions and series of near isogenic lines in which different HMW subunits are incorporated into a common genetic background. The current status of such lines is discussed in detail by Shewry *et al.* (2003b).

The conclusions from these studies can be summarised as follows:

- Subunits encoded by all three genomes (1A, 1B, 1D) may be associated with quality but the magnitude of this effect differs.
- The subunit pair 1Dx5 + 1Dy10 (encoded by chromosome 1D) is associated with the highest quality when compared with the allelic pairs 1Dx2 + 1Dy12, 1Dx3 + 1Dy12 and 1Dx4 + 1Dy12 which are all associated with poor quality.
- The presence of a 1Ax subunit (1Ax1 or 1Ax2*) is always superior to the null (i.e. silent) allele.
- The subunit pair 1Bx17 + 1By18 is generally superior to all other alleles encoded by chromosome 1B (i.e. single 1Bx subunits or 1Bx + 1By subunit pairs).

It has also been possible to combine data from such studies to assign 'quality scores' for individual subunits or subunit pairs, as shown in Table 8.1. Despite being relatively minor components in terms of amount (see below), the HMW subunits have been calculated to account for between about 45% and 70% of the

Table 8.1 Quality scores assigned to individual HMW subunits or subunit pairs. Taken from Payne *et al.* (1987) with permission

Score	Locus		
	Glu-A1	*Glu*-B1	*Glu*-D1
4	-	-	5 + 10
3	1	17 + 18	-
3	2*	7 + 8	-
2	-	7 + 9	2 + 12
2	-	-	3 + 12
1	null	7	4 + 12
1	-	6 + 8	-

variation in breadmaking performance within European wheats (Branlard and Dardevet, 1985; Payne *et al.*, 1987, 1988).

8.3.3 Quantitative variation in HMW subunits

The HMW subunits have been reported to account for up to about 12% of the total grain protein, corresponding to 1–1.7% of the flour dry weight (Nicolas, 1997). However, more detailed analyses show that each individual HMW subunit protein accounts on average, for about 2% of the total, and that variation in subunit number is associated with differences in the total amount of HMW subunit protein (Seilmeier *et al.*, 1991; Halford *et al.*, 1992). For example, Halford *et al.* (1992) compared the total amounts of HMW subunit protein in cultivars with similar HMW subunit compositions but with either the null 1Ax gene (i.e. expressing a total of four subunits) or expressing subunit 1Ax1 or 1Ax2* (i.e. expressing five subunits). The presence of a 1Ax subunit was associated with an increase in total HMW subunit protein from 8.015 to 10.211% (SE = 0.1411) of the total extractable protein. Comparisons of the SDS-PAGE patterns of HMW subunit proteins from different cultivars also indicate that some allelic forms are expressed at higher levels than others (cf. 1Bx and 1By subunits in Fig. 8.4) but these differences have not been studied in detail.

8.3.4 HMW subunits have quantitative and qualitative effects on breadmaking quality

These studies indicate, therefore, that HMW subunits have two types of effect on quality. These are quantitative effects, which are related to differences in gene expression, and qualitative effects, which relate to differences in the structures and properties of allelic subunits. These have provided a basis for attempts to improve breadmaking quality by manipulating HMW subunit composition in transgenic plants.

8.4 The genetic transformation of wheat

The first prerequisite for manipulating plant traits using genetic engineering is the availability of gene(s) that determine the trait of interest, including the protein-coding sequence and regulatory sequences that control the pattern and level of gene expression. These coding and regulatory sequences do not necessarily need to be from the same gene, as discussed below.

Assuming the appropriate gene(s) are available, the transformation of wheat can be divided into three parts: gene delivery, the selection and regeneration of transformed plants, and characterization. These parts will now be discussed in turn.

8.4.1 DNA delivery

The delivery of DNA into plant cells can be achieved either by using a vector system or by direct insertion. Early work on plant transformation focused on the exploitation of *Agrobacterium tumefaciens*, a soil bacterium that naturally infects the roots of many plant species. Infection with *Agrobacterium* is accompanied by the insertion of a small piece of DNA (present on the tumour-inducing or T_i plasmid) into the DNA of the host cells, with the subsequent expression of the genes present on this plasmid leading to cell proliferation and 'crown gall' formation. Replacement of genes present on the native T_i plasmid with foreign genes can result in the latter being inserted into the DNA of the host cell and expressed. *Agrobacterium*-mediated transformation has become the standard system for transformation of dicotyledonous plants but has proved less successful for monocotyledonous plants, many of which are outside the normal host range for *Agrobacterium*. Nevertheless, it is now widely used for rice and maize transformation, and is under development for wheat (Cheng *et al.*, 1997) and barley (see review by Barcelo *et al.*, 2001).

The standard method of DNA delivery to wheat, which is currently used in a number of laboratories worldwide, is direct gene transfer by particle bombardment (Vasil *et al.*, 1992; Weeks *et al.*, 1993). This simple procedure involves coating the DNA onto microscopic gold particles which are then literally shot into cells of the recipient tissues using a 'gun' driven by helium pressure. Although there is inevitably some tissue damage, at least some cells that receive the particles become transformed by insertion of the DNA and can be regenerated to give transformed plants.

8.4.2 Regeneration and selection

It is important that the transformed cells are capable of being regenerated to form a transformed plant and it is therefore necessary to use a highly embryogenic tissue. It is usual with wheat to use immature embryos, isolated from developing seeds, but immature inflorescences can also be used, particularly for durum wheat (Barcelo *et al.*, 2001; Shewry *et al.*, 2003).

Of course, not all of the cells in the bombarded tissue are transformed and regeneration of such a tissue would give rise to a chimaeric plant in which untransformed cells are mixed with those that have been transformed. Consequently, it is necessary to use a selection system in which all non-transformed cells are killed, allowing only transformed cells to regenerate. In practice this leads to the regeneration of homogeneous transformed plants arising from single cells. The most widely used selection system is resistance to the herbicide phosphinothricin (the active ingredient of Basta), conferred by the *bar* gene from the bacterium *Streptomyces*. The *bar* gene is usually present on a separate DNA plasmid to the gene of interest but a high level of co-transformation occurs, meaning that the majority of herbicide-resistant regenerants are also transformed with the gene of interest. In addition to the 'selectible' marker gene, a second 'scoreable' marker gene is also often used, allowing transformation to be confirmed by a simple colorimetric assay. The most widely used is the *uidA* gene from *Escherichia coli*, encoding the enzyme β-glucuronidase which gives a blue product when incubated with the appropriate substrate.

8.4.3 Characterization

It is usual for the regenerated transgenic plants to be heterozygous for the genes of interest. Furthermore, particle bombardment may result in insertion of transgenes at two or more loci which may be unlinked and hence segregate in the T_1 (1st transformed) generation. The first priority is, therefore, to select homozygous transgenic lines and this can be facilitated by 'doubled haploidization' technology, either by *in vitro* culture of microspore-derived anthers or crosses with maize pollen (Zhang *et al*., 1996; Massiah *et al*., 2001). Both approaches result in haploid embryos which need to be 'rescued' from the mother plant, treated with a chromosome duplicating agent to give a homozygous diploid chromosome complement, and cultured *in vitro* to give plantlets.

Further analyses will depend on the gene of interest and use of the transgenic plants (i.e. for research purposes or commercial exploitation) but will include one or all of the following:

- Determination of the numbers of inserted transgenes, their structures (i.e. intact or rearranged), organisation and location in the genome, including analysis of the DNA flanking the transgene inserts.
- Determination of the pattern and level of transgene expression in the plant based on analysis of the corresponding mRNA.
- The pattern and level of accumulation of the transgene product within the plant.
- The inheritance of the transgenes and their stability of expression over several generations in the field and under containment conditions.
- The effect of the transgene on the plant phenotype (development, composition, etc.) and on functional properties.

It is crucial that transgenes should behave in a similar fashion to endogenous genes, without any undesired or unpredictable effect, if transgenic crops are to be acceptable for commercial production.

8.5 Manipulating HMW subunit composition and dough properties

Genes for HMW subunits of bread wheat were initially isolated in the mid-1980s (Forde *et al.*, 1985; Sugiyama *et al.*, 1985; Thompson *et al.*, 1985) and a number of genes are now available, both from cultivated wheat (including the whole gene family from cv. Cheyenne and allelic forms from other cultivars) and from related cultivated (*Triticum timopheevi*) and wild (*T. tauschii*, *Aegilops cylindrica*) species (see Shewry *et al.*, 2003b).

The ready availability of these genes and the clear correlation between the number of expressed genes, the total amount of HMW subunit protein and breadmaking quality of wheat (see above), resulted in HMW subunit composition being the first target for attempts to improve wheat quality by genetic engineering. The first success was reported by Blechl and Anderson (1996) who constructed a gene encoding a hybrid protein comprising the *N*-terminal part of subunit 1Dy10 (residues 1–124) fused to the *C*-terminal part of subunit 1Dx5 (residues 130–848). This strategy was used to allow the novel subunit to be clearly resolved from the endogenous subunits 1Dx5 and 1Dy10 present in the host cultivar (Bobwhite). Accumulation was observed at similar levels to those of the endogenous subunits, but the chimaeric protein was shown to form circular monomeric structures stabilized by head-to-tail disulphide bonds and was not incorporated into glutenin polymers.

Subsequent studies have used genes encoding HMW subunits 1Ax1, 1Dx5 and 1Dy10 to transform either Bobwhite (Altpeter *et al.*, 1996; Blechl *et al.*, 1998; Anderson and Blechl, 2000), commercial Argentinean lines (Alvarez *et al.*, 2000) or near isogenic lines differing in their HMW subunit composition (Barro *et al.*, 1997). In all cases the transgenes were expressed at levels up to and exceeding those of the endogenous HMW subunit genes. However, both Blechl *et al.* (1998) and Alvarez *et al.* (2000) have reported that the expression of HMW subunit transgenes can result in reduced expression of one or all of the endogenous HMW subunit genes, a phenomenon that has been observed previously in transgenic plants (Meyer and Saedler, 1996).

8.5.1 Effects of HMW transgenes on dough functional properties

Analyses of dough functional properties have been carried out using Mixograph and breadmaking tests. The Mixograph is a recording dough mixer that measures energy input. It was originally designed as an alternative to the Farinograph, and is more suited to strong doughs such as those made from North American wheat (Walker and Hazelton, 1996). It is particularly attractive as a tool for research as it is available in versions using 35 g, 10 g and 2 g of flour. It can therefore be used in early generation programmes including the analysis of grain from single plants. A typical Mixograph curve allows the measurement of a number of parameters, the most important being peak resistance (in arbitrary units), mixing time (in seconds) and resistance breakdown (i.e. stability to overmixing, also measured in arbitrary units).

The Bobwhite lines used have the 1BL/1RS chromosome translocation which is associated with the sticky dough character. Hence, the Mixograms from this variety tend to be flatter than those of a strong breadmaking wheat. Vasil *et al.* (2001) showed that field-grown samples of two transgenic lines expressing subunit 1Ax1 had longer mixing times than the control line, while the same two lines and a third line all showed slightly higher loaf volumes (by approx. 5% and 10%). Anderson and Blechl (2000) also showed that the expression of transgenes for subunits 1Dx5 and 1Dy10 in the Bobwhite background resulted in a dramatic increase in mixing time (from about 4 to 17 min) but a decrease in peak resistance.

More typical Mixograms are exhibited by the Argentinean cv. ProInter Federal which was transformed by Alvarez *et al.* (2000) to express the 1Ax1 or 1Dx5 transgenes. Expression of subunit 1Dx5 resulted in a two-fold increase in mixing time and lower resistance breakdown, but the dough failed to develop normally (Alvarez *et al.*, 2001). Other transgenic lines in which suppression of subunit expression occurred had lower mixing times and lower peak resistance, indicating reduced dough strength.

These studies are consistent in that they show that transformation to express additional HMW subunit genes resulted in increased mixing time, which is consistent with increased dough strength, but without a corresponding increase in peak resistance. This indicates that the transgenic subunits had complex effects on gluten structure and properties rather than a straightforward increase in strength. Further insight can be gained from more detailed studies carried out by ourselves in collaboration with the groups of Yves Popineau (INRA, Nantes, France) and Ferenc Békés (CSIRO, Canberra, Australia) (Barro *et al.*, 1997; Rooke *et al.*, 1999; Popineau *et al.*, 2001).

Our work has focused on transformants in two near isogenic lines, L88-6 which expresses five HMW subunits (1Ax1, 1Dx5, 1Dy10, 1Bx17, 1By18) and L88-31 which expresses only subunits 1Bx17 and 1Bx18. Lines B73-6-1 and B72-8-11b have high-level expression of the 1Dx5 transgene in the L88-6 and L88-31 backgrounds, respectively (Fig. 8.5a). Analysis of field-grown material of these lines showed that the transgene had detrimental effects on mixing properties, with dramatic decreases in both mixing time and peak resistance (Fig. 8.5, B, D, E, F). Furthermore, in both cases the flour failed to form a cohesive dough on hydration and mixing, as previously reported for glasshouse grown material of B73-6-1 (Rooke *et al.*, 1999) and by Alvarez *et al.* (2001) for ProInter Federal expressing the 1Dx5 gene. The effect in the L88-31 background was particularly unexpected as this line expresses only two endogenous HMW subunits and hence has poor intrinsic mixing properties (Fig. 8.5B). Contrasting

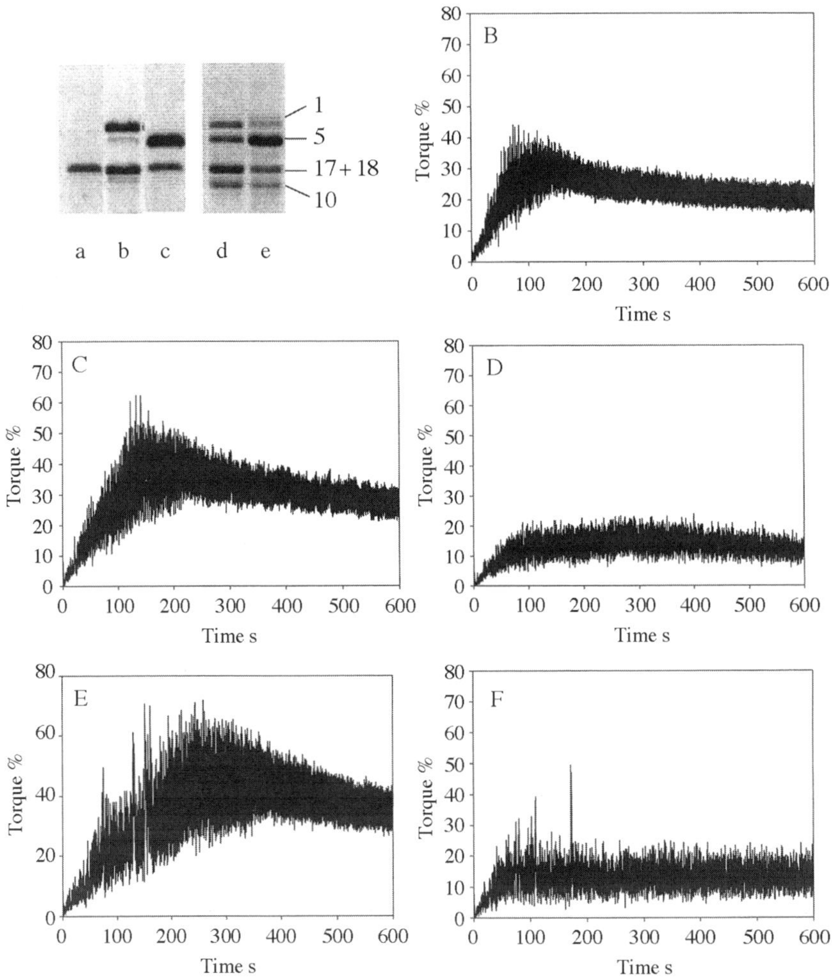

Fig. 8.5 Analysis of the mixing properties of transgenic wheats expressing additional HMW subunits using the 2 g Mixograph. (a) SDS-PAGE of the HMW subunits from a, L88-31: control line (1A null, 1Bx17+1By18, 1D null); b, B72-8-11b: transformed line expressing 1Dx5 subunit transgene in the L88-31 background; c, B102-1-2: transformed line expressing 1Ax1 subunit transgene in the L88-31 background; d, L88-6: control line (1Ax1, 1Bx17+1By18, 1Dx5+1Dy10); e, B73-6-1: transformed line expressing 1Dx5 subunit transgene in the L88-6 background. **B–F** are mixograms of (**B**), L88-31; (**C**), B102-1-2; (**D**), B72-8-11B, (**E**), L88-6; (**F**), B73-6-1. The resistance is given as torque % and the mixing time in seconds. Taken from Popineau *et al.* (2001), with permission.

results were observed when L88-31 was transformed with the subunit 1Ax1 gene (in line B102-1-2, Fig. 8.5C), with the expected increases in mixing time and peak resistance being observed. It is clear, therefore, that the 1Ax1 and 1Dx5 transgenes had fundamentally different effects on mixing properties, even when expressed in the same poor quality background. We have also shown that different effects on breadmaking quality occur, with B73-6-1 and B72-8-11b giving low loaf volumes (approx. 53% and 89% of the values for the respective parents) while B102-1-2 had a loaf volume approx. 10% higher than that of the parent (Darlington *et al.*, 2003).

In order to determine the basis for the different effects of the 1Ax1 and 1Dx5 transgenes, Popineau *et al.* (2001) fractionated flour protein from the transgenic lines using sequential extraction and carried out rheological analyses of gluten fractions. Sequential extraction showed that presence of the 1Dx5 transgene resulted in a massive increase in the proportion of the total protein present as insoluble polymers (i.e. extracted only by 2% (w/v) SDS + 2% (v/v) dithiothreitol (to reduce disulfide bonds) with sonication), from about 11 to 18% in B72-8-11b and from about 13.5 to 29% in B73-6-1. Furthermore, the gluten fractions from B72-8-11b and B73-6-1 had higher viscoelastic plateaux (G_N^o) which were similar to those of gluten fractions which had been modified by treatment with transglutaminase to increase cross-linking. In contrast, B102-1-2 had an increased amount of soluble glutenin polymer (extracted by 2% (v/v) SDS with sonication) and the increase in G_N^o was more modest (i.e. as would be expected for the addition of an endogenous gene).

It was concluded that the subunit 1Ax1 and 1Dx5 transgenes have fundamentally different effects on the structure and properties of gluten. The 1Ax1 gene appears to have a similar effect to that predicted on the basis of comparisons of cultivars and near-isogenic lines differing in numbers of endogenous HMW subunit genes. In contrast, the subunit 1Dx transgene gives rise to highly cross-linked glutenin polymers which do not hydrate normally and result in 'overstrong' type mixing properties. This may relate to the requirement for an appropriate balance of subunits 1Dx5 and 1Dy10, the two proteins forming an allelic pair. A further contributory factor may be the presence in subunit 1Dx5 of a cysteine residue which is absent from all other characterized subunits. It is possible, therefore that this cysteine forms inter-chain disulfide bonds which increase polymer cross-linking in the transgenic lines.

8.6 Future trends: improving bread quality

The initial results described in the preceding section are disappointing in that they have failed to demonstrate the possibility of achieving a real improvement in mixing properties or breadmaking quality in a commercially relevant line. However, they have provided important information which is benefiting more applied studies. In particular, our current efforts are focused on exploiting the subunit 1Ax1 transgene, which occurs as a single subunit (i.e. not as part of an allelic pair) in bread wheat and has been shown to result in improved quality of the L88–31 and Bobwhite lines (Vasil *et al.*, 2001; Popineau *et al.*, 2001). We have therefore generated lines of the commercial cultivars Imp, Canon and Cadenza which express the 1Ax1 transgene (Pastori *et al.*, 2000) and preliminary Mixograph analyses indicate that at least some of these have improved mixing properties (increased dough strength) (unpublished results of G. Pastori, H. Jones, S. Steele, P.R. Shewry, B. Butow and F. Békés).

Previous work has generated only a small number of transgenic lines which has precluded the possibility of relating transgene expression levels to functional properties. It is therefore necessary to generate larger populations in order to select lines with optimal expression levels. It is possible that these levels will be substantially lower than the high expression levels which have been selected in previous work.

It may also be of value to carry out transformation with genes encoding subunits with unusual structures and properties. These could be generated by mutagenesis of currently available genes, or novel genes could be used derived from exotic lines of bread wheat, other cultivated wheat species or wild relatives. Two specific features of HMW structure may be important in this respect.

- **Subunit size**. There is evidence from mixing studies that larger HMW subunit proteins have a greater positive effect on dough strength than smaller subunits (Békés *et al.*, 1995). Genes encoding 1Dx subunits with higher molecular weights than 1Dx2 (1Dx2.2, 1Dx2.2*) (see Fig. 8.4) could be isolated from exotic lines of bread wheat (D'Ovidio *et al.*, 1996). Alternatively, genes encoding modified large and small forms of subunit 1Dx5 have been constructed and expressed in *E. coli* (D'Ovidio *et al.*, 1997) and in transgenic wheat (He *et al.*, 2000). The impact of the latter on dough strength has not yet been determined.
- **Crosslinking**. The number and distribution of cysteine residues can be expected to affect the formation of inter-chain cross-links and hence the structures and properties of the glutenin polymers. For example, the presence of an additional cysteine residue in subunit 1Dx5 may account for its dramatic effect on dough properties in transgenic wheat as discussed above. Mutation could be used to remove or add cysteine residues to the proteins encoded by the currently available genes. We have also recently isolated a novel 1Dy gene encoding a protein containing an additional cysteine residue from the wild relative *Aegilops cylindrica* (Wan *et al.*, 2002). This is again being used for transformation of bread wheat.

Of course, although the HMW subunits are the major group of proteins that affect dough strength, other components also affect breadmaking quality and it may be necessary in the future to identify wider targets for manipulation. We have already transformed pasta wheat with genes encoding low molecular weight subunits to increase dough strength (Tosi *et al.*, 2000) and a similar approach may allow the fine tuning of the structure and properties of bread wheat gluten. The promoters of the HMW subunit genes provide an ideal tool to drive transgene expression in such studies, as they confer specific expression only in the starchy endosperm of the developing grain (Lamacchia *et al.*, 2001). However, at present the HMW subunits remain the most viable target for attempts to improve breadmaking quality.

8.7 Sources of further information and advice

There is a vast volume of literature on the genetic engineering of plants, much of which is highly technical. Barcelo *et al.* (2001) provide a comprehensive and up-to-date account focusing on cereals. Similarly, the reader is referred to reviews by Payne (1987) on the genetics of HMW subunits and their role in determining quality and by Shewry *et al.* (2001, 2003b) on their structures and manipulation. The commercialisation of GM wheats with improved quality is still some way off, so advice must be sought from individual researchers, contacted via their Institute web sites or addresses quoted on publications. Most journals now include e-mail addresses of communicating authors.

8.8 Acknowledgements

IACR receives grant-aided support from the Biotechnology and Biological Sciences Research Council of the United Kingdom.

8.9 References

ALTPETER, F., VASIL, V., SRIVASTAVA, V. and VASIL, I. K. 1996. Integration and expression of the high-molecular-weight glutenin subunit 1Ax1 gene into wheat. *Nature Biotechnology* **14**, 1155–1159.

ALVAREZ, M.L., GUELMAN, S., HALFORD, N.G., LUSTIG, S., REGGIARDO, M.I., RYABUSHKINA, N., SHEWRY, P., STEIN, J. and VALLEJOS, R.H. 2000. Silencing of HMW glutenins in transgenic wheat expressing extra HMW subunits. *Theoretical and Applied Genetics* **100**, 319–327.

ALVAREZ, M.L., GÓMEZ, M., CARRILLO, J.M. and VALLEJOS, R.H. 2001. Analysis of dough functionality of flours from transgenic wheat. *Molecular Breeding* **8**, 103–108.

AMEND, T. 1995. Der Mechanismus der Teigbildung: Vorstoß in den molekularen Strukturbereich. *Getreide Mehl und Brot* **49**, 359–362.

ANDERSON, O.D. and BLECHL, A.E. 2000. Transgenic wheat – challenges and opportunities. In: *Transgenic Cereals* (L. O'Brien and R. Henry eds), AACC, St. Paul, MN, pp 1–27.

BARCELO, P., RASCO-GAUNT, S., THORPE, C. and LAZZERI, P.A. 2001. Transformation and Gene Expression. *Advances in Botanical Research* **34**, 59–126.

BARRO, F., WOKE, L., BÉKÉS, F., GRAS, P., TATHAM, A.S., FIDO, RJ., LAZZERI, P., SHEWRY, P.R. and BARCELO, P. 1997. Transformation of wheat with HMW – subunit genes results in improved functional properties. *Nature Biotechnology* **15**, 1295–1299.

BÉKÉS, F., GRAS, P.W. and GUPTA, R.B. 1995. The effects of purified cereal polypeptides on the mixing properties of dough. In: *45th Australian Cereal Chemistry Conference – Capturing the Benefits of Research for Consumers*. Conference Proceedings, Adelaide, 10–14 September 1995. The Royal Australian Chemical Institute Cereal Chemistry Division, pp. 92–98.

BELTON, P.S., DUCE, S.L. and TATHAM, A.S. 1987. Nuclear magnetic resonance studies of wheat gluten. In: *Cereals in a European Context – First European Conference on*

Food Science and Technology (I.D. Morton ed.) Ellis Horwood Ltd, Chichester. pp. 489–495.

BLECHL, A.E. and ANDERSON, O.D. 1996. Expression of a novel high-molecular-weight glutenin subunit gene in transgenic wheat. *Nature Biotechnology* **14**, 875–879.

BLECHL, A.E., LEE, L.Q. and ANDERSON, O.D. 1998. Engineering changes in wheat flour by genetic transformation. *Plant Physiology* **152**, 703–707.

BRANLARD, G. and DARDEVET, M. 1985. Diversity of grain protein and bread wheat quality. II Correlation between high molecular weight subunits of glutenin and flour quality characteristics. *Journal of Cereal Science* **3**, 345–354.

BURNOUF, T. and BOURIQUET, R. 1980. Glutenin subunits of genetically related European hexaploid wheat cultivars: their relation to breadmaking quality. *Theoretical and Applied Genetics* **58**, 107–111.

CHENG, M., FRY, J.E., PANG, S., ZHOU, H., HIRONAKA, C.M., DUNCAN, D.R., CONNER, T.W. and WAN, Y. 1997. Genetic transformation of wheat mediated by *Agrobacterium tumefaciens*. *Plant Physiology* **115**, 971–980.

CRESSEY, P.J., CAMPBELL, W.P., WRIGLEY, C.W. and GRIFFIN, W.B. 1987. Statistical correlations between quality attributes and grain-protein composition for 60 advanced lines of crossbred wheat. *Cereal Chemistry* **64**, 299–301.

DARLINGTON, H., FIDO, R., TATHAM, A.S., JONES, H., SALMON, S.E. and SHEWRY, P.R. 2003. Milling and baking properties of field grown wheat expressing HMW subunit transgenes. *Journal of Cereal Science* in press.

D'OVIDIO, R., LAFIANDRA, D. and PORCEDDU, E. 1996. Identification and molecular characterization of a large insertion within the repetitive domain of a high-molecular-weight glutenin subunit gene from hexaploid wheat. *Theoretical and Applied Genetics* **93**, 1048–1053.

D'OVIDIO, R., ANDERSON, O.D., MASCI, S., SKERRITT, J. and PORCEDDU, E. 1997. Construction of novel wheat high-M_r glutenin subunit gene variability: modification of the repetitive domain and expression in *E. coli*. *Journal of Cereal Science* **25**, 1–8.

FIELD, J.M., SHEWRY, P.R. and MIFLIN, B.J. 1983. Solubilisation and characterisation of wheat gluten proteins: correlations between the amount of aggregated proteins and baking quality. *Journal of the Science of Food and Agriculture* **34**, 370–377.

FORDE, J., MALPICA, J.M., HALFORD, N.G., SHEWRY, P.R., ANDERSON, O.D., GREENE, F.C. and MIFLIN, B.J. 1985. The nucleotide sequence of a HMW subunit located on chromosome 1A of wheat (*Triticum aestivum* L.). *Nucleic Acids Research* **13**, 6817–6832.

HALFORD, N.G., FIELD, J.M., BLAIR, H., URWIN, P., MOORE, K., ROBERT, L., THOMPSON, R., FLAVELL, R.B. TATHAM, A.S. and SHEWRY, P.R. 1992. Analysis of HMW glutenin subunits encoded by chromosome 1A of bread wheat (*Triticum aestivum* L.) indicates quantitative effects on grain quality. *Theoretical and Applied Genetics* **83**, 373–378.

HE, G.Y., D'OVIDIO, R., ANDERSON, O.D., FIDO, R., TATHAM, A.S., JONES, H.D., LAZZERI, P.A. and SHEWRY, P.R. 2000. Modification of storage protein composition in transgenic bread wheat. In: *Wheat Gluten* (P.R. Shewry and A.S. Tatham eds). Royal Society of Chemistry, Cambridge, pp. 84–87.

HUEBNER, F.R. and WALL, J.S. 1976. Fractionation and quantitative differences of glutenin from wheat varieties varying in baking quality. *Cereal Chemistry* **53**, 258–269.

LAMACCHIA, C., SHEWRY, P.R., DI FONZO, N., FORSYTH, J., HARRIS, N., LAZZERI, P.A., NAPIER, J.A., HALFORD, N.G. and BARCELO, P. 2001. Endosperm-specific activity of a storage protein gene promoter in transgenic wheat seed. *Journal of Experimental Botany*

355, 243–250.

LAWRENCE, G.J., MOSS, H.J., SHEPHERD, K.W. and WRIGLEY, C.W. 1987. Dough quality of biotypes of eleven Australian wheat cultivars that differ in high-molecular-weight glutenin subunit composition. *Journal of Cereal Science* **6**, 99–101.

LEVY A.A., GALILI G. and FELDMAN M. 1988. Polymorphism and genetic control of high molecular weight glutenin subunits in wild tetraploid wheats *Triticum turgidum* var. *dicoccoides*. *Heredity* **61**, 63–72.

MARGIOTTA, B., URBANO, M., COLAPRICO, G., TURCHETTA, T. and LAFIANDRA, D. 1998. Variation of high molecular weight glutenin subunits in tetraploid wheats of genomic formula AAGG. In: *Proceedings 9th International Wheat Genetic Symposium* (A.E. Slinkard ed.) vol. 4 pp. 195–197, University Extension Press, University of Saskatchewan, Saskatoon, Canada.

MASSIAH, A., RONG, H.L., BROWN, S. and LAURIE, S. 2001. Accelerated production and identification of fertile, homozygous transgenic wheat lines by anther culture. *Molecular Breeding* **7**, 163–173.

MECHAM, D.K., COLE, E.W. and NG, H. 1972. Solubility effect of mercuric chloride on the gel proteins of wheat flour. *Cereal Chemistry* **49**, 62–67.

MEREDITH, O.B. and WREN, J.J. 1966. Determination of molecular-weight distribution in wheat-flour proteins by extraction and gel filtration in a dissociating medium. *Cereal Chemistry* **43**, 169–186.

MEYER, P. and SAEDLER, H. 1996. Homology dependent gene silencing in plants. *Annual Review of Plant Physiology and Plant Molecular Biology* **47**, 23–48.

MOONEN, J.H.E., SCHEEPSTRA, A. and GRAVELAND, A. 1982. Use of SDS-sedimentation test and SDS-polyacrylamide gel electrophoresis for screening breeders' samples of wheat for breadmaking quality. *Euphytica* **31**, 677–690.

NICOLAS, Y. (1997) Les prolamines de blé: extraction exhaustive et développement de dosages chromatographiques en phase inverse et de dosages immunochimiques à l'aide d'anticorps anti-peptides. Ph. D. Thesis. University of Nantes.

ORTH, R.A. and BUSHUK, W.A. 1972. Comparative study of the proteins of wheat of diverse baking quality. *Cereal Chemistry* **49**, 268–275.

PASTORI, G.M., STEELE, S.H., JONES, H.D. and SHEWRY, P.R. 2000. Transformation of commercial wheat varieties with high molecular weight glutenin subunit genes. In: *Wheat Gluten* (P.R. Shewry and A.S. Tatham eds). Royal Society of Chemistry, Cambridge, pp. 88–92.

PAYNE, P.I. 1987. Genetics of wheat storage proteins and the effect of allelic variation on bread-making quality. *Annual Review of Plant Physiology* **38**, 141–153.

PAYNE, P.I. and LAWRENCE G.J. 1983. Catalogue of alleles for the complex gene loci, *Glu-A1, Glu-B1, Glu-D1* which code for high-molecular-weight subunits of glutenin in hexaploid wheat. *Cereal Research Communication* **11**, 29–35.

PAYNE, P.I., HOLT, L.M., WORLAND, A.J. and LAW, C.N. 1981. Structural and genetical studies on the high-molecular-weight subunits of wheat glutenin. Part III. Telocentric mapping of the subunit genes on the long arms of the homologous group 1 chromosomes. *Theoretical and Applied Genetics* **63**, 129–138.

PAYNE, P.I., NIGHTINGALE, M.A., KRATTIGER, A.F. and HOLT, L.M. 1987. The relationship between HMW glutenin subunit composition and the breadmaking quality of British grown wheat varieties. *Journal of the Science of Food and Agriculture* **40**, 51–65.

PAYNE, P.I., HOLT, L.M., KRATTIGER, A.F. and CARRILLO, J.M. 1988. Relationship between seed quality characteristics and HMW glutenin subunit composition determined using wheats grown in Spain. *Journal of Cereal Science* **7**, 229–235.

POMERANZ, Y. 1965. Dispersibility of wheat proteins in aqueous urea solutions – a new parameter to evaluate breadmaking potentialities of wheat flours. *Journal of the Science of Food and Agriculture* **16**, 586–593.

POPINEAU, Y., DESHAYES, G., LEFEBVRE, J., FIDO, R., TATHAM, A.S. and SHEWRY, P.R. 2001. Prolamin aggregation, gluten viscoelasticity, and mixing properties of transgenic wheat lines expressing 1Ax and 1Dx high molecular weight glutenin subunit transgenes. *Journal of Agricultural and Food Chemistry* **49**, 395–401.

ROOKE, L., BÉKÉS, F., FIDO, R., BARRO, F., GRAS, P., TATHAM, A. S., BARCELO, P., LAZZERI, P. and SHEWRY, P. R. 1999. Overexpression of a gluten protein in transgenic wheat results in greatly increased dough strength. *Journal of Cereal Science* **30**, 115–120.

SEILMEIER, W., BELITZ, H.-D. and WIESER, H. 1991. Separation and quantitative determination of high-molecular-weight subunits of glutenin from different wheat varieties and genetic variants of the variety Sicco. *Zeitschrift Für Lebensmittel-Untersuchung Und-Forschung* **192**, 124–129.

SHEWRY, P.R. (2002). Wheat gluten proteins. In: *Wheat Gluten Protein Analysis* (P.R. Shewry and G. Lookhart eds) pp. 1–17, American Association of Cereal Chemists, St. Paul, Minnesota, USA.

SHEWRY, P.R., HALFORD, N.G. and TATHAM, A.S. 1992. The high molecular weight subunits of wheat glutenin. *Journal of Cereal Science* **15**, 105–120.

SHEWRY, P.R., TATHAM, A.S. and HALFORD, N. 1999. The prolamins of the triticeae. In: *Seed Proteins* (P.R. Shewry and R. Casey eds) Kluwer Academic Publishers, Dordrecht, pp. 35–78.

SHEWRY, P.R., POPINEAU, Y., LAFIANDRA, D. and BELTON, P. 2001. Wheat gluten subunits and dough elasticity. *Trends in Food Science and Technology* **11**, 433–441.

SHEWRY, P.R., BELL, P., DI FONZO, N., LAMACCHIA, C., TOSI, P., LAZZERI, P. and BARCELO, P. 2003a. Genetic manipulation of durum wheat: application to grain composition and quality. In *Durum Wheat*, in press.

SHEWRY, P.R., HALFORD, N.G., TATHAM, A.S., POPINEAU, Y., LAFIANDRA, D. and BELTON, P. 2003b. The high molecular weight subunits of wheat glutenin and their role in determining wheat processing properties. *Advances in Food and Nutrition Research* **45**, 221–302.

SUGIYAMA, T., RAFALSKI, A., PETERSON, D. and SOL, D. 1985. A wheat HMW glutenin subunit gene reveals a highly repeated structure. *Nucleic Acids Research* **13**, 8729–8737.

THOMPSON, R.D., BARTELS, D. and HARBERD, N.P. 1985. Nucleotide sequence of a gene from chromosome 1D of wheat encoding a HMW glutenin subunits. *Nucleic Acids Research* **13**, 6833–6846.

TOSI, P., NAPIER, J.A., D'OVIDIO, R., JONES, H.D. and SHEWRY, P.R. 2000. Modification of the LMW glutenin subunit composition of durum wheat by microprojectile-mediated transformation. In: *Wheat Gluten* (P.R. Shewry and A.S. Tatham eds), Royal Society of Chemistry, Cambridge, pp. 93–96.

VASIL, V., CASTILLO, A.M., FROMM, M.E. and VASIL, I.K. 1992. Herbicide resistant fertile transgenic wheat plants obtained by microprojectile bombardment of regenerable embryonic callus. *Bio/Technology* **10**, 667–674.

VASIL, I.K., BEAN, S., ZHAO, J., MCCLUSKEY, P., LOOKHART, G., ZHAO, H.-P., ALTPETER, F. and VASIL, V. 2001. Evaluation of baking properties and gluten protein composition of field grown transgenic wheat lines expressing high molecular weight glutenin gene 1Ax1. *Journal of Plant Physiology* **158**, 521–528.

WAINES, J.G. and PAYNE P.I. 1987. Electrophoretic analysis of the high-molecular-weight glutenin subunits of *Triticum monococcum*, *T. urartu* and the A genome of bread wheat (*T. aestivum*). *Theoretical and Applied Genetics* **74**, 71–76.

WALKER, C.E. and HAZELTON, J.L. 1996. Dough rheological tests. *Cereal Foods World* **41**, 23–38.

WAN, Y.F., WANG, D., SHEWRY, P.R. and HALFORD, N.G. 2002. Isolation and characterization of five novel high molecular weight subunit genes from *Triticum timopheevi* and *Aegilops cylindrica*. *Theoretical and Applied Genetics* **104**, 828–839.

WEEKS, J.T., ANDERSON, O.D. and BLECHL, A.E. 1993. Rapid production of multiple independent lines of fertile transgenic wheat (*Triticum aestivum*). *Plant Physiology* **102**, 1077–1084.

ZHANG, J., FRIEBE, B., RAUPP, W.J., HARRISON, S.A. and GILL, B.S. 1996. Wheat embryogenesis and haploid production in wheat × maize hybrids. *Euphytica* **90**, 315–324.

9

Analysing wheat and flour

M. Hajšelová, Consultant, UK and A. J. Alldrick, Campden and Chorleywood Food Research Association, UK

9.1 Introduction

(Where reference is made to a recognised standard or test method, further information can be accessed from the relevant publication or website as detailed in Section 9.7).

Quality has been described as, 'The totality of features and characteristics of a product or service that bear on its ability to satisfy stated or implied needs' (BS 4778–2:1991). Quality is therefore understood to be the sum of attributes rather than a single measurable parameter. In common with other foods, the quality of bread is to a degree a reflection of that of its ingredients. Given that the key ingredient is flour, there is a need for tests which can assess the suitability of wheat to make flour and of the flour to make bread. The quality attributes of a 'good' wheat or flour will largely be determined by three key factors: type of bread (e.g. pan bread, or varietal); manufacturing process (e.g. 'Chorleywood Breadmaking Process', or batch fermentation) and; economics, both in terms of production costs and consumer expectations.

In an ideal world, the breadmaking quality of an unknown wheat or flour might be assessed by a single end-use (baking) test. Recommendations as to how such tests should be constructed have been published (ISO 6820:1985) and a number of test methodologies (e.g. ICC 131; AACC 10-09; AACC 10-10B; AACC 10-11) based on breadmaking ability have been developed. In order to be efficient, such tests must reflect the production environment. Given the different types of product, breadmaking processes and the variations which can occur between bakeries using similar processes, the utility of breadmaking tests, on their own, to determine quality parameters and enforce specifications commercially, is limited. A further disadvantage is that such tests are demanding both in terms of resource (floor-space and equipment) and time to perform. Consequently, a number of quality attributes have been identified, which, taken in combination, give a better prediction of how suitable grain and/or flour are for use in production. These can be viewed arbitrarily as being physical (e.g. specific density, kernel characteristics, damaged grain, size uniformity) or chemical (e.g. moisture and ash contents, protein quantity together with quality, enzymatic activity).

Quality attributes have two principal functions. They assist in predicting how a given raw material will behave in the milling or baking processes and thus are of importance to the plant breeder, miller and baker. Secondly, they underpin the contractual relationship between vendor and purchaser. Consequently they have considerable commercial importance. For example, following the 2002 harvest in the UK, the Home-Grown Cereals Authority (Anon, 2002) identified five broad specifications for wheat, based on: varietal classification; protein content; Falling Number and test weight. Specifications issued by individual purchasers are more specific and, in a number of countries, will also specify the exact wheat variety as well as physical and chemical characteristics.

Given the above, there is a necessity for key quality attributes to be agreed and for the development of common methods for their measurement. In light of their technological and commercial importance, these methods must not only give the required technological information, but also be sufficiently robust to support any commercial decision (e.g. accept or reject at intake). Analytical methods must, therefore, fulfil two criteria:

- They must yield results for any particular sample, which are capable of being repeated within the same laboratory and reproduced in others.
- The methods must be recognised within commercial and regulatory environments.

These requirements are usually expressed in the form of nationally or internationally recognised methods, such as those issued by national or international quality standards bodies (e.g. British Standards Institute (BS) and International Organisation for Standardisation (ISO); relevant scientific bodies (e.g. American Association of Cereal Chemists (AACC) and International Association of Cereal Science and Technology (ICC)) and/or; within the context of an industrial forum, as in the case of the UK (Anderson and Salmon, 2002). The particular quality attribute tested for, and the method used by plant breeders, millers and bakers may differ; however, the ultimate objective – a flour with appropriate breadmaking potential – remains common to all.

This chapter aims to provide an overview of methods most commonly used to assess the quality attributes of wheat and flour related to breadmaking quality. A number of test methods rely on a particular piece of equipment. Where reference is made to such equipment, this is a reflection of what is considered to be common practice and not an endorsement by the authors. Although many analytical methods have relevance to both grain and flour, some are restricted to one material or the other.

9.2 Sample collection and preparation

A pre-requisite of any analytical regime is that the sample to be analysed is not only representative of the total bulk, but also that it is homogeneous. In terms of sampling, any procedure must address two types of 'risk' (discussed by Coker, 1998):

- Producer risk, the probability that the regime will result in rejection of material which actually possesses the desired quality attributes, in other words, meets specification.
- Consumer risk, the probability of accepting material which does not meet specification.

Protocols have been developed which deal with these two risk factors and have been incorporated in appropriate standards (e.g. ISO 13690:1999; ICC 130).

A second factor to be considered in sample handling is, that once collected, the sample must be presented in an appropriate form for analysis. In the case of a number of methods, in particular for cereal quality, it is important to use material of a specified particle size. In the case of whole kernels therefore, specific methods have been developed to achieve this (e.g. Anderson and Salmon, 2002; FTWG 03p; FTWG 04p).

9.3 Grain quality parameters

This section deals with a selection of methods concerned mainly with the general quality of grain, rather than parameters more specifically related to breadmaking performance (discussed in Section 9.4).

9.3.1 Moisture

Grain moisture is important both from a food-safety point of view in managing microbiological contamination and also commercially. The commercial aspects include the redundant nature of water and the fact that many specifications for a particular analyte are based on a set common moisture content. Reference methods (ISO 711:1985; ISO 712:1998) for moisture determination involve measuring the difference in mass following oven drying. However, these methods are time consuming and automated methods based on the electrical properties of the grain-moisture complex or using Near Infra Red (NIR) reflectance have been developed. As in the case of other derivative methods of analysis, moisture meters and other equipment which measure moisture content indirectly, need to be calibrated on a regular basis (ISO 7700–1:1984).

9.3.2 Ash

Ash may be defined as the residue left following the controlled combustion of the test material (ISO 2171:1993). It can be determined in a furnace by incineration at high temperatures (e.g. 900°C, ICC 104/1) or indirectly, for example, using electrical conductivity (ICC 157) or NIR (AACC 08–21).

9.3.3 Besatz

Besatz is the term used to describe that proportion of a grain shipment which differs from the expected norm (ICC 102/1). A non-exhaustive list of besatz includes damaged, sprouting and undersized grains, foreign grains, ergot, chaff, etc. The besatz content of any shipment is estimated by a combination of selective sieving and manual selection.

9.3.4 Bulk density/test weight

The test weight of a grain sample is an indicator of its density and soundness. Generally, the greater the test weight, the better the quality. The term test weight is often replaced by either 'Hectolitre weight (metric)' or, 'Bushel weight (US)'. They are expressed in terms of kilograms per hectolitre or pounds per bushel respectively. Essentially these methods are based on determining the mass of a known volume of grain, using a calibrated measuring container (ISO 7971:1986; ISO 7971–2:1995; AACC 55–10).

9.3.5 Grain hardness

The terms 'hard' and 'soft' relate to the milling quality of the grain. Broadly speaking, the concept of wheat 'hardness' relates to the resistance of the grain to the milling process and the pattern of particle formation. It manifests itself in the granularity of the resultant flour which the kernel produces. Differences between hard and soft wheat are reflected in their sieving properties (Kent and Evers, 1994). Subjectively, hardness was determined by estimating how easy it was to crush a kernel between the teeth (Matz, 1991). Absolute measures for hardness are based on the 'Particle Size Index' of a meal ground and subsequently sieved under defined conditions (Hoseney, 1987). Automated methods have been developed, based either on NIR (discussed by Sanders, 1990; AACC 39–70) or on mechanical destruction (e.g. with the Perten Single Kernel Characterisation System 4100; AACC 55–31; FTWG 20). The latter measures the force necessary to crush a single kernel. In the case of the single kernel characterisation system, hardness is expressed in terms of a 'hardness index', calculated from the mean value for 300 kernels. Both the NIR and mechanical systems require appropriate standards (e.g. those produced by the National Institute of Standards & Technology, Gaithersburg MD 20899) in order to calibrate the equipment.

9.3.6 Varietal identification

In a number of countries, in particular when dealing with home-produced wheats, millers will not only specify quality parameters in terms of, for example,

Falling Number (discussed below) and test weight, but also in terms of variety. A good marker for variety is its protein composition. This is determined genetically and is independent of temporal and environmental conditions. Since the wheat plant is essentially self-pollinating, the protein compositions of different wheat varieties remain constant over a number of generations. The protein composition of a particular wheat or flour sample can be characterised in terms of a profile of the gliadin proteins by polyacrylamide gel electrophoresis (ISO 8981:1993; ICC 143). This in many ways is analogous to a fingerprint. Comparison of the gliadin profile generated with those from known standards, can confirm or refute that a delivered wheat is of the required variety. It can also be employed in identifying material of unknown variety. This technique is sufficiently robust to be applied to both composite samples, as well as individual kernels. In the United Kingdom, millers would generally expect no more than 1 kernel in 50 being of incorrect origin, by this method (nabim, 2002).

9.4 Flour quality: protein

As Kent and Evers (1994) have pointed out, two of the principal characteristics of a good breadmaking flour are:

- sufficient protein, both in terms of content and also gluten performance (elasticity, strength and stability);
- an appropriate diastatic potential, with regards to the amounts of amylase as well as susceptible (damaged) starch present to produce sufficient sugars to support yeast activity.

In attempting to predict breadmaking performance, most test methods relate to the content and functionality of the protein and/or starch present in the grain (flour). These in turn are dictated by wheat variety, agronomic conditions and the general condition of the grain itself. This section concentrates on those analytical methods linked to protein and starch. The protein content of wheat or flour can be viewed from two angles: the total protein content and the amount of functional protein present, where the term ‘functional’ refers to the ability of the protein to produce a dough with the desired rheological properties.

9.4.1 Total protein content

Historically, the protein content of flour and grain has been chemically determined by measurement of its nitrogen content and application of an appropriate conversion factor. Until recently, the reference method usually used was the Kjeldahl method (e.g. AACC 46-10). The original method was not only time consuming, but also involved working with strong acids and digests, which gave off acrid and corrosive fumes. The Kjeldahl method has undergone various modifications, which have reduced both the hazards it presents to the analyst and its adverse environmental impact. These include automated systems, which enable the test to be conducted with smaller samples and less reagent.

A more recent development is the Dumas combustion method (ICC 167; FTWG 19). Like the Kjeldahl method, it involves measuring the total amount of nitrogen present in the sample and applying an appropriate conversion factor to determine how much protein is present. While the Kjeldahl method involves the liberation of nitrogen by acid digestion and its estimation using titrimetric techniques, the Dumas method is based on the liberation of nitrogen following combustion at very high temperatures (850–1100 °C) and its subsequent detection using a thermal conductivity detector. Within most of the United Kingdom milling industry, the Dumas method has now superseded that of Kjeldahl as the reference analytical method for determining protein content. The current minimum protein content specified for good breadmaking wheat in the UK is 13% on a dry matter basis (nabim, 2002).

Secondary analytical methods have also been developed, which are widely used on the basis of speed and convenience; in particular, those based on near infra-red (NIR) reflectance or transmittance. These are rapid instrumental techniques for the analysis of cereals both in the laboratory and on-line. Standard methods for the determination of protein content in both whole grains and flour by NIR have been described (AACC 39-25; AACC 39-11). Essentially, the technique relies upon the fact that specific chemical groups within a molecule (in particular, peptide bonds in proteins and the hydroxyl groups in starch and water) absorb near infra-red light to different degrees, dependent on the wavelength. By exposing a material to a known spectrum of NIR and measuring the amount and wavelength of energy reflected, it is possible, using mathematical processing of the signals, to determine the protein content of a sample. It is important to remember that analysis by NIR is a secondary method and is therefore dependent on calibration against standards derived from a suitable standard reference method (e.g. Dumas).

9.4.2 Protein and dough functionality

The key protein complex which contributes to dough functionality is 'gluten'. This is the visco-elastic substance formed through the interaction between the wheat proteins glutenin and gliadin, wheat lipids and water during mixing (Matz, 1991). The principal questions, which need to be addressed, concern the quantity and quality of the gluten present. From first principles this could be achieved by determining the amino acid composition of gluten. However, it has not proven possible to find essential differences in the amino acid profiles of *good* and *bad* flour (Matz, 1991). Attention has therefore focused on either the quantity of gluten present, or on parameters predictive of how the gluten might behave when processed in a dough.

In terms of simple quantity, the 'Gluten Washing Test' (ICC 106/2) is probably the best known method. Fundamentally, the procedure involves making a dough from flour and a buffered solution of sodium chloride. This is then carefully washed with water or with a specific buffer system. The procedure can be performed manually (AACC 39-10); however, the washing process can be automated, for example, the Perten Glutomatic (ICC 137/1). In this case, the wet gluten is isolated by washing the dough with a buffered sodium chloride solution and the residual liquid adhering to the gluten is removed by centrifugation. The remaining gluten is either weighed or dried and weighed according to whether the wet- or dry-gluten content is required.

A related method is the 'Gluten Index' (ICC 155). This is defined as the percentage of the wet gluten that remains on the inside of a special sieve after centrifugation. This method characterises the gluten as being weak, normal or strong.

Gluten quality can be assessed in a number of ways. It is frequently measured on the basis of how it behaves when suspended in an aqueous environment. These measurements usually involve an assessment of the viscosity, sedimentation and/or the swelling properties of the gluten. The most commonly used test method is the Zeleny sedimentation test (Pinkney *et al.,* 1957; ISO 5529:1992; ICC 116/1). Essentially this involves suspending ground wheat particles in a graduated cylinder containing dilute lactic acid and measuring the amount of sedimented material after a standard time interval. The degree of sedimentation is taken as a measure of the hydration capacity of the flour proteins and by inference a measure of baking quality. A higher gluten content and/or a better gluten quality both give rise to slower sedimentation rates and higher sedimentation volumes. An adaptation of the Zeleny method involving the use of sodium dodecyl sulphate (SDS sedimentation volume) has been described (Axford *et al.,* 1979). This has found use in a number of countries, for example in the United Kingdom (BS 4317-19:1990) and also in the assessment of durum wheat (ICC 151).

As Matz (1993) has pointed out, while many dough properties are reflections of gluten characteristics, most cereal chemists prefer to study dough behaviour and thus measure quality attributes under conditions closer to the production environment. Since the original observations by Jenö von Hankoczy in 1927 (Příhoda *et al.,* 1994), who found a correlation between mixer-energy input and dough consistency, a number of analytical techniques have been developed which measure the performance of doughs made from either a grain meal or flour. While some of the methods simply measure rheological behaviour, others simulate what happens under certain technological processes. They find their use both commercially and in terms of basic cereal science and thus have relevance to plant breeders, millers and bakery technologists. These methods find their utility in a wide range of applications, for example:

- Predicting how a material is going to behave in production. This is of importance at the plant breeding stage in assisting the selection process as well as in supplier evaluation and monitoring (e.g. new harvest).
- As part of quality control checks, to ensure that flour meets specification, both prior to dispatch from the mill and/or receipt by the bakery.

- In product development, to assess either the suitability of a new flour for use in a particular product or the influence of other functional ingredients (e.g. ascorbic acid, enzymes, emulsifiers, etc.) on dough performance.

Probably the oldest and still the most extensively used method is based on the Brabender Farinograph (D'Appolonia and Kunnerth, 1984; ISO 5530:-1:1997). In principle, the farinograph consists of a drive unit which powers rotating mixer blades within a specially-designed, temperature-controlled bowl where the test dough is mixed. The instrument measures the resistance (change of the consistency) of the dough against the rotating mixer blades from the beginning of dough formation and records it (graphically – 'Farinogram') as torque (Brabender Units – BU) versus time. The resistance increases as the dough develops and then decreases as the dough breaks down.

The method can be used to determine the water absorption of a particular flour or ground grain sample and also to predict the processing characteristics of wheat and rye. Water absorption is essentially the amount of water needed to yield a dough with a specified maximum torque. It is particularly useful as a predictor of dough yield within the bakery. If dough mixing is allowed to continue, the resulting 'Farinogram' also supplies information about the potential mixing characteristics of the flour. In addition to water absorption the parameters usually recorded are: development time; stability; degree of softening and mixing tolerance index (MTI).

Variants of the farinograph are used in different countries (e.g. Rheologica Instruments Mixograph; AACC 54-40A). These differ in terms of the one or a number of factors including: mixer blade shape and configuration; temperature control systems; variable mixer speeds and how energy (torque) inputs are measured.

A second approach to measuring the behaviour of gluten in dough is to investigate the doughs' response to stretching. Two of the commonest methods used, involve either the Brabender Extensograph (Rasper and Preston, 1991; ISO 5530-2:1997) or the Chopin Alveograph (ISO 5530-4:2002; ICC 121).

The Brabender Extensograph measures the stretching properties of wheat flour doughs, in particular their resistance to extension and their extensibility. Both of these characteristics depend predominantly on the gluten structure of the dough. Essentially the method involves preparing a dough sample from flour, distilled water and salt under standard conditions in the farinograph. After a defined rest time (originally 45 minutes), the dough is stretched to breaking in the extensograph. If required, the dough is then remoulded, rested and assessed two more times (45 minute intervals). On each occasion, the force exerted is measured and recorded. Alternative protocols based on shorter resting times, similar to those used under production conditions, also exist. The output ('Extensogram') from the assay gives information on:

- *Resistance to extension* – a predictor of flour (gluten) 'strength' – care needs to be used in interpreting these data. While low values are indicative of poor quality, so too are those which are excessively high.

- *Extensibility* – an indicator of elasticity of the dough.
- *Ratio of resistance/extensibility* – predicts the behaviour and stability of the dough and is considered to be related to finished product volume.
- *Area under the curve (energy)* – an estimate of the sensitivity of the dough to processing and handling within the bakery.

The second commonly used method is based on the Chopin Alveograph. This also measures resistance to stretching and extensibility, but does so by blowing air through a thin sheet of dough and measuring the pressure generated manometrically. The dough is mixed at constant water absorption and, as a result, the resistance to stretching is affected by anything that affects water absorption, including damaged starch. Initially, a rapid rise in pressure, as the dough sheet resists expansion, is observed. This is followed by a decline, as the thinning sheet loses strength and stretches, forming a bigger 'bubble'. The volume of the bubble is often a good indication of the baking strength of the flour. Eventually, the bubble bursts, with a sudden and complete loss of the excess pressure being recorded.

Most laboratories record:

- *Peak height* (P) or *tenacity* – this is the maximum pressure achieved in blowing the dough bubble and reflects the stability of the dough.
- *Length of curve* (L) – a measure of extensibility.
- *Area under the curve* (W) or *energy* – corresponds with the work involved in blowing the bubble and is related to the gluten strength.

Some laboratories also record the configuration ratio of P to L.

Although the outputs may appear to be similar, there are important differences between the two methods. This can in part be attributed to the fact that while the two methods measure resistance to extension and the extensibility, they do so in different ways. At the most basic, these arise from the different formulations used for making the doughs and conditions of measurement. The alveograph is valuable in identifying flours of weak to medium strength, and is the instrument most favoured by mills that specialise in soft wheat milling. The alveograph is not recommended for use with extra strong wheat flours, where the extensograph is preferred. As in the case of the farinograph, other systems are also coming into prominence, which include the Chopin Alveo-Consistograph (AACC 54-50).

9.5 Flour quality: starch and other attributes

Starch has two important roles in the breadmaking process: as a substrate for yeast, through its breakdown by amylolytic enzymes to maltose, and secondly, by its contribution to the physical structure of the finished product. Key parameters associated with starch are: the presence of amylolytic enzymes in the grain; its water absorbing properties and the degree to which starch present in the flour acts as a substrate for yeast in the desired dough system. Baking performance is a consequence of the interaction of these factors. Thus assays which reflect this complexity are of more immediate relevance than those which measure either enzymatic activity or starch attributes separately. Analytical techniques, which provide information on a number of parameters simultaneously, include amongst others, the amylograph and Falling Number methods. As discussed below, these methods are based on measuring changes in the viscosity of suspensions during and/or after the starch has been gelatinised.

The Brabender Amylograph (Shuey and Tipples, 1982; ISO 7973:1992) is one of the principal instruments for measuring the gel-forming properties of the starch present in the flour or grain and also the activity of endogenous amylolytic enzymes. Its applicability is not restricted to wheat and it has found use in flours made from rye and rice as well. The equipment comprises a rotating bowl, within which a suspension of either flour or a finely ground meal in water is heated. Into the suspension is placed a measuring probe. The equipment measures the change in viscosity through the heating cycle. The peak viscosity value is a reflection of both the amount of damaged starch and indigenous enzyme activity. The lower the value, the greater the enzyme activity and the more damaged starch present.

The Falling Number method (ISO 3093:1982; ICC 107/1) uses the starch within the kernel as a substrate for endogenous amylases. The starch is rapidly gelatinized when a test tube with the sample (flour or ground meal) suspended in water is inserted into a boiling water bath and stirred with a purpose designed rod. Subsequently, the amylolytic enzymes in the sample start to hydrolyse the starch (liquefaction), reducing its viscosity; the speed of liquefaction being dependent on the enzyme activity. The Falling (Hagberg) Number is measured as a function of the time taken for the stirring rod, once raised, to travel a set distance through the sample. High endogenous amylolytic enzyme activity leads to rapid liquefaction, a speedy breakdown in viscosity, manifesting itself in lower falling number.

The Falling Number method has considerable significance since there is a direct relationship between enzyme activity and finished product attributes (e.g. bread crumb quality, loaf volume, etc.). It is also quick to perform (about 10 minutes). Falling Number and protein content are two of the key determinants in setting commercial specifications for grain and also act as criteria for premium payments and penalty charges.

Other equipment, which operate on similar principles, include among others, the derivation of a 'Stirring Number' using the Newport Scientific Rapid Visco Analyser (RVA, ICC 161; AACC 22-08).

In addition to their use as QC tools, as described above, measurement of starch characteristics and amylolytic enzyme activity has other benefits:

- general flour production control, as well as controlling the blending of grain or flours;
- detection of sprout damage;

- predicting the baking properties of flours (e.g. in product development, determining the need or otherwise for additional malt or fungal alpha amylase).

There are other tests that are routinely used to measure the parameter of starch damage independently. In particular those based on the starch's ability to be degraded by amylolytic enzymes. Until recently, the preferred method in the UK has been based on the method of Farrand (1964). However, this has since been replaced with methods based on enzymic test kits supplied by the Megazyme Company (Anderson and Salmon, 2002; ICC 164).

9.5.1 Process-linked methods

A number of pieces of equipment have been developed, capable of testing dough under similar conditions to those found in the proving stage, for example the Chopin Rheofermentometer. These find their applications in the control of flour quality (e.g. fermentative potential, gluten strength, etc.), as well as yeast performance; determining the effects of approved additives for improving the quality of flour as well as evaluating frozen dough and its keeping quality. Most equipment used for these purposes measure both the amount of carbon dioxide generated over a fixed time together with changes in dough volume (gas retention).

9.6 Conclusion

The ultimate test for the quality of any wheat or flour is its ability to produce a good product under the conditions used by a particular baker. Wheat quality is inherently variable depending on, amongst other things, agronomic and climatic factors. Given the diversity of bakery products in general, as well as baking technologies and their application, it is usually infeasible to mill bespoke flours for each and every customer. It has therefore proven necessary to identify general quality parameters predictive of how the wheat or flour will subsequently perform. These have to be measured using standardised and agreed test methods which are sufficiently robust to support commercial contracts.

9.7 Sources of further information and advice

Further information on the standards or methodological approaches referred to in this chapter can be obtained from the sources below:

AACC American Association of Cereal Chemists (2000) *Approved Methods of the AACC (10th Edition) – Including the 2001 Supplement,* St Paul, American Association of Cereal Chemists

BS	British Standards Institute, www.bsi-global.com
ICC	International Association of Cereal Science and Technology, www.icc.or.at
ISO	International Organisation for Standardisation, www.iso.ch
FTWG	Anderson, C & Salmon, S. (eds) (2002) *Guideline No. 3 Manual of Methods for Wheat and Flour Testing* (3rd edn). Chipping Campden, Campden & Chorleywood Food Research Association Group.

A number of manufacturers have been mentioned where their products are specifically used in an analytical method. It should be remembered that there are also other manufacturers in the market. Further information, concerning the manufacturers of specific equipment or materials mentioned in this chapter, can be found from their websites as listed below.

Brabender OHG	www.brabender.com
Chopin SA	www.chopin-sa.com
Megazyme Ltd	www.megazyme.com
Newport Scientific Ltd Pty	www.newport.com.au
Perten Instruments AB	www.perten.com
Rheologica Instruments AB	www.rheologica.se

9.8 References

ANDERSON, C. and SALMON, S. (2002) *Guideline No. 3 Manual of Methods for Wheat and Flour Testing* (3rd edn) Chipping Campden, Campden & Chorleywood Food Research Association Group.

ANONYMOUS (2002) *mi Quality Update* (18 September 2002), London, Home Grown Cereals Authority.

AXFORD, D.W., MCDERMOTT, E.E. and REDMAN, D.G. (1979) A note on the sodium dodecyl sulphate test for bread-making quality: a comparison with the Pelshenke and Zeleny tests. *Cereal Chemistry*, **56**, 582–584.

COKER, R.D. (1998) Design of sampling plans. In: Sinha, K.S. and Bhatnagar, D. (eds) *Mycotoxins in Agriculture and Food Safety*, New York, Marcel Decker.

D'APPOLONIA, B.L. and KUNERTH, W.H. (eds) (1984) *The Farinograph Handbook (3rd Ed.).* St Paul, *American* Association of Cereal Chemists Inc., 64 pp.

FARRAND, E.A. (1964) Modern bread processes in the United Kingdom with special reference to α-amylase and starch damage. *Cereal Chemistry* **41**, 98–111.

HOSENEY, R.C (1987) Wheat hardness. *Cereals Food World* **32,** 320–322.

KENT, N.L. and EVERS, A.D. (1994) *Kent's Technology of Cereals* (4th edn), Oxford, Elsevier Science, 334 pp.

MATZ, S.A. (1991) *Baking Technology and Engineering* (3rd edn), New York, Van Nostrand Reinhold, 853 pp.

MATZ, S.A. (1993) *Cereals as Food and Feed* (2nd edn), New York, Van Nostrand Reinhold, 751 pp.

NABIM (2002) *Recommended code of practice for mill intake.* www.nabim.org.uk.

PINKNEY, A.J., GREENAWAY, W.T. and ZELENY, L. (1957) Further developments in the sedimentation test for wheat quality. *Cereal Chemistry* **34**, 36.

PŘIHODA, J., KAISROVÁ, A., NOVOTNÁ, D. and HAJŠELOVÁ, M. (1994) *Hodnocení kvality obilovin a výrobků z nich,* Praha, Mlynářské noviny, 111 pp.

RASPER, V.F. and PRESTON, K.R. (eds) (1991) *The Extensigraph Handbook*, St Paul, American Association of Cereal Chemists Inc.

SANDERS, K.F. (1990) Wheat hardness may form basis of future US grain standards. *The Wheat Grower* (Apr) 12–14.

SHUEY, W. and TIPPLES, K. (eds) (1982) *Amylograph Handbook*, St. Paul, American Association of Cereal Chemists, 37 pp.

10

Milling and flour quality

C. Webb and G. W. Owens, Satake Centre for Grain Process Engineering, UK

10.1 Introduction

As the principal component of bread, flour is vitally important in breadmaking and the production of flour by milling was among the first and most important of all industrial operations. The processes of milling and baking have therefore developed side by side throughout history, progressing from domestic to village to large-scale industries. It is appropriate then to present first a brief history of flour milling in terms of its origins, and its development, to the modern roller milling process. Then, separately, flour milling and its characteristics in relation to flour quality are discussed. By the end of the chapter, the reader will appreciate the technologies that are employed in flourmills and the factors influencing flour quality at the milling stage. References are also given that allow the reader to research the entire subject in greater detail if desired.

The prime objective of producing flour is to render cereal grains more accessible as food, by increasing palatability. This is done through a series of size reduction and separation stages. The size reduction enables enzymatic and cooking processes to be carried out more effectively. The separation steps are desirable in order to minimise the indigestible bran components of the grain in the final food product and also to remove the oil-rich germ component. This would otherwise increase the possibility of rancidity occurring and dramatically reduce the storage shelf-life of the flour and hence its palatability. Removal, however, also reduces the nutritional value of the cereal and is therefore optimised as much as possible.

So, flour production consists, principally, of grinding (milling) and sifting (separation) of the grain. In the modern mill, however, much of the process is also taken up by so-called 'cleaning' operations. These are necessary to ensure that only the desired grain is milled. Almost all stages of the process are carried out dry, using mainly mechanical techniques. The quality of the flour produced through the flour milling process will depend on the sequence of operations carried out and the degree of separation achieved. The definition of quality will also vary according to the intended end-use, whether for factory-produced white bread, wholemeal or artisan breads.

10.2 Flour milling

The purpose of flour milling is to isolate the starchy endosperm in as pure a state as possible, uncontaminated by either germ or bran. The process includes the preparatory stages of wheat cleaning and tempering as well as the more obvious steps associated with grinding and sifting. While some of the details of the process are specific to flour milling, many of the basic principles are such that any chemical, mechanical or process engineer would be familiar with them. Roller mills are used for grinding, sieves are used for size separation, air is used for density and drag-based separations and pneumatics are used for conveying. Power consumption, yield and losses are all terms used in flour milling that would be familiar to engineers of most disciplines.

Flour milling is characterised as a process industry, since it transforms a basic raw material, via a primary process, into a product that is of value to a wide range of consumers. The other features of flour milling that place it into the process industry category are that operations are carried out on a large scale and on a continuous basis. Another characteristic of the industry is that the process is performed via a series of interacting unit operations.

10.2.1 The evolution of modern flour milling

The process of flour milling dates back to Egyptian and earlier times. Primitive people ground wheat by using stones to pound the grain and release the edible seeds from their hulls. Later, as shown in illustrations from ancient inscriptions, grain was crushed using a mortar and pestle, with the resulting material being sieved to produce material of greater purity. About five millennia ago the saddle stone appeared, while two and a half thousand years later the lever mill (which was an improvement over the saddle stone process because it combined both shearing and grinding) was introduced. The first rotary mill was developed 2300 years ago. The millstones were first hand operated, then driven by animals and finally driven by waterpower. About a thousand years ago wind flourmills were used which, together with the water-driven flourmills, remained active until the late 19th century. The latter power source allowed the development of the first automatic mills, where all operations between wheat delivery and flour collection were performed automatically.

Millstones dominated the process used to produce flour until the 1870s when roller mills began to supplement them on a large scale, because of the superior

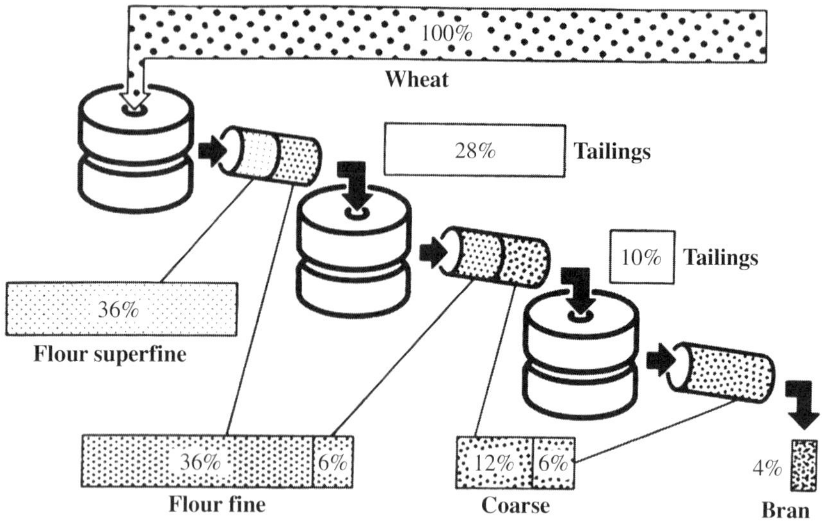

Fig. 10.1 The 'French Process' for flour milling.

flour that could be produced using them. However, the modern flour milling process that was introduced at the same time as the widespread adoption of the roller mills has its origins in what is now known as the 'French Process' (see Fig. 10.1). This process began to emerge in the 16th century and advocated the use of a number of grinding stages with intermediate sieving of the products to produce final products of intermediate grades. The other notable feature was the fact that the stones used were set in such a way as to perform a gentle grinding action and thus maintain the purity of the endosperm being separated.

The flour milling process as it is known today evolved considerably during the years between 1830 and 1870 (Storck and Teague, 1952). During the latter half of the 19th century the present concept of a gradual reduction in the milling of wheat was developed which improved the flour colour (incorrect grinding prevents the effective removal of bran and thereby influences the colour). During this period middling purifiers also appeared which helped the removal of the branny material. At the same time, the first roller mill was also introduced which, instead of stones, used porcelain, and their rolls were usually smooth. Later, break rolls were made of chilled iron. Along with roller mills plansifters and purifiers also appeared in order to improve and handle larger quantities of wheat.

It was a Manchester-based engineer, Henry Simon, who commissioned the first commercial mill using the technologies that dominate the industry today (Simon, 1997). This was the last complete technological revolution that occurred in flour milling. Simon perfected the gradual reduction system, using a large number of process stages in an extension of the French process and making exclusive use of roller mills for grinding. The principles of gentle grinding and intermediate sieving were developed to give the break, purification and reduction systems that are used extensively today.

The important features of this mill were steel roller mills and the gradual reduction system. Prior to these technological developments stone mills and shorter processes were normal. The advantage of the newer, more elaborate process was that higher yields of quality flour could be produced. The older processes employed can only be described as crude and produced typically only 10% high-quality flour from the wheat grain compared with more than 70% in roller milling plants. The remaining flour was of very poor quality and heavily contaminated with bran and germ constituents.

Many variations in the detail of the process flow sheet emerged in the years after the perfection of the gradual reduction system. This was due to the varying requirements of customers and even political directives. For example, in wartime, the British government ordered millers to produce high extraction flour in order to extend supplies and to supplement people's diet with the essential vitamins and minerals found in the outer layers of the wheat berry (Smith, 1938).

Development of flour milling technology since the introduction of roller mills and the gradual reduction system include flow sheet alterations to accommodate newer machine technologies and increased machine capacities as well as the adoption of information-based technologies.

10.2.2 The modern flour milling process

There are a number of aspects of the overall process of flour milling that are particular to it and are not to be found in other industries. Thus a group of terms are used which would be unfamiliar to individuals not acquainted with the process. This section aims to provide a brief introduction to the process and the terms employed to describe aspects of it. Posner and Hibbs (1997) describe the process in great detail and the reader is recommended to consult this book for further insight.

Flour milling is the continuous process by which raw wheat grain (the berry or kernel) is transformed into a form that is of use to the baking and other industries and the domestic consumer. A small portion of production is geared towards wholewheat flours, employing a simplified process flow sheet, but most demand and effort is directed towards the production of white flour.

White flour is the ultimate product of flour milling. The aim of white flour milling is to extract a maximum amount of endosperm from the wheat berry in as pure a form as possible. The outer bran layers become the co-product of the process called wheat feed. Many operations also separate the embryo (germ). This is a high value co-product when a market exists. Where a market for germ does not exist, it is sold for animal feed with the wheat feed bran. These co-products contribute significantly to the financial viability of milling operations.

One of the keys to the success of a flour milling operation is the efficient, economical separation of starchy endosperm from the rest of the berry. The process has developed along very specific lines towards achieving this goal. There is just one accepted manner in which flour is produced globally, known as the gradual reduction system. This is the process of taking the wholewheat berry

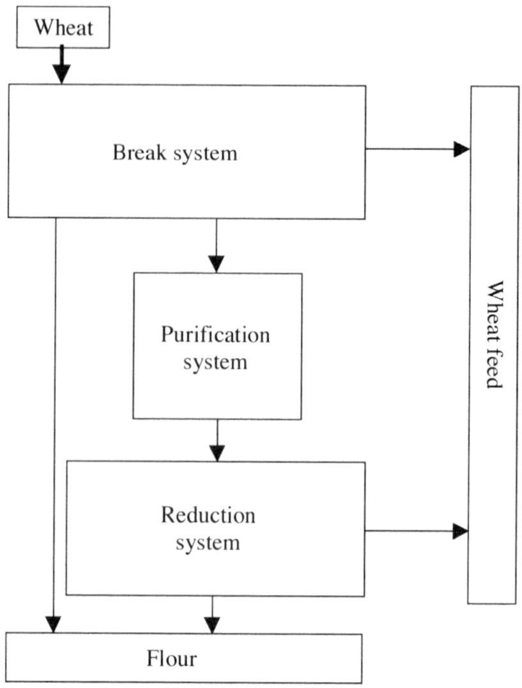

Fig. 10.2 Simple block diagram representation of the flour milling process.

and, via a series of grinding and sieving stages, producing white flour of the desired quality and yield. The gradual reduction system has enabled the production of flours of low ash content and high yield. Specialist, high-quality flours, are produced by extracting high-purity subproducts from within the process.

The flour milling process can be represented by a simple block diagram (Fig. 10.2). There are three principal divisions within the process. These are known as the break system, the purification system and the reduction system. The purification system is not favoured by many millers and may be absent from processes. It is often replaced by what is known as the sizing system. However, the other two blocks are present in all gradual reduction flourmills in operation today.

Preparing the wheat for milling (cleaning and tempering)

Before the wheat reaches the first milling stage, it has to undergo several preliminary operations that ensure the correct performance of the main process. The first of these is wheat cleaning. The grain, having been received at the flourmill, analysed and stored in a silo, must be separated from any contaminants prior to preparation for milling. The contaminants removed during the cleaning process include burnt, immature, spouted or shrivelled grains, other seeds, sand, straw and stones. Many of these impurities can be separated by simple sieving to remove small and large materials, aspirated sieving to remove lighter materials and sieving on inclined screens to separate dense materials such as stones. In the aspirated systems the wheat is subjected to strong air currents that remove the lighter impurities, while with the inclined systems, advantage is taken of the effects of gravity to separate dense from less dense materials as they are shaken along the screen.

Disc separators are used to remove impurities of similar size and cross-section as the wheat grains, such as oats, barley and other seeds. The discs have small pockets on both sides and revolve partly immersed in the grain, picking out particles small enough to enter the pockets and discharging them into troughs between the discs on the downward side. Scourers, consisting of a stationary cylindrical wire or perforated screen, and a few rotating surfaces (beaters) are used to scour off impurities, such as dirt trapped in the crease. Any ferrous metals that might be present are removed in magnetic separators.

After cleaning, the wheat is tempered (or conditioned) prior to be milled. Tempering is one of the most important parts of the flour milling process. In this stage, the moisture content of the grain is increased by adding water and by allowing the grain to sit for a period of time. This conditioning process toughens the kernel and prevents bran powder being formed during grinding, thus simplifying the physical separation of endosperm from the bran. The tempering also affects the way in which endosperm particles are broken in the subsequent milling steps and the extent of mechanical starch damage achieved. However, if the wheat is over-dampened, sifting becomes difficult and the capacity of the mill is immediately reduced. The amount of water added depends both on the existing moisture content of the wheat and the hardness of the grain. Hard (winter or spring) wheat is tempered to 16.5% moisture while soft wheat is usually tempered to 15.0–15.5% moisture. Durum wheat is tempered to a higher level of moisture.

The time required for the grain to reach an even distribution of moisture depends on the hardness of the grain rather than the characteristics of the bran layers. For hard wheat, the time required for tempering varies and can be from 10 to 36 hours, while for soft wheat, the time required is from 6 to 18 hours. Tempering times also depend whether the process is carried out hot or cold, with hot tempering being much the quicker method.

The break system

The preparation being complete, the wheat is now ready to pass to the first break rolls for milling. Referring again to Fig. 10.2, the breaking block or break system is the area of the process where most endosperm separation is achieved. This work is performed principally on roller mills whose surfaces have a saw tooth profile. The rolls run at differential speeds towards each other. The combination of these two attributes in operation mean that, in the first contact with the wheat berry, the grain is split open and a significant amount of material is released which, after sieving, will make its way into the purification and reduction blocks. A small amount of flour is also produced at this stage and is removed before further processing occurs. The coarse endosperm material produced is referred to as semolina and must be further size reduced. The material that remains in the break system after first contact (i.e. material too course to pass to the purification and reduction systems) is presented to a second set of rolls (second break) for further grinding. Again, material is released into the purification and reduction blocks along with a small amount of flour production. This procedure is repeated four or five times until it is deemed no longer worthwhile trying to release further material (because of diminishing returns both in terms of quantity and quality of material released). At this point the remaining material is discharged to the wheat feed stream, the main co-product of flour milling.

The amount of material released at each break passage is limited by adjusting the gap between the rolls. The term 'break release' is often used and represents the percentage of material released at each particular break passage. The aim of the break system is to achieve maximum release of endosperm particles with minimal disintegration of the bran.

The purification system

The purification block contains three machine types, purifiers, roller mills and sifters. Purifiers are machines that separate particles on the basis of differences in size, air resistance and particle specific gravity, simultaneously. The purifier is essentially an inclined sieve that becomes coarser from head to tail, which is oscillated, and through which an air current passes upwards. The heavier particles of endosperm stay on the sieve until they reach openings big enough for them to fall through, while the air currents lift out the lighter branny material and convey them out of the system.

The streams feeding the purification system come from the break system discussed above. They are classified on a size basis prior to entry to the purification system and contain a mixture of pure endosperm and intermediate purity particles. Particles that are pure enough are separated immediately, leave the purification system, and are passed to the reduction system for further processing. The remaining particles are processed on roller mills whose surfaces have a fine saw tooth profile. Further bran and endosperm separation is possible as a result of this grind. Ultimately most of the material that enters the purification system is passed on to the reduction system. The remaining material is sent back to the break system. Generally, purification of the stock obtained after first and second breaks results in almost pure endosperm, while purification of the stock obtained from later breaks results in material containing a higher proportion of bran. Because the aim of this block is the purification of milling streams, almost no flour is produced in the purification system.

The reduction system

The reduction block is the main flour-producing part of the process. It is also the area where another desirable property in flour is manipulated. Mechanical starch damage is induced in order to increase the water-absorbing capacity of the flour, which in turn improves bread yield. The reduction block consists of a series of roller mills and sifters in sequence. Material is transferred from the break and purification blocks to these roller mills principally for size reduction, although the sieving apparatus also removes some remaining impurities. The roller mills used in the reduction system differ from those used in the other blocks in that they are usually smooth surfaced and are operated at lower differential speeds (Scanlon and Dexter, 1986). Material that is not sufficiently reduced in size in a particular grinding pass is sieved out and ground again in a subsequent grinding stage. This process is repeated up to 11 times in what are termed long surface mills.

The starch damage mentioned earlier is achieved through the application of shear and pressure to the starch granules that constitute the endosperm. The shear stresses are applied by virtue of the differential speeds employed by the grinding rolls. The mechanical linkage that supports the grinding rolls applies the pressure to the particles.

Flour release

Flour separation takes place in the plansifters. Any material that passes through the finest sieves is, by definition, flour. All other material is returned to the appropriate point of the system for further processing. Flour is produced at each stage, although the quality of that obtained from the break stages may differ from that produced in the reduction stages (because of contamination with fine bran particles). The result of the iterative grinding and sieving operations described above is the cumulative release of endosperm from the wheat berry, followed by the cumulative release of flour from this material. The actual quantities released at each stage vary widely between particular examples of mills. The operational settings depend on factors as diverse as wheat type, plant operator, customer demands, equipment supplier, geographical location and even tradition.

Typical releases for a flourmill in the British Isles are (NABIM, 1990):

- *A* – Cumulative release from the break system: 88%
- *B* – Rejection from the purification and reduction systems: 10%
- Cumulative flour release (*A*–*B*): 78%

While no two milling plants are the same, the differences between them occur in the intensity with which the processes described above are applied and the particular machine configurations used. This same process has been applied for more than one hundred years with only minimal changes to processing strategy.

10.2.3 Recent developments in flour milling

The most notable process developments in flour milling have been the double grinding of intermediate streams prior to sieving and the debranning of wheat prior to main processing. New machines have been developed to exploit these techniques. These machines and their application are well documented in trade literature (see for example, *World Grain*) and are the subject of numerous patent applications (Forder, 1996; Posner and Hibbs, 1997). The effect of introducing these new concepts into the flow sheet has largely been to reduce the total amount of machinery used.

The double-grinding roller mill

The double-grinding roller mill has been successfully employed in flourmills only in the last decade, even though it was first examined during the original development of the gradual reduction system (Storck and Teague, 1952). A variation, the six-roller mill, has been in use in the malting industry for many years (Bühler, 1981). The configuration of the double-grinding roller mill enables mill stocks to be ground twice without intermediate sifting or grinding. This development reduces sifting and conveying requirements within the process significantly. Considerable capital savings can be made in terms of equipment purchase requirements and the size of building required to house the process.

Debranning

The most radical of recent developments has been the advent of debranning as a unit operation in flour milling (Posner and Hibbs, 1997). This is the process of removing the outer bran layers from the wheat berry prior to the milling process proper. In the PeriTec® process commercialised by the Satake Corporation of Japan, wheat passes through abrasion and friction devices, which remove the outer bran then sequentially the thinner bran layers below the seed coat. A rationalised process flow is possible with such a debranning step installed. The conventional break system is rendered obsolete with only one or two fluted rolls being required. (Fluted rolls refer to the rolls employed in the break system that are finished with a saw tooth profile.) Because of the purity of the semolina produced in such a grinding system, the requirement for a purification system is also significantly reduced. Both these consequences of debranning enable a streamlined process with corresponding savings in building size and running costs. In addition, results are beginning to emerge that suggest that better breadmaking performance can be linked with the use of flours produced from debranned wheat. At this stage, however, it is unclear as to the reason for the improved performance; suggestions include the presence of greater quantities of aleurone cells in flour from debranned wheat and the lower incidence of microbial contaminants, which are normally present in the bran layers of the wheat kernel and which would normally be carried through the conventional milling process.

Pin mills

Pin mills have also been adapted for use throughout the process. Pin mills are a type of comminution device that consists of a revolving rotor with a series of pins attached. The material to be comminuted enters the mill at the centre of the rotor and is thrown towards the periphery by centrifugal forces. The particles impact the pins along this route and are comminuted as a consequence of these impacts. They have been among the key factors in reducing the amount of grinding equipment required in flourmills. The pin mills used are standard machines but the innovation is in the process locations where they have been applied. Pin mills are favoured where starch damage of the bulk flour will not be adversely affected or where flour production rates are relatively low. Pin mills are widely used in plants where starch damage production is not an issue or where it is undesirable.

In general there are two features that newer developments have in common. These are that they appear neither to enhance nor to reduce product quality and cannot be applied universally. However, significant capital and operational savings are possible through even limited application and so they are always worth applying as extensively as possible.

On-line process control

Technology exists today that can measure most process control properties on-line and in real time. This technology centres on the use of NIR (near infra-red) techniques, but X-ray fluorescence and gravimetric methods are also emerging. However, confidence levels in these technologies are not yet sufficient to allow an increase in plant operators' dependence on on-line control for functions other than protein addition and open loop control of other parameters. The main difficulty is the concept of establishing sufficient calibrations that are regularly verified.

However, the work documented by Graf (1994), Fearn and Maris (1991) and others provides examples of the developments that are beginning to bring process plants closer to the goal of producing a mill that can be intelligently controlled. The applications examined by Fearn and Maris were the automatic control of gluten addition to flour and a miniature gravimetric sensor designed to measure the release of material from a milling passage.

Liveslay and Maris (1992) performed a survey on the use of Computer Integrated Manufacturing (CIM) and Programmable Automation (PA) in the milling and baking industries. The conclusions were that:

- the milling and baking industries have a small number of examples of mature CIM developments, namely automation and manning reduction;
- the number of such developments is likely to increase, although slowly;
- much of the industry remains unconvinced about the benefits of CIM and PA.

It is possible to demonstrate some specific benefits that CIM and PA will bring to the industry. The benefits of CIM include quality improvement and enhanced production control. Many CIM processes also have features that enhance hygiene aspects of the process. In addition, data collection and analysis as well as product traceability are built-in features of CIM that are readily exploited by processors to ensure optimum quality for customers and as tools to enhance profitability.

Improving control saves energy and improves product consistency by ensuring key process variables are more stable, thus allowing comfort margins to be reduced. Processes may also be operated closer to optimum values or constraints (Hart *et al.*, 1996). In many cases a simple control system will achieve the desired effect but others require a more sophisticated approach.

10.3 Flour milling and flour quality

Quality is a term that encompasses a large number of parameters with respect to the product delivered from the flourmill to the baker. Essentially, it means satisfying the baker's requirement for consistent performance from the flour. The challenge for millers is to achieve this while maintaining acceptable performance from the mill.

10.3.1 Quality components

Factors influencing flour quality before wheat processing occurs include wheat variety, the presence of impurities, drying and storage conditions before processing and conditioning regimes. Parameters influencing flour quality during processing include the flow sheet employed, the condition of the roller mills and other processing equipment, the settings of the roller mills, atmospheric conditions and the flour divides taken. Obviously, the extraction level of the process has a major impact on the performance of the flour produced. This is because it dictates to a significant extent the amount of bran and mineral contamination of the final flour.

The miller's role is clearly defined given the above influencing factors. He must supply to the customer a product that will satisfy their requirements. In the context of breadmaking this means the miller must supply a flour that will produce a loaf of bread with the correct crumb structure, volume and colour, both inside and out. The miller must also supply flour that will deliver adequate yield, i.e. that has sufficient water absorption.

The miller has many choices to make in order to satisfy these requirements, although these choices are made in an evolutionary style in many mills. In other words, the customer may start with a standard flour from the mill's product range and over time get the miller to change some parameters in order to deliver the required performance. Thus one can see that the process of developing customised flour is very much a two-way process, where the miller and baker are in regular contact.

Producing flour for a particular application begins with the choice of wheat to be used as the basis for the flour. This is usually, but not always, a blend of different wheats. The resulting blend is commonly known as the grist, hence the common phrase 'grist to the mill'. Wheat types dramatically affect flour characteristics, particularly in the context of wheat gluten. For most bread flours North American wheat is considered highly desirable and finds its way into many breadmaking grists. However, because of its cost and the ready availability of affordable vital wheat gluten, the quantity of this grain type used in breadmaking grists in Europe has declined dramatically over recent years. It has been substituted with wheat of lower protein content (and thus cheaper!) and vital wheat gluten. The net product is similar in performance because the flour miller can manipulate many of the properties of the final flour during milling.

Water absorption capacity is a key parameter in the purchase of flour for breadmaking. This is because it is directly linked to bread yield. Water absorption is influenced by wheat variety in the grist but also by the amount and type of grinding performed during milling. Water absorption in flour is manipulated by disrupting the starch granules that form the endosperm of the wheat grain. Roller mills cause more starch damage than pin mills and so are preferred in mills producing breadmaking flours. The mechanism of starch damage is not clearly understood, but it is thought that pin mills cause particle comminution through fracture and impact, while roller mills break particles using compression and shear forces (Moss *et al.*, 1994). The latter mechanism is thought to affect starch granules at a molecular level and thus influence the amount of water absorbed at molecular level.

Colour and ash content of flour are loosely related because both reflect the amount of bran powder present. This bran powder is the portion of the outer layers of the wheat berry that has been ground by the milling process to flour-sized particles. Generally this material is considered undesirable in the flour and is regarded as a contaminant. It is, however, impossible to produce flour without any contamination, so the objective is simply to keep it to a minimum. Acceptable levels are determined through experience to be what the baker customer can tolerate.

This last point is very important because excessive levels of contamination affect the baking characteristics of the flour. However, assuming that the mill is being operated in an efficient manner, the level of bran contamination present in the flour is directly proportional to the flour yield from a given grist. Thus flour with a high level of bran contamination is higher-yield flour. The impact of this is that the flour can be produced at a lower cost per tonne. This should be reflected in the price paid by the baker for the flour. Therefore it is in the interest of the baker to know what level of contamination is acceptable for his or her requirements. He or she will then be in a position to purchase the highest extraction (cheapest) flour that will produce a good product. The consequent saving could be significant.

A point bakers often do not consider in terms of flour quality is the amount of contamination inherent in the wheat used for producing the flour. The immediate impact impurities have is on the ash and colour of the flour, but there can be a number of far more serious consequences; for example, ergot is poisonous and mite contamination can cause allergic reactions.

10.3.2 Manipulating flour quality

Once processing of the wheat into flour begins, there are a number of ways that flour properties and quality can be manipulated. Particle size is manipulated by the amount of grinding and the choice of sieve apertures employed throughout the process (Scanlon *et al.*, 1988). The choices available in the latter area are enormous. There are literally hundreds of individual sieves in the plansifters employed in mills. Various other sieves are employed in drum detachers, bran finishers and redressers. The latter device is primarily used to detect sieving problems upstream of it in the milling process since it is the last processing point during milling and all flour passes through it before passing to the silos. Redressers can therefore be considered one of the critical control points in the milling process.

In practical terms the mill operator will manipulate flour particle size by controlling the extent of grinding on the roller mills and by choosing various aperture size combinations in the sieving equipment. Adjustment of roller mills has a complex effect on the process because adjustments influence the particle size distribution of the material produced and hence the quantities of material passing to other areas of the process. Thus the setting of subsequent machines must also be taken into account.

In practice, it is impossible to consider all the consequences of such adjustment and mill operators generally assess the impact only on the end product. A good example of such a process is the adjustment of a first break roller mill. Increasing the grind on first break releases flour earlier on in the process (Hseih *et al.*, 1980). The water absorption of this flour cannot be further manipulated and so the water absorption of the finished flour is affected. The other consequence of this action is that the material balance of the mill is affected by reduced quantities passing through other sections of the process. This can result in small feeds in other areas of the process, which, in turn, can cause machine damage and poor sieving characteristics in some machines. The net result could be a flour product with significantly different characteristics from the one produced before the adjustment was performed.

Flour divides

Another tool that is used by flour millers to great effect is called flour dividing. This is a simple concept that has been around for almost as long as flour has been produced on a large scale. In a modern mill flour is produced after each processing step, of which there can be 15 or more; therefore a 'straight run' flour might consist of at least 15 component flours, each with slightly different characteristics. In a typical mill these component flours are grouped into three principal groups based on their general characteristics. The highest quality flours will be grouped into flour one, intermediate flours will be grouped into flour two and lower-quality flours will be grouped into flour three. The miller will manipulate the content of these three groups in order to match a baker's requirements. Flour one will be high in endosperm purity and possibly high in starch damage. Flour two will still be high on purity, but with less starch damage. Flour three will have the highest levels of bran contamination and will originate from the latter stages of the milling system and from auxiliary machines such as bran finishers.

Flour one is sometimes taken off separately to form a high-quality product for, say, cake manufacture, but is often used as the primary component of bread flour. Bread flour is generally made up of flour one, flour two and most of flour three combined. However, sending some components to wheat feed or to flour varies the exact composition. When all other options have been exercised, the flour miller will manipulate these streams in order to meet customer specifications. While it is highly undesirable for a miller to send flour to wheat feed, it is an efficient and *instant* way of manipulating product quality. In fact many mill operators analyse the properties of the individual flour streams on a routine basis and know without going onto the mill floor what effect the inclusion or exclusion of certain streams will have on the final flour produced (Flores *et al.*, 1991). This is a very useful tool when reacting to unplanned situations.

The principle of flour dividing has also received attention outside the main milling process. Many new mills constructed today incorporate flour blending in their storage silos. This allows millers to create composite flours by blending different mill products. The resulting composite flours can thus be tailored to match requirements that it would not be possible to satisfy directly in the mill. The idea of flour blending also has advantages for mill operators, because it is possible to mill base flours of a limited number of wheat types and blend these after milling to produce a myriad of different products. Thus product possibilities are extended while production runs on the mill are lengthened.

The divide principle is also widely used in the manufacture of wholemeal and wheatmeal flour. In this instance diverter valves are used to send the bran fractions present in the process directly to product streams or on to further processing. Within the milling process there exist particles of all shapes and sizes and thus it is possible with a little thought to produce almost any particle size distribution desired by bakers. The grinding process can also be controlled to a large extent to further manipulate the particle size distribution of the product.

In the production of wholemeal flour the entire content of the wheat berry is included in the product, while wheatmeal may exclude some parts of the mill product or may even include some extra components. For example, some wheatmeal flour includes extra bran.

10.3.3 Technological developments affecting flour quality

Although the milling process still employs the same principles as it did a hundred years ago it has undergone a major transformation in recent decades. The mills of yesteryear simply would not be capable of producing some of the products in common use today. This is particularly true of the high-starch damage flours required by some of today's plant bakers. There are enormous physical forces required to achieve this level of mechanical damage on the starch granules and older machines are simply not capable of producing the required levels of force.

New mills are designed with hygiene in mind, unlike older mills where producing a white powder was the goal. The grain entering today's mills can be cleaned like never before. There are even machines to scour the surface of the grain to remove microbial and fungal contaminants. Machines installed in modern mills are often self-cleaning and have product hold-up areas reduced to a minimum. After processing, the products are passed through infestation destroyers before storage in the product silos. They are often then subjected to a second treatment prior to dispatch to packing plants or to bulk customers.

It should be noted that the maintenance of high-quality products in milling involves a lot of routine work on the part of the plant operators. An enormous amount of time and effort is invested in preventative cleaning and maintenance to avoid contamination of mill products with biological or mechanical contaminants. In order to further safeguard the product, fumigation of all potential contamination areas is carried out under strict supervision and at regular intervals. As a result of such exhaustive routines and cross-checks, problems with flour contamination are rare in today's environment.

Hygiene considerations have even impacted on the new generations of mill processes where debranning machines have now been adapted for use in flour milling, such as the PeriTec® process referred to earlier. These machines have been employed for many years in the milling of rice for human consumption but their application to flour milling is quite a recent development. Debranning is the process of removing the outer bran layers from the wheat berry by selectively moistening the outer layers and subsequently using friction and abrasion to 'peel' off these layers. The logic behind the approach is two-fold. The outer layers of the wheat berry are porous and so contamination with moulds, fungi and bacteria occurs in these layers. Removal, therefore, influences the microbiological quality of any subsequent products in a positive manner. Secondly the debranning process aims to remove the bran layers from the wheat berry before any comminution is performed. In principle this ensures that the subsequent flour contains less bran and should perform more consistently because of the lack of disruptive contaminants.

While there is a learning curve with all new technologies and there undoubtedly have been and will be some failures, there seems to be growing evidence that this new processing approach is delivering on its promises and is gaining greater acceptance among the milling and baking community. However, given the historical nature of the milling industry, it will probably take decades before such a technology will reach maturity.

Debranning is the most radical of new technologies to affect flour milling and flour quality and promises the most significant advances in those areas. However, other advances have taken their place in the repertoire of flour millers. These include double grinding roller mills, pin mills and on-line process management. The first two developments have served not so much to enhance flour quality as to rationalise the milling process. Nevertheless flour millers have had to adopt their processes to ensure a consistent product having incorporated these technologies. For example pin mills release flour quickly from the process but do not generate as much starch damage as roller mills. Therefore the roller mills present alongside pin mills have to work harder to generate the desired starch damage levels in the final product.

Similarly, double grinding generates high levels of flour release and simultaneously generates high levels of starch damage in endosperm grinding applications. However, the aggressive grinding also grinds down bran particles present and so streams processed by double grinding passages need to be quite pure for processing to be effective and efficient. This can, at times, be a challenge for the miller and can mean that purifiers must be installed in the product streams.

10.3.4 Automation and testing

On-line process management has been the hidden revolution that has transformed the quality of flour being delivered to bakeries today. Never before has it been possible to deliver such a consistent product to customers. Every aspect of the art of milling is now scrutinised in a routine and scientific manner, to the point of exhaustion in some cases. Figure 10.3 gives an illustration of the type of monitoring that computer technology makes possible today.

Each consignment of raw material is tested as it passes from site reception right through to processing. All process operations, from moisture addition to blending, are monitored either manually or, increasingly, automatically. Machine settings can be automatic and are often logged. Improver addition is typically microprocessor controlled and flour blending is often centrally controlled.

Computers and their use in conjunction with standards such as ISO 9000 have resulted in products that are traceable, not just in terms of raw material origins, but also in terms of processing conditions and process settings. With this level of monitoring and control, incidences of non-conformance in flour products are

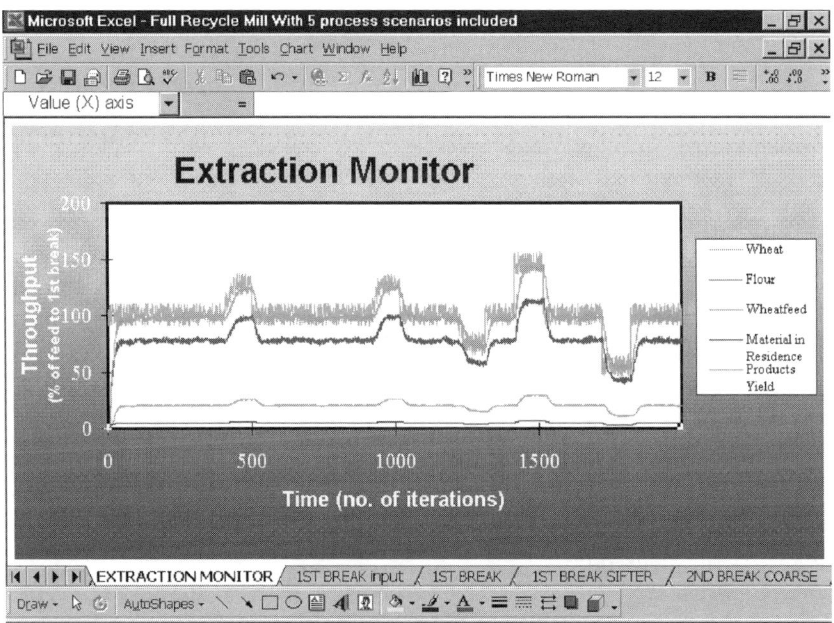

Fig. 10.3 Monitoring output possible with modern computer technology.

exceptionally rare and it is without doubt a major contributory factor to the successful development of the automated plant bakery. These plants simply cannot function without consistently performing flour.

Flour is the product of a natural raw material that is inevitably variable in quality and is therefore itself subject to variability. Flour quality is therefore closely monitored in terms of an extensive list of properties that include, as mentioned earlier, colour, ash, water absorption, speck count, particle size distribution, protein and even wheat variety. Protein content is one of the few quality parameters widely measured in real time and used for automatic control purposes; most other properties can only be measured off-line. However, discrete sample instruments are available to perform rapid tests for most other quality parameters and these are now being adapted for real time automated measurement (Osbourne, 1980; McFarlane, 1992). This will open the door to automated responses to process changes (Reyman, 1992). When this happens, in the not too distant future, the whole definition of high-quality and consistent flour will develop to new levels. Whereas today nothing is known about flour quality between discrete sampling events, in the future all anomalies will be detected and acted upon.

10.3.5 Product delivery

A final area of note in terms of milling and flour quality is the whole area of delivery to the customer. Generally, delivery is the responsibility of the miller who will have the capability to deliver product in a number of pack sizes and also in bulk. The latter option is by far the most desirable from a quality perspective because the mechanical elements involved in the transfer from miller to customer are minimised and there is no question of packaging materials leading to contamination. Moreover the possibilities of waste and contamination are minimised. In other words, everyone is a winner. Indeed, in this era of greater environmental awareness, the removal of packaging from the flour production equation is an enormous step forward, and so the argument for bulk products from a quality and environmental perspective is compelling.

10.4 Milling research

Milling research can be divided into two categories, namely commercial research and academic research. Commercial research tends to be confidential in nature, but the results of this work can be seen in the new products and processes marketed by manufacturers in the field. These developments have been discussed above. In contrast, academic research is well documented and may be discussed extensively.

Engineering research that has been reported to date has tended to be rather experimental in nature with little emphasis placed on understanding the basic principles employed; for example, Moss *et al.* (1980) and Hook *et al.* (1984).

There are, however, notable exceptions to this rule: Nuret and Thielin (1949) and Ruffet (1994) performed considerable amounts of research in the area of mechanical performance of plansifters and roller mills. Owens (2000) worked in the area of overall process modelling and optimisation. He also evaluated the potential of process recycle techniques in flour milling. Even more recently, Campbell and others at UMIST in Manchester, UK, have been working towards complete understanding and a mathematical description of the roller milling operation (Campbell and Webb, 2001; Campbell *et al.*, 2001; Bunn *et al.*, 2001). This new area of research is ongoing (Webb *et al.*, 2000).

10.5 The future of flour milling

The aspects of milling already discussed highlight the way in which this sector of the cereals processing industry has developed to date and may develop in the future. Developments in machine capacity have been dramatic, but progress in this area is sure to slow down due to limits in material properties being reached. However, maximizing utilization of that capacity has moved to the forefront of both commercial and academic research. Some of that effort has resulted in commercial applications, namely the adoption of computer control to facilitate longer operational runs in plants as well as minimizing downtime. Nevertheless, sophisticated control strategies have yet to make a real impact in the sector, despite significant progress in many other industries. It should be simply a matter of time before advanced control systems and optimization algorithms become an important aspect of commercial milling operations.

The recycle research mentioned briefly above (Owens, 2000) could also form the foundations for fundamental changes in the way milling of wheat is carried out and may lead to more streamlined milling processes which may ultimately lead to small-scale integrated milling and baking units. As organizations get larger and raw materials and processes become more regulated this type of proposition becomes more practical.

In addition to the above process developments, external influences will have a significant impact on the manner in which mill processes are operated. For example, new products and product specifications will demand different things from mill processes and necessitate change. These include, among others, product shelf-life requirements, heat treatment processes and food safety concerns.

It is also certain that technologies that are in their infancy today, for example double grinding and debranning, will become normal features of mill processes and gain greater acceptance among the milling community in the future.

To conclude, the future of development in the milling industry is likely to take the form of incremental evolution of those technologies mentioned above along with the introduction, in some form, of the new technologies discussed. Profitability and developments downstream of mills will have a major impact since these will influence decisions about reinvestment and new investments. Competition between millers can also be a catalyst for change as new plants can have competitive advantages over older ones. This, for example, has been one of the major driving forces in the enormous investment seen in recent years in the USA.

10.6 Conclusion

This discussion on milling began with an overview of the history of the process and its recent development. It can be seen from this review that the process has a large historical background and this has had an enormous influence on how mills operate. The other point to note is how little variation exists in the manner in which milling is performed. This means that millers throughout the world produce flour in exactly the same way and experience exactly the same problems. Flour quality is a concept that is at the forefront of mill operators' concerns and development within the industry is guided by end-user requirements. Given this attitude, it is little wonder that the quality, especially the consistency of mill products, has improved significantly in recent times. This evolution looks set to continue as process monitoring and control become more extensive and accepted in every facet of the production process.

10.7 References

BÜHLER, (1981), *Malt Mill, Diagram*, Buhler Ltd., CH9240, Uzwil, Switzerland, Dec, **78**, 7–8.

BUNN, P.J., CAMPBELL, G.M., FANG, C. and HOOK, S.C.W. (2001), On predicting roller milling performance. Part III. The particle size distribution from roller milling of various wheats using fluted rolls, *Proceedings of the 6th World Chemical Engineering Congress,* Melbourne, Australia, September 2001, P2–068.

CAMPBELL, G.M. and WEBB, C. (2001), On predicting roller milling performance. I. The Breakage Equation, *Powder Technology*, **115**(3), 234–242.

CAMPBELL, G.M., BUNN, P.J., WEBB, C. and HOOK, S.C.W. (2001), On predicting roller milling performance. II. The Breakage Function, *Powder Technology*, **115**(3), 243–255.

FEARN, T. and MARIS, P.I. (1991), An application of Box Jenkins methodology to the control of gluten addition in a flour mill, *Applied Statistics*, **40**(3), 477–484.

FLORES, R.A., POSNER, E.S., MILLIKEN, G.A. and DEYOE, C.W. (1991), Modelling the milling of hard red winter wheat: estimation of cumulative ash and protein recovery, *Transactions of the ASAE*, **34**(5), 2117–2122.

FORDER, D.E. (1996), Flour milling process for the 21st century, in Campbell, G.M., Webb, C. and McKee, S.L. (eds) *Cereals: Novel Uses and Processes*. Plenum Press, New York, 257–264.

GRAF, D.O. (1994), Verbesserung Zum Vermahlen Von Kornerfruchten, *Die Muhle + Mischfuttertechnik*, **131**(14), 172–173.

HART, D., CHENG, P., JOHNSTON, J., GOODALL, A. and OGDEN-SWIFT, A. (1996) *Reducing*

Energy Costs in Industry with Advanced Computing and Control, Energy Efficiency Enquiries Bureau, ETSU, Harwell, Didcot.

HOOK, S., BONE, G. and FEARNE, T. (1984) Influence of air temperature and relative humidity on milling performance and flour properties, *Journal of Science in Food and Agriculture*, **35**, 597–600.

HSEIH, F.H., MARTIN, D.G., BLACK, H.C. and TIPPLES, K.H. (1980), Some factors affecting the first break grinding of Canadian wheat, *Cereal Chemistry*, **57**(3), 217–223.

LIVESLAY, W.A. and MARIS, P.I. (1992), Computer integration and programmable automation in the UK milling and baking industries, *Institution of Chemical Engineers Symposium Series*, No. 126, 499–505.

MCFARLANE, I. (1992) Use of in-line sensors and closed-loop control for food manufacturing processes, *Institution of Chemical Engineers Symposium Series*, No. 126, 133–164.

MOSS, R., STENVERT, N., KINGSWOOD, K. and POINTING, G. (1980) The relationship between wheat microstructure and flour milling, *Scanning Electron Microscopy*, III, 613–620.

MOSS, R., MCCORQUODALE, J., BYRNES, T., EGAN, J., LEE, A., REID, A. and STEPHENSON, J. (1994) The effect of mill variables on starch damage and other aspects of flour quality, *Association of Operative Millers Bulletin*, November, 6459–6462.

NABIM, (1990) *Mill Processes (I),* Flour Milling Industry Correspondence Course, 1st Edition, NABIM, London, 37.

NURET, H. and THIELIN, P. (1949) Etude quantitative et qualitative d'un diagramme-type de quintaux par 24 heures, *Bulletin De L'E.N.S.M.I.C.,* No. 113, 300–350.

OSBOURNE, B.G. (1980) Current and future applications of NIR analysis, a paper presented at the technical session of the FMBRA Annual General Meeting, 2 April.

OWENS, W.G. (2000*) Engineering The Flour Milling Process*, PhD Thesis, Satake Centre For Grain Process Engineering, UMIST, Manchester.

POSNER, E.S. and HIBBS, A.N. (1997) *Wheat Flour Milling*, AACC, St Paul, Minnesota, 91–187.

REYMAN, G. (1992) Integrated control development for food industry, *Institution of Chemical Engineers Symposium Series*, No. 126, 165–178.

RUFFET, M. (1994) Break roller mill performances, in B. Godon and C. Wilm, *Primary Cereal Processing*, VCH Publishers, New York, 189–210.

SCANLON, M. and DEXTER, J. (1986) Effect of smooth roll grinding conditions on reduction of hard red spring wheat farina, *Cereal Chemistry*, **63**(5), 431–435.

SCANLON, M., DEXTER, J. and BILIADERIS, C.G. (1988) Particle size related physical properties of flour produced by smooth roll reduction of hard red spring wheat farina, *Cereal Chemistry*, **65**(6), 486–492.

SIMON, B. (1997) *Henry Simon Of Manchester*, Perdue Press, Leicester, 47–58.

SMITH, L. (1938) *Flour Milling Technology*, 2nd edition, Thomas Robinson & Son Ltd, Rochdale, England.

STORCK, J. and TEAGUE, W.D. (1952) *Flour for Man's Bread, A History of Milling,* University of Minnesota Press, Minneapolis.

WEBB, C., CAMPBELL, G.M., PANDIELLA, S.S., OWENS, G.W. and BUNN, P.J. (2001) Engineering the flour milling process, in M. Wooton, I.L. Batey and C.W. Wrigley (eds), *Cereals 2000*, RACI, Melbourne, 28–31.

11

Modifying flour to improve functionality

C. A. Howitt, L. Tamás, R. G. Solomon, P. W. Gras, M. K. Morell, F. Békés and R. Appels, CSIRO Plant Industry, Australia

11.1 Introduction

The free-threshing hexaploid wheat in common use today (*Triticum aestivum* L.) was most likely derived from a hulled hexaploid progenitor that originated from a hybridisation between a tetraploid emmer wheat, (*T. dicoccum* (Schrank) Schübl) and diploid goatgrass (*Aegilops tauschii*) (Harlan, 1992; reviewed in Feldman, 2001). The genetic complement of hexaploid wheat comprises three sets of seven chromosomes (the A, B and D genomes, Sears, 1954). The level of genetic redundancy associated with this genetic composition provides for many different gene combinations that can contribute to final flour processing properties. In order to select the 'right' gene combinations, a significant challenge in modern breeding programs is the ability to efficiently analyse wheat flour functionality in the context of end-product quality.

Figure 11.1 summarises the major steps in the processing of wheat flour. It is evident that the modification of flour to meet the demands of this processing chain requires an integration of knowledge from cereal chemistry and molecular biology, layered on top of market demands for consistency of supply of flour of a certain quality type.

The mixing and baking properties of wheat flour doughs are largely determined by properties of the prolamin seed storage proteins, particularly the glutenin subunit proteins. Within the glutenin subunit proteins, much work has focused on elucidating the features of the high molecular weight glutenin subunit (HMW-GS) proteins that are primarily responsible for variation in the breadmaking quality of wheat cultivars. Apart from the length and sequence motifs of the central repetitive regions (Anderson *et al.*, 1996; Shewry *et al.*, 1992), features of the *N*- and *C*-terminal non-repetitive domains, such as the

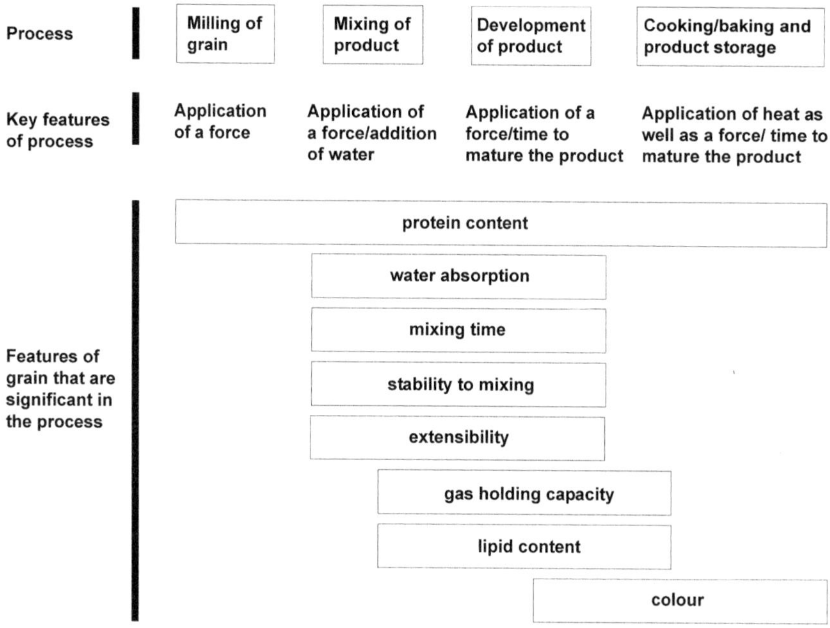

Fig. 11.1 Summary of the flour processing chain.

number of cysteine residues, have been found to be critically important for breadmaking quality (Shewry *et al.*, 1992; Tamás *et al.*, 2002). In this chapter this area of research is discussed in detail with the view to relating experimental observations to ideas about how the dough complex is formed. In addition, some specific areas such as the impact of waxy wheats are considered because starch is a major component of flour and thus a dramatic change in its structure can be expected to have significant consequences on breadmaking quality.

11.2 Definition of some key terms and components

The unique properties of wheat flour to produce the wide variety of end-products available today are derived from the gluten protein interacting with a range of components in the milled wheat grain. The components in the wheat grain that modify the overall performance of the flour include other seed storage proteins, starch, non-starch carbohydrates, lipids, cell wall components and minor components such as phytic acid and vitamins (Table 11.1; Evers *et al.*, 1999; see also Rahman *et al.*, 2000). Among the material added to flour to form dough, water is the most significant and can be viewed as a 'key facilitator' (Given, 1991). Components such as emulsifiers and salt are also added to formulations to modulate how the flour constituents interact with water and hence with each other.

Table 11.1 summarises the types and amounts of the major components in flour (Gras *et al.*, 2001). The figures in the table emphasise the molar excess of

Table 11.1 Estimates of the weight distribution of components in 1 g of wheat flour

Broad category of component (mg/g of flour)	Specific component (mg/g of flour)	μmoles[1]
120 mg protein	39 mg LMW glutenin	1.1
	9 mg HMW glutenin	0.1
	48 mg gliadin	1.5
	24 mg 'globulins'	0.89
710 mg starch	554 mg amylopectin	0.0014
	156 mg amylose	0.39
50 mg minor components[2]	(pentosans, lipids, cell wall components)	Not applicable
120 mg moisture	120 mg water	13333

[1]The figures are approximate and reflect the relative size of the molecules (modified from Gras *et al.*, 2001).
[2]The minor components are described in detail in Evers *et al.* (1999).

water in flour with the standard amount of 12% moisture as well as the large variation in the relative amounts of the components known to be important in contributing to the final properties of the flour. As detailed in later sections each of the specific components listed in Table 11.1 consist of families of molecules that can have different effects on flour properties depending on the particular types of families represented.

In terms of definitions, a number of different mechanical procedures are used to assess flour-processing properties. Although there are too many of these procedures to reproduce a detailed list here, it is necessary to briefly summarise the variables coming from the mixograph since this has been used extensively to assess the ability of proteins to modify flour processing. Small-scale assessments of flour using a 2g Mixograph (a pin-mixer) have been particularly useful in studying the functional properties of individual proteins (Gras *et al.*, 2001). The mixing action develops the dough through successive elongation-rupture-relax cycles of the flour-water mixture (Anderssen *et al.*, 1998; Anderssen and Gras, 2001). Key variables include the time to peak resistance of the dough (related to the strength of the dough), the maximum width of the Mixograph curve (resulting from successive elongation-rupture-relax cycles) related to the extensibility characteristics and the resistance breakdown of the dough to the sensitivity of the dough mixture to overmixing. In experiments where the functional properties of proteins are studied two types of states of the proteins are distinguished, namely, the added state and the incorporated state.

The effect of simply *adding* monomeric glutenin subunits to a flour is to effectively reduce the average molecular weight of the protein in the composite flour, because the relative amounts of polymer and monomer would be shifted towards the monomer. This added state of the protein is distinguished from the incorporated state where through a careful cycle of reduction and oxidation, the protein under investigation is made an intimate part of the dough as it is formed and its respective contributions to dough formation monitored in the Mixograph. Studies of the effects of a range of reductants and oxidants on the functionality of gluten proteins during dough mixing showed that it was possible to effectively destroy dough functionality with a reductant and then recover the functionality by subsequent oxidation (Békés *et al.*, 1994b). Careful selection of the oxidant, its concentration and reaction conditions allowed essentially complete recovery of the original dough mixing properties. This reduction/oxidation procedure has been used to incorporate a wide range of partially purified fractions or individual purified glutenin subunits into the polymeric glutenin phase (Békés *et al.*, 1994a, 1994c; Sapirstein and Fu, 1996; Veraverbeke *et al.*, 1999). The application of the reduction/oxidation procedure to dough preparations produced for extension measurement required different conditions to those for mixing studies, presumably because of continuing slow oxidation of dough components during the long relaxation required before stretching. The differences may offer further insight into the nature of the changes taking place between mixing and extension testing. Both the formulation and the reduction/oxidation conditions have to be modified if the resulting doughs are to be baked, because of the toxicity of dithiothreitol to yeast. Nevertheless, it has been possible to develop protocols in which functionality is maintained (Uthayakumaran *et al.*, 2000a,b).

11.2.1 Water-flour component interactions

The uptake of water by flour follows first-order kinetics (Menkov *et al.*, 2002). Within this overall view of water interacting with flour, greater detail on the nature of the interaction of water with flour components can be followed using real time near infrared (NIR) spectroscopy (Fig. 11.2; Wesley *et al.*, 1998a,b). The measurements at 1160 nm (or 1381 nm in Fig. 11.2) demonstrate a close relationship between water moving into a bound state, defined here as a state that is sufficient to alter the O–H bond stretch characteristics in NIR, and optimal dough formation by the wheat flour. The way this experiment was carried out required the detection of the reflected light at wavelengths including 1160 nm during the course of mixing wheat flour, calculating the second derivative of the reflected spectra and inverting this (Wesley *et al.*, 1998a). As the signal at 1160 nm is monitored with time, it can be assessed against the energy required to mix the flour as it forms dough. The data indicates that a close correlation exists between the formation of dough and a change in the O–H stretch characteristics of water in the mixture. These dynamic measurements are consistent with extensive studies using NMR spectroscopy (Belton *et al.*, 1994; Rugraff *et al.*, 1996; Ruan *et al.*, 1999; Grant *et al.*, 1999) and differential calorimetry (Davies and Webb, 1969; Bushuk and Mehrotra, 1977; Li *et al.*, 1998) indicating the existence of bound and unfreezable water, respectively in dough. The model is that water bound to flour components such protein and starch forms a phase that is distinct from the water that is free in the

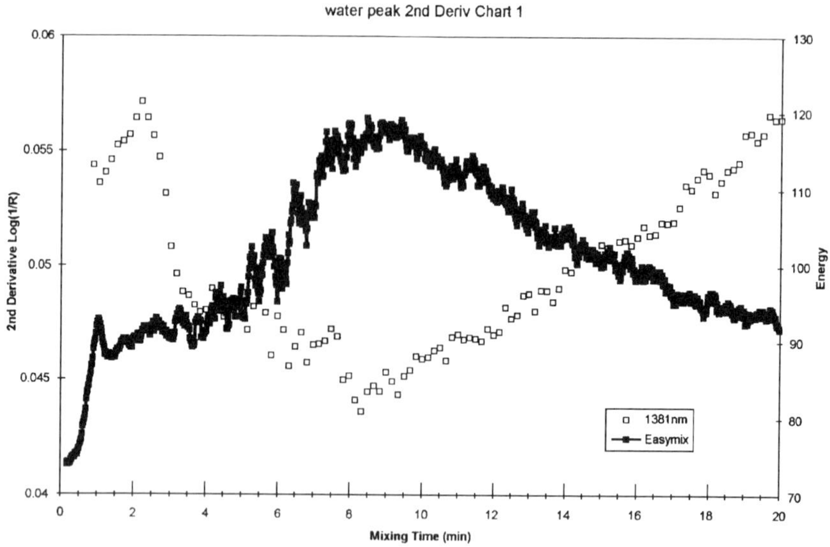

Fig. 11.2 Monitoring of flour mixing (filled symbols, and energy units on right-hand axis) and the NIR spectrum of water (open symbols and left-hand axis) versus time (X-axis). The NIR spectrum of water shows several maxima, including ones at 1160nm and 1381 nm (Curcio *et al.*, 1951) and although the data shown was collected using the 1381 nm wavelength, the 1160 nm wavelength has proven to be more suitable in experiments of this type (see text, Wesley *et al.*, 1998a,b).

mixture. Although the bound water is distinct in terms of its relaxation spectra in NMR spectroscopy and the reduced O–H stretch in NIR spectroscopy, there is still expected to be an exchange between this bound water and free water through diffusion, as well as chemical exchange of protons (Grant *et al.*, 1999).

Implicit in the concept of the importance of bound water is that the formation/redistribution of H-bonds is critical in the process of mixing wheat flour with water and the formation of dough. Although the basis for optimal dough formation is discussed in the following section, the redistribution of H-bonds to facilitate the new intermolecular and intramolecular interactions is discussed here. It is evident from Fig. 11.2 that over-mixing, associated with loss of dough structure, is also associated with release of water from the bound state and emphasises the dynamic state of the dough formation process. Early studies with heavy water (D_2O) replacing normal water indicated the importance of interactions involving water molecules in the formation of dough (Tkachuk and Hlynka, 1968). In the presence of heavy water, the mixing time of the dough was significantly longer and the dough that was formed had a greater Rmax in extension experiments. This is most likely due to a stronger gluten network since these effects were also seen in the mixing of separated gluten. Tkachuk and Hlynka (1968) argued that the effects of exchanging normal water with heavy water reflected the establishment of networks of H-bonding interactions in the mixing process. In the presence of heavy water these networks required a longer time to form but, once established, were stronger. Belton (1999) has argued that extensive networks of H-bonds involving the glutamines in high and low molecular weight glutenin subunit proteins form as flour is mixed with water ('train' model, see Belton, 1999; Gras *et al.*, 2001). The effects of heavy water, in this model, can then be accounted for by the fact that the replacement of deuterium for hydrogen in the H-bonding interactions results in a slower, but eventually stronger, establishment of H-bonds.

The NMR spectroscopic analysis of starch in water has also indicated the presence of bound water (Le Botlan *et al.*, 1998) that interacts with the OH groups of the glucose residues. This was estimated to be 0.17 g of water per g of dry starch. Some of these bound water molecules can be displaced by the binding of proteins such as gliadins and glutenin subunit proteins to the starch, as indicated by real time (dynamic) NIR spectroscopy of starch-water-protein mixtures during the course of heating (Wesley and Blakeney, 2001). The observations with gliadin were particularly striking in demonstrating the release of free water from starch (measured at 1160 nm, see Fig. 11.2) as the mixtures were heated. The available controls were consistent with the interpretation that the protein bound to starch and displaced bound water. The details of how the properties of the gelatinised starch are affected by the bound protein have not been determined.

Other well-defined flour components that bind water are the arabinoxylans (AX). The AXs are cell wall components consisting of linked D-xylopyranose residues with water-binding capacities of up to 0.47 g/g dry matter (reviewed in Shewry and Morell, 2001). Oxidative cross-linking of AX via ferulic acid residues (Garcia *et al.*, 2002) tends to lead to further water-binding capacities either directly or through new AX-protein interactions (Courtin and Delcour, 2002). The water-soluble AXs (WE-AX, Courtin and Delcour, 2002) provide the main contribution, among the AX complement, to the overall water absorption of flour.

11.3 Protein modification and breadmaking quality

Several different models for how the dough complex formed from wheat flour can be viewed have been discussed in the literature (Kasarda *et al.*, 1976; Khan and Bushuk, 1978; Kasarda, 1989; Gao *et al.*, 1992; Gras *et al.*, 2001). From the point-of-view of considering the modification of protein components and the subsequent effects on breadmaking quality, the 'fringed micelle' for dough structure (Fig. 11.3; Gras *et al.*, 2001) provides a reasonable summary of the basic concepts that are crucial for considering the mechanics of possible changes. The diagrammatic representation of the 'fringed micelle' in Fig. 11.3 illustrates the alignment of macromolecular protein complexes (Gras *et al.*, 2000) interfaced with the more extensive amorphous phase that also contains starch granules, unbound water, non-starch carbohydrates and proteins (Gras *et al.*, 2001). The details of the macromolecular protein complex are initially

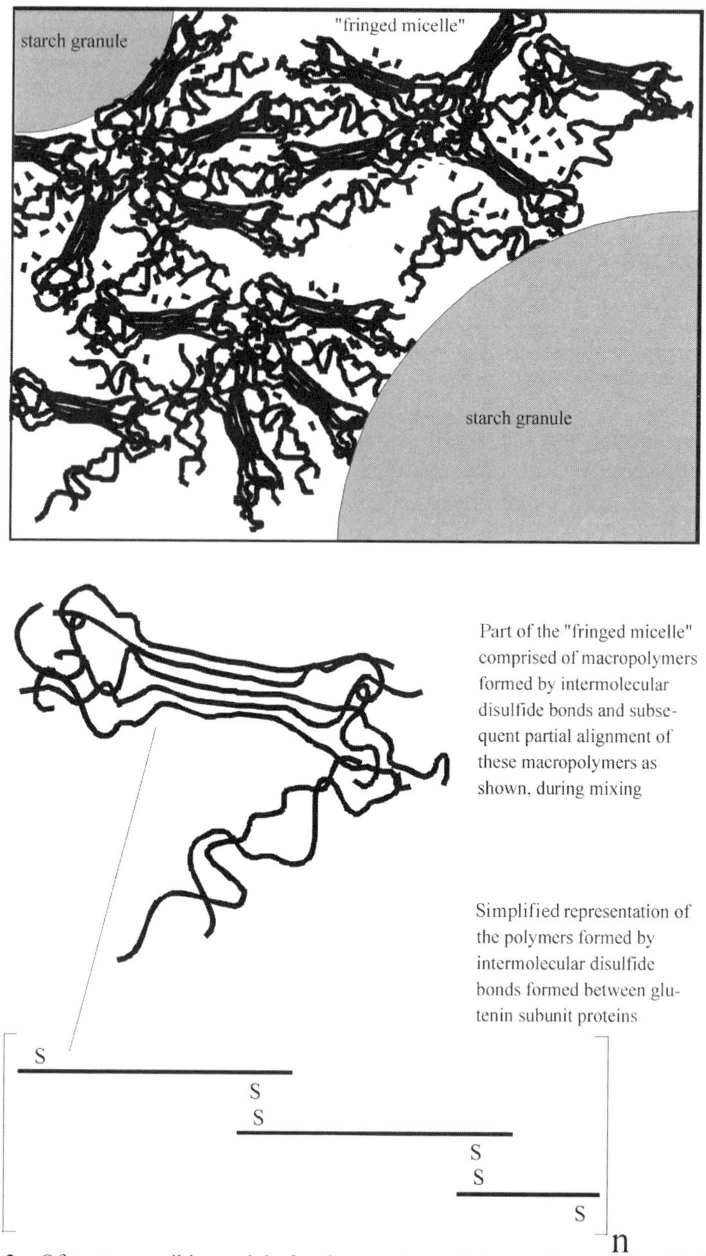

Fig. 11.3 Of many possible models for the structure of dough (Gras *et al.*, 2001), the "fringed micelle", is illustrated. The assembly of the macromolecular complexes involves macro-polymers (black lines) partially aligning and these partially aligned structures then engaging each other through entanglement (top panel). The macro-polymers are comprised of various combinations of glutenin subunit proteins (and possibly other proteins as discussed in the text) joined through disulfide bridges (lower panel).

established during the biological phase of protein accumulation in protein bodies during endosperm development. The timing of the deposition of proteins into protein-bodies and the details of their molecular structure have been reported (Clarke *et al.*, 2001). A key feature of the glutenin subunit protein structure in relation to both their deposition into protein bodies and the formation of the macromolecular protein complexes are indicated in Fig. 11.3, namely, the distribution of disulfide bonds. The protein, protein disulfide isomerase (PDI), is a chaperone involved in transferring glutenin subunit proteins from cytoplasmic sites of synthesis into the lumen of endoplasmic reticulum destined for protein bodies (Gilbert 1997; Ciaffi *et al.*, 1999; Johnson *et al.*, 2001). Consistent with the importance of the distribution of disulfide bonds in assembly macromolecular protein complexes, the enzymatic activity associated with PDI is required for the normal formation of protein bodies *in vivo*, in rice (Takemoto *et al.*, 2002).

A major problem in exploring the relationships between the structure and properties of prolamin storage proteins lies in their sequence diversity within the overall pattern of motifs defining the particular class of prolamin. Attempts to extract information on the relative importance of protein sequence elements in, for example, the HMW-GS class of proteins, have been hindered by the myriad of seemingly minor sequence differences (amino acid substitutions, insertions and deletions) between members of the class. Various workers have attempted to circumvent these problems through the modification of HMW-GS proteins and analysing the properties of the proteins after production in a bacterial expression system. Examples of this include the swapping of N-terminal non-repetitive domains between x- and y-type HMW-GS (Anderson *et al.*, 1996; Shani *et al.*, 1992) to try to elucidate the mechanism of the apparent synergistic effect of x+y pairs on dough properties (Békés and Gras, 1994); the modification of the size of the repetitive domain (D'Ovidio *et al.*, 1997; Andersen *et al.*, 1996; Arêas and Cassiano, 2001) and of the particular repeated motifs present within the domain (Anderson *et al.*, 1996) to determine how molecular size affects HMW-GS properties; and the modification of the cysteine content of a repetitive domain derived from HMW-GS 1Dx5, to remove the contribution of the terminal domains from the functional consequences of incorporating a prolamin into a dough formulation (Buonocore *et al.*, 1998; Tatham *et al.*, 1998). In addition novel proteins such as secalin 2 from rye (Murray *et al.*, 2001) have been introduced into wheat lines (L.Rooke, F. Bekes and R. Appels manuscript in preparation) and the changes in flour processing properties monitored. In general the studies provide a consistent view of the features of proteins that result in modifications to flour processing. The analysis of ANG proteins (section 11.3.1) provides a useful way of reviewing these properties.

11.3.1 The Analogue Glutenin (ANG) proteins

The identification of the functional properties of various motifs within the HMW-GS proteins in determining dough quality parameters provides a good starting point for considering the modification of quality attributes. Utilising a model protein with characteristics similar to the HMW-GS proteins and to express it at high levels in *E. coli* (Tamás *et al.*, 1994), to allow *in vitro* characterisation of its effects on dough properties (Tamás *et al.*, 1998), has proven to be a useful approach. The protein studied extensively by Tamás and colleagues is a modified C hordein, referred to as an Analogue Glutenin (ANG) protein, which has unique short N- and C-terminal domains flanking a long central repetitive region. Although the central repetitive domain of C hordein bears only limited sequence similarity to the homologous domains of the HMW-GS family of proteins, it possesses a similar conformational transition between βI/III turn-rich and polyproline II-like structures in equilibrium in solution (Gilbert *et al.*, 2000). This structure is generally recognised as being related to the unique viscoelastic properties of the glutenin subunit proteins.

ANG proteins with up to two cysteine residues have been expressed in bacteria, purified, and mixing, extension and baking experiments carried out with these proteins either added to, or incorporated into the gluten network of, the dough (Tamás *et al.*, 2002; Tamás *et al.*, 1998). A series of small peptides comprising sequences from either terminal region of the ANG proteins have also been synthesised and tested for their effect on dough quality parameters (Tamás *et al.*, 2002). As discussed in more detail below, these studies have demonstrated the ability of the ANG proteins to modify dough mixing and baking quality parameters by acting as either chain extenders or chain terminators, depending simply on the number of cysteine residues, and also addressed the effects of protein monomer size on the magnitude of the effects observed.

Changes in protein for targeting mixing properties can be considered as affecting either the aligned molecular complexes (see Fig. 11.3) or the amorphous regions. For the aligned regions of the dough complex the positioning of cysteine residues within protein subunits, and hence the location of intermolecular disulfide bonds, can be seen to be crucial. In addition, the repetitive regions, rich in glutamines that can form H-bonds need to be accessible for alignment. Experiments discussed below have shown that major changes in dough structure can occur when novel proteins with unusual locations of cysteines and numbers of repetitive units are introduced into dough.

The effects on mixing, extension and baking parameters of incorporating the various ANG proteins, described in Table 11.2, into dough are shown in Figs 11.4–11.6 (L. Tamás, R. G. Soloman, M. K. Morell, F. Békés, R. Appels, unpublished). The control for these experiments was the oxidation/reduction procedure in the absence of added protein, and further controls were obtained using HMW-GS 1Bx7 and WT-ANG proteins, the latter having no cysteine residues and thus not being able to be incorporated into the glutenin macropolymer structure. The effects of the ANGs were more complex than expected (L. Tamás, R. G. Soloman, M. K. Morell, F. Békés, R. Appels, unpublished), given the previously reported tendency for polypeptides with single cysteine residues to act as chain terminators and those with two cysteine residues to act as chain extenders (Tamás *et al.*, 2002).

Table 11.2 Summary of proteins used in this work

Polypeptide	Abbreviation	Description
HMW-GS 1Bx7	1Bx7	High molecular weight glutenin subunit Glu1Bx7
WT-ANG	WT	Parent molecule from which ANGs are derived ≡ C hordein, genebank accession X60037
ANGCys13Cys231	1N1C	Ser13Cys Pro231Cys mutant of WT-ANG
ANGCys7Cys13	2N	Ser7Cys Ser13Cys mutant of WT-ANG
ANGCys231Cys236	2C	Pro231Cys Thr236Cys mutant of WT-ANG
ANGCys7Cys13Cys 236	2N1C	Ser7Cys Ser13Cys Thr236Cys mutant of WT-ANG
ANGCys7Cys231Cys 236	1N2C	Ser7Cys Pro231Cys Thr236Cys mutant of WT -ANG
ANGCys7Cys13 Cys231Cys236	2N2C	Ser7Cys Ser13Cys Pro231Cys Thr236Cys mutant of WT-ANG

The results for 1N1C (Tamás *et al.*, 2002) indicate the protein provides stronger and more stable dough with improved loaf height (refer to Figs 11.4–11.6). It is evident from these results that the placement of the cysteine residues within the terminal domains result in the sulfhydryl groups being readily available for the formation of inter-chain disulfide bonds. The 2N and 2C proteins have very similar effects on dough properties. Although they have no significant effect on mixing time or extensibility, they appear to weaken the dough to some extent, as seen in increased resistance breakdown and decreased maximum resistance to extension compared to controls (see Figs 11.4 and 11.5). In this, they appear to mimic the effects of the WT-ANG protein, and it is concluded that the closely apposed cysteine residues in these proteins are more likely to form intra-chain disulfide bonds (Fig. 11.7) than to participate in the extension of the glutenin macropolymer chain. In such a situation, the proteins then become part of the pool of monomeric prolamin proteins, resulting in a reduction in the polymeric:monomeric protein ratio (Békés and Gras, 1999). However, the effect of 2N and 2C on loaf height (Fig. 11.6) appears to be inconsistent with their effect on dough mixing and extension properties. Incorporation of these proteins causes a large increase in loaf height, similar to that achieved with the 1N1C protein and with HMW-GS 1Bx7, whereas the WT-ANG has no significant effect on loaf height.

This apparent inconsistency between the baking results and the mixing and extension results for 2N and 2C may be a consequence of differences in the rates of formation or the stability of intra- and inter-molecular disulfide bonds involving the cysteine residues in these two proteins, and may also reflect a greater lability of such bonds during baking. The baking experiment differs from the mixing experiment in a number of ways, most notably in the addition of salt, improver and yeast to the dough formulation and in the timescale of the experiment. It is envisaged that these various factors may combine to increase the relative availability of the cysteine residues in 2N and 2C for participation in

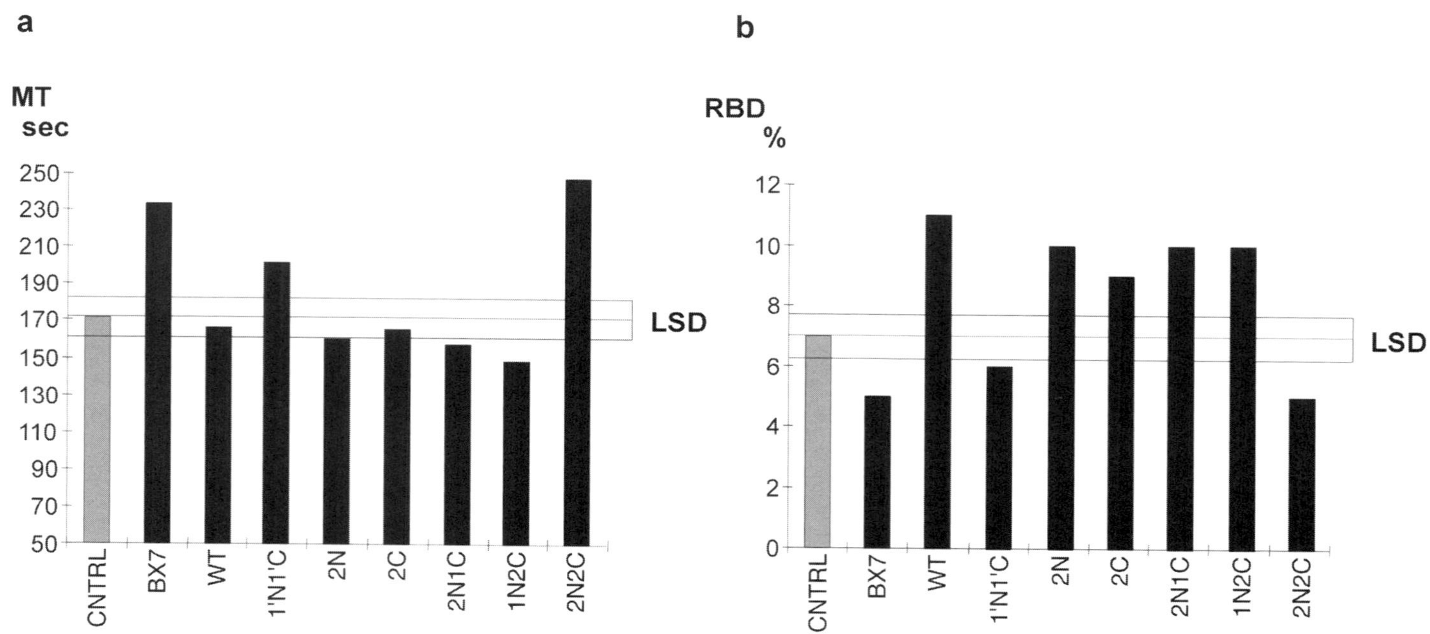

Fig. 11.4 (a) Mixing time (MT) and (b) resistance breakdown (RBD) results upon incorporation of ANG proteins with altered numbers of cysteine residues. The added protein is indicated below the x-axis (Table 11.2). CNTRL = reduction/oxidation mixing in the absence of added protein; BX7 = incorporation of HMW-GS 1Bx7; WT = incorporation with wild-type C hordein. Other signifiers identify the cysteine content of the incorporated ANG protein, as described in the Experimental section. LSD = least significant difference from the control. Mixing tests were conducted with a prototype Mixograph using a modification of the standard method for 35 g of flour scaled down to 2 g size15. The dough formulation comprised flour, water, dithiothreitol (DTT) as reductant, potassium iodate (KIO_3) as oxidant and 60 nm of protein was added where appropriate. Mixing parameters were determined using a modification of a previously reported computer program (Gras *et al.*, 1990). Parameters determined were mixing time (MT) and breakdown in resistance (RBD). The reversible reduction/oxidation procedure for incorporation of purified proteins into the gluten matrix was carried out according to the method developed by Békés *et al.* (1994b). All tests were performed three times. The least significant differences were calculated for $p = 0.005$ using Students t test.

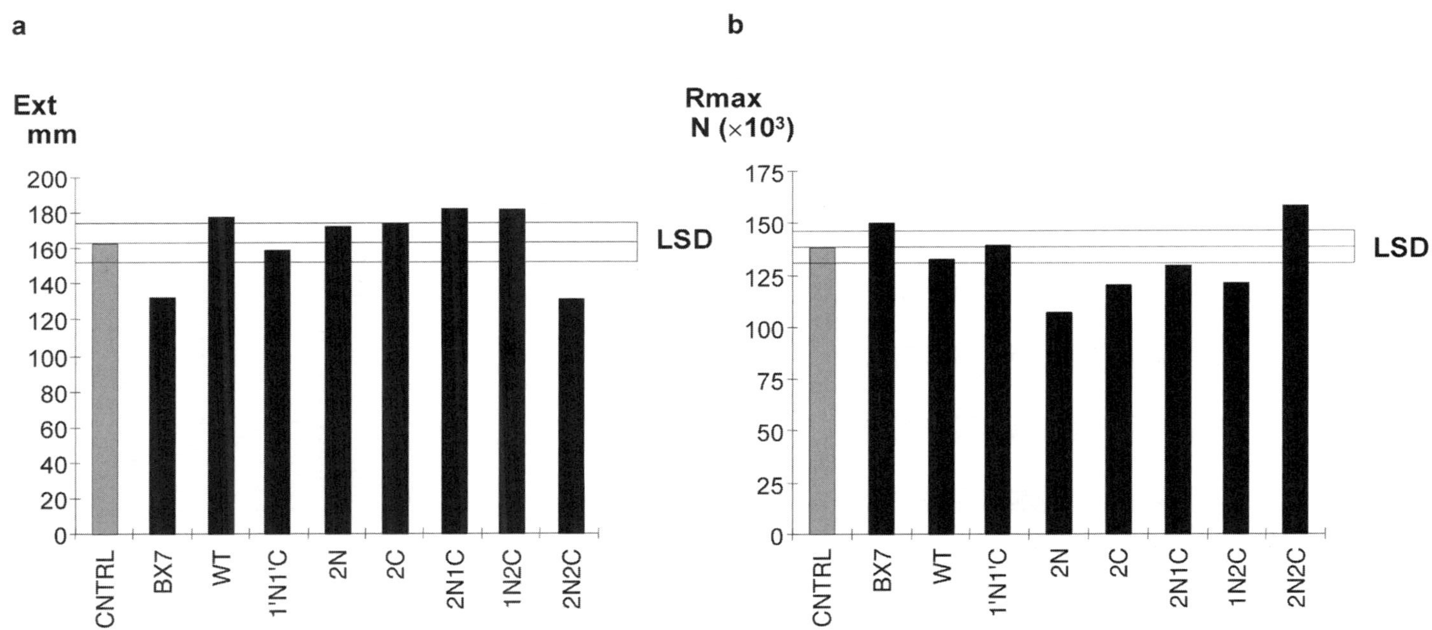

Fig. 11.5 (a) Extensibility (Ext) and (b) maximum resistance to extension (Rmax) results upon incorporation of ANG proteins with altered numbers of cysteine residues. Column signifiers as per Fig. 11.4. A small-scale extension tester was used, providing results for maximum resistance (Rmax) and extensibility (Ext) which are closely related to those from the Brabender Extensograph (Rath *et al*., 1994). Sample preparation and handling methods were as published earlier (Gras *et al*., 1997; Uthayakumaran *et al*., 2000). The dough formulation comprised flour, water, sodium chloride, dithiothreitol (DTT) as reductant, potassium iodate (KIO_3) as oxidant and added protein where appropriate. All tests were performed three times. The least significant differences were calculated for $p = 0.005$ using Students t test.

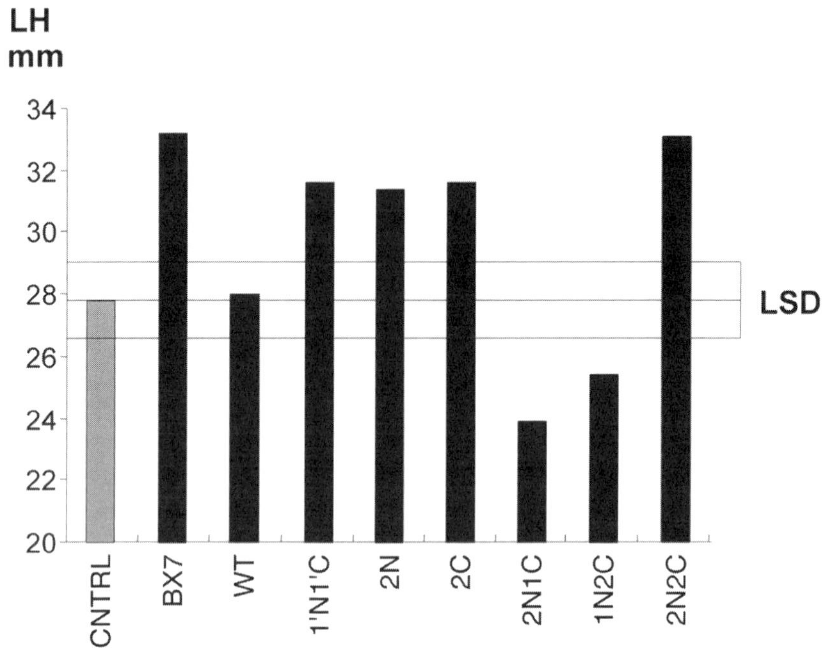

Fig. 11.6 Loaf height (LH) results upon incorporation of ANG proteins with altered numbers of cysteine residues. Column signifiers as per Fig. 11.4. Test baking was carried out using a recently developed procedure for the incorporation of added proteins, employing 2.4 g of dough prepared in the 2 g Mixograph 20. The dough formulation comprised flour, water, sodium chloride, improver, dithiothreitol (DTT) as reductant, potassium iodate (KIO_3) as oxidant and added protein or oligopeptide where appropriate. The improver was obtained from BRI Australia Ltd (Sydney, Australia), and included malted wheat flour, mineral salt, emulsifier, ascorbic acid and amylase enzyme. The height of the loaves was measured, and is used as an approximation to the effect of the incorporated proteins on loaf volume. All tests were performed three times. The least significant differences were calculated for $p = 0.005$ using Students t test.

inter-protein disulfide bonds, the formation of which allows the increase in dough height relative to the control, i.e. relative to the incorporation of WT-ANG protein in the baking experiment. The tendency for 2N and 2C to act in a similar manner to WT-ANG, with respect to mixing and extension experiments, may reflect a kinetic preference for the formation of intramolecular disulfide bonds within these proteins, with such bonds being intrinsically less thermodynamically stable, possibly due to strain, than the intermolecular bonds which may then form by disulfide exchange catalysed by redox compounds and/or yeast metabolites during the extended period available during proofing and baking.

Alternative explanations for the effect of 2N and 2C on loaf height include the possibility that the ANG-2N and ANG-2C proteins remain as monomeric species throughout the baking process and are not incorporated by the mechanism suggested above, or do not incorporate in a chain-extending conformation. In such a scenario, it is possible that their effect on loaf height is obtained through an effect on gas retention, perhaps by modulation of the interaction between the glutenin polymer and lipidic components forming the gas-retaining structure in the dough.

The observation that ANG2N and ANG2C provide an apparent boost to loaf volume without adversely affecting the mixing requirement makes them good candidates for forming the basis of an ingredient capable of reducing mixing requirement but still capable of maintaining or improving loaf volume, and further research in that direction is currently being pursued.

The preferential formation of intramolecular disulfide bonds by the closely apposed cysteine residues of 2N and 2C ANGs is supported by the results obtained for the ANG proteins containing 3 cysteine residues, namely 2N1C and 1N2C (see Table 11.2). The results for these two proteins are much the same as each other, and indicate a dough weakening effect, seen as reduced mixing time and maximum resistance to extension and increased RBD and Ext (Figs 11.4 and 11.5). These are similar to the effects obtained with single-cysteine, chain-terminating ANG proteins such as ANG-1N and ANG-1C (Tamás *et al.*, 2002). These results are consistent with the paired cysteine residues in the terminal domains of these proteins participating in intramolecular disulfide bonds, as hypothesised for the 2N and 2C proteins, leaving the unpaired cysteine to form chain-terminating disulfide bonds (see Fig. 11.7). In this case, the observed reduction in loaf height is consistent with the effects of the proteins on mixing and extension parameters. An alternative explanation for the chain terminating effects of 2N1C and 1N2C proteins involves the formation of head-to-tail intramolecular disulfide bonds, possibly between Cys13 and Cys236 in 2N1C and between Cys7 and Cys231 in 1N2C (see Fig. 11.7). Such circularised proteins would also have a single cysteine residue available to act in chain-terminating intermolecular disulfide bonding (Buonocore *et al.*, 1998). Buonocore and colleagues produced a number of variants of a peptide, derived from the repetitive domain of HMW-GS 1Dx5, containing up to 2 cysteine residues near either terminus (i.e. up to 4 cysteine residues in total). After chemical oxidation of the peptides, they observed bands with increased electrophoretic mobility in SDS-PAGE gels for those peptides containing two or more cysteine residues, and ascribed these bands to forms of the proteins containing head-to-tail intramolecular disulfide bonds. Upon chemical incorporation into doughs, the peptide containing three cysteine residues was also shown to have effects on mixing parameters consistent with a role as a chain-terminator (Buonocore *et al.*, 1998). Such head-to-tail disulfide bond formation has also been reported in a chimaeric HMW-GS formed by replacing the N-terminal non-repetitive domain of subunit 1Dx5 with the homologous domain from its allelic partner, subunit 1Dy10 (Shimoni *et al.*, 1997). Upon expression of this modified glutenin in transgenic wheat, changes were noted in its electrophoretic mobility and in its ability to incorporate into the glutenin macropolymer, with substantial amounts of the intramolecularly-

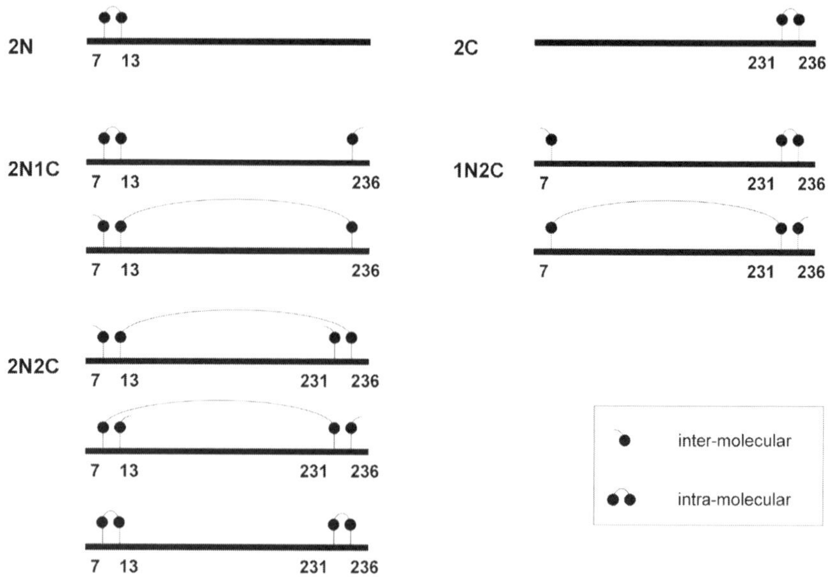

Fig. 11.7 Schematic representation of disulfide-bonding patterns of some of the ANG proteins used in this work. The figure is not meant to be comprehensive of all possible disulfide bonding patterns, but illustrates some of the possibilities discussed in the text.

bonded recombinant protein remaining in the monomeric (gliadin) protein fraction.

The results obtained from the incorporation of 2N2C in flour mixing experiments indicate a substantial strengthening of the dough, equal to or surpassing the effects of HMW-GS 1Bx7 or 1N1C. In this regard, it also mimics the effects of Buonocore *et al.* (1998) 4-cysteine peptide. MT and Rmax are significantly higher than controls, and RBD and Ext are greatly reduced (Figs 11.4 and 11.5). The increase in LH (Fig. 11.6) also mirrors the effect of 1Bx7. However, this implies that at least two of the cysteine sulfhydryls in 2N2C are available for the formation of inter-protein disulfide bonds, given the protein's strong chain-extending properties. Such a result would be inconsistent with an explanation invoking preferential formation of disulfide bonds between the paired cysteine residues in either terminal domain (see Fig. 11.7). However, N-to-C-terminal intramolecular disulfide bond formation is a possibility provided that only one such bond is formed per 2N2C protein chain. This would leave available one cysteine from each terminal domain to provide for chain extending intermolecular disulfide bond formation (Fig. 11.7), satisfactorily explaining the observed effects of the protein incorporated in dough. It is possible, therefore, that although the closely apposed cysteine residues in ANG proteins may form disulfide bonds, such bonds are intrinsically less stable than those formed by cysteine residues from opposite ends of the molecule or those formed between protein chains. This differs slightly from the explanation provided by Buonocore *et al.* (1998), who concluded that cysteine residues placed six residues apart in the protein chain were unlikely to form intramolecular bonds due to them being on opposite sides of the repetitive domain β-spiral structure. This seems more likely in situations where alternative disulfide bonds cannot be formed, for example in 2N and 2C ANG proteins, where the cysteine residues are placed in the non-repetitive region of the ANGs and are probably not in a region of β-spiral secondary structure.

The ability of ANG and glutenin subunit protein with two or more cysteine residues to act as either extenders or terminators depending on whether intramolecular disulfide bonds are formed reflects the dynamic nature of dough formation at a molecular level in response to environmental conditions (additives present in the dough mix). Such bonds alter the effective number of cysteine residues available for intermolecular interactions, causing unexpected effects on dough properties. To focus on the availability of cysteines more closely, L. Tamás, R. G. Soloman, M. K. Morell, F. Békés and R. Appels (unpublished) monitored the effects of the short synthetic peptides (Table 11.3) on mixing parameters (Fig. 11.8). The mixing results indicate that the simple addition of the short peptides causes a dough weakening, measured as reduced mixing time and reduced tolerance to over-mixing (resistance breakdown see Fig. 11.8), consistent with the peptides acting in a manner similar to a gliadin fraction. The exception is peptide #130, which has no effect on mixing parameters upon simple addition. Peptide #130 has a relatively hydrophobic intervening sequence based on an antifreeze protein repeat sequence (De Vries, 1986), rather than on prolamin repetitive domain sequences.

Upon incorporation, most of the peptides act as dough strengthening agents, similar to that seen for the ANG proteins, increasing mixing time and reducing resistance breakdown. Thus the cysteines in these peptides were clearly available for intermolecular disulfide formation. However, the hydrophobic peptide #130 has no effect on mixing parameters and this is indicative of the peptide being unable to form disulfide bonds with the glutenin proteins. It is most likely that this is due to its hydrophobic nature, and the fact that this peptide partitions to a different 'phase' of the dough and does not interact with the gluten phase. Since peptide #130 does not even produce an effect on dough properties when simply

Table 11.3 Aligned sequences of oligopeptides used in the current study, with cysteine residues in bold italic type

Peptide #	Peptide sequence	Inter-cysteine spacing	Molecular weight
125	RQLNP***C***SQELQS----------------***C***IWSMV	6	2122.4
126	RQLNPSSQELQ***C***--------PQQPFPQQ***C***IWSMV	8	3073.5
127	RQLNP***C***SQELQS--------PQQPFPQQ***C***IWSMV	14	3073.5
128	RQLNP***C***SQELQSPGQ-----GQQG--QQ***C***IWSMV	15	3031.4
129	RQLNP***C***SQELQSPQQPFPQQPQQPFPQQ***C***IWSMV	22	4024.6
130	GSNE***C***TAANAAAAAALR***C***GGT	12	1880.0

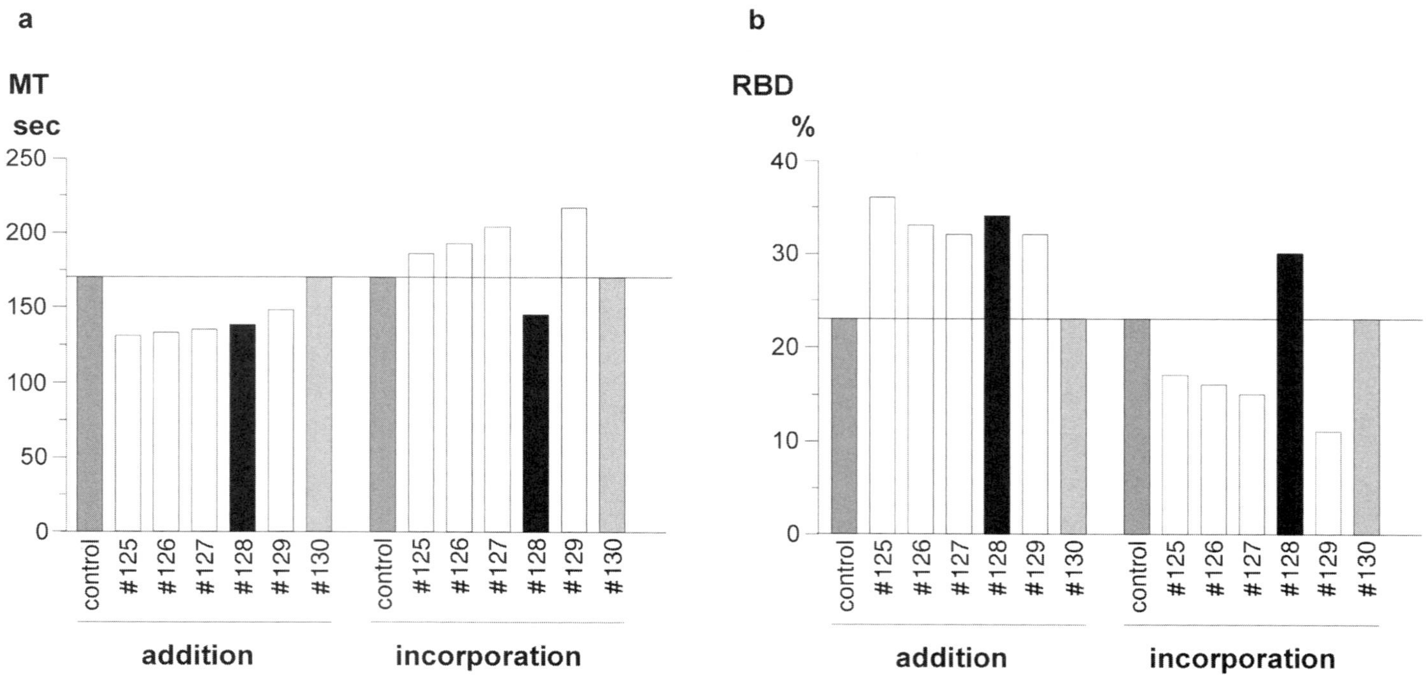

Fig. 11.8 Effect of the small cysteine-containing oligopeptides on dough mixing properties, (a) mixing time (MT) and (b) resistance breakdown (RBD), in addition and incorporation experiments. The addition control is without added peptide. The incorporation control is mixing with oxidation and reduction in the absence of added peptide. See Fig. 11.5 for experimental details. All tests were performed three times. The least significant differences were calculated for $p = 0.005$ using Students t test.

added, it seems that it does not affect the monomeric:polymeric protein ratio in any way. In the incorporation experiments, peptide #128 was another exception in that it failed to form intermolecular disulfide bonds even though the cysteines were present. In this case L. Tamás, R. G. Soloman, M. K. Morell, F. Békés and R. Appels (unpublished) concluded that the presence of a number of glycine residues in the intervening sequence, a motif derived from the repetitive domain of the x-type HMW-GS class of proteins allowed the extra flexibility to preferentially form an intramolecular disulfide bond. In this way peptide #128 was able to act in both addition and incorporation experiments as a gliadin-type molecule, present in the glutenin phase of the dough but not forming disulfide bonds with the glutenin proteins.

The diagrammatic representation of glutenin subunit proteins in Fig. 11.9 provides a summary of the dynamic nature of the interactions responsible for dough formation, illustrated by the above discussions. The net result of the interactions illustrated is to control the type of protein polymer that forms for subsequent participation in the macromolecular complexes in Fig. 11.2.

11.3.2 New domains in glutenin subunit proteins

The ANG proteins described in the preceding section to study, in detail, the nature of the interactions involved in forming dough, and how these can be modified, have also provided a template for investigating the possibility of introducing new functionalities into the gluten complex. Morell and colleagues (Morell *et al.*, 2000) investigated the possibility of altering protein-protein, protein-lipid and protein-starch interactions within the gluten matrix and showed that new surface-active molecules or part of molecules could be introduced into the gluten matrix. A starch-binding domain from *Bacillus circulans* cyclo dextrin glycosyltransferase has been shown to be capable of directing a model protein (luciferase) to starch granules *in vivo*, in transgenic potato (Qin Ji *et al.*, 2003).

Morell and colleagues (Morell *et al.*, 2000) found that modified ANG proteins with the following domains could be produced; the lipid-binding domain of the barley oleosin gene, the lipid-binding regions of the wheat CM16 protein, and the starch-binding domain of the glucoamylase from *Aspergillus niger*. The ANG protein with the starch-binding domain of the glucoamylase from *A. niger* (ANG-SBD, Fig. 11.10) was characterised in most detail and it was demonstrated that ANG-SBD could bind β-cyclodextrin (an indicator for the starch binding activity) and still participate in the formation of a glutenin macro-polymer. Some of the properties of the dough formed with the ANG-SBD containing glutenin subunit protein were unexpected and are currently under further investigation.

Although this area of research is still in its early stages the concept of modifying the properties of the glutenin macro-polymer through the introduction of novel protein domains represents an exciting approach for specifically altering the properties of dough.

a

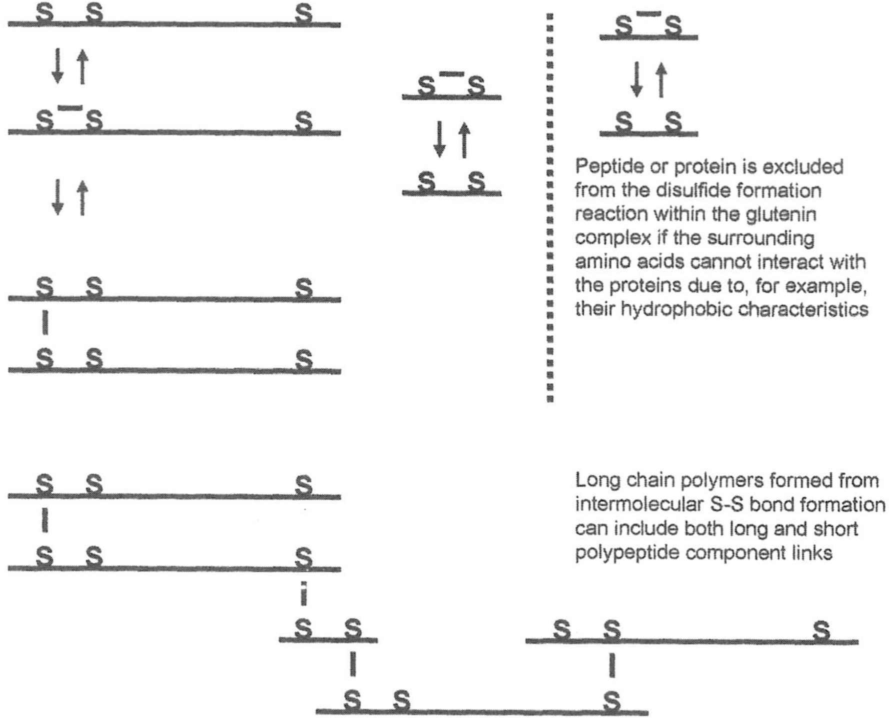

Peptide or protein is excluded from the disulfide formation reaction within the glutenin complex if the surrounding amino acids cannot interact with the proteins due to, for example, their hydrophobic characteristics
Long chain polymers formed from intermolecular S-S bond formation can include both long and short polypeptide component links

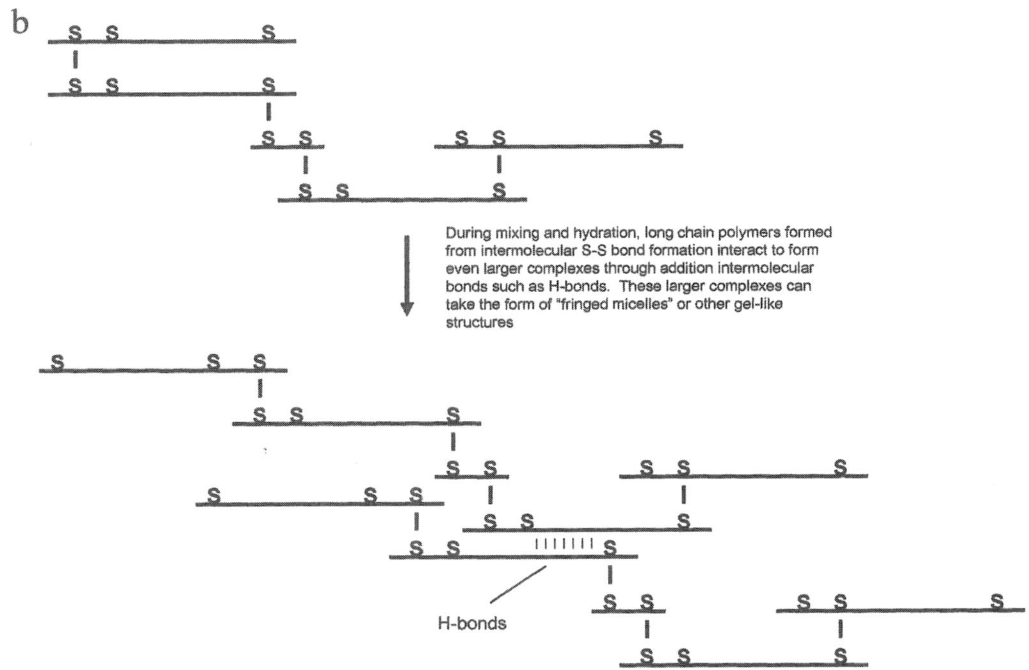

Fig. 11.9 Diagrammatic summary of proteins and peptides interacting, as discussed in the text. (a) Formation of macro-polymers via disulfide bridges and exclusion of certain peptides from these interactions within the gluten complex due to properties of the amino acids separating the cysteine residues. (b) Assembly of macro-polymers into even larger structures as a result of multiple H-bonds, in series, between regions of subunits of the macro-polymers (Belton, 1999).

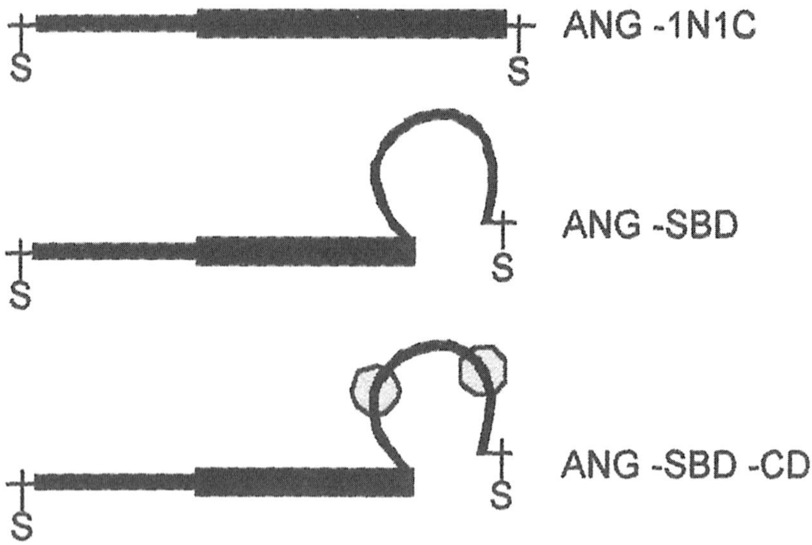

Fig. 11.10 The modification of an ANG protein by the insertion of a new domain, after deleting a region of the repetitive domain in the original protein. The domain inserted is a starch-binding domain (SBD), as described in the text, and, as indicated in the lower panel, this domain is still capable of binding β-cyclodextrin in the ANG-SBD form, to generate ANG-SBD-CD. The 'S' indicates the location of a cysteine residue.

11.3.3 Foam stabilising proteins

A significant feature of dough is that it entraps air as it undergoes the processing summarised in Fig. 11.1. The development of yeast-leavened products from dough is possible because fermentation gas is retained in the form of a foam structure. In addition to the gluten network, Courtin and Delcour (2002) have argued that water extractable arabinoxylans (WE-AXs) also contribute to gas retention through increasing the viscosity of the aqeous phase and the generation of a network of high molecular AXs that helps to stabilise the gluten network. The presence of WU-AXs as distinct particles has been argued to be detrimental because their positioning within the walls of gas bubbles could destabilise the bubbles (Courtin and Delcour, 2002).

The stabilisation of the gluten network to reduce the rupture of gas bubbles within the dough, in the presence of added components, including lipids, that form part of the formulation, can be affected by specific proteins in the flour that are known to be foam stabilising proteins and lipid-binding proteins (Keller *et al.*, 1997; Douliez *et al.*, 2001, 2002). Puroindolins are a well-characterised group of small proteins located near the hardness locus on chromosome 5D that can bind lipids and are highly active in foam stabilisation (Dubriel *et al.*, 1997). Variation in these proteins has been correlated with variation in grain texture (Turnbull and Rahman, 2002), but it is not clear how variation the proteins changes the stability of bubbles formed within the dough during processing.

Low molecular weight classes of prolamin have been identified in wheat flour over a long period of research in this area (Salcedo *et al.*, 1979; Rocher *et al.*, 1996; Anderson *et al.*, 2001; Clarke *et al.*, 2001). A subgroup of these proteins (low molecular weight gliadins, LMGli) has been shown to have strong homology to proteins in barley associated with foam-stabilisation (17 kDa, ε-hordeins, Vaag *et al.*, 2000) and investigated further with regards their properties in dough formation (Clarke *et al.*, 2003). The LMGli proteins contain cysteines that would be expected to participate in intermolecular disulfide formation (Clarke *et al.*, 2003) but are unusual in not carrying the hydrophilic repetitive domains (... QPQQ ...) that characterise glutenin subunit proteins and gliadins. Experiments using pure protein (isolated from bacteria engineered to carry the gene) showed that when it was simply added to flour and mixing carried out to form a dough, a decrease in mixing time and an increase in resistance breakdown (reduced stability to over-mixing) was observed. This was similar to the behaviour of normal gliadin proteins. Under conditions where LMGli was able to incorporate into the gluten macro-polymer, the protein caused an increase in mixing time and maximum bandwidth in the Mixograph at levels comparable to those found for LMW glutenins (Lee *et al.*, 1999). Clarke *et al.* (2003) noted that it was unusual for LMGli to increase both the mixing time and sensitivity to over mixing (resistance breakdown). Furthermore the incorporation of LMGli generated a clear increase in loaf volume in small-scale baking tests and the authors suggested this might have resulted from a slightly greater increase of hydrophobicity of the gluten complex. The increased hydrophobicity would have resulted from the incorporation of LMGli as a subunit lacking the highly hydrophilic repetitive (... QPQQ ...) domain and the authors suggested this could result in improved CO_2 retention during baking.

The characterisation of low molecular weight seed storage proteins such as LMGli provide the basis for accounting in detail for variation in flour processing properties in genetic studies (discussed below, see Section 11.4). Variation in the classical glutenin subunit group of proteins accounts for approximately 60% of the variation found in different wheat cultivars (Eagles *et al.*, 2001, 2002) and it is possible that variation in proteins such as the low molecular weight gliadins will help predict the properties of wheat flour more accurately.

11.3.4 Proteins to modify water structure in frozen dough

The behaviour of water during the freezing of dough, an intervention commonly used in the flour/dough-processing scheme (Fig. 11.1), is critical in determining the final properties of the product formed. Besides the effects of freezing on yeast viability, the freezing/storage of dough results in loss of dough strength and gas retention, and hence deterioration of product quality. The growth of ice crystals during the frozen storage (Räsänen *et al.*, 1998) can result in increased protein polymer concentrations and separation of the gluten matrix from starch granules (Zounis *et al.*, 2002a). These negative effects can be minimised by the addition of oxidants such as ascorbic acid and surface-active agents such as sodium stearoyl-2-lactylate have been added to dough mixtures to help to maintain dough quality during frozen storage (Inoue *et al.*, 1994; Abd El-Hady *et al.*, 1999; Laaksonen and Roos, 2001). In addition, the deployment of flour with very strong gluten (Inoue and Bushuk, 1992) and reduced temperatures during the dough formation (Mathews, 1989; Nemeth *et al.*, 1996; Zounis *et al.*, 2002b) further help to reduce the negative effects of freezing dough.

Antifreeze proteins have, for sometime, been of interest as an alternate means of preventing damage due to ice crystal growth, during frozen storage. A wide range of antifreeze proteins is found in nature to protect life forms against damage due to freezing (De Vries, 1986; Duman and Olsen, 1993). At concentrations of approximately 1g/l, purified proteins bind to nascent ice crystals and reduce the freezing point of the solution during a process referred to as thermal hysteresis. Under conditions of thermal hysteresis the slower growth of ice crystals occurs along the less favourable *c*-axis of the crystal, giving rise to hexagonal bipyramidal crystals (Fig. 11.11). The type 1 AFP (anti-freeze protein) is a particularly well-studied protein from winter flounder (De Vries, 1986; Knight *et al.*, 1991) and synthetic forms of the small protein have been produced in bacteria on a large scale (Solomon and Appels, 1999). The type 1 AFP has been produced on a large scale by Solomon and Appels (1999) and was active in reducing ice crystal growth in controlled experiments (Fig. 11.11) as well as industrial experiments with ice-cream preparations (unpublished

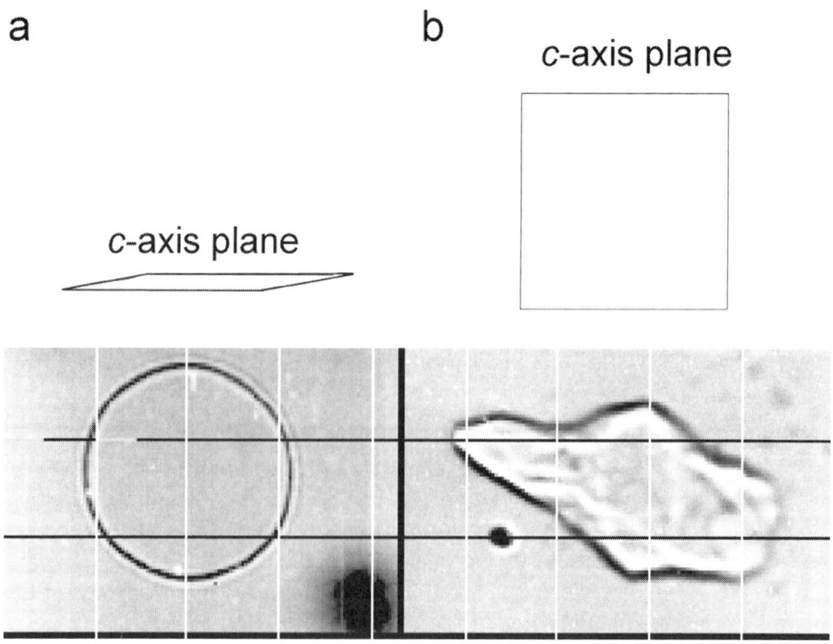

Fig. 11.11 Single ice crystals grown in (a) water or (b) 1.0 g/l recombinant antifreeze protein -1 in water. The approximate orientation of the *c*-axis of the crystal is indicated. Modified from Solomon and Appels (1999).

observations). Preliminary experiments also demonstrated that type 1 AFP provided some amelioration of the negative effects of freezing on dough extensibility (Solomon and Appels, 2001). The production of transgenic wheats with the type 1 AFP gene expressed in the grain are currently being generated and the analysis of flour properties with respect to behaviour in frozen dough experiments will be of interest.

11.4 Genetic modification of flour properties

The analysis of genetic factors underpinning flour-processing properties has largely focussed on defining the functional properties of individual glutenin subunit proteins (reviewed in Shewry and Morell, 2001; Gras *et al.*, 2001) and has provided advances in utilising genetic variation in these proteins for modifying the flour processing properties of wheat cultivars (Payne *et al.*, 1987). Direct genetic evidence for the importance of variation in the HMW and LMW glutenin loci on the long and short arms (respectively) of chromosomes 1A, 1B and 1D is evident in quantitative trait analysis of doubled haploid lines from a cross between wheat lines differing at these loci (Appels *et al.*, 2000; Békés *et al.*, 2002; see also Fig. 11.12). The genetic analyses define quantitative trait loci (QTLs). Eagles *et al.* (2001, 2002) have reported a detailed study of the relationship between variation in glutenin subunit proteins and dough processing properties in Australian wheat lines. The glutenin alleles (defined electropheretically) most common in Australian wheat are GluA3 (null), GluA3b, GluA3c, GluB3b, GluB3c, GluB3h, GluD3a, GluD3b, GluD3c, GluD3d for the LMW glutenin subunit proteins and GluA1a (1), GluA1b (2*), GluA1c (null), GluB1b (7+8), GluB1c (7+9), GluB1e (20), GluB1i (17+18), GluD1a (2+12), GluD1d (5+10), for the HMW glutenins (O'Brien *et al.*, 2001). Variation for these proteins, plus some other minor alleles, indicated that approximately 60% of the variation in maximum resistance of dough and its extensibility could be accounted for. An additional 6% of variation in the dough properties was accounted for when 2-way interactions between loci were taken into account. The latter observation indicated interactions between glutenin subunit proteins occurred to determine the final processing properties of the flour. These observations are consistent with the chemical properties described in the preceding sections and provide a means of modifying processing properties through normal crossing procedures to manipulate the complement of the types of glutenin subunit proteins present in flour. The contribution of the small classes of protein such as LMGli and LTPs (see Section 11.3) has not yet been determined and if variation in these components is taken into account it may be possible to predict variation in dough properties even more reliably.

Interacting with the above proteins is the complement of starch in the grain. Starch is a major component of flour (see Table 11.1) and is particularly important in the context of bread and noodle in defining water uptake by the

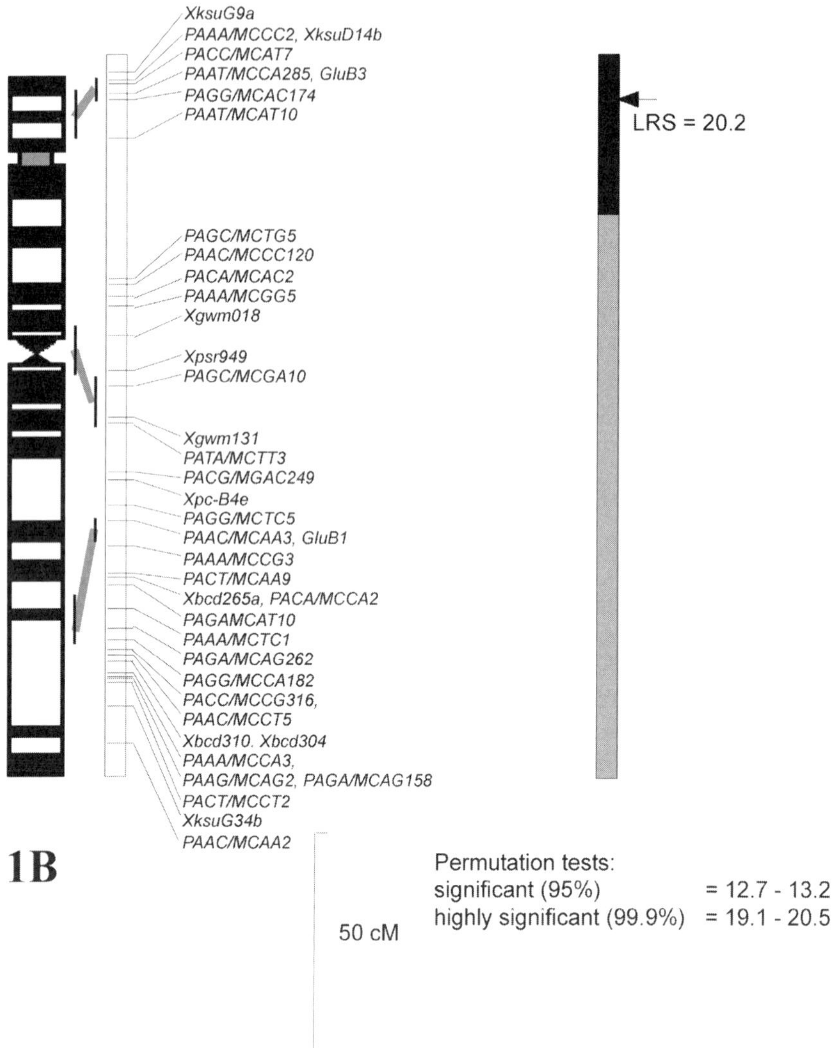

Fig. 11.12 The location of a quantitative trait locus (QTL) for mixing time in wheat. The figure shows a classical schematic of a C-banded chromosome of chromosome 1B of wheat on the far left and next to this the molecular genetic map (Chalmers *et al.*, 2001), with the approximate location of sections of the genetic map within the C-banded or physical map indicated by grey lines. The scale of the molecular genetic map is in centiMorgans (cM). The bar on the far right indicates the association for variation in the mixing time trait among 163 lines in the mapping population derived from the cross Cranbrook × Halberd (grey non-significant and black highly significant). Modified from Békés *et al.* (2001).

starch granules and the degree to which they swell as well as the viscosity of the starch upon heating. Wheat starch is deposited as discrete granules with a bimodal distribution of large (A) and small (B) granules. The minimum between the distributions of A and B granules falls at approximately 10 μm and generally 15–30% of the starch (by weight) falls below this 10 μm benchmark (reviewed in Shewry and Morell, 2001). The starch granules are imbedded in the matrix of the grain and the nature of the interaction defines grain texture (reviewed in Turnbull and Rahman, 2002). The hard grain types result in extensive starch granule damage during milling of the grain and this affects the amount of water that is absorbed by the flour (see Section 11.2.1) during processing. This variation in the hard/soft nature of the grain is a major 'background' that moderates the processing properties of flour, in addition to the changes caused by variation in glutenin subunit protein, as discussed in the previous sections. The genetic analysis of doubled haploid lines from a cross between two lines of wheat that are virtually identical in glutenin protein subunits but differing in the hardness trait indicates most of the variation in flour processing characteristics (defining a QTL) is associated with the hardness (Ha) locus on chromosome 5D (Australian Winter Cereals Molecular Marker Program, R. Appels unpublished). Generally the more pronounced effects coming from variation in the glutenin subunit proteins hide this variation associated with the Ha locus.

Variation in grain hardness alters the water absorption characteristics of flour and this can then alter the processing properties as described in an earlier section (see section 11.3). Related to this is the fact that variation in water-soluble arabinoxylans (WE-AX, see section 11.3) can also alter water absorption characteristics. A significant QTL for variation in WE-AX (especially arabinose: xylose ratio) is located distal to the HMWglutenin subunit protein locus on chromosome 1B (Martinant *et al.*, 1998) and it is possible that this could also provide 'background' variation that would moderate the effects observed for glutenin subunit variation. Minor variation in WE-AX was also associated with chromosomes 4B and 6B.

The molecular genetic analysis of starch properties in wheat has indicated distinct quantitative trait loci associated with granule swelling, starch viscosity and the distribution of starch granule size (Zao *et al.*, 1998; Batey *et al.*, 2001). This level of variation can, again, be genetically controlled (through a null mutation at the granule bound starch synthase locus on chromosome 4A, Batey *et al.*, 2001) and define wheat lines with specific properties for a product such as udon, or white salted, noodles (Crosbie *et al.*, 1992). Further modification in starch properties has been investigated (Båga *et al.*, 1999) but at present the consequences for flour processing properties are not clear. A dramatic change in the processing properties occurs when wheat lines are missing the entire complement of granule bound starch synthases (Yamamori and Quynh, 2000; Kim *et al.*, 2003). The starch of so-called waxy wheats has no amylose, a much reduced gelatinisation temperature and higher water absorption (Kim *et al.*, 2003) and it is evident that the flour processing/baking properties are completely altered from normal wheat. The waxy wheats are now being produced in large quantities with different complements of glutenin subunit proteins in order to find new combinations of genes that can produce novel end products. It is evident that in the area of modifying wheat flour, the advent of waxy wheats provides a good example of the major changes that can occur even without the use of genetic engineering technologies. Modifying gene complexes, rather than just single genes, are clearly important in future considerations for generating wheat lines with new processing properties.

11.5 References

ABD EL-HADY E A, EL-SAMAHY S K and BRÜMMER J-M (1999), 'Effect of oxidants, sodium-stearoyl-2-lactylate and their mixtures on rheological and baking properties of nonprefermented frozen doughs', *Lebensm-Wiss u-Technol*, 32, 446–454.

AMIOUR N, JAHIER J, TANGUY A M, CHIRON H and BRANLARD G (2002), 'Effects of 1R(1A), 1R(1B) and 1R(1D) substitution on technological value of bread wheat', *J Cereal Sci*, **35**, 149–160.

ANDERSON O D, KUHL J C and TAM A (1996), 'Construction and expression of a synthetic wheat storage protein gene', *Gene*, **174**, 51–58.

ANDERSON O D, HSIA C C, ADALSTEINS A E, LEW E J-L and KASARDA D D (2001), 'Identification of several new classes of low-molecular-weight wheat gliadin-related proteins and genes', *Theor Appl Genet*, **103**, 307–315.

ANDERSSEN R S, GRAS P W and MACRITCHIE F (1998), 'The rate-independence of the mixing of wheat flour dough to peak dough development', *J Cereal Sci*, **27**, 167–177.

APPELS R., GRAS P W, CLARKE B C, ANDERSSEN R, WESLEY I and BÉKÉS F (2000) Molecular genetic studies on processing trats of wheat flour. Euphytica **119**, 49–54.

ARÊAS E P G and CASSIANO M M (2001), 'Folding interpenetration in a gliadin model: the role of the characteristic octapeptide motif', *Biophysical Chem*, **90**, 135–146.

BÅGA M, REPELLIN A, DEMEKE T, CASWELL K, LEUNG N, ABDEL-AAL E S, HUCL P and CHIBBAR R N (1999), 'Wheat starch modification through biotechnology', *Starch/Stärke*, **51** (4), 111–116.

BATEY I L, HAYDEN M J, CAI S, SHARP P J, CORNISH G B, MORELL M K and APPELS R (2001), 'Genetic mapping of commercially significant starch characteristics in wheat crosses'. *Aust J Agric Res* **52**, 1287–1296.

BÉKÉS F and GRAS P W (1994), 'Effects of individual HMW glutenin subunits on mixing properties', in *Proc* 5th *Int Workshop on Gluten Proteins*, Association of Cereal Research, Detmold, Germany, 170–179.

BÉKÉS F and GRAS P W (1999), 'In vitro studies on gluten protein functionality', *Cereal Foods World*, **44**, 580–586.

BÉKÉS F, ANDERSON O, GRAS P W, GUPTA R B, TAM A, WRIGLEY C and APPELS R (1994a). The contribution to mixing properties of 1D glutenin subunits expressed in a bacterial system. In *Improvement of Cereal Quality by Genetic Engineering* (R Henry, JA Ronalds, eds), pp. 97–104. Kluger, London.

BÉKÉS F, GRAS P W and GUPTA R B (1994b), Mixing properties as a measure of reversible reduction/oxidation of doughs. *Cereal Chem.* **71**, 44–50.

BÉKÉS F, GRAS P W, GUPTA R B, HICKMAN D R and TATHAM A S (1994c), Effects of 1Bx20 HMW glutenin on mixing properties. *J. Cereal Sci.* **19**, 3–7.

BÉKÉS F, MA W and GALE K (2002). QTL analysis of wheat quality traits. *Acta Agronomy.*

BELTON P S (1999), 'On the elasticity of wheat gluten', *J Cereal Sci*, **29**, 103–107.

BELTON P S, COLQUHOUN I J, GRANT A, WELLNER N, FIELD J M, SHEWRY P R and TATHAM A S (1995), 'FTIR and NMR studies on the hydration of a high-M_r subunit of glutenin', *Int J Biol Macromol*, **17**(2), 74–80.

BUONOCORE F, BERTINI L, RONCHI C, BÉKÉS F, CAPORALE C, LAFIANDRA D, GRAS P, TATHAM A S, GREENFIELD J A, HALFORD N G and SHEWRY P R (1998), 'Expression and functional analysis of M_r 58000 peptides derived from the repetitive domain of high molecular weight glutenin subunit 1Dx5', *J Cereal Sci*, **27**, 209–215.

BUSHUK W and MEHROTRA V K (1977), 'Studies of water binding by differential thermal analysis', *Cereal Chem*, **54**, 320.

CHEN P L, LONG Z, RUAN R and LABUZA T P (1997), 'Nuclear magnetic resonance studies of water mobility in bread during storage', *Lebensmittel Wissenschaft und Technologie (LWT)*, **30**, 178–183.

CIAFFI M, LEE J K, TAMÁS L, GUPTA R B, SKERRITT J and APPELS R (1999), 'The low molecular weight glutenin subunit proteins of primitive wheats. III: the genes from D genome species', *Theor Appl Genet*, **98**, 135–148.

CLARKE B C, PHONGKHAM T, GIANIBELLI M C, BEASLEY H and BÉKÉS F (2003), 'The characterisation and mapping of a family of LMW-gliadin genes: effects on dough properties and bread volume', *Theor Appl Genet*, **106**, 629–635.

CLARKE B C, RHEE S, BÉKÉS F, GALE K, LARROQUE O and APPELS R (2001), 'New markers for mapping: the frequent classes of genes expressed in wheat and corn endosperm tissue', *Aus J Agric Res*, **52**, 1181–1193.

COURTIN C M and DELCOUR J A (2002), 'Arabinoxylans and endoxylanase in wheat flour breadmaking', *J. Cereal Sci*, **35**, 225–243.

CROSBIE G B, LAMBE W J, TSUTSUI H and GILMOUR R F (1992) 'Further evaluation of the flour swelling volume test for identifying wheats potentially suitable for Japanese noodles. *J Cereal Science* **15**, 271–280.

CURCIO J A and PETTY C C (1951), 'The near infrared absorption spectrum of liquid water', *Journal of the Optical Society of America*, **41**(5), 302–304.

D'OVIDIO R, ANDERSON O D, MASCI S, SKERRITT J and PORCEDDU E (1997), 'Construction of novel wheat high-M(r) glutenin subunit gene variability: Modification of the repetitive domain and expression in *E. coli.*', *J Cereal Sci*, **25**, 1–8.

DAVIES R J and WEBB T (1969), 'Calorimetric determination of freezable water in dough', *Chem Ind*, **16**, *1138.*

DE VRIES A L (1986), 'Antifreeze glycopeptides and peptides: interactions with ice and water', *Methods in Enzymology*, **127**, 293–303.

DOULIEZ J P, MICHON T, ELMORJANI K and Marion D (2000), Structure, biological and technological functions of lipid transfer proteins and indolines, the major lipid binding proteins from cereal kernels. *J Cereal Sci* **32**, 1–20.

DOULIEZ J P, JEGOU S, PATO C, LARRE C, MOLLE D and MARION D (2001), Identification of a new form of lipid transfer protein (LTP) in wheat seeds. *J Agric Food Chem.* **49**, 1805–1808.

DUBRIEL L, COMPOINT J-P and MARION D (1997), The interaction of puroindolines with wheat polar lipids determines their foaming properties. *J Agric Food Chem* **45**, 108–116.

DUMAN J G and OLSEN T M (1993), 'Thermal hysteresis protein activity in bacteria, fungi, and phylogenetically diverse plants', *Cryobiology*, **30**, 322–328.

EAGLES H A, BARIANA H, OGBONNAYA F.C, REBETZKE G J, HOLLAMBY G J, HENRY R J, HENSCKE P H and CARTER M (2001), Implementation of markers in Australian wheat breeding. *Aust J Agric Res* **52**, 1349–1356.

EAGLES H A, HOLLAMBY G J and EASTWOOD R F (2002), Genetic and environmental variation for grain quality traits routinely evaluated in southern Australian wheat breeding programs. *Aust J Agric Res* **53**, 1047–1057.

EVERS A D, BLAKENEY A B and O'BRIEN L O (1999), 'Cereal structure and composition', *Aust J Agric Res*, **50**, 629–650.

FELDMAN M (2001), 'Origin of cultivated wheat', in *The World Wheat Book*, eds Bonjean A P and Angus W J, Lavoisier Publishing, London, Paris, New York, pp. 3–56.

GAO L, NG P K W and BUSHUK W (1992), 'Structure of glutenin based on farinograph and electrophoretic results', *Cereal Chem*, **64**, 452–455.

GARCIA R, RAKOTOZAFY L, TELEF N, POTUS J and NICOLAS J (2002), 'Oxidation of ferulic acid or arabinose-esterified ferulic acid by wheat germ peroxidase', *J Agric Food Chem*, **50**, 3290–3298.

GILBERT S M, WELLNER N, BELTON P S, GREENFIELD J A, SILIGARDI G, SHEWRY P R and TATHAM A S (2000), 'Expression and characterisation of a highly repetitive peptide derived from a wheat seed storage protein', *Biochimica et Biophysica Acta*, **1479**, 135–146.

GIVEN P S (1991), 'Molecular behavior of water in a flour-water baked model system', in Levine H and Slade L, *Water Relationships in Food*, New York, Plenum Press, 465–483.

GRANT A, BELTON P S, COLQUHOUN I J, PARKER M L, PLIJTER J J, SHEWRY P R, TATHAM A S and WELLNER N (1999), 'Effects of temperature on sorption of water by wheat gluten determined using deuterium nuclear magnetic resonance', *Cereal Chem*, **76**, 2, 219–226.

GRAS P W, ANDERSSEN R S, KEENTOK M, BÉKÉS F and APPELS R (2001), 'Gluten protein functionality in wheat flour processing: a review', *Aust J Agric Res*, **52**, 1311–1323.

GRAS P W, CARPENTER H C and ANDERSSEN R S (2000), 'Modelling the developmental rheology of wheat-flour dough using extension tests', *J Cereal Sci*, **31**, 1–13.

GRAS P W, ELLISON F W and BÉKÉS F (1997), 'Quality evaluation on a micro-scale', in *Proc Int Wheat Quality Conference*, Steele J L and Chung O K, eds, Grain Industry Alliance, Manhattan, Kansas, 161–171.

GRAS P W, HIBBERD G E and WALKER C E (1990), 'Electronic sensing and interpretation of dough properties using a 35 gram Mixograph', *Cereal Foods World*, **35**, 568–571.

HARLAN J R (1992), 'Origins and processes of domestication', in *Grass Evolution and Domestication* (ed G P Chapman), Cambridge University Press, Cambridge, UK, pp. 159–175.

INOUE Y and BUSHUK W (1992) Studies on frozen doughs II. Flour quality requirements for bread production from frozen dough. *Cereal Chemistry* **69**, 423–428.

INOUE Y, SAPIRSTEIN H D, TAKAYANAGI S and BUSHUK W (1994), Studies on frozen doughs III. Some factors involved in dough weakening during frozen storage and freeze-thaw cycles. *Cereal Chemistry* **71**, 118–121.

JOHNSON J C, CLARKE B C and BHAVE M (2001), 'Isolation and characterisation of cDNAs encoding protein disulfide isomerases and cyclophilins in wheat', *J Cereal Sci*, **34**, 159–171.

KASARDA D D, BERNARDIN J E and NIMMO C C (1976), in *Advances in Cereal Science and Technology*, Pomeranz Y, ed., AACC, St Paul, MN, 277–302.

KASARDA D D, (1989), Gluten structure in relation to wheat quality. In *Wheat is unique* (Y. Pomeranz, ed.) pp. 158–236 (AACC: St Paul, MN).

KELLER R C A, ORSEL R and HAMER R J (1997), 'Competitive adsorption behaviour of wheat flour components and emulsifiers at an air-water interface', *J Cereal Sci*, **25**, 175–183.

KHAN K and BUSHUK W (1978), 'Glutenin: structure and functionality in breadmaking', *Bakers Digest*, **58**, 14–20.

KIM W, JOHNSON J W, GRAYBOSCH R A and GAINES C S (2003) 'Physicochemical properties and end-use quality of wheat starch as a function of waxy protein alleles'. *J Cereal Science* **37**, 195–204.

KNIGHT C A, CHENG C C and DE VRIES A L (1991), 'Adsorption of α-helical antifreeze peptides on specific ice crystal surface planes', *Biophys J*, **59**, 409–418.

LAAKSONEN T J and ROOS Y H (2001), 'Dielectric relaxations of frozen wheat doughs containing sucrose, NaCl, ascorbic acid and their mixtures', *J Cereal Sci*, **33**, 331–340.

LE BOTLAN D, RUGRAFF Y, MARTIN C and COLONNA P (1998), 'Quantitative determination of bound water in wheat starch by time domain NMR spectroscopy', *Carbohydrate Res*, **308**, 29–36.

LEE Y K, BÉKÉS F, GRAS P W, APPELS R and MORELL M (1999), The low molecular weight glutenin subunit proteins of primitive wheats IV. Functional properties of products from individual genes. *Theoretical and Applied Genetics* **98**, 149–155

LI S, DICKINSON L C and CHINACHOTI P (1998), 'Mobility of "unfreezable" and "freezable" water in waxy corn starch by ^{2}H and ^{1}H NMR', *J Agric Food Chem*, **46**(1), 62–71.

MARTINANT J P, CADALEN T, BILLOT A, CHARTIER S, LEROY P, BERNARD M, SAULNIER L and BRANLARD G (1998), 'Genetic analysis of water-extractable arabinoxylans in bread wheat endosperm', *Theor Appl Genet*, **97**, 1069–1075.

MATHEWS J (1989), 'Making frozen dough', *Bakers' Review*, November, 19–20, 22.

MENKOV N D, YANCHEV I G and ZHELYAZKOV (2002), 'Kinetics of wheat flour water uptake', *Nahrung*, **46**(2), 76–77.

MORELL M, TAMÁS L, APPELS R and BÉKÉS F (2000), 'Modified proteins', patent application August 1999, PCT number 00563.

MURRAY F R, SKERRITT J H and APPELS R (2001), A gene from the *Sec2 (Gli-R2)* locus of a wheat 2RS.2BL chromosomal translocation line. *Theor Appl Genet* **102**, 431–439.

NEMETH L J, PAULLEY F G and PRESTON K R (1996), 'Effects of ingredients and processing conditions on the frozen dough bread quality of a Canada western red spring wheat flour during prolonged storage', *Food Res Int*, **29**(7), 609–616.

O'BRIEN L, MORELL M, WRIGLEY C and APPELS R (2001), 'Genetic pool of Australian wheats', in *The World Wheat Book*, Bonjean A P and Angus W J eds, Lavoisier Publishing, London, Paris, New York, 611–648.

PAYNE P I, NIGHTINGALE M A, KRATTINGER A F and HOLT I M (1987), The relationship between the HMW glutenin subunit composition and bread-making quality of British grown wheat varieties. *J Science Food and Agriculture* **40**, 51–65.

QIN JI, VINCKEN J-P, SUURS L C J M and VISSER R G F (2003), Microbial starch-binding domains as a tool for targeting proteins to granules during starch biosynthesis. *Plant Mol Biol* **51**, 789–801.

RAHMAN S, LI Z, BATEY I, COCHRANE M P, APPELS R and MORELL M (2000), 'Genetic alteration of starch functionality in wheat', *J. Cereal Sci*, **31**, 91–110.

RÄSÄNEN J, BLANSHARD J M V, MITCHELL J R, DERBYSHIRE W and AUTIO K (1998), 'Properties of frozen wheat doughs at subzero temperatures', *J Cereal Sci*, **28**, 1–14.

RATH C R, GRAS P W, ZHEN Z, APPELS R, BÉKÉS F and WRIGLEY C W (1994), 'A prototype extension tester for two-gram dough samples', in *Proc 44th Australian Cereal Chem Conference*, Panozzo J F and Downie P G, eds, RACI Cereal Chem Division, North Melbourne, pp 122–126.

ROCHER A, CALERO M, SORIANO F and MÉNDEZ E (1996), 'Identification of major rye secalins as coeliac immunoreactive proteins', *Biochim Biophys Acta*, 1295, 13–22.

RUAN R R, WANG X, CHEN P L, FULCHER R G, PESHECK P and CHAKRABARTI S (1999), 'Study of water in dough using nuclear magnetic resonance', *Cereal Chem*, **76**(2), 231–235.

RUGRAFF Y L, DESBOIS P and LE BOTLAN D J (1996), 'Quantitative analysis of wheat starch-water suspensions by pulsed NMR spectroscopy measurements', *Carbohydrate Res*, **295**, 185–194.

SALCEDO G, PRADA J and ARAGONCILLO C (1979), 'Low MW gliadin-like proteins from wheat endosperm', *Phytochemistry*, **18**, 725–727.

SAPIRSTEIN H D and FU B X (1996), Characterization of an extra-strong wheat: Functionality of 1) gliadin- and glutenin-rich fractions, 2) total HMW and LMW subunits of glutenin assessed by reduction-reoxidation. Proc. 6th International Gluten Workshop, (C W Wrigley, ed.), pp. 302–306, Royal Australian Chemical Institute, North Melbourne, Vic.

SEARS E R (1954), 'The aneuploids of common wheat', *Missouri Agricultural Experimental Research Station Bulletin*, **572**, 1–58.

SHANI N, STEFFEN-CAMPBELL J D, ANDERSON O D, GREENE F C and GALILI G (1992), 'Role of the amino- and carboxy-terminal regions in the folding and oligomerization of wheat high molecular weight glutenin subunits', *Plant Physiol*, **98**, 433–441.

SHEWRY P R and MORELL M (2001), 'Manipulating cereal endosperm structure, development and composition to improve end-use properties', in *Advances in Botanical Research incorporating Advances in Plant Pathology*, Shewry P R, Lazzeri P A and Edwards K J, eds, Academic Press, San Diego, San Francisco, New York, Boston, London, Sydney, Tokyo, **34**, 165–236.

SHEWRY P R, HALFORD N G and TATHAM A S (1992), 'High-molecular-weight subunits of wheat glutenin', *J Cereal Sci*, **15**, 105–120.

SHIMONI Y, BLECHL A E, ANDERSON O D and GALILI G (1997), 'A recombinant protein of two high molecular weight glutenins alters gluten polymer formation in trangenic wheat', *J Biol Chem*, **272**, 15488–15495.

SOLOMON R G and APPELS R (1999), 'Stable, high-level expression of a Type I antifreeze protein in *Escherichia coli*', *Protein Expression and Purification*, **16**, 53–62.

SOLOMON R G and APPELS R (2001), 'Impact of biotechnology on the production of improved cereal varieties'. *Advances in Botanical Research* **34**, 289–300

TAKEMOTO Y, COUGHLAN S J, OKITA T W, SATOH H, OGAWA M and KUMAMARU T (2002), 'The rice mutant *esp2* greatly accumulates the glutelin precursor and deletes the protein disulfide isomerase', *Plant Physiol*, **128**, 1212–1222.

TAMÁS L, BÉKÉS F, GREENFIELD J, TATHAM A S, GRAS P W, SHEWRY P R and APPELS R (1998), 'Heterologous expression and dough mixing studies of wild-type and mutant C hordeins', *J Cereal Sci*, **27**, 15–22.

TAMÁS L, GRAS P W, SOLOMON R G, MORELL M K, APPELS R and BÉKÉS F (2002), 'Chain extension and termination as a function of cysteine content and the length of the central repetitive domain in storage proteins', *J Cereal Sci*, **36**, 313–325.

TAMÁS L, GREENFIELD J, HALFORD N G, TATHAM A S and SHEWRY P R (1994), 'A beta-turn rich barley seed protein is correctly folded in *Escherichia coli*', *Protein Expression and Purification*, **5**, 357–363.

TATHAM A S, GREENFIELD J A, HALFORD N G and SHEWRY P R (1998), 'Expression and functional analysis of M_r 58000 peptides derived from the repetitive domain of high molecular weight glutenin subunit 1Dx5', *J Cereal Sci*, **27**, 209–215.

TKACHUK R and HLYNKA I (1968), 'Some properties of dough and gluten in D_2O^1', Vol 80–87.

TURNBULL K M and RAHMAN S (2002), Endosperm texture in wheat. *J Cereal Sci*, **36**, 327–337.

UTHAYAKUMARAN, S., STODDARD, F.L., GRAS, P.W. and BÉKÉS, F. (2000), Optimized methods for incorporating glutenin subunits into wheat dough for extension and baking studies. *Cereal Chemistry* **77**, 731–736.

VAAG P, BECH L, CAMERON-MILLS V and SORENSEN M (2000), '17 kDa foam protein', *International patent application number*, W0 00/14237.

VERAVERBEKE, W.S., VERBRUGGEN, I.M. and DELCOUR, J.A. (1999), Effects of increased HMW-GS content of flour on dough mixing behavior and breadmaking. *J. Agric. Food Chem.* **46**, 4830–4835

WESLEY I J and BLAKENEY A B (2001), 'Investigation of starch-protein-water mixtures using dynamic near infrared spectroscopy', *J Near Infrared Spectrosc*, **9**, 211–220.

WESLEY I J, LARSEN N, OSBORNE B G and SKERRITT J H (1998a), 'Non-invasive monitoring of dough mixing by near infrared spectroscopy', *J Cereal Sci*, **27**, 61–69.

WESLEY I J, LARSEN N G, OSBORNE B G and SKERRITT J H (1998b), 'Monitoring of dough properties', *PCT patent application number*, W0 98/48271.

YAMAMORI M and QUYNH N T (2000), 'Differential effects of Wx-A1, -B1, -D1 protein deficiencies on apparent amylose content and starch pasting properties in common wheat. *Theoretical and Applied Genetics* **100**, 32–38.

ZAO X C, SHARP P J, CROSBIE G, BARCLAY I, WILSON R, BATEY I L and APPELS R (1998), 'A single genetic locus associated with starch granules and noodle quality in wheat. *J Cereal Science* **27**, 7–13.

ZOUNIS S, QUAIL K J, WOOTTON M and DICKSON M R (2002a), 'Studying frozen dough structure using low-temperature scanning electron microscopy', *J Cereal Sci*, **35**, 135–147.

ZOUNIS S, QUAIL K J, WOOTTON M and DICKSON M R (2002b), 'Effect of final dough temperature on the microstructure of frozen bread dough', *J Cereal Sci*, **36**, 135–146.

12

The nutritional enhancement of wheat flour

C. M. Rosell, IATA-CSIC, Spain

12.1 Introduction

Cereal grains have been used as a basis of the human diet from ancient times. Today, the world cereal production is estimated at around 2086 million tonnes, of which around 583 million tonnes is wheat. Wheat consumption is around 70 kg per capita per year (Fig. 12.1) (FAOSTAT, 2002), with consumption being greatest in Europe. In Africa wheat consumption has decreased to 45 kg per capita per year owing to the high intake of other cereals such as oat, millet and sorghum. Such data help us to appreciate that the wheat provides a great part of the required energy in the human diet. In Europe, wheat and wheat-based foods supply up to 800 calories per person per day (Fig. 12.2), which means that up to 25% of the daily energy intake is provided by wheat and its products. In other developed countries, wheat-based products provide up to 23% of the daily energy intake, while in the developing countries that percentage decreases to 18% owing to the high consumption of rice and other cereals. Wheat is also a cheap source of protein compared with animal proteins. In Europe, wheat supplies up to 25 g of proteins per person per day, which constitutes about 26% of the daily protein intake. Similar values are found in other developed countries, where wheat provides 23% of the daily protein intake, and that value falls to 20% in the developing countries.

Although the consumption of wheat and wheat-based products has decreased in the last decade because of social changes (industrialisation, high per capita income, fast- and takeaway foods) favouring the consumption of fat and proteins from animal sources, in recent years a reversed trend has been observed due to growing nutrition awareness. Therefore, wheat and related products are still considered as good providers of nutrients.

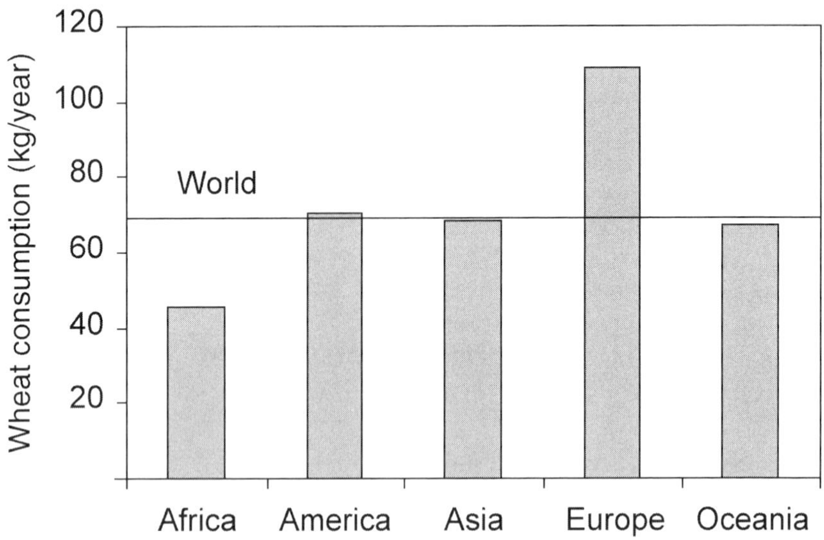

Fig. 12.1 Wheat consumption (kg per capita per year) in the different continents. Source: FAOSTAT, 2002.

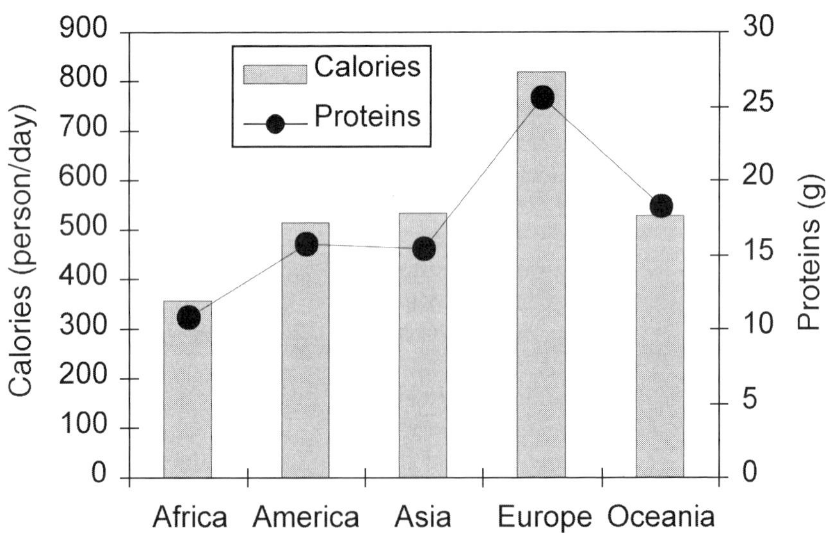

Fig. 12.2 Contribution of wheat and derived wheat product to the daily energy and protein intake. Source: FAOSTAT, 2002.

12.2 The nutritional value of wheat

The consumption of wheat-based products is highly recommended owing to their excellent nutritional profiles and their contribution to the diet:

- complex carbohydrates;
- dietary fibre;
- low content in fat (without containing cholesterol);
- minerals, especially calcium, phosphorus, iron and potassium, B vitamins.

Wheat is considered an important source of energy. It provides between 1220 and 1450 kJ per 100 g of cereal (Fig. 12.3). In addition it is an important supplier of different nutrients necessary for a healthy diet. The complex carbohydrates, mainly starch, are the major components of wheat (61–65%); but it is also an excellent source of dietary fibre (9–12%).

The protein content of wheat is one of the highest of the cereals. It ranges between 10 and 15%, although the nutritional value of the cereal proteins is lower than the animal proteins owing to a deficiency in some essential amino acids, mainly lysine. The fat content is very low (1.7–2.0%) and mainly comprises polyunsaturated fats with an absence of cholesterol, in contrast to the saturated forms present in animal fats.

Wheat is also a source of B vitamins, namely thiamine (B1), riboflavin (B2) and niacin (B6), and minerals. Important quantities of calcium, phos-

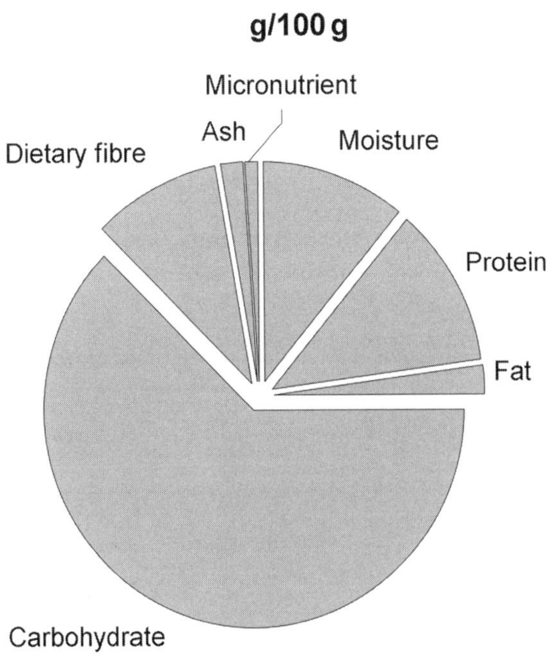

Fig. 12.3 Nutritional value of wheat.

phorus, iron, sodium, magnesium and potassium are found in the aleurone layer of wheat. Special consideration must be paid to phosphorus, since the largest part of the total is not present in an inorganic form, but is found as phytic acid or myo-inositol hexaphosphate. This form is considered to be an anti-nutritional compound because of its adverse effects on the bioavailability of minerals. In the gastrointestinal tract, phytic acid or phytate forms insoluble complexes with multivalent cations, especially Zn^{2+}, Ca^{2+}, Mg^{2+} and Fe^{3+} (Davies and Nightingale, 1975; Erdman, 1979; Zemel and Shelef, 1982; Munoz, 1985), decreasing the bioavailability of those minerals. Therefore, from the nutritional point of view, the major part of phosphorus present in wheat is not readily available and it also promotes a reduction in the bioavailability of the rest of minerals present in wheat and wheat flours.

12.2.1 Effects of processing on the nutritional value of wheat-based products

Wheat is not directly consumed as a grain cereal and processing is generally required to transform the grains into a powder form which may have a modifying effect on the nutritional pattern described above.

The distribution of the nutrients within the wheat grain is not uniform. The germ and the aleurone layers are particularly rich in minerals and vitamins. Therefore, wheat processing, like flour milling, can result in significant change in the nutritional value of the products derived from wheat. The nutrients concentrated in the bran layers and germ will be removed during wheat milling to provide white flour. The data given in Table 12.1 show all of the nutrients during the extraction of the endosperm from the wheat grain. Different extraction rates are shown and it can be seen that only the starch, which is abundant in the endosperm, remains unaffected. The most affected compound is the crude fibre followed by the ash (minerals). Therefore the removal of the outer parts of the wheat grain involves an increase in food acceptance to the detriment of the nutritional value. Obviously, a nutritional

Table 12.1 Effect of milling on the nutritional composition of wheat flours at different extraction rates

Extraction rate (%)	Protein (%)	Fat (%)	Ash (%)	Crude fibre (%)
100	100 (13.8)	100 (2.5)	100 (1.6)	100 (2.2)
90	99.3	68.3	49.0	15.2
80	97.1	57.5	38.7	6.0
70	93.5	46.4	26.5	0.5

Data in parenthesis indicate the composition referred to percentage in grain.
Source: Benedito (1999).

improvement of the wheat flours can be reached by increasing the extraction rate and this is seen in the contrast between white and wholemeal flours.

12.3 Increasing the nutritional value of wheat flour

In order to partially recover the original nutritional value of wheat, and also to use wheat and related products for the delivery of micronutrients, different approaches have been used. In its simplest and well-recognised form it consists of putting back the compounds removed during milling. The enrichment of wheat flour with different nutrients is regulated differently in individual countries. The Economic and Social Department of FAO has detailed a review about the legislative status of wheat flour fortification (Table 12.2; FAO, 2002). There are some countries, such as Canada and the UK, where the fortification of wheat flour is mandatory. Others, such as New Zealand, Finland and Norway have laws prohibiting the addition of micronutrients to wheat flour.

12.3.1 Addition of compounds removed during milling: B-vitamins and minerals

The term 'enriched flour' was born during World War II, when the American government in the interest of winning the war decided to improve the diet of its soldiers by using bread as a carrier for different nutrients. At the beginning, the enrichment consisted of the three major B vitamins (thiamine, riboflavin and niacin) and iron, in spite of little scientific information about the real nutritional improvement likely to be achieved. Later, studies were conducted to verify the benefits of the flour enrichment and the improvement of the nutrient intake. The American Chemical Association (1972) analysed the Food and Drug Administration (FDA) proposal to increase the level of iron in flours, owing to the prevalence of iron-deficiency anaemia in the USA. They concluded that this proposal would be more beneficial rather than constitute a hazard to human health.

The supplementation of B-vitamins in wheat flour is faced with the problem of instability of the enriching compounds during storage. A study carried out comparing the efficiency of the enrichment at the mill or at the bakery showed that the vitamin content decreased by 33.3–58.1% when wheat flour was enriched at the mill and was greater than a 17.0–38.7% reduction observed when the vitamins were added to bread dough (Stepanova *et al.*, 1988).

Hallberg *et al.* (1986) analysed the bioavailability of commercially available iron powders used for flour enrichment. By specific labelling techniques, they found that the relative bioavailability of the iron was unexpectedly low and highly dependent on the meal served (phytates, calcium and polyphenols decrease the bioavailability of iron), arriving at the conclusion that it would be necessary to reconsider the use of elemental iron powders for flour enrichment. Nevertheless, the supplementation of wheat flour with iron is still a common practice adopted in some countries as a solution for iron deficiency. From 1997,

Table 12.2 Legislation in different countries concerning wheat flour enrichment. Values are expressed in mg/kg

Country	Law status	Legislation	Known nutrient deficiences
Angola	–	No	Vit A, I, Fe
Australia	Voluntary	Thiamine: >6.4; B_1: 15, B_2: 11; niacin: 70; B_6: 11; E: 72; Folate: 2.9; Fe: 86; Mg: 2300; Zn: 52	–
Bolivia	–	No	Vit A, I, Fe
Canada	Mandatory	B_1: 4.4–7.7, B_2: 2.7–4.8; niacin: 35–64; Fe: 29–43. The following are voluntary: pantothenic acid: 10–13; B_6: 2.5–3.1; folate: 0.4–0.5; Ca: 1100–1400; Mg: 1500–1900	–
Costa Rica	–	No	I, Fe
Cuba	Voluntary	No	Fe, Ca, Vit B_1, Vit C, Vit A
Ecuador	Voluntary	B_1: >2; B_2: >2.6; niacin: >35.3; Fe: >24; Ca: > 1100	–
El Salvador	–	No	Vit A, I, Fe
Ethiopia	–	Draft standard proposal	Vit A, I, Fe
Finland	Prohibited	Restoration allowed	–
Gambia	–	No	I, Fe, Ca, folate, B_2
Haiti	–	No	–
Honduras	Voluntary	B_1: 4.4, B_2: 2.6; niacin: 35.2; Fe: 28.7; Ca: 1100	Vit A, I, Fe
Hungary	Voluntary	A serving must contain 1/3 of RDA	–
Malta	Voluntary	Mandatory for brown or wholemeal flour: B_1, B_2: 2.4; niacin: 16; Fe: 16.5	
Mauritania	–	No	I, Fe, folate
Morocco	–	No	Vit A, I, Fe
New Zealand	Prohibited	In process of reviewing the legislation	–
Norway	Prohibited	–	–
Pakistan	–	No	B-Vits, I, Fe
Peru	–	No	–
Philippines	Voluntary	Recommend the addition of Fe and B-Vits to 1/3 of RDA	Vits A, B_1, B_2, C, and I, Fe, Ca
South Africa	–	–	–
Sweden	Voluntary	Thiamine-HCl: 4–8; B_2: 1.5-3.0; nicotinic acid: 40–80; piridoxine HCl: 3.5–7.0; Fe: 65-90	–
Switzerland	Voluntary	B_1: 4.4; B_2: 2.0; niacin: 50; Fe: 29	–

Table 12.2 (continued)

Country	Law status	Legislation	Known nutrient deficiences
Tanzania	–	No	Fe, I, folate, Vit A
Turkey	Voluntary	No	Fe, I, Ca, Vits B_2, B_6, C, D
UK	Mandatory	B_1: >2.4; niacin: 16; Fe: 16.5; Ca: 940–1560	–
Uruguay	–	No	Fe, I, Vit D
Vietnam	Voluntary	No	Vit A, I, Fe

the Organización Panamericana de la Salud (OPS) has been supporting the supplementation of iron to the diet by using wheat flour as a vehicle because of its low cost and widespread consumption. In recent years, different countries from South and Central America have implemented measures for reducing the incidence of iron-deficient anaemia after the successful results obtained in Chile, where wheat flour has been fortified for 30 years.

A problem with iron fortification emerges from its low bioavailability, colour and instability during storage. Ferrous sulfate has a high solubility and high bioavailability but often leads to the development of unpleasant colours and flavours owing to reactions with other components of the food matrix. Different formulations of iron, such as the use of ferrous sulfate coated with a layer of crystallised $FeSO_4.7H_2O$ (Deng *et al.*, 1994) or NaFeEDTA (Hurrell *et al.*, 2000) have been developed in order to solve the adsorption problem and to enhance the bioavailability of carbonyl iron, which is highly dependent on the compounds served at the same meal.

Other minerals which have been added to wheat flour are zinc and calcium. Wang and Tang (1995) reported the effect of the fortification with calcium and zinc lactate on the rheological and baking properties of wheat flour. They showed that the supplementation improved the dough expansion without any adverse effect on the quality of the resulting bread, which retained between 87 and 93% of the added calcium and zinc.

12.3.2 Vitamin supplementation of wheat flour

Apart from the addition of compounds removed during milling, there are some studies showing the beneficial effects of using wheat flour as a carrier of other micronutrients such as vitamins and minerals. Vitamin A has been added to wheat flour buns (pandesal) and an increase in the initial serum retinol concentrations was observed in school-age children after a daily consumption of these buns for 30 weeks (Solon *et al.*, 2000). However, the usefulness of adding vitamin A to wheat flour products is limited because of its poor stability in the presence of oxygen or air. The loss of vitamin A during 70 minutes of baking at 200 °C could be greater than 50% (Cakirer and Lachance, 1975).

In recent years some studies have confirmed the role of folates in congenital malformations and the development of chronic diseases in later life such as Alzheimer disease (Czeizel and Dudas, 1992) have encouraged the addition of folic acid to cereal based foods (Rader *et al.*, 2000; Molley and Scott, 2001). Data related to the effectiveness of this fortification programme to ensure that women have the right folate intake to reduce their risk of having pregnancy affected by a neuronal tube birth defect will not be available for several years. However, in the meantime some short-term studies allow an estimate of the potential positive effects of the folate supplementation on the reduction of total homocysteine concentrations (Malinow *et al.*, 1998; Jacques *et al.*, 1999). It is, however, premature to consider that such a reduction would cause a decrease in the incidence of cardiovascular disease. The US government has passed a mandatory regulation covering folic acid fortification, of all flour (Food and Drug Administration, 1996a,b,c). In the European Union no mandatory fortification policy has been adopted in order to regulate the addition of folic acid.

In the USA, the FDA approves the supplementation of four B vitamins (thiamine, niacin, riboflavin and folic acid) and iron to wheat flour, to provide the so-called 'enriched flour', although this practice is not regulated by Federal law and depends on the State legislation. The UK regulations require that the nutrients removed with the bran during the milling must be replaced in all types of flours except wholemeal. White and brown flour must have thiamine, niacin, iron and calcium, although the addition should not be done in amounts that might be harmful to people.

12.3.3 Mineral fortification of wheat flour

During milling a high proportion of minerals, mainly located in the outer layers, is lost resulting in a reduction in the nutritional value of wheat flour. A further supplementation with minerals could be carried out by:

- directly adding minerals, such as iron and calcium;
- modifying the milling process;
- using organic material as a mineral source.

The first point has been summarised earlier. Regarding the second it is easy to understand that if the minerals are located in the outer grain layers, then higher extraction yields will provide nutritionally richer wheat flours, although with poorer breadmaking properties. Modification of the milling process is possible to address this issue.

The wheat milling process consists of grinding and separating various grain components. Grinding is done on break sizing and reduction rolls. The smaller particles of the ground material obtained after grinding are called break middlings, which are a mixture of pure endosperm, endosperm attached to particles of bran and some small particles of bran. One way in which the resulting wheat flour could be enriched could consist of mixing between 5 and 50% middling (from the first and second breaks) with wheat flour (straight, patent, break or clear) to obtain a mineral enrichment of up to 20% with respect to the initial flour content of calcium, zinc, iron, manganese and phosphorus without changing the fermentation and baking characteristics (Maldonado, 2000).

A third way to enrich wheat flour with minerals is by indirectly adding minerals, for example, by blending small amounts of seaweed extracts such as *Laminaria japonica* into wheat flour, to act as a source of iodine (Miki *et al.*, 1993a,b).

12.3.4 Lysine enrichment of wheat flour

Wheat is a widely accepted source of proteins, but compared with animal proteins, the essential amino acid composition of cereal proteins is of considerably lower nutritional value because of the lower content of lysine. In consequence, one of the main purposes of nutrition programmes is to determine the necessary amount of lysine to improve the bioavailability value of the wheat proteins without producing adverse effects. Graham *et al.* (1969, 1971) conducted studies to establish the appropriate amount of lysine to be added in lysine-enriched flours in those areas where wheat flour is the main source of proteins. At least 0.2% of lysine was recommended for wheat flour enrichment in these cases.

The use of fortified flour with up to 0.3% lysine in breadmaking does not modify the organoleptic values for appearance, texture, flavour and taste, and the overall acceptability of the resulting breads (Yasoda-Devi and Geervani, 1979). The nutritional improvement, measured as the relative protein value, was confirmed in adult rats, who showed better growth when fed with lysine-enriched wheat flour (Mekhael *et al.*, 1989).

12.3.5 Nutritional enrichment of wheat flour through the tempering and the reduction of the antinutritional compounds present in wheat

The initial step in the flour milling process is cleaning. After the removal of foreign material, it is common to use a tempering process in order to facilitate the easier removal of the bran layers and to mellow the starchy endosperm. During this step, water diffuses through the pericarp into the endosperm, and this step could be an ideal vehicle for introducing enzymes into the kernel. Haros *et al.* (2002a,b) reported the addition of carbohydrases into the tempering water which gave a wheat flour with better breadmaking performance. The resulting breads had high specific volume and showed reduced staling during storage. The same method can be applied for introducing enzymes, such as phytase, with the aim of reducing the phytate (myo-inositol hexaphosphate) content of wholewheat flour. Phytase is an esterase that catalyses the stepwise hydrolysis of phytates to phosphate and inositol via penta- to monophosphates. The wheat flour obtained from the wheat tempered in the presence of fungal phytase showed lower phytate content than the wholewheat flour tempered with water

Table 12.3 Phytate content in bread dough from wheat flours tempered in the presence of different phytase concentrations

Dosage of phytase (U/100 g kernel)	Phytate content (%)
0	100.0
160	85.9
320	77.8
640	73.7
960	71.1

(Table 12.3). This result has several benefits from the nutritional point of view. The reduction of the phytate content, and in consequence the increase in the phosphorus bioavailability, and an improvement in the adsorption of different cations such as Ca^{2+}, Fe^{3+}, Zn^{2+} and Mg^{2+} (Harland and Harland, 1980; Lopez *et al.*, 2000, 2001). This treatment also yields wheat flour with better breadmaking performance, because the hydrolysis of phytates releases calcium which is necessary for the activation of α-amylase, as a consequence of the hydrolysis of phytates high α-amylase activity is displayed (Haros *et al.*, 2000, 2001a).

12.4 Improving the nutritional value of wholewheat flours

The increasing awareness of the potential benefits of high-fibre diets has promoted a growing interest for the consumption of whole-grain breads and bran breads. A low intake of complex carbohydrates and dietary fibre has been related with a high incidence of several diseases, including coronary heart disease, diabetes, obesity and certain types of cancer. Wholewheat flours are an important source of fibre, complex carbohydrates, proteins, minerals and vitamins. Nevertheless, along with these nutritional benefits, wholewheat flours contain significant amounts of undesirable compounds, such as phytates.

An improvement to the nutritional value of the wholewheat flour can be obtained during the breadmaking process. This improvement comes from the reduction of the anti-nutritional compounds, phytates, in the wholewheat flour. Phytates are hydrolysed by phytases, enzymes that are naturally present in cereals and also present in bacteria, yeast and fungi. During breadmaking, the phytate content decreases as a result of the endogenous phytase's activity, although this reduction is insufficient to lead a significant improvement on the mineral bioavailability.

The extent of phytase activity during bread performance depends on the wheat flour extraction rate, the proofing temperature and time, dough pH, and the amount of yeast (Turk and Sandberg, 1992; Turk *et al.*, 1996; Fernandez *et al.*, 2002). Therefore, it would be possible to control the phytase activity by modifying the process conditions, and in consequence the resulting phytate content (Table 12.4). The addition of malted flour provides extra phytase

Table 12.4 Effect of different breadmaking conditions on both the phytase activity and phytate content of whole-wheat dough

Process variables	Phytase activity (U/g)	Phytate content (mg/g)
Control	24.0	17.9
Malted flour addition (0.15%)	26.5	15.1
Inactive sourdough addition (1.25%)	25.0	15.2
Lactic acid addition (0.13%)	28.0	17.8
Proofing temperature (40 °C)	26.0	14.5
Fungal phytase addition (3000 U/g)	38.0	10.8

activity and leads to a decrease of the phytate content. However, the greatest reduction was observed by increasing the proofing temperature and by adding fungal phytase to the wholewheat flour. Haros *et al.* (2001a) have reported a nutritional improvement produced by the addition of different concentrations of fungal phytase to breadmaking, along with quality benefits such as acceleration of proofing, an increase of bread specific volume and a softness effect on the bread crumb (Haros *et al.*, 2001b).

A different approach for decreasing the phytate content consists of increasing the yeast concentration; although different results have been obtained with this strategy. Harland and Harland (1980) reported a half reduction of the phytate content by increasing the yeast concentration, while Tangkongchitr *et al.* (1981) did not find any additional phytate degradation from modifying the yeast amount.

12.5 Future trends: protein supplementation and fibre enhancement

The simplest way for achieving a nutritional enrichment of wheat flour-based products is through the addition of different ingredients to the flour at the start of the breadmaking process. Numerous ingredients have been added to wheat breads with the main aim of increasing their protein content and some approaches have been made to augment the dietary fibre supply or the complex carbohydrate content. Some of these approaches have been selected for illustrating ways of obtaining the nutritional enhancement of wheat-based products.

12.5.1 Protein supplementation of wheat breads

Different ingredients (animal or vegetal origin) can be used as a source of proteins and added to wheat flour to supplement the protein content of the original wheat flour. Some countries have adopted this strategy for increasing the protein intake of their populations. For instance, in 1983 the Pan American

Health Service developed an enriched-bread with cereal protein concentrates, obtained from cereal by-products after starch extraction. This bread was commercialised as 'Pan de Vida' in Honduras (Hammond, 1983). However, before adopting a policy of supplementing bread with different sources of protein it is necessary to analyse the economic viability of that measure. For example, a study carried out in Ecuador revealed that the addition of soy flour to enhance the nutritional value of wheat bread was not appropriate as a nutrition intervention (Gormely, 1978).

Dried buttermilk and dried skim milk have been added at levels up to 6% to obtain nutritional enrichment of the breads and to improve flavour (Mostafa *et al.*, 1982). A nutritional improvement of wheat breads has also been achieved by adding 1–3% pasteurised sweet cheese whey solids and no change of quality in taste or texture was observed (Guy, 1984). Even fish protein concentrates have been used as a source of proteins, and mainly for increasing the lysine content in wheat breads (Terra *et al.*, 1976). In Morocco, where bread constitutes a staple food, adding 5% fish protein concentrate yielded differences in taste which were detected but the bread was still acceptable (Servais, 1981).

Among vegetable sources, chickpea flour (*Cicer arietinum*), has been added at levels of up to 15% to wheat flour without seriously affecting the resultant bread quality. This practice resulted in an additional increase of 3.5% in the protein content and the chickpea flour also increased the fibre, ash and fat contents in the final breads (Figuerola *et al.*, 1987). The supplementation of 5% edible grade cottonseed flour was recommended to increase the protein content in breads without having a significant deleterious effect on their organoleptic qualities of bread (El-Shaarawy and Mesallam, 1987). Lentil protein concentrates and sweet lupin flour also have been added to enhance the nutritional value of wheat breads (Yanez *et al.*, 1985), but a decrease in the overall acceptability was observed when adding lentil protein concentrates (Khairy *et al.*, 1986). Some reports have described the use of soy flour or soy protein concentrates for raising the protein content of breads, with a simultaneous balance of the essential amino acid composition and an increase in the caloric value (Tsen and Hoover, 1973; Ranhotra *et al.*, 1974).

A comparison study of proteins from different sources (whey protein concentrate, dried skim milk and fish protein concentrate) or direct supplementation with lysine revealed that the most efficient method to enhance the essential amino acid content, mainly lysine and threonine, of wheat breads was through the addition of whey protein concentrates to wheat flour. This was confirmed by rat growth studies (Tsen and Hoover, 1973).

Abd-El-Kader *et al.* (1984) reported a different approach of raising the total protein content by adding up to 15% wheat flour from germinated grains. Such additions led to a nutritional improvement of the bread through an increase in B-vitamins content, the modification of amino acid composition in favour of some essential amino acids, and the reduction of the phytic acid content.

12.5.2 Fibre enhancement of wheat breads

Recently the importance of consuming dietary fibre has increased owing to its relation with the reduction of blood cholesterol levels and the incidence of colon cancer. Cereals are the major source of dietary fibre, and in this case we refer only to wheat because of its high consumption. Wholewheat bread meets the requirements for dietary fibre intake; however, the largest proportion of bread consumed is of a white type (Barber *et al.*, 1983). There have been different approaches to increasing the dietary fibre content of wheat bread; in this section some of them are presented.

Wheat bread can be enriched with dietary fibre by adding wheat bran (Ranhotra *et al.*, 1990; Sidhu *et al.*, 1999), hydrocolloids, such as guar gum, modified celluloses and β-glucans (Pomeranz *et al.*, 1977; Knuckles *et al.*, 1997). However, the addition of such compounds promotes a detrimental effect on bread quality in terms of loaf volume, texture, colour and sometimes taste. Recently, Rosell *et al.* (2001) reported on the use of hydroxypropylmethyl cellulose as a potential additive for breadmaking. It gave better loaf volume, softer breadcrumb and had a retarding effect on bread staling.

In the search for reducing the negative effects of fortified wheat flours on the bread, different approaches have been developed. They include the addition of carbohydrases to degrade the carbohydrate cell walls, which yields an increase in the dough water adsorption and an improvement of the bread quality (Laurikainen *et al.*, 1998; Haros *et al.*, 2002a), and supplementation with different commercial fibres. There are several fibres from different sources on the market, which raise the dietary fibre content of bread with a simultaneous decrease in the detrimental effect associated with the use of fibres. Barber *et al.* (1981) pioneered the addition of different fractions from rice bran to the wheat bread formulation and obtained breads of acceptable quality. Recently, Abdul-Hamid and Siew Luan (2000) reported on the use of deffated rice bran as a source of dietary fibre in breadmaking, but reduced loaf volume and increase of crumb firmness were obtained. Other commercial fibres used in breadmaking include carob, inulin and pea (Jinshui Wang *et al.*, 2002). Their addition to breadmaking led to breads of acceptable quality. Softer crumb was obtained with the addition of carob and pea fibres (Jinshui Wang *et al.*, 2002).

12.5.3 Future developments

Today the term 'functional foods' seems to be old-fashioned; wheat products and mainly wholemeal wheat products are one of the best carriers for different nutritional compounds. In this chapter we have described the use of wheat and its products as vehicles for the delivery of vitamins, minerals, proteins, complex carbohydrates and fibres. It is also possible to use them as vehicles for delivery of ω-fatty acids and probiotic compounds, such as fructooligosaccharides and inulin. Such approaches are likely to be implemented in the immediate future.

12.6 Sources of further information and advice

Food Quality and Nutrition (1977), Downey W K. Dublin, Applied Science Publisher, 565–589.

Fundamentals of Nutrition (1978), Lloyd L E, McDonald B E and Crampton E W. San Francisco, W H Freeman and Company.

Handbook of Food Allergies (1987), Breneman J C. New York, Marcel Dekker Inc.

Nutrición Humana (1979), Anderson L, Dibble M, Mitchell H S and Rynbergen H J. Barcelona, Ediciones Bellaterra.

Nutritional Quality of Cereal Grains: Genetic and Agronomic Improvement (1987), Olson R A and Frey K J. Wisconsin, American Society of Agronomy Inc., Crop Science Society of America Inc, Soil Science Society of America Inc.

Ranhotra G S (1994), 'Wheat: contribution to food supply and human nutrition', in Bushuk W and Rasper V F, *Wheat: Production, Properties and Quality*, Glasgow, Blackie Academic and Professional.

12.7 References

ABD-EL-KADER M A, MORAD M M, EL-BADAWI A A, ASKAR A and OMRAN H T (1984), 'Nutritional value and baking behaviour of wheat flour enriched with germinated wheat, corn or soybeans', *Getreide Mehl und Brot*, **38**(7), 206–210.

ABDUL-HAMID A and SIEW LUAN Y (2000), 'Functional properties of dietary fibre prepared from deffated rice bran', *Food Chem*, **68**, 15–19.

AMERICAN CHEMICAL ASSOCIATION (1972), 'Iron in enriched wheat flour, farina, bread, buns and rolls', *J Am Med Assoc*, **220**(6), 855–859.

BARBER S, BENEDITO C and MARTÍNEZ J (1981), 'Rice bran proteins. II Potential value of rice bran fractions as protein food ingredients', *Rev Agroquim Tecnol Aliment*, **21** (2), 247–258.

BARBER S, BENEDITO C and LLACER MD (1983), 'Dietetic fiber content, sensory attributes of quality and chemical composition of commercial whole wheat flour bread', *Rev Agroquim Tecnol Aliment*, **23**(1), 119–131.

BENEDITO C (1999), 'Cereales y derivados', in Hernandez Rodriguez M and Sastre Gallego A, *Tratado de nutrición*, Madrid, Díaz de Santos, 401–411.

CAKIRER O M and LACHANCE P A (1975), 'Added micronutrients: their stability in wheat flour during storage and the baking process', *Bakers' Digest*, **49**(1), 53–57.

CZEIZEL A E and DUDAS I (1992), 'Prevention of first occurrence of neural tube defects by periconceptional vitamin supplementation', *New England J Med*, **327**, 1832–1835.

DAVIES N T and NIGHTINGALE R (1975), 'The effect of phytate on intestinal absorption and secretion of zinc and wholebody retention of zinc, copper, iron and manganese in rats', *J Nutr*, **34**, 243–247.

DENG S Y, GEN J Q, LIU S G and DENG B (1994), 'Iron enrichment of flour for prevention of iron deficient anaemia. I. Selection and preparation of iron supplement and test on the stability of iron enriched flour', *J Chinese Cereals Oils Assoc*, **1**, 16–22.

EL-SHAARAWY M I and MESALLAM A S (1987), 'Feasibility of Saudi wheat flour enriched with cottonseed flour for bread making', *Zeitschrift für Ernaehrungswissenschaft*, **26**(2), 100–106.

ERDMAN J W (1979), 'Oilseed phytates: Nutritional implications', *J Am Oil Chem Soc*, **56**, 736–741.

FAO (2002), Economic and Social Department. http://www.fao.org/WAICENT/FAOINFO/ECONOMIC/ESN/fortify/reguls.htm

FAOSTAT (2002), *Food Balance Sheet*, 2000, Food Agric Org, UN Rome.

FERNANDEZ A, HAROS M and ROSELL C M (2002), 'Nutritional improvement of whole wheat bread through the phytases activity during breadmaking', *Proceeding of the European Symposium on Enzymes in Grain Processing 3 (ESEGP-3)*, Belgium.

FIGUEROLA R F, ESTEVEZ A A and CASTILLO V E (1987), 'Supplementation of wheat flour with chickpea (*Cicer arietinum*) flour. I. Preparation of flours and their breadmaking properties', *Arch Latinoamericanos Nutricion*, **37**(2), 378–387.

FOOD AND DRUG ADMINISTRATION, USA (1996a), 'Food standards: amendment of standards of identity for enriched grain products to require addition of folic acid: final rule (21 CFR Parts 136, 137 and 139)', *Federal Register*, **61**, 8781–8797.

FOOD AND DRUG ADMINISTRATION, USA (1996b), 'Food standards: food labelling: health claims and label statements: folate and neural tube defects: final rule (21 CFR Part 101)', *Federal Register*, **61**, 8752–8781.

FOOD AND DRUG ADMINISTRATION, USA (1996c), 'Food additives permitted for direct addition to food for human consumption; folic acid (folacin); final rule, (21 CFR Part 172)', *Federal Register*, **61**, 8797–8807.

GORMELY P J (1978), 'Are high protein foods economically efficient? The case of soy fortified wheatflour in Ecuador', *Food Policy*, **3**(4), 280–288.

GRAHAM G G, PLACKO R P, ACEVEDO G, MORALES E and CORDANO A (1969), 'Lysine enrichment of wheat flour: evaluation in infants', *Am J Clin Nutr*, **22** (11), 1459–1468.

GRAHAM G G, MORALES E, CORDANO A and PLACKO R P (1971), 'Lysine enrichment of wheat flour: prolonged feeding of infants', *Am J Clin Nutr*, **24**, 200–206.

GUY E J (1984), 'Evaluation of the bread baking quality and storage stability of 12% soy fortified wheat flour containing sweet cheese whey solids', *Cereal Chem*, **61**(2), 83–88.

HALLBERG L, BRUNE M and ROSSANDER L (1986), 'Low bioavailability of carbonyl iron in man: studies on iron fortification of wheat flour', *Am J Clin Nutr*, **43**(1), 59–67.

HAMMOND N (1983), 'Utilization of wheat protein concentrates in baked products in Central America', *Develop Food Sci,* **5B**, 1069–1074.

HARLAND B F and HARLAND J (1980), 'Fermentative reduction of phytate in rye, white and whole wheat breads', *Cereal Chem*, **57** (3), 226–229.

HAROS M, ROSELL C M and BENEDITO C (2001a), 'The use of fungal phytase to improve breadmaking performance', *J Agric Food Chem,* **49** (11), 5450–5454.

HAROS M, ROSELL C M and BENEDITO C (2001b), 'Phytase as potential bread-making additive', *Eur Food Res Technol*, **213**(4–5), 317–322.

HAROS M, ROSELL C M and BENEDITO C (2002a), 'Improvement of flour quality through carbohydrases treatment during wheat tempering', *J Agric Food Chem*, **50**, 4126–4130.

HAROS M, ROSELL C M and BENEDITO C (2002b), 'Effect of different carbohydrases on fresh bread texture and bread staling', *Eur Food Res Technol*, **215**, 425–430.

HURRELL R F, REDDY M B, BURRI J and COOK J D (2000), 'An evaluation of EDTA compounds for iron fortification of cereal based foods', *Br J Nutr*, **84**(6), 903–910.

JACQUES P F, SELHUB J, BOSTOM A G, WILSON P W F and ROSENBERG I H (1999), 'The effect of folic acid fortification on plasma folate and total homocysteine concentrations'. *New England J Med*, **340**(19), 1449–1454.

JINSHUI WANG, ROSELL C M and BENEDITO C (2002), 'Effect of the addition of different fibres on wheat dough performance and bread quality', *Food Chem*, **79**(2), 231–236.

KHAIRY M, MORSI S, EL-FARRA A A, IBRAHIM N A and HASSAN S A (1986), 'Characteristics of bread made from wheat flour fortified with isolated legume protein concentrates', *Egyptian J Food Sci*, **14**(2), 435–440.

KNUCKLES B E, HUDSON C A, CHIU M M and SAYRE R N (1997), 'Effect of ß-glucan barley fractions in high-fibre bread and pasta', *Cereal Foods World*, **42**(2), 94–100.

LAURIKAINEN T, HARKONEN H, AUTIO K and POUTANEN K (1998), 'Effects of enzymes in fibre-enriched baking', *J Sci Food Agric*, **76**, 239–249.

LOPEZ H W, OUVRY A, BERVAS E, GUY C, MESSAGER A, DEMIGNE C and REMESY C (2000), 'Strains of lactic acid bacteria isolated from sour doughs degrade phytic acid and improve calcium and magnesium solubility from whole wheat flour', *J Agric Food Chem*, **48**, 2281–2285.

LOPEZ H W, KRESPINE V, GUY C, MESSAGER A, DEMIGNE C and REMESY C (2001), 'Prolonged fermentation of whole wheat sourdough reduces phytate level and increases soluble magnesium', *J Agric Food Chem*, **49**, 2657–2662.

MALDONADO A (2000), 'Mineral enhanced bakery products', United States Patent.

MALINOW M R, DUELL P B and HESS D L (1998), 'Reduction of plasma homocyst(e)ine levels by breakfast cereal fortified with folic acid in patients with coronary heart disease', *New England J Med*, **338**, 1009–1015.

MEKHAEL K G, BASSILY N S, MEKHAEL N A and SAID A K (1989), 'Response of adult rats to lysine supplementation', *Nahrung*, **33**(7), 625–630.

MIKI T, FUKATSU M and HARADA S (1993a), 'Processed food made of iodine-enriched wheat flour', European Patent Application, EP 0 394 904 A2, JP89–108970 (19890427).

MIKI T, FUKATSU M, and HARADA S (1993b), 'Processed food made of iodine-enriched wheat flour', United States Patent, US 5 232 728, JP 89–108970 (19890427).

MOLLOY A M and SCOTT J M (2001), 'Folates and prevention of disease', *Public Health Nutr*, **4** (2B), 601–609.

MOSTAFA M K, HAMED A S and FODA Y H (1982), 'Enrichment of wheat flour with dry skim-milk and dry buttermilk and its effect on the baking quality', *Egyptian J Food Sci*, **8**, 33–39.

MUNOZ J M (1985), 'Overview of the effects of dietary fiber on the utilization of minerals and trace elements' in Spiller G, *Handbook of Dietary Fiber in Human Nutrition*, Boca Raton, CRC Press, 193–200.

POMERANZ Y, SHOGREN M, FINNEY K F and BECHTEL D B (1977), 'Fibre in breadmaking effects on functional properties', *Cereal Chem*, **54**, 25–41.

RADER J I, WEAVER C M, and ANGYAL G (2000), 'Total folate in enriched cereal-grain products in the United States following fortification', *Food Chem*, **70**, 275–289.

RANHOTRA G S, LOEWE R J and LEHMANN T A (1974), 'Breadmaking characteristics of wheat flour fortified with various commercial soy protein products', *Cereal Chem*, **51**(5), 629–634.

RANHOTRA G S, GELROTH J A, ASTROTH K and POSNER E S (1990), 'Distribution of total and soluble fiber in various millstreams of wheat', *J Food Sci*, **55**(5), 1349–1351.

ROSELL C M, ROJAS J A and BENEDITO C (2001), 'Influence of hydrocolloids on dough rheology and bread quality', *Food Hydrocolloid*, **15**(1), 75–81.

SERVAIS J P (1981), 'Supplementation of wheat flour by fish protein concentrate in bread making in Morocco', *Rev Fermentation Ind Alimentaries*, **36**(6), 192–198.

SIDHU J S, AL-HOOTI S N and AL-SAQER J M (1999), 'Effect of adding wheat bran and germ fractions on the chemical composition of high-fiber toast bread', *Food Chem*, **67**, 365–371.

SOLON F S, KLEMM R D, SANCHEZ L, DARNTON-HILL I, CRAFT N E, CHRISTIAN P and WEST K P (2000), 'Efficacy of a vitamin A fortified wheat flour bun on the vitamin A status of Filipino schoolchildren', *Am J Clin Nutr*, **72**(3), 738–744.

STEPANOVA E N, SHATNYUK L N, VERZHINSKAYA M F, KOSTYLEVA M G, BUKINA N A, SPIRICHEV V B, SHUKHNOV A F, KOSTEL'TSEVA N N, BISTROVA A I and EMTSEVA I B (1988), 'Effect of method of vitaminization of high grade wheat flour bread on its content on thiamine, riboflavin and niacin', *Voprosy-Pitaniya*, **2**, 67–71.

TANGKONGCHITR U, SEIB P A and HOSENEY R C (1981), 'Phytic acid: its fate during breadmaking', *Cereal Chem*, **58**, 229–234.

TERRA N N, MELLER A C, MUSSOI E and ABREU L E V (1976), 'Supplementation of foods with fish protein concentrate', *Rev Centro Ciencias Rurais*, **6**(2), 115–120.

TSEN C C and HOOVER W J (1973), 'High protein bread from wheat flour fortified with full-fat soy flour', *Cereal Chem*, **50**(1), 7–16.

TURK M and SANDBERG A (1992), 'Phytate degradation during breadmaking: effect of phytase addition', *J Cereal Sci*, **15**, 281–294.

TURK M, CARLSSON N and SANDBERG A (1996), 'Reduction in the levels of phytate during wholemeal bread making; effect of yeast and wheat phytases', *J Cereal Sci*, **23**, 257–264.

WANG S and TANG Y R (1995), 'Fortification of Ca and Zn in wheat flour', *J Zhengzhou Grain College*, **16**(1), 26–34.

YANEZ E, BALLESTER D and VANOVIC D (1985), 'Wheat and oat fortification with sweet lupin flour (*Lupinus albus* cv. *Multolupa*)', *Nutr Rep Int*, **31**(2), 493–499.

YASODA-DEVI M and GEERVANI P (1979), 'Aceptability of fortified wheat products', *Indian J Nutr Dietetics*, **16**(2), 49–51.

ZEMEL M B and SHELEF L A (1982), 'Phytic acid hydrolysis and zinc and iron in whole wheat bread as affected by calcium containing additives', *J Food Sci*, **47**, 535–537.

Part II

Dough and bread quality

13

The molecular basis of dough rheology

P. S. Belton, University of East Anglia, UK

13.1 Introduction

The rheological properties of dough are an essential factor in the determination of bread quality. Doughs that are too strong do not allow proper development of the bubbles and result in the formation of dense, unpalatable loaves of small volume, while doughs that are too weak cannot retain the bubbles and result in large holes in the loaf or in the collapse of the loaf. It is well known that in order to optimise bread quality, mixing must be stopped at the correct level of mechanical input. The actual process called mixing in reality has two separate processes going on within it: one is the homogenisation of the various ingredients of the dough, which is a true mixing process, and the other is the development of dough structure by the mechanics of mixing energy into the system. Although the former is of vital importance, it is a process common to most food preparation processes; it is the latter process that demonstrates the uniqueness of wheat flour dough. As mechanical energy is put into the dough, its resistance to extension increases and then after some critical point decreases again. Optimum bread quality is achieved by choosing to stop mixing at the appropriate point on the mixing curve (usually close to, but not at, the maximum resistance).

When dough is subjected to mechanical perturbation it shows viscoelastic behaviour. That is, the mechanical force applied to the dough results in dimensional changes that are partially but not fully reversed when the force is removed. The observation of a maximum of resistance during the mixing process implies that the dough stores some of the mechanical energy expended as elastic potential energy. This chapter will examine the molecular basis of these phenomena in terms of the molecular behaviour of the dough. It will first of all consider the various factors that can affect dough rheology and then discuss the role of networks in doughs and how they may be formed. A key issue is how such networks can store elastic energy and the particular role of the high molecular weight subunits in dough rheology. When all the factors have been dealt with the question will be asked, 'How much dough rheology can be explained?' The point of this is to test ideas against observations in rigorous way, since there are many discussions in the literature that put forward models of the molecular structure of dough but do not test them against existing data or make any predictions that are testable.

13.2 Factors affecting dough rheology

There are a wide variety of substances added to the dough mix that might be generally classed as processing aids and which may have secondary effects on dough rheology. Stear (1990) gives a useful summary of these. In this chapter, the main concern is those substances that have a primary effect on rheology and throw light on the factors controlling the response to the input of mechanical energy.

The substances of interest are listed below:

- water;
- D_2O (deuterium oxide – heavy water);
- esterifying agents for glutamine residues;
- urea;
- salts;
- agents affecting disulfide bonding;
- the protein subunits present.

Water is of course a prerequisite for making dough: water plasticises dough, and the control of water content is of critical importance in mixing. The actual level of hydration in dough is quite low, typically the level of added water to flour is in the order of 0.6 g of water per gram of flour. Since the intrinsic level of water in the flour is of the order of 14%, the total water is about 0.75 g per gram. If the water is equally partitioned between the components of the flour this will mean that there is about 0.75 g of water per gram of gluten. In molecular terms this means that there will be about 5.5 water molecules per amino acid residue. This represents a highly concentrated protein system. Results reported using nuclear magnetic resonance (NMR) to measure the amount of mobile protein, in a preparation of high-molecular-weight (HMW) subunits of gluten (Belton *et al.*, 1994), indicate that in this region of water to protein ratio, the quantity of mobile material is highly sensitive to water content. This is illustrated in Fig. 13.1. In breadmaking, the water content of dough is thus chosen to be in a region where small changes in water content are likely to make a large change in the behaviour of the proteins.

Changing H_2O to D_2O has the effect of strengthening the dough (Tkachuk and Hlynka, 1968). This must indicate a role for hydrogen bonding in the dough,

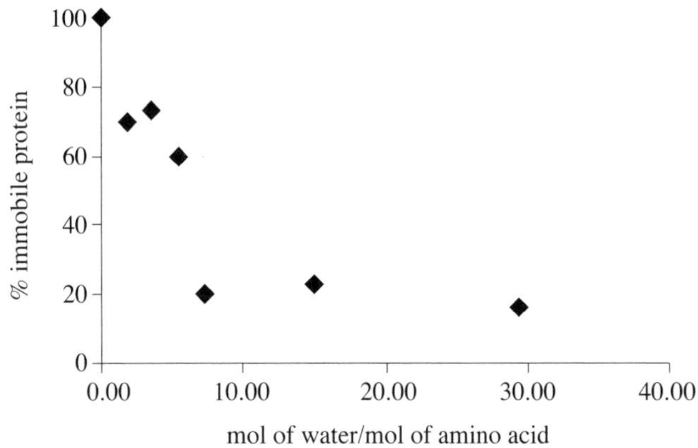

Fig. 13.1 A plot of the variation in the mobile fraction of high molecular weight subunits with water content.

as the hydrogen bonds formed by D_2O are significantly stronger than those formed by H_2O and there seem to be no other significant differences between the two isotopic forms that could have an effect. Whereas strengthening hydrogen bonds strengthens the dough, treatment to esterify glutamines residues, thus removing their hydrogen bonding ability weakens the dough (Beckwith *et al.*, 1963; Mita and Matsumoto, 1981). The indication of this effect is that the glutamine amino side chains are involved in some hydrogen-bonding network that is important in controlling dough rheology. In a similar vein, the weakening effects of urea on dough rheology (Wrigley *et al.*, 1998) have been interpreted as being due to disruption of hydrogen bonding.

Eliasson and Larsson (1993) have reviewed the effects of salts on the behaviour of dough. The addition of sodium chloride to dough influences gas retention, increases the time to optimum dough development and increases the stability of the dough. These effects may arise from a variety of causes not directly linked to the interactions of the proteins. There may be effects on enzymes and yeast; however, more extensive studies have shown that both gluten strength (Preston, 1989) and extractability of proteins (Preston, 1985) are modified by the addition of salts.

For metal chloride salts, the gluten strength is increased with the charge density of the metal ion. Since generally higher charge densities result in a more hydrogen bonded water structure, this may be taken to imply that increasing the hydrogen bonding capacity of the solvent increases the gluten strength. Conversely, the extraction data for a series of sodium salts showed that the greater the capacity of the counter-ion to break down hydrogen bonding structure, the more it facilitated protein extraction. Apart from the obvious effects of shielding electrostatic charge interactions, the role of salts in protein is difficult to understand and much discussion has gone on in the literature. However spectroscopic results on gluten at constant water content (Wellner *et al.*, 2002) indicate that for the series NaCl, NaBr and NaI, increasing counter-ion size, and hence water structure breaking capacity, causes an increase in the amount of beta turn present and the amount of mobile protein present. This result is consistent with those of Preston (1985, 1989).

The role of disulfide linkages in the control of dough rheology is of the utmost importance. If disulfide bonds are reduced by a chemical agent, such as dithiothreitol, a dramatic reduction in dough strength is observed (Wrigley *et al.*, 1998) which is recovered on re-oxidation. The addition of various oxidising and reducing agents that can affect the interchange of disulfide bonds also have major effects (Eliasson and Larsson, 1993). The actual mode of action of the various agents that can affect both the interchange among, and the number of disulfide bonds, is not entirely clear (Eliasson and Larsson, 1993; Weegels *et al.*, 1994). However, their effect is profound. Indeed, the role of disulfide interchange in dough rheology has led Bushuk (1998) to remark that 'The importance of the disulfide interchange reaction in the development and stress relaxation of bread doughs cannot be overemphasised'.

The role of the nature of the various protein subunits in dough rheology and loaf quality has been the subject of intensive research. Gliadins are generally agreed to contribute to the viscous nature of the dough and glutenins to the elastic nature of the dough. Of the glutenins the most important are the HMW subunits even though they only constitute 12% of the total flour proteins or 1–1.7% of the flour dry weight. (For a more detailed discussion see Shewry *et al.*, 2001.)

The way in which rheology develops during mixing has been the subject of much publication; however, very often conditions have not been well defined and the results are subsequently hard to interpret. The work of Gras *et al.* (2000) is, however, a very detailed and well-defined set of experiments using a Mixograph. This may not be the perfect model for commercial, or even domestic mixing practice, but the experiments are extremely useful for testing ideas about the factors affecting dough rheology during the extension, rupture and relaxation occurring sequentially during the mixing process. They observed the following:

- Dough development can be considered as involving in two phases. The first is a hydration stage, assisted by the even distribution of water within the flour by the mixing, and the second is the input of energy through deformation.
- Dough deformation results in the storage of energy in the dough through the modification of molecular structures. In effect mixing results in the storage of elastic energy in the dough.
- Dough resistance reaches a maximum in a Mixograph during mixing. The height of the maximum increases with decreasing water content.
- After resting for an hour all mixed doughs show a decrease in resistance to extension with the number of revolutions of the mixer.
- After resting for an hour all mixed doughs show a decreased degree of extension to break.
- After resting for an hour resistance to extension of the mixed dough decreases with increasing water content.

- After resting for an hour degree of extension to break of the mixed dough increases with increasing water content.

These observations, together with the chemical effects discussed previously, define a set of chemical and mechanical phenomena that a molecular model of dough must seek to explain. In the next sections such a model will be set out and compared with these observations.

13.3 Polymer networks in doughs

On the appropriate time scale viscous materials flow when subjected to a mechanical force, or stress. In contrast, solid materials do not flow: the whole mass is displaced, or, if one part of the sample is fixed in place, the sample is deformed. The deformation is usually termed strain. In the viscous limit flow occurs because the interactions between the molecules are not sufficient to transmit stress throughout the whole of the sample. Solids, on the other hand, have high levels of molecular interaction that ensure transmission of stress. In some solids, for example rubbers, application of mechanical force results in large-scale deformation followed by recovery of the original shape when the force is removed. This phenomenon is known as elasticity. In order to display elastic properties there must be connectivity between at least some of the molecular constituents through out the whole of the material. If it were not so, application of force would result simply in flow.

Dough is viscoelastic; it exhibits both flow and elastic characteristics. This seems to imply both connectivity and non-connectivity in the system, which cannot occur. Two factors need to be considered to understand this: the stress that is applied to the dough and the length and time scales over which perturbation occurs. If the stress applied to the dough is such that the bonds, which create the connectivity of the network within the dough, are disrupted, then flow will result since the network will no longer exist. This may occur at both the level of non-covalent interactions such as hydrogen bonding or at the level of covalent interactions such as disulfide cross-linking. Both of these are discussed further below.

The interacting effects of time and length are illustrated in Fig. 13.2, which shows two junction points connected by two elastic strings. It may be imagined that they are two elastic bands tied together at each end or two elastic molecules chemically joined at each end (Fig. 13.2A). (For this purpose we ignore entropic effects in the molecules.) If the displacement is small then the slack that exists can take it up and no elastic effects will be observed (Fig. 13.2B). If the displacement is larger than the total length of the strings they will be stretched and an elastic effect will be observed (Fig. 13.2C). Thus the length scales of displacement can determine whether or not elastic phenomena are observed. The effects of time scales are illustrated in Fig. 13.3. In Fig. 13.3A the situation is the same as in Fig. 13.2A, only this time there is an equilibrium between a situation

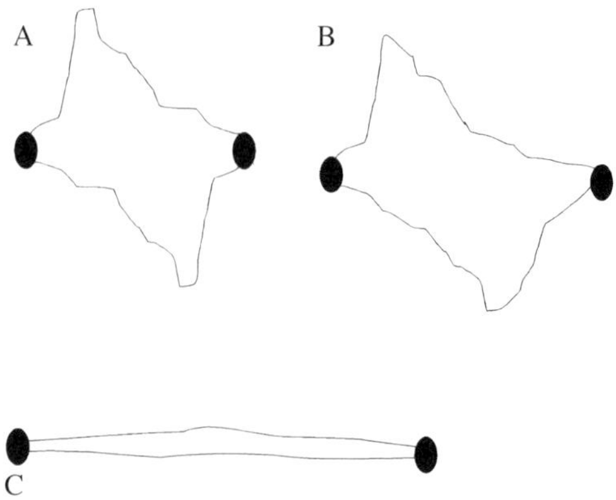

Fig. 13.2 The effects of extension on a network: A unextended, B slightly extended, C fully extended.

in which the junction points are connected (Fig. 13.3A) and one in which they are broken (Fig. 13.3B), The time constant for the equilibrium is τ. When a large displacement is applied in a time short compared to τ, the parts of the system in state A will be end up in state C but those in B will end in D. If the displacement is applied on time scale long compared to τ all the molecules will have had time to get to state B and a net displacement as in state D will have taken place. Further extension will result in a state E. Thus in this case both time- and

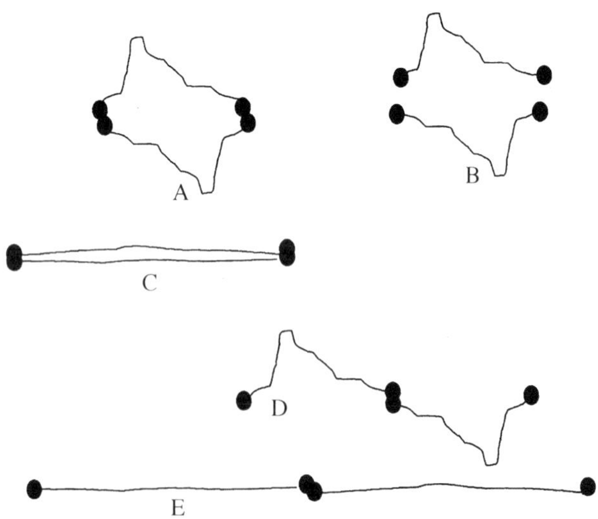

Fig. 13.3 The effects of an equilibrium between cross-linked and uncross-linked polymers undergoing extension: A cross-linked state, B unlinked state, C extended linked state, D extended unlinked state after reformation of the cross-link, E extended re-linked state.

displacement-dependent behaviour will be observed. The observation therefore of connectivity in a system will depend on the applied force, and the time and length scale over which the experiment takes place. Depending on this the system can behave as if it is connected or not connected.

There is no doubt that large networks exist in dough, and that these are related to the mechanical properties of the dough. Evidence comes not only from rheological properties but also from the direct observation of the formation of very large polymeric aggregates in dough called the dough macropolymer (Lindsay and Skerritt, 1999; Shewry *et al.*, 2003). This consists mainly of high- and low-molecular-weight (LMW) glutenin subunits of gluten and is formed by both covalent and non-covalent interactions. By its nature the macropolymer is not easy to extract from the dough. It has been defined (Graveland *et al.*, 1980) as the wheat protein fraction unextractable in 1.5% sodium dodecyl sulfate (SDS). The effect of mixing is to increase the amount of protein that can be extracted in SDS (Graveland *et al.*, 1980). This may imply that the macropolymer is depolymerised during mixing making extraction of protein more easy. However, extractability is not necessarily an indication of the state of the macropolymer; it may be an indication of the accessibility of the proteins since mixing may well create a laminar structure or other perturbations to the homogeneity of the material that simply makes extraction more efficient.

Thus far it has been argued that connectivity, in dough supplied through polymer networks, is required for elasticity. However, the existence of a network is a necessary, but not sufficient, condition for a system to exhibit elasticity. Consider a network of ideal non-extendable strings. The network may be deformed but after deformation there will be no recovery of the original shape since each string in the network will not have changed its length, only its orientation. If the network is made from rubber strips the network may be extended and the individual strips may change their length. In this way a restoring force is created and there will be recovery from the deformation. In the case of doughs there is a further complication to the rheology that not only is the dough elastic but it also changes its resistance to extension during the mixing process. The resistance at first increases and then decreases. An increase in resistance to extension implies that the stress required to extend an element of dough becomes greater. Thus elastic potential energy must be stored in the dough. As the resistance to extension decreases again, the ability of the dough to store elastic potential energy must once more decrease. The challenge is to explain these observations in terms of a molecular model. This will be discussed in the next section.

13.4 The molecular mechanism of energy storage in dough

In order to consider the molecular mechanisms involved in dough rheology attention will be focused on the primary role of the HMW subunits. Other proteins undoubtedly have a role but the model put forward assumes that this role is a secondary one of modifying the HMW interactions. This assumption is justified by the critical role the experimental evidence shows the HMW subunits to play (for a review see Shewry *et al.*, 2003).

High molecular weight subunits have a number of features that must be considered in order to understand the relationship between molecular structure and functional behaviour. They consist of relatively small C and N termini, which contain cysteine residues and are thus able to form disulfide bonds, separated by very long repeat regions of between 480 and 680 residues (Shewry *et al.*, 2001). The repeat regions are very rich in glutamine and proline. The HMW subunits are insoluble in water but, in spite of their high molecular weights, show components with remarkable mobility when examined by NMR (Belton *et al.*, 1994, 1995). As shown in Fig. 13.1 the amount of mobile material is very dependent on the water content. Infrared spectroscopy has demonstrated that two types of structure are present in HMW subunits and model systems, a beta sheet structure and a beta turn structure (Feeney *et al.*, 2003; Belton *et al.*, 1995). The ratio of these depends on the amount of water present. When the material is dry the protein is mainly disordered, as water content increases the amount of beta sheet present increases and reaches a maximum. As water content increases yet further beta sheet declines and is replaced by beta turn.

A model to explain these phenomena has been developed (Belton, 1999; Shewry *et al.*, 2003). It assumes that the glutamine residues in the repeat units of the HMW subunits are able to readily form inter-chain hydrogen bonds with those in neighbouring chains. When the proteins are dry, inter- and intra-chain hydrogen bonds are the only ones possible. Under these circumstances there will be little conformational order as there is no water present to plasticise the system and allow sufficient motion for rearrangement. As water levels increase, molecular motion allows the formation of inter-chain beta sheets. As water levels increase still further there is competition between the water and glutamine for hydrogen bond formation. This results in the replacement of the inter-chain hydrogen bonds by water and thus the replacement of inter-chain beta sheet by a hydrated extended structure which is identified with beta turns. It is also this structure which gives rise to the mobile portion of the NMR signal. This process is shown diagrammatically in Fig. 13.4. The beta sheet regions have been termed 'loops' and the beta turn regions have been termed 'trains' (Belton, 1999) in analogy with the behaviour of polymers at surfaces. At any water content, therefore, there is an equilibrium between sheets and turns. If the sheet to turn ratio is perturbed the tendency will be to restore the equilibrium and minimise the free energy of the system.

The mechanism of elasticity can now be seen by assuming that disulfide cross-links connect the proteins into a network. If a mechanical force is applied the first thing that happens is the distortion of the turn region. This is the most mobile and therefore will be the first to be affected. If the displacement is small there will be only a weak elastic restoring force due to weak polymer–polymer interactions. As the displacement is increased the turns will become straightened out and proximity with other chains will cause the formation of sheet-like

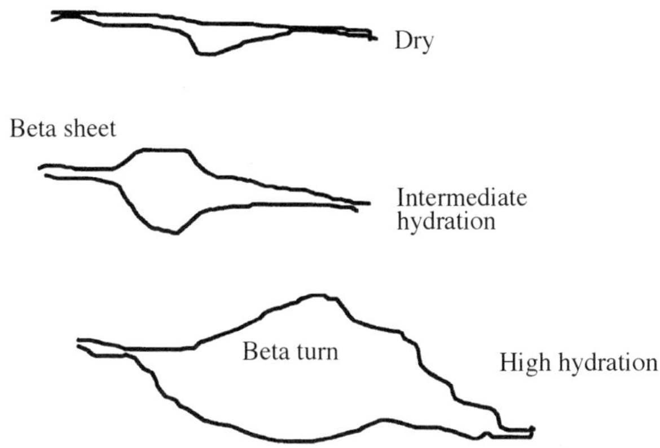

Fig. 13.4 The effects of hydration on the interactions and structure of HMW subunits.

structures. As displacement is increased still further the trains may become ruptured and, finally, there may be breakage of the inter-chain disulfide links. The elastic restoring force in the process is provided by the equilibrium between loops and trains. Perturbation of this will move the system from its free energy minimum to a higher energy state. The elastic restoring force thus comes from the high level of trains formed by extension and the propensity of these to re-form loops, which restore the original dimensions of the system. If the rate of deformation of the dough is faster than the rate of relaxation to its equilibrium state it will get further and further from its equilibrium state and store more and more energy. It will thus become harder and harder to deform the dough as all the loops will be converted to trains and trains will have to be disrupted. At some point the input of energy will be such that the covalent links in the system will be disrupted and breakage will occur. The system will then tend to flow more and resistance to extension will decrease.

A very simple way of visualising this is to consider that the dough consists of uniform polymers in network. When the dough is extended, stretched polymers are created; at the same time the input of mechanical work causes some of the stretched polymers to break, creating broken polymers. Typically, therefore, in a mixer each revolution of the mixer will result in the stretching and breaking of the polymers: after R revolutions of the mixer there will be N_U unstretched polymers, N_S stretched polymers and N_B polymers broken from the network.

The rates of formation and disappearance of the various species can be approximated, for illustrative purposes, by first order rate laws. Thus:

$$dN_s/dR = k_1 N_U - k_2 N_s \quad 13.1$$

$$dN_U/dR = -k_1 N_U \quad 13.2$$

$$dN_B/dR = k_2 N_s \quad 13.3$$

and

$$N_U + N_S + N_B = N_0 \qquad 13.4$$

where k denotes a rate constant.

The number of stretched polymers will depend on both the number of polymers available to be stretched and the rate of polymer breakage; this is reflected in equation 13.1. The rate of disappearance of unstretched polymers is given in equation 13.2. As the system becomes more stretched more breakages will occur as shown in equation 13.3. The resistance to extension ρ will be given by:

$$\rho = E_S N_S + E_U N_U + E_B N_B \qquad 13.5$$

where E denotes a constant.

It is assumed that the storage of the elastic energy in the network will be the most important factor in explaining the resistance curves, but that some energy must be expended to stretch the polymers and that broken polymers will cause some resistance to motion. Thus, $E_S > E_U > E_B$. To a first approximation, therefore, the set of equations 13.1–13.5 represents the build-up of elastic potential energy in the dough.

The details of the shape of the resistance curves with degree of mixing will depend on the rate constants chosen for the equations and the relative values of E. However, as Fig. 13.5 illustrates, the general trends of resistance versus number of revolution observed in Mixograph experiments of Gras *et al.* (2000) are reproduced.

The model suggested so far can account for the elasticity of gluten and the build-up and breakdown of resistance to extension on mixing. It assumes an

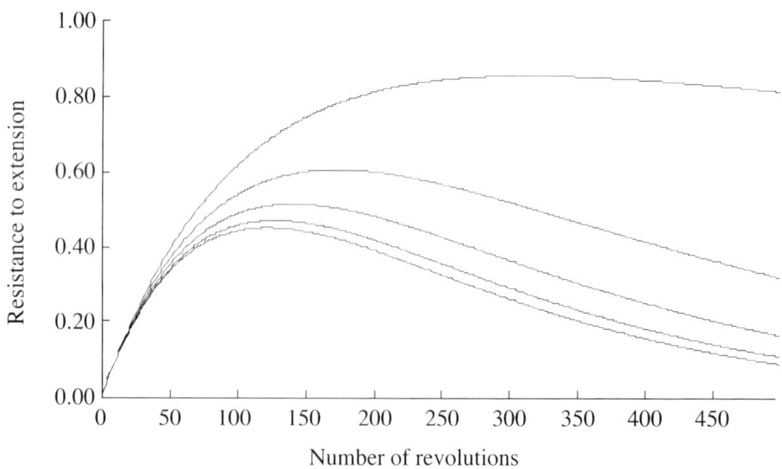

Fig. 13.5 The variation in resistance to extension with the number of Mixograph revolutions calculated for a system of stretched, unstretched and broken polymers. Each curve is the result of different rate constants for stretching and breaking. For details see text.

essentially static network of HMW subunits except for breakages. In a real dough system there is no doubt that disulfide interchange can take place. The question is what is the nature of the change that is likely to take place under extension. Figure 13.3 gives an indication. If disulfide bond breaking takes place before or during the extension process, the tendency is for the polymers to be displaced in the direction of the extending force. Thus the situation shown in Fig. 13.3A will become that shown in Fig. 13.3D, and further extension will result in the situation in Fig. 13.3E. Similarly, the situation shown in Fig. 13.3C will reach that shown in Fig. 13.3E. The extension in Fig. 13.3E is twice that in Fig. 13.3C. The total work required to reach both Figs 13.3C and 13.3E is the same since both result in the extension of two polymers to the same extent. The effect of disulfide exchange during extension is thus to increase extensibility. During mixing the dough piece is moved around in a more or less random manner at the molecular level. Extension will therefore be applied equally in all possible directions so there will be no net orientation of the polymers with respect to some external direction in the system due to disulfide interchange. However, there will be a general extension of the network equally in all directions. Although disulfide interchange can lead to an increase in extensibility, not many polymer displacements will result in a successful reformation of disulfide linkages. There will also be a steady accumulation of broken disulfide linkages, both from this and the direct effects of mechanical force causing bond breakage, although this effect will be to some extent ameliorated by reformation of the bonds. There are thus two opposing tendencies at work, the increases in extensibility due to the disulfide interchange effect and the accumulation of broken disulfide linkages, which will reduce the amount by which the dough can be extended before breakage. Typically doughs with more resistance to extension will accumulate breakages faster because more mechanical energy is put into extend them.

While the role of the HMW subunits is critical in the determination of dough rheology, other proteins have a role to play. The dough macropolymer contains LMW subunits which may extend or terminate chains depending on the number of possible disulfide bonds that they can make (Lindsay and Skerritt, 1999; Shewry *et al.*, 2003). It is also possible that other gluten proteins could block the formation of inter-chain hydrogen bonds if they do not have the correct geometry for hydrogen bond formation. Alternatively if they do possess the correct geometry they might enhance inter-chain hydrogen bond formation. The details of such interactions have yet to be understood. However, the role of gliadins in increasing the viscous component of dough viscoelasticity may be interpreted as blocking the formation of inter-chain hydrogen bonds, resulting in reduced elasticity and hence increased viscosity.

13.5 How much dough rheology can we explain?

The loop and train model asset out above is consistent with a considerable amount of spectroscopic evidence (Shewry *et al.*, 2003) and can explain the effects of components as set out in Section 13.2. The role of water is critical since it determines the ratio of loops to trains and hence the ability of the dough to be extended and to resist extension. If H_2O is replaced by D_2O, hydrogen bond strengths are increased and the force required to extend the HMW subunits by disrupting the beta sheet region is increased. Similarly esterifying the glutamine residues will result in less hydrogen bonding and a less cohesive material. Urea can disrupt hydrogen bonds and hence will have a similar effect to esterification. The molecular effects of salts can be quite subtle and the details of the mechanisms of the interactions of salts with proteins are not completely understood. Salts can affect both hydrogen bonding and protein solubility, both of which will affect the cohesiveness of the system. The role of disulfide bonding is critical both in the interchange mechanism and in the role of covalent links in maintaining the HMW network. Changing the total number of disulfide bonds will affect the degree of extension possible before breakage, either increasing it or decreasing it, depending on an increase or decrease in the total number of bonds. Enhanced interchange of bonds will increase the extensibility of the system as discussed above. Finally, the nature of the proteins subunits present can be understood first in the key role of the HMW subunits and the secondary role of the other gluten proteins. It is clear that some arrangements of amino acids along the repeat region of the HMW subunits will favour the formation of loops to a greater extent than others. Particular subunits may therefore impart more elasticity or resistance to extension than others. Some evidence for this effect has been found in a study of model peptides (Feeney *et al.*, 2003). The peptides contained perfect repeat units when compared with a repeat unit from a 1Dx5 single subunit, which is imperfect. The perfect repeats showed a higher level of beta sheet content, which would offer a greater resistance to extension.

The loop and train model is also consistent with the results of Gras *et al.* (2000). It offers a mechanism for the storage of elastic potential energy in dough and can explain the maximum of resistance reached in the Mixograph. In rested doughs, where it may be assumed that the normal loop to train equilibrium is re-established, the decrease in resistance to extension is explained by the increased number of polymers broken from the network. Similarly the damage to the network explains the decreased distance to break. The effects of water observed can be explained as discussed above by the changes in the loop to train ratio caused by changes to water content.

The loop and train model thus offers a self-consistent explanation of the molecular basis of dough rheology and is of sufficient detail to make testable predictions about the changes that may be expected in rheology as conditions change.

13.6 Future trends

There is still a long way to go in our understanding of the molecular basis of dough rheology. Some of the problems may be formulated in terms of questions that need to be answered:

- How specific is inter-molecular disulfide bonding and what factors control it?
- Are there specific amino acid sequences that favour inter-chain hydrogen bond formation?
- What, in molecular terms, is the role of LMW subunits and other gluten proteins?

Although there has been considerable development mapping of inter-chain disulfide bonds it is not clear how complete this is and to what extent it reflects the true situation in worked doughs (Shewry and Tatham, 1997; Shewry *et al.*, 2003). A greater understanding of this could lead to better control of the rheology through network formation. It might also indicate particular proteins or sequences that were targets for breeding or genetic manipulation.

Evidence from expressed and synthesised peptides (Feeney *et al.*, 2003) indicated that the sequence detail of the repeat regions of the HMW subunits has an effect on the inter-chain hydrogen bonding. This may in part explain the particular contribution that the combination of 1Dx5 and 1Dy10 subunits makes to dough rheology although the possibility of specific disulfide interactions cannot be excluded (Shewry *et al.*, 2003). Further elucidation of the particular requirement will come from the study of expressed proteins and synthesised peptides with specific sequences combined with modelling studies. The final test will be expression of particular combinations of sequences in transgenic wheats, which will allow mixing and baking experiments to be carried out.

The role of the proteins other than the HMW subunits has still to be fully understood. Once again, the use of transgenic lines combined with synthesised and expressed peptides will be of great value. In addition, experiments where specific proteins and peptides are added to dough mixes may also help to understand the role of these proteins.

Undoubtedly, the use of spectroscopic methods of investigation will continue to play an important role. Both NMR and FTIR have proved to be of great value. Recent work, using high-resolution solid-state NMR methods (Alberti *et al.*, 2001; Gil *et al.*, 2001) have begun to show interesting results indicative of differences in interactions between different pairs of HMW subunits. Data obtained also indicate the existence of glutamine residues in different environments on the same subunit. This type of information is likely to be of great value in exploring the fine detail of subunit interactions.

13.7 Sources of further information and advice

There are number of excellent books on breadmaking and gluten. Among those I have found particularly useful are:

HAMER, R J and HOSENEY, R C (1998) *Interactions: Keys to Cereal Quality,* American Association of Cereal Chemists, Minnesota.

ELIASSON A-C and LARSSON K (1993) *Cereals in Breadmaking,* Marcel Dekker, New York

STEAR C A (1990) *Handbook of Breadmaking Technology*, Elsevier Applied Science, London.

A useful series for keeping up to date is the publications arising from the international gluten workshops. The latest was held in 2000 and the Proceedings are published as *Wheat Gluten*, Shewry P R and Tatham, A S (2000) Royal Society of Chemistry, Cambridge.

13.8 Acknowledgement

I would like to thank Bob Anderssen for a critical reading of the manuscript.

13.9 References

ALBERTI, E, HUMPFER, E, SPRAUL, M, GILBERT, S M, TATHAM, A S, SHEWRY, P R and GIL, A M (2001) 'A high resolution ^{1}H magic angle spinning NMR study of a high M_r subunit of wheat glutenin', *Biopolymers*, **58**, 33–45.

BECKWITH, A C, WALL, J S and DIMLER, R J (1963) 'Amide groups as interaction sites in wheat gluten proteins: effects of amide ester conversion', *Arch Biochem Biophys*, **103**, 319–330.

BELTON, P S (1999) 'On the elasticity of wheat gluten', *J Cereal Sci.*, **29**, 103–107.

BELTON, P S, COLQUHOUN, I J, FIELD, J M, GRANT, A, SHEWRY, P R and TATHAM, A S (1994) ^{1}H and ^{2}H NMR relaxation studies of a high M_r wheat glutenin and comparison with elastin', *J Cereal Sci.*, **19**, 115–121.

BELTON, P S, COLQUHOUN, I J, FIELD, J M, GRANT, A, SHEWRY, P R, TATHAM, A S and WELLNER, N (1995) 'FTIR and NMR studies on the hydration of a high M_r subunit of glutenin', *Int J Biol Macromol*, **17**, 74–80.

BUSHUK, W (1998) 'Interactions in wheat doughs', in Hamer, R J and Hoseney, R C, *Interactions: Keys to Cereal Quality*, American Association of Cereal Chemists, Minnesota, p 10.

ELIASSON, A-C and LARSSON, K (1993) *Cereals in Breadmaking*, Marcel Dekker, New York, pp. 261–324.

FEENEY, K A, WELLNER, N, GILBERT, S M, HALFORD, N G, TATHAM, A S, SHEWRY, P R and BELTON, P S (2003) 'Molecular structures and interactions of repetitive peptides based on wheat glutenin subunits depend on chain length', *Biopolymers*, **72**, 123–131.

GIL, A M, ALBERTI, E and SANTOS, D (2001) 'An insight into the structure of foods using ^{13}C and ^{1}H magic angle spinning (MAS) NMR: application to wheat dough', in Webb, G A, Belton, P S, Gil, A M and Delgadillo, I, *Magnetic Resonance in Food Science: A View to the Future*, Royal Society of Chemistry, Cambridge, p. 43.

GRAS, P W, CARPENTER, H C and ANDERSSEN, R S (2000) 'Modelling the developmental rheology of flour dough using extension tests', *J Cereal Sci.* **31**, 1–13.

GRAVELAND, A, BOSVELD, P, LICHTENDONK, W J and MOONEN, J H E (1980) 'Superoxide involvement in the reduction of disulphide bonds of wheat gel proteins', *Biochem Biophys Res Commun*, **93**, 1189–1195.

LINDSAY, M P and SKERRITT, J H (1999) 'The glutenin macropolymer of wheat flour doughs: structure–function perspectives', *Trends Food Sci Tech,* **10**, 247–253.

MITA, A and MATSUMOTO, H (1981) 'Flow properties of aqueous gluten and gluten methyl ester dispersions', *Cereal Chem,* **58**, 57–61.

PRESTON, K R (1985) 'Use of lyotropic salts to study the hydrophobic properties of wheat gluten proteins' in Graveland A and Moonen J H E, *Gluten Proteins, Proceedings of the 2nd International Workshop on Gluten Proteins*, TNO, Netherlands, pp. 207–217.

PRESTON, K R (1989) 'Effects of neutral salts of the lyotropic series on the physical properties of a Canadian red spring wheat flour', *Cereal Chem*, **66**, 144–148.

SHEWRY, P R and TATHAM, A S (1997) 'Disulphide bonds in wheat gluten proteins', *J Cereal Sci*, **25**, 202–227.

SHEWRY, P R, POPINEAU, Y, LAFIANDRA, D and BELTON, P (2001) 'Wheat glutenin subunits and dough elasticity: findings of the EUROWHEAT project', *Trends Food Sci Tech*, **11**, 433–441.

SHEWRY, P R, POPINEAU, Y LAFIANDRA, D, HALFORD, N G TATHAM, A S and BELTON, P S (2003) 'The high molecular weight subunits of wheat glutenin and their role in determining wheat processing properties', *Adv Food Sci Tech*, **45**, pp. 219–302.

STEAR, C A (1990) *Handbook of Breadmaking Technology*, Elsevier Applied Science, London, pp 27–59.

TKACHUK, R and HLYNKA, I (1968) 'Some properties of dough and gluten in D_2O', *Cereal Chem*, **45**, 80–87.

WEEGELS, P L, HAMER, R J and SCHOFIELD, J D (1994) 'Functional properties of wheat gluten', *J Cereal Sci*, **23**, 1–18.

WELLNER, N, BIANCHINI, D, MILLS, E N C and BELTON, P S (2002) 'Effect of selected Hofmeister anions on the secondary structure and dynamics of wheat prolamins in gluten', *Cereal Chemistry*, in press.

WRIGLEY, C W, ANDREWS, J L, BEKES, F, GRAS, P W, GUPTA, R B, MACRITCHIE, F and SKERRIT, J H (1998) 'Protein–protein interactions–essential to dough rheology', in Hamer, R J and Hoseney, R C, *Interactions: Keys to Cereal Quality*, American Association of Cereal Chemists, Minnesota, pp. 17–46.

14

Molecular mobility in dough and bread quality

Y. H. Roos, University College Cork, Ireland

14.1 Introduction

Molecular mobility as a general term is often used to refer to changes that occur at a molecular level and affect time-dependent properties of amorphous food solids. The term in the materials sciences and theories arena relates the mobility of molecules in a system to mechanical and other material properties under observation. In amorphous systems, such as most foods (White and Cakebread, 1966), solids at high concentrations exhibit no particular organised structure typical of equilibrium crystalline states of solids. Such solids are known as 'glasses' independently of whether they are inorganic or organic materials. Glassy materials, in general, are brittle and often transparent structures in which molecular mobility is higher than in a crystalline state at a corresponding environment (Sperling, 1992; Roos, 1995). However, molecular mobility in the glassy state is restricted to vibrations and rotations at a molecular level (Sperling, 1992).

This chapter introduces the concept of molecular mobility in concentrated food systems, such as dough and bread. The emphasis will be on introducing the glass transition concept and physical state effects on time-dependent properties. Glass transition and molecular mobility are important factors affecting the stability and quality of frozen dough and physicochemical changes occurring in baking and in the baked products. Therefore, such quality properties as ice formation and recrystallisation in frozen dough, gelatinisation of starch and non-enzymatic browning in baking as well as retrogradation of starch will be discussed.

14.2 Molecular mobility in dough

14.2.1 Dough formation

A typical plain dough contains flour and water as the main components. Depending on the dough types, other components, such as sugars and shortenings, are added in varying proportions, and salt is added to improve the flavour of the baked products. The composition of the flour, together with water and all additional components, has an impact on various types of molecular mobility in dough and baked products. Much of the research on dough has been done using models to monitor the glass transition and molecular mobility properties of the main flour components (e.g. Bizot *et al.*, 1997; Jouppila and Roos, 1997; Jouppila *et al.*, 1998; Rolee and Le Meste, 1999) or bread (Roudaut *et al.*, 1999a,b). However, it is important to understand that a dough and a baked product formed using the dough is a complex system, as described by Fig. 14.1, in which a large number of changes may take place in mixing, baking and storage.

Dough formation occurs in the mixing of flour and water, as water is sorbed by flour components and the gluten proteins form an elastic protein network (Levine and Slade, 1990). It is likely that water will be sorbed by the native and amorphous gluten proteins and amorphous, non-crystalline parts of starch components. The water content of the hydrated flour is sufficient to allow substantial mobility to hydrated and solubilised molecules in the system. Further increases in molecular mobility would result from heating of the dough and melting of crystalline regions of starch components during gelatinisation.

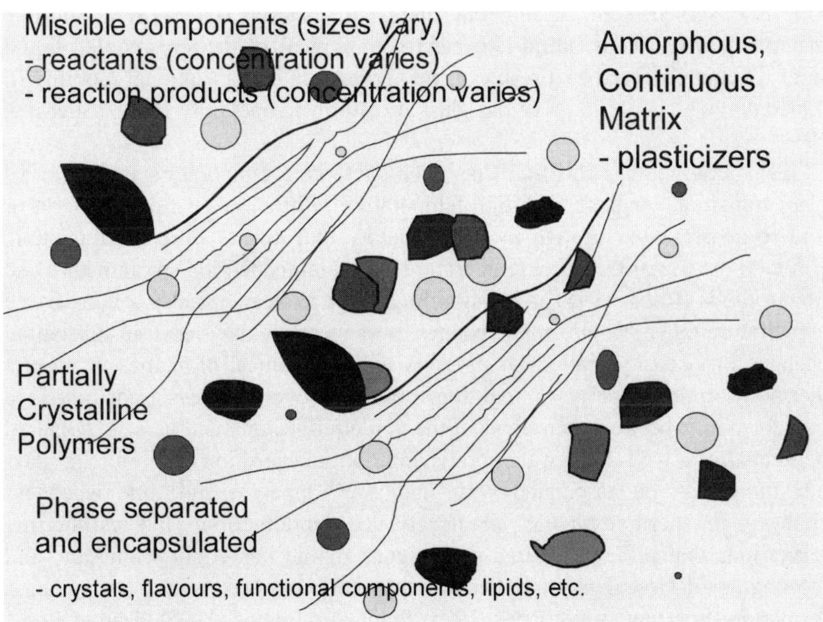

Fig. 14.1 A schematic representation of dough microstructure.

However, all crystalline regions in starch are not melted in gelatinisation and recrystallisation may occur rapidly after baking and during storage (Jouppila *et al.*, 1998).

14.2.2 Glass transition

Glass transition is a property of all amorphous materials, i.e. it occurs when non-crystalline materials undergo a change between the supercooled liquid- and solid-like states. The viscosity of solid-like systems with glassy properties is $>10^{12}$ Pa s (Sperling, 1992), restricting the molecular motions to molecular vibrations and side-chain rotations. Glass transition is often observed in heating of non-crystalline substances with solid characteristics as they gain liquid-like properties over a material-specific temperature range. Most food solids in highly concentrated systems, such as dehydrated and frozen foods, are amorphous and may exist as amorphous solids below their glass transition (Roos, 1995). However, food components in dough and other foods exist in crystalline, partially crystalline, partially amorphous and amorphous states, which all contribute to their properties and stability.

Amorphous, non-crystalline materials are plasticised or softened by temperature and plasticisers (Slade and Levine, 1991; Roos, 1995). As a result of plasticisation, materials may suffer a transformation into a supercooled liquid state over a temperature or plasticiser concentration range, referred to as glass transition. During plasticisation of a solid glass over its glass transition, there is a dramatic change in molecular mobility and corresponding decrease in viscosity. This is observed from an apparent change in material state as the solid-like material gains viscous liquid-like behaviour typical of the supercooled liquid state. This is observed as a result of the appearance of translational mobility of molecules which occurs over the glass transition temperature range (Sperling, 1992).

Each amorphous compound has a characteristic temperature range for its glass transition. A glass transition temperature is often taken from an onset or mid-point of a step change in heat capacity that occurs over the transition, although the use of the onset temperature is recommended for low-moisture and frozen foods (Roos, 1995). Although the change in heat capacity occurs over a temperature range, where small changes in temperature may result in substantial changes in viscosity, molecular mobility and mechanical properties, it gives a thermodynamic property to the transition. However, a glass transition is a transformation between two states of the non-equilibrium materials and it cannot be treated as a well-defined thermodynamic phase transition. Therefore, a glass transition may be associated with numerous apparent and time-dependent changes in thermodynamic quantities, e.g. endothermic and exothermic relaxations that reflect the time dependence of the non-equilibrium state and the degree of 'freezing' of molecules into various glassy states during glass formation (Sperling, 1992; Roos, 1995; Slade and Levine, 1995). Temperatures of the glass transition ranges of biomaterials and foods are particularly dependent on the presence and amount of water, which is the main plasticiser and softener of major hydrogen bonding substances, i.e. carbohydrates, proteins, and other water-soluble or water-miscible compounds (Slade and Levine, 1991; Roos *et al.*, 1996).

The major components of flour are proteins and starch, which in dough-making are mixed with water. Both native and denaturated gluten as well as gelatinised starch exist in an amorphous or partially amorphous state (Levine and Slade, 1990). These flour components are plasticised by water and their glass transitions are affected by the increasing water content (Levine and Slade, 1990; Laaksonen and Roos, 2003). Such plasticisation is found also in seeds (Leopold *et al.*, 1994) and both for gluten (Hoseney *et al.*, 1986) and starch (Zeleznak and Hoseney, 1987). Water-sensitive amorphous systems may exhibit several changes in their physicochemical properties, depending on water content and temperature.

Studies of molecular mobility, thermal behaviour and glass transition of cereal systems with different water contents have used electron spin resonance (ESR) spectroscopy (Rolee and Le Meste, 1999), nuclear magnetic resonance (NMR) spectroscopy (Roudaut *et al.*, 1999a), differential scanning calorimetry (DSC) (Laaksonen and Roos, 2000), dynamic mechanical analysis (DMA/DMTA) (Laaksonen and Roos, 2000) and dielectric analysis (DEA/DETA) (Roudaut *et al.*, 1999b; Laaksonen and Roos, 2000) as the main thermal analytical techniques. These techniques allow determination of a change in heat capacity that occurs over the glass transition (DSC) or a primary relaxation causing a substantial change in mechanical moduli or dielectric properties (Laaksonen and Roos, 2000). NMR methods have also been applied to studies of molecular mobility in dough components and bread (Kalichevsky *et al.*, 1992a; Roudaut *et al.* 1998). Relaxations below the glass transition are known as β- and γ-relaxations while the relaxation occurring over the glass transition is referred to as α-relaxation (Sperling, 1992).

14.2.3 The physical state of dough

Various time-dependent changes resulting from molecular movements and flow can be observed from changes in rheological properties and food texture, and a tendency of component crystallisation (Slade and Levine, 1991; Roos, 1995; Roos *et al.*, 1996). The rates of these changes are dependent on molecular mobility and they occur as a function of the extent of plasticisation that can be related to the temperature above the glass transition ($T - T_g$) (Roos, 1995; Jouppila and Roos, 1997; Jouppila *et al.*, 1998). However, an exact temperature for the glass transition cannot be defined or measured, because of the non-equilibrium nature of the transition and its time dependence, e.g. the temperature range over which the transition occurs is dependent on the time of observation (Slade and Levine, 1995). It should be noted, however, that DSC measurements give relatively consistent values for the T_g, because the observation time cannot be dramatically changed. It may also be assumed that such a dramatic change in observation time may correspond to years of relaxation below an observed T_g.

In dough, gluten and starch exist in their native state, making the dough system extremely complex and evaluation of molecular mobility of the dough components extremely challenging. It may be assumed that water and solutes in mixing of dough are distributed within the gluten and starch polymers. The glass transition temperatures of the gluten proteins decrease as water is added and they reach freezing temperatures at water contents typical of dough (Laaksonen *et al.*, 2001). After mixing, a dough may be considered as an amorphous, plasticised gluten network with dispersed starch granules within an aqueous phase containing dissolved flour components and added solutes, such as salts and sugars. The water molecules are likely to have a high mobility in the system (Roudaut *et al.*, 1999a) and the water plasticisation of the carbohydrate polymers and proteins provide mobility to the non-crystalline phases of these materials.

Among the first studies of glass transitions of gluten and starch were those of Hoseney *et al.* (1986) and Zelesnak and Hoseney (1987), respectively, who showed that wheat gluten and starch exhibit glass transition that is dependent on water content. In water plasticisation, an increasing water content at a constant temperature leads to an increased segmental mobility of the molecular chains in the amorphous regions of glassy polymers. This increased mobility leads to a primary structural relaxation and the glass transition (Levine and Slade, 1990).

The glass transition of cereal proteins has received much attention and several studies have reported glass transitions for the dough components (Nicholls *et al.*, 1995; Noel *et al.*, 1995; Nikolaidis and Labuza, 1996a,b). Cocero and Kokini (1991) studied the thermal and mechanical behaviour of glutenin using DSC and mechanical spectroscopy. Their results showed a correlation between the end-set temperature of the change in heat capacity determined using DSC and the temperature at which the loss modulus, G'', showed a maximum.

The first attempt to estimate T_g for starch and its dependence on water content was reported by van den Berg (1981). The estimate for anhydrous T_g for starch, 151 °C, was used by Marsh and Blanshard (1988) to predict the T_g of starch at various water contents. This result, and those of Zeleznak and Hoseney (1987) obtained by DSC for the determination of T_g of native and pregelatinised wheat starch at several water contents, have established the water plasticisation properties of starch. However, the estimated anhydrous T_g values of starch have varied significantly. The glass transition temperatures of anhydrous starch or gluten have not been determined experimentally owing to high temperatures of the transitions, probable thermal decomposition and other difficulties in detecting glass transition-induced changes in physical properties at low water contents. Experimental T_g data for maltodextrins (starch hydrolysis products) have given more reliable estimates for the T_g of high molecular weight glucose polymers. The estimated T_g of 243 °C for starch was obtained by using the Fox and Flory equation (Roos and Karel, 1991a) that is commonly applied in the polymer science to relate T_g data of homopolymers to their molecular weight. It seems that the anhydrous T_g values for dough components decrease in the following order: starch about 250 °C, dough proteins about 150 °C and water about −135 °C.

Effects of sugars and proteins on the physical state of starch at various water contents have been studied extensively because of their importance as component compounds of most starch-based foods (Slade *et al.*, 1989). Kalichevsky and Blanshard (1992) studied the physical state of mixtures of amylopectin, casein and gluten. Amylopectin–gluten mixtures (1:1) were plasticised by water and showed miscibility after heating with sufficient water. Sugars may also plasticise higher molecular weight carbohydrates and other dough components (Kalishevsky *et al.*, 1992b; Kalischevsky and Blanshard, 1993; Pouplin *et al.*, 1999). However, at high sugar concentrations their glass transition behaviour becomes dominating (Roos, 1995).

The glass transition behaviour of dough is dependent on composition and dough microstructure. Several authors have observed decreasing glass transitions for dough made with added sucrose. For example, decreased glass transitions have been reported by Cherian *et al.* (1995) for wheat gluten films with added sucrose, and Georget and Smith (1996) for wheat flakes with added fructose and sucrose. Laaksonen *et al.* (2001) also found that added sucrose decreased the T_g of dough with different water contents from 5 to 10 °C, over the a_w range of 0.113–0.753, compared with plain dough. These results suggested that sugars may either plasticise amorphous dough solids or exist in separate, carbohydrate-rich regions with limited interaction with dough proteins. Such phase separation of proteins and sugars is likely, as in most protein–carbohydrate food solids the glass transition is mainly affected by the sugar component. Limited miscibility and at least partial phase separation of sugars and proteins has been reported (Kalischevsky *et al.*, 1992b) and such behaviour is typical of dairy powders with amorphous lactose (Jouppila and Roos, 1994).

Our studies have shown that the equilibrium water contents in dough with added sucrose are higher over the whole a_w range when compared with the equilibrium water contents of the plain dough (Laaksonen and Roos, 2003). When NaCl was added, the water contents of dough were slightly higher than with added sucrose. Dielectric relaxations were determined for the different wheat doughs. The α-relaxation peak, taken as the T_g, were highly dependent on frequency. A decrease with increasing water content in T_g due to water plasticisation was fairly linear as a function of water activity. The T_g values of the dough with added sucrose and NaCl were the lowest and they also had the highest water contents at corresponding water activities. In general, the T_g values of dough are lower than those of gluten and starch at corresponding levels of water plasticisation. The lower T_g may result from added sugars and other T_g decreasing components. It should also be noted that yeast and enzyme activity during dough fermentation may decrease the molecular weight of the dough component polymers and thereby the T_g of the dough system.

14.2.4 The physical state of frozen dough

The significant effects of water on the physical state of sugars and other carbohydrates can be explained using state diagrams. State diagrams relate the phase and state transition temperatures of food systems to water content explaining water plasticisation and its effect on material properties (Roos *et al.*, 1996). In a state diagram, the glass transition curve at high water contents approaches the glass transition temperature of water in the vicinity of −135 °C (Roos, 1995; Roos *et al.*, 1996). The Gordon–Taylor equation (Gordon and Taylor, 1952; Roos, 1995) is often used to fit experimental data to allow prediction of the glass transition of food solids at various water contents. A schematic state diagram for the main dough components is shown in Fig. 14.2.

The glass transition behaviour of a freeze-concentrated system, such as frozen dough, differs from that of a system containing liquid water only. The separation of ice in a freeze-concentrated system from food solids increases the amount of dissolved solids in an unfrozen phase and decreases water plasticisation of the unfrozen phase components. In such systems, the solids may become partially freeze-concentrated with an extent of plasticisation corresponding to the unfrozen water content. Frozen systems may also become

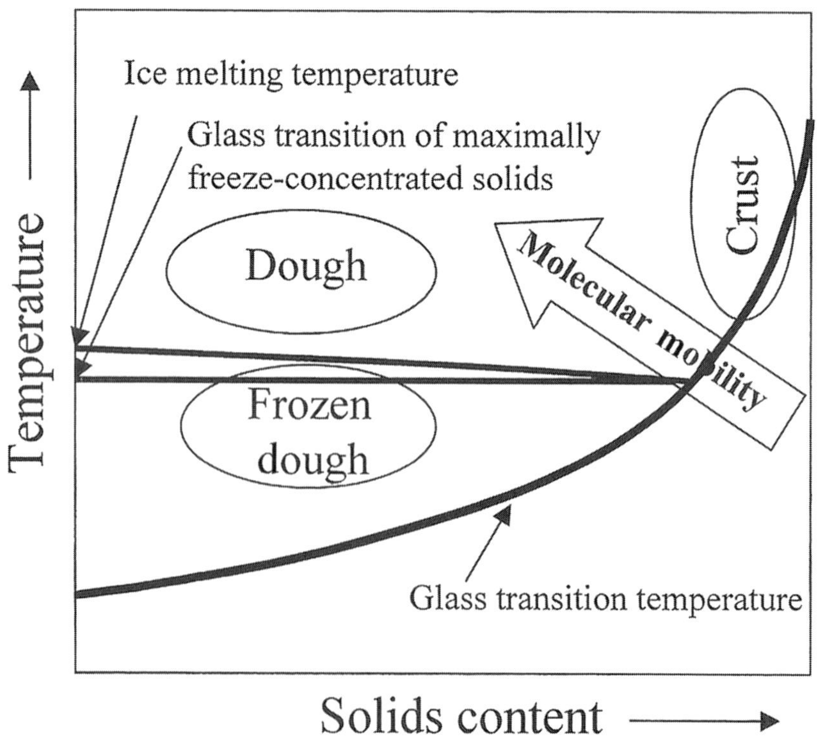

Fig. 14.2 A schematic state diagram explaining molecular mobility in dough.

maximally freeze-concentrated when ice formation occurs at a temperature at which the diffusion of liquid water within a vitrifying solid structure ceases and no further ice forms. The glass transition temperature of the maximally freeze-concentrated system, T_g', and the onset temperature of ice melting, T_m', are, therefore, independent of initial solids concentration (Roos and Karel, 1991b).

As shown by Roos and Karel (1991a) the T_g' and T_m' of food components increase with increasing molecular weight. Therefore, the T_g of gluten/starch systems with maximum freeze-concentration are relatively high. However, in mixtures, the glass transition temperature is a function of the glass transition temperatures and concentration of the miscible components. We have shown that the glass transition of dough decreases with added small molecular weight solutes, such as salts and sugars (Laaksonen and Roos, 2001a,b). Salts and sugars exist also naturally as flour components and the T_g' and T_m' values of a dough system seems to be in the vicinity of −30 °C (Räsänen *et al.*, 1998; Laaksonen and Roos, 2000, 2001a,b).

The glass transition temperature and maximum freeze-concentration in dough is related to the stability of the systems (Levine and Slade, 1990). Although ice recrystallisation rate is likely to increase with increasing temperature above T_m', rates of deteriorative reactions may also increase as a result of rapidly increasing molecular mobility. However, a major concern in dough storage is retention of desired viability of yeast and sufficient yeast activity after freezing. Maximum ice formation causes a high osmotic pressure in cells due to separation of water as ice and consequent extensive freeze-concentration. Maximum ice formation also results in the highest mechanical stress and cellular damage due to expansion as a result of the phase transition of water. Our studies have shown that temperatures above T_g' and T_m' or significantly below these to avoid maximum ice formation are needed during frozen dough storage to maximise the retention of yeast viability (Sieviläinen and Roos, 1996). It is important to note, however, that the conditions for maximum ice formation are affected by added solutes and in some applications added monosaccharides may enhance retention of yeast activity more than disaccharides (Sieviläinen and Roos, 1996). Furthermore, yeast activity in a prefermented dough affects the frozen state transitions of dough and the transition temperatures. According to our studies, the glass transition of a frozen dough decreases dramatically with increasing time of dough fermentation.

14.3 Dough properties in baking

Dough is a complicated polymer and solute system that undergoes several physical changes in baking. Changes occurring in baking are affected by state transitions of the dough components and redistribution of water within the system. These changes continue during storage of baked products, as the molecular mobility is not greatly restricted. However, dough is a concentrated system and the knowledge of effects of glass transition and glass transition related changes on molecular mobility is useful in describing time-dependent properties in baking and storage of baked products.

Schofield *et al.* (1983) showed that free sulfhydryl groups in glutenin were involved in rheological changes that occurred at temperatures between 55 and 75 °C. They postulated that glutenin proteins were unfolded on heating, which facilitated sulfhydryl/disulfide interchange between exposed groups. Similar phenomena occurred in gliadin proteins, which with glutenins belong to the low-molecular-weight (LMW) subunits of wheat gluten proteins (Shewry *et al.*, 1986), after heating at 100 °C. Slade *et al.* (1988) suggested that heating of gluten in the presence of water to above the glass transition temperature allows sufficient mobility, because of thermal and water plasticisation, for the molecules to form a thermoset network via disulfide cross-linking. According to Levine and Slade (1990), thermosetting of gluten polymers in baking is analogous to chemical curing and vulcanisation of rubber. Such thermosetting cannot occur in the glassy state of the protein and it seems that heating of dough and consequent protein denaturation together with thermal and water plasticisation allow thermosetting and formation of the elastic crumb structure.

Cocero and Kokini (1991) reported softening of glutenin above T_g, but the storage modulus, G', did not decrease to the typical value of the rubbery region. They assumed that the high G' value above T_g was due to thermosetting of the protein via formation of the disulfide cross-linking during heating. Kokini *et al.* (1994) reported state diagrams for cereal proteins. These state diagrams explained dough component plasticisation by water and the thermosetting properties of cereal proteins upon heating. Madeka and Kokini (1994) found a large increase in G' above 70 °C, which was assumed to reflect increasing elasticity resulting from the formation of a cross-linked protein network structure during heating of dough.

In dough, starch granules seem to exist separated from the elastic gluten–water phase. An increasing temperature during the baking process results in disruption of starch granules, melting of amorphous regions of amylopectin and release of non-crystalline amylose from the granules (Atwell *et al.*, 1988). The process results in water migration in dough and increases the water content of the starch phase allowing gelatinisation. At the same time viscosity increases. However, gelatinisation in baking is not complete and some of the starch granules remain ungelatinised with crystalline regions of the amylose and amylopectin molecules (Laine and Roos, 1994).

14.3.1 Molecular mobility in baked dough

The most obvious changes following baking involve crystallisation of amorphous amylose and amylopectin. The linear amylopectin molecules released from the starch granules crystallise rapidly (Miles *et al.*, 1984). The branched amylopectin molecules crystallise much slower and exhibit increasing crystallinity with storage of the baked product (Miles *et al.*, 1985). Crystallisation of amylopectin is considered as one of the main changes occurring in starch retrogradation and it is one of the causes for staling of bread. Retrogradation refers to changes that occur in starch paste, gel or starch-containing foods on storage. Starch retrogradation is a temperature- and time-dependent phenomenon, which may also be analysed as a crystallisation phenomenon similar to crystallisation of synthetic polymers (Flory, 1953; Roos and Jouppila, 2002). This agrees with the definition of starch retrogradation (Atwell *et al.*, 1988), stating that retrogradation occurs when molecules comprising gelatinised starch begin to reassociate in an ordered structure, which, under favourable conditions, results in an increase in crystalline order. Hence, the rate of retrogradation of starch is an important measure of molecular mobility in a baked product.

Changes in amylopectin crystallinity may be detected from a change in the appearance of a starch gel or paste, from changes in X-ray diffraction patterns, or from an increasing size of a melting endotherm in DSC scans with increasing storage time (Roos and Jouppila, 2002). The rate of retrogradation is dependent on the ratio of amylopectin and amylose, and on the molecular weight of the starch components. The rate of retrogradation is a function of molecular mobility and the process has similarities with crystallisation of synthetic polymers (Cornford *et al.*, 1964; McIver *et al.*, 1968; Roos and Jouppila, 2002).

The rate of starch retrogradation is dependent on water content. Zeleznak and Hoseney (1986) reported that retrogradation of gelatinised wheat starch with water contents between 20 and 80% (w/w) was slow at low and high water contents. An increase of melting enthalpy during storage at 25 °C showed a maximum at water contents between 50 and 60% (w/w). Levine and Slade (1990) have emphasised the use of the fringed-micelle model in the description of the partially crystalline structure of starch. The model is particularly useful in the description of the partially crystalline structure of starch gels, which contain amorphous regions and microcrystalline regions as junction zones between molecules. According to Levine and Slade (1990) retrogradation is a non-equilibrium polymer crystallisation process in completely amorphous starch–water melts, which proceeds with a rate that is determined by temperature and water content. The slow crystallisation of amylopectin was considered as a nucleation-limited growth process, which occurs above glass transition in a mobile, viscoelastic, fringed-micelle network. The crystallisation mechanism is thermally reversible above melting temperature and applies to both amylose and amylopectin.

Studies on bread staling have often concluded that staling occurs due to changes in the starch fraction, which shows increasing crystallinity as measured by X-ray diffraction (e.g. Cornford *et al.*, 1964; Zobel, 1973). Cornford *et al.* (1964) pointed out that the rate of increase in starch crystallinity increases with decreasing temperature. The crystallisation process in high-polymer systems was stated to be slow just below the melting temperature, to become faster with increasing supercooling, and then to decrease to zero at lower temperatures, where the molecular mobility is insufficient to allow crystallisation. Basically, the principles found to affect time-dependent crystallisation in starch gels apply also to starch retrogradation in bread, which has also been suggested by our results on crystallisation properties of gelatinised corn starch (Jouppila and Roos, 1997; Jouppila *et al.*, 1998).

14.4 Controlling molecular mobility to improve bread quality

14.4.1 Crystallinity of starch

The driving force for crystallisation is determined by the extent of supercooling below the equilibrium melting temperature or the extent of supersaturation (e.g. Roos, 1995). Amorphous food systems can be considered as supercooled and supersaturated materials, which have a temperature- and concentration-dependent driving force for crystallisation. Some of the main factors controlling the rates of crystallisation are probably diffusion of molecules to nucleation sites and their ability to reorient themselves into the crystal lattice structure. In concentrated systems, the diffusion of molecules may become restricted by slow molecular mobility, high viscosity and, thereby, the rates of crystallisation are likely to decrease. Theoretically, crystallisation ceases below the glass transition, as the molecules freeze in the solid, glassy state. Above the T_g, nucleation occurs rapidly, but the growth of crystals occurs slowly owing to the high viscosity and slow diffusion (Slade and Levine, 1995). At temperatures below, but close to the equilibrium melting temperature, T_m, nucleation occurs slowly but crystal growth is fast, as the driving force for nucleation decreases, but the mobility of the molecules increases (Slade and Levine, 1995; Jouppila and Roos, 1997; Jouppila *et al.*, 1998). Therefore, the maximum rate of crystallisation appears between the T_g and T_m of the dough polymer system (Jouppila and Roos, 2002), as illustrated in Fig. 14.3.

The studies of rates of crystallisation in starch have often used starch gels with 50% water or lower as models to agree with water contents of white bread. A well-agreed result is that the rate of crystallisation in gelatinised wheat starch containing 50% solids is higher at refrigeration temperatures than at room temperature (e.g. Colwell *et al.*, 1969; Roos and Jouppila, 2002). This has suggested that staling of bread may occur faster during storage at +4 °C than at room temperature. Colwell *et al.* (1969) found that the rate of crystallisation in gelatinised wheat starch containing 50% solids decreased with increasing storage temperature from 2 to 43 °C. They suggested that crystal formation occurred more rapidly at low temperatures, because the degree of supercooling and, therefore, the driving force for crystallisation was greater.

Our studies (Jouppila and Roos, 1997; Jouppila *et al.*, 1998; Roos and Jouppila, 2002) have suggested that the rate of crystallisation at all water contents is dependent on temperature in addition to the $T - T_g$. The rate constant, k, increased with increasing storage temperature with concomitant decrease in the half-time of crystallisation. The data suggested that the extent of crystallinity or the perfection of the crystalline structure also increased with temperature. Therefore, the rate of crystallisation may depend also on

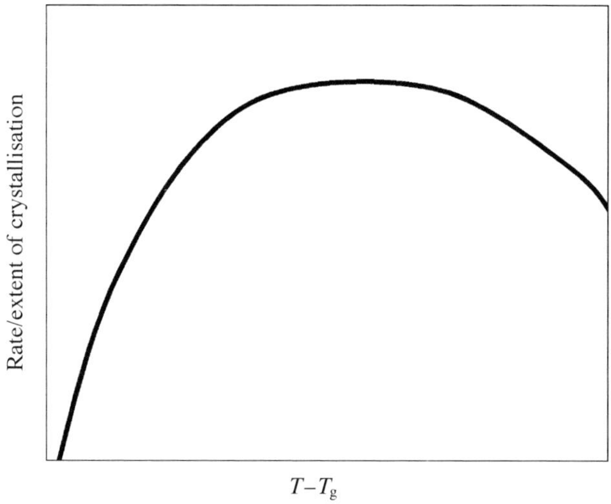

Fig. 14.3 Effect of plasticisation on the rate and extent of crystallisation of amorphous starch.

the crystalline structure formed in addition to the temperature and $T - T_g$ effects.

Our studies have also confirmed that the extent of crystallisation of gelatinised corn starch is a function of the $T - T_g$. The relationship between the extent of crystallinity and $T - T_g$ in crystallised corn starch is parabolic. The parabolic relationship between the extent of crystallisation and $T - T_g$ can be explained using the polymer crystallisation theory, as suggested by Slade and Levine (1991). At low $T - T_g$ conditions crystallisation seemed to occur to a lower extent, because molecular mobility is low. Therefore, the crystal growth is kinetically restricted and slow although nucleation is likely to occur rapidly. Moreover, the crystals formed at the beginning of storage at low $T - T_g$ conditions act as barriers for molecular rearrangements and crystallisation by forming rigid amorphous regions in the vicinity of the crystallites (Flory, 1953). At high $T - T_g$ conditions, crystal growth occurs rapidly and there is the possibility of the formation of larger and more perfect crystallites.

The parabolic relationship between the extent of crystallinity and $T - T_g$ suggested that the maximum extent of crystallisation in gelatinised corn starch occurred at a $T - T_g$ of 87 °C. Therefore, the predicted T_g for gelatinised corn starch containing 60% solids was at 55 °C, suggesting that a maximum extent of crystallisation was obtained at 32 °C. This agreed with the finding of Zeleznak and Hoseney (1986) that a maximum extent of crystallisation at 25 °C occurred in wheat starch gels containing 60% starch for which we predicted the $T - T_g$ to be 80 °C. The extent of crystallisation in wheat starch containing 50% solids has been found to decrease with increasing storage temperature at temperatures above 2 °C (Colwell *et al.*, 1969; Roos and Jouppila, 2002). Our predicted T_g for starch containing 50% solids was −78 °C (Jouppila and Roos, 1997), and, therefore, the predicted $T - T_g$ at 2 °C was 80 °C suggesting that a maximum extent of crystallisation in wheat starch containing 50% solids occurred at a $T - T_g$ of 80 °C or lower.

Starch concentration and the extent of crystallisation at a constant temperature may also be related to $T - T_g$ defined by water plasticisation. We have found that a maximum extent of crystallisation at 4 °C in wheat starch gel containing 50% solids corresponds to crystallisation at a $T - T_g$ of 82 °C (Jouppila and Roos, 1997). However, the temperature at which the maximum extent of crystallinity is produced is likely to be sensitive to product composition and water content, which should be taken into account when results of various systems, e.g. amorphous starch, starch gels, dough and bread, are compared. Starch gels and bread may also be at least partially amorphous and, therefore, more sensitive to water, as water plasticisation probably occurs only within the amorphous regions. It should also be remembered that the crystallinity may depend on the availability of water to hydrate the crystalline structure.

14.4.2 Crust formation and crispness

Crust formation and crispness can be considered as significant factors affecting the sensory properties of baked products. The molecular mobility in dough at early stages of baking is presumably high and affected by thermal and water plasticisation. Dehydration on the dough surface allows the formation of a surface layer with a higher concentration of solids and reactants. This concentration effect with reactant mobility is likely to control the rate of the non-enzymatic browning reaction and development of crust flavour. Furthermore, the reduced water content increases the glass transition temperature of the crust and contributes to the development of the crust crispness (Roos, 1995).

Several studies have investigated effects of glass transition, water activity and other factors on the rates of non-enzymatic browning (Bell *et al.*, 1998; Lievonen *et al.*, 1998). These studies have shown that the reaction rates are low in the glassy state and increase with increasing temperature above the glass transition. This increase may continue until other factors, such as dilution or a decreased temperature, reduce the rate of the reaction. In baking, the dough components and the glass transition of the crust at various stages of baking are probably the main factors affecting browning and flavour development. However, it should be remembered that other factors, such as temperature, pH, reactant concentration, etc., may affect the reaction rate and reaction products (Lievonen and Roos, 2002).

Crust crispness results from the decreased water content and solidification of the surface into the glassy state. Studies of bread crispness have suggested that several factors affect crispness. Obviously, a glassy state of the solids would provide brittleness to the product, but the porous structure of the product and solid, thin pore membranes may significantly contribute to the sensory properties of the product. Our studies have shown that crispness of porous carbohydrate structures typical of, for example, bread and breakfast cereals decreases rapidly around the glass transition of the products (Roos *et al.*, 1998). Roudaut *et al.* (1999a) have found that initial water sorption of white bread increases stiffness before crispness decreases at higher levels of water plasticisation. The mobility of water in the glassy systems, where the mobility of the solids is presumably very low, has been observed to remain high (Roudaut *et al.*, 1999a). This confirms that water molecules in baked systems may diffuse rapidly and water migration during product storage may significantly affect the state of other components during product storage.

14.5 Future trends

Most low-moisture and frozen foods contain amorphous carbohydrates, which exist either in the glassy state or in the more liquid-like supercooled, liquid state. These materials are miscible with water, which is often observed from its plasticising effect. Water plasticisation results in a decrease in the glass transition temperature range and, therefore, at some critical water content or water activity the glass transition occurs at or below ambient temperature. The depression of the glass transition to below ambient temperature is observed from dramatic changes in mechanical properties, decrease in viscosity and enhanced flow properties of the material. These changes suggest increased molecular mobility occurring above the glass transition which may also enhance crystallisation of sugars and carbohydrate polymers, as they approach the equilibrium crystalline state favoured over the non-equilibrium amorphous state. The glass transition seems to control the crystallisation of amorphous carbohydrates, including the apparent increase in crystallinity in retrogradation of starch. The crystallisation, however, is a complicated process affected by dough composition, temperature and water content, and it seems that the glass transition is not the sole factor contributing to the rate and extent of crystallisation of carbohydrates in dough and baked products. However, crystallisation of amorphous components is not likely to occur in glassy carbohydrate systems. Therefore, the knowledge of the glass transition can be used with temperature and water content adjustments to control favourable crystallisation in baking or to reduce detrimental crystallisation of starch carbohydrates during dough processing and storage. Understanding of molecular mobility is extremely important to frozen dough technology and the control of yeast survival and activity. The frozen dough technology with knowledge of molecular mobility and water migration in baking and storage of baked products, including the control of reactions responsible for crust flavour and sensory properties, as well as starch retrogradation are the most important areas of future research of molecular mobility in dough and baked products.

14.6 Sources of further information and advice

Molecular mobility in food systems and baked products has become a significant area of research, as the amorphous state of the systems define rates of changes in texture and contribute to rates of changes in processing and storage. The various phenomena have been discussed in detail by Harry Levine and Louise Slade (e.g. Levine and Slade, 1990; Slade and Levine, 1991; Slade and Levine, 1995). The physical state and molecular mobility-related changes in cereal proteins has been studied extensively by Jozef Kokini (e.g. Kokini *et al.*, 1994). The development of crystallinity in gelatinised starch in relation to molecular mobility in the starch system has been studied by Kirsi Jouppila and Yrjö Roos (Jouppila and Roos, 1997; Jouppila *et al.*, 1998). Many of the recent findings have been included in the book by Kaletunc and Breslauer (2002). Additional further information is available in numerous recent books on bread, water activity and glass transition.

14.7 References

ATWELL, W.A., HOOD, L.F., LINEBACK, D.R., VARRIANO-MARSTON, E. and ZOBEL, H.F. (1988) The terminology and methodology associated with basic starch phenomena. *Cereal Foods World* **33**: 306–311.

BELL L.N., TOUMA, D.E., WHITE, K.L. and CHEN, Y-H. (1998) Glycine loss and Maillard browning as related to the glass transition in a model food system. *J. Food Sci.* **63**: 625–628.

BIZOT, H., LE BAIL, P., LEROUX, B., DAVY, J., ROGER, P. and BULEON, A. (1997) Calorimetric evaluation of the glass transition in hydrated, linear and branched polyanhydroglucose compounds. *Carbohydr. Polym.* **32**: 33–50.

CHERIAN, G., GENNADIOS, A., WELLER, C. and CHINACHOTI, P. (1995) Thermomechanical behavior of wheat gluten films: effect of sucrose, glycerin, and sorbitol. *Cereal Chem.* **72**: 1–6.

COCERO, A.M. and KOKINI, J.L. (1991) The study of the glass transition of glutenin using small amplitude oscillatory rheological measurements and differential scanning calorimetry. *J. Rheol.* **35**: 257–270.

COLWELL, K.H., AXFORD, D.W.E., CHAMBERLAIN, N. and ELTON, G.A.H. (1969) Effect of storage temperature on the ageing of concentrated wheat starch gels. *J. Sci. Food Agric.* **20**: 550–555.

CORNFORD, S.J., AXFORD, D.W.E. and ELTON, G.A.H. (1964) The elastic modulus of bread crumb in linear compression in relation to staling. *Cereal Chem.* **41**: 216–229.

FLORY, P.J. (1953) *Principles of Polymer Chemistry*. Cornell University Press, Ithaca, NY.

GEORGET, D.M.R. and SMITH, A.C. (1996) The effects of sugars on the mechanical properties of processed cereals. *J. Thermal Anal.* **47**: 1377–1389.

GORDON, M. and TAYLOR, J.S. (1952) Ideal copolymers and the second order transitions of synthetic polymers. 1. Non-crystalline copolymers. *J. Appl. Chem.* **2**: 493–500.

HOSENEY, R.C., ZELEZNAK, K. and LAI, C.S. (1986) Wheat gluten: a glassy polymer. *Cereal Chem.* **63**: 285–286.

JOUPPILA, K. and ROOS, Y.H. (1994) Glass transitions and crystallization in milk powders. *J.*

Dairy Sci. **77**: 2907–2915.

JOUPPILA, K. and ROOS, Y.H. (1997) The physical state of amorphous corn starch and its impact on crystallization. *Carbohydr. Polym.* **32**: 95–104.

JOUPPILA, K., KANSIKAS, J. and ROOS, Y.H. (1998) Factors affecting crystallization and crystallization kinetics in amorphous corn starch. *Carbohydr. Polym.* **36**: 143–149.

KALETUNC, G. and BRESLAUER, K. (2002) *Characterization of Cereals and Flours: Properties, Analysis, and Applications*, Marcel Dekker, New York.

KALICHEVSKY, M.T. and BLANSHARD, J.M.V. (1992) A study of the effect of water on the glass transition of 1:1 mixtures of amylopectin, casein and gluten using DSC and DMTA. *Carbohydr. Polym.* **19**: 271–278.

KALICHEVSKY, M.T. and BLANSHARD, J.M.V. (1993) The effect of fructose and water on the glass transition of amylopectin. *Carbohydr. Polym.* **20**: 107–113.

KALICHEVSKY, M.T., JAROSZKIEWICZ, E.M., ABLETT, S., BLANSHARD, J.M.V. and LILLFORD, P.J. (1992a) The glass transition of amylopectin measured by DSC, DMTA and NMR. *Carbohydr. Polym.* **18**: 77–88.

KALICHEVSKY, M.T., JAROSZKIEWICZ, E.M. and BLANSHARD, J.M.V. (1992b) Glass transition of gluten. 1: Gluten and gluten-sugar mixtures. *Int. J. Biol. Macromol.* **14**: 257–267.

KOKINI, J.L., COCERO, A.M., MADEKA, H. and DE GRAAF, E. (1994) The development of state diagrams for cereal proteins. *Trends Food Sci. Technol.* **5**: 281–288.

LAAKSONEN, T.J. and ROOS, Y.H. (2000) Thermal, dynamic-mechanical, and dielectric analysis of phase and state transitions of frozen wheat doughs. *J. Cereal Sci.* **32**: 281–292.

LAAKSONEN, T.J. and ROOS, Y.H. (2001a) Dielectric relaxations of frozen wheat doughs containing sucrose, NaCl, ascorbic acid, and their mixtures. *J. Cereal Sci.* **33**: 331–339.

LAAKSONEN, T.J. and ROOS, Y.H. (2001b) Thermal and dynamic-mechanical properties of frozen wheat doughs with added sucrose, NaCl, ascorbic acid, and their mixtures. *Int. J. Food Properties,* **4**: 201–213.

LAAKSONEN, T.J. and ROOS, Y.H. (2003) Water sorption and dielectric relaxations of wheat dough containing sucrose, NaCl, and their mixtures. *J. Cereal Sci.* **37**: 319–326.

LAAKSONEN, T.J., ROOS, Y.H. and LABUZA, T.P. (2001) Comparisons of the use of desiccators with or without vacuum for water sorption and glass transition studies. *Int. J. Food Properties*, **4**: 545–563.

LAINE, M.J.K. and ROOS, Y. (1994) Water plasticization and recrystallization of starch in relation to glass transition. In *ISOPOW Practicum II, Proceedings of the Poster Session*, A. Argaiz, A. López-Malo, E. Palou and P. Corte (eds.), pp. 109–112. Universidad de las Américas Puebla Cholula, Puebla, México.

LEOPOLD, A.C., SUN, W.Q. and BERNAL-LUGO, I. (1994) The glassy state in seeds: analysis and function. *Seed Sci. Res.* **4**: 267–274.

LEVINE, H. and SLADE, L. (1990) Influences of the glassy and rubbery states on the thermal, mechanical, and structural properties of doughs and baked products. In *Dough Rheology and Baked Product Texture*, ed. H. Faridi and J.M. Faubion. AVI Publishing Co., New York, pp. 157–330.

LIEVONEN, S.M. and ROOS, Y.H. (2002) Nonenzymatic browning in amorphous food models: Effects of glass transition and water. *J. Food Sci.* **67**: 2100–2106.

LIEVONEN, S.M., LAAKSONEN, T.J. and ROOS, Y.H. (1998) Glass transition and reaction rates: Nonenzymatic browning in glassy and liquid systems. *J. Agric. Food Chem.* **46**: 2778–2784.

MADEKA, H. and KOKINI, J.L. (1994) Changes in rheological properties of gliadin as a function of temperature and moisture: development of a state diagram. *J. Food Eng.* **22**: 241–252.

MARSH, R.D.L. and BLANSHARD, J.M.V. (1988) The application of polymer crystal growth theory to the kinetics of formation of the β-amylose polymorph in a 50% wheat-starch gel. *Carbohydr. Polym.* **9**: 301–317.

MCIVER, R.G., AXFORD, D.W.E., COLWELL, K.H. and ELTON, G.A.H. (1968) Kinetic study of the retrogradation of gelatinized starch. *J. Sci. Food Agric.* **19**: 560–564.

MILES, M.J., MORRIS, V.J. and RING, S.G. (1984) Some recent observations on the retrogradation of amylose. *Carbohydr. Polym.* **4**: 73–77.

MILES, M.J., MORRIS, V.J., ORFORD, P.D. and RING, S.G. (1985) The roles of amylose and amylopectin in the gelation and retrogradation of starch. *Carbohydr. Res.* **135**: 271–281.

NICHOLLS, R.J., APPELQVIST, I.A.M., DAVIES, A.P., INGMAN, S.J. and LILLFORD, P.J. (1995) Glass transition and the fracture behaviour of gluten and starches within the glassy state. *J. Cereal Sci.* **21**: 25–36.

NIKOLAIDIS, A. and LABUZA, T.P. (1996a) Use of dynamic mechanical thermal analysis (DMTA): Glass transitions of a cracker and its dough. *J. Thermal Anal.* **47**: 1315–1328.

NIKOLAIDIS, A. and LABUZA, T.P. (1996b) Glass transition state diagram of a baked cracker and its relationship to gluten. *J. Food Sci.* **61**: 803–806.

NOEL, T.R., PARKER, R., RING, S.G. and TATHAM, A.S. (1995) The glass-transition behaviour of wheat gluten proteins. *Int. J. Biol. Macromol.* **17**: 81–85.

POUPLIN, M., REDL, A. and GONTARD, N. (1999) Glass transition of wheat gluten plasticized with water, glycerol, or sorbitol. *J. Agric. Food Chem.* **47**: 538–543.

RÄSÄNEN, J., BLANSHARD, J.M.V., MITCHELL, J.R., DERBYSHIRE, W. and AUTIO, K. (1998) Properties of frozen wheat doughs at subzero temperatures. *J. Cereal Sci.* **28**: 1–14.

ROLEE, A. and LE MESTE, M. (1999) Effect of moisture content on thermomechanical behavior of concentrated wheat starch-water preparations. *Cereal Chem.* **76**: 452–458.

ROOS, Y.H. (1995) Phase Transitions in Foods. Academic Press, San Diego, CA.

ROOS, Y.H. and JOUPPILA, K. (2002) Plasticization effect of water on carbohydrates in relation to crystallization. In *Characterization of Cereals and Flours: Properties, Analysis, and Applications*, G. Kaletunc and K. Breslauer (eds). Marcel Dekker, New York. In press.

ROOS, Y. and KAREL, M. (1991a) Water and molecular weight effects on glass transitions in amorphous carbohydrates and carbohydrate solutions. *J. Food Sci.* **56**: 1676–1681.

ROOS, Y. and KAREL, M. (1991b) Amorphous state and delayed ice formation in sucrose solutions. *Int. J. Food Sci. Technol.* **26**: 553–566.

ROOS, Y.H., KAREL, M. and KOKINI, J.L. (1996) Glass transitions in low moisture and frozen foods: effects on shelf life and quality. *Food Technol.* **50**(11): 95–108.

ROOS, Y.H., ROININEN, K., JOUPPILA, K. and TUORILA, H. (1998) Glass transition and water plasticization effects on crispness of a snack food extrudate. *Int. J. of Food Properties* **1**: 163–180.

ROUDAUT, G., VAN DUSSCHOTEN, D, VAN AS, H., HEMMINGA, M.A. and LE MESTE, M. (1998) Mobility of lipids in low moisture bread as studied by NMR. *J. Cereal Sci.* **28**: 147–155.

ROUDAUT, G., MAGLIONE, M., DUSSCHOTEN, D. and LE MESTE, M. (1999a) Molecular mobility in glassy bread: a multispectroscopy approach. *Cereal Chem.* **76**: 70–77.

ROUDAUT, G., MAGLIONE, M. and LE MESTE, M. (1999b) Relaxations below glass transition temperature in bread and its components. *Cereal Chem.* **76**: 78–81.

SCHOFIELD, J.D., BOTTOMLEY, R.C., TIMMS, M.F. and BOOTH, M.R. (1983) The effect of heat on wheat gluten and the involvement of sulphydryl–disulphide interchange reactions. *J. Cereal Sci.* **1**: 241–253.

SHEWRY, P.R., TATHAM, A.S., FORDE, J., KREIS, M. and MIFLIN, B.J. (1986) The classification and nomenclature of wheat gluten proteins: a reassessment. *J. Cereal Sci.* **4**: 97–106.

SIEVILÄINEN, E. and ROOS, Y.H. (1996) Cryopreservation of baker's yeast in frozen dough. ISOPOW VI, 2–8 March, 1996, Santa Rosa, CA.

SLADE, L. and LEVINE, H. (1991) Beyond water activity: recent advances based on an alternative approach to the assessment of food quality and safety. *Crit. Rev. Food Sci. Nutr.* **30**: 115–360.

SLADE, L. and LEVINE, H. (1995) Glass transitions and water-food structure interactions. *Adv. Food Nutr. Res.* **38**: 103–269.

SLADE, L., LEVINE, H. and FINLEY, J.W. (1988) Protein-water interactions: Water as a plasticizer of gluten and other protein polymers. In *Protein Quality and the Effects of Processing*, D. Phillips and J.W. Finlay (eds). Marcel Dekker, New York, pp. 9–124.

SLADE, L., LEVINE, H. and FINLEY, J.W. (1989) Protein–water interactions: water as a plasticizer of gluten and other protein polymers. In *Protein Quality and the Effects of Processing*. R.D. Phillips and Finley, J.W. (eds). Marcel Dekker, New York, pp. 9–124.

SPERLING, L.H. (1992) *Introduction to Physical Polymer Science*. New York: John Wiley and Sons.

VAN DEN BERG, C. (1981) *Vapour sorption equilibria and other water-starch interactions; a physico-chemical approach*. Doctoral thesis, Agricultural University, Wageningen, the Netherlands.

WHITE, G.W. and CAKEBREAD S.H. (1966) The glassy state in certain sugar-containing food products. *J Food Technol.* **1**: 73–82.

ZELEZNAK, K.J. and HOSENEY, R.C. (1986) The role of water in the retrogradation of wheat starch gels and bread crumb. *Cereal Chem.* **63**: 407–411.

ZELEZNAK, K.J. and HOSENEY, R.C. (1987) The glass transition in starch. *Cereal Chem.* **64**: 121–124.

ZOBEL, H.F. (1973) A review of bread staling. *Baker's Dig.* **47**(10): 52–61.

15

The role of water in dough formation and bread quality

A. Schiraldi and D. Fessas, University of Milan, Italy

15.1 Introduction

Once wheat flour is mixed with water to prepare a dough, the system undergoes amazing macroscopic changes: the flour powder forms an apparently homogeneous sticky wet paste that, under suitable kneading, absorbs the residual flour powder and slowly turns into a rubber-like dough. These transformations take place during 5 minutes of mixing and can be satisfactorily monitored with a standard Farinograph that also allows a rough characterization of the flour on the basis of a few empirical parameters, namely: the time required to attain a maximum mixing torque, the value (in arbitrary units, or Brabender units) of the maximum torque, and the 'length' of its decay after reaching the maximum value. Although very practical for routine applications, this approach cannot reveal the mechanism of the dough-making process at the molecular and supra-molecular level and therefore cannot be of help to those who aim at enhancing the dough quality through recipe modifications, mechanical stress–relaxation treatments, leavening and baking conditions.

To follow such rheological changes various authors using a wide range of experimental approaches have carried a number of investigations. Some evaluation methods, such as calorimetry and traditional thermal analyses and rheology, allow detection of macroscopic properties, such as heat capacity, transition enthalpy, thermodynamic stability, viscoelastic behavior (Schiraldi *et al.,* 1999). Others, such as nuclear magnetic resonance (NMR), mainly relaxometry (Hills, 1999; Leung *et al.,* 1983; Vittadini, 1998; Kim, 2001), and magnetic resonance imaging (MRI) (Hall *et al.*, 2001) and near infrared reflectance (NIR) (Wesley *et al.,* 1998; Alava *et al.,* 2001) can provide information about phenomena that involve single molecules or relatively small clusters of molecules; these techniques are usually very sensitive and specific for single components of the dough, although the relevant experimental outputs can be affected to some extent (that depends on the specific technique) by other dough components. Whenever the target of the investigation is water, all these approaches produce evidence of many coexisting states of water molecules within a bread dough (Kim, 2001; Kou *et al.*, 2002). These states can be singled out and split from one another because of the different relaxation times related to the short-range mobility, as in the case of NMR investigations, or, to a minor extent, because of changes in the dough domains where water has the highest mobility, as in the case of calorimetry, other thermal analyses and rheological investigations. Most of these methods allow a continuous monitoring of the water partition in the course of breadmaking.

It is, however, a matter of fact that different experimental techniques have their own perspective and therefore enhance the detection of some phenomena while being unable to shed light over many others. An integrated picture may indeed be achieved only when, or if, all the pieces of information drawn from various kinds of investigations can be matched together, following some main guideline that allows one to accomplish the final arduous task, namely, to achieve a comprehensive interpretation of what is currently observed by a baker who is preparing a bread dough. In other words, the understanding of phenomena at the microscopic and mesoscopic level coming from sophisticated investigations must be 'scaled up' to provide a clear explanation of the overall macroscopic behavior of a dough.

A very powerful tool, drawn from the experience of those who studied water solutions, is the 'watery eye' (Bockris and Reddy, 1973), since water, which is ubiquitous in the dough, can be used as a reliable probe for the changes that take place in the various steps of the dough preparation. A series of questions pave the way toward a comprehensive view of the evolution of the system:

- Where is water primarily conveyed once it comes in contact with the flour powder?
- Which molecules or molecular aggregates compete for water?
- What is the result of this competition?
- How does the partition of water within the system correlate with the state of the dough before and after leavening, and after baking?

15.2 Dough as a disperse system

The bread dough is a foam because of the air cells trapped in the mixing process. Foams are, by definition, disperse systems; however, bread dough, like many other food products, is much more complex a system than a simple dispersion of bubbles in an aqueous medium. It can indeed be referred to as a hierarchy of dispersed gaseous, liquid and solid phases, which can be dramatically modified by changes in temperature and moisture, as well as by mechanical stresses.

The state of dispersion implies the presence of domains which are kept apart from one another because of the overall medium hindrance that opposes the formation of bulk phases layered according to the relevant densities. If some excess water is added, such a layering can be actually achieved by ultra-centrifugation that simply accelerates the change of a finely dispersed system into separated layers of bulk phases (Larrson and Eliasson, 1996). One can also recognize that a bread dough is indeed formed by several aqueous phases, each rich in a given component of the dough – gluten, starch granules, globular proteins, non-starch carbohydrates, etc. This evidence supports the description of a bread dough as a metastable dispersed system with a huge inter-phase surface across which water can move from one phase to another (Tolstoguzov, 1997).

The main parameter governing this phase separation is the difference between excluded volumes of polymer components of the dough: the resulting immiscibility reflects the so-called thermodynamic incompatibility between different polymers, such as proteins and polysaccharides (Grinberg and Tolstoguzov, 1997). The separated aqueous phases form fine water-in-water emulsions (1–5 μm droplets) where water exchanges are of an osmotic nature (Tolstoguzov, 2000), that is they are driven by gradients of water chemical potential. In a standard dough these droplets can undergo a partial coalescence giving rise to domains that behave like hydrophobic bodies separated by an aqueous inter-phase layer that hosts most of the amphiphilic compounds, like fats and globular proteins (Sahi, 1994; Gan *et al.,* 1999). These compounds consequently play a crucial role in stabilizing size and distribution of the air cells that directly affect the volume and the texture of the final baked product (Gan *et al.,* 1990).

Phase separation in flour dough, experimentally verified by Larsson and Eliasson (1996), is in line with the expected thermodynamic incompatibility between starch carbohydrates and pentosans (Zasyskin *et al*., 1997) and between these and flour soluble proteins (Grinberg and Tolstoguzov, 1997). Gliadin and glutenin fractions of wheat flour are not miscible with albumins, globulins, starch and non-starch polysaccharides (Grinberg and Tolstoguzov, 1997; Tolstoguzov, 2000). It was recently found (Fessas and Schiraldi, 2003) that aggregation of gliadins which takes place at a lower temperature when their concentration is increased, occurred at a lower temperature in the presence of arabinoxylans, as is indeed expected in the case of a phase separation produced in aqueous mixtures of incompatible polymers.

The formation of a 3 D network inhibits the attainment of a true thermodynamic equilibrium between the co-existing phases (Tolstoguzov, 1997) of a dough, since the partition of water is hindered by the local viscosity. However, when some energy is transferred to the system through mechanical stirring, the hindrance is partially overcome since the polymer chains tend to align along the main shear direction: the more ordered structure allows an easier interaction between polymer chains (e.g., increased number of hydrogen bonds) and between polymers and water (Sahi, 1994; Tolstoguzov, 1997). If the temperature is below the threshold of starch gelatinization (50–60 °C), stirring makes starch granules behave like ball-bearing bodies that actively contribute to reshaping the gluten domains that eventually form layers that wrap around the starch granules (Tolstoguzov, 2000). Electron microscopy images of a freshly prepared unleavened dough indeed show an apparently continuous protein phase spotted with homogeneously dispersed starch granules (Parker *et al.,* 1990).

Phase separation and dispersion at the start of mixing a dough undergo modifications during the rest after mixing, during proofing and, above all, during baking. The phenomena that take place at room temperature or below the temperature threshold of starch gelatinization are mainly attributable to displacements of water, which depend on and/or affect the strains and the strain relaxations within the dough, including those related to the expansion of the gas phase during proofing.

Further changes occur when starch undergoes gelatinization and the gluten phase stiffens like a thermosetting polymer (Slade and Levine, 1995). The picture of starch granules wrapped up by a soft gluten phase changes into a starch gel (where the original contours of granules can be still recognized; (Willhoft, 1973) interrupted by a thermoset gluten network (Tolstoguzov, 2000). Since amylose and amylopectin are incompatible with each other in aqueous medium (Kalichevsky and Ring, 1987) the number of dispersed phases therefore increases after starch gelatinization. The real number of separated and dispersed phases within a dough above the threshold of starch gelatinization may therefore exceed that of the phases separated by means of ultra-centrifugation. Accordingly a huge expansion of the inter-phase region has to be expected: this might indeed be referred to as a major 'component' of bread dough and bread crumb.

15.3 Water displacements in dough

A large water fraction of the system has a rather high mobility according to the relevant T_2 relaxation time determined in NMR investigations (Kim, 2001; Kou *et al.,* 2002) and is responsible for large value of relative humidity (RH) of the dough (above 95%) (Czuchajowska and Pomeranz, 1989). Another water fraction behaves like a structure component of the starch gel and the gluten network: this fraction is hard to remove and does not contribute to the overall RH of the system (Schiraldi and Fessas, 2003).

Each of these water fractions is indeed split into several families of molecules that specifically interact with the various components of the dough. Those that are less tightly bound can be exchanged and therefore should be supposed to flip from one environment to another, mainly under the effects of short-range forces. Because of the phase separation and dispersion described above, these displacements take place across phase boundaries and therefore depend on the extension of the separation surfaces and the viscosity of the inter-phase regions. For this reason water migration between dough phases is directly affected by low molecular-mass compounds, such as simple sugars and oligosaccharides, lipids and surfactants (Sahi, 1994; Tolstoguzov, 1997) and, as in any colloidal aqueous system, salts, the role of which is related to their position in the lyotropic series. However, even larger effects are produced by polymer additives (Fessas and Schiraldi, 1998).

Many of these 'minor' ingredients are often added to a standard dough recipe with the aim of improving the quality of the final baked product; however, they play a specific role mainly in one or two steps of the breadmaking process, i.e. when they can really affect the relevant water displacements. It is therefore expedient to review the main steps of the breadmaking process describing them with the eyes of water.

15.3.1 Dough mixing

The proportions of the various populations of water molecules can change with mixing and during the rest period after mixing (Fessas and Schiraldi, 2001a; Schiraldi, 2002). For example, changes that take place after mixing encompass one to two hour span and therefore prove that a steady partition of water between different dough phases is not a fast process. The long-range displacements of water within a dough can be monitored with MRI sophisticated equipments (Hills, 1998; Hall *et al.*, 2001) whereas local motions can be revealed by means of NMR relaxometry (Kou *et al.*, 2002). Relaxation of nuclear spins is related to the interaction between a given kind of spin and its close environment (spin–lattice) and to the interactions between individual spins (spin–spin), with relaxation times T_1 and T_2, respectively. Both T_1 and T_2 are related to the local molecular mobility: the larger the relaxation time, the higher the mobility of the molecules that host the relaxing nuclear spin (although this is not necessarily true for T_1 in a glassy environment).

According to Richardson *et al.* (1986) the ^{1}H spin–spin relaxation signal recorded from dough samples with T_2 in the millisecond range is mainly related to mobile water molecules that interact with flour components. Protons of water molecules tightly bound to the substrates or protons of starch polysaccharides and gluten moieties have T_2 in the range of microseconds. By using a CPMG (Carr–Purcell–Meiboom–Gill) pulse sequence with intervals of 100–200 μs, one can single out signals from the more mobile protons (Ruan and Chen, 1998). The Free Induction Decay (FID) of the spin–spin relaxation of a given kind of nucleus can be reliably described with an exponential law. Because of the various kinds of water molecules, it is a reasonable expectation that the overall FID detected from a dough sample can be split in several exponential components, each with its own T_2 (Yong-Ro, 2001): such a split should, however, take into account cross-over processes and therefore cannot be fully reliable. With this limitation in mind, a deconvolution of the overall FID from a dough in three exponential components with T_2 in the 2–5, 9–18, and 50–200 ms range, respectively (Kim, 2001), seems likely. This means that at least three main kinds of water molecules can be distinguished because of differences of their local molecular mobility.

Both the relative FID intensity, which is roughly proportional to the number of protons of a given type, and the relevant T_2 value change during mixing and during rest after mixing (Kim, 2001). In particular, the mobility of water in a dough decreases with mixing and the partition of water over the three main families of mobile molecules is modified as the number of those with intermediate mobility becomes larger at the expenses of that of the other two populations (Kim, 2001). If the 'loops-and-trains' model of the gluten structure proposed by Belton (1999) is accepted, then the above findings can be interpreted as the result of a decrease of the number of 'loops' (protein–water hydrogen bonds) and an increase of the number of 'trains' (protein–protein hydrogen bonds) following the stretching of the kneaded dough. Over-mixed dough samples are more sticky, in spite of the fact that gluten is expected to become weaker: this behavior, which looks like that of a thixotropic gel with a water content that may be increased by over-mixing, can been tentatively interpreted as the result of a change in the gluten structure (Shewry *et al.,* 2000).

When the dough is allowed to rest after mixing, it relaxes, coming back to the initial proportion of 'loops' and 'trains': this produces a decrease of the number of protons with intermediate mobility. Since the relevant T_2 also decreases during rest (Kim, 2001) one may conclude that the mobility of this kind of water molecules decreases, possibly because their interactions with the gluten become stronger (Kim, 2001). This conclusion is in line with the finding that the glass transition temperature, T_g, of the dough increases with resting (Piazza and Schiraldi, 1997).

The above picture of mobile water molecules in a dough can be improved by describing the dough as a system formed by several micro domains exchanging water molecules between one another (Kou *et al.,* 2002). In this approach the NMR relaxation data are used to define a continuous distribution of molecular mobility around some mean values of T_2 that characterize the interaction of water with single dough components: two main broad T_2 domains are accordingly recognized (Kou *et al.,* 2002).

A much simpler experimental evidence can be obtained with thermo-gravimetry (TG) investigations. The TG-DSC (differential scanning calorimetry) instruments available today combine TG with DSC and provide both outputs, namely, heat flux, HF (dQ/dt), and mass loss (m) versus $T(t)$, the latter being also in the derivative form, DTG (dm/dt). The HF/DTG ratio allows a simple check of the enthalpy drop related to the mass loss, since HF/DTG = $dQ/dm = \Delta_{vap}H$ (in J g^{-1} units). In the case of dehydration processes the evaluated enthalpy drop is always close to 2.2 J g^{-1} (Schiraldi and Fessas, 2003) which is the vaporization enthalpy of pure water. This evaluation cannot be of help in separating out the enthalpies of the different water fractions of a given food system, since the differences between them are some order of magnitude smaller than the vaporization enthalpy. However one can recognize different states of water because of the different temperatures at which water is released during an experimental run.

In the case of wheat flour dough, the DTG record shows multiple peaks and/or shouldered peak that can be deconvoluted in main Gaussian components, each relevant to a fraction of the dough moisture (Fessas and Schiraldi, 2001a). An approximate picture can be achieved by splitting the DTG signal into a couple of main gaussian components, with maximums below and well above 100 °C, respectively (Fessas and Schiraldi, 2001a). The latter can be referred to as structural water, in the sense that it is an element of the structure of the substrate. It has been recognized that this fraction is mainly linked to the gluten component (Fessas and Schiraldi, 2001a). The relevant DTG peak is relatively narrow and accounts for about 15% of the overall moisture content of the dough. The temperature of the maximum of this DTG peak moves toward lower temperatures when the overall dough moisture is increased, and toward higher temperatures when the dough is left at rest after mixing (Fessas and Schiraldi, 2001a). No substantial changes are instead observed for the position of the first Gaussian component, which mainly depends on sample mass and heating rate in the same way as a diffusion-limited process that obeys Fick's law (Fessas and Schiraldi, 2001a).

It must be noted that the release of water that takes place in the course of the TG run strongly depends on the starting conditions of the system; this means that mixing and kneading (which are carried out at room temperature) produce the water partition in the two main fractions described above. These fractions are not substantially modified during the TG run in spite of important changes that take place, namely, starch gelatinization and gluten reticulation. This is because, in the experimental conditions of a TG run, both processes imply short-range displacements of water: in other words, starch and gluten mainly involve the next neighboring water molecules to undergo gelatinization and networking, respectively.

The overall result of such a behavior is that the DTG trace can be used as a record of the water partition attained before the experimental run. This is why the DTG traces of samples from freshly mixed dough and from dough allowed to rest for a couple of hours are different. Since such structure relaxations mainly affect the gluten component of the dough (starch in a mixed dough is still in the state of dispersed wetted granules), the part of the DTG trace where the relevant changes are detected is the high temperature peak (Fig. 15.1).

15.4 Dough proofing and baking

During proofing a disperse gas phase is produced because of the leavening process. The air trapped within the dough in the course of the previous mixing allows the growth of gas cells: without these early dispersion of air throughout the dough, the leaven CO_2 would be quickly lost and the final loaf volume would be rather poor. Gas pressure and dough overall viscosity, which are

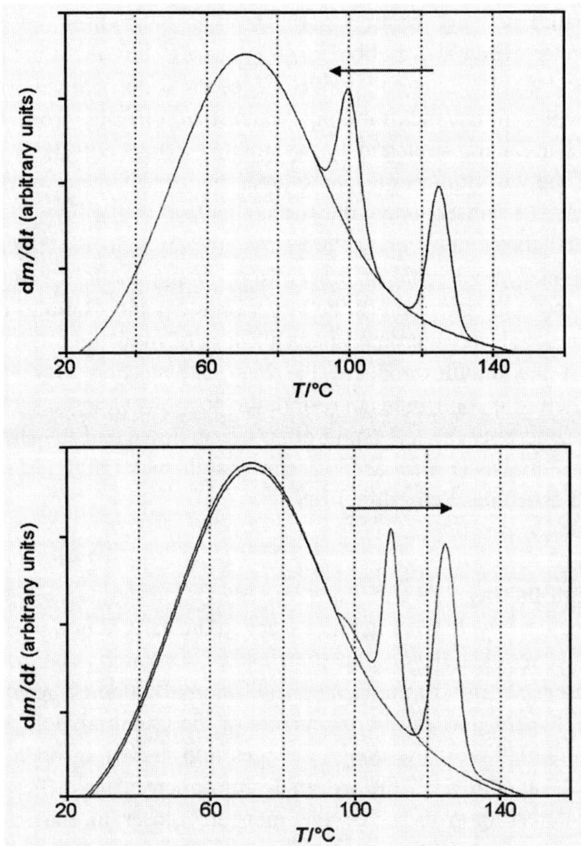

Fig. 15.1 Schematic representation of the DTG records from wheat flour. Top: the high temperature peak shifts toward lower temperature with increasing the overall dough moisture. Bottom: the high temperature peak shifts toward lower temperature when the dough is over-mixed (an 'intermediate' shoulder appears between the main components of the signal, but it comes back to the starting position after a two hour rest).

mainly related to water vaporization and gluten ripening (attained during mixing), respectively, govern the growth of gas bubbles within the dough (Mitchell *et al.,* 1998).

The gas cells are lined by an aqueous layer that hosts gas–liquid and liquid–liquid surfactants, such as globular proteins (flour albumins and globulins) and polar lipids (Gan *et al.,* 1990; MacRitchie, 1976). The leaven CO_2 is partially dissolved in this liquid layer and contributes to the overall internal pressure of the bubbles. This contribution is substantial in the early phase of proofing and during the oven spring (the early phase of baking) but becomes practically negligible when the water vapor pressure increases with increasing temperature: at 70 ºC the pressure in the bubbles is mostly related to water vapor (Mitchell *et al.,* 1998).

A gluten film network hosting starch granules surrounds the bubbles and becomes thinner when the bubble size increases, but the thickness of the aqueous layer does not change that much, since the water that vaporizes into the bubbles is replaced by the imbibing water that diffuses from neighboring regions, which are being stretched by the bubble expansion itself. Because of the expansion of the gas–liquid inter-phase area, the surfactant to water mass ratio decreases and, as a consequence, the surface tension of the lining aqueous film increases. The balance between gas pressure and compliance of the surrounding matrix determines the bubble size according to the Laplace principle: small bubbles tend to disappear whereas the larger ones expand. Bubbles closer to the surface of the dough loaf may pop out with loss of moisture.

Rupture of the bubbles reduces the overall extension of the inter-phase regions and therefore contributes to restore the previous water to surfactant mass ratio. At the same time the leavening effects drop down and the stiffness of the bubble walls increases because of the starch gelatinization and gluten reticulation (Fessas and Schiraldi, 1998).

15.4.1 Dough baking

An enhancement of the leavening takes place in the early stages of baking (oven spring), but it cannot last beyond the threshold of the starch gelatinization which causes coalescence and rupture of many internal bubbles resulting in the formation of cavities that are the precursors of the crumb alveoli.

The foamy structure of the dough changes into that of an open sponge that releases water vapor toward the external atmosphere. The diffusion of water follows an interesting path, since the moisture closer to the surface freely evaporates, while that of the loaf core experiences two generalized forces, namely, the concentration gradient, which sustains the water migration toward the surface, and the temperature gradient which has the opposite effect. The competition between these forces produces different water displacement according to the distance from the loaf surface. The innermost internal regions see water moving toward the loaf core, as the temperature gradient prevails, whereas in the regions closer to the surface water displacements are directed outward.

Because of the temperature gradient, starch gelatinization does not occur simultaneously throughout the dough loaf, but starts first in the more external regions and progresses if a sufficient water supply is available. It is almost (see below) exhausted after 5 minutes in the crust region, where the system experiences the highest temperature, and after 15 minutes in the under-crust of the loaf (Zanoni *et al.*, 1991). In the innermost internal regions, the temperature never attains 100 °C and the moisture content remains large enough to sustain a complete starch gelatinization in about 20 minutes (Zanoni *et al.*, 1991).

The overall water content of the system decreases during baking (from about 45 down to about 35% w/w) since much water vapor is released into the external atmosphere. This loss of water is sustained by a Fickian diffusion (Fessas and

Schiraldi, 2001a) with a rate that goes through a maximum at about 80 °C and then decreases, since the diffusion of water through the starch gel is slower than in the not yet gelatinized system.

A careful determination of starch gelatinization in a bread dough was simulated with *ad hoc* DSC investigations (Fessas and Schiraldi, 2000) carried out with pierced pans so as to allow water vaporization from the sample. The extent of starch gelatinization was assessed after a previous heat treatment, during which the moisture content of the sample decreased because of the vaporization. A TTT (Temperature, Time, Transformation) diagram was finally defined (Fessas and Schiraldi, 2000) that predicts an incomplete starch gelatinization in the water poorest regions of the dough.

The RH of the bread crumb at room temperature remains above 95% (Schiraldi *et al.,* 1996) which means that, even after baking, a fraction of water is relatively free to evaporate. Beside this fraction, which accounts for about 85% of the total crumb moisture (Fessas and Schiraldi, 2001a) there is some 'structure' water that can be delivered only at temperatures above 100 °C (Fessas and Schiraldi, 2001a).

The proportion between these water fractions changes during bread aging (in sealed bags), since some more mobile molecules are displaced toward sites where they can be more tightly fixed. The water displacement can be viewed either macroscopically, as the result of the crumb-to-crust concentration gradient (Piazza and Masi, 1995), or at the molecular level, as the temperature, T_{vap}, at which the more tightly bound fraction is released, increases with aging (Schiraldi and Fessas, 2003); this means an increase of the relevant configuration entropy, $R \ln(a_W)$, as long as

$$T_{vap} \approx \frac{\Delta_{vap}H}{A + R \ln a_w}$$

where $\Delta_{vap}H$ and a_W stand for vaporization enthalpy and water activity, respectively, at T_{vap}, and A is a constant value of about 100 J mol^{-1} K^{-1}, and R is the universal gas constant.

Water activity of bread crumb decreases with aging (Schiraldi *et al.,* 1996) while moisture migrates toward the crust (Piazza and Masi, 1995). Because of this partial dehydration and the concurrent amylopectin retrogradation, the bread crumb becomes firmer and harsher. Any ingredient that can release water reduces the rate of dehydration (Piazza *et al.,* 1996; Schiraldi and Fessas, 2001) and the overall bread staling.

15.5 Dough freezing

Frozen dough preparations are nowadays rather common in industrial bakery since they are used as long-lasting raw materials for easy-to-transport and ready-to-produce freshly baked foods that meet the desire of the consumers. A storage temperature as low as −25 °C is believed adequate to preserve the dough from microbial spoilage at the cost of minor physical damage of the yeast cells. A deeper inspection of the system, however, reveals that a real stability of the system would be attained only at temperatures below −30 °C (Laaksonen and Roos, 2000). Beside the storage temperature, other major issues of frozen dough preparations are the cooling and the thawing rates, since both of them are related to the growth of ice crystals which are detrimental for the overall performance of the material.

The dream of a breadmaker should be the preparation of a fully amorphous frozen dough, as the result of an infinitely fast quenching of the fresh product, which may be fully recovered with an equally quick thawing. This dream seems far from becoming true, but it is not completely foolish since frozen dough is largely amorphous: any decrease of its crystalline fraction, namely, ice, would be a step forward.

The formation of ice is a matter of competition between the rate of action of the driving forces that sustain ice nucleation and growth and the rate of cooling and/or thawing that do not allow the system to fully relax toward its thermodynamically stable state. However, much of the dough water does not crystallize, since the relevant phase becomes too viscous to allow nucleation and growth of a solid phase. This is indeed the result of a freeze–concentration process that subtracts water from liquid phases. The process produces strains within the dough structure, which, if the cooling treatment is sufficiently rapid, cannot relax. The details of this mechanism in a multi-component and multi-phase system, like a bread dough, remain largely unexplained.

It is well known that molecular displacements are almost totally hindered when the temperature of the system falls below the threshold of the so-called 'glass transition' which actually encompasses a temperature range where the viscosity of the system undergoes a change of several orders of magnitude. Usually this temperature range can be singled out by means of thermo-mechanical analysis (TMA) and dynamic mechanical analysis (DMA) and/or calorimetric (DSC and MTDSC) investigations carried out on reheating the frozen system (Laaksonen and Roos, 1996). A detailed description of the experimental findings relevant to the glass transition is not in the scope of the present chapter, but the reader can easily get adequate information from the literature (Levine and Slade, 1989; Roos, 1995; Goff *et al.* 2002).

The main concern here is the role of water during the freezing process. Water molecules migrate toward the ice embryos, which behave as seeds, leaving the original liquid phases that become more concentrated and more viscous. When no further displacement of water molecules is allowed from a given phase, namely when the temperature approaches the 'local' glass transition threshold, no further ice growth takes place in the surrounding regions. However, the process can still be in progress in the vicinity of other aqueous phases which have not yet reached their own glass transition. This means that, at the nano- and meso-scale level, no ice front can be actually recognized (Cornillon *et al.,* 1995) as the process is indeed jeopardized throughout the system. NMR relaxometry and MRI can be of help in as much as the signal intensity in the ms range decreases following the trend of the number of liquid-like protons, allowing a rough evaluation of the freezable *vs.* the unfreezable water (Cornillon *et al.*, 1995; Kerr *et al.*, 1998; Hills, 1999).

It can be also argued that some phases, where water molecules can be still easily displaced, can supply solvent toward those that tend to attain the low mobility condition earlier. Such a process would require some hours of annealing (Räsänen *et al.*, 1998) well above the lowest glass transition temperature (the long lasting storage of frozen dough preparations at −25 °C might meet this requirement), or a very slow cooling treatment. This water displacement would make the phases hosting some residual liquid water attain the same viscosity level at which ice growth is hindered almost simultaneously. This could explain why a frozen dough shows a single maximum freeze–concentration glass transition (at $T = T'_g$(Laaksonen and Roos 2000)). The relevant solute concentrations (in % w/w units) can nonetheless be rather far apart, depending on the molecular mass of the solutes: a diluted aqueous solution of a polymer can be as viscous as a concentrate aqueous solution of simple sugars at the same temperature (Fessas and Schiraldi, 2001b).

15.6 Future trends

The present knowledge of physics and chemistry of breadmaking will certainly be improved through a better understanding of the role of inter-phases, across which exchanges of water take place with a rate that depends on the concentration of low-molecular-mass solutes and water soluble proteins and non-starch carbohydrates. The inter-phase extension in a bread dough is rather large and can be affected by the conditions of breadmaking, such as duration and strength of mixing, rest after mixing, proofing, and the thermal history experienced during baking. As a result of different mechanical and thermal histories, different breads can be obtained from the same dough recipe.

Different flours or flour blends may, however, require different breadmaking procedures, because of their specific composition. One should take into account the role of each ingredient and the interactions between ingredients, including the competition for the available water.

A better understanding of these issues should reduce the presently wide attitude to look for ‘magic’ additives to perfect breadmaking, which is a physical rather than chemical process.

15.7 Sources of further information and advice

For more detailed information on this subject, including interpretations and models that sometimes differ from those presented in this chapter, the reader is addressed to some important sources, including:

- *Interactions: The Key to Cereal Quality*, R.J. Hamer and R.C. Hoseney, Eds., AACC Publ., St. Paul, Minnesota (1998).
- *Baked Goods Freshness*, R.E. Hebeda and H. Zobel, Eds., Marcel Dekker, Inc. Publ., New York, Basel, Hong Kong (1996).
- *Bread Staling*, Y. Vodovotz and P. Chinachoti, Eds., CRC Press, Boca Raton, FL (2001).
- *Starch and Starch Containing Origins – Structure, Properties and New Technologies*, V.P. Yuryev, A. Cesaro, W. Bergthaler, Eds., Nova Science Publisher, NY (2002).
- *Magnetic Resonance Imaging in Food Science*, B.P. Hills, John Wiley & Sons, Inc., NY (1998).
- *Bubbles in Food*, G.M. Campbell, C. Webb, S.S. Pandiella and K. Niranjan Eds., Eagan Press, St. Paul, MN, USA (1999).
- *Amorphous Food and Pharmaceutical Systems*, H. Levine, Ed., The Royal Society of Chemistry, Cambridge (2002).

15.8 References

ALAVA, J.M., MILLAR, S.M. and SALMON, S.E. (2001) The determination of wheat breadmaking performance and bread dough mixing time by NIR spectroscopy for high speed mixers. *J. Cereal Sci.*, **33,** 71–81.

BELTON, P.S. (1999) On the elasticity of wheat gluten. *J. Cereal Sci.*, **29**, 103–107.

BOCKRIS, J.O.M. and REDDY, A.K.N. (1973) *Modern Electrochemistry*, Plenum Press, New York.

CORNILLON, P., ANDRIEU, J., DUPLAN, J-C. and LAURENT, M. (1995) Use of NMR to model thermophysical properties of frozen and unfrozen model food gels. *J. Food Eng.*, **25**, 1–19.

CZUCHAJOWSKA, Z. and POMERANZ, Y. (1989) Differential scanning calorimetry, water activity, and moisture contents in crumb center and near-crust zones of bread during storage. *Cereal Chem.*, **66**, 305–309.

FESSAS, D. and SCHIRALDI, A. (1998) Texture and staling of wheat bread crumb: effects of water extractable proteins and 'pentosans'. *Thermochim. Acta*, **323**, 17–26.

FESSAS, D. and SCHIRALDI, A. (2000) Starch gelatinization kinetics in bread dough: DSC investigations on 'simulated' baking processes. *J. Therm. Anal. Calorim.*, **61**, 411–423.

FESSAS, D. and SCHIRALDI, A. (2001a) Water properties in wheat flour dough I: classical thermogravimetry approach. *Food Chem.*, **72**, 237–244.

FESSAS, D. and SCHIRALDI, A. (2001b) State diagrams of arabinoxylan-water binaries. *Thermochim. Acta*, **370**, 83–89.

FESSAS, D. and SCHIRALDI, A. (2003) Interactions between arabinoxylans and wheat proteins, *Food Chem.*, submitted.

GAN, Z., ANGOLD, R.E., WILLIAMS, M.R., ELLIS, P.R., VAUGHAN, J.G. and GALLIARD, T. (1990) The microstructure and gas retention of bread dough. *J. Cereal Sci.*, **12**, 15–24.

GAN, Z., VAN DER GRAAF, J., LEONARD, S.A., BROOKER, B.E., PARKER, M.L. and SCHOFIELD, J.D. (1999) The role of wheat proteins and polar lipids in the stabilisation of the foam structure of dough, in *Bubbles in Food*, G.M. Campbell, C. Webb, S.S. Pandiella and K. Niranjan, Eds., Eagan Press, St. Paul, MN, USA, pp. 89–93.

GOFF, H.D., MONTOYA, K. and SAHAGIAN, M.E. (2002) The effect of microstructure on the complex glass transition occurring in frozen sucrose model systems and foods, in *Amorphous Food and Pharmaceutical Systems*, H. Levine, Ed., The Royal Society of Chemistry, Cambridge, pp. 145–157.

GRINBERG, V.Y. and TOLSTOGUZOV, V.B. (1997) Thermodynamic incompatibility of proteins and polysaccharides in solutions. *Food Hydrocolloids*, **11**, 145–158.

HALL, L-D, AMIN, M.H.H., EVANS, S., NOTT, K.P. and SUN, L. (2001) Quantitation of diffusion and mass transfer of water by MRI, in *Water Science for Food, Health, Agriculture and Environment*. Z. Berk, R.B. Leslie, P.J. Lillford and S. Mizrahi, Eds., Technomic Publ., Lancaster, PA, USA, pp. 255–271.

HILLS, B.P. (1998) *Magnetic Resonance Imaging in Food Science*, John Wiley & Sons, Inc., New York.

HILLS, B.P. (1999) NMR studies of water mobility in foods, in *Water Management in the Design and Distribution of Quality Foods*. Y.H. Roos, R.B. Leslie and P.J. Lillford, Eds., Technomic Publ. Co., Lancaster, PA, USA, pp. 107–131.

KALICHEVSKY, M.T. and RING, S. (1987) Incompatibility of amylose and amylopectin in aqueous solution. *Carbohydrate Res.*, **162**, 323–328.

KERR, W.L., KAUTEN, R.J., MCCARTHY, M.J. and REID, D.S. (1998) Monitoring the formation of ice during food freezing by magnetic resonance imaging. *Lebensm. Wiss. u. Technol.*, **31**, 215–220.

KIM, Y.-R. (2001) *Physicochemical properties of hard wheat flour dough influenced by processing*, PhD Thesis, Purdue University.

KOU, Y., ROSS, E.W. and TAUB, I.A. (2002) Microstructural domains in foods: effect of constituents on the dynamics of water in dough, as studied by magnetic resonance spectroscopy, in *Amorphous Food and Pharmaceutical Systems*. H. Levine, Ed., The Royal Society of Chemistry, Cambridge, pp. 48–58.

LAAKSONEN, T.J. and ROOS, Y.H. (2000) Thermal, dynamic-mechanical and dielectric analysis of phase and state transitions of frozen wheat doughs. *J. Cereal Sci.*, **32**, 281–292.

LARSSON, H. and ELIASSON, A.-C. (1996) Phase separation of wheat flour dough studied by ultra-centrifugation and stress relaxation. *Cereal Chem.*, **73**, 18–31.

LEUNG, H.K., MAGNUSON, J.A. and BRUINSMA, B.L. (1983) Water binding of wheat flour doughs and breads as studied by deuteron relaxation. *J. Food Sci.*, **48**, 95–99.

LEVINE, H. and SLADE, L. (1989) Influences of the glassy and rubbery states on the thermal, mechanical and structural properties of doughs and baked products, in *Dough Rheology and Baked Product Texture*. H. Faridi and J.M. Faubion, Eds., Van Nostrand Reinhold/AVI, New York, pp. 157–330.

MACRITCHIE, F. (1976) The liquid phase of dough and its role in baking. *Cereal Chem.*, **53**, 318–326.

MITCHELL, J.R., FAN, J-T. and BLANSHARD, J.M.V. (1998) Simulation of bubble growth in cereal systems, in *Bubbles in Food*, G.M. Campbell, C. Webb, S.S. Pandiella, K. Niranjan, Eds., Eagan Press, St. Paul, MN, USA, pp. 107–112.

PARKER, M.L., MILLS, E.N.C. and MORGAN, M.R.A. (1990) The potential of immunoprobes for locating storage proteins in wheat endosperm and bread. *J. Sci. Food Agric.*, **52**, 35–45.

PIAZZA, L. and MASI, P. (1995) Moisture redistribution throughout the bread loaf during staling and its effect on mechanical properties. *Cereal Chem.*, **72**, 320–325.

PIAZZA, L. and SCHIRALDI, A. (1997) Correlation between fracture of semi-sweet hard biscuits and dough viscoelastic properties. *J. Texture Studies*, **28**, 523–541.

PIAZZA, L., SCHIRALDI, A., BRENNA, O. and VITTADINI, E. (1996) Structure and properties of bread dough and crumb. *J. Thermal Analysis*, **47**, 1339–1360.

RÄSÄNEN, J., BLANSHARD, J.M.V., MITCHELL, J.R., DERBYSHIRE, W. and AUTIO, K. (1998) Properties of frozen wheat doughs at subzero temperatures. *J. Cereal Sci.*, **28**, 1–14.

RICHARDSON, S.J., BAIANU, I.C. and STEINBERG, M.P. (1986) Mobility of water in wheat flour suspensions as studied by ^{1}H and ^{17}O NMR, *J. Agric. Food Chem.*, **34**, 17–23.

ROOS, Y.H. (1995) *Phase Transitions in Foods*. Academic Press, New York, USA.

RUAN, R.R. and CHEN, P.L. (1998) *Water in Foods and Biological Materials. A Nuclear Magnetic Resonance Approach*, Technomic Publ. Co, Lancaster, PA, USA.

SAHI, S.S. (1994) Interfacial properties of the aqueous phases of wheat flour doughs. *J. Cereal Sci.*, **20,** 119–127.

SCHIRALDI, A. (2002) Water partition in starch products: thermophysical methods and nuclear magnetic resonance applications, in *Starch and Starch Containing Origins – Structure, Properties and New Technologies*, V.P. Yuryev, A. Cesaro, W. Bergthaler, Eds., Nova Science Publisher, NY, USA, pp. 287–295.

SCHIRALDI, A. and FESSAS, D. (2001) Bread staling: an overview, in *Bread Staling*, Y. Vodovotz and P. Chinachoti, Eds., CRC Press, Boca Raton, FL, pp. 1–17.

SCHIRALDI, A. and FESSAS, D. (2003) Classical and Knudsen Thermogravimetry to check States and Displacements of Water in Food Systems. *J. Therm. Anal. Cal.*, **71**, 221–231.

SCHIRALDI, A., PIAZZA, L. and RIVA, M. (1996) Bread staling: a calorimetric approach. *Cereal Chem.*, **73**, 32–39.

SCHIRALDI, A., PIAZZA, L., FESSAS, D. and RIVA, M. (1999) Thermal analyses in foods and food processes, in *Handbook of Thermal Analysis and Calorimetry*, Vol. 4 *From Macromolecules to Man*, chapter 16, R. Kemp, Ed., Elsevier Publ., Amsterdam, pp. 829–921.

SHEWRY, P.R., POPINEAU, Y., LAFIANDRA, D. and BELTON, P.S. (2000) Wheat glutenin subunits and dough elasticity: findings of the EUROWHEAT project. *Trends Food Sci. Technol.*, **11**, 433–441.

SLADE, L AND LEVINE, H. (1995) Glass transition and water–food interactions. *Adv. Food Nutr. Res.*, **38**, 103–269.

TOLSTOGUZOV, V.B. (1997) Thermodynamic aspects of dough formation and functionality. *Food Hydrocolloids*, **11**, 181–193.

TOLSTOGUZOV, V.B. (2000) Food as dispersed system. *J. Therm. Anal. Calorim.*, **61,** 397–409.

VITTADINI, E. (1998) *Water mobility in heterogeneous systems as examined by 1H, 2H and 17O NMR*. PhD Thesis, Univ. of Massachusetts at Amherst.

WESLEY, I.J., LARSEN, N., OSBORNE, B.G. and SHERRITT, J.H. (1998) Non invasive monitoring of dough mixing by near infrared spectroscopy. *J. Cereal Sci.*, **27**, 61–69.

WILLHOFT, E.M.A. (1973) Mechanism and theory of staling of bread and baked goods, and associated changes in the textural properties. *J. Texture Stud.*, **4**, 292–322.

ZANONI, B., SMALDONE, D. and SCHIRALDI, A. (1991) Starch gelatinization in chemically leavened bread baking. *J. Food Sci.*, **56**, 1702–1706.

ZASYSKIN, D.V., BRAUDO, E.E. and TOLSTOGUZOV, V.B. (1997) Multicomponent biopolymer gels. *Food Hydrocolloids*, **11**, 159–170.

16

Foam formation in dough and bread quality

P. Wilde, Institute of Food Research, UK

16.1 Introduction

One of the principal components, which has a major influence on the textural and sensory attributes of bread, is the gas phase (Campbell and Mougeot, 1999). Gas makes up more than 70% of the final volume of bread, and the main challenge is controlling the gas phase volume. The main components of the flour, such as the gluten and starch, clearly have a massive influence on the properties of the final product, but the presence of gas cells within the bread is essential. The size and number of gas cells can vary greatly between bread types, and will change the sensory properties of the bread. For example, a standard tin loaf is characterised by numerous small gas cells, whereas baguette-type bread typically has fewer, larger gas cells. The formation and stability of gas cells or bubbles in dough during the baking process are therefore critical factors for the overall quality and sensory attributes of bread. This chapter will introduce the main principles underlying the formation and stability of foams in general, followed by a discussion of the importance of the different components and processes. Some techniques which have been applied to cereal systems will also be described.

A foam is defined as a suspension of gas bubbles in a continuous liquid phase. Sometimes the liquid phase may be solidified (in the case of bread) to 'freeze' the foam structure in place. Liquid foams are unique structures in that by combining two fluids of low viscosity (e.g. air and water), one can create a self-supporting structure with increased viscoelastic properties. The problem with foams is that they are intrinsically unstable. Pure liquids are incapable of supporting a foam, and require surface active molecules to form a film around the bubbles to stabilise them against coalescence. The obvious example is the addition of detergent to a running bath. Bubbles are generated in the absence of detergent, but only in the presence of a detergent are they stabilised to form a foam. There are two fundamental stages in the lifetime of a foam: formation and stability.

16.2 Foam formation

Foam is formed when air bubbles are incorporated into the liquid phase upon agitation; in breadmaking, this occurs during the mixing stage. Once bubbles are incorporated, they may be broken up further during mixing or agitation, thus reducing the mean bubble size, and increasing the number of bubbles. The size of the bubbles is usually very important, as this can influence how the gas cells grow during proving, and ultimately the texture of the final product. The size of the gas cells created is generally limited by three main factors: (1) energy input during mixing; (2) continuous phase viscosity; and (3) surface tension.

During foam formation, the amount of energy used (i.e. speed and time of mixing) can influence the bubble size distribution (Hanselmann and Windhab, 1998). Greater mixing energies will incorporate more gas, and will also be able to break up existing bubbles more effectively. Figure 16.1 shows how a bubble is broken in a shear field. In a greater shear field, the bubble experiences larger dispersion forces trying to break it up, until eventually it does break up. In a similar way, the rheology of the continuous phase can influence the shear field in such a way to increase the dispersion forces experienced by the bubbles. Increasing the ratio of the continuous phase : dispersed phase viscosities increases the transfer of energy to the bubble. Hence higher-viscosity systems tend to produce smaller bubbles, for a given mixing speed. This has been

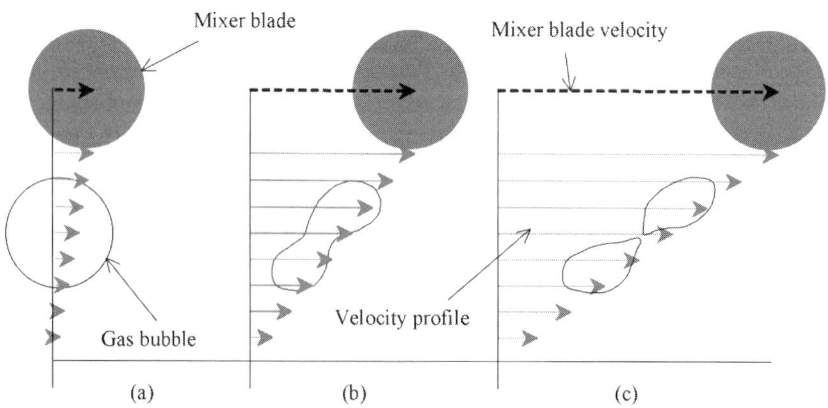

Fig. 16.1 Bubble break-up during mixing. (a), (b) and (c) represent increasing mixer speed. At low speed (a), the velocity gradient (shear field) across the bubble is small, resulting in little or no deformation. Bubble becomes longer and more deformed as mixer speed, hence the velocity gradient across the bubble increases. Finally, the shear field becomes large enough to tear the droplet apart (c).

observed in both foam (Koczo and Racz, 1991) and emulsion systems (Pandolfe, 1981). This is also related to the energy input of the system, as higher-viscosity phases will require more energy to maintain a given mixing speed.

Surface tension is an important factor, particularly for foams formed in low-viscosity liquids. Creating a foam or emulsion involves the formation of many small bubbles or droplets. This results in a huge increase in interfacial area of the system. The energy barrier to this surface expansion is the surface free energy or surface tension. The surface or interfacial tension is defined as the force per unit length acting on a surface. Therefore a lower surface tension will require less energy to produce a specific increase in surface area. Foam formation processes, such as whipping or mixing, occur on short time scales ($\ll 0.1$ s). As more interfacial area is produced, rapid adsorption is required to occupy the newly created surface.

Hence, it has been shown that proteins that lower the surface tension rapidly are most efficient at creating foams (Graham and Phillips, 1976; Kalischewski and Schurgerl, 1979; Kim and Kinsella, 1985; Lorient *et al.,* 1989; Nakai and Li-Chan, 1993). In the case of proteins, the rate of change of interfacial tension is heavily influenced by the molecular weight and hydrophobicity. Smaller molecular weight proteins will diffuse more quickly to the interface and often result in increased functionality (Grunden *et al.*, 1974). Hydrophobic proteins, if soluble, will lower the surface tension more effectively, and improve functionality (Horiuchi *et al.*, 1978; Kato and Nakai, 1980; Townsend and Nakai, 1983; Slack and Bamforth, 1983; Mitchell, 1986).

A parameter that is specific to breadmaking is the pressure during mixing (Campbell *et al.*, 1998). By mixing at different pressures, it is possible to control the level of oxidation required for the development of the gluten network. Variations in pressure can also control the development of the gas cells. Increased pressure during mixing allows high levels of gas to be incorporated, to accelerate the oxidation process, but it results in large gas cell sizes in the final dough. If a partial vacuum is subsequently applied during mixing, the resulting dough has much smaller gas cells. This results in complete, rapid oxidation and a fine crumb structure. By using different pressure regimes, breads of very different gas cell structures can be produced from the same basic formulation. For example, the very open gas cell structure of 'baguette' style bread can be created using increased pressure through the entire mixing process (Cauvain *et al.*, 1999).

It is well known therefore that energy input, viscosity, surface tension and mixing pressure can influence the bubble sizes created during foam formation. Specifically, during the breadmaking process, it is not clear how significant the different factors are. Clearly, the mixing speed and pressure have a large impact on bubble size and structure, but the balance between the surface tension and dough rheology is not clear and requires further investigation.

Once the gas cells have been created, they act as nucleation sites for CO_2 produced during yeast fermentation. The gas cells then expand, causing the dough to rise. The structure of the dough is now dependent on the foam stability processes to maintain the overall desired gas cell structure. The main foam stability processes will now be discussed.

16.3 Foam stability

Gas bubbles are generally around a thousand times less dense than the aqueous phase, so they tend to migrate rapidly upwards. This induces rapid drainage of liquid downwards. Hence the bubbles come into close contact very quickly. For a foam to remain stable, each gas cell must remain as a separate discrete entity for as long as possible. There are three main instability mechanisms which control the overall structure of a foam: drainage, coalescence and disproportionation.

16.3.1 Foam drainage

Foam drainage is simply the amount or rate at which the continuous aqueous phase drains out of the foam structure. This is most important for low-viscosity fluids, where hydrodynamic forces allow rapid flow of the liquid between the bubbles. The flow of liquid is mainly influenced by the bubble size, the properties of the bubble surface, the density and rheology of the aqueous phase (Haas and Johnson, 1967). The bubble size is important, mainly because it defines the size, and number of the gaps between bubbles (plateau borders), which contain most of the liquid in the foam. Therefore a smaller bubble size distribution results in slower drainage and a more dense foam. The properties of the bubble surface can influence the flow of liquid past them. A rigid, elastic surface will slow down the flow of liquid and hence drainage, whereas a fluid surface will allow more rapid fluid flow and hence more rapid drainage (Wilde, 1996). The biggest influences are the density and rheology of the aqeuous phase. Foam drainage is mediated by gravity, so higher-density liquid will drain more rapidly, and because the other issue is fluid flow between bubbles, fluids with a higher viscosity will drain more slowly (Haas and Johnson, 1967). In the case of bread dough, the rheological properties of the dough are such that drainage is not really an issue on a bulk scale. That is, the viscoelastic modulus of most doughs is high enough to prevent movement against gravity, of the individual gas cells, over the time scale of the normal breadmaking process. Without this movement, drainage will not take place.

16.3.2 Coalescence

However, during the later stages of proving, the bubbles will increasingly make contact, and may not be separated by the gluten network. Then the drainage properties of the thin liquid film which is thought to separate the bubbles at this stage, becomes very important for the stability of the adjoining gas cells.

Pure liquids (such as water) are not capable of supporting a foam. By agitating water, bubbles will be formed, determined by the parameters discussed above.

However, the bubbles will soon collapse, or coalesce with each other, until the foam is completely destroyed. Therefore, to stabilise the foam, a protective, adsorbed layer of molecules is formed around the bubbles. This layer helps prevent coalescence, and hence allows the formation of a foam that may be stable for long periods, depending on the surface properties of the adsorbed layer.

Surface charge, dynamics and rheology of adsorbed layers are thought to be determinant factors for stabilising foams against coalescence. A prerequisite for coalescence to occur is that the surfaces of neighbouring bubbles must be close enough for the thin film between them to rupture. Therefore, to prevent close approach, large repulsive forces may be employed. These are usually electrostatic or steric forces. In foams, the bubbles are forced into close proximity, and the long-range electrostatic and steric forces are no longer effective. In this situation, two main stability mechanisms have been identified and are visualised in Fig. 16.2:

- The **viscoelastic** mechanism requires the lamellae to be stabilised by adsorbed layers with considerable interfacial elasticity. The proteins or polymers forming this adsorbed layer effectively form a 2D gel at the surface, with equivalent viscoelastic properties to a 3D polymer gel. If a lamella is deformed, the elastic adsorbed layer stretches and deforms. As long as the deformation is within the elastic limit, and they remain intact, the two adsorbed layers will return to their original position, thus forming a physical barrier between the gas cells, preventing coalescence.
- The **Gibbs–Marangoni** mechanism relies on a highly fluid adsorbed layer of surfactant, detergent or emulsifier. Deformation of a lamella will cause local thinning and deplete the local surfactant concentration. The molecules will naturally migrate to the depleted area to reduce the concentration gradient. The flow of the surfactants will drag interlamellar fluid to the thinner region and hence restore it to its original thickness.

The problem with many food systems is that in addition to protein, they often contain surfactants, in the form of lipids, detergents, emulsifiers, etc. These are either present as endogenous components or added as functional improvers. The problem is that these low-molecular-weight surfactants are often more surface active (i.e. can achieve lower surface tensions) than proteins, and will therefore compete for interfacial area (**competitive adsorption**). There has been much interest in the effect of competitive adsorption of surfactants and proteins on functionality, particularly in the field of food colloids, where this effect is most commonly found. It is known that adding surfactants to protein stabilised foams will have a detrimental effect on stability (Halling, 1981; Lee and Tynan, 1988; Pearce *et al.*, 1991; Dickinson *et al.*, 1993; Chen *et al.*, 1993). Figure 16.3 shows how the competitive adsorption is thought to destabilise foam lamellae. In fact surfactants are often utilised in this way as antifoaming agents (Lee and Tynan, 1988). Surfactants effectively adsorb in the protein adsorbed layer, and reduce the surface viscoelasticity which is vital for the stability of protein foams (MacRitchie, 1978; Halling, 1981; Dickinson, 1992; Wüstneck *et al.*, 1996). In

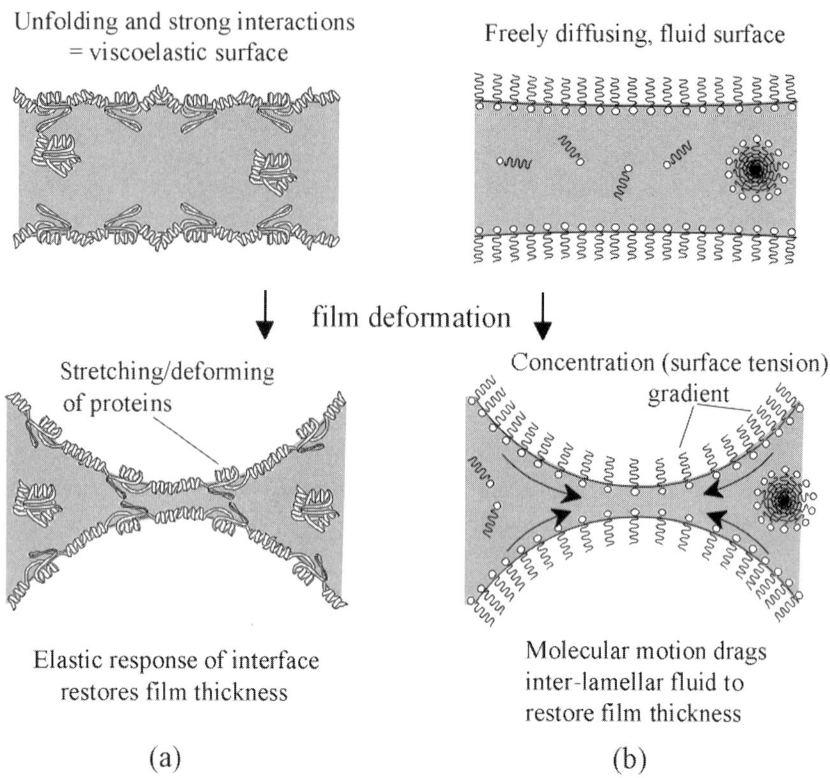

Fig. 16.2 Foam film stabilisation mechanisms. (a) Viscoelastic mechanism employed by proteins: Adsorbed proteins unfold and interact to form viscoelastic surfaces which are resistant to deformation. (b) Gibbs–Marangoni mechanism used by surfactants. Surfactants respond to changes in surface concentration by rapid migration to regions of low concentration. This movement drags associated solvent to thinner regions of the film and restores the film thickness.

contrast, the presence of the remaining protein hinders the diffusion of the surfactant (Clark *et al.*, 1994a), thus the surfactant is unable to stabilise using the Gibbs–Marangoni mechanism. The result is a less stable foam. Recent advances in surface imaging techniques have revealed the structure of mixed protein surfactant adsorbed layers (Mackie *et al.*, 1999). These studies show that the surfactant forms domains within the protein network, thus weakening it. The discrete domains restrict diffusion of the surfactant within the domains, limiting their effectiveness.

16.3.3 Disproportionation

The final instability mechanism is disproportionation, which is the transport of gas between bubbles (Garrett, 1993). The driving force is the Laplace pressure within bubbles generated by the surface tension on the bubble surface. Surface

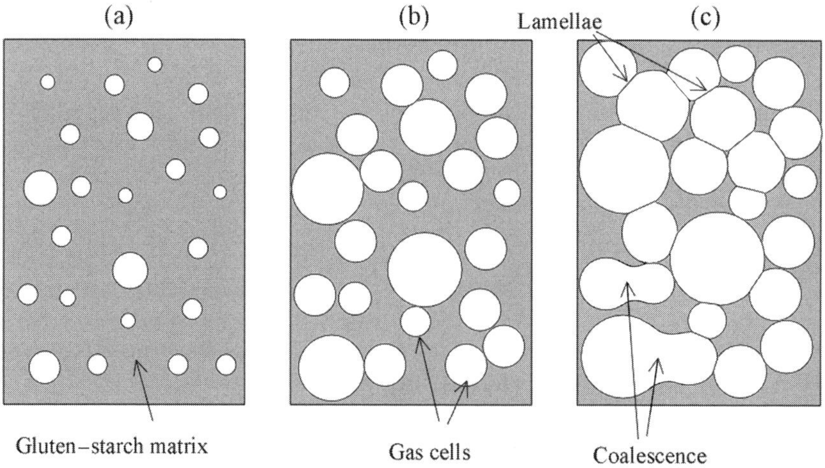

Fig. 16.3 Bubble evolution during proof. (a) Post-mixing, the gas cells are separated and supported by the gluten–starch matrix. (b) Mid-way through proof, the gas cells grow and begin to make contact. (c) End of proof, gluten–starch matrix is excluded from between some bubbles, leaving thin liquid lamellae. Unstable lamellae result in coalescence.

tension acts all around the bubble surface, effectively trying to contract the bubble, keeping it spherical. This contraction of the bubble induces a small pressure increase inside the bubble, which is known as the Laplace pressure, and is defined as:

$$\Delta P = \frac{2\gamma}{r}$$

where ΔP is the Laplace pressure difference between the inside and outside of the bubble, γ is the surface tension and r is the radius of the bubble. Hence, ΔP is greater in smaller bubbles. Therefore, when two bubbles come into close proximity, there will be a net transfer of gas from the smaller bubble to the larger one. Eventually the smaller bubbles completely disappear, and the whole foam structure becomes more coarse with time. The rate of transfer and hence foam coarsening will depend on the bubble size difference, and the solubility of the gas. Other factors that may be important are the surface elasticity (Lucassen, 1981) and the surface density (Quoc *et al.*, 2002).

Surface elasticity is the ability of a surface to resist deformation. It has been shown semi-theoretically that if the surface elastic modulus was greater than $\gamma/2$, then disproportionation would be stopped (Lucassen, 1981), as the elastic properties of the surface of the bubble would resist the expansion or shrinkage of the bubbles. However, recent experimental work has cast doubts on this hypothesis (Dickinson *et al.*, 2002).

The other barrier to disproportionation is the permeability of the surface layer. Increasing the packing density of surfactants on a surface can dramatically decrease the permeability of the adsorbed layer to oxygen (Quoc *et al.*, 2002). Bread doughs are more complex, because it is not simply the case of transport of gas between the bubbles, as gas (CO_2) is continually being produced by the yeast or leavening agents, and the net transport is into the bubbles from the surrounding aqueous phase. Therefore it is still a debatable point whether disproportionation actually occurs in dough systems. Recent studies suggest that the Laplace pressure can play an effect, such that the rate of gas transport into the smaller bubbles is slower, because the internal pressure is higher. Depending on the partial pressures of CO_2 in the surrounding phase, the smaller bubbles may stop growing, and in fact may disappear (Shimiya and Nakamura, 1997; Shah *et al.*, 1998).

16.4 Surface active dough components

When considering foams, there are three main classes of surface active material in food systems: polymers, lipids and emulsifiers. They differ in their molecular properties and surface behaviour, and as discussed above, they can often behave antagonistically at the surface.

16.4.1 Polymers

The main surface active polymers in bread, and most other food systems, are proteins. Proteins come in a wide variety of forms, molecular weights and secondary structures (Schofield, 1994). To be surface active, the protein must generally be soluble in the aqueous phase from which the foam is formed in order for it to adsorb to the bubble surface. This is because most insoluble proteins will aggregate in solution, making the hydrophobic side-chains unavailable to stabilise the bubble surface. However, it may be possible for less soluble proteins to be transported to the surface through the mechanical action of mixing (Ornebro *et al.*, 2000). The surface activity of wheat proteins has not been widely studied (Ornebro *et al.*, 2000); however general fractions such as albumins, gliadins, glutenins and globulins have been investigated (Keller *et al.*, 1997) and shown some surface activity. Other specific proteins that show some surface activity include the puroindolines (Kooijman *et al.*, 1998; Douliez *et al.*, 2000; Biswas *et al.*, 2001) and lipid transfer proteins (Subirade *et al.*, 1995; Kooijman *et al.*, 1998). Some carbohydrates, particularly those with modified or hydrophobic substitutions, have been used as foam active ingredients in the food industry. In breadmaking, there may be some starch or cellulose fragments which may show some surface activity, but there is no evidence for this. Pentosans, however, have been linked with improved crumb structure (van Hamel *et al.*, 1993), and one theory is that they are surface active and help stabilise the gas cells against coalescence (Courtin and Delcour, 2002). Pentosans have been shown to interact with proteins at a surface and strengthen the interactions between them, and hence increase the stability of protein foams (Sarker *et al.*, 1998).

16.4.2 Lipids

Wheat also contains significant levels of lipids of various types (phospholipids, sphingolipids, fatty acids, sterol lipids, glycosyl-glycerides, etc.) with a range of fatty acid compositions (Morrison, 1983). Many of these lipids are very surface active, and will compete with proteins to stabilise the bubble surface (Dubreil *et al.*, 1997; Gan *et al.*, 1995; Keller *et al.*, 1997). The surface and foam stabilising properties of wheat lipids have not been studied extensively, but the main trends can be related to other model studies of protein–lipid mixtures. In general, the foaming properties of the non-polar lipids similar to those present in wheat are not particularly good. To be foam active, lipids and surfactants are required to be able to diffuse rapidly at the interface and interact with the aqueous phase to be able to stabilise foam lamellae through the Gibbs–Marangoni mechanism (Fruhner *et al.*, 2000). Modification of the lipids by lipase (Castello *et al.*, 1998) has been found to improve the breadmaking properties (Weegels and Hamer, 1992), suggesting that the foaming activity of the intrinsic lipids has considerable room for improvement. Not enough is known about which wheat lipids are foam positive, and their levels in dough. In addition, not all lipids will be available, as many are known to be bound by various components in the gluten–starch matrix (Morrison, 1994). Nevertheless, lipids are thought to have a great influence on the stability of gas cells in dough (MacRitchie, 1983; Gan *et al.*, 1995). It is thought that, as discussed above, that the lipids competitively adsorb with the protein and destabilise the protein surface of the gas cells.

16.4.3 Emulsifiers

Another class of lipid-like components important to breadmaking are the emulsifiers. Emulsifiers such as DATEM (di-acyl tartaric esters of mono-glycerides) and SSL (sodium stearoyl lactylate) are added to the formulation, and are known to improve loaf volume and stability of the dough during proving (Mettler and Seibel, 1993; Carson and Sun, 2000). Again, little is known about the underlying mechanism, but it is thought that they replace the lipids at the surface, and perhaps even the proteins also, and stabilise the gas cells against coalescence, as they are more foam active than the wheat lipids owing to their detergent-like structure.

In summary, therefore, the various mechanisms and basic principles of foam formation stability have been introduced, together with the main components of dough which are thought to play a role in the stabilisation of the gas cells. Bread dough itself is very different from food systems that have conventionally been used to study foam formation and stability. Therefore, many of the fundamental mechanisms underlying the stabilisation of the gas cells in bread are still poorly understood. Therefore the following sections will discuss current opinion, rather than undisputed facts, regarding the stabilisation of gas cells in dough.

16.5 The aqueous phase of dough and foam formation

It is important for baking quality that, during certain stages of the breadmaking process when the gas cells come into close contact, the gas cells remain stable. Otherwise, gas cell coalescence and collapse of the dough will occur. Figure 16.3 shows how the gas cells increase in size and come into contact during proving. It is thought that the aqueous phase in dough is the main source of surface-active material that stabilises the gas cells during these critical stages. Does the aqueous phase in dough actually exist? And if so, what form is it in? The gluten–starch matrix absorbs most of the water added to dough. Ultracentrifugation will separate a liquid phase or dough liquor from the matrix, but this does not necessarily mean that there is an aqueous phase stabilising the gas cells. However, it has been found that no gas is retained by dough containing less than 35% water, gas cell stability improves with added water (MacRitchie, 1976). Microscopy has also suggested the existence of thin aqueous films separating gas bubbles in dough (Gan *et al.*, 1990, 1995). The stability and surface properties of aqueous extracts of dough have been related to baking quality (Sahi, 1994). Here we will discuss how it is envisaged that the aqueous phase of dough may play a role in creating and stabilising the gas cells in dough.

16.5.1 Physical aspects

The mechanisms of foam formation were discussed earlier. The main parameters contributing to bubble formation are the energy input, the viscosity of the dough and the surface tension. The role of surface tension during mixing is difficult to determine, as it impossible to measure the surface tension directly on a freshly created interface of the dough itself. It is thought that in bread doughs, the viscoelasticity of the dough is mainly responsible for the entrapment of gas cells during mixing (Fig. 16.3a). Some evidence for this comes from measuring dough density during proofing. Changing the surface tension of the dough by adding emulsifiers does not appear to affect the amount of gas incorporated during whipping (Campbell *et al.*, 2001). The surface tension was also thought to have an effect on dough rheology, but the effect is probably very small, unless there is a very high gas phase volume (Bloksma, 1981). Therefore, it would appear that the surface properties of the aqueous phase will only have limited impact on the number and size of the bubbles during mixing.

In order to create and stabilise foams in the conventional sense, the foam active components should be soluble and free to diffuse to the surface of the newly created bubbles. Bread dough is a complex, multi-component viscoelastic network. The physical properties and morphology of the network change with time and processing. The structure of the network will have a profound impact on the ability of surface active species to diffuse to surfaces. In simple aqueous solutions, the relationships between molecular weight, diffusion coefficients and hence adsorption rates and foam properties are reasonably well established

(Grunden *et al.*, 1974). However, the gluten–starch network will have a large effect on the motion of surface active molecules, as it is more likely to restrict molecular diffusion. This will result in a much larger emphasis on molecular weight, as the larger proteins will be unable to pass through the network in the same way as the low-molecular-weight lipids and surfactants. However, the role of mixing in assisting the adsorption of less soluble components is not clear. It is thought that during the wetting and hydration process during mixing, surface active material may be able to spread at the surface of bubbles which come into contact with insoluble particles (Ornebro *et al.*, 2000). Therefore, the less soluble proteins in the gluten matrix may well play a role stabilising the surface of the gas cells.

Interactions and binding are known to affect rates of adsorption, such that adsorption of ligands which interact with a non-adsorbing molecule can be reduced or prevented. The starch–gluten matrix is the major component of dough, hence the potential for it to bind surface active material, and prevent adsorption, is huge. It is known that significant amounts of lipid are affected this way (Morrison, 1994). The binding of lipids by soluble proteins such as puroindolines and lipid transfer proteins has been studied (Wilde *et al.*, 1993; Dubriel *et al.*, 1997; Tassin-Moindrot *et al.*, 2000). It has been found that these interactions can prevent lipids from destabilising protein foams (Clark *et al.*, 1994b). It is possible, therefore, that these 'lipid-binding' proteins may play an important role in reducing the destructive effects of lipids.

The main structure in conventional liquid foams that prevents coalescence between gas cells, is the thin liquid lamellae which separate gas bubbles. As discussed in the introduction, the stability of these lamellae is critical for the lifetime of the foam (Fig. 16.3c). During mixing, gas cells are entrapped in the gluten–starch matrix, and are kept as discrete cells, separated by the matrix. Here the interfacial stability mechanisms have no impact, as the stability of the bubbles is ensured by being immobilised by the dough matrix. During proving, the gas cells expand and eventually may come into contact as shown in Fig. 16.3. It has been shown that it is possible that by this stage the bubbles may no longer be separated by the gluten–starch matrix, but are stabilised by a thin, liquid lamella (Gan *et al.*, 1990, 1995). The stability of the gas cells is then determined by the interfacial properties of the components adsorbed to them, as described in the introduction. The competition between the proteins, lipids and other surface active components in the dough will determine whether the lamellae are stabilised either by the viscoelastic or Gibbs–Marangoni mechanisms, or whether the competition is antagonistic and results in an unstable lamella. This will be discussed in more detail in the next section.

In the absence of the gluten–starch matrix from the lamellae, the liquid is able to drain more freely. As the liquid drains from the lamellae, they become thinner, and once the thickness approaches 100 nm, the stabilisation mechanisms (viscoelastic or Gibbs–Marangoni) determine whether the lamellae remains intact or not. Therefore for an unstable system, the rate of drainage, or at least, the time before the lamella reaches a critical thickness, is important for the

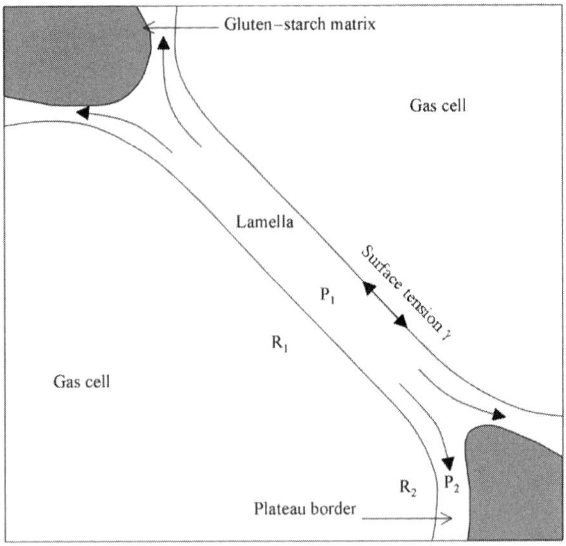

Fig. 16.4 Drainage mechanism from foam lamella. Surface tension forces increase pressure inside gas cell. Radius of curvature in plateau border (R_2) is smaller than in lamella (R_1). Hence the pressure in the border (P_2) is less than pressure in lamella (P_1), resulting in a net flow of liquid from lamella into plateau border (arrows).

stability of the gas cells. The physical situation for a lamella in bread dough is very different from one in a free draining foam. In dough, the lamella is held in place by an annulus of gluten–starch matrix, and therefore the liquid has nowhere to drain. The drainage mechanism here is the disjoining pressure. The disjoining pressure is the reduction in pressure below a curved surface caused by the surface tension as shown in Fig. 16.4. Therefore, if the edge of the lamella (plateau border) has a small radius of curvature relative to the centre, the disjoining pressure will be lower than in the centre, therefore liquid will flow with the pressure gradient away from the centre of the lamellae, to the outside, causing local thinning, which may cause instability.

The flow of the liquid is influenced by physical properties of the liquid and the surface. The viscosity of the liquid will affect drainage, such that increased viscosity of the aqueous phase may slow down the flow of liquid through the lamellae (Haas and Johnson, 1967). However, this can be counteracted by higher liquid density which will increase drainage from (non-horizontal) lamellae. The surface properties will also influence drainage. A viscoelastic surface will resist drainage, because, any drainage process will change the curvature of the lamellae, and hence the surface area (Brierley *et al.*, 1996). Surfaces with a high viscoelastic modulus will resist the change in surface area, and may decrease or halt drainage. Fluid, mobile surfaces may initially oppose liquid flow through the Gibbs–Marangoni mechanism, but ultimately, they cannot physically resist drainage. Mobile, surfactant stabilised lamellae can drain over an order of magnitude more quickly than lamellae stabilised by viscoelastic interfaces.

Disproportionation is the other main cause of instability in foam systems. As explained earlier, this is the transport of gas from small to large bubbles due to higher internal Laplace pressure in smaller bubbles (Garrett, 1993). Disproportionation in dough ought to be quite simple, as the bubbles are immobilised by the gluten–starch matrix, and the gas (CO_2) has a high solubility in the aqueous phase. However, the complicating factor is the fact that the gas is being produced continually, and there is a net transport of gas into the bubbles to make the dough rise, whereas in conventional liquid foams, the total volume of gas is fixed. In addition, the bubbles have to grow against the resistance of the viscoelasticity of the gluten–starch matrix. In this sense, disproportionation does not occur in the same way that it does in conventional liquid foams. In fact some researchers suggest that disproportionation does not occur in dough at all. Some evidence for this comes from comparing fermented and unfermented doughs. Disproportionation has been found to occur in unfermented doughs, resulting in a loss of smaller bubbles and a growth of the larger bubbles (Shimiya and Nakamura, 1997). In fermented doughs, where CO_2 was being produced, the smaller bubbles neither shrunk nor grew. This was probably because the higher Laplace pressure in the smaller gas cells merely balanced the increased CO_2 production, resulting in growth of the larger gas cells only. However, during the early stages of baking, when fermentation ceased, the smaller bubbles rapidly disappeared, presumably through disproportionation. Therefore, while disproportionation may not occur during normal proofing, the effects of surface tension and Laplace pressure influences the development of the bubble size distribution. The growth of the bubbles in dough has been modelled, illustrating the importance dough elasticity (Shimiya and Nakamura, 1997), CO_2 production (Shah *et al.*, 1998) and surface tension (Shimiya and Nakamura, 1997; Shah *et al.*, 1998).

To summarise, various physical principles can influence the creation, growth and stability of the gas cells in dough. Some of these physical processes can be modelled and even predicted and controlled. However, one of the limitations in this field is that it is not clear how, or indeed which, of the individual components in dough affect these particular physical properties. Therefore we can only discuss the limited research findings to date, and speculate as to how the different components may influence the behaviour of gas cells in dough.

16.6 Dough composition and foam stability

Despite much research interest in the gas cells in dough, it is still not known what components stabilise the surface of gas cells. The principal reason is that it is extremely difficult to remove intact gas cells from the dough. In liquid foams, it is relatively simple to allow the liquid in the foam to drain away, and analyse the components on the surface of the bubbles. This has been used successfully to identify certain proteins important in stabilising beer foam (Sorenson *et al.*, 1993). Many components have been studied, but no definitive evidence obtained. Therefore, considering the physical processes involved we will discuss the likely roles of the main components in bread dough, and review the supporting evidence.

16.6.1 Proteins

The protein content of wheat grain used for breadmaking varies usually between 9 and 15% dry weight, and the correlation between protein content and breadmaking quality is very clear (Pomeranz, 1987; He and Hoseney, 1992; Tronsmo *et al.*, 2002). However, the majority of the breadmaking quality is contributed through the elastic properties of the gluten network. Gluten proteins (gliadins and glutenins) constitute about 85% of flour proteins, and their structure and interactions are responsible for the development of the extensibility and elasticity in doughs (Schofield, 1994). The ability of the gluten network to support the gas cells is very important, as a weak gluten network will not support the gas cells, and the cell walls will fracture more easily (i.e. at an earlier stage of proofing), reducing loaf volumes (He and Hoseney, 1992). Gliadins have been associated with foaming properties of gluten (Mita *et al.*, 1978), and specific gliadins have also been linked with loaf volumes (van Lonkhuijsen *et al.*, 1992). The remaining 15% of flour proteins are classified as albumins (60%) and globulins (40%), and are more soluble than the gluten proteins in water or salt solutions. The albumins and globulins make up a whole range of proteins (Schofield, 1994) expressed in the wheat grain for a wide variety of physiological roles. Many of these proteins are surface active, and specific proteins such as puroindoline, and lipid transfer proteins (LTP) have been associated with foam stabilising properties (Wilde *et al.*, 1993; Sorenson *et al.*, 1993; Dubriel *et al.*, 1997).

The elastic gluten–starch matrix is responsible for supporting and stabilising discrete gas cells during mixing and early stages of proofing. However, at later stages of proofing and the early stages of baking, when the gas cells come into contact, the surface properties of the gas cells are responsible for their stability (Figs 16.2 and 16.3). It is thought that proteins play a major role in stabilising the gas cells. The protein content of dough is very high, so there is more than enough protein to stabilise the many gas cells created during breadmaking. Estimates vary, but there is probably a 50-fold excess of protein to stabilise the gas cells in bread. However, the question remains, which proteins? They can be split into soluble and non-soluble proteins. One assumption is that the proteins must be soluble to be able to adsorb and stabilise the bubbles. And this is indeed the case in liquid foams. However, bread dough is a concentrated, protein–polysaccharide matrix, so normal diffusive adsorption may play a limited role. Mixing-assisted adsorption is probably more important, which may assist the adsorption of highly insoluble gluten proteins (Ornebro *et al.*, 2000). Soluble proteins such as puroindolines and lipid transfer proteins (LTPs) have been studied, and found to have significant surface activity (Kooijman *et al.*, 1997; Douliez *et al.*, 2000; Biswas *et al.*, 2001) as have the soluble fractions of albumins, globulins and gliadins (Paternotte *et al.*, 1994; Keller *et al.*, 1997; Ornebro *et al.*, 2001). Insoluble proteins such as glutenins can be spread at surfaces (Balla *et al.*, 1997, 1998), supporting the idea that they may be spread at the surface during mixing.

Proteins tend to form more stable foams when they form a strong elastic film on the surface of interfaces (Mitchell, 1986). Aqueous extracts of dough have also been shown to form a strong elastic interface (Sahi, 1994), which varied among varieties of differing baking quality, suggesting a role for proteins in stabilising the gas cells. However, the role of other components such as lipids are also important, as are component interaction, which will be discussed in more detail later.

16.6.2 Lipids

There are many types of lipids in wheat (Morrison, 1983), and their diversity stems from their physiological role. Within the grain, half of the lipids are contained in the endosperm. About 40% of endosperm lipids are contained within the starch granules. These are strongly associated with the starch, and are not thought to influence breadmaking quality. The germ and aleurone contain most of the rest of the lipids. The lipids are often classified into polar lipids (PoL) and non-polar or neutral lipids (NL). The main neutral lipids of interest in flour are trigycerides, diglycerides, free fatty acids and sterol esters. Most of the neutral lipids are found in the germ and aleurone, and most of the polar lipids are found in the endosperm. The main polar lipids in wheat are the glycolipids such as monoglycosyldiglyceride (MGDG) and diglycosyldiglyceride (DGDG), and the main phospholipid types such as phosphatidylethanolamine (PE), phosphatidylcholine (PC), phosphatidylglycerol (PG) and phosphatidylinositol (PI). The fatty acid composition of the lipids is dominated by linoleic (C18 : 2) the rest comprising mainly oleic (C18 : 1) and palmitic (C16 : 0). It is well known in other foam studies, as discussed earlier, that surface and foam activity of lipids will depend on lipid type and structure (MacRitchie, 1983; Keller *et al.*, 1997). The distribution of different lipids throughout the grain will mean that lipid concentration and composition in flour will be influenced by variety, environmental conditions and milling stream (Morrison and Barnes, 1983; Matsoukas and Morrison, 1991; Sahi, 1994; Morrison, 1994). Lipids are generally more surface active than proteins, but again this will depend on solubility. The solubility of the lipids will decrease with longer, more saturated chain lengths, and they will tend to form aggregates, which are slower to adsorb. This again assumes that the adsorption is diffusive, but if mixing-assisted adsorption is found to be significant, then the less soluble lipids may well play a significant role in the surface properties of the gas cells. The structure of lipids is important, as the headgroups need to interact with the aqueous phase to form stable foams (Fruhner *et al.*, 2000). Therefore neutral lipids such as fatty acids, triglycerides and alcohols do not form stable foams, whereas more amphipathic lipids with larger or more polar headgroups, such as monoglycerides and lyso-phospholipids can form stable foams

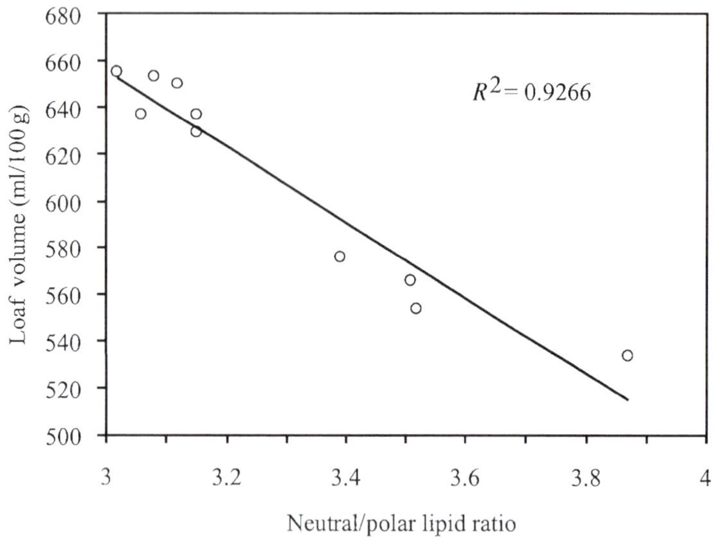

Fig. 16.5 Effect of weight ratio of neutral : polar lipids on the loaf volume of 10 Greek breadwheats. The solid line is a linear regression fit. (Data from Matsoukas and Morrison, 1991.)

(MacRitchie, 1983). It has been seen that the ratio of polar to neutral lipids can be important as shown in Fig. 16.5 (Matsoukas and Morrison, 1991). It is clear from the literature that lipids are very important for breadmaking quality; this is thought to be principally concerned with their role in stabilising gas cells (MacRitchie, 1983; Matsoukas and Morrison, 1991; Gan *et al.*, 1995). However, it is commonly thought that the stability of gas cells in dough and bread is mainly determined by the interaction of lipids and proteins (Gan *et al.*, 1995; Ornebro *et al.*, 2000).

16.6.3 Protein–lipid interactions

As discussed earlier, proteins and lipids stabilise foams by very different and antagonistic mechanisms. Thereby, competition between lipids and proteins to stabilise foams can often result in destabilisation (Clark *et al.*, 1994b; Wilde, 1996; Sarker *et al.*, 1998). Reconstitution experiments have shown that removing lipids can sometimes increase loaf volume, then addition of low levels of lipid can destabilise the foam (Mita *et al.*, 1977), followed by a recovery in loaf volume or foam stability at high lipid contents (MacRitchie and Gras 1973; MacRitchie, 1983). Figure 16.6 shows how the presence of lipids can influence loaf volume. Neutral lipids are usually more detrimental to foam, and do not recover the foam stability or loaf volume. Polar lipids can still be destructive, but at higher concentrations they can stimulate recovery of the functionality. This is probably because they contain some lipids that are foam active. In addition, the ratio of neutral to polar lipids in flour have been strongly correlated with loaf volume (Matsoukas and Morrison, 1991).

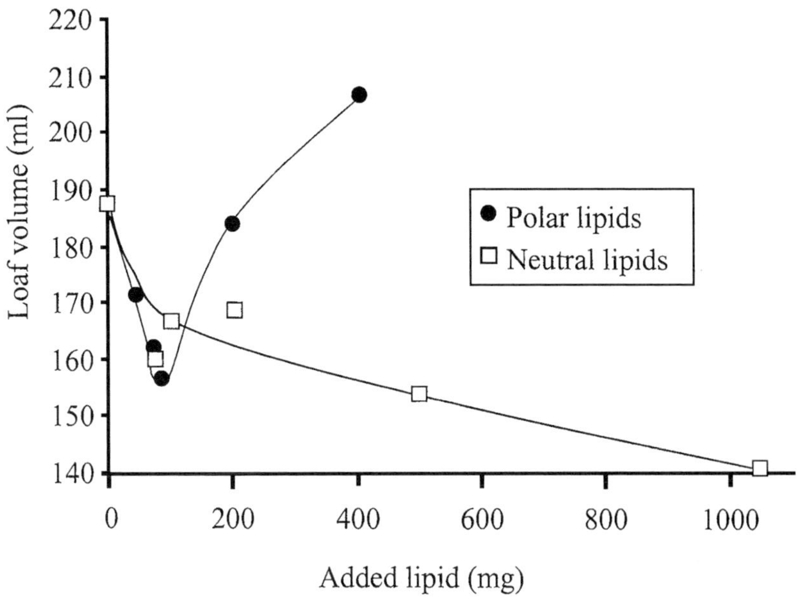

Fig. 16.6 Influence of added lipids on loaf volume. The minimum observed for the polar lipids corresponds to an unstable protein : lipid interface. The polar lipids appear to be more foam positive than the neutral lipids, which show no recovery in loaf volume. (Data from MacRitchie and Gras, 1973.)

Lipid binding proteins have created much interest in recent years. In non-bread systems, their ability to bind lipids has been exploited to prevent lipid-induced destabilisation of protein foams (Wilde *et al.*, 1993; Clark *et al.*, 1994b; Dubriel *et al.*, 1997). These proteins include puroindolines and LTPs from wheat. Therefore it is thought that these proteins may play a role in limiting lipid damage. Studies to date have shown that they can prevent lipid destabilisation in model systems, but their ability to achieve this in real baked products is still not conclusive (Dubriel *et al.*, 1997; Douliez *et al.*, 2000; Igrejas *et al.*, 2001).

16.6.4 Polysaccharides

Polysaccharides are not generally surface active, although some modified polysaccharides have been found to have surface active and foam stabilising properties. The main non-starch polysaccharide components in wheat flour that have had the most influence on baking quality are the arabinoxylans or pentosans. Their water holding capacity has been known to affect dough rheology, extensibility and texture (van Hamel *et al.*, 1993; D'Appolonia and Rayas-Duarte, 1994). Loaf volumes have also been improved in the presence of pentosans (Delcour *et al.*, 1991; van Hamel *et al.*, 1993). Pentosans with ferulic acid moieties have been shown to electrostatically cross-link proteins at the air–water interface (Sarker *et al.*, 1998). This was found to strengthen the adsorbed layer, reduce drainage rates in lamellae and make foams more resistant to lipid-induced destabilisation. Propylene glycol alginate (PGA) was also found to have a similar effect (Sarker and Wilde, 1999), and is used extensively to improve the stability of beer foam. Therefore certain pentosans may have a role in improving the functionality of the proteins which stabilise the gas cells. Some studies have involved the addition of hydrocolloids such as guar gum and modified cellulose (Mettler and Seibel, 1993), but the improvements observed were not associated directly with improved stabilisation of the gas cells. Acetylated pectins in pumpkin powder added to flour were thought to be responsible for improved gas cell stability in the dough (Ptitchkina *et al.*, 1998). However, the protein content of a pectin preparation was found to be responsible for much of its functional properties (Akhtar *et al.*, 2002).

16.6.5 Added ingredients

A whole range of added improvers and functionality enhancers are added to bread to improve baking quality. Most of them are involved in improving the development of the gluten–starch matrix; however, there are some that are added specifically to improve the stability of the gas cells.

Emulsifiers such as SSL and DATEM have been used for many years to improve the loaf volume and crumb texture of bread (Weegels and Hamer, 1992; Kokelaar *et al.*, 1995; Keller *et al.*, 1997; Carson and Sun, 2000). The emulsifiers are thought to improve the stability of gas cells in dough, particularly during the later stages of proof when many of the gas cells are thought to be stabilised by thin liquid films (Gan *et al.*, 1995). In the absence of these emulsifiers, some breads are particularly sensitive to handling at the end of proof, such that knocking or vibration can cause the loaf to collapse. In the presence of these emulsifiers, this is not the case. The specific mechanism by which they achieve this improvement has not been demonstrated, but surface measurements have shown that they are capable of inserting into protein films more readily than wheat lipids (Keller *et al.*, 1997). Therefore it is likely that they shift the stability regime from a less stable mixed lipid–protein system over to a surfactant-dominated system, as described earlier.

As discussed earlier, the different varieties of lipids in flour have different foam activities (MacRitchie, 1983). The balance of foaming and non-foaming lipids is likely to be influenced by wheat variety, environmental conditions and processing conditions (including milling) (Morrison and Barnes, 1983). Lipases have been used to hydrolyse di- and tri-acyl lipids into monoglycerides and lysolipids (Castello *et al.*, 1998); this has increased loaf volumes and crumb structure, and has been offered as a replacement for emulsifiers (Weegels and Hamer, 1992; Monfort *et al.*, 1999). The effect is probably based on the fact that the hydrolysis products are more soluble and surface active and foam active than the substrate lipids. These products then act in a way not dissimilar to emulsifiers such as DATEM or SSL.

Bakery fats are often added to improve functionality and have been observed to have a similar effect to some emulsifiers. There was some evidence that they may stabilise the gas bubble interface (Brooker, 1996), but this has never been fully investigated.

16.7 Processing stages and foam stability

We have discussed in various sections critical aspects of gas cell formation and stability, and the importance of various stages of the breadmaking process have been mentioned. However, it is probably useful to summarise the important factors in each of the processing stages here to bring all this information together.

16.7.1 Mixing

The mixing process is critical as the gas cells are created here. The size of the gas cells created is mainly dependent on the energy input during mixing (Hanselmann and Windhab, 1998), which includes the rheological properties of the dough. The surface tension does not seem to play a role in such a viscous system (Campbell *et al.*, 2001). The size of the bubbles is very important, as this determines whether they grow or not during the proving process (Shimiya and Nakamura, 1997; Shah *et al.*, 1998). The other important role of mixing is thought to be the assisted adsorption of material. The structure and viscosity of the gluten–starch matrix are likely to hinder adsorption of water-soluble components. The mixing process, during hydration of the dough, will bring both soluble and insoluble components to the surface, and hence may assist adsorption (Ornebro *et al.*, 2000). This will heavily influence the composition of the gas cell surface throughout the process, and will therefore influence the stability of the gas cells during the later stages of breadmaking.

16.7.2 Proving

Proving involves the transfer of gas from the bulk, into the gas cells. The surface tension of individual gas cells increases the pressure inside the gas cells. This effect is greater in smaller gas cells, such that a critical gas cell radius is found, which is dependent on the surface tension and the elasticity of the dough. Below this critical radius, the gas cells do not grow, and eventually disappear altogether. Above this radius, the gas cells will increase in size during proving as expected (Kumagai *et al.*, 1991; Shimiya and Nakamura, 1997; Shah *et al.*, 1998). Hence the initial gas cell size distribution is critical for the final crumb structure. It is pointless trying to create too many small gas cells, as these do not grow, and will not contribute to the final crumb structure.

During the later stages of proof, the gas cells achieve close contact, and if the gluten films separating them fail, then the gas cell stability becomes dependent on the surface properties of the liquid films (Gan *et al.*, 1995). Whether the surfaces are stabilised by proteins, lipids or emulsifiers, or a mixture, will determine whether the gas cells remain stable.

16.7.3 Baking

It is thought that many of the smaller bubbles disappear during the early stages of baking. The reason is not totally clear: some workers believe it is due to disproportionation (Shimiya and Nakamura, 1997), others think it may be due to coalescence (Hayman *et al.*, 1998b). During proving, the smaller bubbles did not disappear because they reached equilibrium with gas production in the aqueous phase. During early baking, gas production ceases, therefore there is an excess pressure in the smaller bubbles, and they lose gas to the larger bubbles, as has been observed in unleavened bread (Shimiya and Nakamura, 1997).

The early stages of baking are similar to the later stages of proof. Gas cells stabilised by thin liquid lamellae will be dependent on the surface properties of the adsorbed layer to remain stable (Gan *et al.*, 1995). The gas cells must remain stable until the gluten and starch matrix gels, essentially 'freezing' the gas cell structure in place. After this the gas cell walls can rupture, and the bread converts from a foam to an open sponge structure, so the gas can freely escape, and collapse will not occur during cooling.

16.8 Analytical techniques

I will now review the main techniques and approaches that have been used to study the surface properties of dough systems. The main problem is that most surface techniques require liquid or low-viscosity systems. Therefore the first part of this section will look at approaches that have been used to extract or prepare samples to gain information about the surface properties of dough.

16.8.1 Sample extraction and preparation

The approach used when studying the surface properties of liquid foam systems is fairly straightforward. The surface properties can be directly measured on the planar surface of the liquid itself, as it is assumed that the largely diffusive adsorption process at the planar interface simulates the process of adsorption to the bubble surface during foaming. This enables the determination of which species are adsorbing in a complex mixture. Another approach is to create a foam, and allow it to drain extensively, then characterise the composition of the remaining foam (Sorenson *et al.*, 1993). Unfortunately bread dough is too viscous to make direct measurements at the surface. The viscosity also suggests, as discussed earlier, that purely diffusive adsorption cannot account for the surface properties, and that the mixing process is also important for driving material to the gas cell surface (Ornebro *et al.*, 2000). Therefore, the only direct measure of the surface properties is on the bubble surface *in situ* in the dough. Until a method is developed that can do this, the only way to access surface information in doughs is to extract surface active material from the flour or the dough and study them separately.

The simplest protocol is to extract components from the flour. A simple water extraction will extract the albumins and the presence of salt will extract the globulins. The less water-soluble gluten proteins can be extracted using alcohol (gliadins) and acid/alkali solutions (glutenins) (Pomeranz, 1987; Schofield, 1994). These extracts can be further purified into the different gliadins and glutenins. Further extraction procedures can be used for other specific proteins. Puroindolines (PIN) and lipid transfer proteins (LTP) have been investigated in relation to breadmaking quality owing to their lipid binding ability. PINs are purified from a detergent extract of flour (Blochet *et al.*, 1993; Douliez *et al.*, 2000) and they are present at higher concentrations in soft milling wheats (Igrejas *et al.*, 2001). LTPs are purified from a KCl extraction of flour (Desormeaux *et al.*, 1992). The surface properties of these proteins have been studied, in particular their interactions with lipids (Kooijman *et al.*, 1998; Douliez *et al.*, 2000; Biswas *et al.*, 2001). In turn, flour lipids can be extracted using either polar or non-polar solvents to extract polar and non-polar lipids, and further purified into separate components (Barnes, 1983).

The employment of flour extracts is a useful approach for investigating the surface activities of specific flour components. However, the effect of dough hydration and processing causes a wide variety of changes within the dough matrix, which may influence composition and functionality. Therefore, another method is to extract material from the dough. These are more direct methods of accessing the surface properties of dough, but they still have limitations. The simplest method is to make a dough, and go through the normal mixing procedure. Then two methods have been used to prepare solutions. The first is to freeze dry the dough and pulverise, then solubilise the powder in aqueous solution for study (Kokelaar and Prins, 1995). The second method is to ultracentrifuge the dough ($1–2.10^5$g) for 30–60 minutes (Sahi, 1994; Dubreil *et al.*, 1998). This produces a straw-coloured viscous liquid. The surface properties of the dough liquor have been correlated with lipid content and baking performance. This still does not take into account the role of mixing and adsorption of insoluble components, and uses diffusive adsorption of the components to measure surface properties. However, it is the nearest method there is for looking at the components in dough that play an active role at the surface of the gas cells.

16.8.2 Surface techniques

The main techniques used to study the surface properties of wheat components are surface tension, surface rheology and ellipsometry. These approaches have been reviewed recently (Ornebro *et al.*, 2000). Some other techniques and approaches that have been used will also be discussed.

Surface tension

The vast majority of studies measuring the surface tension of wheat or dough components have used the Wilhelmy plate method (Weser, 1980). The principle is simple: a small flat plate (length L, thickness d) is lowered and placed in contact with the surface, the meniscus of the liquid pulls down on the plate, with a force equal to the surface tension (γ) multiplied by the wetted perimeter ($= 2L + 2d$) of the plate. The plate is normally made of platinum or ground glass, and if pre-wetted, the contact angle between the plate and the liquid is zero, so no compensation for the contact angle need be made. The plate is connected to a microbalance or load cell, and the force exerted by the surface of the liquid is measured. The measurements are normally made in a large square container known as a Langmuir trough. The size of the trough ensures a very flat surface for the measurement, and surface barriers enable the surface to be compressed or expanded to study the compression properties of the surface. Material can be adsorbed from the solution in the trough, or can be spread on the surface, either from solution or from a dry powder (Eliasson *et al.*, 1991; Balla *et al.*, 1998).

The spreading method has been used for many insoluble proteins and lipids, and is important for studying the surface properties of insoluble components which may stabilise the gas cells as a result of mixing, as discussed earlier. The trough design also enables sequential measurements. For example, protein or lipid components could be spread on the surface of the trough, then other components can be added underneath the surface to see if they can adsorb in the pre-spread film, or even displace it. This has given very important information regarding the interaction between proteins and lipids at interfaces. For example, it has been shown that certain protein fractions can lower the surface tension in lipid films, even when the surface tension of the spread lipid film is lower than the protein can reach alone. This confirms that these proteins can bind to adsorbed lipids, and insert into the film. This may also provide useful information about the physiological role of these proteins and their interactions with membranes.

Another technique that is becoming more popular is the pendant drop technique (Ambwani and Fort, 1979). It is useful because it uses very little sample volume, in the order of 1 ml rather than 1000 ml in the case of a Langmuir trough. This is more practical when dealing with small volumes of material such as dough liquor or purified proteins. The principle is simple, by using a syringe with a flat tipped needle, a droplet of the sample liquid is partially expelled, until it hangs from the tip of the needle (pendant). The tear shape of the droplet is a result of the balance of forces acting on the drop. Gravity is pulling the droplet down, elongating it, whereas the surface tension is trying to contract the droplet into a spherical shape. By measuring the size and shape of the drop, the surface tension can be calculated using the derived Young–Laplace equation (Ambwani and Fort, 1979). With the advent of high-speed computing, this technique is now very popular, as the image capture and analysis can be done in real time. The disadvantage to the technique is that sequential spreading and adsorption measurements are not possible.

Surface rheology
The rheological properties of surfaces are very sensitive to surface composition. Surface tension is a useful measure of surface occupancy, and can detect some differences in molecular composition; however, systems with identical surface tensions may have very different surface rheological properties. This is particularly important when studying mixed protein : lipid surfaces, because most proteins form a viscoelastic surface, and lipids on the whole do not. Two approaches are generally used, surface shear rheology and surface dilatational rheology.

Surface shear rheology is a direct method of measuring the rheological properties of an interface. The technique normally uses an oscillatory method to determine the elastic and viscous moduli. The analogy with bulk rheology is very close, as surface rheology is simply measuring the viscoelasticity of a two-dimensional film, rather than a three-dimensional bulk material. The geometry is usually a bicone or ring, which is located at the surface, and oscillated at controlled stresses or strains, the resultant stress or strain and phase angle are measured as in bulk rheology. Adsorbed protein films have similar viscoelastic characteristics to bulk protein networks, so similar analysis can be used. Surface shear measurements are very sensitive to interactions between adsorbed molecules, hence the presence of even low levels of lipids that disrupt the protein–protein interactions, can have a dramatic effect on the surface shear rheology (Sahi, 1994), thus giving an insight into the detrimental effect that lipids may have on the stability of gas cells in dough.

The other approach is surface dilatational rheology. This involves oscillatory changes in surface area (strain) and measuring the resultant change in surface tension (stress). This is an indirect measure of the rheological properties of the surface, but it is very sensitive to the composition of the surface, particularly for mixed protein–lipid systems. The strength of the technique is that the surface tension is being measured simultaneously, so by plotting the data in terms of the surface tension, a unique fingerprint of the system is obtained (Paternotte *et al.*, 1994). The precise contributions of the various constituents to the surface composition and physical properties may then be determined. Original dilatational measurements were performed on a Langmuir trough, where surface barriers would oscillate backwards and forwards, changing the surface area. Another approach is the ring trough method (Kokelaar *et al.*, 1991, 1995) where a circular barrier is raised and lowered vertically through the surface, causing isotropic expansion/compression of the surface.

Ellipsometry
This technique is used to measure the adsorbed amount of material to a surface. It uses a principle whereby an elliptically polarised light beam is reflected off a surface (De Feijter *et al.*, 1978). The ellipticity of the reflected beam is dependent on the angle of incidence, refractive index and the optical properties of the substrate. If the refractive index values are known, then it is possible to calculate the thickness of the adsorbed layer, to very high resolution. The

technique is very sensitive to the position of the surface, so adsorption to solid surfaces is technically simpler. Therefore most of the measurements on adsorption of wheat proteins using this technique have been on solid surfaces. However, using hydrophobic surfaces can simulate adsorption to a bubble surface. The strengths of the technique are the high resolution, such that competition between different proteins can show one protein interacting with or displacing another, imparting important information regarding how different wheat proteins are likely to interact at the gas cell surface (Ornebro *et al.*, 2001).

Other techniques
The other approaches worth mentioning are potential attempts to measure the surface properties of the dough directly. First is the use of contact angle measurements which are used to measure the interfacial tension between a solid and a liquid. By depositing a drop of liquid with known surface tension, onto a solid surface, the angle which the droplet contacts the solid surface is dependent on the surface tension of the surface. Therefore, this may be a method to determine surface tension in the dough itself. However, the surface of the gas cell may not effectively be solid. If the liquid film hypothesis is correct, then there is probably a thin hydrated layer on the surface of the dough bubble, which will interfere with this measurement. The other approach is to measure the pressure of bubbles in dough. The pressure of a bubble in an elastic medium is the sum of the Laplace pressure imposed by the surface tension, and the elastic resistance of the medium. One attempt was made to extract the surface tension component from alveograph measurements (Hayman *et al.*, 1998a), but if the Laplace pressure is calculated for these bubbles, then the values are below the resolution of the instrument. If it were possible to measure the pressure inside a much smaller bubble, and account for the viscoelasticity of the dough matrix, then it may be possible to measure the surface properties of gas cells *in situ*.

16.9 Future trends

As I have explained in the previous sections, there are great technical difficulties studying the mechanism of stabilisation of the gas cells in dough. There is still much work to do applying some of the above approaches to gradually increase our knowledge of the gas cell stability and dough functionality. There are two main areas for future development: one is to increase our understanding of the underlying mechanisms, and the other is to use this knowledge to develop biotechnological approaches to improving dough functionality.

16.9.1 Understanding the mechanisms

Future research should focus in two specific, but not mutually exclusive, areas. In the past, much research in this field has, understandably, been largely empirical. With new techniques and approaches, research has become more focused on the underlying mechanisms. The two main areas should be: (1) understanding how individual components stabilise the gas cell surface; and (2) how the physical properties of the gas cell surface influence stability and breadmaking quality. Using information from these areas will tell us exactly how certain proteins or lipids influence baking quality and may help develop future quality markers. Focusing on the physical properties of the surface will help develop techniques that can be used by the baking industry to predict quality and functionality. For example, probe microscopy has recently allowed significant advances in our knowledge of protein–surfactant interfaces, and the same methodology could be applied to dough systems to help us understand the individual roles of the proteins and lipids, and help develop new strategies to improve quality.

16.9.2 Improving functionality

Improved understanding about how certain proteins or lipids stabilise (or destabilise) gas cells, and the impact on breadmaking quality, will allow informed strategies to be developed. The strategies can cover the entire process from breeding, growing and storage through to milling, mixing and baking. Advances in molecular biology can assist breeders and biotechnologists to develop new varieties targeted at specific functional properties. Complementary knowledge of component interactions, and process engineering may simply allow modification of the production process to target specific functional properties.

The future of research in this area still contains many unanswered questions and technical challenges. There is still an enormous amount of work to do before we really start to understand exactly what is happening at the surface of the gas cells in dough.

16.10 Sources of further information and advice

- *Lipids in Cereal Technology*, Barnes P J (ed.), New York, Academic Press, 1983.
- *Wheat: Production, Properties and Quality*, Bushuk W and Rasper V F (eds.), London, Chapman and Hall, 1994.
- *Modern Cereal Science and Technology*, Pomeranz Y, New York, VCH Publishers, 1987.
- *Wheat: Chemistry and Technology*, Pomeranz Y (ed.), St. Paul, MN, American Association of Cereal Chemists, 1988.
- *Handbook of Cereal Science and Technology*, Kulp K and Ponte, J G (eds.), New York, Marcel Dekker, 2000.

16.11 References

AKHTAR M, DICKINSON E, MAZOYER J and LANGENDORFF V (2002), 'Emulsion stabilizing properties of depolymerized pectin', *Food Hydrocolloids*, **16**(3), 249–256.

AMBWANI D S and FORT JR T (1979), 'Pendant drop technique for measuring liquid boundary tensions', in Good R J and Stromberg R R, *Surface and Colloid Science*, vol. 2, New York, Plenum, 93–119.

BALLA A, BLECKER C, RAZAFINDRALAMBO H and PAQUOT M (1997), 'Interfacial properties of wheat gluten films from flours with different breadmaking qualities', *Sci Aliments*, **17**, 271–278.

BALLA A, RAZAFINDRALAMBO H, BLECKER C and PAQUOT M (1998), 'Interfacial properties of gluten monolayers spread on various chloride salt solutions. Effects of electrolytes, salt concentrations, and temperature', *J Agric Food Chem*, **46**, 3535–3539.

BARNES P J (1983), *Lipids in Cereal Technology*, London, Academic Press.

BISWAS S C, DUBREIL L and MARION D (2001), 'Interfacial behaviour of wheat puroindolines: Study of adsorption at the air–water interface from surface tension measurement using Wilhelmy plate method', *J Colloid Interface Sci*, **244**, 245–253.

BLOCHET J E, CHEVALIER C, FORET E, PEBAY-PEYROULA E, GAUTIER M-F, JOUDRIER P, PÉZOLET M and MARION D (1993), 'Complete amino acid sequence of puroindoline, a new basic and cystine-rich protein with a unique tryptophan-rich domain, isolated from wheat endosperm by Triton X-114 phase partitioning', *FEBS Lett*, **329**, 336–340.

BLOKSMA A H (1981), 'Effect of surface-tension in the gas–dough interface on the rheological behavior of dough', *Cereal Chem*, **58**, 481–486.

BRIERLEY E R, WILDE P J, ONISHI A, HUGHES P J, SIMPSON W J and CLARK D C (1996), 'Fundamental studies on the influence of ethanol on the foaming properties of beer protein fractions', *J Sci Food Agric*, **70**(4), 531–537.

BROOKER B E (1996), 'The role of fat in the stabilisation of gas cells in bread dough', *J Cereal Sci*, **24**, 187–198.

CAMPBELL G M and MOUGEOT E (1999), 'Creation and characterisation of aerated food products', *Trends Food Sci Technol*, **10**, 283–296.

CAMPBELL G M, RIELLY C D, FRYER P J and SADD P A (1998), 'Aeration of bread dough during mixing: effect of mixing dough at reduced pressure', *Cereal Foods World,* **43**(3), 163–167.

CAMPBELL G M, HERRERO-SANCHEZ R, PAYO-RODRIGUEZ R and MERCHAN M L (2001), 'Measurement of dynamic dough density and effect of surfactants and flour type on aeration during mixing and gas retention during proofing', *Cereal Chem*, **78**, 272–277.

CARSON L C and SUN X S (2000), 'Breads from white grain sorghum: effects of SSL, DATEM, and xanthan gum on sorghum bread volume', *Appl Eng Agric*, **16**, 431–436.

CASTELLO P, JOLLET S, POTUS J, BARET J L and NICOLAS J (1998), 'Effect of exogenous lipase on dough lipids during mixing of wheat flours', *Cereal Chem*, **75**(5), 595–601.

CAUVAIN, S P, WHITWORTH, M B and ALAVA, J M (1999) The evolution of bubble structure in bread doughs and its effect on structure. In (Eds. G M Campbell, C Webb, S S Pandiella and K Niranjan) *Bubbles in Food*, St. Paul, MN, USA, AACC.

CHEN J, DICKINSON E and IVESON G (1993), 'Interfacial interactions, competitive adsorption and emulsion stability', *Food Structure*, **12**(5), 135–146.

CLARK D C, MACKIE A R, WILDE P J and WILSON D R (1994a), 'Differences in the structure and dynamics of the adsorbed layers in protein stabilised model foams and emulsions', *Royal Soc Chem Faraday Discussion,* **98**, 253–262.

CLARK D C, WILDE P J and MARION D (1994b), 'The protection of beer foam against lipid-induced destabilisation', *J Inst Brew*, **100**, 23–25.

COURTIN C M and DELCOUR J A (2002), 'Arabinoxylans and endoxylanases in wheat flour bread-making', *J Cereal Sci*, **35**(3), 225–243.

D'APPOLONIA B L and RAYAS-DUARTE P (1994), 'Wheat carbohydrates: structure and functionality', in Bushuk W and Rasper V F, *Wheat Production, Properties and Quality*, London, Chapman and Hall, 107–127.

DE FEIJTER J A, BENJAMINS J and VEER F A (1978), 'Elipsometry as a tool to study the adsorption behaviour of synthetic and biopolymers at the air–water interface', *Biopolymers*, **17**, 1759–1772.

DELCOUR J A, VANHAMEL S and HOSENEY R C (1991), 'Physicochemical properties of rye nonstarch polysaccharides II Impact of a fraction containing water soluble pentosans and proteins on gluten–starch loaf volumes', *Cereal Chem,* **68**, 72–76.

DESORMEAUX A, BLOCHET J-E, PEZOLET M and MARION D (1992), 'Amino acid sequence and conformation from infrared and Raman spectroscopy of a wheat non-specific lipid transfer protein. Role of disulphide bridges and phospholipids in the stabilisation of the alpha-helix structure', *Biochim Biophys Acta*, **1121**, 137–152.

DICKINSON E (1992), 'Structure and composition of adsorbed protein layers and the relationship to emulsion stability', *J Chem Soc Faraday Trans,* **88**(20), 2973–2983.

DICKINSON E, OWUSU R K and WILLIAMS A (1993), 'Orthokinetic destabilization of a protein-stabilized emulsion by a water-soluble surfactant', *J Chem Soc Faraday Trans*, **89**(5), 865–866.

DICKINSON E, ETTELAIE R, MURRAY B S and DU Z P (2002), 'Kinetics of disproportionation of air bubbles beneath a planar air–water interface stabilized by food proteins', *J Colloid Interface Sci*, **252**(1), 202–213.

DOULIEZ J P, MICHON T, ELMORJANI K and MARION D (2000), 'Structure, biological and technological functions of lipid transfer proteins and indolines, the major lipid binding proteins from cereal kernels', *J Cereal Sci*, **32**, 1–20.

DUBREIL L, COMPOINT J P and MARION D (1997), 'Interaction of puroindolines with wheat flour polar lipids determines their foaming properties', *J Agric Food Chem*, **45**, 108–116.

DUBREIL L, MELIANDE S, CHIRON H, COMPOINT J P, QUILLIEN L, BRANLARD G and MARION D (1998), 'Effect of puroindolines on the breadmaking properties of wheat flour', *Cereal Chem*, **75**, 222–229.

ELIASSON A C, SILVERIO J and TJERNELD E (1991), 'Surface-properties of wheat flour-milling streams and rheological and thermal-properties after hydration', *J Cereal Sci*, **13**, 27–39.

FRUHNER H, WANTKE K D and LUNKENHEIMER K (2000), 'Relationship between surface dilational properties and foam stability', *Colloids Surfaces A – Physicochem Eng Aspects*, **162**, 193–202.

GAN Z, ANGOLD R E, WILLIAMS M R, ELLIS P R, VAUGHAN J G and GALLIARD T (1990), 'The microstructure and gas retention of bread dough', *J Cereal Sci*, **12**, 15–24.

GAN Z, ELLIS P R and SCHOFIELD J D (1995), 'Gas cell stabilization and gas retention in wheat bread dough', *J Cereal Sci*, **21**, 215–230.

GARRETT P R (1993), 'Recent developments in the understanding of foam generation and stability', *Chem Eng Sci*, **48**(2), 367–392.

GRAHAM D E and PHILLIPS M C (1976), 'The conformation of proteins at the a/w interface and their role in stabilising foams', in Akers, R J, *Foams,* London, Academic Press, 195–215.

GRUNDEN L P, VADEHRA D V and BAKER R C (1974), 'Effects of proteolytic enzymes on the functionality of chicken egg albumin', *J Food Sci*, **39**, 841–843.

HAAS P A and JOHNSON H F (1967), 'A model and experimental results for drainage of solution between air bubbles', *I & EC Fundamentals,* **6**(2), 255–263.

HALLING P J (1981), 'Protein-stabilized foams and emulsions', *CRC Crit Rev Food Sci Nutr,* October, 155–203.

HANSELMANN W and WINDHAB E (1998), 'Flow characteristics and modelling of foam generation in a continuous rotor/stator mixer', *J Food Eng*, **38**(4), 393–405.

HAYMAN D, SIPES K, HOSENEY R C and FAUBION J M (1998a), 'Factors controlling gas cell failure in bread dough', *Cereal Chem*, **75**, 585–589.

HAYMAN D, HOSENEY R C and FAUBION J M (1998b), 'Bread crumb grain development during baking', *Cereal Chem*, **75**, 577–580.

HE H and HOSENEY R C (1992), 'Effect of the quantity of wheat-flour protein on bread loaf volume', *Cereal Chem*, **69**, 17–19.

HORIUCHI T, FUKISHIMA D, SUGIMOTO H and HATTORI T (1978), 'Studies on enzyme-modified proteins as foaming agents: effect of structure on foam stability', *Food Chem*, **3**(3), 35–42.

IGREJAS G, GABORIT T, OURY F X, CHIRON H, MARION D and BRANLARD G (2001), 'Genetic and environmental effects on puroindoline-a and puroindoline-b content and their relationship to technological properties in French bread wheats', *J Cereal Sci*, **34**, 37–47.

KALISCHEWSKI K and SCHURGERL K (1979), 'Investigation of protein foams obtained by bubbling', *Colloid Polymer Sci,* **257**, 1099–1110.

KATO A and NAKAI S (1980), 'Hydrophobicity determined by a fluorescent probe method and its correlation with surface properties of proteins', *Biochim Biophys Acta,* **624**, 13–20.

KELLER R C A, ORSEL R and HAMER R J (1997), 'Competitive adsorption behaviour of wheat flour components and emulsifiers at an air-water interface', *J Cereal Sci*, **25**, 175–183.

KIM S H and KINSELLA J E (1985), 'Surface activity of food proteins: relationships between surface pressure development, viscoelasticity of interfacial films and foam stability of bovine serum albumin', *J Food Sci,* **50**, 1526–1530.

KOCZO K and RACZ G (1991), 'Foaming properties of surfactant solutions', *Colloids Surfaces,* **56**, 59–82.

KOKELAAR J J and PRINS A (1995), 'Surface rheological properties of bread dough components in relation to gas bubble stability', *J Cereal Sci*, **22**, 53–61.

KOKELAAR J J, PRINS A and DE GEE M (1991), 'A new method for measuring the surface dilational modulus of a liquid', *J Colloid Interface Sci*, **146**, 507–511.

KOKELAAR J J, GARRITSEN J A and PRINS A (1995), 'Surface rheological properties of sodium stearoyl-2–lactylate (SSL) and diacetyl tartaric esters of mono (and di) glyceride (DATEM) surfactants after a mechanical surface-treatment in relation to their bread improving abilities', *Colloids Surfaces A – Physicochem Eng Aspects*, **95**, 69–77.

KOOIJMAN M, ORSEL R, HAMER R J and BEKKERS A C A P (1998), 'The insertion behaviour of wheat puroindoline-a into diacylgalactosylglycerol films', *J Cereal Sci*, **28**, 43–51.

KUMAGAI H, LEE B H, KUMAGAI H and YANO T (1991), 'Critical radius and time course of expansion of an isolated bubble in wheat-flour dough under temperature rise',

Agric Biol Chem, **55**, 1081–1087.
LEE J C and TYNAN K J (1988), 'Antifoams and their effects on coalescence between protein-stabilised bubbles', *Proc 2nd International Conf on Bioreactor Fluid Dynamics*, London, Elsevier Applied Science, 353–377.
LORIENT D, CLOSS B and COURTHAUDON J-L (1989), 'Surface properties of the bovine casein components: relationships between structure and foaming properties', *J Dairy Res,* **56**, 495–502.
LUCASSEN J (1981), 'Dynamic properties of free liquid films and foams', in Lucassen-Reynders E H, *Anionic Surfactants*, New York, Marcel Dekker, 217–265.
MACKIE A R, GUNNING A P, WILDE P J and MORRIS V J (1999), 'The orogenic displacement of protein from the air/water interface by surfactant', *J Colloid Interface Sci*, **210**, 157–166.
MACRITCHIE F (1976), 'The liquid phase of dough and its role in baking', *Cereal Chem*, **53**, 318–326.
MACRITCHIE F (1978), 'Proteins at interfaces', in Anfinsen C B, Edsall J T and Richards F M, *Advances in Protein Chemistry* Vol 32, New York, Academic Press, 283–326.
MACRITCHIE F (1983), 'Role of lipids in baking', in Barnes P J, *Lipids in Cereal Technology*, London, Academic Press, 11–32.
MACRITCHIE F and GRAS P W (1973), 'The role of flour lipids in baking', *Cereal Chem*, **50**, 292–302.
MATSOUKAS N P and MORRISON W R (1991), 'Breadmaking quality of 10 Greek breadwheats: 2 — relationships of protein, lipid and starch components to baking quality', *J Sci Food Agric*, **55**, 87–101.
METTLER E and SEIBEL W (1993), 'Effects of emulsifiers and hydrocolloids on whole wheat bread quality – a response-surface methodology study', *Cereal Chem*, **70**, 373–377.
MITA T, NIKAI K, HIRAOKA T, MATSUO S and MATSUMOTO H (1977), 'Physicochemical studies on wheat protein foams', *J Colloid Interface Sci,* **59**(1), 172–178.
MITA T, ISHIDA E and MATSUMOTO H (1978), 'Physicochemical studies on wheat protein foams. II Relationships between bubble size and stability of foams prepared with gluten and gluten components', *J Colloid Interface Sci*, **64**(1), 143–153.
MITCHELL J R (1986), 'Foaming and emulsifying properties of proteins', in Hudson B J F, *Developments in Food Proteins – 4,* London, Elsevier, 291–338.
MONFORT A, BLASCO A, SANZ P and PRIETO J A (1999), 'Expression of LIP1 and LIP2 genes from *Geotrichum* species in baker's yeast strains and their application to the bread-making process', *J Agric Food Chem*, **47**(2), 803–808.
MORRISON W R (1983), 'Acyl lipids in cereals', in Barnes P J, *Lipids in Cereal Technology*, London, Academic Press, 11–32.
MORRISON W R (1994), 'Wheat lipids: structure and functionality', in Bushuk W and Rasper V F, *Wheat Production, Properties and Quality*, London, Chapman and Hall, 128–142.
MORRISON W R and BARNES P J (1983). 'Distribution of wheat acyl lipids and tocols into millstreams', in Barnes P J, *Lipids in Cereal Technology*, London, Academic Press, 149–164.
NAKAI S and LI-CHAN E (1993), 'Recent advances in structure and function of food proteins, a QSAR approach', *Crit Rev Food Sci Nutr,* **33**(6), 477–499.
ORNEBRO J, NYLANDER T and ELIASSON A C (2000), 'Interfacial behaviour of wheat proteins', *J Cereal Sci*, **31**, 195–221.
ORNEBRO J, NYLANDER T, ELIASSON A C, SHEWRY P R, TATHAM A S and GILBERT S M (2001),

'Adsorption of the high molecular weight glutenin subunit 1Dx5 compared to the 58–kDa central repetitive domain and alpha-gliadins', *J Cereal Sci*, **34**, 141–150.

PANDOLFE W D (1981), 'Effect of dispersed and continuous phase viscosity on droplet size of emulsions generated by homogenisation', *J Dispersion Sci Technol*, **2**, 459–474.

PATERNOTTE T A, ORSEL R and HAMER R J (1994), 'Dynamic interfacial behavior of gliadin-diacylgaloctosylglycerol (MGDG) films – possible implications for gas-cell stability in wheat-flour doughs', *J Cereal Sci*, **19**, 123–129.

PEARCE R J, MARSHALL S C and DUNKERLEY J A (1991), 'Reduction of lipids in whey protein concentrates by microfiltration – effect on functional properties', *Int Dairy Fed special issue 9201– New applications of membrane processing*, 118–129.

POMERANZ Y (1987), *Modern Cereal Science and Technology*, New York, VCH Publishers.

PTITCHKINA N M, NOVOKRESCHONOVA L V, PISKUNOVA G V and MORRIS E R (1998), 'Large enhancements in loaf volume and organoleptic acceptability of wheat bread by small additions of pumpkin powder: possible role of acetylated pectin in stabilising gas-cell structure', *Food Hydrocolloids*, **12**, 333–337.

QUOC P N, ZITHA P L J and CURRIE P K (2002), 'Effect of foam films on gas diffusion', *J Colloid Interface Sci*, **248**(2), 467–476.

SAHI S S (1994), 'Interfacial properties of the aqueous phases of wheat-flour doughs', *J Cereal Sci*, **20**, 119–127.

SARKER D K and WILDE P J (1999), 'Restoration of protein foam stability through electrostatic propylene glycol alginate-mediated protein-protein interactions', *Colloids Surfaces B-Biointerfaces*, **15**, 203–213.

SARKER D K, WILDE P J and CLARK D C (1998), 'Enhancement of protein foam stability by formation of wheat arabinoxylan-protein cross-links', *Cereal Chem*, **75**(4) 493–499.

SCHOFIELD J D (1994), 'Wheat proteins: structure and functionality in milling and breadmaking', in Bushuk W and Rasper V F, *Wheat Production, Properties and Quality*, London, Chapman and Hall, 73–106.

SHAH P, CAMPBELL G M, MCKEE S L and RIELLY C D (1998), 'Proving of bread dough: Modelling the growth of individual bubbles', *Food Bioproducts Processing*, **76**, 73–79.

SHIMIYA Y and NAKAMURA K (1997), 'Changes in size of gas cells in dough and bread during breadmaking and calculation of critical size of gas cells that expand', *J Texture Studies*, **28**, 273–288.

SLACK P T and BAMFORTH C W (1983), 'The fractionation of polypeptides from barley and beer by hydrophobic interaction chromatography: the influence of their hydrophobicity on foam stability', *J Inst Brew*, **89**, 397–401.

SORENSON S B, BECH L M, MULDBJERG M, BEENFELDT T and BREDDAM K (1993), 'Barley lipid transfer protein 1 is involved in beer foam formation', *MBAA Technical Quarterly*, **30**, 136–145.

SUBIRADE M, SALESSE C, MARION D and PEZOLET M (1995), 'Interaction of a nonspecific wheat lipid transfer protein with phospholipid monolayers imaged by fluorescence microscopy and studied by infrared-spectroscopy', *Biophys J*, **69**, 974–988.

TASSIN-MOINDROT S, CAILLE A, DOULIEZ J P, MARION D and VOVELLE F (2000), 'The wide binding properties of a wheat nonspecific lipid transfer protein – Solution structure of a complex with prostaglandin B-2', *Eur J Biochem*, **267**(4), 1117–1124.

TOWNSEND A and NAKAI S (1983), 'Relationships between hydrophobicity and foaming characteristics of food proteins', *J Food Sci*, **48**, 588–594.

TRONSMO K M, FAERGESTAD E M, LONGVA A, SCHOFIELD J D and MAGNUS E M (2002), 'A study of how size distribution of gluten proteins, surface properties of gluten and dough mixing properties relate to baking properties of wheat flours', *J Cereal Sci*, **35**, 201–214.

VAN HAMEL S, CLEEMPUT G, DELCOUR J A, NYS M and DARIUS P L (1993), 'Physicochemical and functional-properties of rye nonstarch polysaccharides. 4 The effect of high-molecular-weight water-soluble pentosans on wheat-bread quality in a straight-dough procedure', *Cereal Chem*, **70**, 306–311.

VAN LONKHUIJSEN H J, HAMER, R J and SCHREUDER C (1992), 'Influence of specific gliadins on the breadmaking quality of wheat', *Cereal Chem*, **69**, 174–177.

WEEGELS P L and HAMER R J (1992), 'Improving the bread-making quality of gluten', *Cereal Foods World*, **37**(5), 379–395.

WESER C (1980), 'Measurement of interfacial tension and surface tension general review for the practical man', *GIT Fachzeitschrift Laboratorium*, **24**, 642–648.

WILDE P J (1996), 'Foam measurement by the microconductivity technique: an assessment of its sensitivity to interfacial and environmental factors', *J Colloid Interface Sci*, **178**, 733–739.

WILDE P J, CLARK D C and MARION D (1993), 'The influence of competitive adsorption of a lysopalmitoyl phosphatidylcholine on the functional properties of puroindoline, a lipid binding protein isolated from wheat flour', *J Agric Food Chem*, **41**, 1570–1576.

WÜSTNECK R, KRÄGEL J, MILLER R, WILDE P J and CLARK D C (1996), 'Adsorption of surface-active complexes between β-casein β-lactoglobulin and ionic surfactants and their shear rheological behaviour', *Colloids Surfaces A: Physicochem Eng Aspects,* **114**, 255–265.

17

Bread aeration

G. M. Campbell, Satake Centre for Grain Process Engineering, UK

17.1 Introduction

The distinctive appeal of bread derives from its aerated structure, made possible by the unique ability of wheat gluten proteins to retain gases produced by yeast fermentation. In the past the fermentation process by which this structure was created was mysterious and magical, probing at the fundamental but unknown forces of life itself, at the same time fascinating but slightly sinister. Leavening was associated with corruption, leading to a recurrent suspicion that perhaps the superbly palatable yeast-leavened bread was less wholesome and nutritious than its unrisen counterpart. This belief took various forms, from the pronouncements and warnings of physicians and health evangelists, to the production of alternative breads that avoided biological fermentation, to the attempted prohibition of yeast-raised bread. Quoting Elizabeth David (1977):

> The interdependence between the grain and the yeast, between bread and fermenting liquor, was certainly established ... in the earliest times and has persisted throughout history. This circumstance, together with its very mystery, accounts perhaps for the curiously ambivalent attitude towards leavened bread and leaven generally as expressed in both the Old Testament and the New; 'a symbol of silent pervasive influence, usually of that which is corrupt'. ... In more recent times, one of the factors making for the success of Dr Dauglish's famous aerated bread was that being unfermented it was believed by many people to be more wholesome than ordinary bread. A ferment indicated a kind of decay, a corruption, and was therefore suspect.

However, public opinion down the centuries is persistently in favour of yeast-leavened bread, even against legal machinations. Elizabeth David again:

> In seventeenth century France, the Paris Faculty of Medicine spent months debating the question of beer leaven and whether the bakers should be permitted to use it, eventually coming up with the solemn decision that it was injurious to health and should be forbidden. (Apart from a decree passed by the Paris parliament and shortly afterwards rescinded, nobody took much notice. The ancient spontaneous leaven and sourdough systems were all right for bread made from coarse brown meal, but for the fine light bread liked by the gentry, a good barm was needed.)

The allure of well-risen bread conquers all. To the Roman masses, bread was famously equalled in appeal only by circuses, and much of Rome's political and economic organisation was fuelled by the need to secure the wheat supplies (Atkin, 1995). Socrates stated 'No man qualifies as a statesman who is entirely ignorant of the problems of wheat' (Evans, 1993), and Athens, before the rise of Rome, had similarly built her military policy around the need to protect the sea trade routes for the import of grain (Atkin, 1995). In more recent times, developing nations have endangered their economies through the excessive import of wheat to make the raised bread demanded by their citizens, and riots have ensued following increases in bread prices. The ready availability of raised bread has blinded the modern Western world to the unique and special quality of this most charismatic of foods. We are used to seeing food exotica presented to us in curry houses and trendy sushi bars; but recently in the Far East, in an enterprising cultural reversal, the equivalent 'bread boutique' has been established, to bring this wondrous alien food to peoples accustomed to consuming their cereals in a more intact, unexpanded, less delicate form (Anon, 2002).

The enduring appeal of bread is in the palatable texture and attractive appearance derived from its aerated structure, and empires have been organised, decrees ignored, economies threatened and venturesome business schemes established in the pursuit of this. Yet despite five thousand years of bread production and over a century of scientific research, the mysteries underlying the creation and control of this aerated structure to a large extent remain. Gas production and retention are the keys to breadmaking, but gases were identified by Chamberlain in 1979 as 'the neglected bread ingredients'. Any ambition to master breadmaking at the scientific level must encompass understanding of the gas phase behaviour throughout the process.

While the term 'bread' is used generically, there is a myriad variety of different breads, reflecting local and national culture and practice, and frequently identified by distinctive aerated structures, or by the specific shapes that the rising process allows the loaf to adopt. Also, bread is not the only aerated baked product. Matz (1960) proposed: 'There are probably innumerable schemes for classifying bakery products, but the most sensible way to categorize

Table 17.1 Classification of selected cereal-based food products on the basis of the leavening action. (Note that some names are used variously in different regions or communities, and that recipes and formulations vary, so this classification is approximate and indicative rather than prescriptive or authoritative. It is also incomplete, in particular not listing the innumerable varieties of breads, cakes and biscuits.)

Yeast-leavened	Chemically leavened	Air/steam-leavened	Unleavened
Bagels, baguettes, baps, barley bread†, breadsticks, brioche, ciabatta bread, corn bread†, croissants, buns, crackers, crumpets, Danish pastry, dumplings, extruded products, focaccia bread, gluten-free breads, grissini, muffins, naan breads, oatcakes, pain au levain, pancakes, pikelets, pitta breads, pizza bases, potato bread†, pretzels, quiche pastry, rice bread†, rolls, rusks, rye bread, Sally Lunn, scones, simnel cakes, sourdough breads, steamed breads, stollen, toast, yeast cakes, yeasted wheat breads (too numerous to list separately, although a few common or representative examples have been named)	Biscuits, cakes, cookies, doughnuts, expanded extruded products, mandelbrot, soda breads, waffles	Angel food cake, biscuits, chapatis, choux pastry, cornflakes, crispbreads, éclairs, extruded breakfast cereals, flat breads, lucchi, matzos, parathas, piadine, popcorn, poppadoms, profiteroles, puff pastry, rice crispies, soufflés, sponge cake, tortillas, vol au vents, wafers, Yorkshire puddings	Cous cous, pie pastry, pasta, porridge, rice

† Note that breads such as potato bread, rice bread and barley bread generally contain significant proportions of wheat flour.

them for the bakery engineer or cereal chemist would seem to be on the basis of the source of leavening gas'; he then classified bakery foods into yeast-leavened, chemically leavened, those that are air-leavened and unleavened. In fact, most cereal-based food products, not just baked, derive their palatable textures from aeration to some extent, including cakes, crumpets, biscuits, pikelets, pizza bases, Yorkshire puddings, profiteroles, cornflakes, rice crispies, extruded breakfast cereals, wafers, waffles, soufflés and, of course, popcorn. Even flat breads are significantly aerated. Table 17.1 shows the classification of cereal-based foods on the basis of Matz's categories (recognising that his air-leavened products are actually mainly leavened by steam entering bubble nuclei that initially contain air, and that air and steam similarly contribute to the yeast- and chemically leavened goods). But leavened bread encompasses most of the issues of food aeration, has the greatest market volume, and is the best understood in terms of gas phase behaviour. This chapter therefore describes the historical development, current status and future direction of bread aeration studies, but with an eye on the application of the principles of bubble and gas phase behaviour to other cereal-based products.

17.2 The development of bread aeration studies

The gas cells in bread are so clearly its defining feature that bakery research scientists have found it impossible to interpret or apply their findings without substantial speculation about bubble behaviour. However, studying the bubbles in bread directly is difficult, in part owing to the difficulty of developing and applying appropriate experimental techniques and interpreting the results. For this reason, within the vast corpus of bread research literature, relatively few studies have been reported that investigate the gas phase behaviour within the process directly. It is universally recognised that gas retention in wheat flour doughs is attributable to the unique rheology of the gluten proteins, so rheological studies of bread doughs abound, but their interpretation in terms of bubble creation and control is still vastly constrained.

The direct study of gas phase behaviour in bread starts with the pioneering work of JC Baker and MD Mize in the late 1930s and 1940s (Baker and Mize, 1937, 1939a,b, 1941, 1942, 1946; Baker, 1941). These workers demonstrated that the origin of gas cells in bread is the bubbles created in the dough during mixing, that aeration coincides with dough development, and that the gas composition in the headspace during mixing directly affects dough development. Their seminal papers are still a rich source of inspiration. Mechanical dough development had been discovered by Swanson and Working a few years earlier (1926), but Baker and Mize's discovery that the use of oxidants reduced the power required to develop doughs mechanically allowed scale-up of this new approach. This led to the development by Baker of the Do-Maker, the first commercial mechanical dough development (MDD) system. This continuous system was followed by others in the USA, the UK and Australia, and by the batch Chorleywood Breadmaking Process (CBP). (Baker's gas-related studies also included the maturation of flour with nitrogen trichloride, described, on the occasion of Baker's award of the AACC's Osborne Medal in 1945, as 'the most important contribution of the science of cereal chemistry to mankind'.)

The Chorleywood Breadmaking Process was developed at the British Baking Industries Research Association (BBIRA) based at Chorleywood in the UK (subsequently absorbed into the Flour Millers and Bakers Research Association, FMBRA, and now part of Campden & Chorleywood Food Research Association, CCFRA). It was introduced into the UK in the early 1960s, and is now responsible for the majority of the bread consumed in this country. It has also been adopted widely in Australia, New Zealand, South Africa and elsewhere. In the USA sponge-and-dough bread processes are preferred, which give the finer, higher-volume loaves preferred by American consumers.

Its use of a high-speed batch mixer made the CBP more suitable to small bakeries than the less flexible continuous systems. However, it was soon in demand for plant bakery use as well, requiring scale-up of the original mixer. This presented a problem, that on scale-up, loaf quality deteriorated, the cause of which was diagnosed as excessive aeration of the dough during mixing. (Thus, although bubbles in the dough are required in order to produce good bread, too many are detrimental.) Reference to the earlier work of Baker and Mize prompted rectification of this problem through the application of partial vacuum during mixing, and this became a distinguishing feature of the CBP. As a result of this, numerous dough aeration studies were, and still are, carried out in the UK, such that the greatest understanding of bread aeration has been gained on CBP-type processes.

Oxidants (or 'improvers') are an essential feature of mechanical dough development, facilitating the alignment of glutenin proteins that defines dough development. For many years potassium bromate and azodicarbonamide were the preferred oxidants, but their ban in most countries in the 1990s left bakers with only ascorbic acid (Vitamin C) as an allowable oxidant in breadmaking. Ascorbic acid is in fact an anti-oxidant, exerting its improving effect through conversion to dehydro-ascorbic acid via reaction with atmospheric oxygen. But mixing under partial vacuum, as practised in the CBP, limits the availability of atmospheric oxygen and hence the effectiveness of ascorbic acid. This once again put aeration under the spotlight in the UK, and was addressed by the introduction of pressure-vacuum mixing (APV Corporation Ltd, 1992; Cauvain, 1994). In this system, doughs are initially mixed under positive pressure, to increase oxygen availability. Then, towards the end of mixing, vacuum is applied, to reduce the air content to the level required to give the desired bread texture. This new technology has given bakers additional versatility to produce a range of bread textures through manipulation of the mixing pressure regime, but once again demands an even more sophisticated understanding of aeration processes throughout breadmaking, obtained with appropriate measurement tools and techniques.

17.2.1 Methods for studying bread aeration

The pioneering work of Baker and Mize drew its insightful conclusions based on two key measurement tools: examination of baked loaves and measurement of dough densities, combined with inventive dough preparation and processing techniques based mainly on mixing under vacuum and under various gases. These techniques have been widely applied in subsequent studies and remain predominant today. Baker and Mize (1939a,b) also investigated gas phase behaviour during baking in an electrical resistance oven, and related pressure and loaf volume changes to transitions and events during baking. Other techniques that have been applied to bread aeration studies include: examination of photographs of frozen part-proved doughs; scanning electron and light microscopy; vacuum expansion of doughs; measurement of CO_2 production and retention; proof height; measurement of volume changes and gas release during baking; measurement of bubble size distributions in doughs; measurement of interfacial properties of dough components; bubble inflation rheometry; X-ray tomography of proving doughs; measurement of ultrasound propagation; image analysis of bread crumb; texture analysis of bread crumb. (In addition, studies on related systems such as cake batters have provided analogies with bread dough aeration.)

Alongside these techniques have been interpretations, sometimes flawed, of the observations in terms of the behaviour of the gas phase during processing. Bubble behaviour in bread doughs is complex, and measurements and observations must be integrated into mechanistic models for the behaviour of the gas phase throughout the breadmaking process. The understanding required to describe gas phase behaviour throughout breadmaking is still far from complete. The next section describes aeration processes throughout breadmaking, acknowledging that different breads are likely to behave differently, that many causes and effects are multifaceted, and that our current knowledge is limited and is mostly derived from studies of white sandwich bread.

17.3 Aeration during breadmaking: mixing

The essential operations to manufacture raised bread are mixing, proving and baking, and each of these contributes to the aeration of the loaf. Mixing entrains bubbles into the dough; proving inflates these bubbles with carbon dioxide gas generated by yeast fermentation; and baking transforms the foam structure containing discrete bubbles into a sponge of interconnected gas cells, and sets the structure. The following paragraphs briefly describe current understanding of the gas phase behaviour at each stage. The aim of this section is not to present a critical review or a comprehensive treatise. Rather, the purpose is to establish sufficient understanding that the opportunities for further studies that would contribute to improving bread quality become evident.

In no-time processes such as the CBP, the bubble structure created in the dough mixer directly affects baked loaf structure and texture. The nitrogen gas

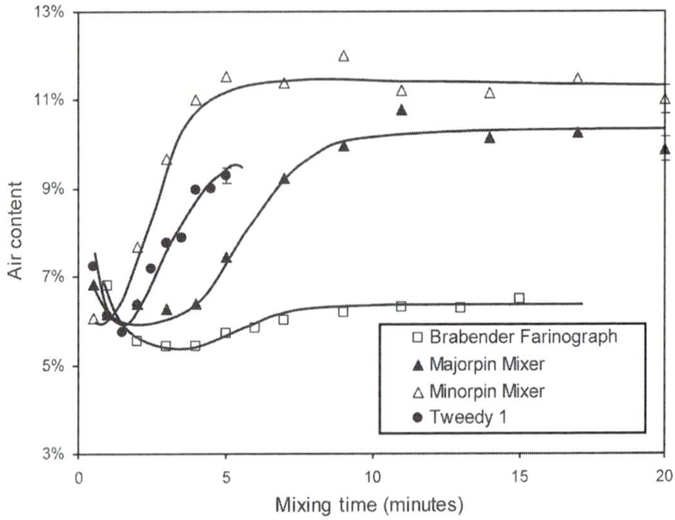

Fig. 17.1 Change in air content during the course of mixing in four laboratory-scale mixers.

in the air forms bubbles that act as nucleation sites for the carbon dioxide gas generated during proving. Oxygen in the air contributes to oxidation of the dough and hence dough development, both directly and through interaction with ascorbic acid, and at the end of mixing is rapidly removed from the bubbles by yeast metabolism. Baker and Mize (1937) showed that mixing doughs under high vacuum eliminated these nucleation sites, preventing retention of CO_2 gas and resulting in loaves of low volume and poor structure. Dunn and White (1939) later found the same result for cake batters.

Four factors are important with regard to aeration of the dough during mixing: the gas content of the dough, the rate of turnover of gas during mixing, the distribution of the gas in terms of the bubble size distribution, and the gaseous composition of the bubbles. Figure 17.1 illustrates the changing air contents throughout the course of mixing for several laboratory-scale mixers, showing how air is initially removed from the dough, then entrained as the gluten network develops, before achieving a steady state. For doughs mixed in air or nitrogen, the gas content is proportional to the mixing pressure and depends on the dough formulation and on the mixer design, operation and scale. Typical air contents for doughs mixed at atmospheric pressure are 5–10%. Increasing the mixing pressure gives a more open bread crumb grain. Doughs based on strong flours entrain less air than those based on weak flours; however, the formulation and processing factors affecting the quantity of gas initially entrained into doughs have not been established comprehensively.

Although it is clear that the initial air content of the dough is important in affecting baked loaf texture, it is not precisely clear why. The quantity of gas produced by yeast fermentation and by evaporation of water during baking is much greater than that initially entrained in the dough, and the majority of the bubbles initially entrained combine through coalescence. Thus the detailed mechanism by which the initial air content translates into altered bread structure is not yet resolved.

The air in dough is distributed in bubbles varying widely in size, from below 30 μm up to several millimetres in diameter, with an average size of around 100 μm, and with around 30–100 bubbles per mm^3. Mixing at different pressures changes the number of bubbles per unit volume but has relatively little effect on their size distribution (Campbell *et al.*, 1998). High-speed mixing, as practised in the CBP, appears to give smaller average bubble sizes than slower speed mixing. Measuring bubble size distributions in dough is difficult and time-consuming, and the effect of only a very few factors regarding dough formulation and mixer design and operation have been studied to date.

The air content and bubble size distribution depend on the balance between the rates of entrainment and disentrainment of air during mixing and on bubble break-up. Knowledge of the rate of disentrainment affects the timing of the application of vacuum during pressure-vacuum mixing and hence effective use of this technology. The turnover of air during mixing also affects oxygen availability. Chamberlain (1979) noted a 'suspicion that the availability of oxygen varies from one mixer to another' and that in studies with oxygen-enriched headspace atmospheres in the mixer, the oxygen 'clearly was not gaining access to the dough in sufficient quantities'. Marston's (1986) work on the delayed provision of oxygen in the mixer headspace indicated that the resultant effect was not homogeneously distributed throughout the dough in the time available. This is the result of the slow turnover of air during mixing, as demonstrated by Campbell and Shah (1999). As well as the need to understand better the factors determining the gas content and its static distribution in the dough at the end of mixing, there is clearly a need to understand the dynamics of aeration during mixing.

Most research on dough mixers is carried out on laboratory-scale mixers, while most commercial breadmaking uses larger mixers. The fluid mechanics of mixers change on scale-up, affecting dough development and aeration, and compromising the interpretation of results obtained at laboratory scale. Scale-up studies related to dough mixing would give a sound basis for transferring research results and new product development to plant-scale operations, but few such studies have been reported.

Several workers over the decades have mixed doughs under varying headspace gas compositions, either in an attempt to understand aeration behaviour, or with a view to establishing commercial processes. Baker and Mize (1937) showed that doughs do not develop in the absence of oxygen during mixing. Chamberlain and Collins (1979) established that at the end of mixing, residual oxygen was rapidly removed by yeast activity, and that nitrogen was thus the gas that made up the initial bubble nuclei. Mixing under pure oxygen gave bread of low volume and poor structure similar to that obtained from mixing in a high vacuum, while mixing in a 60 : 40 oxygen : nitrogen atmosphere gave results similar to mixing at half atmospheric pressure, as the partial pressure of nitrogen is similar in both cases. Marston (1986) presented evidence that exclusion of oxygen at the beginning of mixing, followed by mixing in air for the latter half of mixing, improved bread compared with mixing in air throughout mixing. Providing an oxygen-enriched atmosphere for the latter half of mixing enhanced the benefit further, as did blowing air over the dough during mixing. His results demonstrated 'some possibilities for lessening dependence on other additives by controlling the atmosphere above dough during mixing'. Todd *et al.* (1954) patented the mixing of dough in an oxygen-enriched atmosphere at around atmospheric pressure, while Suntheimer (1961) presented a patent in which dough was mixed under a CO_2 atmosphere, benevolently to 'aid in reducing the strain on the yeast as a producer of carbon dioxide'. Despite these, modified atmosphere mixing is not yet practised commercially. Spooner (1999) encourages a renewed investigation into the possibilities of this approach.

Mixing creates the bubble size distribution that will be inflated by proving. This inflation depends on the initial size, number and gas composition of the bubbles. In principle, altering the mass transfer dynamics during proving, by altering gas composition during mixing, ought to allow the creation of distinctive gas cell structures in the baked loaf. The envelope of possibilities in this area has not yet been explored fully. Figure 17.2 shows the gas contents of doughs mixed in nitrogen and carbon dioxide atmospheres at different pressures. The increased gas solubility of carbon dioxide at higher pressures

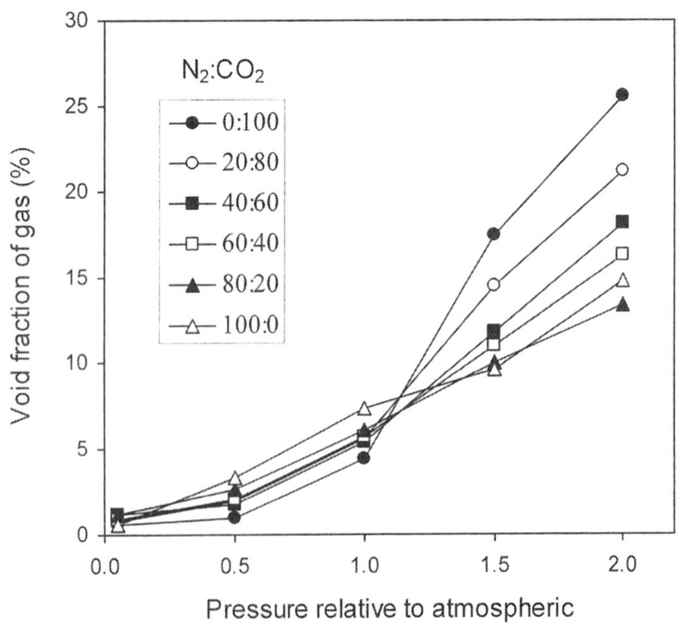

Fig. 17.2 Effect of gas composition (proportions of nitrogen and carbon dioxide) in the headspace atmosphere during mixing on the aeration of the dough.

causes very large gas contents in the dough when the pressure is released. Potentially this could alter bubble growth and coalescence during proving and baking and hence give rise to novel bread textures, although maintaining oxygen availability during mixing is still essential.

Mixing is the critical control point in no-time breadmaking processes. It is the most complicated of operations in the process, and the one that has undergone the greatest transformations in the last century. The creation and control of bread structure and quality start with aeration of the dough during mixing, and there is much scope for applying new techniques to define and exploit its contribution more precisely.

17.4 Aeration during proving

Although mixing is the most critical and complex operation in breadmaking, proving is still the heart of breadmaking and its defining operation. Proving links the bubble size distribution created in the mixer to the gas cell distribution apparent in the baked loaf, through the dynamics of CO_2 generation by yeast and its mass transfer into bubbles, and of bubble coalescence. Without understanding this link, knowledge of aeration of the dough during mixing is impotent.

Two related examples serve to illustrate some of the current mysteries of proving. Firstly, in the CBP, mixing under a partial vacuum gives fewer bubbles in the dough, but results in a finer gas cell structure (i.e. more gas cells) in the baked loaf. Secondly, there are many fewer gas cells in the baked loaf than bubbles in the dough; 99% of the bubbles do not survive to become gas cells. Thus, while aeration during mixing directly affects baked loaf architecture, the baked loaf is not simply an expanded version of the bubble structure created in the mixer. Studies based primarily on examination of baked loaves are hence notoriously difficult to interpret. The link between aeration during mixing and the baked loaf structure is a complex competition for the available CO_2 gas that depends on relative bubble sizes and partial pressures, on the rate of CO_2 generation and its partitioning between the aqueous and gas phases, and on the extensive coalescence of bubbles during the later stages of proving and early stages of baking.

Bubble inflation during proving is a diffusive mass transfer operation, described by the equation

$$Q = kA(C - C^*) \qquad [17.1]$$

where Q is the molar rate of transfer of CO_2 into a bubble, k is the mass transfer coefficient, A is the surface area of the bubble, C is the concentration of CO_2 in the aqueous dough phase, and C^* is the concentration of CO_2 in equilibrium with the gaseous CO_2 concentration in the bubble (Shah *et al.*, 1998). The latter is related to the partial pressure of CO_2 in the bubble via Henry's Law. Larger bubbles have the advantage of lower internal pressures, hence C^* is lower, favouring mass transfer. However, smaller bubbles have higher mass transfer coefficients, compensating for this, such that large and small bubbles tend to grow much the same rate. This is evident in real time images of proving doughs obtained using X-ray tomography. Disproportionation (the growth of large bubbles at the expense of smaller ones, due to their lower internal pressures) is not applicable to actively inflating bubbles; although called upon from time to time, it is largely irrelevant in no-time breadmaking processes, except perhaps to bubbles very near the surface of the proving dough which can lose gas to the atmosphere. (Studies that purport to demonstrate disproportionation in no-time doughs are, in the author's opinion, reporting artefacts of the measurement system that only allows observation of bubbles near the surface of thin samples of dough. Observed losses of gas are to the atmosphere rather than to larger bubbles, and would not occur in the bulk of a larger dough piece. Disproportionation may, however, be relevant to doughs from bulk fermentation and sponge-and-dough processes, and to frozen doughs.) The cause of the greatly reduced numbers of gas cells in the bread compared with the number of nitrogen bubbles in a MDD dough is primarily extensive coalescence during proving and baking. These coalescence events occur extremely rapidly within the dough, and are hence difficult to observe and study. Doughs with good strain hardening properties resist coalescence, resulting in finer crumb structures in the baked loaf (van Vliet *et al.*, 1992; Dobraszczyk and Roberts, 1994).

The above picture is complicated by the distribution of bubble sizes in the dough, which is dominated by very large numbers of small bubbles and their very high surface area for mass transfer. It is further complicated by the rate of CO_2 generation by the yeast and its subsequent partitioning between the liquid phase of the dough and the bubbles. Spatial variations in CO_2 concentration throughout the dough piece also occur, as near the surface of the proving dough, CO_2 is lost to the atmosphere, reducing its concentration and hence the rate of growth of the bubbles near the surface. These factors can in principle be modelled mathematically, with model validation provided by experimental results from measuring the growth of the dough piece and loss of gas and, more recently, X-ray tomography (Whitworth and Alava, 1999; Whitworth, 2002). Such models, fully implemented, would allow the growth of the initial bubble size distribution to be explored. This would facilitate understanding of the effect of changing the initial number, sizes and gas composition of the bubbles, and would give guidance as to how specific desirable bread textures could be created by suitable control of the mixing operation.

At this point the rheology of the dough becomes critical. In mechanical dough development processes, the primary purpose of mixing is to develop the dough rheology, such that bubbles resist coalescence and remain discrete during proving and baking. Studies by Dobraszczyk (1997, 1999) have demonstrated that dough rheology as characterised by bubble inflation rheometry correlates well with bubble failure strain and with baked loaf volume from flours of differing breadmaking quality. The correlations are improved when studies are carried out at temperatures of 55 °C, as experienced by the dough during the early stages of baking. This work has confirmed that strain hardening of wheat flour doughs is the property that confers their unique bubble stability, as proposed by van Vliet *et al.* (1992). Although the assumptions currently employed to derive strain hardening parameters using this method have been criticised (Charalambides *et al.*, 2002a,b), the technique could be extended to a much wider range of dough formulations and ingredients, to quantify the contribution they each make to improving bread quality. Then, mechanistic and quantitative models are needed, to allow these fundamental rheological measurements to be integrated into refined and testable predictions of breadmaking performance.

As well as gluten rheology, the surface rheology of bubbles is also important. MacRitchie (1976) proposed the liquid film hypothesis, that towards the end of proving, discontinuities develop in the thin gluten film between bubbles. However, coalescence is retarded by the presence of a thin liquid film containing water-soluble surface active components. Good breadmaking performance requires not only good gluten quality, but also the contribution from surface active materials to contribute that little extra bubble stability that distinguishes a good breadmaking flour from a poor one. Örnebro *et al.* (2000) reviewed the interfacial behaviour of wheat proteins. Gan *et al.* (1995) reviewed the evidence in support of the liquid film hypothesis, but it remains speculative, and a research priority is to confirm it as a realistic framework of understanding. If sound, it gives a basis for specifying the action of bread ingredients, particularly surfactants, more precisely. Either way, it is clear that both bulk and surface rheological effects contribute in some way to bubble stability and bread aeration.

Precise measurements of dough bulk and surface rheology can currently only be related qualitatively and in vague terms to final bread structure and texture. This line of research is also in danger of being swamped by the endless identification of chemical species contributing, uniquely or otherwise, to bread quality. A breakthrough is needed to unify the vast and growing body of information of these diverse constituents, and to allow it to be integrated into mechanistic, predictive and quantitative models of the creation of the aerated state of bread.

Bread quality can be improved through effective application of an increasing range of ingredients, including surfactants (emulsifiers), enzymes and bakery fat. Current understanding of their action is extremely qualitative and limited, although many are believed to exert at least part of their influence during proving. Focusing on the behaviour of the gas phase throughout breadmaking, and developing and applying an appropriate repertoire of measurement tools and techniques, offers a coherent framework for understanding the action of these ingredients and hence for their effective application.

A curiosity of the CBP is the need for high melting point bakery fat, the absence of which gives characteristic blisters on the surface of the dough at the end of proving, and little or no oven spring (Chamberlain *et al.*, 1965; Williams and Pullen, 1998). Most studies of the action of bakery fats have concentrated on the baking stage and the behaviour of fats at baking temperatures (Baker and Mize, 1939b, 1942; Brooker, 1996), but clearly fat has effects at proving temperatures in the CBP, and the fat effect relates to the bubble behaviour. Bell *et al.* (1981) confirmed that vacuum expansion of doughs, which mimicked baking expansion but avoided the application of heat, demonstrated volume improvements from high melting point fats; however the reason for the improvement is still not certain. Williams and Pullen (1998) commented 'It is a source of great frustration to bakers and cereal scientists alike that ... the fundamental understanding of the "fat effect" in breadmaking still remains unclear.' Given the priority throughout the food industry of reducing fat contents in foods, it would be helpful to obtain a more complete insight into the action of fats in the CBP and other breadmaking processes.

Bran is increasingly recognised as an important component of a healthy diet; increasing bran consumption is arguably the single most promising dietary factor that could increase the health of the population. Bread is clearly an appropriate vehicle for incorporation of bran into the diet, but bran has a deleterious effect on bubble stability (Gan *et al.*, 1989), limiting the appeal of wholemeal and bran-enriched breads. Further research on the interaction of bran components with bubbles may lead to enhanced appeal and consumption of healthy, fibre-rich breads.

At the end of proving, the risen dough contains about 70–75% gas by volume, and is extremely delicate. It now remains to convert this highly expanded mass into a digestible and stable loaf, suitable for consumption and conforming to aesthetic expectations.

17.5 Aeration during baking

The baking process is a relative newcomer in cooking terms (McGee, 1984). It is a sophisticated process that requires the facility to enclose the food items in the dry heat environment of an oven. An oven is a permanent structure that requires the settled lifestyle that historically was enabled by the construction of cities. Many artisan and commercial bakers insist that the oven is the most critical component in the production of really good bread (Collister and Blake, 2000; Calvel, 2001), and that baking is 'the most important step in breadmaking' (Pyler, 1973). Baking causes a complex series of physical, chemical and biochemical transformations, the spatial arrangement and sequencing of which are critical. The elaborate processes operating within baking, by which a light, attractive, palatable, shelf-stable and readily digestible loaf is delivered, have been described in detail elsewhere (e.g. Pyler, 1973; Wiggins, 1998).

From the perspective of aeration, baking contributes additional leavening action and transforms the foam structure at the end of proving into a sponge structure containing a porous interconnected network of fine gas cells separated by thin walls. Wall thickness and the size, shape and orientation of the gas cells are defining factors regarding the specific bread variety and the evaluation of its quality and functionality, along with loaf volume, which gives a much cruder indication of bread quality.

The key physical phenomena underpinning the contribution of baking to bread aeration are that heat transfer takes place from the outside of the loaf to the inside, that dissolved gases are less soluble as temperature increases, that water and ethanol evaporate into bubbles as temperature increases, and that setting of the gluten–starch matrix through partial gelatinisation and denaturation causes and coincides with bubble rupture events by which leavening gases are lost.

During baking, six phenomena occur that contribute to bread aeration:

1. The increasing temperature causes increased yeast activity for a while, until the yeast is killed at about 50 °C. This can be considered a high-temperature extension of proving, and contributes additional leavening action known as 'oven rise'.
2. The gases in the dough undergo thermal expansion as the temperature rises.
3. Carbon dioxide, ethanol and other components dissolved in the aqueous phase of the dough come out of solution into bubbles. In addition, water evaporates into steam which enters and inflates the bubbles. These phenomena, in combination with the thermal expansion of the gases, contribute further volume increases to the dough known as 'oven spring'. The terms 'oven rise' and 'oven spring' are often used interchangeably, but for the sake of precision should be used to describe different phenomena occurring during baking. Oven rise and oven spring combine to bring the final gas volume of the baked loaf to around 75–85%, dependent on the bread variety.
4. The formation of a crust on the outside of the loaf increases the pressure within the baking dough and encourages coalescence of bubbles and the loss of fine structure (Hayman *et al.*, 1998b). Coalescence is a low-temperature event between bubbles separated by still viscous dough, which causes loss of bubble numbers and coarsening of the bubble structure, but does not result in a loss of gas. Surface active agents in the liquid film (if this hypothesis is correct) contribute to stability against coalescence and help maintain the fine gas cell structure.
5. The starch–gluten matrix sets, through gluten denaturation and starch gelatinisation, causing rupture of bubble walls and creating interconnections that render the crumb porous and bring about the loss of leavening gases and their replacement by air. (In the limited water environment of baking dough, the extent of denaturation and gelatinisation is limited.) Rupture is a high-temperature phenomenon that probably coincides with setting of the matrix (Bloksma, 1990a,b) and causes loss of gas, but leaves the interconnected gas cells distinct, if no longer discrete. Many of the ingredients in the dough formulation are believed to contribute to bread structure and texture by delaying this setting, so that bubbles remain discrete and stable for a little longer during baking, giving a finer crumb grain and increasing loaf volume.
6. Firming of the cell walls, which are now the defining components of the mechanical properties of the aerated structure.

Bloksma (1990a) estimated that evaporation of water contributes more than half of the oven spring increase, with evaporation of carbon dioxide contributing about a quarter, and thermal expansion of gases contributing about 15%. Hood and Lowe (1948) found the relative contributions to expansion during baking of cakes depended on the type of fat used. For cake formulations containing baking powder and made with butter, they estimated the increases from thermal expansion of air, water evaporation and carbon dioxide evaporation to be around 8, 32 and 60%, respectively.

Hayman *et al.* (1998a), using an optimised straight dough process, reported that the aerated structure apparent on micrographs of doughs prepared from flours of different breadmaking performance appeared similar at the end of proving, but that differences appeared within the first 12 minutes of baking. Whitworth (2002), by contrast, reported that in CBP breads differences could be observed by the end of proof using X-ray tomography, concluding that differences in texture from flours of differing breadmaking quality originate during proof and not just during baking. The apparently contradictory observations may in part reflect the different breadmaking systems being studied.

Tomography is a tool that is beginning to yield new insights into the evolution of the dough during proving (Whitworth and Alava, 1999; Whitworth, 2002). Combined with image analysis of the final loaf, this gives clues to some of the phenomena occurring during baking. However, direct tomography of doughs during baking has not yet been reported and would complement the proving studies, as would tomographic studies of mixing. In addition, quantitative and mechanistic models are required to interpret the qualitative observations enabled with this powerful tool. Magnetic resonance imaging (MRI) potentially offers greater resolution than X-ray tomography, while electrical capacitance, resistance and impedance tomography offer cheaper and more portable and accessible systems.

The result of baking is a structure defined by a relatively solid outer crust and a soft, delicate crumb comprising cell walls that surround gas cells and determine the mechanical properties of the loaf. The internal and external appearance, compressibility and fracture mechanics of the loaf are key factors determining its aesthetic appeal, apparent freshness and performance during consumer handling. The factors within the breadmaking process, described above, that create the aerated structure are understood to only a limited extent; our understanding of the relation of this structure to the texture of the loaf is even more circumscribed. The mechanical properties of baked crumb structures of both bread and cakes have been studied under compression or tension or both by a number of workers (Platt and Kratz, 1933; Platt and Powers, 1940; Chen *et al.*, 1994; Swyngedau and Peleg, 1992). Promising results based on the material science of cellular foams have been reported (Keetels *et al.*, 1996; Attenburrow *et al.*, 1989; Scanlon *et al.*, 1997, 2000; Zghal *et al.*, 2001; Liu and Scanlon, 2002), in which mechanical properties such as Young's modulus and critical fracture stress have been related to structure, as defined by bulk density, using relations of the form

$$\frac{\sigma}{\sigma_w} = \left(\frac{\rho}{\rho_w}\right)^n \qquad [20.2]$$

where σ and σ_w are a given mechanical property of the foam and of the cell walls, respectively, and ρ and ρ_w are the bulk density of the foam and of the continuous matrix, respectively. The exponent n depends on the particular mechanical property under consideration, the architecture of the foam, and the thermo-rheological state of the matrix. The measured properties correlate well with sensory assessments, completing the link between structure and perceived texture. Image analysis can be applied to refine these relationships further (Zghal *et al.*, 1999). Further such studies, combining image analysis, texture analysis and tomography, are required to complete the picture of the creation of the food world's most enchanting texture. Similarly, the creation of the crust itself, often a defining characteristic of bread, has been studied very little.

Baked loaves are accessible for evaluation and analysis, and numerous innovative techniques have been applied to study and understand the baking process. Gas production and retention during proving have also been studied extensively, and the advent of tomography is beginning to probe the mysteries of proving with greater scrutiny. (Moulding has not been discussed here explicitly; this operation also affects bubbles through degassing and coalescence and through constraining their growth during proving (Cauvain *et al.*, 1999; Whitworth and Alava, 1999), and has much scope for greater understanding leading to improved bread quality.) Mixing is, however, where the baker has most control of bread quality, including aeration. The bubble structure initially created in the mixer, and the rheology developed therein that is the key to gas retention, render the evolution of the dough during proving and baking largely deterministic. As noted above, without qualitative and quantitative models of proving and baking, knowledge of aeration during mixing is impotent; however, mixing remains the critical control point in breadmaking, and the point where current knowledge is most constrained and where opportunities for improving the breadmaking process are greatest.

17.6 Future trends

On the basis of the historical development of breadmaking, will greater understanding of the gas phase behaviour lead to improved quality? The CBP is the process that has demanded the most investigation into gas phase behaviour, but its detractors would claim, rightly or wrongly, that it delivers bland, insipid, unsatisfying loaves and has denied a generation access to really good bread. In this respect, bread is a victim of its own success, that the technologies that have simplified the process and reduced the price of bread so greatly have also reduced the status of bread to an undervalued staple. Bread remains for many the staff of life, a staple component of the diet – essential, but largely taken for granted. However, increasingly bread is taking on an alternative status, of a luxury item, an opportunity for exotic variety in the diet, even a status symbol, particularly in the production and consumption of specialty and artisan breads. The report of an artisan loaf retailing for nearly £10 (*Daily Mail*, 2 September 2002) illustrates the esteem that really good bread commands. In recent years there has been an explosion of bread varieties on the market, and increased interest in bread quality and variety, encouraged by the prevalence of domestic breadmakers. Thus bread fulfils a dual role in the shopping basket, both as a staple for everyday consumption within the context of a weekly or fortnightly shopping trip, and as a treat or a dinner party fashion accessory. The future of commercial breadmaking is thus to continue to deliver ever more convenience at ever reduced price, but even more, it is to aspire to the variety, quality and appeal delivered by artisan breads.

And will further studies of gas phase behaviour play a role in this? Firstly, it is germane that many specialty and artisan breads have distinctive bread structures, often with eye-catchingly large holes that are central to their appeal. Secondly, the tools and techniques that have been developed to study gas phase behaviour in primarily white sandwich breads could, without prejudice, be applied to elicit the secrets of the more romantic breads. Other areas to which the above approaches could be profitably applied include frozen doughs and the production of gluten-free breads, both of which are compromised by the inferior textures of their respective products. Then there are the numerous other breadmaking systems beyond the Chorleywood Bread Process, each with its own subtleties regarding the creation and control of the aerated structure of bread. Beyond bread, there is the rest of the baking section of the supermarket to be addressed; the bubbles in cakes, crumpets, biscuits, etc. have also been neglected and are bursting to be noticed.

17.7 Conclusions

In the context of improving bread quality, this chapter has contended that greater understanding of gas phase behaviour throughout breadmaking will be central to this ambition. As Chamberlain (1979) noted, 'The better measurement and manipulation of the cheapest and most abundant bread ingredient may lead to significant advances in the management of bread quality'. Creation of the initial bubble structure in the dough mixer is key. Linking this structure via proving and baking to final bread structure and texture allows the initial aeration requirements from the mixer to be specified, and the exploitation of rheological and surface active components to be mastered. A framework of understanding of the entire breadmaking process in terms of aeration then serves as the basis for understanding ingredient contributions and formulation–process interactions. Rising stars for research have been identified:

- effects of dough formulation and mixer design and operation on aeration during mixing;

- aeration dynamics during mixing, to improve oxygen availability and effective exploitation of pressure-vacuum mixing;
- modified atmosphere mixing;
- scale-up of dough mixers;
- degassing and coalescence during moulding;
- models of growth, coalescence and rupture of bubbles during proving and baking;
- confirmation of the liquid film hypothesis;
- interactions of bran components with bubbles, and the creation of healthy yet palatable breads;
- effects of fat, surfactants and enzymes;
- tomography during mixing and baking, in addition to further proving studies;
- mechanics of crust formation;
- combined tomographic, textural and image analysis studies, and relating baked structure to texture;
- integrative, accessible, computer-based models of bread aeration incorporating bulk and surface rheology, coalescence and rupture, and heat and mass transfer effects;
- effect of dough freezing and frozen storage on gas phase behaviour and baked loaf structure;
- creation of artisan-like breads through informed manipulation of the bubble structure;
- extension of experimental techniques beyond white sandwich bread to other breads, including gluten-free breads, and to other baked goods.

The physicist and gastronome Nicholas Kurti stated 'It is a sad reflection on our civilisation that while we can and do measure the temperature in the atmosphere of the planet Venus, we do not know what goes on in our soufflés' (Kurti and Kurti, 1988). The direct study of gas phase behaviour that has breathed new life into bread research in recent years may yet encompass this commendable aspiration, to the benefit of manufacturers and consumers of baked goods.

17.8 Acknowledgements

The author is grateful to past and current students and colleagues, including Nyuk Ling Chin and Peter Martin whose results are reported here, and to the Satake Corporation of Japan and the Biotechnology and Biological Sciences Research Council for research funding.

17.9 Further reading

BLOKSMA, A.H. (1981) Effect of surface tension in the gas-dough interface on the rheological behaviour of dough. *Cereal Chemistry*, **58**, 481–486.

BROOKER, B.E. (1993) The stabilisation of air in cake batters the role of fat. *Food Structure*, **12**, 285–296.

BROOKER, B.E. (1995) The role of fat in bread doughs. *Scanning*, **17**, 81–82.

CAMPBELL, G.M., RIELLY, C.D., FRYER, P.J. and SADD, P.A. (1991) The measurement of bubble size distributions in an opaque food fluid. *Trans. IChemE, Part C: Food and Bioproducts Processing*, **69**, 67–76.

CAMPBELL, G.M., RIELLY, C.D., FRYER, P.J. and SADD, P.A. (1993) Measurement and interpretation of dough densities. *Cereal Chemistry*, **70**, 517–521.

CAMPBELL, G.M., RIELLY, C.D., FRYER, P.J. and SADD, P.A. (1999) Reconstruction of bubble size distributions from slices. In Campbell G.M., Webb C., Pandiella S.S. and Niranjan K., *Bubbles in Food*, Eagan Press, St. Paul, MN, USA, pp. 207–220.

CAMPBELL, G.M., WEBB, C., PANDIELLA, S.S. and NIRANJAN, K. (1999) *Bubbles in Food*, Eagan Press, St. Paul, MN, USA.

CAMPBELL, G.M., HERRERO-SANCHEZ, R., PAYO-RODRIGUEZ, R. and MERCHAN, M.L. (2001) Measurement of dynamic dough density, and the effect of surfactants and flour type on aeration during mixing and gas retention during proving. *Cereal Chemistry*, **78**, 272–277.

CAUVAIN, S.P. (1998) Breadmaking processes. In Cauvain SP and Young LS, *Technology of Breadmaking*, Chapman and Hall, London, UK, pp. 18–44.

CAUVAIN, S.P. and YOUNG, L.S. (1998) *Technology of Breadmaking*, Chapman and Hall, London, UK.

CHAMBERLAIN, N. (1975) Advances in breadmaking technology. In, *Bread: Social, Nutritional and Agricultural Aspects of Wheaten Bread*, Spicer, A. (Ed.), Applied Science Publishers, London, UK, pp. 259–272.

CHAMBERLAIN, N., COLLINS, T.H. and ELTON, G.A.H. (1962) The Chorleywood Bread Process. *Bakers Digest*, **36**, 52–53.

DE CINDIO, B. and CORRERA, S. (1995) Mathematical modelling of leavened cereal goods. *Journal of Food Engineering*, **24**, 379–403.

DOBRASZCZYK, B.J., CAMPBELL, G.M. and GAN, Z. (2001) Bread: a unique food. In, *Cereals and Cereal Products: Chemistry and Technology*, Dendy D.A.V. and Dobraszczyk B.J., Aspen Publishers Inc., Maryland, USA, pp. 82–232.

ELIASSON, A-C. and LARSSON, K. (1993) *Cereals in Breadmaking*. Marcel Dekker Inc., New York, USA.

GAN, Z., ANGOLD, R.E., WILLIAMS, M.R., ELLIS, P.R., VAUGHAN, J.G. and GALLIARD, T. (1990) The microstructure and gas retention of bread dough. *Journal of Cereal Science*, **12**, 15–24.

HANDLEMAN, A.R., CONN, J.F. and LYONS, J.W. (1961) Bubble mechanics in thick foams and their effects on cake quality. *Cereal Chemistry*, **38**, 294–305.

HAYMAN, D., SIPES, K., HOSENEY, R.C. and FAUBION, J.M. (1998c) Factors controlling gas cell failure in bread dough. *Cereal Chemistry*, **75**, 585–589.

HIBBERD, G.E. and PARKER, N.S. (1976) Gas pressure–volume–time relationships in fermenting doughs. I. Rate of production and solubility of carbon dioxide in dough. *Cereal Chemistry*, **53**, 338–345.

JUNGE, R.C., HOSENEY, R.C. and VARRIANO-MARSTON, E. (1981) Effect of surfactants on air incorporation of dough and crumb grain of bread. *Cereal Chemistry*, **58**, 338–342.

KNIGHTLY, W.H. (1988) Surfactants in baked foods: current practice and future trends. *Cereal Foods World*, **33**, 405–412.

KOKELAAR, J.J. and PRINS, A. (1995) Surface rheological properties of bread dough components in relation to gas bubble stability. *Journal of Cereal Science*, **22**, 53–61.

KROG, N. (1981) Theoretical aspects of surfactants in relation to their use in breadmaking. *Cereal Chemistry*, **58**, 158–164.

MAHDI, J.G., VARRIANO-MARSTON, E. and HOSENEY, R.C. (1981) The effect of mixing atmosphere and fat crystal size on dough structure and bread quality. *Bakers Digest*, **55**, 28–36.

POMERANZ, Y. and CZUCHAJOWSKA, Z. (1993) Gas formation and gas retention. II. Role of vital gluten during baking of bread from low-protein or fiber-enriched flour. *Cereal Foods World*, **38**, 504–510.

SAPIRSTEIN, H.D., ROLLER, R. and BUSHUK, W. (1994) Instrumental measurement of bread crumb grain by digital image analysis. *Cereal Chemistry*, **71**, 383–391.

SWANSON, C.O. and WORKING, E.B. (1926) Mechanical modification of dough to make it possible to bake bread with only fermentation in the pan. *Cereal Chemistry*, **3**, 65–83.

WILLIAMS, A. (1975) *Breadmaking: the Modern Revolution*. Hutchinson Benham, London, UK.

ZANONI, B., PIERUCI, S. and PERI, C. (1993) Study of the bread baking process. I. A phenomenological model. *Journal of Food Engineering*, **19**, 383–398.

ZANONI, B., PIERUCI, S. and PERI, C. (1994) Study of the bread baking process. II. Mathematical modelling. *Journal of Food Engineering*, **23**, 321–336.

17.10 References

ANONYMOUS (2002), http://www.breadtalk.com.

APV CORPORATION LTD (1992) UK Patent GB 2 264 623A.

ATKIN, M. (1995) *The International Grain Trade*, 2nd Edition. Woodhead Publishing Ltd., Cambridge, UK.

ATTENBURROW, G.E., GOODBAND, R.M., TAYLOR, L.J. and LILLFORD, P.J. (1989) Structure, mechanics and texture of a food sponge. *Journal of Cereal Science*, **9**, 61–70.

BAKER, J.C. (1941) The structure of the gas cell in bread dough. *Cereal Chemistry*, **18**, 34–41.

BAKER, J.C. and MIZE, M.D. (1937) Mixing doughs in vacuum and in the presence of various gases. *Cereal Chemistry*, **14**, 721–734.

BAKER, J.C. and MIZE, M.D. (1939a) Effect of temperature on dough properties, I. *Cereal Chemistry*, **16**, 517–533.

BAKER, J.C. and MIZE, M.D. (1939b) Effect of temperature on dough properties, II. *Cereal Chemistry*, **16**, 682–695.

BAKER, J.C. and MIZE, M.D. (1941) The origin of the gas cell in bread dough. *Cereal Chemistry*, **18**, 34–41.

BAKER, J.C. and MIZE, M.D. (1942) The relation of fats to texture, crumb and volume of bread. *Cereal Chemistry*, **19**, 84–94.

BAKER, J.C. and MIZE, M.D. (1946) Gas occlusion during dough mixing. *Cereal Chemistry*, **23**, 39–51.

BELL, B.M., DANIELS, D.G.H. and FISHER, N. (1981) Vacuum expansion of mechanically developed doughs at proof temperature: effect of shortening. *Cereal Chemistry*, **58**, 182–186.

BLOKSMA, A.H. (1990a) Rheology of the breadmaking process. *Cereal Foods World*, **35**, 228–236.

BLOKSMA, A.H. (1990b) Dough structure, dough rheology and baking quality. *Cereal Foods World*, **35**, 237–244.

BROOKER, B.E. (1996) The role of fat in the stabilisation of gas cells in bread dough. *Journal of Cereal Science*, **24**, 187–198.

CALVEL, R. (2001) *The Taste of Bread*. Aspen Publishers, Inc., Maryland, USA.

CAMPBELL, G.M. and SHAH, P. (1999) Entrainment and disentrainment of air during bread dough mixing, and their effect on scale-up of dough mixers. In, Campbell GM, Webb C, Pandiella SS and Niranjan K, *Bubbles in Food*, Eagan Press, St. Paul, MN, USA, pp. 11–20.

CAMPBELL, G.M., RIELLY, C.D., FRYER, P.J. and SADD, P.A. (1998) Aeration of bread dough during mixing: the effect of mixing dough at reduced pressure. *Cereal Foods World*, **43**, 163–167.

CAUVAIN, S.P. (1994) New mixer for variety bread production. *European Food and Drink Review*, Autumn, 51–53.

CAUVAIN, S.P., WHITWORTH, M.B. and ALAVA, J.M. (1999) The evolution of bubble structure in bread doughs and its effect on bread structure. In, Campbell G.M., Webb C., Pandiella S.S. and Niranjan K., *Bubbles in Food*, Eagan Press, St. Paul, MN, USA, pp. 85–88.

CHAMBERLAIN, N. (1979) Gases – the neglected bread ingredients. In, *Proc. 49th Conf. Brit. Soc. Baking*, pp. 12–17.

CHAMBERLAIN, N. and COLLINS, T.H. (1979) The Chorleywood Bread Process: the role of oxygen and nitrogen. *Bakers Digest*, **53**, 18–24.

CHAMBERLAIN, N., COLLINS, T.H. and ELTON, G.A.H. (1965) The Chorleywood Breadmaking Process: Improving effects of fat, *Cereal Science Today*, **10,** 415–432.

CHARALAMBIDES, M.N., WANIGASOORIYA, L., WILLIAMS, J.G. and CHAKRABARTI, S. (2002a) Biaxial deformation of dough using the bubble inflation technique. I. Experimental. *Rheologica Acta*, **41**, 532–540.

CHARALAMBIDES, M.N., WANIGASOORIYA, L. and WILLIAMS, J.G. (2002b) Biaxial deformation of dough using the bubble inflation technique. II. Numerical modelling. *Rheol. Acta*, **41**, 541–548.

CHEN, P., WHITNEY, L.F. and PELEG, M. (1994) Some tensile characteristics of bread crumb. *Journal of Texture Studies*, **25**, 299–310.

COLLISTER, L. and BLAKE, A. (2000) *Country Bread*, Octopus Publishing Group, London, UK.

DAVID, E. (1977) *English Bread and Yeast Cookery*, Penguin Books, London, UK.

DOBRASZCZYK, B.J. (1997) Development of a new dough inflation system to evaluate doughs. *Cereal Foods World*, **42**, 516–519.

DOBRASZCZYK, B.J. (1999) Measurement of biaxial extensional rheological properties using bubble inflation and the stability of bubble expansion in bread doughs. In, Campbell G.M., Webb C., Pandiella S.S. and Niranjan K., *Bubbles in Food*, Eagan Press, St. Paul, MN, USA, pp. 173–182.

DOBRASZCZYK, B.J. and ROBERTS, C.A. (1994) Strain hardening and dough gas cell-wall failure in biaxial extension. *Journal of Cereal Science*, **20**, 265–274.

DUNN, J.A. and WHITE, J.R. (1939) The leavening action of air in cake batter. *Cereal Chemistry*, **16**, 93–100.

EVANS, L.T. (1993) *Crop Evolution, Adaptation and Yield*, Cambridge University Press, Cambridge, UK.

GAN, Z., ELLIS, P.R., VAUGHAN, J.G. and GALLIARD, T. (1989) Some effects of non-endosperm components of wheat and of added gluten on wholemeal bread microstructure.

Journal of Cereal Science, **10**, 81–91.

GAN, Z., ELLIS, P.R. and SCHOFIELD, J.D. (1995) Mini review: gas cell stabilisation and gas retention in wheat bread dough. *Journal of Cereal Science*, **21**, 215–230.

HAYMAN, D., HOSENEY, R.C. and FAUBION, J.M. (1998a) Bread crumb grain development during baking. *Cereal Chemistry*, **75**, 577–580.

HAYMAN, D., HOSENEY, R.C. and FAUBION, J.M. (1998b) Effect of pressure (crust formation) on bread crumb grain development. *Cereal Chemistry*, **75**, 581–584.

HOOD, M.P. and LOWE, B. (1948) Air, water vapour and carbon dioxide as leavening gases in cakes made with different types of fat. *Cereal Chemistry*, **25**, 244–254.

KEETELS, C.J.A.M., VAN VLIET, T. and WALSTRA, P. (1996) Relationship between the sponge structure of starch bread and its mechanical properties. *Journal of Cereal Science*, **24**, 27–31.

KURTI, N. and KURTI, G. (1988) *But the Crackling is Superb: An Anthology on Food and Drink by Fellows and Foreign Members of The Royal Society*. Institute of Physics Publishing, London, UK.

LIU, Z. and SCANLON, M.G. (2002) Understanding and modelling the processing-mechanical property relationship of bread crumb assessed by indentation. *Cereal Chemistry*, **79**, 763–767.

MACRITCHIE, F. (1976) The liquid phase of dough and its role in baking. *Cereal Chemistry*, **53**, 318–326.

MARSTON, P.E. (1986) Dough development for breadmaking under controlled atmospheres. *Journal of Cereal Science*, **4**, 335–344.

MATZ, S.A. (1960) *Bakery Technology and Engineering*, AVI Publishing Co. Inc., Westport, CT, USA.

MCGEE, H. (1984) *On Food and Cooking: The Science and Lore of the Kitchen*, Harper Collins Publishers, London, UK.

ÖRNEBRO, J., NYLANDER, T. and ELIASSON, A.C. (2000) Interfacial behaviour of wheat proteins. *Journal of Cereal Science*, **31**, 195–221.

PLATT, W. and KRATZ, P.D. (1933) Measuring and recording some characteristics of test sponge cakes. *Cereal Chemistry*, **10**, 73–90.

PLATT, W. and POWERS, R. (1940) Compressibility of bread crumb. *Cereal Chemistry*, **17**, 601–621

PYLER, E.J. (1973) *Baking Science and Technology*, Siebel Publishing Company, Chicago, USA.

SCANLON, M.G., FAHLOUL, D. and SAPIRSTEIN, H.D. (1997) A measure of fracture toughness of bread crumb. *Cereal Chemistry*, **74**, 612–613.

SCANLON, M.G., SAPIRSTEIN, H.D. and FAHLOUL, D. (2000) Mechanical properties of bread crumb prepared from flours of different dough strength. *Journal of Cereal Science*, **32**, 235–243.

SHAH, P., CAMPBELL, G.M., MCKEE, S.L. and RIELLY, C.D. (1998) Proving of bread dough: modelling the growth of individual bubbles. *Trans IChemE, Part C: Food and Bioproducts Processing*, **76**, 73–79.

SPOONER, T.F. (1999) Controlled atmosphere: Could bakers better control mixing results by changing headspace gases in the mixer? *Baking & Snack*, February, 96–102.

SUNTHEIMER, F.J. (1961), British Patent 931637.

SWYNGEDAU, S. and PELEG, M. (1992) Characterisation and prediction of the compressive stress-strain relationship of layered arrays of spongy baked goods. *Cereal Chemistry*, **69**, 217–221.

TODD, J.P., HAWTHORN, J. and BLAIN, J.A. (1954), British Patent 771367.

VAN VLIET, T., JANSSEN, T., BLOKSMA, A.H. and WALSTRA, P. (1992) Strain hardening of dough as a requirement for gas retention. *Journal of Texture Studies*, **23**, 439–460.

WHITWORTH, M.B. and ALAVA, J.M. (1999) The imaging and measurement of bubbles in bread doughs. In, Campbell GM, Webb C, Pandiella SS and Niranjan K, *Bubbles in Food*, Eagan Press, St. Paul, MN, USA, pp. 221–231.

WHITWORTH, M.B. (2002) Development of bubble structure in bread dough. *Grain and Feed Milling Technology*, **112**(6) 10–11.

WIGGINS, C. (1998) Proving, baking and cooling. In Cauvain S.P. and Young L.S., *Technology of Breadmaking*, Chapman and Hall, London, UK, pp. 120–148.

WILLIAMS, A. and PULLEN, G. (1998) Functional ingredients. In, Cauvain S.P. and Young L.S., *Technology of Breadmaking*, Chapman and Hall, London, UK, pp. 45–80.

ZGHAL, M.C., SCANLON, M.G. and SAPIRSTEIN, H.D. (1999) Prediction of bread crumb density by digital image analysis. *Cereal Chemistry*, **76**, 734–742.

ZGHAL, M.C., SCANLON, M.G. and SAPIRSTEIN, H.D. (2001) Effects of flour strength, baking absorption and processing conditions on the structure and mechanical properties of bread crumb. *Cereal Chemistry*, **78**, 1–7.

18

Measuring the rheological properties of dough

B. J. Dobraszczyk, The University of Reading, UK

18.1 Introduction: dough rheology and bread quality

Within the cereal science community, there is a widespread conviction that the rheological properties of dough are related to baking quality, mainly due to a long tradition of subjective manual assessments of dough rheology prior to baking; for example the practice among bakers of kneading and stretching the dough by hand to assess its quality. Although this is a very subjective method of measuring rheology, it gives us an indication of the sort of rheological measurements we should be making in order to predict baking performance.

Gluten is the major protein in wheat flour doughs, responsible for their unique viscoelastic behaviour during deformation. It is now widely accepted that gluten proteins are responsible for variations in baking quality, and in particular it is the insoluble fraction of the high-molecular-weight (HMW) glutenin polymer which is best related to differences in dough strength and baking quality among different wheat varieties (Weegels *et al.*, 1996; MacRitchie and Lafiandra, 1997). However, the exact molecular mechanisms responsible for this variation still remain unclear, largely because information about the molecular size and structure of this fraction is inaccessible by conventional polymer size characterisation techniques such as gel permeation chromatography (GPC) and size exclusion high-performance liquid chromatography (SE-HPLC) due to its insolubility. Therefore other techniques sensitive to differences in polymer molecular weight distribution (MWD) and structure are necessary.

From studies in the field of polymer physics, it is widely accepted that molecular size, structure and MWDs of polymers are intimately linked to their rheological properties and ultimately to their performance in various end-use applications. Modern polymer physics concepts relate the molecular size and structure of polymer melts and concentrated solutions to their rheology and end-use performance, and can give quantitatively accurate predictions of the structure and interactions of long-chain branched polymers from relaxation mechanisms and large deformation extensional flow (Doi and Edwards, 1986; McLeish and Larson, 1998). Rheological measurements are increasingly being used as rapid, sensitive indicators of changes in polymer molecular structure and end-use quality traits and, since gluten can be viewed as a polymer, an understanding of the rheological properties of gluten in relation to its molecular structure and its functional behaviour during the whole breadmaking process would therefore be highly desirable in relation to understanding the physical mechanisms responsible for variations in breadmaking quality. However, many of the rheological tests on doughs reported in the literature are inappropriate because they do not use deformation conditions appropriate to baking, and are quite often insensitive to changes in molecular weight and structure. The rheological techniques commonly used to assess dough quality and their relationship to baking quality will be reviewed in this chapter.

Gluten proteins comprise a highly polydisperse system of polymers, classically divided into two groups based on their extractability in alcohols: gliadins and glutenins. The gliadins are single-chain polypeptides with molecular weight (MW) ranging from 2×10^4 to 7×10^4, while the glutenins are multiple-chain polymeric proteins in which individual polypeptides are thought to be linked by inter-chain disulphide and hydrogen bonds to give a wide MWD ranging between 10^5 to well beyond 10^8 (Carceller and Aussenac, 2001). It is only recently that it has been possible to measure and quantify the high-molecular-weight (HMW) glutenin polymers known to be responsible for variations in breadmaking quality using techniques such as dynamic light scattering (Wahlund *et al.*, 1996; Stevenson and Preston, 1996; Egorov *et al.*, 1998; Southan and MacRitchie, 1999), and their conformation and structure by techniques such as X-ray and electron scattering (Thomson *et al.*, 1999), nuclear magnetic resonance (NMR) spectroscopy (Alberti *et al.*, 2002a,b), hydrodynamic studies in solution (Field *et al.*, 1987) and atomic force microscopy (AFM) (Humphris *et al.*, 2000). These show that gluten has a bimodal MW distribution which roughly parallels the classical division based on solubility into gliadins and glutenins. Glutenins have an extended rod-like structure *ca.* 50–60 nm in length made up of a regular spiral structure, termed β-spirals (Shewry *et al.*, 2002), and are branched (Humphris *et al.*, 2000). The polymer MW distributions and structures can vary between different wheat varieties and can also be modified by environmental and growing conditions, suggesting that both genetic and agronomic conditions could be varied to improve the breadmaking quality of wheat.

18.1.1 Rheology and bubble wall stability during proof and baking quality

Baking is about the growth and stability of bubbles: their size, distribution, growth and failure during the baking process will have a major impact on the final quality of the bread, both in terms of its appearance (texture) and final volume. It is considered that the limit of expansion of these bubbles is related directly to their stability, owing to coalescence and the eventual loss of gas when the bubbles fail. The rheological properties of the bubble walls will therefore be important in maintaining stability against premature failure during baking, and also in relation to gas cell stabilisation and gas retention during proving and baking, and thus to the final structure and volume of the baked product (Dobraszczyk *et al.*, 2001). Hence, baking may in reality be governed by failure processes: the failure of the bubble walls at large deformations. It is increasingly being recognised in the polymer literature that extensional strain hardening is essential for most large deformation processing situations, such as blow moulding, fibre drawing or stretching and in foam formation, where the deformation is essentially biaxial extension around the expanding gas cells. Polymer size needs to be above a critical entanglement MW to give rise to strain hardening, which is also enhanced by long chain branching and a broad, bimodal polymer MWD. Strain hardening has been observed in doughs, and has been shown to be related to the failure of gas cell walls measured under large deformation biaxial extension (Dobraszczyk and Roberts, 1994; Dobraszczyk, 1997). Therefore it is expected that strain hardening is an important mechanism in providing greater stability of bubble walls against early coalescence during proof and better gas retention during baking expansion (Dobraszczyk *et al.*, 2003). The measurement of strain hardening in doughs and its relevance to baking quality will be reviewed in Section 18.6.

18.2 The role of rheology in quality control

Rheology is the study of the flow and deformation of materials. A controlled, well-defined deformation or strain is applied to a material over a given time and the resulting force response is measured (or vice versa) to give an indication of material parameters such as stiffness, modulus, viscosity, hardness, strength or toughness of the material. The general aims of rheological measurements are:

- to obtain a quantitative description of the materials' mechanical properties;
- to obtain information related to the molecular structure and composition of the material;
- to characterise and simulate the material's performance during processing and for quality control.

Rheology can be used as an aid in process control and design, and as a tool in the simulation and prediction of the materials response to the complex flows and deformation conditions often found in practical processing situations which can be inaccessible to normal rheological measurement.

For example, it is difficult to access dough during mixing, sheeting, proving and baking without interrupting the process or disturbing the structure of the material. We therefore have to predict the range of conditions the dough experiences during a given process and then extrapolate from rheological measurements made under simple, well-defined laboratory conditions, often via modelling simulations using Computational Fluid Dynamics (CFD) software (Scott and Richardson, 1997). For example, mixing and sheeting of doughs has been simulated (Love *et al.*, 2002; Binding *et al.*, 2002), and bubble growth during proof of bread doughs and expanded cereal extrudates has been modelled through the simulation and growth of bubbles (Fan *et al.*, 1994; Shah *et al.*, 1999)

Rheology can be related to product functionality: many rheological tests have been used to attempt to predict final product quality such as mixing behaviour, baking performance, mouthfeel and texture, quality of biscuits, pasta, pastry, etc. This is based on the ideas of structural engineering analysis of materials, where small-scale laboratory measurements of mechanical properties have successfully been extrapolated to the behaviour of large engineered structures such as bridges, buildings, pressure vessels etc., resulting in the idea that controlled tests on well-defined small samples of food in the laboratory can be related to the larger, more complex multi-component situations found in practical processing conditions.

Humankind has always had an intuitive feel for rheological testing, for example in tactile and visual assessments of material properties such as hardness, stiffness, flexibility and viscosity and their relation to end-use quality characteristics. People often intuitively assess the quality of solid foods by gently squeezing them, or liquid viscosity is assessed by gently rotating the liquid in its container, and indeed these sort of tests are often applied on the factory floor as a crude measure of quality. These intuitive assessments gradually became formalised into quantitative descriptions of material properties by scientists such as Newton, Boyle, Pascal, Hooke, Young and Cauchy (Tanner and Walters, 1998). Since then rheology has grown rapidly as a science and contributed to a number of applications such as colloids, suspensions and emulsions, polymer processing, extrusion and polymer modelling. Recent developments in polymer rheology have established a quantitative link between the molecular size and structure of polymers to their rheology and end-use performance (de Gennes, 1979; Doi and Edwards, 1986). Rheological measurements are increasingly being used as rapid, sensitive indicators of polymer molecular structure and predictors of end-use performance (Marin and Montfort, 1996) and are being applied to bread doughs as indicators of the gluten polymer molecular structure and predictors of its functional behaviour in breadmaking.

18.3 Rheological tests

Rheological tests attempt to measure the forces required to produce given controlled deformations, such as squashing (compression), bending or pulling apart (tension), and to present them in such a way as to be independent of sample size, geometry and mode of testing. They measure a well-defined property, such as stress, strain, stiffness or viscosity. A small test piece of the material is usually deformed in a controlled way, normally on a motor-driven machine, and the force is measured as well as the distance moved or displacement of the object. The force is then usually plotted against the displacement to give a force–displacement curve. Normally for stiff materials we would divide the force and displacement by the *original* sample dimensions to obtain stress (force/cross-section area) and strain (displacement/original dimension), because the changes in sample dimensions are small and uniform; this allows us to remove the sample size as a variable. Many food materials are not stiff and undergo large deformations in practice, where the geometry often changes in a non-uniform and unpredictable manner giving large and non-uniform stresses and strains along the sample. For example, dough thins out non-uniformly when stretched, in common with many polymers, giving rise to large stresses and strains not correctly calculated by the conventional method of dividing by original sample dimensions. In this case, it is necessary to normalise by *actual* change in dimensions during deformation, in which case sample dimensions should be measured locally and independently by using contact extensometers or non-contact techniques such as laser, video or photographic techniques. For materials that flow under normal measurement time scales, stress is normally divided by strain rate (strain/time of strain application) to give viscosity.

Rheological properties should be independent of size, shape and how they are measured; in other words, they are universal, rather like the speed of light or density of water, which do not depend on how much light or water is being measured or how it is being measured. It would be comforting to know that the stiffness of bread or viscosity of dough measured in a laboratory in Reading will be the same measured in any laboratory in the world, even if they are measured using different tests, sample sizes or shapes. The whole point of the rheological approach is that the properties that are measured are reproducible and can be compared between different samples, test sizes and shapes, and test methods.

18.3.1 Rheological test methods

There are many test methods used to measure rheological properties. It is not possible to describe all the available testing geometries here, and the reader is referred to general reviews of rheology (Ferry, 1980; Barnes, 1989; Whorlow, 1992), rheological testing of foods (Sherman, 1970; Carter, 1990; Rao and Steffe, 1992; Dobraszczyk and Vincent, 1999; van Vliet, 1999) and cereal products (Bloksma and Bushuk, 1988; Faridi and Faubion, 1986, 1990; Muller, 1975). It is common to categorise rheological techniques according to the type of strain imposed: e.g. compression, extension, shear, torsion, and also the relative magnitude of the imposed deformation, e.g. small or large deformation. The main techniques used for measuring cereal properties have traditionally been divided into descriptive empirical techniques and fundamental measurements.

18.4 Descriptive rheological measurements

Within the cereals industry there has been a long history of using descriptive empirical measurements of rheological properties, with an impressive array of ingenious devices such as the Penetrometer, Texturometer, Consistometer, Amylograph, Farinograph, Mixograph, Extensigraph, Alveograph, various flow viscometers and fermentation recording devices, reviewed by Muller (1975) and Shuey (1975) (Table 18.1). Empirical tests are easy to perform and are often used in practical factory situations, providing data that are useful in evaluating performance during processing and for quality control. The instruments are often

Table 18.1 Rheological methods used for cereal products

Method	Products	Property measured
Empirical methods		
Mixers:	Dough	Mixing time/torque
Farinograph		Apparent viscosity
Mixograph		
Reomixer		
Extensigraph	Dough	Extensibility
TAXT2/Kieffer rig	Dough, gluten	Extensibility
Alveograph	Dough, gluten	Biaxial extensibility
Amylograph RVA	Pastes, suspensions	Apparent viscosity Gelatinisation temp.
Consistometer	Sauces, fillings	Apparent viscosity
Flow cup	Fluids, sauces, batters	Apparent viscosity
Falling ball	Fluids	Apparent viscosity
Flow viscometers	Fluids, pastes	Apparent viscosity
Fermentometers	Dough	Height, volume
Penetrometers	Semi-solid foods, gels	Firmness, hardness
Texturometer, TPA	Solid foods	Texture, firmness
Fundamental methods		
Dynamic oscillation		
Concentric cylinders	Fluids, pastes, batters	Dynamic shear moduli
Parallel plates	doughs	dynamic viscosity
Tube viscometers:		
Capillary	Fluids	Viscosity
Pressure, extrusion	Sauces, pastes, dough	Viscosity
Pipe flow		In-line viscosity
Transient flow:		
Concentric cylinders	Semi-solid viscoelastic	Creep relaxation
Parallel plates	material	moduli and time
Extension:	Solid foods, doughs	
Uniaxial, biaxial		Extensional viscosity
TAXT2 dough inflation system		strain hardening
Lubricated compression		

RVA = Rapid Visco-Analyser, Newport Scientific Pty., Australia.
TPA = Texture Profile Analysis.

robust and capable of withstanding demanding factory environments, and do not require highly skilled or technically trained personnel. Simply because they do not provide data in fundamental units does not mean that these tests are worthless: indeed, they have provided a great deal of information on the quality and performance of cereal products such as consistency, hardness, texture, viscosity, etc. However, these measurements are not strictly 'rheological' tests since:

- the sample geometry is variable and not well defined;
- the stress and strain states are uncontrolled, complex and non-uniform; and
- it is not possible to define any rheological parameters such as stress, strain, strain rate, modulus or viscosity.

Therefore, these tests are purely descriptive and dependent on the type of instrument, size and geometry of the test sample and the specific conditions under which the test was performed. For example, empirical tests have been used to characterise the behaviour of bread doughs during processing, such as the Farinograph and Mixograph. Many of these are used as 'single point' tests, where a single parameter is often arbitrarily selected from a whole range of data acquired during the test as, for example, in selecting the peak torque from a mixing trace and then using this to correlate with performance. This neglects a large part of the recorded data, and is appropriate only to the set of conditions under which that test was performed and is generally not applicable to any other deformation conditions. Since dough experiences a wide range of conditions of stress states and strain rates during processing and baking, and the rheological properties of dough are dependent both on time and strain, there is often a discrepancy between such single point type tests and actual performance on the plant, where conditions of strain and strain rate may be poorly defined and very different from those in the laboratory test. While this may give the illusion of a 'scientific' test by being performed on a machine (frequently with a computer attached), and may give satisfactory correlations with a textural or processing parameter, it is impossible to compare results between different testing machines, or to extrapolate the results to other deformation conditions.

Most food materials are viscoelastic and therefore their properties depend on how quickly the test is performed (the strain rate or frequency). This is important in many aspects of dough processing: if the dough is deformed quickly, such as in mixing or sheeting, then the rheological properties of the dough will be very different if measured at the typically slower rates of deformation found in conventional testing machines. Alternatively, during processing dough will experience strains very different in magnitude and nature than those generally available in a rheological test. Many food processes operate under extensional flow, while most rheological tests on foods are performed in shear. Tests under only one particular set of conditions of rate, temperature and strain will almost certainly not be applicable to another set of deformation conditions. What is necessary is to define the set of deformation conditions that the food endures in practice and perform tests under similar conditions.

18.5 Fundamental rheological tests

Fundamental rheological tests measure well-defined physical properties independent of size, shape and how they are measured, and can be used for process design calculations and to model complex processing situations not amenable to direct measurement. Problems encountered with such fundamental tests are: complex instrumentation which is expensive, time consuming, difficult to maintain in an industrial environment and requires high levels of technical skill; often inappropriate deformation conditions; difficulty in interpretation of results; and slip and edge effects during testing.

The main types of fundamental rheological tests used in cereal testing are: (i) dynamic oscillation, (ii) creep and stress relaxation, (iii) extensional measurements (iv) flow viscometry (Table 18.1).

18.5.1 Dynamic oscillation measurements

Adapted from techniques developed for measuring viscoelastic properties of polymer melts and concentrated solutions (Ferry, 1980), this is one of the most popular and widely used fundamental rheological techniques for measuring cereal doughs and batters. It has the advantage of a well-developed theoretical background, readily available instrumentation, and simultaneous measurement of elastic and viscous moduli, while the non-destructive nature of the test enables multiple measurements to be performed as temperature, strain or frequency are varied.

Disadvantages of the dynamic oscillation method are that the deformation conditions are often inappropriate to practical processing situations, because they are carried out at rates and conditions very different from those experienced by the dough during processing or baking expansion. For example, rates of expansion during proof and oven rise in bread doughs have been calculated between 5×10^{-3} s^{-1} to 5×10^{-4} s^{-1}, compared with measuring rates in rheological tests several orders of magnitude greater (Bloksma, 1990). Dynamic oscillation measurements are usually made over a relatively narrow window of frequencies, which are not relevant to the strain rates observed during fermentation and baking, and which do not reflect the segmental motions and interactions of the HMW glutenin polymers thought to be responsible for baking performance. The frequency range of most conventional oscillatory shear rheological tests on doughs is limited to a part of the plateau range, at which the HMW polymers are entangled and the value of which is insensitive to changes in the MWD (Graessley, 1974, 1982).

Figure 18.1 shows the effect of increasing MW on the dynamic shear modulus for a series of narrow MW linear polystyrene polymer melts. As the MW increases, a plateau in modulus begins to appear, which increases in width as the MW increases further. The plateau represents the effect of entanglements, which at a certain polymer size, effectively lock the polymer structure into a temporary 3D network with a fixed modulus, the height of which is independent

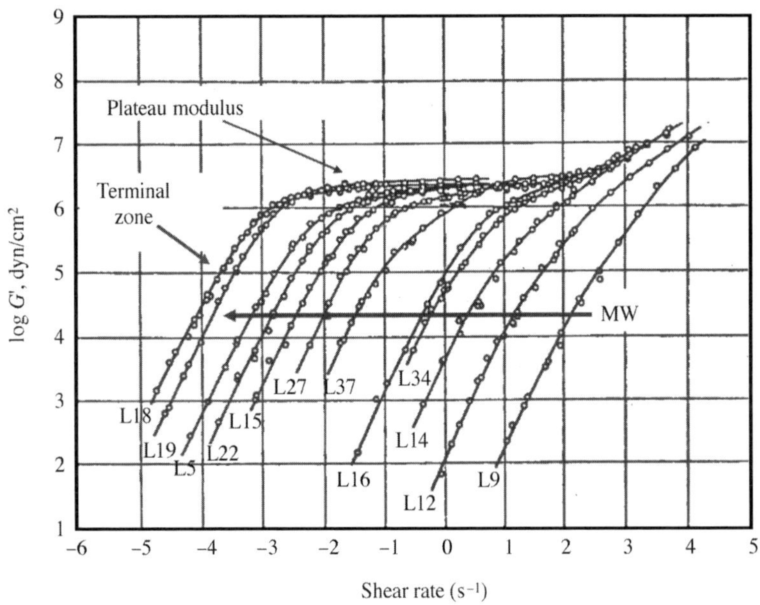

Fig. 18.1 Effect of increasing MW on dynamic shear modulus for a series narrow MW linear polystyrene polymer melts (from Onogi, S., Masuda, T. and Kitagawa, K. (1970) *Macromolecules* **3**: 109–116).

of MW. At some stage the polymer network begins to disentangle, and the modulus starts to decrease rapidly into the terminal zone, where the polymer chains are free to move about and act as a viscous liquid. The larger the polymer, the longer it stays entangled, and therefore the wider the frequency range over which the plateau remains. Thus it is the width of the plateau, or the point at which it descends into the terminal zone (sometimes known as the sol–gel transition in concentrated solutions) which defines the MW of the polymer. Unfortunately, most rheological measurements on dough and gluten have been performed in the plateau region, which is the region most insensitive to differences in MW.

Owing to its wide MW distribution, the plateau region for gluten will be very wide, and the crucial point at which it descends into the terminal zone is inaccessible by normal frequency sweep measurements because of the long times necessary to access this region. It is possible to extend the frequency range slightly by using time–temperature superposition, or by changing concentration of water (Masi *et al.*, 1998). For branched HMW polymers and wide polymer weight distributions such as gluten, the plateau becomes less distinct and the shear modulus is less than the linear polymer equivalent. The effect of polymer chain branching on shear and extensional viscosity has been modelled by McLeish and Larson (1998) (Fig. 18.2). An increase in the number of branches increases strain hardening and extensional viscosity, and decreases shear viscosity. Thus, extensional rheological properties appear to be more sensitive to

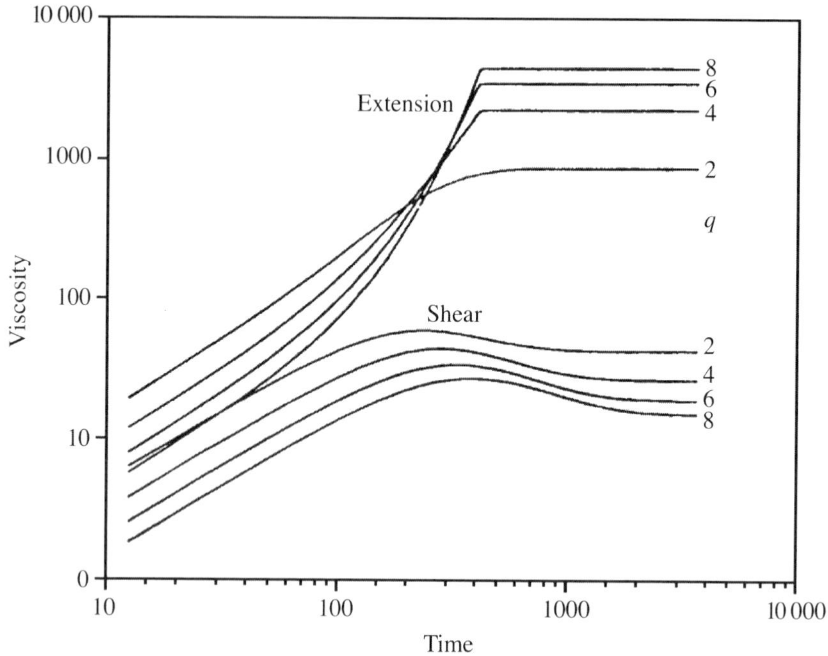

Fig. 18.2 Modelling the effect of polymer chain branching on shear and extensional rheology (q = no. of branch points) (from: MacLeish, T.C.B. and Larson, R.G. (1998) *Journal of Rheology* **42**: 81–110).

changes in MW, polymer entanglements and branching than dynamic shear properties. This is supported by the data in Table 18.2 which show poor correlations between small deformation dynamic rheological properties and baking performance, but much better relationships with large deformation extensional and relaxation properties.

Conventional oscillatory shear rheological tests usually operate in the linear region at small strains in the order of up to 1%, while strain in gas cell expansion during proof is known to be in the region of several hundred per cent (Amemiya and Menjivar, 1992). Furthermore, most rheological tests are carried out in shear, while most large-strain deformations in dough (i.e. sheeting, proof and baking) are extensional in nature. It is not widely recognised within the cereals science community that measurements in one deformation field such as shear are not always directly applicable to a different deformation such as extension. Polymer melt fluid dynamics show that very different results are possible in shear than in extension due to the different physical effects these deformations can have on networks of molecules (Graessley, 1974, 1982; Cogswell, 1981; Padmanabhan, 1995). From extensional studies on long-chain high-molecular-weight polymer melts it is known that entirely different rheological properties are obtained in shear than in tension, especially if the polymer chains are branched. For example, the elongational viscosity of low-

Table 18.2 Correlations between rheological properties and baking performance

Rheological parameter	Baking parameter	Correlation/ discrimination	Reference
Small deformation shear oscillation			
G' 1Hz (dough)	Volume	0.15 ($n = 48$)	Autio *et al.* (2001)
G' slope		0.72	
G' 10 Hz (wet gluten)	Volume	−0.85 ($n = 27$)	Schober *et al.* (2002)
	Form ratio (W/H)	0.65	
G' 1 Hz (gluten)	Volume	ns ($n = 20$)	Tronsmo *et al.* (2002a)
	Form ratio (*H/W*)	0.69	
tan δ	(*H/W*)	−0.71	
G' 1 Hz	Loaf height	−0.64 ($n = 8$)	Uthayakumaran *et al.* (2002)
tan δ		ns	
G'	Volume	ns ($n = 4$)	Safari Ardi and Phan-Thien (1998)
Large deformation			
Biaxial extensional strain hardening	Volume	0.89 ($n = 20$)	Tronsmo *et al.* (2002b)
	Form ratio	0.80	
Max. uniaxial extensional viscosity	Loaf height	0.81	Uthayakumaran *et al.* (2002)
Biaxial strain hardening	Volume	0.92–0.97 ($n = 6$)	Dobraszczyk (1997) Dobraszczyk *et al.* (2003)
Biaxial strain hardening	Volume	Good	Wikstrom and Bohlin (1999a,b)
Biaxial strain hardening	Volume	Good	Kokelaar *et al.* (1996)
Relaxation creep	Volume	Good 0.94 ($n = 23$)	Wikstrom and Eliasson (1998); Wang and Sun (2002)
Relaxation	Quality	Good	Safari-Ardi and Phan-Thien (1998)
Shear relaxation Shear viscosity	Quality	Good	Amemiya and Menjivar (1992)

density (branched-chain) polyethylene melts increases with both strain and strain rate (strain hardening), while the shear viscosity decreases with strain and strain rate (shear-thinning), giving widely different values in final viscosities between elongation and shear (Fig. 18.3a). For doughs, shear and elongational viscosities at low strains are similar, but at higher strains they diverge and the elongational viscosity rises steeply to give a value two orders of magnitude higher at failure (Fig. 18.3b). This increased strain hardening is attributed to the stretching of molecules between entanglement of long-chain or branched molecules during extensional flow, giving rise to the observed steep increase in viscosity at larger strains (Cogswell, 1981; McLeish and Larson, 1998; Munstedt *et al.*, 1998).

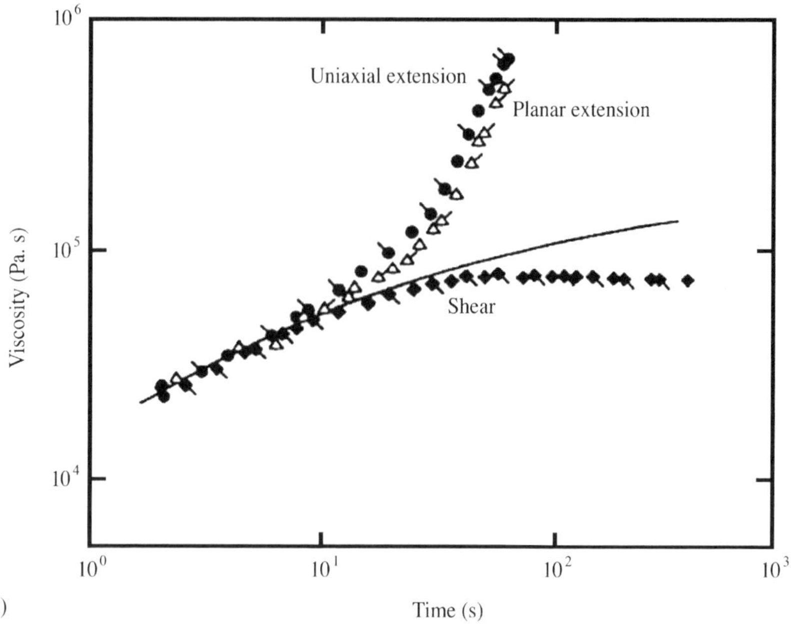

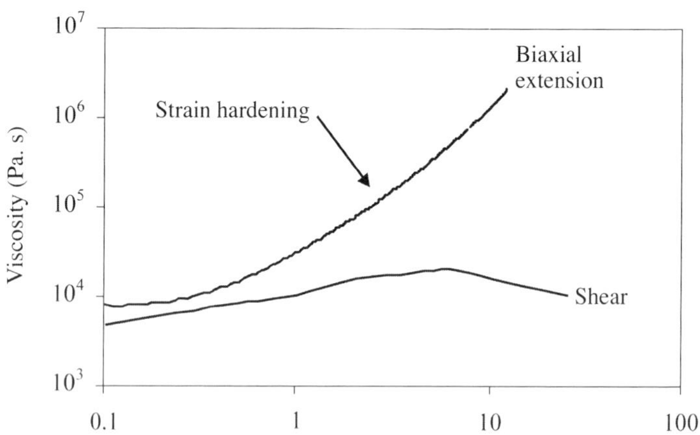

Fig. 18.3 (a) Shear and extensional viscosity of LDPE (low-density polyethylene) at 125 °C at constant strain rate ($0.05\ s^{-1}$) (from: MacLeish, T.C.B. and Larson, R.G. (1998) *Journal of Rheology* **42**: 81–110). (b) Large deformation shear and biaxial extensional viscosities for dough at constant strain rate ($0.1\ s^{-1}$).

Therefore, many of the rheological tests reported in the food literature are inappropriate because they do not measure the system under the appropriate deformation conditions. Small deformation dynamic shear rheological tests are frequently quoted in the cereals literature and applied almost indiscriminately to any system regardless of whether it is valid or not.

18.5.2 Creep and relaxation measurements

In stress relaxation measurements, deformation is held constant and the force response is measured, while in creep the stress is held constant and the deformation is measured. Schofield and Scott Blair (1937) first measured the creep response of dough and obtained values for elastic moduli and viscosity. Bloksma and Bushuk (1988) surveyed the experimental results from the literature for stress relaxation measurements for a number of doughs. None of the curves showed the exponential decay typical of a single relaxation time, but correspond to a decay typical of a broad spectrum of relaxation times. This shows a broad distribution of relaxation processes is responsible for the relaxation process within dough, related to the wide molecular weight distribution of gluten.

Many authors have shown that a slower (longer) relaxation time is associated with good baking quality (Bloksma, 1990; Launay, 1990; Wang and Sun, 2002), with relaxation time relatively independent of water, mixing time or temperature. Stress relaxation measurements on dough and gluten in shear showed that the relaxation behaviour of dough could be described by two relaxation processes (Bohlin and Carlson, 1980): a rapid relaxation over 0.1 to 10 s and a slower process occurring over 10 to 10 000 s. Measurements of large-deformation creep and shear stress relaxation properties were found to be useful in discriminating between different wheat varieties of varying quality, and were found to be closely associated with baking volume (Safari-Ardi and Phan-Thien, 1998), and strength of durum wheat varieties (Edwards *et al.*, 1999, 2001). At small strain amplitudes (0.1%), doughs with different baking quality showed no differences in relaxation behaviour, but at a range of large strains (up to 29%), their creep and relaxation behaviour was closely correlated with the baking behaviour of dough (Safari-Ardi and Phan-Thien, 1998).

Doughs exhibited a characteristic bimodal distribution of relaxation times, with the second peak at longer times clearly discriminating between cultivars with varying strength and quality, which reflects the differences in the MW distribution of glutenin polymers (Rao *et al.*, 2001). The second relaxation peak is related to the entanglement properties of HMW glutenin polymers, and has been shown to be directly related to the insoluble fraction of the HMW glutenins (Li *et al.*, 2003). Relaxation properties of doughs relate well to MW distribution and particularly to entanglements of HMW glutenin polymers and may be useful as a rapid method of discriminating variations in MW distribution between cultivars which vary in baking quality.

18.5.3 Extensional techniques

There are many types of extensional flow measurements, including: simple uniaxial tension, fibre wind-up or spinning, converging flow, capillary extrusion, opposed jets, lubricated compression and bubble inflation (Macosko, 1994). There are several methods that have been used to measure the rheological properties of dough in extension: simple uniaxial extension, where dough is stretched in one direction; and biaxial extension, where the dough is stretched in two opposing directions, which can be achieved either by compression between lubricated surfaces or by bubble inflation (Bagley and Christianson, 1986; Huang and Kokini, 1993; Dobraszczyk, 1997).

Most of the tests described previously have involved relatively small deformations. Many food materials undergo large deformations in practice during processing and eating, and many of these have a large extensional component. For example, extensional flows are important in mixing and sheeting of pastry and dough, converging and diverging flow such as in extrusion and pumping, spreading of soft solids such as butter, cheeses and pastes, and expansion of bubbles in foams such as bread dough, cakes and heat-extruded snacks. Unfortunately, most tests in flow are carried out in shear under small deformations, mainly because most conventional viscometers operate in shear, because the equipment is readily available and the technique well established. However, measurements carried out in shear and using small deformations do not provide information about a material's behaviour under large extension. Therefore there is an obvious need to perform measurements under conditions relevant to those experienced by the material in actual practice.

Uniaxial extension

One of the oldest and most widely used test methods to measure a material's properties is the uniaxial tensile test. A strip of material is clamped at both ends and pulled apart at a fixed rate in a suitable testing machine, and the force measured at the same time as the displacement of the object. The force is generally plotted against the displacement (extension) to give a force–extension curve. Tensile tests may produce an approximately uniform extension of a sample provided necking does not occur. Normally the force and extension are divided by the original sample dimensions to obtain stress and strain, and allows removal of the sample geometry as a variable, but for doughs undergoing large extensional deformation the actual change in dimensions must be measured or calculated. The slope of the stress–strain curve then gives the elastic modulus or stiffness. Many test methods attempt to measure the uniaxial extensional properties of doughs, such as the Simon Research Extensometer, Brabender Extensigraph, Stable Micro Systems Kieffer dough and gluten extensibility rig, but none of these gives rheological data in fundamental units of stress and strain, because the sample geometry is not defined, dimensions change extensively and non-uniformly during testing, and it is therefore impossible to define any rheological parameters such as stress, strain, strain rate, modulus or viscosity.

Studies on the fundamental uniaxial extensional rheological properties of doughs have been carried out by many workers (Schofield and Scott Blair, 1932; Tschoegl *et al.*, 1970a,b; Rasper, 1975; Uthayakumaran *et al.*, 2000). Some of the earliest attempts to characterise the fundamental rheological properties of dough were in a series of uniaxial extensional measurements by Schofield and Scott Blair (1932) who stretched a cylinder of dough floated on a mercury bath and measured the elongation and force. Plastic and elastic components of deformation were resolved and viscosity and elastic modulus were calculated. They showed that the rheological behaviour of dough is non-linear with strain and strain rate, i.e. elastic modulus and viscosity varies with both rate and strain. Tschoegl *et al.* (1970a,b) measured the large extension properties of doughs by extending a ring of dough suspended in a liquid of density equivalent to that of the dough between two hooks at constant deformation rates until rupture. The stress–strain curves showed considerable strain hardening, and strain and stress at rupture was considerably less for poor quality flours than good quality flours.

Biaxial extension

In biaxial extension a sample is stretched at equal rates in two perpendicular directions in one plane, as in an expanding spherical balloon. The most widely used methods for measuring biaxial extension properties of food materials have been inflation techniques and compression between flat plates using lubricated surfaces, which produce purely extensional flow provided no friction occurs (Chatraei *et al.,* 1981).

Inflation was first used as an empirical technique to measure wheat gluten and bread dough extensibility in the 1920s (Hankoczy, 1920; Chopin, 1987). This method was later developed to access rheological parameters (Launay *et al.,* 1977) and further developed to measure the fracture and biaxial extensional rheological properties of wheat doughs and glutens, and to assess the quality of wheat flour doughs (Dobraszczyk and Roberts, 1994; Dobraszczyk, 1997). The major advantage of this test is that the deformation closely resembles practical conditions experienced by the cell walls around the expanding gas cells within the dough during proof and oven rise, i.e. large deformation biaxial extension. Extensional rheological properties can be measured at large strains up to failure and low strain rates, and the gripping problems normally associated with uniaxial tests can be minimised. Extensional rheological properties of wheat doughs have been measured using lubricated compression and bubble inflation (van Vliet *et al.,* 1992; Huang and Kokini, 1993; Dobraszczyk and Roberts, 1994; Kokelaar *et al.*, 1996; Wikström and Bohlin, 1999a,b). Differences in extensional strain hardening between varieties of different baking quality were found to relate to baking quality, with good breadmaking varieties showing greater strain hardening and extensional viscosity (Dobraszczyk *et al.*, 2003; Uthayakumaran *et al.*, 2002).

18.6 Baking quality and rheology

Rheological studies on doughs related to baking have mostly been performed in small deformation shear oscillation at room temperature. Such dynamic rheological measurements on doughs have been investigated in many studies (Hibberd and Wallace, 1966; Hibberd and Parker, 1975a,b; Abdelrahman and Spies, 1986; Smith *et al.*, 1970; Dreese *et al.*, 1988). Viscous (G'') and elastic

(G') moduli for dough are measured over a range of frequencies. Elastic properties predominate over viscous properties, and the moduli are slightly frequency dependent, which is typical of a cross-linked polymer network.

No convincing relationship has been established between dynamic rheological properties and baking performance (Faubion and Faridi, 1985; Dreese *et al.*, 1988, Khatkar *et al.*, 1995; Kokelaar *et al.*, 1996; Hayman *et al.*, 1998; Autio *et al.*, 2001). Abdelrahman and Spies (1986) compared two flours of different baking quality and measured lower values of elastic (storage) modulus (G') for the higher baking quality flour. Similarly, many reports have found lower values of G' corresponded with better baking quality (Weipert, 1988, 1992; Amemiya and Menjivar, 1992; Schober *et al.*, 2002). Others, however, have found that a higher value of G' for glutens and doughs relates to better baking performance (Attenburrow *et al.*, 1990; Janssen *et al.*, 1996), see Table 18.2. Small deformation dynamic rheological measurements have provided useful information about the properties of isolated glutens and their sub-fractions, but similar measurements on whole dough systems have not shown any significant differences (Khatkar *et al.*, 1995). It has been shown earlier that these conflicting results arise because most of these tests are carried out at rates and deformation conditions very different from those experienced by the dough during baking expansion, and also because dynamic rheological parameters in the plateau region are generally insensitive to differences in molecular weight of polymers. However, these parameters are highly sensitive to changes in polymer concentration and diluents such as water, which are almost never kept constant in rheological experiments on doughs. Most dynamic rheological tests on doughs and glutens have been carried out over a relatively narrow frequency range in the plateau zone, the value of which is known to be insensitive to changes in MW. If, as is generally accepted, the structure and size of large MW glutenin polymers are responsible for the variations in breadmaking performance between different wheat varieties, then it is to be expected that measurements of the plateau modulus will not be good indicators of baking performance and are therefore not appropriate to baking quality, as seen in Table 18.2.

During proof and baking the growth and stability of gas bubbles within the dough determines the expansion of the dough and therefore the ultimate volume and texture of the baked product (He and Hoseney, 1991). The limit of expansion of these bubbles is related directly to their stability, owing to coalescence and the eventual loss of gas when the bubbles fail. The rheological properties of the expanding bubble walls will therefore be important in maintaining stability in the bubble wall and promote gas retention (Dobraszczyk *et al.*, 2001). The relevant rheological conditions around an expanding gas cell during proof and baking are biaxial extension, large strain and low strain rate. Any rheological tests which seek to relate to baking performance should therefore be performed under conditions similar to those of baking expansion, such as large strain biaxial extension and low strain rates. Methods such as bubble inflation and lubricated compression potentially offer the most appropriate method for measuring rheological properties of doughs. The major advantage of these tests is that the deformation closely resembles practical conditions experienced by the cell walls around the expanding gas cells within the dough during proof and oven rise, i.e. large deformation biaxial extension and can be carried out at the low strain rates and elevated temperatures relevant to baking (Dobraszczyk *et al.*, 2003).

Extensional rheology is sensitive to polymer chain branching and entanglement interactions between HMW polymers at large deformations (McLeish and Larson, 1998). The theory is simple and relatively well developed, and it generally provides good correlations with the relevant large deformation processing quality parameters (Table 18.2). The disadvantages are that there is no single well-defined and accepted method for extensional measurement, with many different methods being used depending on the type and viscosity of the material being studied. The tests often use large amounts of material and are destructive. The measurement of extensional flow is often difficult, because the deformation is large and non-uniform and it is therefore impossible to calculate strain directly from the machine displacement, requiring the direct measurement or calculation of changes in sample dimensions, often at high speeds.

The failure of gas cell walls in doughs has been shown to be directly related to the elongational strain hardening properties of the dough measured under large deformation biaxial extension (Dobraszczyk and Roberts, 1994). Strain hardening is shown as an increase in the slope of the true stress–strain curve with increasing extension, giving rise to the typical J-shaped stress–strain curve observed for highly extensible materials (Fig. 18.4). Strain hardening in doughs is thought to arise mainly from stretching of polymer chains between points of entanglement in the larger glutenin molecules (Fig. 18.5) which gives rise to the increasing viscosities observed at large strains (Singh and MacRitchie, 2001). Under extensional flow, entangled polymers exhibit strain hardening which is enhanced for polymers with a broad MW distribution, particularly a bimodal distribution (Watanabe, 1999), and branching (McLeish and Larson, 1998). It is therefore expected that the broad bimodal MW distribution and branching typical of gluten will result in enhanced strain hardening and a bimodal distribution of relaxation times.

Gas cell wall failure in expanding dough bubbles has been predicted using the Considere criterion for instability in extension for polymers (Dobraszczyk and Roberts, 1994; Wikström and Bohlin, 1999b). This criterion states that the stability in extension of a viscoelastic material is guaranteed provided the strain is less than that at which a maximum occurs in the force–extension plot and defines a critical strain beyond which failure is inevitable on further extension. Uniform extension of a viscoelastic membrane during inflation is guaranteed, provided the strain does not reach this maximum. Beyond this critical value of strain the criterion states that the material cannot be extended homogeneously and instead undergoes a dynamic failure event (MacKinley and Hassager, 1999). During large extension of materials, plastic strain is uniform throughout the

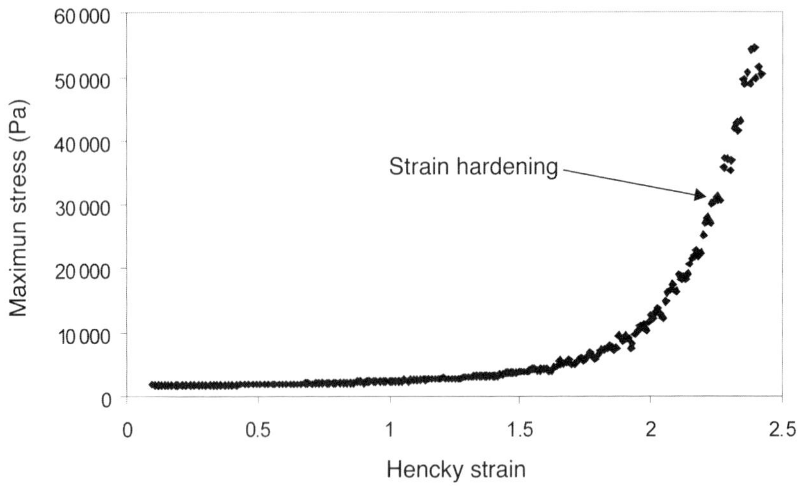

Fig. 18.4 Typical J-shaped stress–Hencky strain curve in biaxial extension for a dough bubble inflated at 50 °C and constant strain rate ($0.1\,s^{-1}$). Bubble inflation using SMS Dough Inflation System, maximum stress and Hencky strain calculated for bubble wall polar region.

sample up to the point of maximum force. Beyond this point, force begins to decrease and it is at this point that localised and non-uniform plastic deformation begins to occur. The cross-sectional area begins to change in a non-uniform way, and a neck or localised constriction forms, which can either stabilise or propagate in an unstable manner to failure. If the cross-section at any point is slightly less than elsewhere or there are any irregularities in the sample when the force is increasing in plastic flow, the stress will increase locally. While the force is increasing the deformation will be stable, i.e. any local constrictions are self-arresting. In contrast, under a decreasing force the deformation is no longer

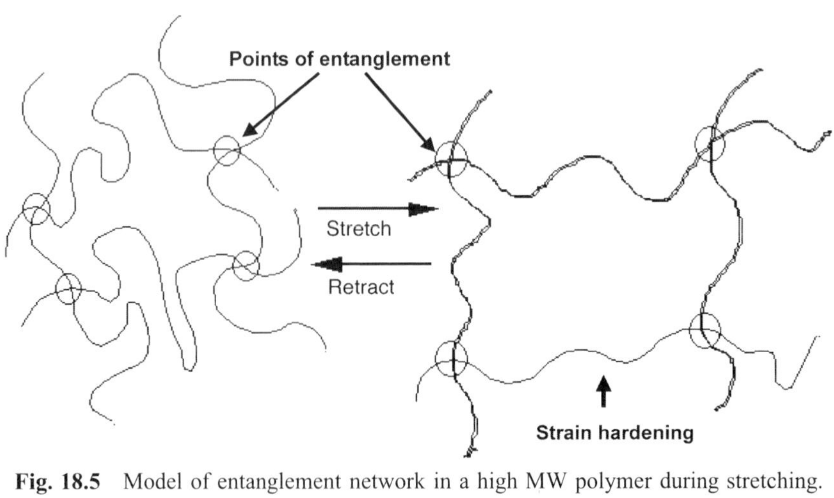

Fig. 18.5 Model of entanglement network in a high MW polymer during stretching.

stable, leading to the formation and cumulative increase of necking and eventual failure. Hence the force maximum defines a point of instability in tension, beyond which fracture is inevitable. The occurrence of strain hardening (sometimes called work hardening) in a material stabilises any regions of incipient localised thinning that could lead to unstable necking and eventual fracture during high extensions, and can allow much larger extensions before rupture than would otherwise be possible.

Recent work has shown that bread doughs exhibit strain hardening under large extensional deformations, and that these extensional rheological properties are important in baking performance (Kokelaar *et al.*, 1996; Dobraszczyk, 1997; Wikström and Bohlin, 1999b; Dobraszczyk *et al.*, 2003). Strain hardening allows the expanding gas cell walls to resist failure by locally increasing resistance to extension as the bubble walls become thinner, and appears to provide the bubbles with greater stability against early coalescence and better gas retention. It is expected therefore that doughs with good strain hardening characteristics should result in a finer crumb texture (e.g. smaller gas cells, thinner cell walls and an even distribution of bubble sizes) and larger baked volume than doughs with poor strain hardening properties. It has been shown that good breadmaking doughs have good strain hardening properties and inflate to larger single bubble volume before rupture, while poor breadmaking doughs inflate to lower volumes and have much lower strain hardening.

Loaf volume for a number of commercial white flour doughs has been related directly to the failure strain and strain hardening properties of single dough bubbles measured at elevated temperatures in biaxial extension (Dobraszczyk *et al.*, 2003). Strain hardening and failure strain of cell walls were both seen to decrease with temperature, with cell walls in good breadmaking doughs remaining stable and retaining their strain hardening properties to higher temperatures (60 °C), while the cell walls of poor breadmaking doughs became unstable at lower temperatures (45 °C–50 °C) and had lower strain hardening (Fig. 18.6). Strain hardening measured at 50 °C gave good correlations with baking volume, with the best correlations achieved between those rheological measurements and baking tests which used similar mixing conditions.

As predicted by the Considere failure criterion, a strain hardening value of 1 defines a region below which gas cell walls become unstable, and discriminates well between the baking quality of a range of commercial flour blends of varying quality (Fig. 18.7). It is suggested that changes in strain hardening and bubble stability in this temperature region are important in relation to bubble coalescence. When strain hardening falls below a value of 1, bubble walls are no longer stable and coalesce rapidly, resulting in loss of gas retention and lower volume and texture. Bubble walls with good strain hardening properties remain stable for longer during baking, allowing the bubbles to resist coalescence and retain gas for much longer. Strain hardening in poorer breadmaking varieties starts to decrease at much lower temperatures, giving earlier bubble coalescence and release of gas, resulting in lower loaf volumes and poorer texture.

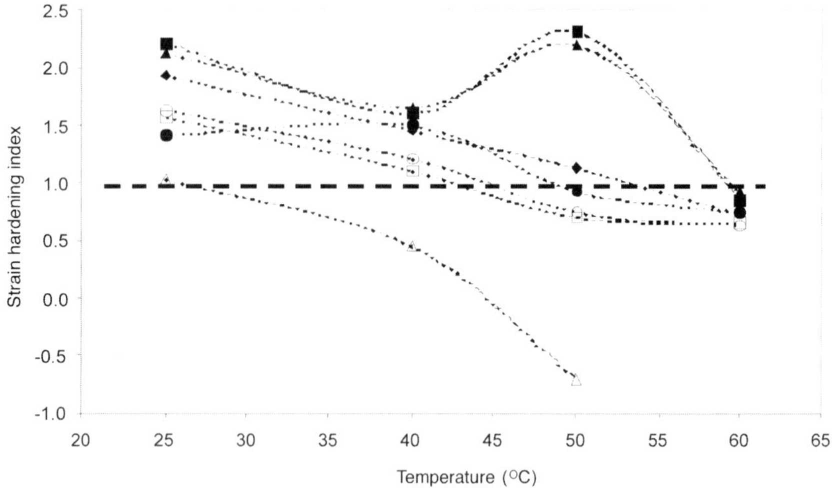

Fig. 18.6 Mean bubble cell wall strain hardening values ($n = 6$) for a number of wheat varieties inflated at constant strain rate ($0.1\,s^{-1}$) at various temperatures on the SMS Dough Inflation System. The dashed line at strain hardening index = 1.0 defines the region below which expanding bubble walls become unstable. ■ = Pillsbury, ▲ = Soissons, ● = Rialto 1999. ◆ = Hereward, ❍ = Rialto 1998, ❐ = Charger, Δ = Riband.

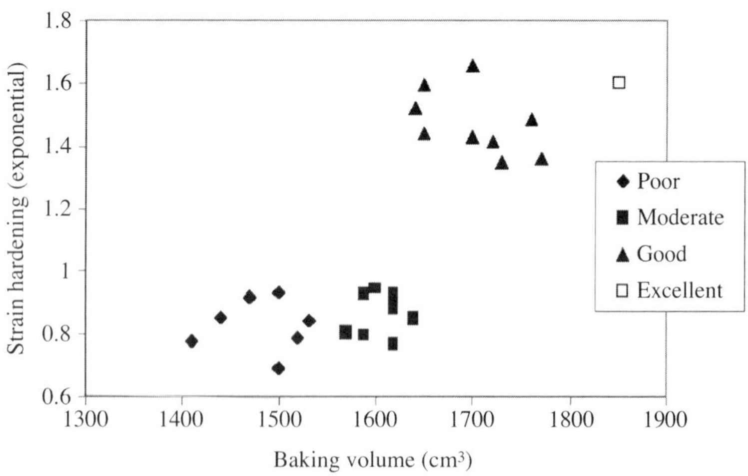

Fig. 18.7 Discrimination of baking performance for a range of commercial flour blends of varying baking quality using mean bubble cell wall strain hardening properties obtained at 50 °C and constant strain rate ($0.1\,s^{-1}$).

18.7 Future trends

The realisation that a small change in polymer structure at the very highest end of the size spectrum can lead to a large change in rheological properties has spurred the search for rheological techniques that will be sensitive to such changes in doughs. It has been recognised that the conventional techniques of small deformation shear oscillation used for doughs over a limited frequency range are insensitive to changes in MW and structure, and that extensional rheology is much more sensitive to polymer chain branching and entanglement interactions between HMW polymers at large deformations. With increasing development of molecular models relating changes in polymer structure to rheological properties, it may soon be possible to predict the structure of the insoluble glutenin polymer fractions known to be responsible for variations in baking quality from rheological and spectroscopic information. Development of imaging techniques such as AFM for biological systems will also allow direct visualisation of molecular structure of these polymers as they are stretched, and spectroscopic characterisation of gluten polymer in the stretched state will reveal the dynamic interactions of the various polymer fractions.

18.8 References

ABDELRAHMAN. A. and SPIES, R. (1986) Dynamic rheological studies of dough systems, in: *Fundamentals of Dough Rheology*, eds. H. Faridi and J.M. Faubion, American Association of Cereal Chemists, St. Paul, MN, USA, pp. 87–103.

ALBERTI, E., GILBERT, S.M., TATHAM, A.S., SHEWRY, P.R., NAITO, A., OKUDA, K., SAITO, H. and GIL, A.M. (2002a). Study of wheat high molecular weight 1Dx5 subunit by ^{13}C and ^{1}H solid state NMR: II. The roles of non-repetitive terminal domains and length of repetitive domain. *Biopolymers-Biospectroscopy section*, **65**(2), 158–168.

ALBERTI, E., GILBERT, S.M., TATHAM, A.S., SHEWRY, P.R. and GIL, A.M. (2002b) Study of wheat high molecular weight 1Dx5 subunit by ^{13}C and ^{1}H solid state NMR spectroscopy: I. The role of covalent crosslinking. *Biopolymers-Biospectroscopy section*, **67**(6), 487–498.

AMEMIYA, J.I. and MENJIVAR, J.A. (1992) Comparison of small and large deformation measurements to characterise the rheology of wheat flour doughs. *Journal of Food Engineering*, **16**, 91–108.

ATTENBURROW, G.E., BARNES, D.J., DAVIES, A.P. and INGMAN, S.J. (1990) Rheological properties of wheat gluten. *Journal of Cereal Science*, **12**, 1–14.

AUTIO, K., FLANDER, L., KINNUNEN, A. and HEINONEN, R. (2001) Bread quality relationship with rheological measurements of wheat flour dough. *Cereal Chemistry*, **78**, 654–657.

BAGLEY, E.B. and CHRISTIANSON, D.D. (1986) Response of commercial chemically leavened doughs to uniaxial compression, in: *Fundamentals of Dough Rheology*, eds. H. Faridi and J.M. Faubion, American Association of Cereal Chemists, St. Paul, MN, USA, pp. 27–36.

BARNES, H.A., HUTTON, J.F. and WALTERS, K. (1989) *An Introduction to Rheology*, Elsevier.

BINDING, D.M., COUCH, M.A., SUYATHA, K.S. and WEBSTER, J.F.E. (2002) Experimental and numerical simulation of dough kneading in filled geometries, accepted *Journal of Food Engineering*.

BLOKSMA, A.H. (1990) Rheology of the breadmaking process. *Cereal Foods World*, **35**, 228–236.

BLOKSMA, A.H. and BUSHUK, W. (1988) Rheology and chemistry of dough, in: *Wheat Chemistry and Technology II*, (Y. Pomeranz, ed.), AACC, St. Paul, MN, USA.

BOHLIN, L. and CARLSON, T.L.G. (1980) Dynamic viscoelastic properties of wheat flour doughs: dependence on mixing time. *Cereal Chemistry*, **57**, 175–181.

CARCELLER, J.L. and AUSSENAC, T. (2001) Size characterisation of glutenin polymers by HPSEC-MALLS. *Journal of Cereal Science*, **33**, 131–142.

CARTER, R.E. (1990) *Rheology of Food, Pharmaceutical and Biological Materials with General Rheology*, Elsevier Applied Science, London.

CHATRAEI, S.H., MACOSKO, C.W. and WINTER, H.H. (1981) Lubricated squeezing flow: a new biaxial extensional rheometer. *Journal of Rheology*, **25**, 433–443.

CHOPIN, M. (1987) Relations entre les proprietes mecaniques des pates de farines et la panification, in: *The Alveograph Handbook*, eds. H. Faridi, V.F. Rasper and B. Launay, American Association of Cereal Chemists, St. Paul, MN, USA.

COGSWELL, F.N. (1981) *Polymer Melt Rheology*, George Godwin Ltd, London.

DE GENNES, P.G. (1979) *Scaling Concepts In Polymer Physics*, Cornell University Press, Ithaca.

DOBRASZCZYK, B.J. (1997) Development of a new dough inflation system to evaluate doughs. *Cereal Foods World*, **42**, 516–519.

DOBRASZCZYK, B.J. and ROBERTS, C.A. (1994) Strain hardening and dough gas cell-wall failure in biaxial extension. *Journal of Cereal Science*, **20**, 265–274.

DOBRASZCZYK, B.J. and VINCENT, J.F.V. (1999) Measurement of mechanical properties of food materials in relation to texture: the materials approach. In *Food Texture: measurement and perception* (A.J. Rosenthal, ed.), Aspen Publishers, MD, USA.

DOBRASZCZYK, B.J., CAMPBELL, G.M. and GAN, Z. (2001) Bread – a unique food, in: *Cereals & Cereal Products: Technology & Chemistry*, eds. B.J. Dobraszczyk and D.A. Dendy, Aspen Publishers, Gaithersburg, MD, USA.

DOBRASZCZYK, B.J., SMEWING, J., ALBERTINI, M., MAESMANS, G. and SCHOFIELD, J.D. (2003) Extensional rheology and stability of gas cell walls in bread doughs at elevated temperatures in relation to breadmaking performance, *Cereal Chemistry*, **80**, 218–224.

DOI, M. and EDWARDS, S.F. (1986) *The Theory of Polymer Dynamics*, Oxford University Press, Oxford.

DREESE, P.C., FAUBION, J.M. and HOSENEY, R.C. (1988) Dynamic rheological properties of flour, gluten and gluten–starch systems. I. Temperature-dependent changes during heating. *Cereal Chemistry*, **65**, 348–353.

EDWARDS, N.M., DEXTER, J.E., SCANLON, M.G. and CENKOWSKI, S. (1999) Relationship of creep-recovery and dynamic oscillatory measurements to durum wheat physical dough properties. *Cereal Chemistry*, **76**, 638–645.

EDWARDS, N.M., PERESSINI, D., DEXTER, J.E. and MULVANEY, S.T. (2001) Viscoelastic properties of durum wheat and common wheat dough of different strengths. *Rheologica Acta*, **40**, 142–153.

EGOROV, T.A., ODINTSOVA, T.I., SHEWRY, P.R. and TATHAM, A.S. (1998) Characterisation of high MW wheat glutenin polymers by agarose gel electrophoresis and dynamic light scattering. *FEBS Letters*, **434**, 215–217.

FAN, J., MITCHELL, J.R. and BLANSHARD, J.M.V. (1986) A computer simulation of the dynamics of bubble growth and shrinkage during extrudate expansion, *Journal of Food Engineering*, **23**, 337–356.

FARIDI, H. and FAUBION, J.M. (1986) *Fundamentals of Dough Rheology*, American Association of Cereal Chemists, St. Paul, MN, USA.

FARIDI, H. and FAUBION, J.M. (1990) *Dough Rheology and Baked Product Texture*, AVI Van Nostrand Reinhold, New York.

FAUBION, J.M. and FARIDI, H. (1985) Dough rheology: its benefits to cereal chemists, in: *Rheology of Wheat Products*, ed. H. Faridi, American Association of Cereal Chemists, St. Paul, MN, USA, pp. 1–9.

FERRY, J.D. (1980) *Viscoelastic Properties of Polymers*, John Wiley and Sons.

FIELD, J.M., TATHAM, A.S. and SHEWRY, P.R. (1987) The structure of a high M_r subunit of durum wheat (*T. durum*) gluten. *Biochemical Journal*, **247**, 215–221.

GRAESSLEY, W.W. (1974) The entanglement concept in polymer rheology. *Advances in Polymer Science*, **16**, 1–179.

GRAESSLEY, W.W. (1982) Entangled linear, branched and network polymer systems – molecular theories. *Advances in Polymer Science*, **47**, 67–117.

HANKOCZY, J. (1920) Apparat für Kleberverwertung. *Zeitschrift für Gesamte Getreidewissenschaft*, **12**, 57.

HAYMAN, D., SIPES, K., HOSENEY, R.C. and FAUBION, J.M. (1998) Factors controlling gas cell failure in bread dough. *Cereal Chemistry,* **75**, 85–89.

HE, H. and HOSENEY, R.C. (1991) Gas retention of different cereal flours. *Cereal Chemistry*, **68**, 334–336.

HIBBERD, G.E. and PARKER, N.S. (1975a) Dynamic viscoelastic behaviour of wheat flour doughs, part IV. Non-linear behaviour. *Rheologica Acta*, **14**, 151–157.

HIBBERD, G.E. and PARKER, N.S. (1975b) Measurement of the fundamental rheological properties of wheat flour doughs. *Cereal Chemistry*, **52**, 1–23.

HIBBERD, G.E. and WALLACE, W.J. (1966) Dynamic viscoelastic behaviour of wheat flour doughs, part 1. Linear aspects. *Rheologica Acta*, **5**, 193–198.

HUANG, H. and KOKINI, J.L. (1993) Measurement of biaxial extensional viscosity of wheat flour doughs. *Journal of Rheology*, **37**(5), 879–891.

HUMPHRIS, A.D.L., MCMASTER, T.J., MILES, M., GILBERT, S.M., SHEWRY, P.R. and TATHAM, A.S. (2000) Atomic force microscopy (AFM) study of interactions of HMW subunits of wheat glutenin. *Cereal Chemistry*, **77**, 107–110.

JANSSEN, A.M., VAN VLIET, T. and VEREIJKEN, J.M. (1996) Rheological behaviour of wheat glutens at small and large deformations. Comparisons of two glutens differing in breadmaking potential. *Journal of Cereal Science*, **23**, 19–31.

KHATKAR, B.S., BELL, A.E. and SCHOFIELD, J.D. (1995) The dynamic rheological properties of glutens and gluten sub-fractions from wheats of good and poor bread making quality. *Journal of Cereal Science*, **22**, 29–44.

KOKELAAR, J.J., VAN VLIET, T. and PRINS, A. (1996) Strain hardening and extensibility of flour and gluten doughs in relation to breadmaking performance. *Journal of Cereal Science*, **24**, 199–214.

LAUNAY, B. (1990) A simplified non-linear model for describing the viscoelastic properties of wheat flour doughs at high shear strains. *Cereal Chemistry*, **67**, 25–31.

LAUNAY, B., BURE, J. and PRADEN, J. (1977) Use of the Chopin Alveograph as a rheological tool. I. Dough deformation measurements. *Cereal Chemistry*, **54**, 1042–1048.

LI, W., DOBRASZCZYK, B.J.. and SCHOFIELD, J.D. (2003) Stress relaxation behaviour of wheat dough and gluten protein fractions. *Cereal Chemistry*, **80**, 333–338.

LOVE, R. J., HEMAR, Y., MORGENSTERN, M. and MCKIBBIN, R. (2002) Modelling the Sheeting of Wheat Flour Dough. 9th Asian Pacific Confederation of Chemical Engineering Congress APCChE 2002 and 30th Annual Australasian Chemical Engineering Conference CHEMECA 2002, Christchurch, New Zealand.

MACKINLEY, G.H. and HASSAGER, O. (1999) The Considere condition and rapid stretching of linear and branched polymer melts. *Journal of Rheology*, **43**, 1195–1212.

MCLEISH, T.C.B. and LARSON, R.G. (1998) Molecular constitutive equations for a class of branched polymers: the pom-pom model. *Journal of Rheology*, **42**, 81–110.

MACRITCHIE, F. and LAFIANDRA, D. (1997) Structure–function relationships of wheat proteins, in: *Food Proteins and Their Applications*, eds. S. Damodaran and A. Paraf, Marcel Dekker, New York, pp. 293–323.

MACOSKO, C.W. (1994) *Rheology: Principles, Measurements and Applications*, VCH Publishers Inc., New York.

MARIN. G. and MONTFORT, J.P. (1996) Molecular rheology and linear viscoelasticity, in: *Rheology for Polymer Melt Processing*, Elsevier Science, B.V. Amsterdam.

MASI, P., CAVELLA, S. and SEPE, M. (1998) Characterisation of dynamic viscoelastic behaviour of wheat flour doughs at different moisture contents, *Cereal Chemistry*, **75**, 428–432.

MULLER, H.G. (1975) Rheology and the conventional bread and biscuit making process, *Cereal Chemistry*, **52**, 89r–105r.

MÜNSTEDT, H., KURZBECK, S. and EGERSDÖRFER, L. (1998) Influence of molecular structure on rheological properties of polyethylenes. Part II. Elongational behaviour, *Rheologica Acta*, **37**, 21–29.

PADMANABHAN, M. (1995) Measurement of extensional viscosity of viscoelastic liquid foods. *Journal of Food Engineering*, **25**, 311–327.

RAO, M.A. and STEFFE, J.F. (1992) *Viscoelastic Properties of Foods*, Elsevier Applied Science, New York.

RAO, V.K., MULVANEY, S.J., DEXTER, J.E, EDWARDS, N.M. and PERESSINI, D. (2001) Stress–relaxation properties of mixograph semolina–water doughs from durum wheat cultivars of variable strength in relation to mixing characteristics, bread- and pasta-making performance. *Journal of Cereal Science*, **34**, 215–232.

RASPER, V.F. (1975) Dough rheology at large deformations in simple tensile mode. *Cereal Chemistry*, **52**, 24r–41r.

SAFARI-ARDI, M. and PHAN-THIEN, N. (1998) Stress relaxation and oscillatory tests to distinguish between doughs prepared from wheat flours of different varietal origin. *Cereal Chemistry*, **75**, 80–84.

SCHOBER, T.J., CLARKE, C.I. and KUHN, M. (2002) Characterization of functional properties of gluten proteins in spelt cultivars using rheological and quality factor measurements. *Cereal Chemistry*, **79**, 408–417

SCHOFIELD, R.K. and SCOTT BLAIR, G.W. (1932) The relationship between viscosity, elasticity and plastic strength of soft materials as illustrated by some mechanical properties of flour doughs. *I. Proceedings of the Royal Society of London A*, **138**, 707–718.

SCHOFIELD, R.K. and SCOTT BLAIR, G.W. (1937) The relationship between viscosity, elasticity and plastic strength of a soft material as illustrated by some mechanical properties of flour dough, IV. The separate contributions of gluten and starch. *Proceedings of the Royal Society of London A*, **160**, 87–94.

SCOTT, G. and RICHARDSON, P. (1997) The application of computational fluid dynamics in the food industry. *Trends in Food Science and Technology*, **8**, 119–124.

SHAH, P. CAMPBELL, G.M., DALE, C. and RUDDER, A. (1999) Modelling bubble growth during proving of bread dough. In *Bubbles in Food* (G.M. Campbell, C. Webb, S.S. Pandiella and K. Niranjan, eds.), American Association of Cereal Chemists, St. Paul, MN, USA.

SHERMAN, P. (1970) *Industrial Rheology: with particular reference to foods, pharmaceuticals, and cosmetics*, Academic Press, London.

SHEWRY, P.R., HALFORD, N.G., BELTON, P.S. and TATHAM, A.S. (2002) The structure and properties of gluten: an elastic protein from wheat grain. *Philosophical Transactions of the Royal Society of London B*, **357**, 133–142.

SHUEY, W.C. (1975) Practical instruments for rheological measurements on wheat products. *Cereal Chemistry*, **52**, 42r–81r.

SINGH, H. and MACRITCHIE, F. (2001) Application of polymer science to properties of gluten. *Journal of Cereal Science*, **33**, 231–243.

SMITH, J.R., SMITH, T.L. and TSCHOEGL, N.W. (1970) Rheological properties of wheat flour doughs III. Dynamic shear modulus and its dependence on amplitude, frequency, and dough composition. *Rheologica Acta*, **9**, 239–252.

SOUTHAN, M. and MACRITCHIE, F. (1999) Molecular weight distribution of wheat proteins. *Cereal Chemistry*, **76**, 827–836.

STEVENSON, S.G. and PRESTON, K.R. (1996) Flow field-flow fractionation of wheat proteins. *Journal of Cereal Science*, **23**, 121–131.

TANNER, R.I. and WALTERS, K. (1998) *Rheology: An Historical Perspective*, Elsevier Science B.V., Amsterdam.

THOMSON, N.H., MILES, M.J., POPINEAU, Y., HARRIES, J., SHEWRY, P. and TATHAM, A.S. (1999) Small angle x-ray scattering of wheat seed-storage proteins: α-, γ- and ω- gliadins and the high molecular weight (HMW) subunits of glutenin. *Biochimica et Biophysica Acta*, **1430**, 359–366.

TSCHOEGL, N.W., RINDE, J.A. and SMITH, T.L. (1970a) Rheological properties of wheat flour doughs I. Method for determining the large deformation and rupture properties in simple tension. *J. Science of Food and Agriculture*, **21**, 65–70.

TSCHOEGL, N.W., RINDE, J.A. and SMITH, T.L. (1970b) Rheological properties of wheat flour doughs II. Dependence of large deformation and rupture properties in simple tension on time, temperature and water absorption, *Rheologica Acta*, **9**, 223–238.

TRONSMO, K.M., MAGNUS, E.M., FAERGESTAD, E.M. and SCHOFIELD, J.D. (2002a) Relationship between gluten rheological properties and hearth loaf characteristics, *accepted by Cereal Chemistry*.

TRONSMO, K.M., FAERGESTAD, E.M., SCHOFIELD, J.D. and MAGNUS, E.M. (2002b) Wheat protein quality in relation to baking performance evaluated by the Chorleywood Bread Process and a hearth bread baking test, in: *Wheat Gluten Proteins: Analysis of properties in relation to baking performance*, Ph.D thesis, Department of Chemistry and Biotechnology, Agricultural University of Norway, submitted to *Journal of Cereal Science.*

UTHAYAKUMARAN, S., NEWBERRY, M., KEENTOCK, M., STODDARD, F.L. and BEKES, F. (2000) Basic rheology of bread dough with modified protein content and glutenin to gliadin ratios. *Cereal Chemistry*, **77**, 744–749.

UTHAYAKUMARAN, S., NEWBERRY, M. P., PHAN-THIEN, N. and TANNER, R. (2002) Small and large strain rheology of wheat gluten. *Rheologica Acta*, **41**, 162–172.

VAN VLIET, T. (1999) Rheological classification of foods and instrumental techniques for their study, in: *Food Texture: measurement and perception*, (A.J. Rosenthal, ed.), Aspen Publishers, MN, USA.

VAN VLIET,T., JANSSEN, A.M., BLOKSMA, A.H. and WALSTRA, P. (1992) Strain hardening of dough as a requirement for gas retention. *Journal of Texture Studies*, **23**, 439–460.

WANG, F.C. and SUN, X.S. (2002) Creep recovery of wheat flour doughs and relationship to other physical dough tests and breadmaking performance. *Cereal Chemistry*, **79**, 567–571.

WATANABE, H. (1999) Viscoelasticity and dynamics of entangled polymers. *Progress in Polymer Science*, **24**, 1253–1403.

WEEGELS, P.L., HAMER, R.J. and SCHOFIELD, J.D. (1996) Critical review: functional properties of wheat glutenin. *Journal of Cereal Science*, **23**, 1–18.

WEIPERT, D. (1988) The benefits of basic rheometry in studying dough rheology. *Cereal Chemistry*, **67**, 311–317.

WEIPERT, D. (1992) Descriptive and fundamental rheology in a new light. *Cereal Foods World*, **37**, 15–24.

WAHLUND. K.G., GUSTAVSSON. M., MACRITCHIE. F., NYLANDER. T. and WANNERBERGER. L. (1996) Size characterisation of wheat proteins, particularly glutenin, by asymmetrical Flow Field-Flow Fraction. *Journal of Cereal Science*, **23**, 113–119.

WHORLOW, R.W. (1992) *Rheological Techniques*, 2nd edn, Ellis Horwood, Chichester, UK.

WIKSTRÖM, K. and BOHLIN, L. (1999a) Extensional flow studies of wheat flour dough. I. Experimental method for measurements in contraction flow geometry and application to flours varying in breadmaking performance. *Journal of Cereal Science*, **29**, 217–226.

WIKSTRÖM, K. and BOHLIN, L. (1999b) Extensional flow studies of wheat flour dough. II. Experimental method for measurements in constant extension rate squeezing flow and application to flours varying in breadmaking performance. *Journal of Cereal Science*, **29**, 227–234.

WIKSTRÖM, K. and ELIASSON, A-C. (1998) Effect of enzymes and oxidizing agents on shear stress relaxation of wheat flour dough, *Cereal Chemistry*, **75**, 331–337.

19

Controlling dough development

S. Millar, Campden and Chorleywood Food Research Association, UK

19.1 Introduction

The process of mixing is a critical element of any breadmaking process as it is at this stage where blending and hydration of the flour components occur, bubble structure is initiated and where the development of the gluten proteins is commenced. For no-time doughs such as are typically produced in modern plant bakeries, dough mixing assumes an even more important role as in the absence of other mechanisms within which dough maturation may occur (such as bulk fermentation), the mixing element is wholly responsible for delivering a dough having the appropriate structure and rheological properties for further processing. Within this chapter a description of a generic dough mixing curve is used as basis from which the changes occurring with the dough system during development will be discussed. Although a number of other chapters within this book deal specifically with the underlying chemical and rheological changes that contribute to our understanding of dough development, an overview of some of the most pertinent is reproduced here to provide a context for subsequent assessment of the practicalities of control of processing industrial. The main mechanisms currently available for the assessment and control of dough properties are then reviewed with reference to the different types of mixer system that may be encountered. Finally, recent developments that are likely to have a significant impact in the future both in terms of understanding the mixing process more completely and as potential techniques for ultimate on-line use are described.

19.2 Dough rheology during mixing

As highlighted in the introduction to this chapter, dough development is a key requirement for the production of bread. It is important to recognise that this requirement is essentially distinct from the processes of blending ingredients, inclusion of air and initiation of bubble structure, which are also important aspects of dough mixing. Nevertheless each element interacts with others in the process and this should be recognised even where one aspect, in this case dough development, is dealt with specifically. It is useful, therefore, to begin at a reasonably common point from which the fundamental principles of mixing and ultimately dough development may be drawn. While recognising that there are many types of mixing process that interact with the subsequent stages of bread production, the concept of dough development allied to the changing rheological properties of dough as a result of physical manipulation is recognised virtually the world over. A useful, general introduction to this is shown in Fig. 19.1.

This type of curve is generated using a mechanical mixing device which moves the dough around in a bowl by means of some sort of agitation device, whether blade or pin in form. The movement of the dough is generated either by moving the bowl in relation to the agitator or vice versa. In the case of the example given here, a Mixograph, the dough is contained in a circular, flat-bottomed bowl which rotates at a constant speed. The dough in turn is also moved and, as it does so, it contacts the pins which are rigidly attached to an arm suspended over the bowl. The dough moves against these pins and as it is moved past them it produces a strain on the pins which is recorded. Even at this elementary level, some key aspects of dough quality may be highlighted. The first of these is that the dough moves around and past the pins. From this it is clear, therefore, that the dough is not a solid with the limited properties of deformation associated with such materials. In other words, the dough has a

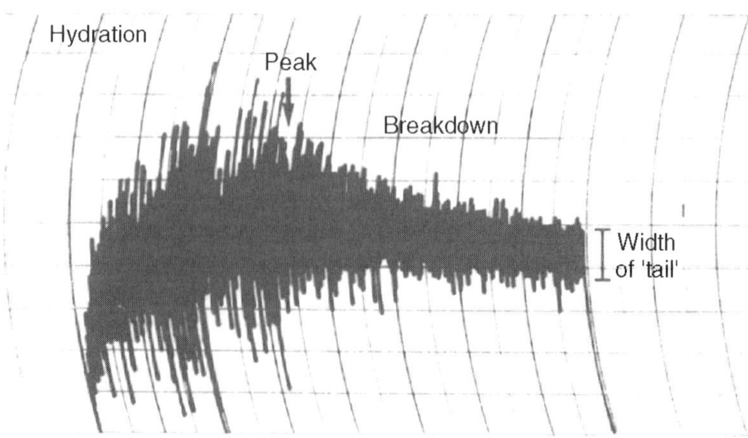

Fig. 19.1 Generalised Mixograph curve (reproduced with permission from Hoseney, 1986).

capacity to flow. However, if one compares dough with its main liquid component, i.e. water, it is clear that the force required to make dough flow is much greater. In other words, dough is viscous. However, dough rheology may not be described solely in terms of its viscosity. As mentioned above, dough exhibits resistance both to being moved around the mixing bowl and to being stretched within the mixing bowl. Dough rheology, therefore, has an elastic component or a tendency to recoil from stretching. This property interacts closely with dough viscosity such that sufficiently large-scale deformation causes an overall decrease in resistance to deformation. Nevertheless, the elastic component of a dough's rheology is critical to the properties required for breadmaking.

When flour and water are introduced to each other, the first requirement is that the flour should be hydrated. This is a prerequisite for subsequent operations as water is required as a medium within which molecules and enzymes became mobile and through which the many complex interactions that ultimately result in bread may take place. It is obvious that this must commence relatively soon after the onset of any mixing process. What is less clear is the way in which the various flour components hydrate, the way in which this water is distributed and redistributed through mixing and when hydration gives way to the next process which, for the purposes of this description, is gluten development. It is clear that gluten formation and development can occur only in the presence of water (compare the properties of wheat endosperm with dough) but it is less clear when this commences. The reason for this is that the initial part of the mixing curve (Fig. 19.1) shows an increase in resistance associated with the transition from dry flour and water to the viscoelastic dough. This increase in resistance continues over a period up to a maximum. Following this, resistance again decreases. There are two key aspects of this. The first is the overall resistance, i.e. the overall height of the curve. The second is the time taken under constant mixing conditions to reach the peak resistance. This is a fundamental property of individual flours which is generally thought to relate to dough strength. Doughs produced using strong flours generally require longer times to reach peak resistance. In this context, 'strong' is usually taken to be a description of the elasticity of the dough mixed. The more elastic the dough, the stronger it is generally perceived to be. It should be noted that while good elasticity is generally desirable for breadmaking, dough strength is not an absolute predictor of bread quality, but it is rather an indicator of the mixing regime required to optimise the quality of the bread to be produced. However, it is generally true that weak flours which are unsuited to breadmaking tend to have mixing curves where shorter times are required to attain maximum resistance.

As long as flour with good breadmaking potential is used, the dough produced tends to be at its most suitable for handling and breadmaking around the time at which resistance is at a maximum. Prior to this, the dough is considered to be underdeveloped, i.e. the development phase is still ongoing. Beyond this point, the dough suffers what is usually referred to as breakdown

(Stauffer, 1998) and becomes more extensible, less elastic and stickier, making it more difficult to handle and with a noticeable loss in bread quality.

The implications of this for controlling dough development are the following:

- There is a point at which the quality of dough and ultimately bread produced will be optimised.
- Changes in dough rheology may be used as an indicator of this point.
- To control dough development effectively a means of quantifying or indicating these changes is required.

19.3 Dough development

Wheat flour doughs are remarkably complex systems and while much has been researched, our understanding of the changes occurring during dough development is as yet incomplete. Nevertheless, there are a number of areas of scientific work that allow an understanding of the molecular basis of dough development. It is important to recognise that although this discussion will primarily focus on wheat flour proteins, there are a number of other factors that, while not covered here, are nonetheless important. These include, but are not limited to, the level of damaged starch produced during milling, the concentration of water-soluble and non water-soluble arabinoxylans, the physical characteristics of the starch granules, the milling extraction rate and so on. Nevertheless, it is generally accepted that the key component in wheat's unique ability to form viscoelastic doughs is wheat protein, specifically those fractions which form gluten when hydrated.

Wheat proteins comprise at least 70 different proteins (Belton, 2002) but for the purposes of assessing their functionality, they may be classified into four main groups on the basis of their solubility in various liquid media (Osborne, 1907). These groups are detailed in Table 19.1. While all of these fractions are important to the wheat plant either as metabolic (the albumins and globulins) or storage (gliadins and glutenins) proteins, it is these latter groups (those that form gluten on hydration) that confer the viscoelastic properties of wheat flour doughs. Gliadins are made up of monomers and are thought to confer extensibility to doughs while the glutenins form the polymeric fraction and confer elasticity. Both high-molecular-weight and low-molecular-weight

Table 19.1 Osborne fractions of plants generally and wheat proteins specifically

Solubility	Plant protein fraction	Wheat protein fraction
Water soluble	Albumins	Albumins
Saline soluble	Globulins	Globulins
Aqueous alcohol soluble	Prolamins	Gliadins
Remaining insoluble fraction	Glutelins	Glutenins

glutenins exist, with the former being the better characterised and more directly associated with dough properties in the work to date as discussed elsewhere.

Early workers trying to develop models linking gluten structure to function recognised the importance of the cysteine residues contained in the glutenin fraction. It was reasoned that the structure of gluten had to be extensively cross-linked to produce the resistance to extension typical of wheat flour doughs and the formation of disulfide bonds between and across individual glutenin units was proposed as the means by which this occurred (Schofield, 1986). These disulfide bonds were thought to emanate from the thiol groups present on the cysteine residues in gluten proteins. Evidence for this was supplied by the fact that oxygen was shown to be a key requirement for dough development. This is demonstrated by mixing studies in which doughs are underoxidised (Baker and Mize, 1941). In addition to this work, the action of oxidative improvers such as dehydro-ascorbic acid (the active improver formed when ascorbic acid is exposed to oxygen under aqueous conditions), potassium bromate or azodicarbonamide also indicates that oxidative conditions are required during mixing for dough development to occur (Cauvain and Collins, 1994).

This model for gluten structure did not provide clear mechanisms for the differences observed between different varieties of wheat and it was in the 1980s that workers demonstrated the importance of allelic variation in the assessment of breadmaking potential. Work by Peter Payne and his colleagues at Plant Breeding International (Payne *et al.*, 1987) showed that there were particular combinations of the glutenin subunits expressed on the 1A, 1B and 1D chromosomes that gave the best potential for good breadmaking. Scores were assigned to all the different subunit combinations and the proportion of the variation in baking performance was estimated to be of the order of 47–60%. This work resulted in very significant interest in this area with many groups subsequently studying this in great detail.

Another aspect of the research into gluten quality which has been extensively studied is that relating to polymer theory of how large molecules interact. This indicates that for polymers to demonstrate elasticity, the molecular weight distribution of the system must cover very large (and in the context of gluten this could be >1 MDa) molecular weights (MacRitchie, 1999). The problem with assessing this sort of structure is that some modification of the material must be made to ensure that the largest molecules are either solubilised or suspended sufficiently to allow for their analysis. Of the techniques used, perhaps the most common is size exclusion high-performance liquid chromatography of the sodium dodecyl sulfate (SDS) soluble proteins using ultrasonication as a means of solubilising the larger polymers (Morel *et al.*, 2000). More recently, however, field flow fractionation has also been applied to the problem with promising results (Schofield, 2000).

Although the importance of the formation of disulfide bonds in dough structure is widely accepted, recently authors have highlighted the importance of other systems that appear to contribute to wheat flour doughs' characteristic rheology. The first of these, hydrogen bonding, has long been recognised as a factor due to the changes in dough rheology observed when D_2O is substituted for H_2O (Tkachuk and Hlynka, 1968). More recently, however, spectroscopic techniques (about which there will be more later) have led to the development of the so-called loop and train model (Belton, 1999). In this, it is postulated that individual glutenin subunits interact with one another by disulfide bonds at the ends of the subunits and hydrogen bonds along the repeat region. The interaction between the repeat regions of adjacent subunits leads to zones of trains where the molecules are closely associated and loops where water is bound to one or both of the subunits. Extension of the system pulls the loops straight such that the loops disappear and trains are formed. Relaxation of the system allows the loop regions to reappear.

Another hypothesis for interactions between glutenin molecules has recently been developed by Katherine Tilley and colleagues at Kansas State University. In this, potential for the formation of dityrosine cross-links has been highlighted, particularly as a function of processing and improver addition (Tilley *et al.*, 2001). While it is clear that this work is at a relatively early stage, it is also apparent that the role of these bonds in dough development is as yet incompletely understood and that further work is warranted in this area.

Another theory as to the changes occurring during mixing has been advanced by Carl Hoseney (Hoseney, 1986). This is of particular interest owing to the way in which it has been developed and explained with reference to the generalised mixing trace above. In this case, it is proposed that the increase in resistance up to the peak is due simply to hydration of the flour proteins. As they gradually become hydrated, the amount of mobile water decreases up to the point at which the proteins are fully hydrated (the peak). Following this point, irreversible changes in the gluten polymers occur, which result in an inability to return the dough, even on resting, to its state prior to the torque peak. Although there is evidence that the process is more complicated than this model, it is interesting to re-visit this hypothesis in the context of some of the more recent spectroscopic studies on dough mixing (see Chapters 5 and 13 and later in this chapter).

19.4 The effects of mixer type

In addition to differences in mixing action, a key distinction between mixers is the intensity of their mixing action, i.e. how quickly the mixer 'works' the dough. In order to apply the principles discussed above to the process of dough mixing as carried out commercially, it is important to understand differences between the mixer types used. In terms of intensity, mixers can be thought of as falling into one of four broad groups: low, medium, high and very high.

19.4.1 Low-intensity mixers

A good example of such a system is the Artofex mixer, as used for the production of traditional bread types or doughs which will undergo bulk fermentation, particularly in continental Europe. Here the dough is worked in a very gentle fashion, giving a similar effect to mixing by hand. The mixing times are very long (typically ~30 minutes) and this means that the margin of error on this time is large and thus that the exact mixing time is less critical for the subsequent process than other factors such as the dough temperature or fermentation time. In fact, the mixing rate is so low that overmixing of the dough is not thought to be an issue. However, it is also highly unlikely that such a mixing regime could produce high-quality doughs in the absence of subsequent dough maturation, by fermentation for example.

19.4.2 Medium-intensity mixers

Spiral type mixers are typical of this category. Their energy input for a given mixing time at their high speed is relatively high when compared with an Artofex but the mixing times in the development phase (rather than the slow speed incorporation phase) are of the order of 10 minutes. Although these mixers may be used for processes where bulk fermentation or sponge systems are used (i.e. where other mechanisms contributed to dough development), they may also be used in systems where fermentation is limited to proof of moulded pieces immediately prior to baking. This indicates, therefore, that the levels of gluten development achievable can be sufficient to produce doughs in the absence of long fermentation steps.

19.4.3 High-intensity mixers

These include the mixing systems that tend to be popular in North America such as the horizontal bar mixer. Here, mixing times are shorter than spiral (typically 6–8 minutes) although other aspects of the process will also be different. As an example, sponge and dough processes are common in the United States and so even with higher-intensity mixers, gluten development as a result of other parts of the process may be important. Nevertheless, this should be tempered with an appreciation of the high protein content and elastic nature of the gluten for many North American wheat varieties which will promote longer mixing times in most cases.

19.4.4 Very high-intensity mixers

In this case, mixers are used as the mechanism of developing gluten structure and initiating bubble structure in no-time doughs. The ability of high mixer speeds to produce rapid gluten development was first noted by Swanson and Working (1926). The first commercial application of this was the Do-Maker system from the 1950s although it was the Chorleywood Bread Process (CBP) developed in the latter part of this decade that combined all the elements required to allow high-quality bread to be produced from much shorter processes than those used at that time (Chamberlain *et al.*, 1961). Since then, the generic term mechanical dough development has been adopted to cover similar processes. A key element of making bread in this way is the delivery of a fixed amount of work to the dough in a short period. In the case of CBP, this was stipulated to be 39.6 $kJ\,kg^{-1}$ (11 $W\,h\,kg^{-1}$) within a period of 2–5 minutes and this figure has stood the test of time well in the UK where the process was developed. It is important to note, however, that the level of work required can vary significantly where wheats of different gluten properties occur. In Australia and New Zealand, for example, work inputs as high as 61.2–72 $kJ\,kg^{-1}$ (17–20 $W\,h\,kg^{-1}$) may be encountered because of the high protein content and strength of many of the wheat varieties used for breadmaking in these countries (Cauvain and Collins, 1994; Gould, 1998).

One of the elements that has arisen through continued development of the CBP has been the use of pressure-vacuum mixing. The effect of a partial vacuum applied during mixing in Tweedy type mixers on bread crumb texture to promote fine and uniform crumb textures had been recognised for some time. However, the development of an over-pressure stage at the beginning of mixing was introduced in the 1990s to ensure that there was significant oxygen available to the dough during the early stages of mixing to promote good gluten development and to ensure that the action of oxidising improvers was maximised (Collins, 1993). The effect of this process on crumb texture is described in Chapter 17 but it is important to mention the practice in this context as it is another means of affecting dough development and ultimately bread quality which is available to the modern plant bakery.

19.5 Controlling dough development

While it is clear from the preceding sections that there are undoubtedly major chemical and rheological changes that take place in dough during mixing, the majority of the methods currently used for the control of the mixing process do not rely on such fundamental measurements. Nevertheless, it is common to use as least one of the methods listed below in order to ensure that the process is under control:

- Mixing to a fixed time.
- Mixing to a fixed energy input.
- Mixing to a fixed dough temperature (or, more correctly, temperature rise).
- Mixing to a fixed dough consistency.

Of these, the first two are probably in widest use although the others have been applied. In addition, the development of new spectrometers capable of recording spectra very rapidly now allows the changes in chemistry to be assessed and this will be covered in later sections.

19.5.1 Mixing to a fixed time

Generally speaking, mixing to a fixed time is applied in processes where longer mixing times using lower-intensity mixers are normal. It is effective because the longer mixing time means that there is more margin for error such that differences in flour quality do not result in the need for greatly different mixing times. However, the more important point to make is that typically this sort of process does not rely solely on the mixing stage for gluten development. Rather, it is usual to include some sort of bulk fermentation stage. This stage allows the gluten to be developed, appropriate gas cell structures to result and serves to 'even up' any differences between flour types. Where higher-intensity mixers are used, mixing to a fixed time becomes impractical because the margin for error becomes less in absolute terms, i.e. a 1 minute difference in mixing time has a proportionately greater effect when the total mixing time is 3 minutes rather than 10 minutes. At higher mixer speeds the interaction between flour types and the mixer becomes more acute such that differences in mixing time of at least 1 minute for an average of, say, 3 minutes may be realistic. Also, high-intensity mixers are typically used for the production of no-time doughs and so there is not a bulk fermentation step during which the dough properties may be changed. In such cases, it is more usual to mix to a fixed energy input.

19.5.2 Mixing to a fixed energy input

This input may be chosen to relate to a particular rheological state in the dough as determined using a mixer torque curve as shown in Fig. 19.2 or it may be a set level which is known to be effective for the range of flour types encountered as in CBP.

Another aspect of the duration of mixing for high-intensity mixers is the effect of differing mixer speeds or rates of work input on the quality of the dough and ultimately bread produced. Although high-intensity no-time dough processes such as the CBP tend to rely on a fixed work input as a means of controlling dough mixing, the time over which this work input must be delivered has always been stipulated (Chamberlain *et al.*, 1967). Optimum results were achieved in the original development work for a fixed work input of 39.6 $kJ\,kg^{-1}$ (11 $W\,h\,kg^{-1}$) when this work was delivered within 2–4 minutes (although an upper limit of 5 minutes was accepted as the absolute limit for production of bread of acceptable quality). This study also demonstrated that total work input affected not only bread quality but also the rate at which it was carried out. In fact, the variation in mixer rate was found to be most marked for strong flours with higher rates of mixing for strong flours of Canadian origin being a means of adjusting for situations where a reduced total level of work input was delivered.

There have been a number of other studies investigating the effect of variation in rate of work input on bread quality for high intensity mixing regimes. Work at the Spillers Research and Technology Centre in Cambridge,

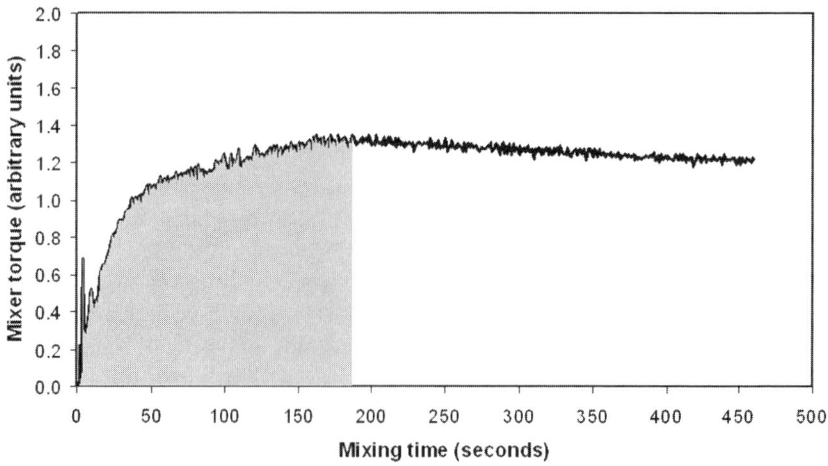

Fig. 19.2 Typical mixer torque curve with energy input indicated.

England, demonstrated the striking effect of increasing mixer speed on the shape of the torque traces generated using a 'Compudomixer', an instrument based on a 300 g Brabender Farinograph bowl but where the mixer speed was variable over a wide range (Frazier *et al.*, 1975). Work in this area was also carried out in the early 1970s at the Canadian Grain Commission's Grain Research Laboratory (GRL) in Winnipeg. Kilborn and Tipples (1972a) worked extensively in this area using Canadian wheats of differing grades under standard CBP conditions. This work also demonstrated the importance of differing rates of work input on the shape of the mixer torque traces generated. Critically, however, they also demonstrated that for high-intensity mixers run at reduced speeds there came a point at which the rate of work input was below a critical point required for dough development to occur. This was especially marked for one particular sample of flour which possessed extremely strong gluten characteristics. A higher mixer speed than normal was required to obtain typical development curves with this sample. As a result of this finding, these workers included an additional stipulation for high-intensity mixing, stating that not only must a certain level of work input be achieved but that this must be produced above a critical mixer speed. Although this was notable because of the particularly strong flours with which this group worked, the original development of CBP and the time limit within which the work was to take place led to effectively the same conclusion albeit by a different mechanism.

Another aspect of mixing control which resulted from the work at Spillers Ltd (Frazier *et al.*, 1975) and GRL (Kilborn and Tipples, 1972b) was the development of mixing systems for which a constant rate of work input could be maintained. As the resistance to mixing increases and the torque on the mixer blades increases, an increase in the rate of work input results. To counteract this, both groups developed systems whereby the mixer speed was automatically adjusted to the consistency of the dough, starting at high speeds during early development and gradually reducing as the dough resistance increased and then increasing again as resistance reduced following the attainment of peak resistance.

19.5.3 Mixing to a fixed dough temperature

The rise in dough temperature during mixing is a natural consequence of the work expended in moving the dough about the mixing chamber and is easily understood for those who have mixed dough by hand. Nevertheless, it should be recognised that where the losses of energy from a mixer are constant, then the temperature rise of the dough relates to the energy used to mix the dough (Levine, 2002). Thus, where the temperature of the ingredients is controlled and the loss of energy from the mixer is the same for each mix, then the temperature at the end of mixing may be used as a control mechanism. In bread dough mixing, the final dough temperature is usually controlled as it has a direct effect both on how the remaining parts of breadmaking (such as proof) proceed as well as how the dough will pass through the moulding equipment. Where high energy levels are required to develop the dough, therefore, the ingredients and the mixing bowl will be subject to cooling to ensure that the large temperature rise associated with the high-intensity mixing does not produce a dough at the end of mixing that will be difficult to pass through the plant or that will prove too quickly.

During the 1960s and 1970s, much work was carried out at the former Flour Milling and Baking Research Association investigating the processes associated with the production of semi-sweet biscuits or cookies (Wade *et al.*, 1965). For these products some gluten development is desirable to ensure that the dough withstands sheeting but over-tough or elastic gluten properties are avoided to ensure that the cut biscuits do not pull back to an oval shape after cutting. This work showed that while the work input during mixing was an important aspect of ensuring that the required changes took place, mixing to a final temperature in the range 37.8–43.3°C was also important to deliver optimum dough quality for subsequent processing. This was to some extent dictated by other changes that were temperature related such as solubilisation of the sucrose in these recipes. Nevertheless, this work is relevant to the study of bread dough mixing as it shows first of all the potential of the use of temperature rise as a control mechanism. Secondly, it also demonstrates that temperature rise has many consequences for the underlying chemistry of the system. Most scientists and technologists will be aware of the increased rate of many reactions at higher temperatures and the changes in rheology that occur for materials when their temperature is changed. This should serve to alert us to the possibility that the increased temperature of the dough itself is an important factor in the process of working the dough as well as the energy that is required to raise the temperature.

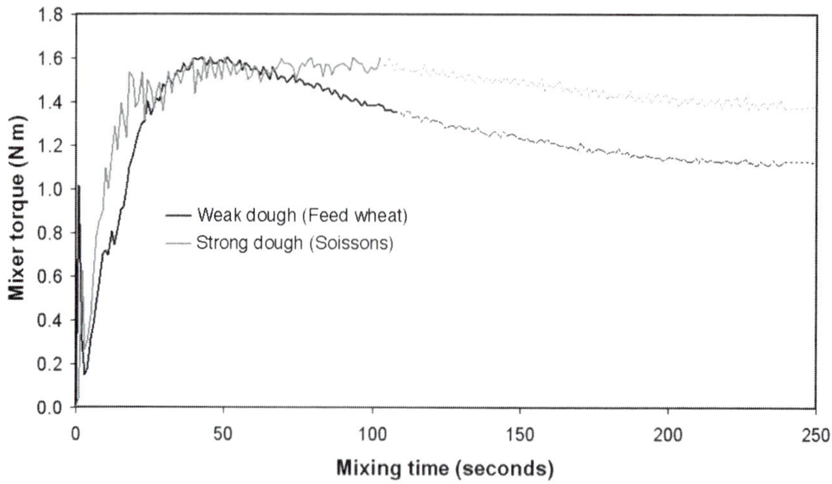

Fig. 19.3 Differing torque curves for two doughs given the same energy input during mixing.

19.5.4 Mixing to a fixed dough consistency

Given the marked differences between doughs from different flour types in terms of mixer torque traces and in light of the importance of dough rheology in the baking process generally, it is perhaps surprising that relatively few control mechanisms rely directly on assessment of the development of dough rheological properties (such as when using a torque curve) on a commercial scale. Although mixing to fixed energy input can result in such control where the flour properties change little, it clearly cannot where the flour properties are very different, as the same amount of energy will result in arrival at very different positions on a mixing curve, as shown in Fig. 19.3.

The fact that one of the key skills required of bakers has always been an ability to judge the consistency of dough by manual manipulation, and adjust the process accordingly, makes the relative lack of control mechanisms based on rheological parameters even more surprising. However, there have been attempts to effect control of the mixing process on this basis. Work at the Spillers Research and Technology Centre in Cambridge, England, during the early 1970s led to the development of a means of controlling mixing on the basis of dough rheological properties. At this time, the use of high-speed mixers had become widespread in the UK and so mixing to a fixed energy input was commonplace. In addition to the problems described above in relating this energy to a point on a development curve, the variability in flour properties also could mean differences in water absorbing capacity. When fixed levels of water were used for a given production run, therefore, variations in dough consistency were observed. Where doughs were too tight (insufficient water added), bread yield was reduced (Russell Eggitt, 1975) and shorter mixing times were required for equivalent energy input (water acting as plasticiser). The converse led to

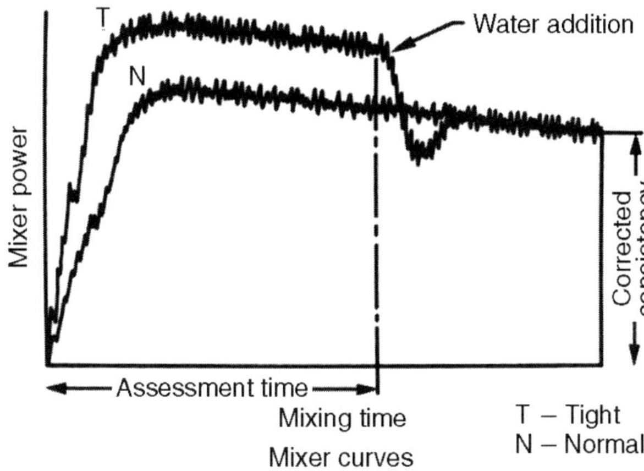

Fig. 19.4 Power consumption curves illustrating the effect of adding water to tight dough on its consistency (reproduced from Russell Eggitt, 1975).

doughs that were too slack, that were difficult to process and that required longer mixing times.

To counteract these problems, Spillers Ltd developed an instrument called the Dough Consistency Controller or DCC (Morison, 1970). With this system, control of dough consistency could be effected on each individual batch of dough rather than adjusting for the next batch on the basis of the properties of the final dough from the current one. The dough's consistency was assessed at a fixed mixing time which was shorter than the total mixing time. At this point the dough consistency could be adjusted by the addition of water to obtain the desired consistency of the dough. Of course, the removal of excess water is impossible and the addition of extra flour impractical. As a result, therefore, each mix was started with a reduced amount of water (about 3% below optimum) such that the doughs were always too tight (Frazier *et al.*, 1975). An appropriate amount of water was then added at the end of the assessment time in order to arrive at the desired consistency at the end of mixing (Fig. 19.4).

19.6 Emerging methods for controlling dough development

19.6.1 Ultrasound

More recently, there has been new interest in the assessment of dough rheological properties through the use of ultrasound techniques (Létang *et al.*, 2001). This work is still at an early stage but other workers (Salazar *et al.*, 2002) have since demonstrated that differences in flour properties may be related to differences in ultrasound velocity through doughs mixed to different levels. It is clear, therefore, that this technique has potential for further development.

19.6.2 Spectroscopic techniques

The use of techniques allowing the underlying chemistry of the dough system to be studied directly rather than inferred from rheological changes is a relatively recent development due primarily to the lack of suitable instrumentation being previously available. Nevertheless, since the 1990s there have been a number of studies where individual dough components have been studied using spectroscopic approaches. Of particular interest have been studies where nuclear magnetic resonance (NMR) spectroscopy has been used to understand changes in water availability (Shewry *et al.*, 2001) and those where Fourier transform infrared (FT-IR) spectroscopy has been used to study protein conformation, particularly of the high-molecular-weight subunits (Popineau *et al.*, 1994; Belton, 1999). This work is described in greater detail in Chapter 13 but a summary of the findings is also relevant to the subsequent sections of this chapter.

Studies using NMR have demonstrated that there appear to be regions of high polymer mobility in mixtures of water and individual high-molecular-weight subunits. The proportion of these regions is affected by the degree of hydration of the polymer chains glutenin but it is also affected by physical manipulation of the polymers. As the proteins are extended, the mobility of the system decreases and then increases again during subsequent relaxation. In addition, FT-IR studies have demonstrated that extension of glutenin subunits results in a maximisation of β-sheet confirmation at the expense of β-turn which is more apparent when the glutenins are in a relaxed state. Drawing on both sets of observations, a hypothesis for the role of hydrogen bonds in the elastic nature of the polymeric fraction of the wheat protein fraction has been proposed, loosely entitled the 'loop and train' model (Belton, 1999). Within this it is thought that adjacent polymers have zones where they are 'unzipped', i.e. not adjacent and it is thought that these are associated with mobile polymer regions. Extension causes these areas to disappear as the polymer chains are aligned next to one another and become 'zipped' together by hydrogen bonds. Relaxation leads to the re-formation of 'unzipped' loop regions among the 'zipped' train regions. Peak resistance has been proposed as being the point at which the maximum proportion of the glutenin polymers are in the zipped state and subsequent loss of resistance has been attributed to breakdown of the polymer structure, perhaps through cleavage of covalent bonds (Belton, 1999).

One of the interesting aspects of this work is the concept of alignment of the gluten polymers. This is a phenomenon that may be understood by any baker describing the properties of optimally mixed dough. Descriptors such as smooth, cohesive and extensibility are often used (Calvel *et al.*, 2001) as opposed to the lumpy appearance of dough during the early stages of mixing. Thus it may be inferred that some sort of linear molecular arrangement occurs during dough development. This has been confirmed at the microscopic level by studies such as that described by Amend and Belitz (1990) where linear arrangements of gluten were clearly seen as development occurred. Thus it appears that spectroscopic, microscopic and practical baking experience all point to an essentially linear basis for an optimally developed gluten structure.

Until recently, however, the instrumentation available to researchers was such that assessment of the changes occurring in full recipe doughs could not be directly monitored. This changed in the later 1990s with the advent of a new generation of near-infrared (NIR) spectrometers which allowed spectra to be acquired extremely rapidly and thus to allow an assessment of the changes occurring during dough mixing to be followed. The first published work in this area was carried out by Ian Wesley and co-workers at BRI Australia Ltd (Wesley *et al.*, 1998). In this study a Perten DA-7000 NIR spectrometer using a stationary grating and a system of diode arrays to allow parallel processing of entire spectra (rather than scanning through a spectrum) was used. Rather than calibrating the NIR data with some appropriate reference technique, as is common practice in many other aspects of cereals assessment, the technique was used in a more classical sense and it was the fundamental changes in the spectrum that were recorded. The technique was applied to mixers having different actions including pin, z-blade and spiral configurations and in the majority of cases, mixing curves could be plotted based on changes in the NIR spectra.

Since then two other groups, Campden and Chorleywood Food Research Association (CCFRA) in England (Millar *et al.*, 1999) and the American Institute of Baking in Manhattan, Kansas, have also used similar instrumentation to extract information from mixing doughs. The former group has carried out a significant amount of work on the fundamental understanding and the subsequent application of the technique. Early work assessed the changes occurring in the NIR spectra by means of principal components analysis (PCA) in which both an overall picture of the sort of mixing curves which could be obtained as well as the underlying spectral basis for these changes. This showed that two spectral regions, 1125–1180 nm and 1375–1525 nm contained information which could be used to plot mixing curves on the basis (Fig. 19.5). Subsequent work demonstrated that the former gave the most consistent results for a range of different flour types, possibly because the region at longer wavelengths contained information from the first overtone band of O–H stretch and the first overtone band of N–H stretch. The peak at 1160 nm has previously been assigned to water (Carcio and Petty, 1951) and had been used by the Australian workers in their initial work. The fact that these regions appear to change during dough mixing is particularly interesting the context of the work described above in which the contribution of hydrogen bonding to the phenomenon of dough elasticity was noted.

Subsequent work at CCFRA demonstrated that the mixing curves that could be derived from plotting NIR data derived from the region 1125–1180 nm related to differences in gluten properties both in terms of the performance expected for a given variety and in relation to methods of analysing the rheology of the gluten fraction such as the elastic modulus of the gel protein fraction (Alava *et al.*, 2001). In addition to this, the mixing time as predicted from the

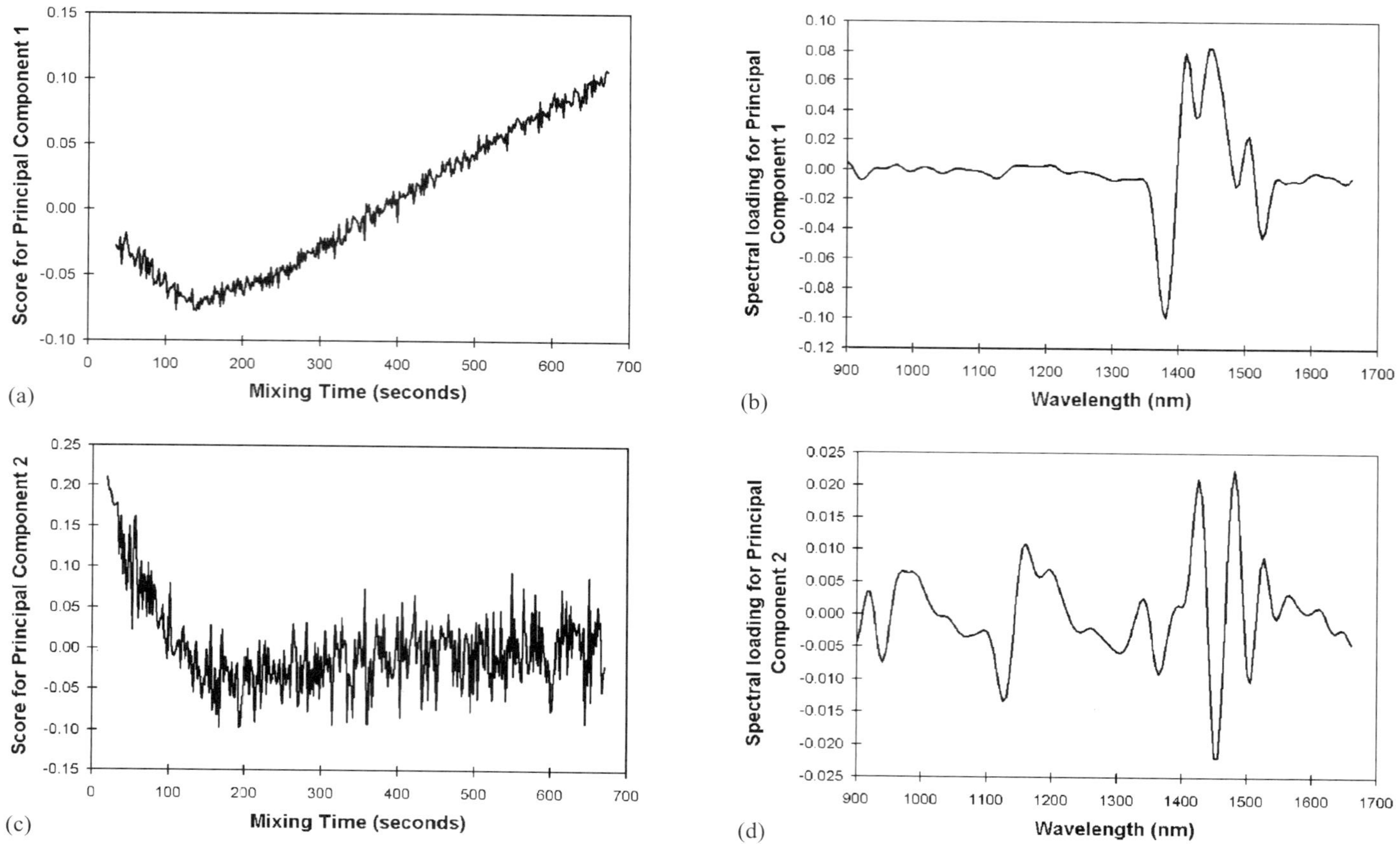

Fig. 19.5 Principal components analysis of NIR data collected from dough produced in a DoCorder using flour milled from Hereward wheat (reproduced from Alava *et al.*, 2001).

NIR data (or the time taken to attain a turning) was found to be affected by systematic changes in formulation in a similar way to mixer torque. In particular, increasing dough water levels led to longer mixing times and vice versa and the addition of a reducing agent such as L-cysteine hydrochloride led to much reduced mixing times (Millar *et al.*, 2001). Although the overall direction of the changes with differing ingredient levels was effectively the same as those observed for the time taken to reach peak mixer torque, in each case the NIR mixing times was approximately 20% longer than this time. This is interesting in the context of how many bakers use dough rheological properties to 'predict' optimum mixing times. It is generally accepted that mixing to a point 110–120% of the time taken to reach peak torque results in bread of superior quality than earlier or later mixing times.

Having established that the technique was capable of responding to differences in mixing performance for different flours, the workers at CCFRA then moved to baking trials. Here doughs were mixed to different times, processed in a standard CBP system and the bread quality assessed by means of volume by seed displacement and crumb texture using image analysis. These studies showed (Millar, 2000) that the finest crumb texture as indicated by lower mean crumb cell areas by image analysis was found at a relatively early stage during mixing. This point was shown to be that at which the peak mixer torque was attained. However, the greatest loaf volume was shown to occur at a later point. This makes sense in light of the practice of mixing longer than the time taken to reach peak mixer torque as described above. It was interesting to note, however, that the NIR turning point occurred at a point intermediate to the finest crumb texture and greatest volume. It appears, therefore, that the changes in dough chemistry to which NIR is sensitive tell a slightly different story from that obtained from rheological techniques and thus that this technique has an application in the study of dough properties and in developing our understanding of this fundamental element of breadmaking. It also suggests that ultimately there may be a commercial application for an NIR-based system.

In the context of the effects of rate of work input on mixing performance described above, the most recent work reported (Millar *et al.*, 2001) is interesting as different mixer speeds have been used to assess how spectroscopic and rheological measures are affected. Mixing using a dough based on a flour produced from a UK-grown batch of Soissons wheat (having very strong gluten properties) showed that gluten development did not occur at very low mixer speeds using both mixer torque and NIR data. The quality of the bread thus produced was markedly inferior to that produced at higher mixer speeds (Fig. 19.6). Although the overall quality of bread produced from a flour having very weak gluten properties (milled from feed wheat) was poor, there appeared to be no detrimental effect of mixing at low mixer speeds because the rate of mixing was above the critical value required for this wheat type.

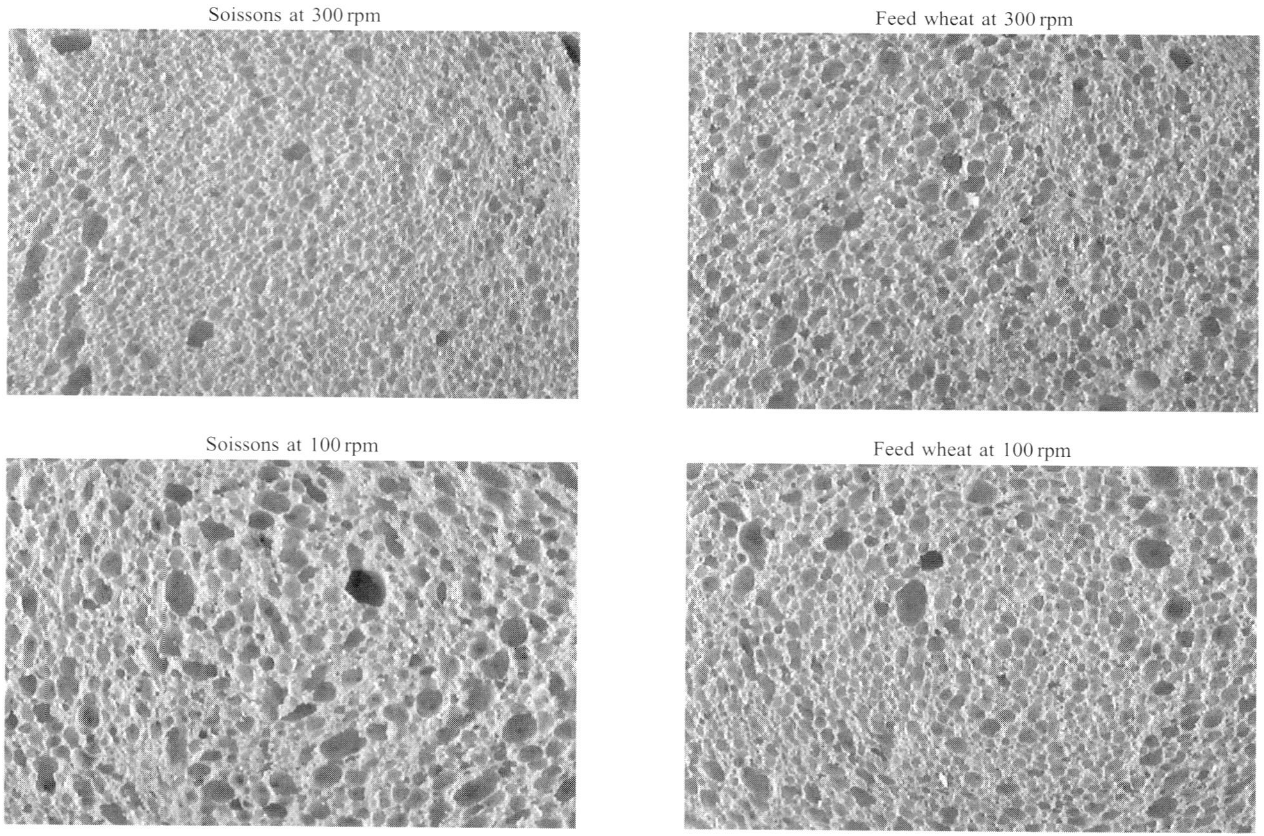

Fig. 19.6 Images of crumb texture of bread produced using flours from Soissons and feed wheat when doughs were mixed at different mixer speeds (reproduced with permission from Millar and Alava, 2002).

19.7 Future trends

Perhaps the most striking element of the mechanisms used to control mixers currently is that the majority are based on parameters that may be measured relatively easily (e.g. time, energy input) rather than the parameters widely accepted in small-scale studies of dough quality (e.g. time to peak mixer torque). The introductory sections of this chapter were written to summarise some of the main theories and assessment techniques relating to dough and gluten development and it is clear in the sections that followed that these are still some way from being adopted more globally for mixer control. This is understandable, as in many cases current practice gives perfectly acceptable results most of the time and the parameters measured are relatively easy to handle. However, it is also clear that this situation relies on either a consistent level of flour properties where high-speed mixers are used or the effect of additional processes such as fermentation where lower-speed mixers are used. Were one to consider systems where a high level of flexibility were required then it is likely that current control systems would be deficient. This is illustrated by the developments highlighted toward the end of the chapter and it is in these areas that future advances are anticipated. The use of ultrasound has been applied to the assessment of both gas bubble size distributions and gluten dough rheological properties resulting from gluten development. This work is interesting as it marks an attempt to assess the rheological properties of doughs in fundamental terms using systems that may ultimately have at-line or even on-line possibilities.

The use of the spectroscopic techniques described is also an area where further development is likely. The work to date has been interesting in the way that depending on the technique used either fundamental studies of model systems or global changes in full formulation systems have been attempted. This is perhaps a good example of how differing levels of scientific, technical and practical understanding may be combined to develop more complete models of complex systems such as wheat flour doughs. As bakeries change across the world to shorter time systems, there will also be increased need for control systems to ensure that the wide range of ingredient qualities encountered may be managed by the entire process and particularly during mixing.

It would appear, therefore, that the future will see an increased need for on-line control of dough development, most likely increasingly within the context of short breadmaking processes and that the natural variation inherent in the raw material, process and final products will ensure that a move to more specific control mechanisms based on fundamental dough properties will be evident.

19.8 Sources of further information and advice

19.8.1 Other books

There are many books covering the relevant areas of cereal and bakery science and the following are a selection that cover both scientific and practical approaches.

- *Wheat: Chemistry and Technology*. American Association of Cereal Chemists, Inc., 3340 Pilot Knob Road, St. Paul, MN, USA.
- *Principles of Cereal Science and Technology*. American Association of Cereal Chemists, Inc., 3340 Pilot Knob Road, St. Paul, MN, USA.
- *The Technology of Breadmaking*. Plymbridge Distributors Ltd, Estover Road, Plymouth, PL6 7P2, UK.
- *Bakery Problems Solved*. Woodhead Publishing Ltd, Abington Hall, Abington, Cambridge, CB1 6AH, UK.

19.8.2 Journals

- *Journal of Cereal Science*. Elsevier Science Ltd., Langford Lane, Kidlington, Oxford, OX5 1GB, UK.
- *Cereal Chemistry*. American Association of Cereal Chemists, Inc., 3340 Pilot Knob Road, St. Paul, MN, USA.
- *Cereal Foods World*. American Association of Cereal Chemists, Inc., 3340 Pilot Knob Road, St. Paul, MN, USA.

19.8.3 Professional organisations

- American Association of Cereal Chemists, Inc., 3340 Pilot Knob Road, St. Paul, MN, USA.
- International Association for Cereal Science and Technology (ICC), Marxergasse 2/Mezzanine floor, P.O. Box 47, A-1033 Vienna, Austria.

Both organisations organise regular conferences dedicated to cereal science issues.

19.8.4 Research organisations

These organisations are sources both of information and in many cases of specialised training in aspects covering the entire range of cereal science applications.

- The American Institute of Baking, 1213 Bakers Way, PO Box 3999, Manhattan, Kansas 66505–3999, USA.
- BRI Australia Ltd, PO Box 7, North Ryde, NSW 1670, Australia.
- Campden and Chorleywood Food Research Association, Chipping Campden, Gloucestershire, GL55 6LD, United Kingdom.
- TNO Nutrition and Food Research, Utrechtseweg 48, P.O. Box 360, 3700 AJ Zeist, The Netherlands.

19.9 References

ALAVA, J.M., MILLAR, S.J. and SALMON, S.E. (2001). The determination of wheat breadmaking performance and bread dough mixing time by NIR spectroscopy for high speed mixers. *Journal of Cereal Science*, **33**, 71–81.

AMEND, T. and BELITZ, H.D. (1990). The formation of dough and gluten – a study by scanning electron microscopy. *Zeitschrift für Lebensmittel Untersuchung Und Forschung*, **190**, 401–409.

BAKER, J.C. and MIZE, M.D. (1941). The origin of the gas cell in bread dough. *Cereal Chemistry*, **18**, 19–34.

BELTON, P.S. (1999). On the elasticity of wheat gluten. *Journal of Cereal Science*, **29**, 103–107.

BELTON, P.S. (2002). Structure function relationships in high molecular weight subunits of gluten. *52nd RACI Cereal Chemistry Conference*, 8–12 September, Christchurch, New Zealand.

CALVEL, R. WIRTZ, R.L. and MACGUIRE, J.J. (2001). *The Taste of Bread*. Aspen Publishers, Inc., Gaithersburg, Maryland, USA.

CARCIO, J.A. and PETTY, C.C. (1951). The near infrared absorption spectrum of liquid water. *Journal of the Optical Society of Am*erica, **41**, 302–304.

CAUVAIN, S.P. and COLLINS, T.H. (1994). Mixing, moulding and processing of bread doughs. In: (ed. A. Gordon) *Baking Industry Europe 1995*. Sterling Publications, London, UK, pp. 41–43.

CHAMBERLAIN, N., COLLINS, T.H. and ELTON, G.A.H (1961). The Chorleywood Bread Process. *British Baking Industries Research Association, Report No. 59*. July 1961.

CHAMBERLAIN, N., COLLINS, T.H. and ELTON, G.A.H (1967). The Chorleywood Bread Process: effect of rate of dough mixing. *Flour Milling and Baking Research Association, Report No. 1*. March 1967.

COLLINS, T.H. (1993). The Chorleywood Bread Process: pressure/vacuum mixing. *Chorleywood Digest No. 130*, September, 94–97.

FRAZIER, P.J., DANIELS, N.W.R. and RUSSELL EGGITT, P.W. (1975). Rheology and the continuous breadmaking process. *Cereal Chemistry*, **52**(3), 106r–130r.

GOULD, J. (1998) Baking around the world. In: (eds. S.P. Cauvain and L.S. Young) *Technology of Breadmaking*, Blackie Academic & Professional, London, UK, pp. 197–213.

HOSENEY, R.C. (1986). *Principles of Cereal Science and Technology*. American Association of Cereal Chemists Ltd, St. Paul, MN, pp. 213–214.

KILBORN, R.H. and TIPPLES, K.H. (1972a). Factors affecting mechanical dough development. I. Effect of mixing intensity and work input. *Cereal Chemistry*, **49**, 34.

KILBORN, R.H. and TIPPLES, K.H. (1972b). Factors affecting mechanical dough development. II. Implications of mixing at a constant rate of energy input. *Cereal Chemistry*, **4**, 48.

LÉTANG, C., PIAU, M., VERDIER, C. and LEFEBVRE, L. (2001). Characterisation of wheat-flour–water doughs: a new method using ultrasound. *Ultrasonics*, **39**, 133–141.

LEVINE, L. (2002). Energy balances in dough mixers. *Cereal Foods World*, **47**(7), 334–338.

MACRITCHIE, F. (1999). Wheat proteins: characterization and role in flour functionality. *Cereal Foods World*, **44**, 188–193.

MILLAR, S.J. (2000). Mixed Fractions. *European Baker*, May/June 2000.

MILLAR, S.J. and ALAVA, J.M. (2002). *Predicting Baked Product Processing Requirements and Quality using NIR Spectroscopy*. Campden & Chorleywood Food Research Association Research Summary Sheet No. 2001–15.

MILLAR, S.J., ALAVA, J.M. and SALMON, S.E. (1999). Using near infrared spectroscopy to assess dough development during mixing. In: (eds A.M.C. Davis and R. Giangiacomo) *Proceedings of the 9th International Conference of Near Infrared Spectroscopy*, Verona.

MILLAR, S.J., ALAVA, J.M. and SALMON, S.E. (2001). NIR spectroscopy as a means of understanding changes occurring during dough mixing. *86th AACC Annual Meeting*, 14–18 October 2001, Charlotte.

MOREL, M-H, DEHLON, P., AUTRAN, J.C., LEYGUE, J.P. and BAR-L'HELGOUAC'H, C. (2000). Effects of temperature, sonication time, and power settings on size distribution and extractability of total wheat flour proteins as determined by size-exclusion high-performance liquid chromatography. *Cereal Chemistry*, **77**, 685–691.

MORISON, P.P. (SPILLERS LTD) (1970). Improvements in or relating to mixers. British Patent No. 1,319,297.

OSBORNE, T.B. (1907). *The Proteins of the Wheat Kernel*. Publication 84, Carnegie Institute, Washington D.C.

PAYNE, P.I., NIGHTINGALE, M.A., KRATTIGER, A.F. and HOLT, L.M. (1987). The relationship between HMW glutenin subunit composition and the bread-making quality of British-grown wheat varieties. *Journal of the Science of Food and Agriculture*, **40**, 51–65.

POPINEAU, Y., BONENFANT, S., CORNEC, M. and PÉZOLET, M. (1994). A study by infrared spectroscopy of the conformations of gluten proteins differing in their gliadin and glutenin compositions. *Journal of Cereal Science*, **20**, 15–22.

RUSSELL EGGITT, P.W. (1975). Dough consistency control. *Bakers Digest*, **49**(3), 31–35.

SALAZAR, J., ÁLAVA, J.M., SAHI, S.S., TURÓ, A., CHÁVEZ, J.A. and GARCÍA, M.J. (2002). Ultrasound measurements for determining rheological properties of flour-water systems. In: (eds D.E. Yuhas and S.C. Scheider). *Proceedings of the 2002 IEEE Ultrasonics Symposium*, Munich, 8–11 October 2002, pp. 855–858.

SCHOFIELD, J.D. (1986). Flour proteins: structure and functionality in baked products. In: (eds J.M.V. Blanshard, P.J. Frazier and T, Galliard). *Chemistry and Physics of Baking*. Royal Society of Chemistry, London.

SCHOFIELD, J. D. (2000). Redox reactions in gluten and dough. Wheat proteins where are we now? Flair-Flow Workshop. London.

SHEWRY, P.R., POPINEAU, Y., LAFIANDRA, D. and BELTON, P. (2001). Wheat glutenin subunits and dough elasticity: findings of the EUROWHEAT project. *Trends in Food Science and Technology*, **11**, 433–441.

STAUFFER, C.E. (1998) Principles of dough formation. In: (eds. S.P. Cauavin and L.S. Young) *Technology of Breadmaking*, Blackie Academic & Professional, London, UK, pp. 262–295.

SWANSON, C.O. and WORKING, E.B. (1926). Mechanical modification of dough to make it possible to bake bread with only the fermentation in the pan. *Cereal Chemistry*, **3**, 65.

TILLEY, K.A., BENJAMIN, R.E., BAGOROGOZA, MOSES OKOT-KOTBER, B., PRAKASH, O. and KWEN, H. (2001). Tyrosine crosslinks: molecular basis of gluten structure and function. *Journal of Agricultural and Food Chemistry*, **49**, 2627–2632.

TKACHUK, R. and HLYNKA, I. (1968). Some properties of dough and gluten in D_2O. *Cereal Chemistry*, **45**, 80–87.

WADE, P., BOLD, E.R. and HASTINGS, W.R. (1965). Investigation of the mixing process for hard sweet biscuit doughs: Part III. Effect of mixing conditions on the finished biscuits. *British Baking Industries Research Association, Report No. 80*. May 1965.

WESLEY, I.J., LARSEN, N., OSBORNE, B.G. and SKERRITT, J.H. (1998). Non-invasive monitoring of dough mixing by near infrared spectroscopy. *Journal of Cereal Science*, **27**, 61–69.

20

The use of redox agents

H. Wieser, German Research Centre of Food Chemistry, Germany

20.1 Introduction

Wheat is unique among cereals in its ability to form a cohesive viscoelastic dough, when flour is mixed with water. Wheat dough retains the gas produced during fermentation or by chemical leavening and this gives a leavened loaf of bread after baking. The major factor accounting for variation in bread quality, particularly in loaf volume and crumb structure, is the protein content of the flour; the relation within a variety is essentially linear between 8 and 18% (Pomeranz, 1988). For different varieties, however, the level and slope of the regression lines for bread volume and protein content differ significantly, indicating differences between varieties in protein quality. Therefore, both a certain quantity and qualitative characteristics of protein are needed in wheat flour to produce a bread of high quality. Whereas protein quantity can be determined with good accuracy by different techniques, protein quality is difficult to define and is dependent on both genotype and growing conditions of wheat.

It is commonly accepted that structures and quantitative compositions of gluten proteins govern dough properties and bread quality and account for the differences between wheat varieties. According to their solubility in aqueous alcohols, gluten proteins can be divided into the soluble, chiefly monomeric, gliadins and the insoluble aggregated glutenins. Only an optimal combination of gliadins and glutenins gives a gluten of desirable properties in dough and bread. Among covalent and non-covalent bonds in gliadins and glutenins, disulfide (SS) bonds play a key role in dough and bread structure, forming cross-links within and between protein chains. In particular, the formation of very large glutenin polymers by inter-chain SS bonds is important determining dough strength and loaf volume (Weegels *et al.*, 1996). Various studies have substantiated that the oxidation of free thiol (SH) groups accompanied by the formation of large protein aggregates and the SH/SS interchange accompanied by the depolymerisation of glutenin polymers are crucial reactions during the breadmaking process (Stauffer, 1990). It has been known for a long time that these reactions can be manipulated by the addition of redox agents, resulting in baked products of desired quality; such redox agents are widely used by millers, bakers, producers of bread improvers and baking industry. This chapter is concerned with the redox state in flour, the redox reactions during processing, and the effects of redox agents and is restricted primarly to wheat flour, dough and yeast-leavened bread.

20.2 The redox state in flour

20.2.1 Cysteine

The amino acid cysteine and its SS form cystine are the main target for most redox reactions in flour and dough. Cysteine is readily converted to cystine even under mild oxidative conditions, equally reduction of cystine to cysteine is easily performed by means of reducing agents (equation 20.1).

$$2R{-}CH_2{-}SH \underset{+2H}{\overset{-2H}{\rightleftharpoons}} R{-}CH_2{-}S{-}S{-}CH_2{-}R \qquad (20.1)$$

Cyst(e)ine amounts to about 2.5 g per 100 g of flour proteins. This value corresponds to a range of 20–25 μmol per g flour with a protein content of 10–12% (Grosch and Wieser, 1999). About 98–99% of total cyst(e)ine is integrated into proteins; the residue is present in small peptides or as free amino acid. The quantitative determination of accessible free SH groups in a series of mature flours from different wheat cultivars resulted in a range of 0.66 to 1.07 μmol per g (Ewart, 1988). According to these data it can be assumed that only about 5% of total flour cyst(e)ine have free SH groups, whereas the major portion is oxidised to SS bonds. Flour milled under nitrogen has a significantly higher SH content than flour milled under atmospheric oxygen. The amount of free SH groups in freshly milled flour is significantly higher compared with mature flour and decreases continuously during storage under atmospheric air. During dough mixing the concentration of SH groups decreases very quickly, down to about 50% of the original value (Sullivan *et al.*, 1963) and intense SH/SS interchange reactions occur. The reactivity of single SH or SS groups, however, is different and depends on the size and mobility of the molecules, of which the groups are a part, and on the three-dimensional availability of the groups. SS bonds are covalent bonds having an energy of 205 kJ/mol; they are usually not broken at room temperature except by chemical reactions, e.g. by reducing agents. Under mechanical stress (high-speed mixing, sonification) and at high temperature (baking) they can be broken, forming reactive thiyl radicals.

20.2.2 Gluten proteins

When water is added to flour and mixed, the water-insoluble proteins hydrate and develop gluten, a cohesive viscoelastic mass in which starch, added yeast and other dough components are embedded. The three-dimensional network of gluten is responsible for the rheological properties and gas retention of dough which makes production of light leavened products possible. Gluten proteins account for about 80% of total flour proteins. Both covalent bonds (disulfide bonds) and non-covalent interactions (hydrogen, ionic and hydrophobic bonds) between gluten proteins play an important role in contribution to the functional properties of gluten. As mentioned above, gluten proteins can be divided into two main fractions according to their solubility in aqueous aliphatic alcohols, e.g. 60% ethanol: the soluble gliadin and the insoluble glutenin fraction. Both fractions differ diametrically in their contribution to the rheological properties of dough. Gliadins are cohesive but with low elasticity and contribute to the viscosity and extensibility of dough; glutenins are both cohesive and elastic and control dough strength and elasticity. Gliadins are considered as acting as 'plastifier' or 'solvent' for glutenin aggregates (Belitz *et al.*, 1986).

Gliadin and glutenin fractions consist of numerous components which can be classified into a few protein types according to related amino acid compositions and sequences and molecular weights (Table 20.1) (Grosch and Wieser, 1999). The gliadin fraction contains mainly monomeric proteins, with rather low molecular masses (M_r) ranging from 28 000 to 55 000. They belong to four different types, the ω5, ω1,2, α- and γ-gliadins. The glutenin fraction comprises aggregated proteins with M_rs from about 10^5 up to several millions that are linked by inter-chain SS bonds. After reduction of SS bonds the resulting glutenin subunits show a solubility in aqueous alcohols similar to gliadins. According to structural differences glutenin subunits can be classified into the

Table 20.1 Classification and characteristics of gluten proteins (Grosch and Wieser, 1999)

Type	%[a]	M_r[b]	Cys[c]
Gliadins			
ω5	3–6	44–55 000	0
ω1,2	4–7	36–44 000	0
α	28–33	28–35 000	6
γ	23–31	31–35 000	8
Glutenins			
x HMW	4–9	83–88 000	4–5
y HMW	3–4	67–74 000	7
LMW	19–25	32–39 000	8

[a] Percentage according to total gluten proteins.
[b] Derived from complete sequences or MALDI-TOF (matrix-assisted laser desorption/ionization time-of-flight mass spectrometry) analysis.
[c] Number of cysteine residues.

x- and y-type of high-molecular-weight (HMW) subunits ($M_r = 67-88\,000$) and the low-molecular-weight (LMW) subunits ($M_r = 32-39\,000$). The amino acid sequences of LMW subunits are homologous with those of α- and γ- gliadins in sequence divisions III and V (Fig. 20.1).

Amount and proportions of the different gluten protein types depend on wheat varieties and growing conditions, but it can be generalised that α-, γ-gliadins and LMW subunits belong to the major and ω- gliadins and HMW subunits to the minor protein components of gluten (Table 20.1). Numerous studies demonstrated that the total amount of gluten proteins, the ratio of gliadins to glutenins, the ratio of HMW subunits to LMW subunits and the presence of specific glutenin subunits determine rheological dough properties and bread quality. In particular, the amount of SS bonded glutenin aggregates with the highest molecular masses (gel protein, glutenin macropolymer) have been considered to be of special importance (Weegels *et al.*, 1996), and this is the fundamental background for the relevance of redox systems within the breadmaking process.

Usually, gliadins are in a monomeric state having either no cysteine (ω-gliadins) or six (α-gliadins) and eight (γ-gliadins) cysteine residues, respectively, forming intra-chain SS bonds. Owing to point mutations of the amino acid sequences, a small number of gliadins have an odd number of cysteine residues and are linked to glutenins by an inter-chain SS bond. Direct SS bond determination demonstrated that six and eight cysteine residues, respectively, in the C-terminal domains (sequence divisions III and V) of α-, γ-gliadins and LMW subunits form homologous intra-chain SS bonds (Fig. 20.1) (Koehler *et al.*, 1993; Müller and Wieser, 1997). These bonds are concentrated in a relatively small area and contribute to a compact three-dimensional structure. Two further cysteine residues located in the N- and C-terminal domains (divisions I and IV) are unique to LMW subunits and responsible for the aggregative character of this protein type. Probably for steric reasons these cysteine residues are not able to form an intra-chain SS bond; they form inter-chain bonds with cysteines in other LMW subunits, HMW subunits and even γ-gliadins. HMW subunits also are involved in both intra-chain and inter-chain SS bonds (Shewry and Tatham, 1997); in most hypothetical gluten models they form the backbone of gluten aggregates by linear inter-chain disulfide bonds between the N- and C-terminal domains (divisions A and C). Furthermore, it has been suggested that LMW subunits branch from the HMW subunit polymers also by disulfide bonds (Graveland *et al.*, 1985). Various studies have indicated a close relationship between the presence and amounts of specific HMW subunits and dough strength or breadmaking quality. Subunits 5 and 10 are often present in varieties with strong dough and high baking quality and subunits 2 and 12 are associated with dough weakness and poor baking quality. Subunits 5 and 10 are better able to polymerise into very large macromolecules (MacRitchie, 1992).

During the whole process of breadmaking, the state of the formation of large glutenin aggregates is characterised by three competitive reactions: (1) the

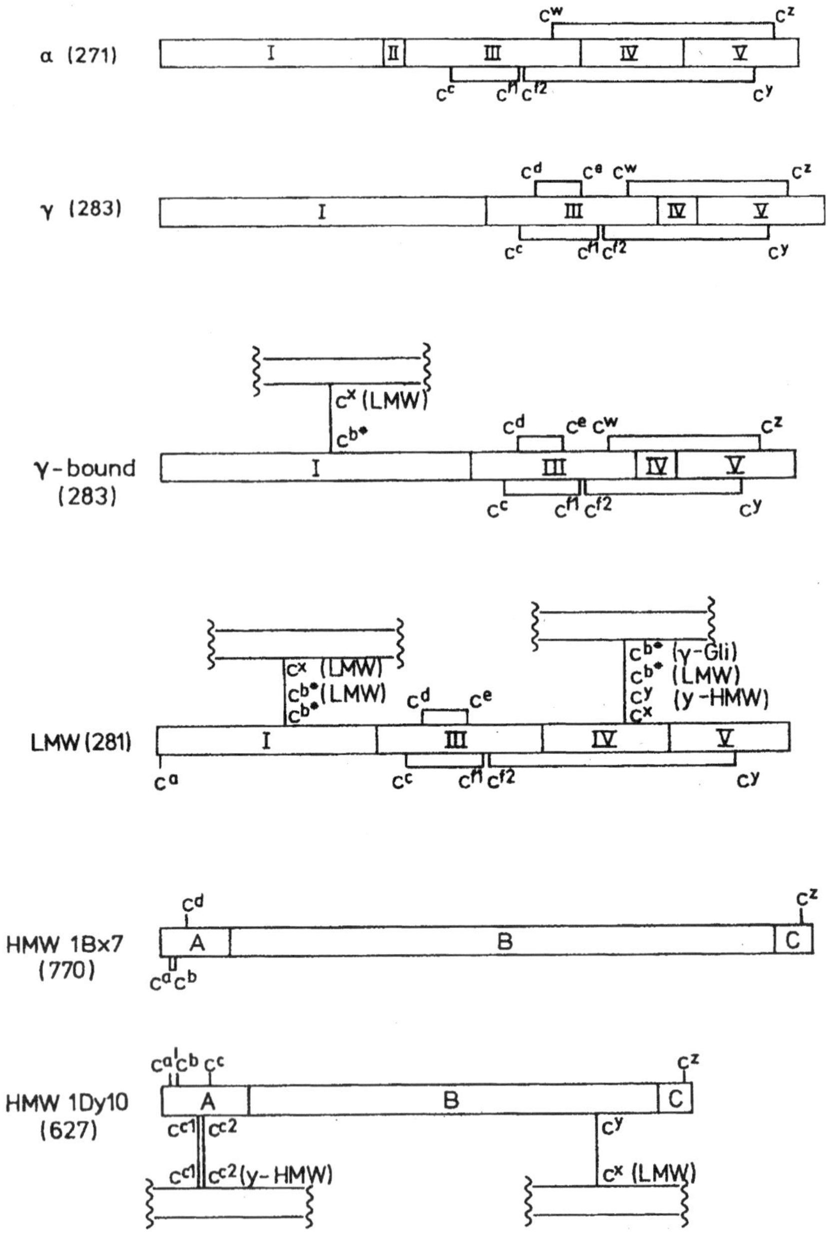

Fig. 20.1 Disulfide structures of gluten proteins (nomenclature of cysteine residues according to Koehler *et al.*, 1993).

oxidation of free SH groups which support the formation of large aggregates; (2) the presence of 'terminators' (LMW thiol compounds such as glutathione, gliadins with an odd number of cysteines) that stop polymerisation of glutenins; and (3) SH/SS interchange reactions between glutenins and LMW thiol compounds that depolymerise the aggregates. The overriding importance of SS bonds can be readily demonstrated by the addition of reducing agents weakening dough and of SH-blocking and oxidising agents strengthening dough. Some of the inter-chain SS bonds have been shown to be particularly important in maintaining the gluten structure: a reduction of only 3-4% of the SS bonds with mercaptoethanol causes a depolymerisation of 80% of the HMW gluten proteins (Seeger and Belitz, 1981). About 3–4% of total cysteines in wheat flour is present in form of protein-bound free SH groups. By means of derivatisation with a fluorescent agent several α- and γ-gliadins and LMW subunits were found to carry free SH groups (Antes and Wieser, 2000). Their positions within the amino acid sequences are not distributed randomly, but are strongly directed to specific cysteine residues which have been identified as being C^z of α- and γ-gliadins, C^{b*} and C^c of γ-gliadins and C^x of LMW subunits (Fig. 20.1). The location of SH groups does neither depend on the cultivars investigated nor on the presence of oxygen during milling and storage of flour.

20.2.3 LMW thiol compounds

In the wheat kernel LMW thiols are mainly located in the germ and aleurone cells, but occur also in white flour. Quantification of LMW thiols and their reaction products in flour and dough is difficult due to their relatively low concentrations. However, sensitive and reliable methods based on RP-HPLC (reversed phase-high-pressure liquid chromatography) of derivatised compounds are now available which allow the determination of both amounts and reaction mechanisms (Sarwin *et al.*, 1992; Schofield and Chen, 1995). The LMW compounds represent only a small proportion ($\approx$10%) of total free SH groups in flour, the major portion ($\approx$90%) is present in proteins. Nevertheless, LMW thiols diffuse more rapidly in dough than proteins and, therefore, are expected to be chemically more reactive as well as rheologically more effective (Bloksma and Bushuk, 1988). Among LMW thiols glutathione (GSH) is the major component, the concentrations of others such as cysteine and the dipeptides derived from GSH (Glu-Cys and Cys-Gly) are much lower. For example, 100 nmol GSH, 17 nmol Glu-Cys, 5 nmol Cys-Gly and 13 nmol CSH have been determined in 1 g of a Dark Northern Spring flour (0.78% ash) (Hahn *et al.*, 2000). Apart from free GSH the oxidised form GSSG and the protein-bound form GSSP exist. The proportions of these three forms can vary over a wide range and are strongly dependent on the wheat cultivar. The analysis of 11 freshly milled flours resulted in 18–81 nmol GSH, 12–27 nmol GSSG and 70–150 nmol PSSG per g flour (Chen and Schofield, 1995). The levels of free GSH and total GS in flour increase with the ash content, which corresponds to the extraction grade of the flours (Table 20.2) (Sarwin *et al.*, 1992). The content of

Table 20.2 Levels of GSH and total GS in wheat flours in relation to the ash content (Sarwin *et al.*, 1992)

Wheat-class/cultivar	Ash (%)	GSH (nmol/g flour)	Total GS (nmol/g flour)
CWRS	0.54	16	172
	0.71	35	348
	1.44	60	575
DNS	0.59	41	175
	0.78	100	345
	1.57	215	657
Maris Huntsman	0.55	20	185
	0.68	94	273
	1.73	210	435

LMW thiols is affected by the presence of oxygen during milling. It has been shown that the GSH content of flour milled under nitrogen is 50% higher than that of flour milled under air. During the storage of flour, LMW thiol content is significantly lowered; the rate of decrease depends on the wheat cultivar and storage temperature. While the loss of GSH in one flour was only 7% in 60 days at 20 °C, the changes in another flour were more dramatic as GSH fell to 57% of the initial value during the first 10 days then remaining constant to 40 days of storage (Chen and Schofield, 1996). The fall in GSH content has been accompanied by an increase in loaf volume. During dough mixing a rapid oxidation of GSH occurs; about 45% of the initial GSH has been found to be depleted in the first minute (Mair and Grosch, 1979). The reaction proceeds and after a 10 minute rest, no GSH could be detected.

Numerous studies have demonstrated the large weakening effect of LMW thiols on dough; examples are given in Table 20.3. This effect has been interpreted by SH/SS interchange reactions with gluten proteins (PSSP) combined with the depolymerisation of gluten polymers and the formation of protein/GSH mixed disulfides (PSSG) (equation 20.2)

$$\text{PSSP} + \text{GSH} \longrightarrow \text{PSH} + \text{PSSG} \qquad (20.2)$$

GSSG has also been demonstrated to weaken dough, but to a lesser extent than GSH. It has been proposed that GSSG undergoes a SH/SS interchange reaction with protein SH groups (PSH) to form PSSG and GSH the latter depolymerising glutenins according to equation 20.2 (Grosch and Wieser, 1999). Based on analytical data it can be estimated that about 9000 nmol of PSSP of gluten proteins are present compared with about 50 nmol of GSH. This extreme disproportion of reactants suggests that the attack of GSH on gluten proteins is strongly directed, otherwise an effect of GSH on dough properties could scarcely be expected. For the identification of specific binding sites of GSH in gluten polymers, wheat flour was mixed with water containing ^{35}S-labelled GSH as tracer (Hüttner and Wieser, 2001a). The distribution of radioactivity in the gliadin and glutenin fractions obtained from dough indicated that the major

Table 20.3 Influence of increasing amounts of added GSH and CSH on dough rheology (Kieffer *et al.*, 1990)[a]

Thiol (μmol/g)	*R*	*E*
GSH		
0.5	43	104
1.0	35	106
2.5	26	112
5.0	15	143
CSH		
0.5	46	110
1.0	40	106
2.5	26	119
5.0	18	124

[a] Data from extensigrams; *R* = maximum resistance to extension (mm), *E* = extensibility (mm).

portion of GSH is bound to glutenins. Isolation and identification of radioactive peptides from digested glutenin demonstrated that 91% of GSH were linked to those cysteine residues of LMW subunits that were found to form intermolecular protein/protein SS bonds (C^{b*} and C^{x} in Fig. 20.1) (Hüttner and Wieser, 2001b). Because these SS bonds have been proposed to be responsible for the aggregative nature of LMW subunits, the results obtained are in accordance with the effect of GSH on the rheological properties of dough. Oxidants such as potassium bromate, azodicarbonamide and dehydro-ascorbic acid, the oxidation product of ascorbic acid, have been proposed to act as bread improvers through oxidation of GSH to GSSG.

20.3 Redox reactions during processing

20.3.1 Flour maturing

Flour is the most important and basic ingredient in breadmaking. Freshly milled flour is sometimes referred to as 'green' flour. This implies that some sort of ripening or maturing of flour must take place before it makes a strong dough with good volume potential in a bakery. Maturing occurs naturally if the flour is stored under ambient conditions for one to two months. During this time some oxidation and hydrolysis of the flour lipids occur and the concentration of free SH groups decreases. For example, the GSH level of freshly milled flour (cv. Mercia) fell from 149 to 85 nmol/g during the first 10 days of storage (Chen and Schofield, 1996). However, it is uneconomical today to store flour until natural maturation. Frequently, flour is less than a week old when it is moved out of the mill to the baker. For these reasons oxidants are applied directly to the flour at the mill to enhance maturation. Where the laws permits it, azodicarbonamide and chlorine dioxide are the main oxidants in use (Stauffer, 1990). The main reaction appears to be the oxidation of free SH groups to SS moieties.

20.3.2 Dough mixing

The mixing of dough has several functions. First, the ingredients – flour, water, yeast, salts and other additives – are transformed into a homogeneous mass and the system appears to become less wet and sticky. Second, gluten proteins become hydrated, and are organised in such a way that they impart to the dough cohesive viscoelastic properties and the desired gas retention. And third, air is occluded, forming the nuclei of the gas cells that grow in size during fermentation. Dough development during mixing is the most critical step in the conversion of flour into bread. Optimal water absorption and mixing time are major factors for dough development and are mainly influenced by the genotype of wheat, type of flour, formula, temperature and type and speed of mixer. To produce optimum bread quality, dough must be mixed to the maximum resistance to extension (dough development time). The rheological properties of the developed dough (cohesivity, viscosity, elasticity) are important for breadmaking; they determine the behaviour of dough during mechanical handling (e.g. dividing, rounding, moulding) and they affect the quality of the end product (loaf volume, crumb structure).

It is commonly accepted that gluten proteins play a key role for dough development and properties. Studies on the relationship between rheological dough properties (development time, maximum resistance to extension, extensibility) and the quantities of single gluten protein types have revealed that development time is highly correlated with the x-type HMW subunits and total HMW subunit content (Wieser and Kieffer, 2001). The maximum resistance is strongly dependent on both quantity of total glutenins and the ratio of gliadins to glutenins, and extensibility is mainly influenced by the ratio of gliadins to glutenins. The correlations of LMW and HMW subunits are similar, but if equal amounts of protein are considered the effect of HMW subunits is twofold. The contribution of x-type HMW subunits is much higher than that of y-type subunits. The different roles of gliadins and glutenins and the importance of HMW subunits for the rheological properties are supported by a series of addition and incorporation experiments (e.g. Bekes *et al.*, 1994a; Schropp and Wieser, 1996; Antes and Wieser, 2001a). Molecular oxygen is known to be essential for optimal gluten development; mixing a dough in the presence of air results in gluten with stronger rheological characteristics than gluten from dough mixed under nitrogen (Dirndorfer *et al.*, 1986). The proportion of HMW protein aggregates is much higher in dough mixed under air and it has been postulated that additional SS cross-links between gluten proteins are formed. Differences between good and poor wheat cultivars are only evident when dough preparation is carried out under air.

During dough mixing intense SH/SS interchange reaction occurs. GSH decreases rapidly, whereas GSSG and PSSG increase (Grosch and Wieser, 1999). In contrast to GSH, CSH increases during mixing and it has been proposed that this is caused by an SH/SS interchange reaction of GSH with CSSC. In order to obtain optimum rheological dough characteristics, a certain amount of SH/SS interchange between LMW thiols and gluten proteins is thought to be necessary, owing to an optimal rearrangement of partially reduced gluten proteins (Bushuk, 1985). This procedure has also been used for the incorporation of glutenin subunits into dough by addition of small amounts of cysteine followed by fast oxidation with iodate (Bekes *et al.*, 1994b). However, too much interchange will lead to a gluten matrix that is weaker than desired for good loaf volume. Various authors have observed that the extractability of glutenin with solvents such as acetic acid and SDS solution increases during mixing (Weegels *et al.*, 1996). Different mechanisms have been proposed to explain this. It has been suggested that the size of protein aggregates decreases either by physical separation and/or by breaking of non-covalent or SS bonds. In an opposed view, the increase in extractability has been thought to be merely an effect of conformational rearrangement of polymers. In summary, the changes in extractability still require investigation. As mentioned above, dough should be mixed to maximum resistance. Continued mixing past the optimum leads to different results depending on the flour being mixed. Some flours resist overmixing and resistance remains with little apparent change. Others show a decrease of resistance and the dough becomes wet and sticky. This is generally spoken of as the dough being broken down or overmixed. The mechanism suggested is that during mixing the gluten proteins become stressed, a few SS bonds are broken and glutenins are partially depolymerised. α-, β-unsaturated carbonyl compounds, such as fumaric or ferulic acid have been shown to be involved in the reactions (Hoseney and Rogers, 1990).

Modifying the redox state of gluten proteins has very important effects on rheological dough properties. Such modification may occur naturally through reaction with atmospheric oxygen and endogenous SH compounds such as glutathione or by the addition of redox agents. For example, dough can be developed much faster at low mixing speed by adding minute quantities of a reducing agent such as cysteine; afterwards the dough can be stabilised by addition of a slow-acting oxidising agent such as potassium bromate (so-called chemical dough development or activated dough development). In processes employing mechanical dough development, bulk fermentation can be replaced by intense mechanical energy input to a dough (Cauvain, 2001). The inclusion of rather high amounts of oxidising agents (e.g. bromate plus L-ascorbic acid) led to the development of 'no-time' systems of breadmaking (e.g. the Chorleywood Bread Process).

20.3.3 Fermentation

Independent of the kind of fermentation (e.g. straight dough, sponge and dough or liquid ferment process), the objectives of fermentation are to bring the dough to an optimum condition for baking. As with most of the processing steps an optimum level of fermentation depends on many factors, which include flour strength, enzymatic activity of the flour, formulation, yeast activity and the type of product desired. Regular bread doughs contain 1.5–2.0% of compressed yeast on flour basis. Too little yeast results in slow fermentation, sticky dough, and poor crumb grain and texture; too much yeast results in an overly porous dough and a bread that stales rapidly (Pomeranz, 1987). During fermentation the yeast converts fermentable sugar into CO_2 and ethanol. Both dissolve in the aqueous phase of the dough. When the aqueous phase is saturated with CO_2, the gas evaporates from there to the gas cells. Ethanol, which is more soluble in water than is CO_2, hardly evaporates during fermentation. Yeast does not create new bubbles in a dough system, therefore air must be incorporated during mixing to provide pre-existing bubbles. The gas cells in the dough become larger, as more gas is produced and growth of gas cells changes from slow to rapid after *ca.* 25 min of fermentation. The viscoelastic dough will flow to equalise the pressure created by the additional gas and as a result the dough will expand. The oxygen in dough is rapidly consumed by the yeast as fermentation starts. Thereafter, the fermentation is anaerobic unless oxygen is added to the system (i.e. by remixing). As CO_2 is dissolved in the aqueous phase, the pH decreases from about 6.0 to 5.0.

During fermentation dough development is continued, that is the dough becomes drier, less sticky, and much more elastic, and gas retention is improved (Hoseney and Rogers, 1990). Fermentation with yeast causes the dough to go from having a large viscous-flow component to one that is elastic. This trend is the same that can be found when oxidants are added to dough; thus, yeast has clearly an oxidising effect. Yeast contains considerable amounts of GSH. In freshly compressed yeast, GSH is retained in the intact cells and only such minute amounts are liberated into the dough that do not influence dough rheology. Water extracts from dried yeast, however, have been shown to contain relatively high amounts of GSH (11–24 μmol per g dry mass) (Hahn *et al.*, 2000). Studies on dough rheology have demonstrated that the maximum resistance is strongly lowered by the addition of water extracts from dried yeast.

20.3.4 Baking

When the fermentation process is finished the dough is baked in an oven. Temperature increases in the interior up to 100 °C and in the crust above 100 °C. Important physical, physico-chemical and biochemical processes changes take place during baking in the crumb and in the crust (Bushuk, 1998; Weegels and Hamer, 1998). The dough expands by approximately 50% (‘oven spring’) due to a further production of CO_2, until yeast is inactivated at about 50 °C, and to the evaporation of water, CO_2 and ethanol. The foam structure in which the gas cells are self-contained is changed into a sponge structure with interconnected gas cells. The predominantly viscous dough is transformed into the elastic crumb and the crisp crust. When the reactive SH groups in the dough are removed in the early oven phase by a slow-acting oxidising agent such as potassium bromate, the SS bonds provide the required stability in the protein matrix until the loaf structure is ‘set’ by the gelatinisation of the starch and the thermal denaturation of the gluten proteins. Above 60 °C dough viscosity increases rapidly as a result of the gelatinisation and swelling of starch, and gluten proteins are denaturated involving changes in both non-covalent and SS bonds. The so-called browning reaction is characterised by both caramelisation of sugars and interaction between sugars and protein material (Maillard reactions) and imparts a deep colour to the crust.

The functional and chemical changes of gluten proteins during heat denaturation have been studied by numerous model experiments (Weegels and Hamer, 1998). Moderate heating of dough leads to increased resistance to extension and reduced extensibility (Belitz *et al.*, 1986). With increasing temperature, heating time and moisture content gluten proteins become less soluble in disaggregating solvents such as urea or SDS. When gluten is heated with a moisture content exceeding 20% the total SH content is decreased. In SDS extracted proteins from heated and unheated gluten, the SH content remains equal, whereas the ratio of SH groups to SS groups in SDS unextracted proteins decreased dramatically.

In baked bread the extractability of gliadins with 60% ethanol under non-reducing conditions is strongly reduced (Wieser, 1998). α- and γ-gliadins containing intra-chain SS bonds are much more affected than S-deficient ω-gliadins. The amount of proteins present in the insoluble glutenin fraction is strongly increased owing to covalently bound gliadins. Under reducing conditions and at increased temperature both gliadins and glutenin subunits are almost completely extractable and the protein patterns of RP-HPLC are similar to those of the corresponding flour fractions. The results suggest that most of the cross-links between gliadins and glutenins induced by heating are reducible SS bonds. The fact that the functionality of heat-denatured gluten can be restored, at least partially, by dough mixing in the presence of reducing agents, also supports the involvement of reactions involving SS bonds in the denaturing process (Stenvert *et al.*, 1981).

20.4 Redox agents: oxidants and reductants

Redox agents are frequently used by millers, bakers and baking industry to optimise dough and bread properties. The use of redox agents in baking confers technological advantages that are necessitated mainly by the demand for economic production in modern breadmaking processes. In former times bakers adapted the timing of mixing and fermentation as well as baking conditions to the variations in flour properties and yeast activity to obtain the desired quality of baked foods. In today's highly automated and controlled production process such a high degree of flexibility is not available, or is available only at an unacceptable cost. Therefore, additives are used to adjust flour characteristics to the process, to improve product quality and to reduce costs. Among additives, redox agents are an important group; their effect is mainly directed to the SH/SS systems of flour and dough. The following is concerned primarily with the oxidising and reducing agents of commercial importance for modifying gluten protein structure.

20.4.1 Effect of oxidants

The main effect of oxidants is twofold. Firstly, to accelerate flour maturing that formerly was obtained by ageing the flour during 1 to 2 months' storage. Some oxidants (e.g. acetone peroxide) also have a bleaching effect on flour by the oxidation of pigments. Secondly, oxidants can improve the rheological properties and the gas retention abilities of dough. They shorten mixing time, lower the energy input in dough mixing and the time required for the dough to mature and bread quality is significantly increased, with larger loaf volume and improved crumb structure. The use of oxidants is most permissive in the United States and very restrictive in the European Union, which permits only ascorbic acid. In countries, where the law permits it, the treatment with oxidants is an established practice. They can be added to flour at the mill or incorporated into a dough conditioner mix and then added at the bakery. The amount of oxidants used varies with the type of oxidant, flour and end product; usually it is about 10–70 mg per kg flour. With mechanical dough development, much larger amounts, 100 mg/kg and more, are used. The main reaction appears to be the oxidation of free SH groups to SS moieties to yield an increased number of cross-links between proteins and to minimise SH/SS interchange reactions. The gluten proteins and LMW thiols of flour are thought to be the main target for this oxidation. In consequence, the oxidation requirement (the amount of oxidant needed to produce the best loaf of bread) is related to total protein and gluten content and SH/SS ratio (Pomeranz, 1987). Because all of the reactions of oxidants with SH groups involve hydrogen (equation 20.1), the oxidative potential in aqueous systems such as dough will be pH dependent and increases as the pH of the dough decreases, e.g. during fermentation. In addition, certain other oxidative reactions may also contribute to the overall effect; oxidation of flour lipids and oxidative dimerisation of feruloyl residues attached to flour pentosans are two examples. Apart from gaseous oxygen, the most important oxidants are L-ascorbic acid (after oxidation to dehydryascorbic acid by atmospheric oxygen), potassium bromate, potassium iodate, azodicarbonamide and acetone peroxide. They can be applied as single additives or in combination; when more than one oxidant is used the results are sometimes difficult to predict beforehand. For convenience and accuracy of weighing the oxidants are made up in pre-weighed portions either as an ingredient of bread improvers (1–2%), as tablets or in water-soluble pouches.

20.4.2 Oxygen

Oxygen from the air affects flour during milling and storage and is mainly responsible for its natural maturing; the SH content of flour is significantly decreased during a storage of days. Atmospheric oxygen that is incorporated into dough during mixing affects dough properties considerably; changes can be observed immediately after mixing, the resistance to extension is increased and extensibility decreased. Oxygen also bleaches the dough. The disappearence of SH groups when flour is mixed in air or oxygen has been well established

(Stauffer, 1990). During mixing and resting in the absence of oxygen, the SH content of dough gradually increases or remains constant (Grosch, 1986). Oxygen mixed into the dough disappeared during processing and is probably consumed by yeast fermentation. Thus, a fermented dough is essentially an anaerobic system. Two other reactions of oxygen, which may influence dough properties, are the oxidation of flour lipids and the oxidative gelatinisation of water-soluble flour pentosans.

20.4.3 Halogenates

The relatively high oxygen availability of halogenates decisively contributes to their effect as oxidising agents. The halogenates used for breadmaking include iodate and bromate as their potassium salts. Potassium bromate was first recommended as a bread improver in 1916 to increase loaf volume and improve texture. Since then bromate and to a lesser degree iodate have been widely used and subjected to extensive research. The effects of both halogenates are based on the oxidation of free SH groups in dough increasing the strength of gluten. In contrast to ascorbic acid, both halogenates do not require molecular oxygen to exert their improving effect. The major difference between iodate and bromate is the fast action of iodate and the slow action of bromate. Under normal conditions, potassium bromate has essentially no effect during mixing, and about half of the added bromate is still present after mixing and a 4 h rest (Bushuk and Hlynka, 1960). Bromate affects dough rheology during fermentation and baking. Because bromate reacts faster at lower pH, its effect is accelerated during fermentation.

In contrast to bromate, iodate is completely consumed during mixing (Bloksma, 1964); consequently the improving effect of iodate is limited to the dough mixing process. Because of the higher oxidation potential ($IO_3^- = +1.09$ V, $BrO_3^- = +0.61$ V) and the faster action of iodate, fewer equivalents of iodate than bromate are required to bring the rheological properties of dough to the level that is the optimum for breadmaking. For bromate the optimum level of addition is in the range of 20–50 mg per kg flour and, for iodate, 10–20 mg per kg. Both halogenates are added either at the mill or later in improver formulations by the baker; bromate is also applied in combination with azodicarbonamide. The initial rate-limiting reduction of halogenate to halogenite by thiol groups proceeds rapidly for iodate and slowly for bromate (equation 20.3) (Tkachuk and Hlynka, 1961). Following this step, reduction of halogenites to halides and oxidation of SH to SS groups occurs as a fast stage (equation 20.4). The stoichiometry of the oxidation of thiol to disulfide by both halogenates is 6; several studies on dough, however, demonstrated that the extent of reactions varied and were less than stoichiometric (Stauffer, 1990). Characteristic differences between the effects of bromate and iodate have also been shown by re-oxidation experiments with isolated LMW and HMW subunits of glutenin (Schropp *et al.*, 1995; Antes and Wieser, 2001b). The results demonstrated that HMW subunits can be slowly re-oxidised with bromate to polymers with molecular weights up to several millions, whereas re-oxidation with iodate proceeded much faster and led to lower proportions of polymerised proteins. Obviously more intra- and fewer inter-molecular disulfide bonds were formed by re-oxidation with iodate compared with bromate. In contrast, LMW subunits formed similarly high proportions of polymers by re-oxidation with both bromate and iodate. Further studies on re-oxidised HMW subunits have demonstrated a significant dependence of polymer size on halogenate concentration (Veraverbeke *et al.*, 2000). At low concentrations, the size of polymers increase more by iodate than by bromate. However, high concentrations of iodate negatively affect the degree of polymerisation; a similar observation has not been made with bromate.

$$BrO_3^- + 2RSH \xrightarrow{slow} BrO_2^- + RSSR + H_2O \quad (20.3a)$$

$$IO_3^- + 2RSH \xrightarrow{fast} IO_2^- + RSSR + H_2O \quad (20.3b)$$

$$BrO_2^- + 4RSH \xrightarrow{fast} Br^- + 2RSSR + 2H_2O \quad (20.4a)$$

$$IO_2^- + 4RSH \xrightarrow{fast} I^- + 2RSSR + 2H_2O \quad (20.4b)$$

20.4.4 L-Ascorbic acid

The improver effect of L-ascorbic acid (AA) on the rheological properties of dough was firstly observed in 1935. It was found that small amounts of AA caused a pronounced increase in dough strength with the consequence that the loaf volume was up to 20% higher than that of the control. Since then AA has been widely used as bread dough improver, mainly in countries where bromate is not permitted. Clearly, AA is a most acceptable additive in bread since it is a vitamin. AA, although itself a reducing agent, exerts an effect similar to that of oxidants. Actually, the effective reagent is not AA, but its oxidation product, namely dehydro-L-ascorbic acid (DA). AA is oxidised in dough by atmospheric oxygen under the influence of both a heat-labile enzyme (ascorbic acid oxidase) and heat-stable catalysts (heavy metal ions). D-Ascorbic acid and a number of related compounds are oxidised in the same way. The rate of AA oxidation depends on the amount of oxygen mixed into the dough, e.g. it is significantly increased when the rotation of the mixer is increased.

The improving effect of DA is based on a rapid removal of endogenous GSH which otherwise would cause dough weakening by SH/SS interchange with gluten proteins (Grosch and Wieser, 1999). Addition of AA accelerates the loss of GSH in dough; for example, only 20% of the GSH level in flour has been found in dough, compared with 44% in the experiment without AA (Hahn *et al.*, 2000). This reaction proceeds at an appreciative speed only with L-threo-DA and dehydroreductic acid, but not with D- or erythro-compounds. The stereochemistry of the improver effect has suggested that an enzyme is involved in the reaction of L-threo-DA with GSH. This enzyme was identified as glutathione dehydrogenase, which occurs predominantly in the germ of wheat, but there are also appreciable activities in flours. The enzyme is specific for GSH and does not accept cysteine as an H-donor (Boeck and Grosch, 1976). It is less specific towards the H-acceptor, L-threo-DA and dehydroreductic acid are the best substrates followed by the two erythro-DAs, whereas D-threo-DA is the worst one (Walther and Grosch, 1987). The chemical basis of the improver effect of AA is summarised in the following equations:

$$2AA + 1/2O_2 \rightarrow 2DA + H_2O \quad (20.5)$$

$$DA + 2GSH \rightarrow AA + GSSG \quad (20.6)$$

First, AA is oxidised to DA by oxygen present in the dough (equation 20.5). The reaction is catalysed enzymatically and non-enzymatically and depends on the amount of oxygen mixed into dough. Thus, the rate of AA oxidation in dough is correlated with the speed of the mixer. In the subsequent reaction, endogenous GSH is withdrawn from the SH/SS interchange with gluten proteins which starts immediately on dough mixing (equation 20.6). Accordingly, the content of free GSH during dough mixing with AA decreases much faster than without AA.

AA is used as a bread improver, as its addition to dough causes an increase in loaf volume and an improvement in crumb structure. It belongs to the slow-acting oxidising improvers: the reaction rate is described as being quick during the initial reaction and then longer lasting because of its cyclical nature (equations 20.5 and 20.6) (Stauffer, 1990). The effectiveness is roughly equivalent to bromate on a weight basis, and the two oxidants when used together are synergistic. AA differs from other dough improvers by requiring exogenous oxygen for its effect. Unlike the effects of halogenates the effect of AA in a resting dough may be greatly reduced by the presence of yeast, which makes the system anaerobic. AA is typically applied in a range of 50–75 mg per kg flour. It is a characteristic feature of AA that the additional level is not critical and it is virtually impossible to over-treat doughs unlike the situation with other oxidants (e.g. azodicarbonamide, iodate).

20.4.5 Azodicarbonamide

The use of azodicarbonamide, $H_2N{-}CO{-}N{=}N{-}CO{-}NH_2$ (ADA) for maturing flour was patented in 1959. The optimum addition level for flour is between 10 and 20 mg per kg flour. ADA is one of the fastest oxidants used as a dough improver in breadmaking reacting within a few minutes after flour and water are mixed (Tsen, 1964). The action of ADA is similar to that of iodate, but even faster. It reduces the SH content of dough and thereby is converted into biurea (equation 20.7). The reaction scheme shows that one molar equivalent of ADA is required to oxidise two equivalents of thiols.

$$-N{=}N- + 2RSH \longrightarrow RSSR + -NH{-}NH- \quad (20.7)$$

Its action in dough is to form a cohesive, dry dough that can tolerate high water absorption. Mixing times are shortened even more than with an equal weight of iodate, energy input in dough mixing is lowered and dough properties are improved by increasing resistance to extension. Resulting breads have a better loaf with respect to texture and volume, in particular in combination with bromate. Over-treatment with ADA is characterised by grey, streaky crumb of poor volume resulting from a tight, extensible dough. This has to be countered in commercial improver mixer by blending with other oxidants and enzymes.

20.4.6 Peroxides

Hydrogen peroxide is presumed to be the active compound produced by dough-improving peroxides. The mechanism by which hydrogen peroxide has its effect is not known in detail. Acetone peroxide (AP) was introduced as a flour bleach and improver in 1961. AP is used on a carrier of starch containing mainly monomeric acetone (about 90%) with the remainder as the acyclic dimer and some residual hydrogen peroxides. AP is a very fast-acting oxidant and rapidly reduces the SH content of dough and increase dough resistance. In contrast to other improvers AP is active in the dry flour and completes its action within 24 h and has a good tolerance to over-treatment but poor storage properties (Fitchett and Frazier, 1986). To attain the same effect in dough, more equivalents of AP than of iodate or azodicarbonamide are required (Tsen, 1964). The typical dose is around 25 mg per kg flour. Calcium peroxide increases the water absorption in the dough and produces a dry, elastic dough with improved machining characteristics at the divider and moulder (Tiekelmann and Stehle, 1991). This material is often combined with enzyme-active soy flour. The recommended dose is in the range of 20–35 mg per kg.

The use of enzymes instead of chemical oxidants is an interesting option for improving the breadmaking performance of dough. Glucose oxidase found in a number of fungal sources is known to strengthen dough and to improve loaf volume and results in a drying of the dough (Vemulapalli *et al.*, 1998). It is expected that glucose oxidase will attain widespread use as a flour improving agent (Hamer, 1995). Replacing successful oxidants such as bromate will require combination with other enzymes, e.g. xylanase. Glucose oxidase catalyses the oxidation of glucose by consuming atmospheric oxygen and produce gluconic acid and hydrogen peroxide (equation 20.8). In contrast to calcium peroxide glucose oxidase is readily soluble in water and has been reported to have stability for at least one year when stored at 2–4 °C.

$$\beta\text{-D-glucose} + O_2 + H_2O \rightarrow \text{D-gluconic acid} + H_2O_2 \qquad (20.8)$$

20.4.7 Lipoxygenase

The addition of small amounts of enzyme-active flours made from soybeans or fava beans to wheat dough leads to bleaching of carotenoid pigments, increases the mixing tolerance, improves the dough handling properties and may increase the bread volume. It has been demonstrated that type II lipoxygenase is responsible for the improver action and the bleaching effect (Grosch, 1986). Lipoxygenase catalyses the oxidation of unsaturated fatty acids to their corresponding monohydroperoxides. In contrast to endogenous wheat flour lipoxygenase, soy and fava bean type II lipoxygenases release peroxy radicals which co-oxidise carotenoids and other flour constituents. The mechanism of lipoxygenase as a dough improver is still unexplained. The effect could only be demonstrated in the presence of lipid and air. It has been postulated that lipid radical intermediates formed by lipoxygenase combine with the activated double bond of water-soluble flour components (e.g. fumaric acid, ferulic acid) (Hoseney *et al.*, 1980). Accordingly these components are prevented from reacting with thiyl radicals of gluten proteins that are formed by the homolytic cleavage of SS bonds during mixing. The action of lipoxygenase can lead to undesirable flavours in bread; therefore, the amount of enzyme-active flours is restricted to approximately 1% flour weight.

20.4.8 Reductants

For economic reasons L-cysteine and sodium metabisulfite are the two reductants in practical use today; additionally a preparation of inactive dry yeast that contains a relatively high amount of GSH is on the market. They react with SS bonds in dough, breaking them with concominant reduction to SH groups. The overall effect is to reduce the average molecular weight of glutenin protein aggregates (Stauffer, 1990). The use of L-cysteine as a reductant was first patented in 1962. It decreases dough development time and dough stability and helps in moulding and shape forming, such as with rolls and baps, without structural damage. It increases loaf volume of strong cultivars of the North American types, since prior to baking the gas trapped within the dough can develop a more spongy dough. Furthermore, L-cysteine is used as a dough softener for biscuits. The first step of the reaction with gluten proteins is an SH/SS interchange freeing one protein unit and leaving an SS bond between protein and cysteine (equation (20.9a). This SS moiety may interact with another cysteine leaving the second protein SH group free and giving the oxidised form of cystine (equation 20.9b). L-Cysteine is used in the hydrochloride form which shows higher solubility in water and is applied in levels between 30 and 70 mg per kg flour. Over-treatment should be avoided as this produces a sticky dough which is difficult to handle (Fitchett and Frazier, 1986).

$$\mathrm{CSH + RSSR' \rightarrow RSH + CSSR'} \qquad (20.9a)$$

$$\mathrm{CSH + CSSR' \rightarrow R'SH + CSSC} \qquad (20.9b)$$

The effect of sodium metabisulfite is similar to that of L-cysteine leading to reduced water and energy requirements during mixing, increased extensibility of the dough, and better shape of the resulting product. Compared with L-cysteine the reaction, however, takes a slightly different course (Fitchett and Frazier, 1986). The metabisulfite ion ($S_2O_5^{2-}$) is hydrolysed by water to bisulfite

(HSO_3^-), which reacts with protein SS groups by interchange, freeing a SH group on one protein but leaving a thiolsulfate ester on the other protein (equations 20.10a and b). This ester is hydrolysed by water to give a free SH group on the protein and a sulfate ion (SO_4^{2-}) (equation 20.10c). Sodium metabisulfite is used as a biscuit/pastry dough softening agent in concentrations of about 200 mg per kg flour. It is cheaper than cysteine, but the latter has been found to be more effective and is more acceptable as a food additive, being a naturally occurring amino acid. In combination with oxidising improvers such as bromate and ascorbic acid, cysteine and sodium metabisulfite are used for the so-called activated dough development (ADD), as an alternative to high-speed mechanical dough development. The ADD process is especially useful for the manufacture of fruited breads, when excessive mechanical mixing would damage the texture and appearance of the fruit (Fitchett and Frazier, 1986).

$$S_2O_5^{2-} + H_2O \rightarrow 2HSO_3^- \quad (20.10a)$$

$$HSO_3^- + RSSR' \rightarrow RSSO_3^- + R'SH \quad (20.10b)$$

$$RSSO_3^- + H_2O \rightarrow RSH + SO_4^{--} + H+ \quad (20.10c)$$

20.5 Future trends

The current trend in the reduction of permitted redox agents in breadmaking has reached its limits in many parts of the world. In the countries of the European Union only ascorbic acid, a vitamin, and cysteine, a naturally occurring amino acid, are allowed. The USA still permits a number of agents such as potassium bromate, but they have an uncertain future because of questions concerning safety of use. The baking industry is faced worldwide with the major challenge of finding replacement of 'chemical' additives by more 'natural' additives. Enzymes appear to be promising candidates for solving these problems. For a long time the use of exogenous enzymes was limited to the addition of amylase in malt products for the adjustment of flour properties. Today, a wide range of enzymes produced specially for breadmaking is offered. They can be prepared from existing natural sources, such as plants and especially microorganisms, or they are the product of the fermentation of genetically modified organism. Glucose oxidase is an oxidising agent that has an effect similar to chemical oxidants such as bromate, especially in combination with xylanase. Some other peroxidases and lipoxygenase are of potential value in the replacement of chemical oxidisers and could be of great technological and commercial value. Transglutaminases catalyse the acyl-transfer reaction between lysine and glutamine residues of gluten proteins, thereby forming inter-chain cross-links. By addition of transglutaminase to flour dough properties and bread quality can be positively influenced similar to the actions of chemical oxidants. Proteinases catalyse the partial hydrolysis of gluten proteins forming weaker doughs; they are, therefore, proposed as a suitable alternative to reducing agents such as cysteine or metabisulfite.

In the future, the design of novel enzymes efficiently produced by microbial fermentation may allow optimal operations during the breadmaking process. However, because the mechanism of the action of single enzymes or of their combinations is not well understood, an integrated approach combining chemical analysis, rheological experiments and baking trials will be necessary to give answers to complicated questions.

20.6 Sources of further information and advice

ASH M and ASH I (2002) *Handbook of Food Additives* (2nd edition), Synapse Information Resources, New York, USA.

BLANSHARD JMV, FRAZIER PJ and GALLIARD T (1986) *Chemistry and Physics of Baking*, Royal Society of Chemistry, London, UK.

HAMER RJ and HOSENEY RC (1998) *Interactions: The Keys to Cereal Quality*, American Association of Cereal Chemists, St. Paul, MN, USA.

HOSENEY RC (1986) *Principles of Cereal Science and Technology*, American Association of Cereal Chemists, St. Paul, MN, USA.

KULP K and PONTE JG (2000) *Handbook of Cereal Science and Technology*, Marcel Dekker, NewYork.

OWENS G (2001) *Cereal Processing Technology*, Woodhead Publishing Limited, Cambridge, UK.

POMERANZ Y (1987) *Modern Cereal Science and Technology*, VCH Publishers, Weinheim, Germany.

POMERANZ Y (1988) *Wheat: Chemistry and Technology*, Vol. I and II, American Association of Cereal Chemists, St. Paul, MN, USA.

STAUFFER CE (1990) *Functional Additives for Bakery Foods*, AVI Book, Van Nostrand Reinhold, New York, USA.

20.7 References

ANTES S and WIESER H (2000), 'Quantitative determination and localisation of thiol groups in wheat flour', in Shewry PR and Tatham AS, *Wheat Gluten*, Cambridge, Royal Society of Chemistry, 211–214.

ANTES S and WIESER H (2001a), 'Effects of high and low molecular weight glutenin subunits on rheological dough properties and breadmaking quality of wheat', *Cereal Chem*, **78**, 157–159.

ANTES S and WIESER H (2001b), 'Reoxidation behavior of wheat and rye glutelin subunits', *Cereal Chem*, **78**, 8–13.

BEKES F, ANDERSON O, GRAS PW, GUPTA RB, TAM A, WRIGLEY CW and APPELS R (1994a), 'The contribution to mixing properties of 1D HMW glutenin subunits in a bacterial system' in Henry RJ and Ronalds RA, *Improvement of Cereal Quality by Genetic Engineering*, New York, USA, Plenum Publishing Corp., pp. 97–103.

BEKES F, GRAS PW and GUPTA RB (1994b), 'Mixing properties as a measure of reversible reduction and oxidation of doughs', *Cereal Chem*, **71**, 44–50.

BELITZ H-D, KIEFFER R, SEILMEIER W and WIESER H (1986), 'Structure and function of gluten proteins', *Cereal Chem*, **63**, 336–341.

BLOKSMA AH (1964), 'Oxidation by potassium iodate of thiol groups in unleavened wheat flour doughs', *J Sci Food Agric*, **15**, 83–94.

BLOKSMA AH and BUSHUK W (1988), 'Rheology and chemistry of dough', in Pomeranz Y, *Wheat: Chemistry and Technology*, Vol. II, St. Paul, MN, USA, American Association of Cereal Chemists, 131–217.

BOECK D and GROSCH W (1976), 'Glutathione – dehydrogenase of wheat flour. Purification and properties', *Z Lebensm Unters Forsch*, **162**, 243–251.

BUSHUK W (1985), 'Rheology: theory and application to wheat flour doughs', in Faridi H, *Rheology of Wheat Products*, St. Paul, MN, USA, American Association of Cereal Chemists, 1–26.

BUSHUK W (1998), 'Interactions in wheat dough', in Hamer RJ and Hoseney RC, *Interactions: The Key to Cereal Quality*, St. Paul, MN, USA, American Association of Cereal Chemists, 1–16.

BUSHUK W and HLYNKA I (1960), 'The bromate reaction in dough. I. Kinetic studies', *Cereal Chem*, **37**, 141–150.

CAUVAIN SP (2001), 'Breadmaking', in Owens G, *Cereal Processing Technology*, Cambridge, UK, Woodhead, 204–230.

CHEN X and SCHOFIELD JD (1995), 'Determination of protein-glutathione mixed disulfides in wheat flour', *J Agric Food Chem*, **43**, 2362–2368.

CHEN X and SCHOFIELD JD (1996), 'Changes in glutathione content and breadmaking performance of white wheat flour during short-term storage', *Cereal Chem*, **73**, 1–4.

DIRNDORFER M, KIEFFER R and BELITZ H-D (1986), 'Veränderungen an Kleber verschiedener Weizensorten durch Oxidation', *Z Lebensm Unters Forsch*, **183**, 33–38.

EWART JAD (1988), 'Thiols in flour and breadmaking quality', *Food Chem*, **28**, 207–218.

FITCHETT CS and FRAZIER PJ (1986), 'Action of oxidants and other improvers', in Blanshard JMV, Frazier PJ and Galliard T, *Chemistry and Physics of Baking*, London, UK, Royal Society of Chemistry, 179–198.

GRAVELAND A. BOSVELD P, LICHTENDONK WJ, MARSEILLE JP, MOONEN JHE and SCHEEPSTRA AA (1985), 'A model for the molecular structure of the glutenins from wheat flour', *J Cereal Sci*, **3**, 1–16.

GROSCH W (1986), 'Redox systems in dough', in Blanshard JMV, Frazier PJ and Galliard T, *Chemistry and Physics of Baking*, London, UK, Royal Society of Chemistry, 155–169.

GROSCH W and WIESER H (1999), 'Redox reactions in wheat dough as affected by ascorbic acid', *J Cereal Sci*, **29**, 1–16.

HAHN B, SARWIN R and GROSCH W (2000), 'Determination of low molecular weight thiols in wheat flours and doughs', in Schofield JD, *Wheat Structure, Biochemistry and Functionality*, Cambridge, UK, The Royal Society of Chemistry, 235–241.

HAMER RJ (1995), 'Enzymes in the baking industry', in Tucker GA and Woods LF, *Enzymes in Food Processing*, Glasgow, UK, Blackie Academic and Professional, 190–222.

HOSENEY RC and ROGERS DE (1990), 'The formation and properties of wheat flour doughs', *Crit Rev Food Sci Nutr*, **29**, 74–93.

HOSENEY RC, RAO H, FAUBION J and SIDHU JS (1980), 'Mixograph studies. IV. The mechanism by which lipoxygenase increases mixing tolerance', *Cereal Chem*, **57**, 163–166.

HÜTTNER S and WIESER H (2001a), 'Studies on the distribution and binding of endogenous glutathione in wheat dough and gluten. I. Distribution of glutathione in Osborne fractions', *Eur Food Res Technol*, **213**, 329–334.

HÜTTNER S and WIESER H (2001b), 'Studies on the distribution and binding of endogenous glutathione in wheat dough and gluten'. II. Binding sites of endogenous glutathione in gluten, *Eur Food Res Technol*, **213**, 460–464.

KIEFFER R, KIM J-J, WALTHER C, LASKAWY G and GROSCH W (1990), 'Influence of glutathione and cysteine on the improver effect of ascorbic acid stereoisomers', *J Cereal Sci*, **11**, 143–152.

KOEHLER P, BELITZ H-D and WIESER H (1993), 'Disulphide bonds in wheat gluten: further cystine peptides from high molecular weight (HMW) and low molecular weight (LMW) subunits of glutenin and from γ-gliadins', *Z Lebensm Unters Forsch*, **196**, 239–247.

MACRITCHIE F (1992), 'Physicochemical properties of wheat proteins in relation to functionality', *Adv Food Nutr Research*, **36**, 1–87.

MAIR G and GROSCH W (1979), 'Changes in glutathione content (reduced and oxidised form) and the effect of ascorbic acid and potassium bromate on glutathione oxidation during dough mixing', *J Sci Food Agric*, **30**, 914–920.

MÜLLER S and WIESER H (1997), 'The location of disulphide bonds in monomeric γ-type gliadins', *J Cereal Sci*, **26**, 169–176.

POMERANZ Y (1987), 'The art and science of bread making', in Pomeranz Y, *Modern Cereal Science and Technology*, Weinheim, Germany, VCH Publishers, 220–257.

POMERANZ Y (1988), 'Composition and functionality of wheat flour components', in Pomeranz Y, *Wheat: Chemistry and Technology*, Vol II, St. Paul, MN, USA, American Association of Cereal Chemists, 219–370.

SARWIN R, WALTHER C, LASKAWY G, BUTZ B and GROSCH W (1992), 'Determination of free reduced and total glutathione in wheat flour by an isotope dilution assay', *Z Lebensm Unters Forsch*, **195**, 27–32.

SCHOFIELD JD and CHEN X (1995), 'Analysis of free reduced and free oxidised glutathione in wheat flour', *J Cereal Sci*, **21**, 127–136.

SCHROPP P, BELITZ H-D, SEILMEIER W and WIESER H (1995), 'Reoxidation of high molecular weight subunits of glutenin', *Cereal Chem*, **72**, 406–410.

SCHROPP P and WIESER H (1996), 'Effects of high molecular weight subunits of glutenin on the rheological properties of wheat gluten', *Cereal Chem*, **73**, 410–413.

SEEGER R and BELITZ H-D (1981), 'Zusammenhänge zwischen der Reduktion von Disulfidbindungen und der Molekulargewichtsverteilung bei Weizenkleber', *Z Lebensm Unters Forsch*, **172**, 182–184.

SHEWRY PR and TATHAM AS (1997), 'Disulphide bonds in wheat gluten proteins', *J Cereal Sci*, **25**, 207–227.

STAUFFER CE (1990), *Functional Additives for Bakery Foods*, New York, USA, AVI Book, Van Nostrand Reinhold, 1–40.

STENVERT NL, MOSS R and MURRAY L (1981), 'The role of dry vital wheat gluten in breadmaking. Part I. Quality assessment and mixer interaction', *Bakers Digest*, **55**, 6–12.

SULLIVAN B, DAHLE LK and SCHIPKE JH (1963), 'The oxidation of wheat flour. IV. Labile and nonlabile sulfhydryl groups', *Cereal Chem*, **40**, 515–531.

TKACHUK R and HLYNKA I (1961), 'Some improving effects of halogenates and their reduction intermediates in dough', *Cereal Chem*, **38**, 394–398.

TIEKELMANN RF and STEHLE RE (1991), 'Higher-assay grade of calcium peroxide improves properties of dough', *Food Technol*, **45**, 106–112.

TSEN CC (1964), 'Comparative study of reactions of iodate, azodicarbonamide, and acetone peroxides in simple chemical systems and in dough', *Cereal Chem*, **42**, 86–97.

VEMULAPALLI V, MILLER KA and HOSENEY RC (1998), 'Glucose oxidase in breadmaking system', *Cereal Chem*, **75**, 439–442.

VERAVERBEKE WS, LARROQUE OR, BEKES F and DELCOUR JA (2000), '*In vitro* polymerization of wheat glutenin subunits with inorganic oxidizing agents. I. Comparison of single-step and stepwise oxidations of high molecular weight glutenin subunits'. *Cereal Chem*, **77**, 582–588.

WALTHER C and GROSCH W (1987), 'Substrate specifity of the glutathione dehydrogenase (dehydroascorbate reductase) from wheat flour', *J Cereal Sci*, **5**, 299–305.

WEEGELS PL and HAMER RJ (1998), 'Temperature-induced changes of wheat products', in Hamer RJ and Hoseney RC, *Interactions: The Key to Cereal Quality*, St. Paul, MN, USA, American Association of Cereal Chemists, 95–130.

WEEGELS PL, HAMER RJ and SCHOFIELD JD (1996), 'Functional properties of wheat glutenin', *J Cereal Sci*, **23**, 1–18.

WIESER H (1998), 'Investigations on the extractability of gluten proteins from wheat bread in comparison with flour', *Z Lebensm Unters Forsch*, **207**, 128–132.

WIESER H and KIEFFER R (2001), 'Correlations of the amount of gluten protein types to the technological properties of wheat flours determined on a micro-scale', *J Cereal Sci*, **34**, 19–27.

21

Water control in baking

S. P. Cauvain and L. S. Young, Campden and Chorleywood Food and Research Association, UK

21.1 Introduction: water composition and properties

Perhaps because of its abundance in nature (over 60% of the surface of our planet is covered with water and we encounter it as rain, snow, in streams, rivers and lakes), the special properties of water (Table 21.1) are often overlooked. A large number of chemical, physical and biochemical reactions in baking depend on the presence of water.

Chemically water comprises 2 atoms of the gas hydrogen combined with 1 atom of the gas oxygen to give the compound formula H_2O with a molecular weight of 18 (see Table 21.1). Even though pure water is a compound of two gases at temperatures above 0 and below 100 °C at standard pressure (1 bar) it exists as a liquid. Many compounds of a similar molecular size to that of water are gases at 20 °C. When the temperature falls below 0 °C pure water turns to ice and when the temperature rises above

Table 21.1 Properties of water

Freezing point	0 °C
Boiling point	100 °C
Specific heat of fusion (freezing)	336 kJ Kg^{-1}
Specific heat of vaporisation (boiling)	2268 kJ Kg^{-1}
Thermal conductivity (20 °C)	0.6 W m^{-1} K^{-1}
Expansivity (20 °C)	$2 \times 10^{-4} K^{-1}$
Molecular weight	18
Solubility of nitrogen (at 28 °C)	14 $cm^3 l^{-1}$
Solubility of oxygen (at 28 °C)	27 $cm^3 l^{-1}$
Solubility of carbon dioxide (at 28 °C)	650 $cm^3 l^{-1}$

100 °C it exists as steam. These transitions are very important in the manufacture of baked products since we may use temperatures below 0 °C as an aid to delay processing and we raise the temperature above 100 °C as part of the heat-setting (baking) of bread.

Because of its structure the electrostatic charges within the water molecule are not equally distributed. The oxygen nucleus has a positive charge of 8 while the hydrogen nuclei each only have positive charges of 1 so that in water there is migration of the negative charge from the hydrogen nuclei in the direction of the oxygen nucleus. The uneven electronic charge causes the water molecule to behave as a weak dipole which attracts other water molecules. A three-dimensional structure forms in water because of the electrostatic charges and this is the basis of the *hydrogen bond*. The existence of hydrogen bonding significantly contributes to the ability of water to take part in the reactions which are important in bread baking.

In the liquid state a few molecules are dissociated into ions. It is generally considered that hydrogen ions (H^+) do not exist in the 'free' state but are attached to water molecules forming hydroxonium ions, H_3O^+. This situation can be described as:

$$2H_2O = H_3O^+ + OH^-$$

The equation is reversible so that in water the numbers of molecules which dissociate equal the numbers re-associating. The hydroxonium ion can readily form ionic solutions, e.g. with sodium chloride.

It is only in water that the maximum potential for hydrogen bonding can be realised because of its equal and opposite pairs of positive and negative charges. As a consequence water has a much higher *specific heat capacity* in comparison with other substances. This means that water can absorb large amounts of heat by comparison with other liquids for the same rise in temperature. These large quantities of heat required to raise the temperature of water make a significant contribution to the design of the heating and cooling processes commonly used in the manufacture of bakery products.

Because of the presence of hydrogen bonding part of any heat transmitted to water is used to break inter-molecular bonds, leaving the remainder to increase the temperature by increasing the molecular kinetic energy. If we begin to heat pure water its molecular kinetic energy continues to increase and the temperature rises to 100 °C. At this temperature a large supply of energy is required to break the mass of hydrogen bonds present in the water in order to vaporise it and turn it to steam. The transition from water to steam at 100 °C requires considerable energy for no change in temperature and the heat required is described as *latent heat of vaporisation*.

At any given pressure, the presence of dissolved substances raises its boiling point above 100 °C. The attractive forces existing between ionic and polar compounds play their part in lowering the saturated vapour pressure in a solution. The reduction in vapour pressure depends on the proportion of solute and its nature. Solutes also affect the freezing point of the solution and depress the freezing point of pure water below 0 °C. In solutions that undergo cooling, when the freezing point is reached some of the water molecules form ice crystals and separate from the liquid. Thereafter during cooling the solute concentration in the solution increases until all of the water molecules have turned to ice and the ice crystals already formed become embedded in a matrix of much smaller crystals.

While many substances dissolve in water they cannot do so *ad infinitum*. Continued additions of the substance leads to a point when there will be undissolved substance and the solution is said to be *saturated*. The solubility of a substance increases as the temperature of the water increases and vice versa. In the manufacture of fermented products we are dealing with complex mixtures of many substances so that we cannot expect that the solubility processes are as straightforward as those described above.

Gases are only partially miscible with liquids and the mass of a gas that will dissolve in water depends on the nature of the gas, the temperature and the pressure of the gas in contact with water. In baking we are interested in the 'neutral' gases (nitrogen and oxygen) which do not dissociate to ions in solution. Such gases are only slightly soluble in water. An essential gas in the formation of the structure of baked products is carbon dioxide derived either from fermentation by yeast or chemical leavening agents. Unlike nitrogen or oxygen, carbon dioxide is readily soluble in water. The solubility of gases in water increases as the temperature falls but the solubility increase for carbon dioxide is far greater than for nitrogen or oxygen, which has significant implications for the production of retarded and frozen fermented dough (Cauvain 1998b).

In addition to the true solution we may encounter other important relationships between water and ingredients used in breadmaking, such as the following:

- *Suspensions* in which particles never ionise in the solvent. Some materials may form weak bonds with water molecules which considerably increase the length of time that the material will remain suspended. For example, when flour and water are shaken together the mixture will retain its 'milky' appearance for many hours because of hydration of the proteins and starch.
- *Colloidal suspensions* in which the particles are so finely divided and dispersed in the solvent that they are invisible to the naked eye. In a simple two-phase colloidal system one of the components is dispersed throughout the other in the form of small particles or droplets. Of particular interest in baking are the *hydrophobic* (water-fearing) and the *hydrophilic* (water-loving) suspensions. The colloidal particles in a hydrophobic colloidal system show no tendency to combine chemically with the dispersing medium, or to absorb it, but they may carry positive or negative electrical charges. When water is the dispersing medium the electrical charge originates from the adsorption of hydrogen ions (H^+), or may be attributed to hydroxyl ions (OH^-) adsorbing on to the colloidal particles. Which of these two ions is adsorbed depends on the nature of the colloidal particles.

- *Emulsions* are two-phased systems in which one (the *disperse*) phase is suspended as small droplets in the second (*continuous*) phase. The suspended particles are usually too small to be seen with the naked eye but are not always small enough to be considered as colloids. Following the act of dispersion there is a very large increase in the surface area between the two components of the system, and for the emulsion to be stable there must be a reduction in interfacial tension and an increase in adhesion. *Emulsifiers* work by forming a 'bridge' between the two phases; in effect they form a double surface, one for each component.
- *Gels* occur when a colloid sets to a semi-solid with no visible liquid present. The solid particles of the colloid join together to form a three-dimensional network trapping water in the spaces of the net. A three-dimensional network is formed in protein gels based on protein–protein (*aggregation*) and protein–solvent interactions. Large quantities of water can be held in the gel by as little as 1% solids even though the distance between the individual macromolecules can be very large. When protein gels are heated water is lost from the gel and the protein undergoes random aggregation, or *coagulation*. This particular process plays a major role in the formation of the (relatively) rigid structures which are formed during baking.
- Water is also a good solvent for substances that do not ionise but have a marked hydrogen bonding potential. These include *polar compounds* which contain hydroxyl (OH), carboxyl (COOH), carbonyl (C=O) and amino (NH_2) groups. All of these groups occur in the ingredients that are commonly used in the manufacture of bakery products and many of these play key roles in the formation of product structures and textures.

21.2 Hygrometry and water hardness

The molecules of a mass of water standing with its surface exposed in a closed container at a given temperature are in constant motion and a number of them will have sufficient energy to escape into the atmosphere above the liquid surface. In the still conditions of a closed container a similar number of molecules to those that have escaped return to the surface to rejoin the liquid and an equilibrium is reached. If water vapour molecules above the liquid are swept away by air movement the equilibrium of the system is disturbed. In such circumstances more water will evaporate from the liquid surface in order to try to restore the previous state of equilibrium and if this process continues for long enough then all of the water in the container will evaporate. Dehydration processes play significant roles in the processing of bread dough, e.g. proving, retarding and baking.

For any given set of conditions the water-vapour content of air can be measured. This measurement is referred to as the *humidity*, or *absolute humidity*, of the air and expressed as mass per unit volume. Under a given set of conditions a given volume of air can hold only a given mass of water vapour and the air is said to be *saturated*. Within a closed vessel where water is in equilibrium with its vapour then that part of the pressure that may be attributed to bombardment by the vapour molecules is known as the *saturated vapour pressure* (SVP). Saturated vapour pressure varies with temperature, with SVP rising as the temperature increases. *Relative humidity* (expressed as a percentage) is defined as the ratio of a mass of water vapour present in a given volume of air, to the mass that would be present if the sample were saturated, if the temperature and total pressure remain the same.

The rate at which moisture evaporates from products is affected by the relative humidity difference between products and the surrounding atmosphere. Commonly, moisture losses are minimised in bread production in order to prevent quality losses. For example, the RH of the atmosphere in the prover is maintained close to that of the dough product in order to limit evaporation and the formation of a hard, dry skin on the surface.

21.2.1 Water hardness

Rainwater from the more remote parts of the earth is normally the purest form of water. It contains in solution small amounts of oxygen, nitrogen and carbon dioxide absorbed from the atmosphere, along with traces of ammonium nitrate formed by electrical discharges in the atmosphere. Other dissolved gases may be present, for example oxides of sulfur, which are derived from the combustion products of fossil fuels at levels that depend on geographical location.

Rainwater as such is not commonly used in baking. More usually water is drawn from reservoirs and wells. Desalinated seawater may be used in some parts of the world. Rainwater will dissolve soluble rocks, such as chalk and calcium carbonate, and the presence of dissolved carbon dioxide makes the water slightly acid (a weak solution of carbonic acid is formed, bringing the pH below 7.0). Water runs off the land into reservoirs or percolates through the soil and rock layers to collect at the aquifer levels which supply wells. During its passage it continues to dissolve minerals, the quantity and type of which largely depend on the geographical location concerned.

The dissolved minerals confer special properties to the water usually recognised by its ability to form, or not, a lather when agitated with soap (Brownsell *et al*., 1989). Water which readily forms a lather is classified as *soft* and that which does not as *hard*. The relative hardness of water is expressed in *degrees of hardness*, one degree of hardness being the equivalent of 10 ppm calcium carbonate in water. Hardness in water comes mainly from calcium and magnesium bicarbonates and sulfates. The hardness coming from the bicarbonates can be removed by boiling and so is usually referred to as *temporary* while that arising from the sulfates cannot be removed by boiling and is therefore considered *permanent*.

21.3 The water absorption capacity of flour

In the manufacture of bread the production of dough with unvarying consistencies is desirable so that process conditions and product quality can be optimised. Achieving this depends to a large extent on being able to work with a known ratio of water to flour and other recipe ingredients. However, it is inevitable that variations occur in the water absorption capacity of flours and it is these changes that millers and bakers seek to measure and predict. The water absorption capacity of the flour is affected by the following:

- Moisture content – lower moisture gives a higher water absorption capacity and vice versa.
- Protein content – lower protein gives a lower water absorption capacity and vice versa.
- Flour grade colour, ash or bran level – higher values for all three properties give higher water absorption capacities and vice versa.
- Water-soluble proteins (pentosans) – higher values give higher water absorption capacity and vice versa.
- Damaged starch – higher values give higher water absorption capacity and vice versa.
- Enzymic activity (amylase) – greater activity gives lower water absorption capacity and vice versa.

There have been numerous attempts to develop mathematical models to predict water absorption from measured flour properties (e.g. Farrand, 1969; Dodds, 1971; Cauvain *et al.*, 1985) but it is still common practice to make a 'direct' measurement based on assessment of flour performance in some form of dough mixing test. The basis of such tests is to mix a dough and measure aspects of its rheology, either during mixing or afterwards. The 'correct' water absorption is identified when the dough meets a predefined rheological condition (e.g. resistance to deformation or viscosity) set by calibration with an expert assessor, the latter having judged the 'correct' dough consistency based on a sensory evaluation and experience in dough processing.

A common method for determining flour water absorption is based on the Brabender Farinograph (e.g. CCFRA FTP, 1991) which assesses a dough with known ratios of flour to water by recording the resistance encountered by the movement of the mixing blades during mixing. Initially there is little resistance from the mixture but as the proteins hydrate and form gluten the resistance increases. The results are recorded graphically and the operator is required to add sufficient water to reach a predetermined height on the graph. In the UK it has become the accepted practice to mix to a maximum viscosity on the 600 line of the Farinograph trace, though in other parts of the world the 500 line is still commonly used. Such differences in assessment technique arise from using different breadmaking processes and dough handling methods.

21.4 Dough formation

Lyophilic colloidal systems are characterised by the dispersed phase showing an affinity for the dispersing medium. In the case of water the colloidal particles are hydrated. The particles of a lyophilic colloid may carry electrical charges, though some pure lyophilic colloids are stable even when uncharged. The sign of a charge on a lyophilic colloid depends more on the environment than the nature of the colloid. Where the media is acidic, lyophilic colloids are often positively charged, and in alkali environments they are often negatively charged. Many hydrophillic colloidal systems occur naturally and the hydration of wheat proteins and starch during dough mixing are important examples. The glutenin and gliadin proteins in wheat flour combine with water to form the gluten protein network that is critical for the retention of air and carbon dioxide gas in the dough during breadmaking. As well as undergoing hydration it is necessary to impart energy to a flour–water mix in order to develop a gluten structure.

The gluten proteins present in wheat flour are embedded in the flour particles along with the other flour components, mainly starch granules. When an excess of water comes into contact with the flour particles there is a gradual uptake of water. The precise nature of the gluten protein–water interactions is complex and unclear. Bernadin and Kasarda (1973) observed flour particles under the microscope 'exploded' when brought into contact with excess water and strands of protein were rapidly expelled into the aqueous phase. Subjecting the mixture to linear stresses caused the protein strands to stretch. The rate of hydration of the protein strands proceeds more rapidly when there is an excess of water which, of course, is not the case in breadmaking, and so clearly this is not the sole mechanism for gluten formation. Stauffer (1998), in reviewing the principles of dough formation in breadmaking, described the protein in flour as existing 'as a flinty material' which softened during hydration. This description agrees with observations of Hoseney *et al.* (1986) who considered that as water is taken up by the wheat gluten proteins they pass through a glass transition stage changing from a hard glassy material to a soft rubbery one.

A proportion of starch granules present in flour are damaged and they absorb five times more water than undamaged granules. The absorption of water by starch is necessary for the *gelatinisation* process which occurs on heating. Damaged and gelatinising starch granules are susceptible to attack from alpha-amylase enzymes. Approximately 2–3% of the weight of flours comprises a mixture of water-soluble proteins and pentosans and they have a significant water absorbing capacity (Stauffer, 1998). The fibrous bran also absorbs water during mixing but more slowly than other flour components.

Water addition has an impact on the dispersion, dissolution and hydration of ingredients (see Table 21.2). It also plays a role in the transfer of energy to the dough and air incorporation during mixing. The rate of transfer of energy to the dough is related to the viscosity of the dough and energy is transferred when the viscosity is higher, that is less water has been added. There is little practical value in making stiff doughs for breadmaking because dough with sub-optimal

Table 21.2 Water movement in breadmaking

Process step	Action/contributes to
Mixing	dispersion, dissolution, hydration, adsorption competition for water between ingredients
Moulding	rheological changes
Proving	rheological changes, diffusion, condensation
Baking	diffusion, expansion, escape
Cooling	migration, escape
Shelf-life	migration in/out/through, spoilage (moulds), staling

water additions is suited to subsequent processing and frequently yields impaired final product qualities. The incorporation of air and the creation of small gas bubbles are critical to the development of bread cell structures (Cauvain, 1998a). The role that water plays in these aspects of the mixing process is largely that of facilitating the development of a suitable gluten network in the dough. Much of the expansion of the gas bubble structures in bread doughs occurs post-mixing, especially in the proof and baking stages, and depends on the gas retention properties of the dough (Cauvain, 1998a) which, in turn, depends on the dough development that has been achieved. A gluten structure with the 'correct' rheological properties is a critical element in the retention of gas and the expansion of fermented dough.

The control of final dough temperature is vitally important to consistent production and product quality and so it is common to control the final dough temperature at the end of mixing. In its simplest form control of the water temperature can be achieved by blending together quantities of hot and cold water until the required water temperature is achieved. Ice may be added to cool the dough-making water and in some cases it may be added directly to the mixer. The latent heat effect as ice converts to water can have a significant effect on final dough temperature. However, Campos *et al.* (1996) found that a uniform mixture of ice and flour which was warmed later gave a homogeneous but underdeveloped dough. This observation confirms the importance of controlling the hydration processes to 'optimum' dough development. A heat of hydration occurs with all flours (Wheelock and Lancaster, 1970). However, the effect of this heat of hydration on temperature rise is small compared with the heat which comes from energy transfer during mixing.

21.4.1 Water in brews and sponges

Some breadmaking processes include a pre-fermentation stage in which some of the recipe ingredients other than flour are mixed together and allowed to stand. If water-based they are known as 'brews'. Other processes have been evolved in which part of the flour and water used in dough-making are mixed together as a preliminary fermentation stage, commonly referred to as a 'sponge'. In these two-stage processes there is commonly a fermentation period before blending with the remaining dough ingredients at a second mixing stage (Cauvain, 1998a). The initial consistency of the sponge may be made somewhat softer by adding water at levels above the determined flour water absorption level. The lower dough viscosity facilitates a more rapid expansion of the dough during fermentation and increases the water activity of the sponge.

The control of the pH and TTA (total titratable acidity) are important in the sponge or brew so that the correct flavour is developed (Sutherland, 1989). This is because the optimum conditions for the activity of different microorganisms vary and if left uncontrolled then flavour development will also vary. In this context the effect of water hardness must be considered. As discussed above, the hardness of water varies according to geological and processing conditions and the presence of calcium carbonate in water will act as buffer and restrict the degree to which pH of the brew or sponge will fall during storage. This buffering effect may be so marked that it becomes necessary to use softened water or add a suitable acid to lower the pH.

21.4.2 Rheology, dough processing and water level

There is clearly an optimum water level for each of the flours used in bread dough-making. This is usually set according to the required dough viscosity and other rheological properties which yield the 'right' product qualities. Optimum dough water levels are set by bakers according to their ability to process the dough into its required sizes and shapes with a minimum of effort and damage to dough properties. If the dough has too little water it will be 'stiff' (i.e. have a high viscosity), and it will be difficult to change the shape of the dough during dividing, handling and moulding. In contrast, if the added water level is too high in the dough it will be 'soft' (i.e. have a low viscosity), and while it will be easier to change its shape during moulding, it will lose its shape after moulding and flow during proof. When the dough is held in a pan this may be acceptable but flow with freestanding breads is detrimental to final product quality.

During fermentation dough softens as the gluten network relaxes, and the longer the fermentation time, the softer the dough becomes. It is common practice to compensate for increased dough softening by reducing the level of water addition to the dough during mixing. In dough-making processes that do not have a fermentation period after mixing and before dividing, the dough would have a much firmer consistency and so it is common practice for levels of water addition with no-time dough to be higher than that seen with bulk fermentation processes. This extra water in the dough is required to ensure that dough consistencies are similar for all breadmaking processes by the time of dividing.

The application of a partial vacuum towards the end of dough mixing when using the Chorleywood Bread Process produces doughs that are firmer and drier to the touch (Cauvain, 1998a). This effect is usually compensated for in the bakery through the addition of extra water. The lower the pressure during mixing, the greater the quantity of extra water needed in order to adjust the

Table 21.3 Comparison of added water levels with different dough making processes (same flour)

Process	Water addition (% flour weight)
1 h bulk	57
4 h bulk	55
No-time, spiral mixer	58
CBP	60
CBP with partial vacuum	62
CBP with pressure	58

dough consistency. Typical levels of water addition to the same flour processed by different breadmaking processes are given in Table 21.3.

In dough processing the dividing and moulding operations subject the dough to a variety of stresses and strains depending on the equipment used. The major stresses occur in the moulding stages where the shape of individual pieces undergoes considerable change, and it is in these stages that the effects on differences in dough consistency are most often observed. Doughs that lack water are 'tight' and offer greater resistance to deformation. They exhibit greater elasticity and reduced extensibility and they have greater susceptibility to 'damage' during dividing and shaping. The dough surface will tear and rupture leading to quality losses (see Fig. 21.1). There are limited opportunities to

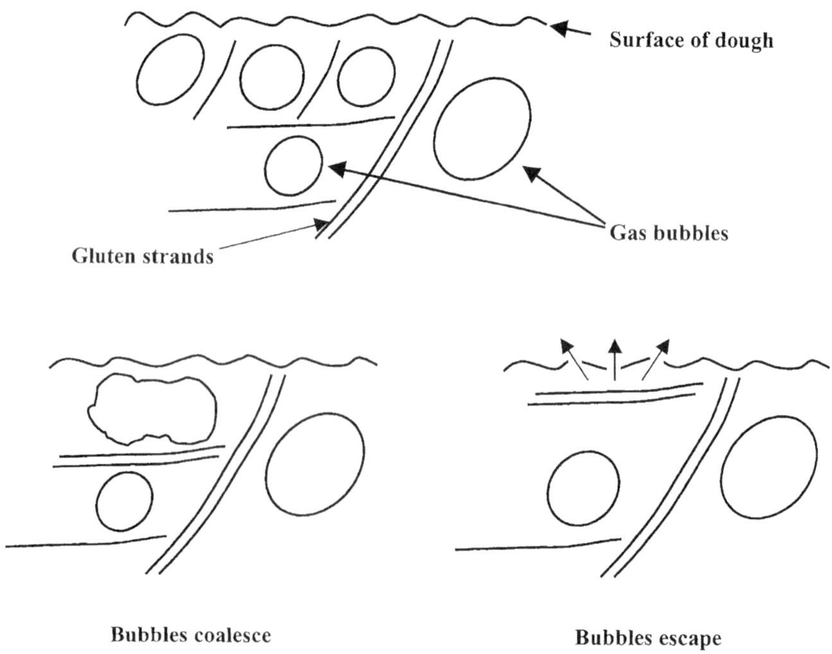

Fig. 21.1 Damage to gas bubble structures in bread dough during moulding.

compensate for such deficiencies in dough rheology during post-mixing processing operations and this increases the importance of judging flour water absorption capacity and optimising added water levels during mixing.

In practice the terms softness and stickiness are often confused. It is possible to have doughs that are soft but not sticky and vice versa, or both. Dough stickiness is most often observed when dough is subjected to shearing rather than compressive forces. Many opportunities for the shearing of dough occur in post-mixing processes, especially during dividing, rounding and moulding operations.

21.5 Proving and baking

After mixing and forming, dough is subject to proof and its structure is expanded to a suitable size before it goes into the oven. If a dough piece stands for any length of time it is likely to lose water from the exposed surface and dry out. If the moisture loss is excessive this may lead to the phenomenon that bakers refer to as 'skinning'. The expansion of the dough piece will be restricted and the bread will have an unacceptable crust character and lack volume. In addition to the effects of the relative humidity of the atmosphere surrounding the dough piece, the velocity of the air over the dough piece, the surface area of the dough piece and the water activity of the dough piece also influence skinning.

Most bread doughs contain few ingredients other than flour, water, salt and yeast. The proportion of water in the dough is such that the relative humidity of the solution within the dough is high, typically above 95%. The relative humidity within a bread dough will fall as the proportion of soluble materials in the formulation (usually sugars) increases but seldom falls below 90%. This being the case the dough relative humidity is almost invariably higher than that of the air in the bakery and so moisture is inclined to evaporate from the dough surface to the surrounding atmosphere. By placing the dough in a prover it is possible to control the temperature of the air in the cabinet and its relative humidity.

Successful proving demands that the relative humidity of the air in the cabinet should be close to but slightly lower than that of the dough. In practice this means around 80–85% relative humidity at about 40 °C. At the start of proving dough pieces gain weight and later they lose it. Wiggins (1998) reported that there was a nett weight loss from dough during proof, typically about 3 g from an 880 g dough piece.

A specialised dough-making process where water control is particularly important because it uses low temperatures for the prolonged dough storage is retarding. The process refers to the specialised refrigerated storage of unproved, yeasted dough pieces for between 16 and 72 h, to allow bakers to 'time-shift' their production (Cauvain, 1992). During the retarding period the activity of the yeast in the dough slows down a considerable extent and if the storage temperature falls low enough the dough may be held in what approximates to a state of suspended activity (Cauvain, 1998b).

The relative humidity of fermented dough pieces is around 90–95% and is much higher than the relative humidity of the air in a retarder at loading. Soon after entering the retarder the dough pieces lose moisture and continue to do so throughout the whole of the storage period. The longer the storage period, the greater the moisture loss. Unlike the situation in dough provers, it is not practical to introduce water vapour into a retarder to limit relative humidity differentials.

The greater the surface area of the dough piece during storage in the retarder, the greater the potential for weight loss through increased evaporation. This means that pieces with large surface areas lose water faster than smaller ones. The surface area is roughly related to the weight of the dough piece and its shape. Thus, a 450 g dough piece in a pan has a much smaller exposed surface area than the same weight of dough which has been formed into a baguette shape. Expansion of dough pieces during retarded storage occurs because of yeast fermentation and may continue to around −5 °C. With this increase in dough piece volume comes an increase in surface area and with it increased evaporation which may lead to greater skinning.

21.5.1 Contribution of steam to crust formation

Water as steam may be deliberately introduced into the oven chamber to raise the humidity and influence final product character. This use of steam adds to the product appearance through the formation of a *gloss* on the surface and influences the texture and eating qualities of the crust. When dough pieces are first placed in the oven their surface temperature is low, typically below 45 °C and with the introduction of steam there is some condensation on the surface of the piece. This excess of water combines with the starch and with the action of the enzymes present and encourages the formation of dextrins and sugars (mainly maltose). As the temperature begins to rise, the surface dries and the enzymic activity ceases but the dextrins and sugars remain and colour as baking proceeds. The dextrins are mainly responsible for the formation of the surface gloss. The whole process is very short because the crust temperature climbs rapidly above the boiling point of water.

In addition to changing the appearance of the product, steaming affects product crustiness; the thicker the gelatinised starch layer that is formed, the greater will be the cracking of the crust on the baked product. This cracking occurs when the product leaves the oven and begins to cool. The crumb of the product has a higher moisture content and is more flexible than the lower moisture content crust and so readily contracts as the temperature of the air within the cells falls. The crust, being less flexible, does not contract as fast or to the same degree as the crumb and the strains placed on the crust by attached portions of crumb become such that splits begin to occur in the surface.

Other key uses for steam during breadmaking include the following:

- The production of part-baked breads in which proved doughs are baked just sufficiently to inactivate the yeast and enzymes, and set the structure with a minimum of crust coloration and moisture loss.
- The production of rye bread doughs in which starch gelatinisation plays a much greater role in structure formation than in wheat breads.
- Chinese steam bread in which the product should have a white crust colour, smooth surface and shiny skin. Baking is carried out in an atmosphere of steam until the structure has become set. The high level of water which condenses on the surface of the heated dough pieces enourages starch gelatinisation but discourages the usual Maillard browning reactions.
- Bagels, which are distinguished by being formed into a ring shape and proved for a short time (typically 15 to 20 min) before they are immersed in boiling water or a boiling water/sugar solution for about 10 to 15 seconds. The boiling water treatment encourages the formation of a surface gloss.

21.5.2 Water and the foam to sponge conversion during baking

Baking is a process of heat gain and moisture loss. The transition from relatively fluid dough to the rigid structure in bread during baking – the foam to sponge conversion – requires the overall loss of water from the product and its redistribution, both on the macroscopic and microscopic scales within the structure. It is because of the loss of water that the dough matrix moves from a glassy to a rubbery state. Water plays a major role as a plasticiser in this context because of its low molecular weight and the transitions invite parallels with changes in synthetic polymers.

Along with the physicochemical changes occurring within the baking dough it is necessary to consider the expansion forces created by the gases which are present in the dough. These come from three sources:

- *Thermal gas expansion* – as the temperature in the dough increases trapped gas bubbles expand according to the gas laws.
- *Release of carbon dioxide* – provided by the yeast, the release of carbon dioxide occurs at relatively low temperatures (yeast is inactivated by 55 °C) but it continues to make a contribution to thermal expansion.
- *Water vapour* – as the temperature in the dough increases some of the water is turned to vapour and adds to the expansion of the matrix. The expansion contribution of the steam depends in part on the water level of the dough, where and how it is held in the matrix and the rate at which heat is input. Bloksma (1990) calculated values for the theoretical expansion of bread during baking and considered that 60% of the volume increase in the oven, that is the difference between proof volume and baked volume (oven spring), came from the steam which was generated.

In its unbaked form bread dough is a complex *foam* in which the gas bubbles (cells) are intact and discrete from one another (see Fig. 21.2). It is because they are intact that they are able to receive carbon dioxide gas from yeast

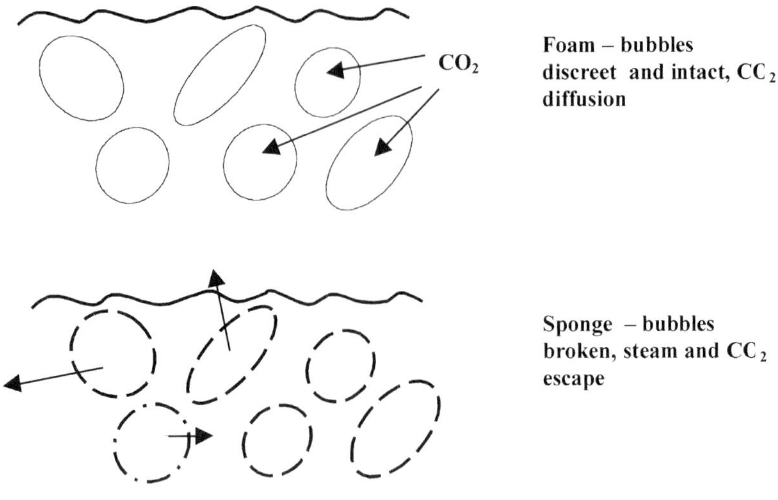

Fig. 21.2 Conversion of foam to sponge during baking.

fermentation which enables the dough to expand (Cauvain, 1998b). The 'surface' of the gas bubbles is stabilised by a combination of gluten and other ingredients, for example, fat and emulsifiers (Williams and Pullen, 1998). During baking the bubble-stabilising components in dough gradually lose control. Fat and emulsifiers melt by 65 °C and the internal bubble pressure becomes so great that bubbles begin to become buoyant but are restrained by the gluten network. The gluten network loses water to the gelatinising starch and loses much of its extensibility. Once the gluten protein structure sets there is no bubble stabilising mechanism left and the gases are free to escape from the baking dough. At this point the foam becomes a *sponge*, that is all of the cells are interconnected and vapours and liquids may freely move through the matrix.

The foam to sponge transition makes the matrix permeable to vapours and liquids. Because of the temperature gradient that is set up during bread baking, the surface of the dough piece reaches the sponge stage long before the centre and so therefore quickly becomes permeable. This permeable surface is crust and through it the water vapour generated during baking escapes. Wiggins (1998) described the heat transfer mechanism of baking dough in terms of evaporation front and heat pipe effects. Evaporation of moisture can occur only once the liquid boiling point has been reached. The low concentration of soluble materials in bread dough means that the boiling point is not much higher than 100 °C.

When dough pieces first enter the oven the water present is largely uniformly distributed throughout the matrix. The proteins and starch form the bulk of the water-absorbing materials and so most of the changes during baking are related to them. As the temperature of the dough rises the starch granules begin to swell (from about 45 °C) and later gelatinise (from about 60 °C). As the temperature increases the polymers in the starch granules vibrate. This breaks the intermolecular bonds and allows hydrogen bonding sites in the granules to engage more water. With more heat input there is a complete loss of crystallinity. To achieve this transition the starch granules need more water than is usually held in their mass and there is a loss of water from the gluten network to the gelatinising starch. At this time the gelatinised starch takes over the structure supporting role from the gluten network in the dough. There is also an increase in enzymic activity and amylase enzymes attack the starch granules, and, in doing so, increase their ability to absorb water. Further action by the amylases breaks down the starch granules and ultimately releases water within the dough matrix.

During baking the rheological character of bread doughs undergoes profound change between 55 and 75 °C (Dreese *et al.*, 1988). This temperature range encompasses the starch gelatinisation at the lower end and protein coagulation at the upper end. The viscosity of the dough at this stage increases by many orders of magnitude; however, the product is not 'baked' at this point since if it is removed from the oven it cannot support itself and will collapse under the influence of gravity. Practical observations have shown that it is necessary to continue heating many types of bread products to achieve a core temperature of between 92 and 96 °C before a sufficiently rigid 'loaf' structure is formed.

21.5.3 Moisture movement during cooling

During cooling there are moisture losses as water evaporates from the product surface because of the temperature differential between the still warm product and the colder air to which it is typically exposed. Variations in air flow rates and temperatures across the product affect the rate at which it cools and loses moisture. As might be expected moisture losses are higher for higher air flow rates. Wiggins (1998) suggested a 20–25 g weight loss for typical UK plant bread during cooling. The humidity of the air used in the cooler may be increased in an attempt to reduce moisture losses from the product. The movement of water throughout the whole of the breadmaking process is summarised in Table 21.2.

Freezing is applied to many bakery products, both unbaked and baked, in order to retain their organoleptic properties for extended periods of time. As the temperature in the product matrix falls below 0 °C the water that is present begins to freeze and forms ice crystals. Bakery products are complex systems in which the starch and protein polymers and low-molecular-weight solutes (e.g. sodium chloride) are dispersed or dissolved in the aqueous phase. Freezing therefore takes place under non-equilibrium conditions so that the normal phase change from liquid to solid is not observed in the product. Thus, the behaviour of the aqueous phase during freezing, and thawing, is significantly affected by the product formulation and the degree to which the water is 'bound' within the product matrix. The term 'bound water' has been used to refer to water in a product that does not freeze when the matrix temperature falls below 0 °C. Referring to water as bound indicates that it is loosely held and may diffuse out of the matrix, usually over relatively long periods of time. This '*non-freezable*' water remains relatively mobile compared with the mobility of water within ice.

21.6 Water activity and the shelf-life of bread

The principles of water activity (Cauvain and Young, 2000) and glass transition are used to explain the role of water in determining the quality of baked products. In the rubbery state the polymer is more mobile than in the glassy state, where the viscosity of the system is much higher and the mobility of all molecules is severely restricted. The quality of bread owes much to the special properties of the proteins and starch that are present in wheat flour. In the key transformations that occur in the baked loaf the plasticising and solvent roles of water are critical. The organolepetic qualities of all baked products change during storage and many of the changes result in a loss of product quality commonly embraced in the general term of '*staling*' (Pateras, 1998; Cauvain and Young, 2000). These changes are most often linked with the movement of water both within and out of the product matrix.

Bread falls into the intermediate moisture range of foods. It has the highest water levels of virtually all baked products and is characterised by a relatively high moisture content in the baked crumb and a lower moisture content in the crust. Water plays a major 'lubricating' role when the product is eaten and because of this the product moisture content has a profound effect on the perception of quality, whether the product has been freshly made or stored. Within limits, the higher the moisture content, the fresher the bread will be perceived by the consumer. The influence of a higher crumb moisture content is seen when bread crumb is compressed with the fingers: the higher the moisture content the easier it will be to deform the crumb and the softer it feels. Too much water and the crumb may be easily deformed but it may not recover to the shape that it was before compression. This combination of easy compression with good recovery is commonly assessed by the 'squeeze' test frequently carried out by consumers at the point of purchase, especially when the bread is cold on the store shelf and the direct link between product warmth and freshness has been lost.

When baking is finished the crusts of all breads have lower moisture contents than those of their respective crumbs. This lower moisture content makes a significant contribution to the crisp eating character of the crust. Gradually, depending on the storage conditions, the moisture held within the crumb migrates to the areas of lower moisture content closer to and at the crust surface, causing the latter to lose its crisp eating character and become soft. Such effects are most readily observed in hearth breads and baguette, where the softening of the crust detracts from the product character as the formerly crisp eating crust assumes a 'chewy' character. Pan breads, on the other hand, actually benefit from this moisture migration phenomenon during storage since a crisp crust is largely undesirable with such products when they are sliced or eaten.

21.6.1 The impact of water activity on microbial shelf-life

Water is the most 'mobile' of all the ingredients that go to make up the finished baked product. Consequently it is involved in many of the factors that affect the quality of the finished product, not least its microbial shelf-life. Most microbes are killed by the baking process. However, once the product has left the oven it falls prey to contamination from the atmosphere, from handling operations, such as cooling, finishing and wrapping, and all contribute to the microbial load. In many of the spoilage mechanisms that concern bakery foods, individual cells must feed before they can begin the process of multiplication that eventually leads to product spoilage. If the water in the product is 'held' by the ingredients then it is not available for use by the moulds. Water activity is a measure of a product's ability to 'lock up' moisture, thus restricting its availability to moulds.

Moisture in different ingredients is held in different ways with different strengths of attraction between the water and the other molecules constituting the ingredient. In bread products, with their high moisture contents and their lack of materials that will bind water, e.g. sugar and salt, the water activity is typically in the range 90–98%. The availability of water in a product is a powerful controlling mechanism on the potential for microbial growth and is often used as a predictor of which organisms may flourish on a given food (Cauvain and Young, 2000) and how quickly a product can be spoiled by such microbial activity.

There is a direct relationship between the water activity (or Equilibrium Relative Humidity, ERH) of a product and the time for which it will remain free of mould growth. This relationship was established in 1968 by scientists working at the Flour Milling and Baking Research Association, Chorleywood (Cooper *et al.*, 1968), using a representative selection of baked products stored at various temperatures. They observed the moment in time at which mould colonies first became visible. This relationship proved robust and later the storage temperature range was extended (Cauvain and Seiler, 1992). Figure 21.3 illustrates the relationship between ERH and mould-free shelf-life (MFSL) for a range of storage temperatures. In simple terms, the lower the storage temperature for a product of given ERH, the longer will be its MFSL. The lower the ERH, the longer the product resists attack by microbes. Consequently if a product ERH can be lowered, its MFSL can be extended. For products with ERHs above 93%, e.g. most breads, reducing the ERH by a relatively small amount, say 1%, causes very little change in MFSL. However, in the range 85 to 70% reducing ERH by 1% causes a much greater increase in shelf-life. In practical terms this means that there is little opportunity for reducing the ERH of bread products without significantly changing their character, e.g. making them sweeter by sugar addition.

In bread, moulds are the most common form of microbial spoilage. These cause the production of colonies of coloured mycelia and sporangia. The colours range from the pale green of *Aspergillus glaucus* through the pinks of *Neurospora sitophila* and the black of *Aspergillus niger* (Pateras, 1998). One bacterium that is not killed by the baking process is *Bacillus subtilis*. In warm humid weather conditions are ripe for the spores to germinate and to grow, forming the characteristic stringy brown mass with the odour of ripe pineapple.

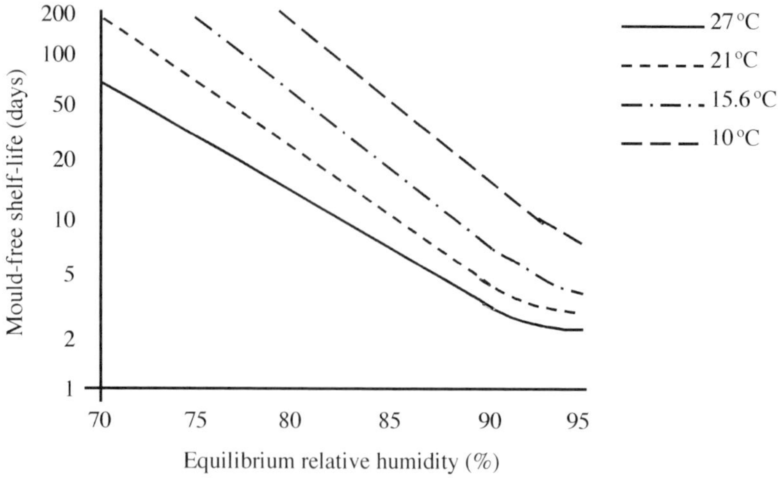

Fig. 21.3 Relationship between product ERH and MFSL (Source: Cauvain amd Seiler, 1992).

As the spoilage continues the crumb is degraded and can be stretched into the long silky threads typical of 'rope'. This typically occurs when the ERH is above 95%.

There are other means of extending MFSL. In some instances a wrapping material that allows some loss of water from the product may be used to control microbial growth or maintain product eating quality. An example of such an effect is the use of perforated wrapping films with crusty products to assist in keeping the crisp eating quality of the crust for as long as possible. Preservatives such as calcium propionate are often included in formulations for shelf-life extensions and occasionally surface spraying with potassium sorbate can give additional days of MFSL; see Table 21.4. However, levels of preservative need to be declared on the label and their presence can often be detected by smell. Modified atmosphere packaging (MAP) with carbon dioxide can also be employed with the advantage that it can increase the shelf-life without affecting the bread's aroma, flavour or appearance.

21.7 Future trends

The fundamental properties of water are hardly likely to undergo any significant change; however, we still have much to learn about the role that water plays in the dough-making process. As the nature of the main raw material, the flour, changes, so do the interactions between it and the water. This is probably most true for the hydration and foam to sponge conversion processes. In both cases the mobility of water is a significant determinant in the changes that occur. A greater understanding of the initial hydration process and the contribution of

Table 21.4 Anti-mould agents commonly used for bread

Preservative	Recommended level of use (% flour weight)
Propionic acid	0.1
Calcium propionate	0.2
Sodium propionate	0.2
Sodium dispropionate	0.2

water to dough rheological changes will have profound effects on the design and operation of bakery plant and equipment. While water is perhaps the least studied of bread ingredients it must be given equal importance to that of any other ingredient if bread quality is to continue to improve.

21.8 Sources of further information and advice

There are a number of books that discuss the general properties of water in foods. Useful general texts include *Water and Food Quality* (1989) Ed. T.H. Hardman. Elsevier Science Publishers Ltd., Barking, UK.

The International Symposium on the Properties of Water (ISOPOW) series provide a wide range of fundamental and applied texts, e.g. *The Properties of Water in Food, ISOPW 6* (1998) Ed. D.S. Reid. Blackie Academic & Professional, London, UK.

A specific bakery-related text is *Bakery Food Manufacture & Quality: Water Control & Effects* (2000) S.P. Cauvain and L.S. Young. Blackwell Science, Oxford, UK.

Specific texts on water activity and product shelf are numerous in the scientific literature. For those involved with assessing product shelf-life sources include *Moisture Sorption: Practical Aspects of Isotherm Measurement and Use* (1984). L.N. Bell and T.P. Labuza. AACC, St. Paul, MN, USA.

Practical calculation of water activity and prediction of mould-free shelf-life can be obtained using ERH CALC™ available from Campden & Chorleywood Food Research Association, UK, www.campden.co.uk

21.9 References

BERNADIN, J.E. and KASARDA, D.D. (1973) Hydrated protein fibrils from wheat endosperm. *Cereal Chemstry*, **50**, 529–536

BLOKSMA, A.H. (1990) Rheology of the breadmaking process. *Cereal Foods World*, **35**, 228–236

BROWNSELL, V.L., GRIFFITH, C.J. and JONES, E (1989) *Applied Science for Food Studies*, Longman Scientific & Technical, Harlow, UK.

CAMPOS, D.T. STEFFE, F.F. and NG, P.K.W. (1996) Mixing flour and ice to form undeveloped dough. *Cereal Chemistry,* **73**, 105–107.

CAUVAIN, S.P. (1992) Retarding, *Bakers' Review,* March, 22–23.

CAUVAIN S.P. (1998a) Breadmaking processes. In *Technology of Breadmaking* (Eds. S.P. Cauvain S.P. and L.S. Young), Blackie Academic & Professional, London UK, pp. 45–80.

CAUVAIN, S.P. (1998b) Dough retarding and freezing. In: *Technology of Breadmaking*, (Eds. S.P. Cauvain and L.S. Young), Blackie Academic & Professional, London, UK, pp. 149–179.

CAUVAIN, S.P. and SEILER, D.A.L. (1992) Equilibrium relative humidity and the shelf life of cakes, *FMBRA Report No. 150*, CCFRA, Chipping Campden, UK.

CAUVAIN S.P. and L.S.YOUNG (2000) *Bakery Food Manufacture & Quality: Water Control & Effects,* Blackell Sciences, Oxford, UK.

CAUVAIN, S.P., DAVIES, J.A. and FERN, T (1985) Flour characteristics and fungal alpha-amylase in the Chorleywood Bread Process. *FMBRA Report No. 12,* CCFRA, Chipping Campden, UK

CCFRA FTP (1991) CCFRA *Flour Testing Panel Methods Handbook*, Method No. 0004, Determination of rheological properties of doughs using a Brabender Farinograph. CCFRA, Chipping Campden, UK.

COOPER, R.M., KNIGHT, R.A., ROBB, J. and SEILER, D.A.L. (1968) The equilibrium relative humidity of baked products with particular reference to the shelf life of cakes. *FMBRA Report No. 19, CCFRA*, Chipping Campden, UK.

DODDS, INITIALS (1971) Damaged starch determination in wheat flours in relation to dough water absorption. *Starke*, **23**, 23–27.

DREESE, P.C., FAUBION, J.M. and HOSNEY, R.C. (1988) Dynamic rheological properties of flour, gluten-starch doughs. II. Effects of various processing and ingredients changes. *Cereal Chem.* **65,** 354–359.

FARRAND, E.A. (1969). Starch damage and alpha-amylase as bases for mathematical models relating to flour water-absorbtion United Kingdom, *Cereal Chemistry*, **46**, 103–116.

HOSENEY, R.C., ZELEZNAK, K. and LAI, C.S. (1986) Wheat gluten: a glassy polymer. *Cereal Chemistry*, **63,** 285–286.

PATERAS, I. (1998) Bread spoilage and staling. In: *The Technology of Breadmaking* (Eds. S.P. Cauvain and L.S. Young), Blackie Academic & Professional, London, UK, pp. 240–261.

STAUFFER, C.E. (1998) Principles of dough formation. In: *Technology of Breadmaking* (Eds. S.P. Cauvain and L.S. Young), Blackie Academic & Professional, London, UK, pp. 262–295.

STREET C.A. (1971) *Flour Confectionery Manufacture*, Blackie Academic & Professional, London, UK.

SUTHERLAND, W.R. (1989) Hydrogen ion concentration (pH) and total titratable acidity tests. *American Institute of Baking Research Department Bulletin, No. XI,* May, **5**.

WHEELOCK, T.D. and LANCASTER, E.B. (1970) Thermal properties of wheat flour. *Starke,* **22**, 44–48

WIGGINS, C. (1998) Proving, baking and cooling. In: *The Technology of Breadmaking* (Eds. S.P. Cauvain and L.S. Young), Blackie Academic & Professional, London, UK, pp. 128–148.

WILLIAMS, A. and PULLEN, G. (1998) Functional ingredients. In: *The Technology of Breadmaking* (Eds. S.P. Cauvain and L.S. Young), Blackie Academic & Professional, London, UK, pp. 45–50.

22

Improving the taste of bread

R. L. Wirtz, Consultant, USA

22.1 Introduction

The majority of people alive today in Westernised civilisations do not think of bread as a staple food. There are, of course, a variety of reasons for the long-term decline in per-capita bread consumption in North America and Western Europe. Some are economic – it is virtually a rule of thumb that both individuals and societies will gradually shift from reliance on a high proportion of grain-based foods to increased consumption of higher-protein foods such as meat and fish as the level of prosperity increases. To a great extent this is simply a reflection of the relative affluence of North American and Western European societies in comparison to much of the rest of the world's population. It has also become apparent in North America and Western Europe that there will be a gradual increase in the proportion of these foods that are processed or fully prepared. The unfortunate health consequences of this long-term trend are becoming especially apparent in the USA, where a large proportion of the population is significantly overweight. The full impact on the nation's health of a switch from grain-based foods, such as bread, to the consumption of high-protein/high-fat foods is not yet fully understood, but there is no question that the incidence of heart disease, diabetes, and other nutritionally related diseases is rising, especially in the USA (Spake and Marcus, 2002).

There are also historical and political reasons for the decline in bread consumption, such as the government-sponsored promotion of mixed flour pan breads in both the USA and UK during the Second World War and the period of wheat shortage that existed for some years following the end of hostilities. This was called the 'American Loaf' in promotional materials in the USA and the 'National Loaf' in the UK. Advances in the science and technology of food production have also played a role, in that the mass production of bakery foods has favored the rapid manufacture of highly standardized food units, to the detriment of the existence of variety and regional specialties.

A phenomenon that also has had a negative impact on the consumption of bread over time has been the cyclical phenomenon of food faddism since the late 19th century, including savage attacks on bread by self-proclaimed dietary experts or 'cult' leaders. While researchers and government-sponsored programs in several Western countries have proclaimed the benefits of the consumption of complex carbohydrates, including grains and bread, other groups have enacted virulent campaigns against bread in general and manufactured bread in particular. Perhaps the most well-known proponent of high-protein, low-carbohydrate diets today is Dr Robert Atkins, whose diet program eschews the consumption of carbohydrates of almost every type. The Atkins diet has proven to be effective in the short term, but also has significant health risks. This and similar diets have convinced many, many people that bread consumption is unhealthy (Aschoff, 2001).

One other important reason for the decline in per-capita bread consumption is the degradation of the taste of bread attributed to modern manufacturing methods and the speed of large-scale bread production (Blanshard *et al.*, 1986). Pomeranz and Shellenberger (1971) noted more than a quarter century ago that:

> A sentiment widely expressed is that bread flavor today is not as good as it once was; yet – despite complaints about the soft, high-volume, bland white bread characteristic of the modern commercial bakery – a large proportion of consumers in Western countries select that type of bread.

The question of bread taste or flavor has received considerable attention in both the industry and popular press over the last decade, and the re-emergence of traditional European hearth products and specialty variety breads as important market segments is in part a reflection of the importance of bread flavor to today's consumer. The degradation of bread taste has been a popular theme among some segments of the public press since the early 1900s and the advent of 'machine-made' bread. However, as Pomeranz (1987) points out, it has been only since the development of liquid and gas chromatographic testing in combination with advanced analytical methods such as nuclear magnetic resonance, UV spectroscopy, and other high-tech tools that it has been possible to examine this question scientifically in such great detail.

Continued research on bread flavor, combined with improvements in the instrumentation and procedures used to conduct this research, has shown that bread flavor is even more complex than had been thought in the past. Pomeranz and Shellenberger (1971) noted that researchers in the mid-1960s had identified more than 70 components of bread flavor (Pomeranz and Shellenberger, 1971). By the mid-1970s more than 200 bread aroma components proceeding from yeast fermentation had been identified, leading some experts to conclude that it may be ultimately very difficult to use a systematic, analytical approach to improve bread taste and flavor through yeast quality improvement (Blanchard *et al.*, 1986).

One other approach to the evaluation of bread flavor has involved the use of sensory analysis by taste panels. Numerous researchers have used this method to study consumer taste preferences, often with intriguing results. One group of French researchers (Rolland *et al.*, 1978) for example, found in the late 1970s that consumers seemed to be more interested in the consistency and texture of bread than in taste, and theorized that improvements in taste would not be sufficient to stimulate higher bread consumption.

Other researchers have noted the effect of trends or cycles on the relative popularity and thus the level of consumption of various types of bread, and taste has certainly been one of the factors involved in these cycles, although not the only one. It is also true that both taste and the perceived health benefits of grain consumption have been foremost among the reasons for the astounding growth in popularity of traditional, multigrain and whole grain breads over the past ten years in both the USA and the UK (Anon, 2000).

One of the most influential figures in this revival in the popularity of traditional hearth and variety breads is Calvel (2001) who feels so strongly about the importance of the flavor of bread that he has written an entire book about it. To date, his text remains the only comprehensive work to focus solely on this subject. However, in the years since the book was published in France in 1990, research in this field has continued to expand our knowledge of the components of bread flavor and aroma and the relationship to consumer taste preference. One of the results of this research has been a re-affirmation of the contribution of traditional methods of bread production to the complex flavor profile of bread.

22.2 Elements of bread flavor

Humans are equipped to detect only four – or perhaps five – different tastes. Because of the structure of the human tongue, we can detect only sweet, sour, bitter and salty tastes, but some researchers feel that there are separate taste receptors for monosodium glutamate. What we call flavor in a popular sense is really a set of highly complex organoleptic qualities that include taste, aroma, mouthfeel (texture, masticability, moistness or dryness, degree of staling, etc.), and which depends to a considerable extent on individual preferences and sensibilities. Taste or flavor perceptions and preferences can also be influenced by ethnic or social group custom, and there is a considerable body of evidence that preferences may also be influenced by geographic region, socio-economic class, sex and age (Stevens *et al.,* 1984). There is empirical evidence that age factors may play an important role in food choice and food intake in the elderly and that taste perceptions may also be more affected by the age and sex of the individual than the actual taste and flavor of the food (Rolls, 1999). This may account in part for the often-repeated statement that bread is not as good tasting as it was in the past. As the average age of the population of most Western countries continues to move upward, enhancement of bread flavor to meet the taste perceptions and expectations of this group of consumers should become more important to bakers and food producers.

Researchers long ago identified a number of factors that contribute to the production of flavor in bread. In her review, Martínez-Anaya (1996) notes that early research indicates that the basic ingredients in breadmaking, primarily flour, are thought to make only a small direct contribution to bread flavor, but that thermal reactions, including non-enzymatic browning and caramelization, are both among the direct contributors to bread flavor. Pyler (1988) also states that the development of flavor in the crumb is dependent to a great extent on the non-enzymatic and non-oxidative thermal browning reactions that occur in the crust of bread (Pyler, 1988). Pomeranz (1987) concurs that the importance of both crust formation and browning and Maillard reactions are extremely important to the production of bread flavor, and notes that: 'Neither a normally fermented dough baked without crust formation nor an improperly fermented dough baked with crust formation has acceptable flavor.'

22.2.1 Heat-dependent reactions in the production of bread flavor

The two thermo-chemical reactions recognized as the primary origins of bread flavor occur in distinctively different fashions: **caramelization** involves the transformation, into brown complex polymers, of the 2–3% of several types of residual sugars that are present in the dough after fermentation. If there is not enough residual sugar in the dough, crust coloration will not occur properly. Caramelization begins when the simple sugars in the outer dough layer begin to melt between 130 and 140 °C (Roussel and Chiron, 2002), continues as the crust surface reaches a temperature above 150 °C, and proceeds through a complex sequence of reactions (Hodge, 1967) up to about 206 °C (Stear, 1990), resulting in the synthesis of various oligosaccharides ranging from sweet through bitter to intensely bitter in taste. Pomeranz (1987) considered that caramelization may begin at temperatures as low as 135 °C.

The second type of flavor-producing thermo-chemical reactions are classified as **melanoidon** or **Maillard** reactions (named in honor of Louis Camille Maillard, the French biochemist who first identified non-enzymic browning) and begin to form temperatures around 100 °C. These involve highly complex interactions between reducing sugars (such as glucose, fructose or maltose), amino acids (such as lysine) and proteins or peptides (Eliasson and Larsson, 1993), resulting in the formation of highly colored aldehyde-amine polymers and heterocyclic nitrogen compounds (Hodge, 1953). Stear (1990) notes that numerous bread aroma compounds, including carbonyl compounds such as aldehydes and ketones, are formed as the crust temperature passes through the range of 100–180 °C. It is the diffusion of some of these substances from the crust to the crumb during cooling and storage that is responsible for much of bread flavor. The volatility of these same carbonyl compounds permits their gradual dissipation from the crumb and crust and a gradual loss of flavor from the loaf, concurrent with the other mechanisms involved in the staling process (Pomeranz, 1987).

22.3 Ingredients and flavor: flour and water

It is universally accepted that wheat flour is the most important ingredient in breadmaking, and many standardized tests have been developed around the world to determine the breadmaking quality of flour. However, other than noting that flour contributes a pleasant 'nutty' or 'wheaty' taste to the end product, relatively little seems to have been written about the taste of flour. This is especially true of white flour, which may often be rather bland in taste. As any wheat farmer can discern, however, there are certainly differences in wheat kernels from one variety to another, from one season to another, and even between wheat grown in fields that are not widely separated. These differences are caused by such things as wheat variety, climate, growing and harvest conditions, soil fertility and mineral content, and many other variables. It is a very common practice for wheat farmers to chew a few kernels of the grain from time to time as it is harvested, in a very personal and unscientific 'test' of wheat quality. Wheat farmers and millers are among the few to experience the unique taste of unprocessed wheat.[1] Because of the blending of different wheats at the mill to obtain a flour with desirable qualities for breadmaking or for the production of other bakery products, the taste variables tend to disappear.

North Americans are fortunate to have large areas of land that are suitable for the production of high-quality wheat. The extent of North American wheat lands, combined with the efficiency of wheat producers and of the American harvesting, collection, storage and transportation networks, have combined to ensure that in most years there will be an adequate supply of wheat and flour of good breadmaking quality. The North American supply chain, moreover, is normally effective in avoiding many of the problems that occur elsewhere in the world, and which may have significant effect on the flavor of bread. Exceptions to this situation do occur however, and may include the following:

- High mold counts and/or aflatoxins, which occur when the grain is harvested or held under moisture conditions that are higher than optimal.
- Off-flavors caused by excessive heating of the grain when it is being dried or held in storage.

[1] Author's note: This completely unscientific test method would seem to be widely practiced among millers as well as farmers. On several occasions, while touring flour mills with groups of millers from other countries, I have observed millers who spoke no common language sharing a few grains of wheat and indicating approval of the taste and odor of the unprocessed grain with nods and smiles. Is there some type of bond of brotherhood between wheat producers and millers due to their shared knowledge and experience?

rs absorbed by the wheat or flour in improper storage or during portation, i.e. from fuel or lubricating oils, agricultural chemicals, other agricultural crops, improperly cleaned transport containers.
- Oxidative rancidity, which occurs most frequently in wholemeal flours, but may also occur in white flour under improper storage conditions.
- Animal contaminants, including rodent hair and excreta, insect fragments, etc.
- Various weed seeds or other vegetable contaminants, some of which may be toxic.
- Contamination with ergot (*Claviceps purpurea*).

Most or all of these problems would ordinarily be detected through standardized mill and bakery testing and corrected or the contaminated material discarded before the grain was milled or the flour was used by bakeries or other end-users. However, with the current trend towards organic growing and processing of grain and the practice among some 'natural' bakers and consumers of buying grain directly from the grower or from another agent such as an agricultural co-op and milling it immediately before use, there is an increased possibility that contaminants or improper handling practices could affect not only the taste but also the sanitation and safety of the finished baked product. One related problem, which has begun to occur more frequently due in part to the use of minimally processed or 'organic' grains, is the occurrence of 'rope' in bread, particularly in those products that contain a proportion of rye flour.

22.3.1 Contribution of water to bread flavor

Water seldom has any direct effect on the taste of bread. However, water is basic to the breadmaking process in that it hydrates the flour and supplies moisture to the starch and proteins, enabling the creation of a dough mass. Water is also necessary to allow the dough to ferment, and to supply the humidity necessary for enzymatic activity (Calvel, 2001). Water is functionally irreplacable because of the many different roles that it plays throughout the breadmaking process, and many of these functions are absolutely critical to the creation of bread taste. Samuel Matz (1972) concurs that:

> the quality of water used as an ingredient can have greater effects on bakery products than is generally recognized. The amount and types of dissolved minerals and organic substances present in the water can affect the flavor, color, and physical attributes of the finished baked goods as well as the machining of doughs.

The native properties of water and its role in the physical, chemical, biophysical and biological processes involved in baking have been examined in considerable depth by Cauvain and Young (2000) while Pyler (1988) has discussed impurities and solutes, water treatment and filtration, water softening, contaminants, effects on fermentation, and water-related processing considerations and problems.

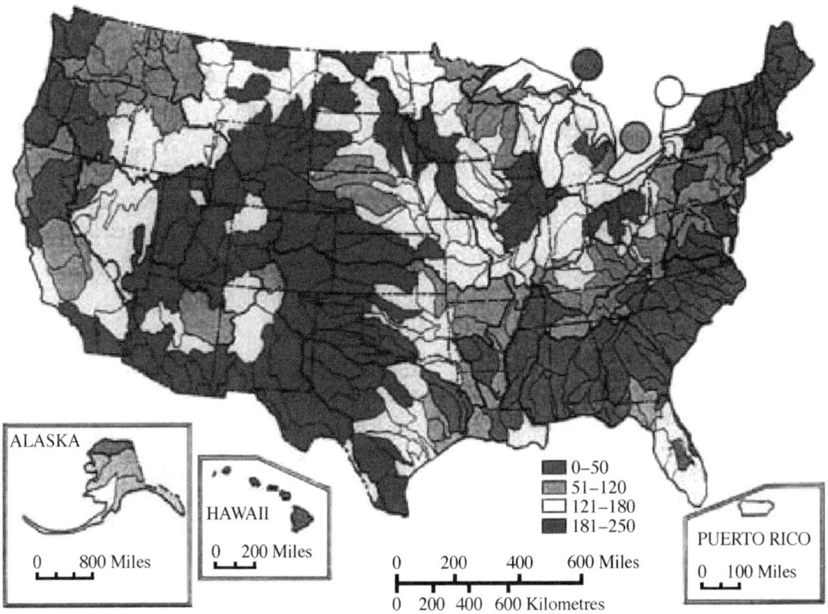

Fig. 22.1 Concentration of hardness as calcium carbonate (mg/l). US Source is USDA, given at: http://water.usgs.gov/owq/hot.html. Hardness figures for Canada available at: http://www.cwqa.com/html/issues.html.

Water used in bread production will in most instances be from municipal water supplies that are regularly subjected to testing according to accepted standards and regulations. In the USA, the National Primary Drinking Water Regulations are published by the Federal Government as 40 °CFR Part 141 and 142.[2] Legal oversight for the drinking quality of water from springs or deep wells is also under the jurisdiction of the Environmental Protection Agency.[3] Drinking water quality standards and guidelines for Canada are under the jurisdiction of Health Canada/Santé Canada.[4] In all cases, water used for bread production should meet the applicable drinking water standards, whether directly from the supplier or following appropriate treatment within the bakery (Fig. 22.1).

As noted previously, water that is potable seldom has any direct effect on bread taste, but water that is high in some types of dissolved salts can have an inhibitory effect on fermentation. According to Matz (1939) and Pyler (1988) calcium oxide, calcium carbonate, calcium sulfate, magnesium chloride,

2 Author's note: These are currently accessible over the Internet at: http://www.epa.gov/safewater/regs/cfr141.pdf and http://www.epa.gov/safewater/regs/cfr142.pdf. These and other safe water regulations are under the jurisdiction of the Environmental Protection Agency, and may be accessed at http://www.epa.gov/safewater/regs/.html#cfr.

3 Accessible at: http://www.epa.gov/safewater/gwr.html

4 Information on these Canadian guidelines are currently accessible at http://www.hc-sc.gc.ca/ehp/ehd/bch/water_quality.htm

magnesium oxide, and sodium bicarbonate all fall into this class. Minute amounts of vanadium or cadmium also are known to inhibit fermentation (Matz, 1939). Calvel (2001) notes that 'water that is too soft reduces dough cohesiveness, while water that is too hard reduces extensibility.' Pyler (1988) cites a 1936 study by C.S. Pickering that concluded:

- Soft waters will generally yield soft and sticky doughs, because the gluten-tightening effect of minerals is absent ...
- Hard waters retard fermentation by toughening the gluten too much ...
- Alkaline waters tend to reduce the fermentation rate and, therefore, require an increase in fermentation time, unless acidification is applied.

22.4 Ingredients and flavor: yeast and lactic acid fermentation

Baker's yeast, which is classed as *Saccharomyces cerevisiae*, is the primary fermentative agent in bread production. Chiron (1994) notes that in Europe the use of beer yeast for bread dough fermentation was practiced as long ago as the first century AD, but during the Dark Ages the practice seems to have been forgotten until the establishment of beer breweries in the northern part of France in the 15th century. Beer brewers began to market beer foam for bread fermentation during the 1770s, and by 1780 distilling companies in Schiedam, in Holland were producing yeast intended solely for that purpose. By the early 19th century bakers in England, Austria and North America (Stauffer, 1990)[5] normally used yeast for everyday production of bakery products, but most French bakers clung to production based on *levain*, or sourdough (Roussel and Chiron, 2002). Calvel (2001) attributes the use of prepared baker's yeast in France to a Baron Zang, who is said to have introduced the Viennese method of *poolish* sponge breadmaking to Paris in the early 1840s. This was a revolution in French bread production, since it allowed the baker to save some of the time and effort previously needed for elaboration of sometimes unpredictable sourdough cultures, and permitted production of a lighter, more crusty loaf. Because of these advantages, the *poolish* method was practiced widely in France until the 1920s, when it began to be supplanted by the simpler straight dough method that was in common use virtually everywhere else in Europe and North America.

Yeast produces an alcoholic fermentation, which is an anaerobic conversion of sugars into carbon dioxide gas and ethanol. These sugars include both those that are directly fermentable as well as those that are produced as the result of the hydration of the starch in flour. In a typical flour, from 1.5% to 2.0% of the dry weight of flour is made up of sugars, including glucose, fructose, sucrose,

[5] Stauffer notes that in the USA, Charles Fleischman began producing baker's yeast commercially in 1868, and ten years later National Distilling (Red Star Yeast) began production of this key ingrediant. Numerous companies were in business by the year 1990.

maltose, and short chain pentose and glucose sugars (Roussel and Chiron, 2002). These simple sugars are normally consumed by the yeast at the beginning of fermentation, and under normal conditions this usually takes place very shortly after the flour is hydrated. Stear (1990) states that within very few minutes after mixing all of the sucrose present in the dough will have been converted into glucose and fructose, and that fermentation of those sugars will then commence immediately. Maltose will be fermented only after the other two sugars have nearly disappeared. In order for fermentation to continue, enzyme activity must convert damaged starch in the flour into fermentable sugars. These conversion processes occur very rapidly during and following the mixing step, and ultimately reach surprising levels. Pomeranz and Finney (1975) report, for example, that the amount of maltose in dough increases from 10 to 15 times the level present at the beginning of mixing. This occurs as a result of the action of the action of α-amylase on available damaged starch.

Martínez-Anaya (1996) notes that:

> Three sources of enzymes can be considered in breadmaking: those existing in flour, those associated with the metabolic activity of yeasts and lactic acid bacteria, and those intentionally included in formulation. Enzymes are of paramount importance in generating bread flavor. Their more or less intensive action can result in positive or negative effects on desirable bread flavor ... their main role is to produce precursors direct or indirectly (the most likely) related to flavor-forming processes.

The enzymatic process involved is extremely complex, and yields a large number of fermentation products, many of which are involved in the generation of bread flavor and aroma. For example, Pyler (1988) states that under anaerobic conditions yeast will convert more than 95% of the glucose it consumes into carbon dioxide and ethanol. Although much of the ethanol is lost in baking, enough remains to serve as a direct, rather than indirect, flavoring agent. Matz (1972) points out that ethanol is an important flavor element, and that its gradual loss after baking is one of the reasons for the loss of flavor that accompanies staling. Only a small amount of ethanol is produced by aerobic fermentation. Stear (1990) notes that:

> The yeast cell is also capable of synthesizing amino acids and protein with the help of inorganic nitrogen sources and carbohydrates. Glucose serves not only as an energy source, but also as a source for intermediate metabolic products, e.g. pyruvic acid is a precursor for such amino acids as alanine, valine, and leucine, and also pentoses.

Generally the taste of yeast itself is not detectable in bread unless the amount of yeast used is greater than 2.5% based on the weight of flour. When yeast is used beyond that level, especially in white flour hearth breads, its flavor becomes more discernible as the amount of yeast used increases. Whether this is objectionable or not depends to a great extent on consumer preferences, but it is generally not a desirable taste attribute (Calvel, 2001).

22.4.1 Contribution of lactic acid fermentation to bread flavor

In addition to yeast fermentation, the action of lactobacilli in the dough has a significant effect on the generation of flavor in bread. It has long been recognized that sourdough breads have a much more distinct flavor profile than breads fermented with yeast. Schieberle (1996) identified 14 intense aroma compounds in sourdough bread, and attributed the most characteristic aroma of sourdough wheat bread crust to the compound 2-acetylpyrroline. Other researchers have shown that bread aroma is significantly enhanced by higher levels of free amino acids in the crumb and especially in the crust (Thiele *et al.*, 2002). Thiele *et al.* (2002) point out that sourdough fermentation produces higher levels of amino acids in the dough, while yeast fermentation actually reduces the level of amino acids. Their work determined the amino acid content of dough by high-performance liquid chromatography (HPLC), and found that ornithine, methionine, phenylalanine, leucine, isoleucine and valine were especially important to bread flavor. Some researchers feel that lactobacilli present in yeast and flour may be associated with flavor development even in yeast leavened breads (Blanshard *et al.*, 1986).

22.5 Processing and flavor: mixing, fermentation and baking

Calvel (2001) has commented at length on the study by Richard-Molard *et al.* (1978) which compared breads made with baker's yeast with those produced from the *levain* and *poolish* methods. This same study also compares the effect of different mixing methods (conventional *vs.* intensive mixing) on the production of various volatile acid fractions in three different types of bread. The *poolish* and *levain* methods produced acetic acid levels that were noticeably higher than the straight dough, which is to be expected. However, the intensified mixing method also had the effect of producing higher levels of acetic acid and of several other acids than the bread made by the conventional mixing method. This may be due in part to the greater frictional heat generated by intensified mixing, which has the effect of stimulating enzymatic activity (Roussel and Chiron, 2002). This is one of the reasons why it is important to control dough temperature during the mixing process.

Calvel (2001) advocates the reduction of total mixing times by at least 15% through the use of the autolysis rest period, which allows the flour to fully hydrate before mixing is continued. Autolysis involves the slow-speed premixing of flour and water, followed by a rest period before the addition of other ingredients and the resumption of mixing. The period of autolysis generally ranges between 15 and 30 minutes but may extend for a longer time without affecting the quality of the dough. The majority of professional bread recipes in the *Taste of Bread* include a recommendation for an optimum autolysis period. This is a simple concept, but one that works extremely well in practice, and is helpful in reducing the amount of dough oxidation and thus in preserving the flavor compounds in bread.

Table 22.1 Conventional mixing with 0.7% bean flour. Quantities expressed in parts per million in relation to the crumb. (From: Calvel, R. (2001) *The Taste of Bread*. Reprinted by permission)

	Straight dough (2% yeast level)	Poolish style sponge (00.7% yeast level)	Naturally leavened starter sponge
Acetic acid	53	105	970
Propionic acid	0.50	0.54	0.68
Isobutyric acid	1.30	1.12	0.32
Butyric acid	0.18	0.29	0.16
Isovaleric acid	0.51	0.48	0.43
Valeric acid	0.13	0.29	0.16
Capriotic acid	0.84	1.06	0.58

Table 22.2 Intensified mixing with 0.7% bean flour. Quantities expressed in parts per million in relation to the crumb. (From: Calvel, R. (2001) *The Taste of Bread*. Reprinted by permission)

	Straight dough (2% yeast level)	Poolish style sponge (00.7% yeast level)	Naturally leavened starter sponge
Acetic acid	55	123	1093
Propionic acid	0.70	0.70	0.90
Isobutyric acid	4.70	1.44	0.77
Butyric acid	0.26	0.18	0.38
Isovaleric acid	1.80	0.56	0.55
Valeric acid	0.16	0.16	0.24
Capriotic acid	1.10	1.00	0.80

Calvel (2001) comments on the conclusion of Richard-Molard *et al.* (1978) that the acetic acid may serve as an intensifier or 'carrier' for other flavor components, although in truth the Professor is not fond of high levels of acetic acid in the dough or bread. He also notes that the longer fermentation of the *poolish* method must certainly have had an effect in the level of acid production, and thus in the taste of the bread.

22.5.1 Effects of fermentation methods on bread taste

Although he does not share the outright condemnation of the straight dough method expressed by the Richard-Molard group, Calvel does state that:

> we must preserve the use of fermentation for French bread, and we should always be aware that the method of dough leavening has an important relationship to taste, since in many cases it determines fermentation time. However, the taste is influenced to an even greater degree by the intensification of mixing and consequent over oxidation of the dough, and it may be greatly damaged by these factors.

Calvel's preference for longer fermentation periods is expressed numerous times throughout the *Taste of Bread* and his earlier works. He realizes that many bakers accomplish an *artificial* dough maturation by overmixing in conjunction with the use of oxidants, such as the lipoxygenase in soya flour which is a common additive in French flours, and by adding a small amount of ascorbic acid. Although this artificial maturation of the dough can produce attractive loaves with large volume, Calvel feels that both the eating quality and aroma of the bread is much inferior to that produced by longer fermentation methods (Fig. 22.2).

22.5.2 Effects of baking on bread taste

As we have noted, several of the processes that produce bread flavor are dependent on the application of heat. The heat of baking serves to evaporate most of the fermentation generated ethanol from the loaf, although a portion of it will remain in the crumb. Both caramelization and Maillard reactions occur as the loaf surface heat rises to 230–250 °C, and the interior of the loaf may reach 100 °C or slightly higher.

In order for the best flavor to develop, it is important that the crust be adequately browned. Calvel (2001) notes that:

> Such a crust will possess the optimal combination of true bread aroma – a subtle amalgam of toast, hazelnut, and frying odors – while avoiding the excessive coloration that results in a bitter taste. Proper bread color must not be excessive, but should be just slightly darker than golden yellow, with a discreet orange tint that borders on light brown ... To avoid an insipid crust taste, it is extremely important that the crust color not be lighter than golden brown. If this occurs, there has been an inadequate Maillard reaction, associated with incomplete caramelization, both of which will noticeably affect the aroma and taste of the bread.

It is unfortunate that many customers have become accustomed to see properly colored as being 'too dark' or 'almost burned'. This is a problem in some parts of Europe and South America, and has been especially so in the Southeastern USA where clients have traditionally had a fondness for soft, lightly tinted breads and baking powder biscuits. This situation is changing as immigrants from other parts of the USA and from Europe have begun to demand higher-quality traditional hearth and variety type breads.

22.6 Innovations in bread flavor

The discussion of bread taste included above relates primarily to traditional methods, production practices and ingredient characteristics. However, bread production and bread taste are currently the subject of a great deal of scientific research and practical product development, and have been since the very

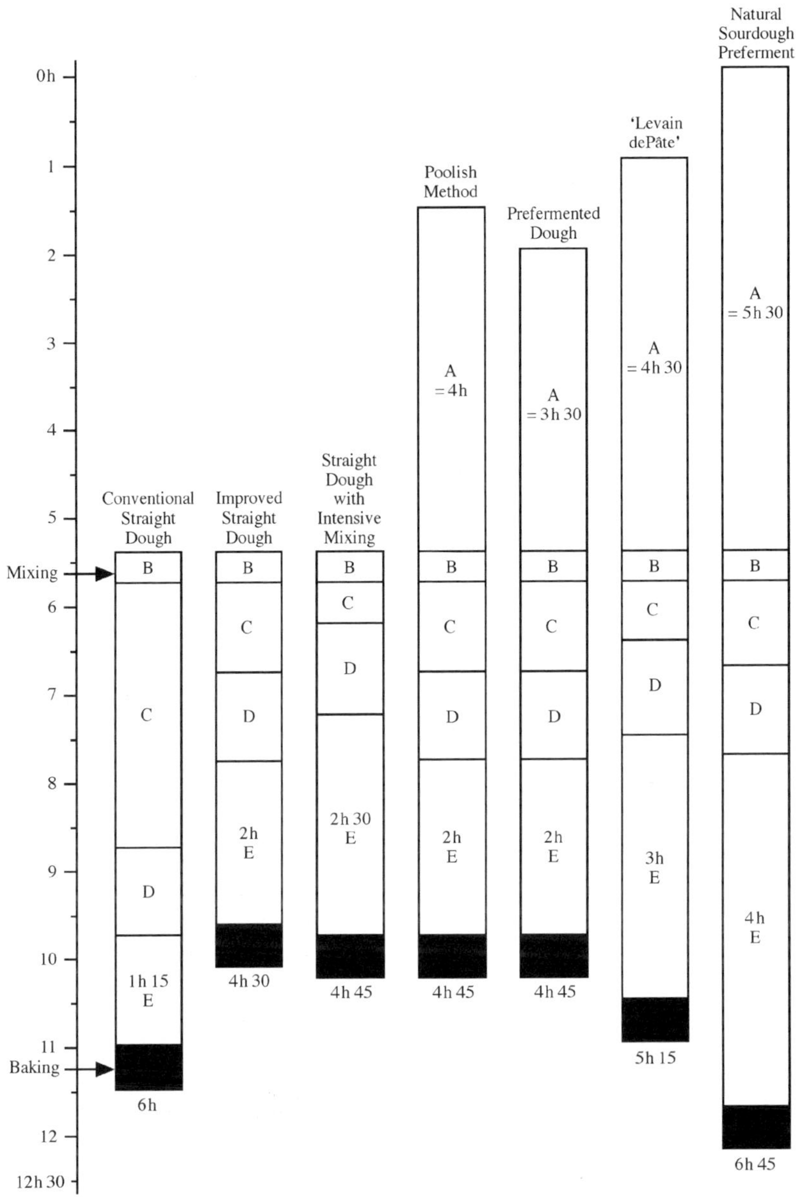

A = Preferment; B = Mixing; C = First or 'bulk' fermentation (note that bulk fermentation times are in direct relation to the duration and speed of mixing); D = Dividing, weighing, forming; E = Proofing.

Fig. 22.2

earliest efforts of Pasteur. This research is international in scope, ranging from investigations on the bread production properties of new strains of yeast or bacteria, a variety of extracts, flavoring agents, and other non-typical ingredients to the use of new and innovative methods of mixing, baking, freezing, preserving or packaging. The present section is by no means a comprehensive examination of this field of endeavor.

While bread fermentation research has most often focused on the traditional bread yeast, *Saccharomyces cerevisiae,* there have been numerous efforts to explore alternative methods of fermentation and flavor enhancement. For example, McKinnon *et al.* (1996) examined the use of 13 different types of wine yeasts in bread production. Although four yeasts were selected for final evaluation, Flor Sherry yeast was found to yield the bread with the most distinctive taste and aroma. Gélinas *et al.* (1995) also studied bread production from pre-fermented dairy products, and found that the *rhamnosus* subspecies of *Lactobacillus casei* yielded good bread with a fermentation period lasting from 16 to 24 hours, especially when used with wholewheat flour. Bread produced by this method was found to have a higher diacetyl and ethanol content and more lactic acid than that made with *Saccharomyces cerevisiae,* and it was determined that dried fermented dairy ingredients at a 1–2% level can be used as flavor enhancers in short fermentation breadmaking processes or at the 10% level as a sourdough base.

Loiez nee Hannette *et al.* (1998) have isolated, hybridized and cultivated several new strains of yeast for commercial bread production. The patent for these genetically modified strains containing only pure yeast DNA has been assigned to Lesaffre et Cie, Paris, France. These modified yeast strains exhibit improved enzymatic activity (maltase-permease, maltase and invertase), have a high multiplication yield and good nitrogen assimilation. Furthermore, they are resistant to drying and have good glucose fermentation activity. A unique feature of these yeasts is that they are fully as capable of rapid fermentation on low sugar or sugar-free dough as yeasts that are specially adapted for that purpose, but that they are also equally as capable of rapid fermentation of high sugar (more than 15% by weight) doughs as those normally used for sugar-containing doughs. No claims have been made for improvement of bread taste by use of these yeasts, but it is interesting to speculate that techniques similar to those used in the development of these new 'synthetic' yeasts might also be useful in developing breadmaking yeasts with the capability of producing improved bread flavor profiles.

Ehret (1998) has developed a method of constituting an 'all natural' biological culture consisting of a specific strain of baker's yeast (*Saccharomyces cerevisiae steineri* DSM 9211) and one or more suitable lactic acid bacteria. The patent has been assigned to Agrano AG of Allschwil, CH. The object of this method is to create an improved fermentation culture that produces bread products comparable in taste and texture to those resulting from traditional artisan or home 'sourdough' processes, without the need to use chemical bread improvers or to use standardized admixtures of pure cultures of yeasts and bacteria, such as is normally done in current commercial practice to create 'reconstructed' sponges. The culture obtained through this process is of the 'batch-fed' type, which allows control of the culture fermentation rate. This culture is equally suitable for commercial production or home/artisan breadmaking. Under commercial production conditions, it may be used immediately after being drawn from the fermentation tank. It may also be stored at a temperature of 3 °C for up to 21 days without loss of viability after having been concentrated by filtration or centrifugation. The yeast in this culture, *Saccharomyces cerevisiae steineri* DSM 9211, was isolated from a home-made sponge that produced bread of excellent organoleptic quality.

The process described in the patent permits a degree of control of the metabolism of the yeast or other microorganisms that make up the culture, and thus regulates the concentration of lactic and acetic acids in the culture and in the final bread product. The culture medium is composed of wheat kernels ground shortly before use, ground wheat germ, yeast autolysate, sea salt, water and an alpha-amylase solution. Two different media are prepared for use by mixing in a bioreactor, then are heated and cooled before the addition of an identical alpha-amylase solution, amyloglucosidase, papine and pancreatine. The yeast noted above is mixed into the culture medium, to which the other microorganisms are added in a predetermined order to adjust the organoleptic qualities of the sponge and therefore of the final product. Bread produced from this culture is said to have excellent taste.

Other methods of improving the flavor of bread include the development of various types of flavoring agents or quality improvers. Lazzari (2000) describes a method for production of a substance derived from baker's yeast that is composed of 45–60% polysaccharides by weight, 35–45% proteins, peptides and amino acids, and 3–5% nucleic acids and nucleotides. The process consists of treating a quantity of *Saccaromyces cerevisiae* yeast with a strong acid, cooking for a specified period, separating the solids from the supernatent by centrifuging or by filtration, suspending the solids in a solution of sodium chloride for 12–15 hours, separating the solid component from the suspension and washing with water, and re-suspending the solids in an alkaline hydroxide solution for a brief period before neutralizing the supernatent with concentrated hydrochloric acid. The solid component is then dried to obtain the improving agent. This product is mixed into bread dough at a rate between 0.5 and 5% by weight of the dough to achieve the desired effects. It is said to improve the volume, softness, shelf-life and organoleptic properties of the bakery product, especially bread or rolls. Baking loss was less than for the control products, while specific volume was 10% higher for the product containing the improving agent. Mold-resistant shelf-life under test conditions was double the control product, and staling or hardening of the test product occurred five times slower than for the control. Taste panel tests conducted on the product indicated that the taste was considerably improved, while the typical bread aroma was also more pronounced than for the control. Patent rights for this product are assigned to Farmint Group Holding SA.

A process for improving the 'roasty' aroma of bakery products has been developed by Bel Rhlid (2002). The product obtained using the process described is a heterocyclic compound identified as 2-acetyl-2-thiazoline or 2-AT, which was identified as long ago as 1971 as a constituent of beef broth. This substance has never been found in bakery foods, but the process described in this patent permits the production of a dried, food grade powder product with an intense popcorn or bread-crust like aroma. The product is especially useful as a surface coating for partially baked (parbaked) and frozen bread, rolls or filled products such as pizza rolls or calzone. Taste panel testing indicated that both bread-like aroma and flavor were enhanced by the use of the coating, and that the heavy tomato flavor of pizza rolls or calzone without the product coating was balanced or slightly masked by the roasted notes of the bread flavor and aroma. The flavoring composition is prepared by centrifuging a specified quantity of commercial cream yeast, discarding the supernatant, and re-suspending the material obtained in a specified amount of a buffer composed of sodium bicarbonate–sodium carbonate to obtain a pH of 9.8. The composition is kept at a pH of 9.8 by the use of an automatic pH-stat device throughout subsequent processing by the metered addition of sodium hydroxide as required. Clysteamine, ethyl-L-lactate, and D-glucose are added at stated points in the 24-hour production process. The resulting liquid may be mixed with a small amount of starch and formula water and diluted up to a 100 g (water) to 1 g (product solids) ratio before being applied to the surface of the prebaked food product. Alternatively, the substance derived from the production process may be treated by vacuum drying to obtain a powder that is remixed with water to the 100 : 1 ratio before use. Best results are obtained by reheating the coated bakery food in a conventional oven rather than in a microwave.

Ishigaki and Iizuka (2001) developed a natural quality improver for bread production that provides bread with desirable eating qualities, fine texture, good volume and shape, good flavor and excellent, long-lasting aroma. The product is based on one or more fermentation products or by-products, including malt from beer-making, germinated hulled rice, sake lees or beer lees. The fermentation product is treated by drying according to commonly used methods, providing that the flavor and fragrance of the fermentation product is retained. For ease of handling the dried fermentation product should be produced in the form of granules. The bread improver also contains added natural vitamins, specifically pantothenic acid or biotin, and mevalonolactone or mevalonic acid. In addition, it also contains a lactic acid fermentation product that is based on enzymatically digested soybean powder that is subsequently inoculated with *Propionibacterium shermanii* and yeast. Finally, it may contain other natural food ingredients such as cystine, sucrose, trehalose, maltose, and enzymes, or additional components such as lactic acid bacteria, yogurt, vegetable oils, margarines, etc.

By preference, artificial chemical products should not be included in the bread improver. The bread improver should be added to the cereal flour component at the rate of 0.01 to 0.2 parts of the improver to 100 parts of the flour component by weight. It may be used in both sponge and straight dough production. A further advantage is that the sponge may be utilized as a base fermentation culture and conserved by refrigeration or freezing without loss of the advantages conferred by the improver. It may also be possible to shorten baking time up to one-half without loss of eating quality and loaf conformation.

Bread and bread products made through the use of this improver have good volume, a thin crust, fine crumb cell structure, a soft mouthfeel and excellent flavor. Flavor retention is enhanced throughout the shelf-life of the baked product. An added advantage is that since the baked product contains an abundance of 'natural' vitamins, proteins and minerals, it may have some marketing possibilities as a health-enhancing product.

Si (2001) has developed and patented an application in baking of a premix composition containing the laccase enzyme. Using laccase is said to result in better machinability of the dough, and to increased volume, better crumb structure, improved softness, and better flavor and freshness to the finished baked product. Although laccase has been used previously in the paper and wood pulp industries, this is the first recorded application to bakery foods. Among other stated advantages is the ability to develop a bakery premix that will produce baked goods acceptable to the consumer from poor quality baking flour or composite flours or meals, including corn flour or meal, rye flour or meal, soy flour, sorghum meal or flour, and potato flour or meal. This premix should also include conventional baking ingredients such as milk powder, gluten, emulsifiers, fat, an oxidant, an amino acid such as cysteine, a sugar or sugars, salts, another enzyme or enzymes, or wheat flour or composite flours or starch. Such a premix would be suitable for use both in large-scale commercial or retail baking, may be used in the production of many types of bakery foods, and may be used for yeast-leavened or chemically leavened products, and is suitable for use in both fresh and frozen doughs. Similar claims are made in other enzyme patents assigned to the same company, and it is reasonable to assume that many enzyme patents, from this or other companies, will be somewhat similar in scope and application (Xu, 2001).

Two patents that take very different approaches to improving bread quality and taste than any of the forementioned are the processes developed by Kato (2001) and Spencer (2001). The Kato (2001) patent is intriguing in that it describes the effect of electrolysis of normal city water with a pH of 7.02 to produce acidic water with a pH of 2.85 and alkaline water of pH 6.95. The process further describes the application of the differences in the treated water to vary and enhance the quality of bakery products without the use of any of the additives commonly utilized in baked foods processing. The patent maintains that flour-based foods such as bread, Chinese buns, pizza, rice crackers and spring rolls when made with acidic water are superior in crispness and mouthfeel to those same products made with conventional city water. This is attributed to the oxidative function of the acidic water on the gluten fractions in the flour. Flour-based foods such as doughnuts, cakes and pancakes made with alkaline water are more plump and moist. This is due to the higher level of hydroxyl ions in the alkaline water, which allows better hydration and starch swelling and consequently better starch pasting. The overall effect is better moisture retention in the dough. Sensory testing on baked bread made using acidic water, alkaline water, and a recombined water showed that desired qualities of the baked product, including taste and other organoleptic qualities, could be significantly modified by the type of water used in kneading.

The patent by Spencer (2001) focuses on the control of the Maillard browning reaction in baked foods. The interesting factor in this patent is the use of one or more 'noble' gases at an elevated pressure to enhance and control the browning reaction. The patent points out that there was not previously any direct means of control of the rate of Maillard reaction in food production, and that previously it was believed that atmospheric composition had no effect on this reaction. The noble gases used in the present instance include argon, xenon, krypton, neon and helium, but not radon owing to its radioactivity and consequent danger to health. Helium is not used owing to the difficulty of containing it, although it is known to accelerate browning reactions. The process involves exposing the bakery product to a concentration of noble gases at least 10% by volume of the atmosphere surrounding the product, at a pressure ranging up to 3 atmospheres. The exposure of the product to the gases may take place during general processing (mixing, blending, storage), to include frozen storage or baking. Rate of reaction may vary according to temperature and to other parameters, including the other components of the processing atmosphere. Argon may increase the speed of the reaction by up to 50%, while the use of nitrogen will tend to inhibit browning. Browning can also be inhibited, if desired, by the use of the proper gas mixture. Since taste of a baked product may be significantly enhanced by Maillard browning in combination with caramelization, it is possible that the exercise of these factors may allow the control of taste to a greater degree than previously has been possible. While all of these innovative approaches to enhancing or improving bread flavor are interesting, their general applicability to the production of bakery foods may depend on a number of factors, including the willingness of bakers to gamble on the consumer acceptance of such non-traditional products and approaches to production. The American consumer especially may be cautious in accepting the use of new ingredients and food production technologies, especially for such basic products as bread, rolls and other traditional bakery foods. However, the offering of products of evident high quality, combined with appropriate doses of consumer education and advertising, should encourage bakers and other food producers to continue to test new and innovative products in the marketplace.

22.7 References

ANON (2000) Industrial imitators, *European Baker*, Issue 39 (September/October 2000), p. 6.

ASCHOFF, S. (2001) 'Fat busters or folly?' *St. Petersburg Times*, October 23, p. 3D.

BEL RHLID, R. (2002) Process for the preparation of flavoring compositions and use of these compositions in bakery products. United States Patent 6,432,459.

BLANSHARD, JVM, FRAZIER, PJ and GALLIARD, T. (1986) *Chemistry and Physics of Baking,* Royal Society of Chemistry, London, p. 129.

CALVEL, R. (2001) *The Taste of Bread.* Aspen Publishers, Inc./Kluwer Academic, Gaithersburg, USA.

CAUVAIN, S.P. and YOUNG, L.S. (2000) *Bakery Food Manufacture and Quality: Water Control and Effects.* Blackwell Science, Oxford, UK.

CHIRON, H. (1994) The effects of technological changes on the characteristics of French bread. *Inds. Cer.*, April, 5–12.

EHRET, A. (1998) Panification ferment containing *Saccharomyces cerivisiae steineri* DSM 9211 and lactic acid bacteria. United States Patent 5,849,565.

ELIASSON, A.C. and LARSSON K, (1993). *Cereals in Breadmaking.* Marcel Dekker, Inc., New York, USA.

GELINAS, P., AUDET, J., LACHANCE, O. and VACHON, M. (1995) Fermented dairy ingredients for bread: effects on dough rheology and bread characteristics. *Cereal Chemistry*, **72**, 151–154.

HODGE, J.E. (1953) 'Chemistry of browning reactions in model systems. *Agr. Food Chem.*, **1**(15), 928–943.

HODGE, J.E. (1967) 'Nonenzymatic browning reactions.' *Chemistry and Physiology of Flavors.* AVI Publishing Company, Westport, USA, pp. 465–491.

ISHIGAKI, R. and IIZUKA, R. (2001) Quality improver for use in producing bread. United States Patent 6,183,787.

KATO, A (2001) Preparation method of dough for flour foods. United States Patent 6,326,048.

LAZZARI, F. (2000) Product based on polysaccharides from bakers' yeast and its use as a technological coadjuvant for bakery products. United States Patent 6,060,089.

LOIEZ NEE HENNETTE, A. (1998) Strains of bread-making yeast, a process for obtaining same, and the corresponding fresh and dry new yeast. United States Patent 5,741,695.

MARTíNEZ-ANAYA, M.A. (1996) 'Enzymes and bread flavor.' *Journal of Agricultural and Food Chemistry*, **44**(9), 2469–2480.

MATZ, S. (1972) *Bakery Technology and Engineering.* AVI Publishing Co., Westport, USA.

MCKINNON, C.M., GÉLINAS, P. and SIMARD, R.E. (1996) Wine yeast preferment for enhancing bread aroma and flavour. *Cereal Chemistry*, **73**(1), 45–50.

POMERANZ, Y. (1987) *Modern Cereal Science and Technology.* VCH Publishers, New York, USA, p. 342.

POMERANZ, Y. and FINNEY, K.F. (1975) Sugars in breadmaking. *Bakers' Digest*, **49**(1), 20–27.

POMERANZ, Y. and SHELLENBERGER, J.A. (1971) *Bread Science and Technology.* AVI Publishing Company, Westport, USA, p. 181.

PYLER, E.J. (1988) *Baking Science & Technology.* Third Edition. Sosland Publishing Company, Merriam, USA, p. 756.

RICHARD-MOLARD, D, NAGO, M.C. and DRAPERON, R. (1978) Rôle de la fermentation. Influence du mode de panification sur la composition en constituants volatiles de la mie du pain. *Bulletin des Anciens Elèves de l'Ecole de Meunerie* (ENSMIC). No. 289.

ROLLAND, M. F. CHABERT, C. and SERVILLE, Y. (1978) Le choix des pains. *Annales de la Nutrition et de l'Alimentatio,* **32**(6), 1285–1299.

ROLLS, B.J. (1999) Do chemo-sensory changes influence food intake in the elderly? *Physiology & Behavior,* **66**(2) 193–197.

ROUSSEL, P. and CHIRON, H. (2002) *Les pains français: évolution, qualité, production.* Maé-Erti, Vesoul, France.

SCHIEBERLE, P. (1996) Intense aroma compounds – useful tools to monitor the influence of processing and storage on bread aroma. *Advances in Food Science,* **18,** 237–244.

SI, J.A. (2001) Use of laccase in baking. United States Patent 6,296,883.

SPAKE, A. and MARCUS, M.B. (2002) A fat nation. *U.S. News & World Report.* **133**(7) August 19, p. 40.

SPENCER, K.C. (2001) Method of controlling browning reactions using noble gases. United States Patent 6,274,185.

STAUFFER, C.E. (1990) *Functional Additives for Bakery Foods.* Van Nostrand Reinhold, New York, USA, p. 199.

STEAR, C.A (1990) *Handbook of Breadmaking Technology*. Elsevier Applied Science, New York, USA, p. 542.

STEVENS, J.C., BARTOSHUK, L.M. and CAIN, W.S. (1984) Chemical senses and aging: Taste versus smell. *Chemical Senses*, **9**(2), 67–179.

THIELE, C., GANZLE, M.G. and VOGEL, R.F. (2002) Contribution of sourdough lactobacilli, yeast, and cereal enzymes to the generation of amino acids in dough relevant for bread flavor. *Cereal Chemistry*, **79**(1), 45–51.

XU, F. (2001) Methods for using dehydrogenases in baking. United States Patent 6,306,445.

23

High-fibre baking

K. Katina, VTT Biotechnology, Finland

23.1 Introduction

The importance of dietary fibre (DF) has been demonstrated in many studies (Newman *et al.*, 1989; Anderson and Siegel, 1990). A typical Western diet contains less than 20 g/day whereas the recommended daily intake is 25–30 g (Drehner, 1987). Thus, at the moment most people eat too little fibre and these low levels of dietary fibre in Western diet contribute to a long list of ills, ranging in severity from dental caries through haemorrhoids to obesity, colorectal cancer and coronary heart disease. For a long time, dietary fibre has been recognised as having health benefits. The effects of fibre may be mechanical (increased faecal weight, decreased transit time, etc.) or physiological (binding of bile acids and salts, production of short chain fatty acids). Fibre may reduce the incidence of gall bladder disease (by reducing bile saturation); it may help control diabetes (by rendering glycaemic dietary components unabsorbable) and has been shown to reduce symptoms of diverticular disease. Soluble fibres exert a hypolipidaemic effect and have been shown to reduce severity of experimental atherosclerosis. There is also epidemiological evidence suggesting that fibre exerts a protective effect in human cardiovascular disease. High-fibre diets also have a protective effect against colon cancer, although the mechanism of action is not clear.

During the last few decades, identified health benefits of fibre have created a growing interest in increasing the fibre content of various foods and the number of fibre sources offered for use in formulating these high-fibre foods have grown almost exponentially. However, when fibre in some form is used in baked products, it is necessary to make adjustments to various process parameters in order to obtain high-quality, high-fibre products which are acceptable to the majority of consumers.

23.2 Sources of fibre in baking

Dietary fibre can be described as the edible parts of plants or analogous carbohydrates that are resistant to digestion and absorption in the human small intestine with complete or partial fermentation in the large intestine. Dietary fibre includes polysaccharides, oligosaccharides, lignin and associated plant substances. Dietary fibre promotes beneficial physiological effects, such as laxation, and/or blood cholesterol attenuation, and/or blood glucose attenuation (McCleary and Prosky, 2001). In considering dietary fibre three categories are used: insoluble dietary fibre (IDF); soluble dietary fibre (SDF); and total dietary fibre (TDF), the sum of IDF + SDF. Insoluble and soluble forms have somewhat different functional characteristics that influence physiological effects and the way they are used in baking. Total dietary fibre is important in the calculation for caloric reduction in foods formulated with these materials. Important sources for dietary fibre include cereals, potatoes, pulses and nuts, vegetables, fruits and berries. Some common fibre sources and proportion of SDF and IDF are presented in Table 23.1.

The main categories of IDFs are cellulose, hemicellulose, amyloids, lignin, resistant starch and Maillard reaction products. Cellulose, located in plant cell walls, is the major component of IDF. It is a β-1,4-linked polyglucan of varying chain lengths. Chemically, cellulose is very similar to amylose, the main difference being the spatial orientation of the bonds that link the successive glucose molecules. In cellulose the molecules are linked in such way that macromolecule forms a straight rod. A cellulose chain contains between 300 and 15 000 glucose units having a molecular weight between 50 $\times 10^3$ and 2.5 $\times 10^6$ Da. The main source of cellulose is wood. Highly purified cellulose (alpha cellulose) is the raw material for food and pharmaceutical industries. Also, cellulose makes up a major portion of IDF in most of the natural fibres being sold to bakers today (Stauffer, 1993).

Table 23.1 Composition of fibre ingredients (according to Stauffer, 1993)

Fibre source	IDF	SDF	Water +ash	Protein	Fat	Carbo-hydrate
Alpha cellulose	95.0	0.0	5.0	0.0	0.0	0.0
Soy fibre	56.5	21.1	8.5	12.0	0.2	2.5
Oat fibre	93.8	0.0	0.0	6.0	0.2	0.0
Pea fibre	80.8	4.6	6.5	2.0	0.5	4.6
Apple fibre	69.0	1.7	1.8	9.1	3.1	15.3
Sugar beet fibre	57.0	25.0	9.7	8.0	0.3	0.0
Processed wheat fibre	90.0	0.0	9.0	1.0	0.0	0.0
Hard wheat bran	40.0	3.0	17.0	15.0	5.0	20.0
Oat bran	11.7	10.5	11.7	20.1	6.0	38.4
Corn bran	87.0	1.0	5.0	3.8	1.0	2.2
Barley bran	64.7	2.9	8.1	17.7	6.6	0.0
Rice bran	23.0	2.0	19.0	14.5	20.0	21.5
Soy hulls	61.1	9.4	10.7	12.2	0.6	6.0

Hemicellulose is a rather complex mixture of polymers, in which the main chains consist of xylans, glucomannans and galactans and side-chains consist of galactose, arabinose and various hexuronic acids. Hemicellulose is a generic term, covering a wide variety of non-cellulose polysaccharides obtained from plant cell walls. In cereal grains, and in wheat and rye flour, the pentosans (both soluble and insoluble) are hemicelluloses. Arabinoxylan is the main component of insoluble pentosans from these flours, while arabinoxylans of lower molecular weight, arabinogalactan and xylans having other side-chains (e.g. glucuronic acid) are found among the soluble pentosans (Stauffer, 1993).

Amyloids are xyloglucans which are located in the cell walls of peas, beans, soybeans and other legumes. The linear main chain is the same as cellulose, but D-xylose, D-galactose and D-fructose are attached at intervals. They are relatively minor component of IDF, except for those fibre ingredient sources derived from legume cell walls.

Lignin is described as the 'glue' that holds together the bundles of cellulose and hemicellulose in mature plant cell walls. It is an amorphous, aromatic hydrocarbon polymer, formed by condensation of phenylpropane subunits. In most of the fibre sources used in bakeries, lignin represents generally 5–10% of TDF. Lignin itself is structurally complex and properly speaking is not one compound but rather a whole class of plant cell wall materials. It contributes relatively little to the water-binding capacity of fibres. On the other hand, lignin might bind bile acids and cholesterol in the gut.

Resistant starch (RS) is formed during the baking process as the amylose part of the gelatinized starch (formed during baking) retrogrades (crystallizes) during cooling. A low-molecular-weight fraction of the amylose crystallizes into tightly knit bundles that are resistant to digestion by alpha-amylase (Sievert *et al.*, 1991). The amount formed depends on the nature of original starch (ratio of amylopectin to amylose) and the baking procedure used. Thus, RS is an artefact of processing and in that sense, not a 'real' dietary fibre. However, RS is largely inert and not digested or absorbed in the human gut and hence a legitimate part of IDF.

Generally, soluble dietary fibres can be described as gums. They may be either endogenous (present as a normal component of another formula ingredient) or exogenous (a relatively pure gum, added to the formula). Gums are made up of long polysaccharide chain with numerous side branches of sugars or oligosaccharides. Frequently the sugar units include carboxylic acids such as D-glucoronic, D-mannuronic or D-galacturonic acids. The highly branched structure contributes to water solubility and anionic gums in the presence of cations, such as Ca^{2+}, often form gels. Gums are sometimes used in dough or batter as a processing aid and also for improved moisture retention in the finished product. Since the amounts used for such purposes are generally rather low (typically 0.1–0.25%, flour basis) their contribution to TDF of the final product is rather small. However, some ingredients, such as oat bran, are used in significantly larger amounts (10–30%, flour basis) in baked products and thus the soluble fibre content may be rather a large part of TDF of the final product (Stauffer, 1993).

The most important endogenous gums are beta-glucan and pectic substances. Beta-glucan is a polyglucose chain in which the linkages are β-1,4 and β-1,3 in varying ratios. Glucans with higher molecular weights and a greater proportion of β-1,4 linkage tend to be insoluble in water and hence assay as IDF; the lower molecular weight species, and those with more β-1,3 linkages are soluble and register as SDF. High contents of beta-glucans are found in oats and barley.

Pectins include a wide variety of materials based upon poly-α-1,4-D-galacturonic acid, and containing side-chains of D-galactose, D-arabinose, D-xylose, D-rhamnose and D-glucose. Commercial pectin may have either a higher (high-methoxy pectin, HMP) or a lower (low-methoxy pectin, LMP) degree of esterification. The major commercial sources of pectin are citrus fruit, apples, pears and sugar beet pulp.

Most exogenous gums are of plant origin, although bacterial polysaccharides (xanthan, gellan) and synthetic derivates (cellulose gums) are also used in the baking industry. Arabic, tragacanth, ghatti and karaya gums are exudates obtained from trees and shrubs. Agar, alginic acid, carrageenan and furcelleran are water extracts of various seaweed species. Locust bean gum and guar gum are obtained from the carob tree. Bacteria often form a certain amount of gum when grown in biofermentors. Examples are xanthan gum and gellan gum. Synthetic gums are also available for baking industry: for example, cellulose substituted with carboxymethyl, methyl and hydroxypropyl plus methyl groups is available for baking applications.

23.3 Problems in high-fibre baking

The most common source of dietary fibre in baking is cereal bran, especially wheat bran (Seibel, 1975). The use of barley bran derived from huskless or de-husked barley or oat bran derived from husked oat kernels is also becoming more widespread thanks to their high SDF content, in particular their beta-glucan content. However, additions of cereal bran, especially in such amounts that health benefits can be expected, cause severe problems in bread quality. According to Seibel (1983) addition of fibre cause following technological changes:

- increases in dough yield;
- a moister and shorter dough;
- decreased fermentation tolerance;
- lower volume;
- a crumb that is tense and non-elastic; and
- flavour changes depending on type of fibre and bread type.

In principle, it is possible to use other fibrous material such as wheat fibres, orange or apple fibres, dietary fibre products derived from carrots, beets and potatoes to enrich cereal products with dietary fibre. However, the problems are the same as for cereal dietary fibres, or even worse due to high water absorption capacity of the some of fibre products mentioned.

In general, additions of fibre increase water absorption of dough. The particle size of the fibre affects water absorption; large particles absorb water more slowly than smaller ones. Also, the ratio of IDF and SDF affects on the ratio of water absorption (Haseborg and Himmelstein, 1988). According to Voit (1989) fibre type and amount of fibre significantly influence dough water absorption and dough development time. If the fibre type used has high water absorption potential, then the dough development time is longer. According to Krishnan *et al.* (1987) the development time of dough is longer if coarser wheat bran is utilized. The volume of bread decreased by 20% if the amount of the bran was 15% and the particle size was small. Also, the mixing time was reduced, especially if fine bran was used.

When dough containing a fibre is first mixed, clean-up is extremely rapid, the fibre seems to rapidly absorb nearly all of the free water, and the initial impression is that dough requires more water. After a few minutes of further mixing the dough relaxes, becomes cohesive and uniform, and develops elastic properties and handling properties similar to a standard dough. This point approximates the optimum mixing time. High-fibre doughs have limited tolerance to over-mixing, and care is necessary to develop a thin, elastic gluten sheet and to avoid over-mixing (Stauffer, 1993). Depending on the fibre source, high-fibre doughs are often sticky and soft or relatively stiff (Haseborg and Himmelstein, 1988).

23.3.1 Volume and texture

Getting acceptable loaf volume with high-fibre breads is difficult to accomplish. Bran supplementation usually weakens the structure and baking quality of wheat dough and decreases bread volume and the elasticity of the crumb. Wheat bran at a substitution level of 20% has been reported to decrease loaf volume by 19% according to Salmenkallio-Marttila *et al.* (2001). The firmness of fresh fibre-enriched bread has been reported to be 41% higher compared with the bread without bran (Laurikainen *et al.*, 1998). It has been suggested that the deleterious effects of fibre addition on dough structure are due to the dilution of the gluten network, which in turn impairs gas retention rather than gas production. This has been detected in a microscopic examination in which major differences between the crumb structure of control and fibre-containing breads were detected (Pomeranz *et al.*, 1987). The crumb structure of wheat breads was composed of thin sheets and filaments which were essentially absent in fibre-enriched breads. However, this well-known effect of bran inclusion in reducing the volume of bread cannot be explained simply in terms of dilution of gluten forming proteins (Gaillard and Gallagher, 1988). According to Gan *et al.* (1992) the bran materials in expanded dough appear to disrupt the starch–gluten matrix and also restrict and force gas cells to expand in a particular dimension. This greatly distorts the gas cell structure and may contribute to the resultant crumb morphology which is an important element of crumb texture. Thus, supplementation of dietary fibre requires changes in processing techniques for production of baked goods with good quality for consumers.

The particle size distribution of bran is important in determining the texture of baked products; for example, fine bran added to flour in CBP baking produces a more dense crumb than in loaves baked from flour to which coarse bran is added. Loaf texture influences consumer acceptability and there is evidence that the physiological effects of bran as dietary fibre differ between fine and coarse bran. Particle size can also play a role in determining the stability of bran containing flour (Gaillard and Gallagher, 1988).

23.3.2 Flavour

There are limits to enriching baked products with cereal bran, as it has detrimental effect on sensory characteristics, appearance, mouth-feel, flavour and chewing characteristics when certain concentrations are exceeded. Enriching foods with high levels of untreated cereal bran is therefore problematic. Various processing methods, such as extrusion cooking and fermentation of bran, have been utilized to modify flavour properties of cereal bran (Meuser, 2001; Salmenkallio *et al.*, 2001). In spite of the possibilities available for modifying the requirement profiles of dietary fibre products, the fact remains that most of them are only suitable for certain applications. Many dietary fibre products are unsuitable for particular applications as they are either not neutral in flavour or have distinctive colour. Neutrality of flavour in particular is a major criterion determining the range of applications. It is shown that, from technological point of view, fibres derived from wrinkled peas, lupins and wheat straw have major advantages over cereal brans (Meuser, 2001). However, the suitability of selected fibre for particular application is always dependent on the characteristics of the product in question and the addition level of the fibre.

23.4 Improving the quality of high-fibre bread

Addition of fibre always necessitates adjustments to other process parameters, notably in the amount of water added to the dough, formulation, mixing and fermentation times, and handling and baking techniques. These adjustments are discussed in the following sections.

23.4.1 Recipe and process adjustments

According to Lai *et al.* (1989b) the largest volume of wheat bread supplemented with fibre is obtained by increasing the amount of dough water as much as possible without increasing stickiness of the dough. The exact amount of water depends on the type of fibre and the fibre enrichment level. Also, additions of margarine or bakery shortenings (up to 3% flour weight) will increase volume of the bread. Additions of fibre tend to increase final fermentation (proofing) time which can be counteracted by adding 1% more yeast in straight dough processes.

The processing of high-fibre doughs differs in some respects from that for regular doughs. If a sponge method is used, the sponge should be slightly unfermented. The gluten should be split, with half of it going into the sponge and the other half being added at the dough mixing state. If a straight dough method is used, intermediate fermentation should be decreased, perhaps by as much as half (Stauffer, 1993).

23.4.2 Pre-treatment of fibre

Reducing the particle size of the fibre decreases its hydration properties and this effect is dependent on the fibre type used (Mongeau and Brassard, 1982). Probably this is due to changes in fibre particles absorbing water during milling. Milling decreases swelling power and water absorption (Auffret *et al.*, 1994). Particle size of the wheat bran does not affect water absorption of dough. Doughs supplemented with fine bran have shorter mixing times compared with those supplemented with coarse bran. Mixing stability is, however, better with coarse bran. Use of coarse bran also produces bigger volume and finer structure to the bread compared with fine bran.

Chemical (acid, base, ethanol) and enzymatic (amylase, protease) modification of wheat bran have been used to improve functional properties of wheat bran (Rasco *et al.*, 1991). These treatments enhance the dietary fibre content of wheat bran to 63–78%. However, the mixing and baking properties of such modified ingredients are adversely affected. Optimal mixing time increased and loaf volume and crumb grain decreased for yeast raised breads (sponge and dough method) containing the treated fibre ingredients relative to wheat bran at the same flour substitution level (10–20% by weight).

23.4.3 Use of baking aids and enzymes

The poor baking performance which may arise from fibre enrichment can be compensated for by adding dry gluten into the flour (Seibel and Brummer, 1991). If the amount of added fibre is over 10%, a flour protein content of around 16% is recommended and the amount of water used should be sufficient to hydrate both the added gluten and fibre components properly (Stear, 1990). The oxidation requirements for a high-fibre dough are much higher than those for a similar standard dough. Ascorbic acid seems to be particularly effective, and 150–200 ppm is routinely used. Where allowed, potassium bromate is used at the 30–50 ppm level, and azocarbonamide (ADA) at up to 25 ppm helps to maintain strength during proofing (Stauffer, 1993). According to Sosulski and Wu (1988) potassium bromate and sodium stearoyl lactylate (SSL) improve the properties of high-fibre wheat breads without the need for the addition of dry gluten. Potassium bromate increases loaf volume and SSL improves texture and gives a more uniform pore distribution. Gums, particularly the high-viscosity types such as guar, locust bean, xanthan, or sodium carboxymethylcellulose, are frequently added to high-fibre formulations at levels between 0.1 and 0.25% flour weight. These gums are supposed to improve gas retention and oven-spring, but the reported effects of gums are contradictory (Stauffer, 1993).

Additions of enzymes improve high-fibre bread quality. According to Haseborg and Himmelstein (1988) additions of two commercial hemicellulases improved dough machinability and oven-spring. Also, bread specific volume was improved and the staling rate decreased. Similar results were obtained when mixtures of xylanases were used (Laurikainen *et al.*, 1998). However, the use of xylanases also increased dough stickiness and softness. Thus, a careful balance is needed when xylanases are used in order to obtain improved bread quality without sacrificing dough machinability. The effectiveness of xylanases in improving bread volume is contributed to result in the redistribution of water from the pentosan phase to the gluten phase. The increase in gluten volume fraction gives the gluten more extensibility, which eventually results in a better oven-spring.

Amylases and proteases also improve the volume of high-fibre breads and soften the crumb structure (Moss, 1989). Furthermore, combinations of amylase and calcium oxide have improved high-fibre bread baking quality (Rasco *et al.*, 1991). The ability of fungal alpha-amylase to improve bread volume has been explained on the basis of increased oven-spring and dough gas holding capacity. Alpha-amylase probably lowers starch gelatinization viscosity which makes possible better oven-spring (Cauvain and Chamberlain, 1988). Thermo-stable alpha-amylase is effective at decreasing bread staling rate by partly hydrolysing starch and producing dextrins (Maleki *et al.*, 1972; Martin *et al.*, 1991). The functionality of proteases in high-fibre baking is based on their ability to shorten the mixing time, improve dough handling properties, improve gas-holding capacity, soften the crumb structure and improve flavour (Kulp, 1993).

23.4.4 Pre-ferments

Sour-dough fermentation is used to improve the texture, volume and keeping quality of breads (Brümmer and Unbehend 1997). Lactic acid bacteria cause acidification of the dough, proteolysis of gluten and moderate hydrolysis of starch. This modification of dough components may affect the physicochemical changes occurring during bread shelf-life (Corsetti *et al.*, 1998). The flour type, temperature, water content and type and amount of added yeast and bacteria all affect the acidification of the dough and bread quality (Brümmer and Unbehend, 1997; Rouzaud and Martínez-Anaya, 1997). Flour type has been shown to be the main factor in the acidification of sour doughs, with most acid produced in sour doughs made from wholemeal flour (Salovaara and Valjakka 1987; Hansen and Hansen, 1994).

Pre-soaking of wheat bran for 1 hour has been reported to have beneficial effects on loaf volume (Lai *et al.*, 1989a,b). Hydration of bran (15 minutes in an excess water at 10 or 97 °C) before addition to the dough increased loaf volume and improved bread quality in wheat bread containing 12% bran (Nelles *et al.*,

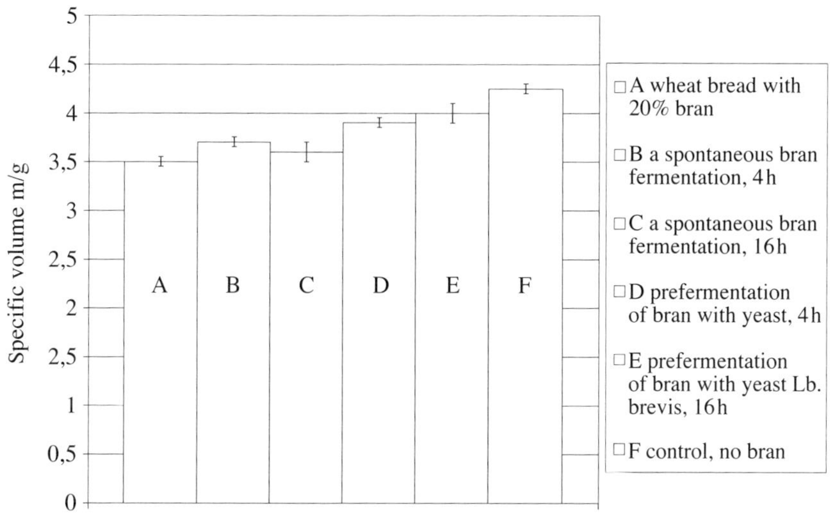

Fig. 23.1 Effect of bran pre-fermentation on the volume of bread supplemented with 20% bran.

1998). Nelles *et al.* (1998) proposed several possible mechanisms for the observed improvement: improved hydration of all brown flour components, lipoxygenase activation and a washing out of free reduced-glutathione. Longer soaking times (4–16 h) or intermediate temperatures (25–30 °C) have not improved loaf volume (Salmenkallio-Marttila *et al.*, 2001). Also, as the bran layer of the kernel is always enriched with various microbes, such as *Bacillus subtilis*, soaking of the bran in conditions which favour the growth of microbes has to be done with care in order to avoid any microbiological risks.

Pre-fermentation of bran with yeast or in particular, with yeast and lactic acid bacteria, improves loaf volume (see Fig. 23.1) and crumb softness (see Fig. 23.2) according to Salmenkallio-Marttila *et al.* (2001). Thus, sour dough technology can also be utilised effectively in high-fibre baking. The effectiveness of fermentation is assumed to be due to activity of endogenous enzymes of flour, especially the amylases and proteases, as well as the enzymes produced by yeast and lactic acid bacteria. The acids produced during fermentation lower the pH of the dough, thereby affecting the enzyme activity and gluten characteristics. The pH optima of carbohydrate-degrading enzymes, such as amylase, pentosanase or cellulase, vary widely (3.6–5.6) depending on wheat variety and germination status (Fox and Mulvihill, 1982). Proteinases and peptidases in flour are active at pH levels ranging between 4 and 9, depending on the substrate. The rapid drop in pH level in sour dough can cause reduced amylolytic activity, whereas the more gradual fall in pH level in spontaneously fermented dough permits further starch degradation (Wehrle and Arendt, 1998). Modifications in starch during fermentation alter the gelatinization characteristics of the starch granules (Siljeström *et al.*, 1988; Eynard *et al.*, 1995).

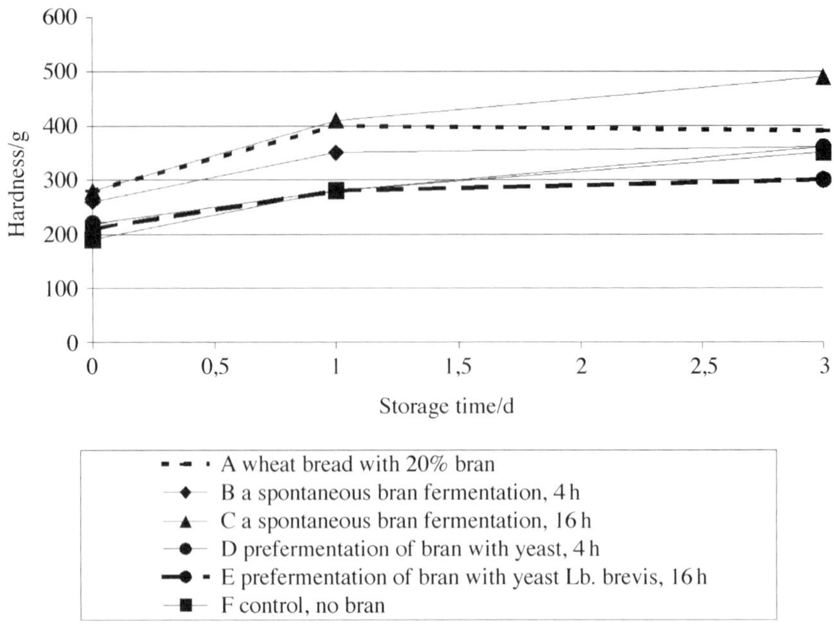

Fig. 23.2 Effect of bran pre-fermentation on crumb firmness of bread supplemented with 20% bran.

23.5 Future trends

There is growing amount of evidence that high-fibre diets, especially those containing cereal fibre, have definite health benefits in reducing risk of chronic diseases such as diabetes, cancer and coronary heart disease. A high-fibre intake is also positively related to control of obesity and physical gastrointestinal tract disorders. As the life expectancy of the population is growing in Western countries, the health-related issues, such as functional foods and individual well-being, will certainly be very appealing topics for the majority of the population also in future decades. Consumers will be interested in buying functional cereal products which will promote their heath and help them avoid being over-weight. High-fibre cereal products will be thus asked for and undoubtedly bought. However, as always with food items, the major criteria for consumer acceptability are good flavour and texture in cereal products. In other words, consumers expect functional cereal products such as high-fibre breads to have at least similar good quality attributes as in standard wheat bread.

Traditionally, introducing fibre into baked products in such amount that health benefits can be expected have meant unavoidable deterioration of quality, both in terms of the flavour and texture of cereal products. However, recent studies of the role of fibre in baking have shown that there are now several options for bakers which overcome problems related to high-fibre baking. Fibre ingredients with improved technological and flavour properties have been developed during the last decade. Also, by employing different process and recipe modifications in high-fibre baking, it is possible to produce high-quality, high-fibre cereal products with definite health benefits and consumer acceptability.

23.6 References

ANDERSON, J. W. and SIESEL, A. E. (1990) Hypocholesterolemic effect of oat products. In Furda, I. and Brine, C. *New Developments of Dietary Fibre: Physiological and Analytical Aspects*, New York, Plenum Press, pp. 17–36.

AUFFRET, A., RALET, M.-C., GIULLON, F., BARRY, J.-L. and THIBAULT, J.-F. (1994) Effect of grinding and experimental conditions on the measurement of hydration properties of dietary fibres, *Lebensmittel Wissenchaft und Technologie*, **27**, 166–172.

BRÜMMER, J-M. and UNBEHEND, G. (1997) Einflussfaktoren auf Weizenvorteige und deren Auswirkungen auf die Brotqualität. *Getreide Mehl Brot*, **51**, 345–349.

CAUVAIN, S.P. and CHAMBERLAIN, N. (1988) The bread improving effect of fungal alfa-amylase. *Journal of Cereal Science*, **8**, 239–248.

CORSETTI, A., GOBBETTI, M., BALESTRIERI, F., PAOLETTI, F., RUSSI, L. and ROSSI, J. (1998) Sourdough lactic acid bacteria effects on bread firmness and staling. *Journal of Food Science*, **63**, 347–351.

DREHNER, M. L. (1987) *Handbook of Dietary Fibre – An Applied Approach*, New York, Marcel Dekker, pp. 2–3.

EYNARD, L., GUERRIERI, N. and CERLETTI, P. (1995) Modifications of starch during baking: studied through reactivity with amyloglucosidase. *Cereal Chemistry*, **72**, 594–597.

FOX, P. F. and MULVIHILL, D. M. (1982) Enzymes in wheat, flour, and bread. In *Advances in Cereal Science and Technology*, Vol. 5, St. Paul, MN, Am. Assoc. Cereal Chem., pp 107–156.

GAILLARD, T. and GALLAGHER, D. M. (1988) The effects of wheat bran particle size and storage period on bran flavour and baking quality of bran/flour blends. *Journal of Cereal Science*, **8**, 147–154.

GAN, Z., GALLIARD, T., ELLIS, P. R., ANGOLD, R. E. and VAUGHAN, J. G. (1992) Effect of the outer bran layers on the loaf volume of wheat bread. *Journal of Cereal Science,* **15**, 151–163.

HANSEN, Å. and HANSEN, B. (1994) Influence of wheat flour type on the production of flavour compounds in wheat sourdoughs. *Journal of Cereal Science*, **19**, 185–190.

HASEBORG, E. and HIMMELSTEIN, A. (1988) Quality problems with high-fibre breads solved by use of hemicellulase enzymes. *Cereal Foods World,* **33**, 419–422.

KRISHNAN, P.G., FAUBION, J.M. and HOSENEY, R. C. (1987) Effect of commercial oat bran on the characteristics and composition of bread, *Cereal Chemistry*, **64**(1), 55–58.

KULP, K. (1993) Enzymes as dough improvers. In Kamel, B. S. and Stauffer, C. E. *Advances in Baking Technology*, Glasgow, Blackie Academic & Professional, pp. 153–178.

LAI, C. S., DAVIS, A. B. and HOSENEY, R. C. (1989a) Production of whole wheat bread with good loaf volume. *Cereal Chemistry*, **66**, 224–227.

LAI, C. S., HOSENEY, R. C. and DAVIS, A. B. (1989b) Effects of wheat bran in breadmaking. *Cereal Chemistry*, **66**, 217–219.

LAURIKAINEN, T., HÄRKÖNEN, H., AUTIO, K. and POUTANEN, K. (1998) Effects of enzymes in fibre-enriched baking. *Journal of Science and Food Agriculture*, **76**, 239–249.

MCCLEARY, B. and PROSKY, L. (2001) *Advanced Dietary Fibre Technology*, Blackwell Science, Oxford, UK.

MALEKI, M., SCHULZ, A. and BRUMMER, J. M. (1972) Uber das Altbackenwerden von Brot. II. Frischhalteeffekt von Bakterien-, Pilz und getreide-Alpha-Amylasen. *Getreide, Mehl- und Brot*, **26**, 221–224.

MARTIN, M. L., ZELEZNAK, K. J. and HOSENEY, R. C. (1991) A mechanism of bread firming. I Role of starch swelling. *Cereal Chemistry*, **68**, 498–503.

MONGEAU, R. and BRASSARD, R. (1982) Insoluble dietary fibre from breakfast cereals and brans: bile salt binding and waterholding capacity in relation to particle size. *Cereal Chemistry*, **59**, 413–417.

MEUSER, F. (2001) Technological aspects of dietary fibre. In McCleary, B. V. and Prosky, L. *Advanced dietary fibre technology,* Blackwell Science, Oxford, UK, pp 248–267.

MOSS, R. (1989) Wholemeal bread quality-processing and ingredient interactions, *Food Australia*, **41**, 694–697.

NELLES, E. M., RANDALL, P. G. and TAYLOR, J. R. N. (1998) Improvement of brown bread quality by prehydration treatment and cultivar selection of bran. *Cereal Chemistry*, **75**, 536–540.

NEWMAN, R. K., NEWMAN, C. W. and GRAHAM, H. (1989) The hypocholesterolemic function of barley β-glucans. *Cereal Foods World*, **34**, 25–51.

POMERANZ, Y., SHOGREN, M., FINNEY, K. F. and BECHTEL, D. B. (1987) Fiber in breadmaking – effects of functional properties. *Cereal Chemistry*, **54**, 24–41.

RASCO, B. A., BORHAN, M., YEGGE, J. M., LEE, M. H., SIFFRING, K. and BRUINSMA, B. (1991) Evaluation of enzyme and chemically treated wheat bran ingredients in yeast-raised breads. *Cereal Chemistry*, **68**, 295–299.

ROUZAUD, O. and MARTÍNEZ-ANAYA, M. A. (1997) Relationships between biochemical and quality-related characteristics of breads, resulting from the interaction of flour, microbial starter and the type of process. *Zeitschrift fur Lebensmittel Untersuchung und Forschung,* **204**, 321–326.

SALMENKALLIO, M., KATINA, K. and AUTIO, K. (2001) Effect on bran fermentation on quality and microstructure of high-fibre wheat bread. *Cereal Chemistry,* **78**(4), 429–235.

SALOVAARA, H. and VALJAKKA, T. (1987) The effect of fermentation temperature, flour type, and starter on the properties of sour wheat bread. *International Journal of Food Science and Technology*, **22**, 591–597.

SEIBEL, W. (1975) Kleie fur die menschlighe Ernährung. *Die Muhle und Mischfuttertechnik*, **112**, 669–970.

SEIBEL, W. (1983) Anreching von Brot und Backverhalten, *Getreide Mehl und Brot*, **12**, 377–379.

SEIBEL, W. and BRUMMER, J-M. (1991) Ballaststoffe und Backverhalten. *Getreide Mehl und Brot*, **45**, 212–216.

SIEVERT, D., CZUCHAJOWSKA, Z. and POMERANZ, Y. (1991) Enzyme-resistant. III. X-ray diffraction of autoclaved amylomaize VII starch and enzyme-resistant starch residues. *Cereal Chenistry*, **68**(1), 86–91.

SILJESTRÖM, M., BJÖRCK, I., ELIASSON, A-C., LÖNNER, C., NYMAN, M. and ASP, N. G. (1988) Effects on polysaccharides during baking and storage of bread – *in vitro* and *in vivo* studies. *Cereal Chemistry*, **65**, 1–8.

SOSULSKI, F. W. and WU, K. K. (1988) High fibre breads containing field pea hulls, wheat, corn, and wild oat brans. *Cereal Chemistry*, **65**, 86–191.

STAUFFER, C. E. (1993) Dietary fibre: analysis, physiology and calorie reduction. In Kamel, B. S. and Stauffer, C. E. *Advances in Baking Technology*, Glasgow, Blackie Academic & Professional, pp. 371–396.

STEAR, C. A. (1990) *Handbook of Breadmaking Technology*, London, Elsevier Science Publishers Ltd, pp. 30–31, 524–525.

VOIT, M. (1989) Dietary fibre addition and the water absorption of bakery doughs, Food Ingredients Europe. Conference Proceedings, Paris, 27–29 September 1989, pp. 186–198.

WEHRLE, K. and ARENDT, E. K. (1998) Rheological changes in wheat sourdough during controlled and spontaneous fermentation. *Cereal Chemistry*, **75**, 882–886.

24

Mould prevention in bread

N. Magan, M. Arroyo and D. Aldred, Cranfield University, UK

24.1 Introduction: the problem of mould in bread

Bread in various shapes and sizes is a staple food world-wide. In the UK annual bread consumption is in the region of 41.5 kg per person (Anon, 1992). Bread in the UK is light in texture, almost exclusively made from wheat flour and leavened by yeast fermentation. In other European countries including Germany, Denmark, Sweden and many Eastern European countries, rye breads made by sourdough fermentation processes and with a low pH are popular. In other parts of the world unleavened breads are more common.

Bread is a highly perishable product. The three most common forms of bread deterioration are staling, moisture loss and microbial spoilage (Seiler, 1984; Legan, 1993). Losses due to mould spoilage are difficult to quantify. However, a conservative estimate of 1% would result in losses of £20 million in the UK alone every year. Across Europe, the economic costs would be in the region of 5–10 times greater.

The reason why moulds are important spoilage organisms in bread is that this food matrix has a relatively high moisture content and water activity (a_w = 0.94–0.97) with a pH of about 6. These bread properties are conducive to germination and growth of any moulds contaminating the bread during or post-production. The bread most prone to spoilage by moulds is sliced, prepacked and wrapped bread. This type of product provides moist cut surfaces for moulds to grow on and wrapping prevents moisture loss allowing a humid atmosphere to form around the loaf. More than 90% of contamination of bread with moulds occurs during cooling, slicing or wrapping operations (Spicher, 1980). Prior to this stage the heating regimes employed in baking mean that most contaminants are eliminated (Legan, 1993; Roessler and Ballenguer, 1996).

Thus contamination is predominantly by fungal spores being deposited from the bakery environment, from flour dust and by introduction from the outside atmosphere. Many different filamentous mould species have been implicated in mould spoilage of bread including *Penicillium, Aspergillus* species, *Cladosporium* species, *Mucorales* and *Neurospora* (Legan, 1993). Tolerance to a wide range of environmental conditions, and their predominantly mycelial growth habit enable them to colonise food products rapidly producing a battery of enzymes to utilise the food matrix. Table 24.1 compares the growth of some spoilage moulds in relation to environmental factors on bread analogues in the laboratory. This shows that growth of the mould species was generally faster at pH 6 than 4.5 and relative growth rate was decreased as the a_w was decreased, regardless of pH. *Eurotium repens* was the fastest growing species, followed by the mycotoxigenic *Aspergillus ochraceus* and *Penicillium verrucosum* strains. Other spoilage *Penicillium* species grew more slowly.

The predominance of *Penicillium* may partially be due to their ability to grow over a wide range of temperatures and water availabilities, and the profuse production of spores which are ubiquitous in the atmosphere. They predominate at cooler temperatures. However, at 22–24 °C there is a 50% reduction in *Penicillium* contamination, and in warmer climates *Aspergillus* and *Eurotium* species prevail. Some of the moulds contaminating bread are also able to produce toxic secondary metabolites. Thus, for a prolonged shelf-life of bread the control of contamination with spoilage moulds and their toxins is critical. This chapter will examine the current methods for mould control and their limitations, consider the developing new methods of mould control and examine possible future trends.

Table 24.1 Effect of environmental factors on colonisation (diametric growth rate, mm day^{-1}) of bread analogues by spoilage moulds at 25 °C (Arroyo and Magan, unpublished data)

	pH 4.5 Water activity			pH 6.0 Water activity		
	0.93	0.95	0.97	0.93	0.95	0.97
Eurotium repens	2.8	2.6	4.6	4.1	7.5	9.3
Penicillium verrucosum (M450)	0.8	1.3	1.4	1.0	1.8	1.9
P. verrucosum (PV3)	0.8	1.0	1.6	1.1	2.2	2.3
Aspergillus ochraceus	1.5	2.3	2.7	1.8	2.8	3.3
Penicillium coryolophilum	0.2	0.4	0.7	0.7	1.2	1.6
Penicillium roquefortii	1.1	1.7	2.2	0.8	1.3	1.4

24.2 Current techniques for mould control and their limitations

Control of mould spoilage in bakery products can be achieved in various ways. Generally, this is achieved by (a) restricting the access of the spoilage mould to the product, (b) inactivating the fungal material and (c) inhibiting growth of the fungus. However, if the fungus gains access to the product the objective turns into controlling its activity and growth on the food itself. For inactivating or inhibiting fungal growth in foods, several physical, chemical and biological measures can be taken.

The most common way to prevent or control mould growth in foodstuffs is by the use of anti-fungal agents. Anti-fungal agents are chemical substances that when added to foods tend to prevent or retard food spoilage by moulds. In practice, most are fungistatic and not fungicidal. Thus they stop germination and growth when present, but growth may occur from untreated pockets. Fungicidal compounds are more effective as they destroy the spoilage moulds directly. Of course the concentration of the anti-fungal agent to a large extent impacts on the safe shelf-life of a food product. Smid and Gorris (1999) suggested that ideally any anti-microbial substance should inhibit microorganisms in their initial lag phase of growth and not in the exponential log phase, since in the latter the necessary dosages of the agent would be too high and would most likely adversely affect the quality of the food (Smid and Gorris, 1999).

At the present time mould spoilage of bread is generally prevented by the addition of food grade preservatives such as propionic, sorbic and acetic acids and their salts. These acids are Generally Recognised As Safe (GRAS) (Liewen and Marth, 1985; Binstok *et al.*, 1998). Sorbic acid, for example, has a half-life in the body of about 40–110 min and in normal conditions it is completely oxidised to CO_2 and H_2O (Liewen and Marth, 1985). However, their potassium, sodium or calcium salts are the forms commonly used because of their higher water solubility and easier handling than their corrosive acids. The status and use of these materials are controlled in many countries by legislation that limits the type and concentration of preservatives that may be used (Table 24.2). This can be affected by the bread type or by other factors such as whether or not the bread is wrapped, and the type of wrapping used. In England and Wales, for example, sorbates are not permitted in all breads but they are permitted for flour confectionery goods at levels of up to 1000 ppm (Seiler, 1984). In other countries such as Germany, Italy and the Netherlands, sorbic acid and its salts are approved in certain types of bread. In the USA high-volume loaves similar to those produced in the UK have successfully been preserved using a spray of potassium sorbate applied immediately after baking (Killian and Krueger, 1983).

In the UK, as in many other countries, propionates are the chemical anti-microbial generally used to control moulds as well as bacterial (*Bacillus* spp.) spoilage of bread (Legan, 1993). Propionates are used mainly as their sodium, potassium or calcium salts because, although more expensive, they are less corrosive and easier to handle than the liquid acid. Their use is permissible at levels

Table 24.2 Different allowable preservatives which can be used in bread and bakery products under EU and UK regulations (89/107/EEC; 95/2/EC; 97/77/EC; SI 1995 No. 3187)

Potassium sorbate	(E202)	Regulated for bread
Sodium benzoate	(E221)	(1000–3000 ppm max. dose)
Calcium propionate	(E282)	
Propyl paraben	(E216)	Regulated for cosmetics; as a food additive with organic acids
Butylhydroxyanisole (BHA)	(E320)	Regulated for foods
Butylhydroxytoluene (BHT)	(E321)	(100–200 ppm max. levels)
Octyl gallate	(E311)	
Essential oils	(None)	No specific regulations

of not more than 0.3% (w/w) of propionic acid equivalent (Anon, 1984). Furthermore, propionates have little or no effect against yeast (Sauer, 1977) which make them highly suitable to control mould spoilage in yeast-raised bread.

But the use of weak acids in bread also has disadvantages. The low absolute efficacy of the propionates means that relatively high concentrations are needed in order to keep bread and other baked goods free of moulds for more than a few days (Lück and Jager, 1997). They are also fungistats and thus not lethal to spores, but delay the capacity for germination and growth. Also at such concentrations serious losses in volume and adverse effects on odour and flavour can occur. Using 0.2% calcium propionate, for example, a reduction of 5–10% of loaf volume occurs in commercial-scale baking because it reduces the yeast activity and alters the dough rheology. Sorbates have even greater adverse effects (Legan, 1993).

Incorporation of calcium propionate into bread at up to 0.3% concentration at pH 4.5 and 0.93–0.97 a_w can effectively control a range of spoilage moulds in bread. However, at sub-optimal concentrations some stimulation was observed, and at pH 6, practically no control was achieved against *E. repens*, *P. verrucosum*, *A. ochraceus*, *P. coryolophilum* and *P. roqueforti* (Arroyo and Magan, unpublished data). Sorbic acid and its salts are among the most thoroughly investigated of all preservatives. Studies with 0.3% (w/w) potassium sorbate on bread have also shown effective control of spoilage moulds at pH 4.5 and 0.93–0.97 a_w with the exception of *P. roquefortii*. However, at pH 6.0 practically no control was achieved with the recommended 0.3% treatment. Furthermore, when calcium propionate concentration was reduced to 0.03% stimulation of *E. repens*, and *P. coryolophilum* and *P. roqueforti* was observed. Recent studies with sponge cakes treated with sodium benzoate or calcium propionate at up to 0.3% concentration and 0.80–0.90 a_w and pH 6 or 7.5 also showed that four *Eurotium* species were only effectively controlled at pH 6 and 0.80–0.85 a_w. Over all other conditions growth was not significantly controlled (Guynot *et al.*, 2002).

Weak acids are lipophilic acids that penetrate the cell membrane in the undissociated form. When the undissociated acid enters the cell a higher pH environment is encountered and the molecule dissociates, resulting in the release of charged anions and protons which cannot cross the plasma membrane. The high solubility, low taste and toxicity of weak organic acids make them highly suitable to be used in bread and bakery product preservation (Ray and Bullerman, 1982; Davison and Juneja, 1990). The pH of the environment and solubility of the acid often determines the foods in which these acids may be effectively used (Ray and Bullerman, 1982). In fact, because of their low pK_a value (4.19–4.87), these substances are effective anti-microbial materials in low pH substrates since this condition favours the uncharged, undissociated state of the molecule which is freely permeable across the membrane.

Alternatives to chemical preservation include destroying or damaging the mould spores that gain access to the surface during the cooling and wrapping processes. This can be achieved using UV light, IR or microwave irradiation. These types of procedures have been used for sourdough bread in continental Europe (Seiler, 1984) but, although effective, their bad publicity and the increase demand for minimally processed and 'fresh' products from the consumers limit their use. Furthermore, UV irradiation for example, does not penetrate the product so mould spores inside the loaf would not probably be affected.

Several other strategies have been followed to limit the rate of mould growth including:

- reformulating of recipes, e.g. by reducing the water availability but without adversely affecting the eating quality of the product or causing changes in volume, shape and texture;
- using novel ingredients such as raisin and prune juice concentrate that inhibit fungal growth (Sanders, 1991);
- using modified atmospheres (MAP) or other packaging techniques (Weng *et al.*, 1999; Abellana *et al.*, 2000).

MAP techniques are based on the fact that moulds are oxygen dependent and highly sensitive to carbon dioxide (Portier *et al.*, 1989; Farber, 1991), thus achieving environments with low O_2 and high CO_2 contents will protect the wrapped food against aerobic spoilage moulds and other microorganisms. Further studies have been focused on the development of polymers that contain the preservative as an active packaging material: in this case the diffusivity of the preservative from the packaging to the food becomes very important (Han and Floros, 1998).

While organic acids and their salts are the most commonly used preservatives in bakery products, there is pressure from EU Directives to reduce the use of these and instead use more natural preservatives such as anti-oxidants and essential oils either alone or in combination with packaging systems to stabilise shelf-life. Any reduction in the concentration of existing preservatives would result in significantly shorter shelf-life and more rapid moulding of bread, especially wrapped, cut varieties. Consumers are demanding more natural bakery products with a minimum of preservatives. Thus the question arises as to whether alternative, more 'natural' additives would be effective, and economically acceptable and feasible.

24.3 Developing new methods for mould control

24.3.1 Synthetic antioxidants

Phenolic-derived antioxidants have been screened for their possible anti-microbial efficacy. Butylated hydroxytoluene (BHT), butylated hydroxyanisole (BHA), propylgallate (PG) and 2-tert-butylhydroxybenzoic (TBHQ) are among them. Studies have been conducted using a range of anti-oxidants with different environmental factors to simulate conditions of bread. For example, screening of the inhibitory effect of anti-oxidants, such as butylated hydroxytoluene (BHT), butylated hydroxyanisole (BHA), *n*-propyl gallate, propyl paraben, octyl gallate at 0.95 a_w and pH 6. This has demonstrated that common bread contaminants such as *Cladosporium herbarum, Penicillium coryolophilum, P. verrucosum* and *Aspergillus ochraceus* were completely inhibited by propyl paraben, BHA and octyl gallate at 500 ppm. Importantly, these anti-fungal agents even at lower concentrations extended the lag phase prior to growth which is indicative of extending shelf-life. Kubo *et al.* (2001), comparing the anti-fungal activity of three gallates, propyl (C3), octyl (C8) and dodecyl (C12), found that octyl gallate was the only active compound against four different fungal genera with a minimum inhibitory concentration (MIC) of 25 ppm. Recent studies have shown that mycotoxigenic spoilage fungi such as *Fusarium* species and *Penicillium verrucosum* are significantly inhibited by both parabens and BHA, and in some cases inhibit mycotoxin production. Interestingly, combinations of anti-oxidants were sometimes more effective at lower concentrations than individual ones, indicating some synergy of efficacy (Etcheverry *et al.*, 2002; Cairns and Magan, 2003; Reynoso *et al.*, 2002; Torres *et al.*, 2003).

Owing to their high pK_a value (8.5), parabens are effective over a wider range of pH (3–8). Antimicrobial activity of parabens is related to the length of the ester group of the molecule. As additives, parabens are applied as alkali solutions or as ethanol or propyl glycol solutions in fillings for baked goods, fruit juices, marmalades, syrups, preserves, olives and pickled sour vegetables (Belitz and Grosch, 1999).

24.3.2 Essential oils (ESOs)

In the last few decades great interest has emerged in the possible use of plant extracts and essential oils for food preservation. Essential oils are mostly derived from spices, i.e. dried aromatic products, obtained from different parts of the plant such as leaves (e.g. rosemary, sage), flowers (e.g. clove), bulbs (e.g. garlic, onion) or fruits (e.g. pepper, cardamon) (Shelef, 1983; Deena and Thoppil,

2000). Extracts and essential oils of many of these plants are now being screened for their anti-microbial effectiveness. *Aspergillus flavus*, one of the most toxigenic foodborne fungi which can contaminate flour used for bakery product production, has been reported to be inhibited by some of these plant derivatives. For example, Dwivedi and Dubey (1993), studying the anti-fungal activity of several umbelliferous plant essential oil against *Aspergillus* species found an important fungistatic effect of *Trachyspermum* seed essential oil at relatively low concentrations (< 500 ppm). Azzouz and Bullerman (1982) established clove and cinnamon as the strongest anti-fungal agents against *Penicillium* and *Aspergillus* species. However, many of these studies neglected to simulate realistic environmental factors appropriate to the specific bakery product.

In some cases (Salmeron *et al.*, 1990) stimulation of growth of the same *Aspergillus* species has been demonstrated when extracts of thyme and oregano were incorporated into nutritive media. Recently, very useful studies were carried out by Lopez-Malo *et al.* (2002) on dose–response curves for *A. flavus* in relation to vanillin, thymol, eugenol, carvacrol and citral or potassium sorbate/sodium benzoate. This showed that over a 60 day period it was possible to predict sensitivity of the mould to each ESO compound and compare this with organic acids. This showed that *A. flavus* had a higher sensitivity to thymol, eugenol, carvacrol, potassium sorbate and sodium benzoate at pH 3.5 than to vanillin or citral. MICs varied from 200 ppm for the organic acids to 1800 ppm for citral.

Screening a range (20+) of different plant essential oils for their activity against four spoilage moulds, *A. ochraceus, C. herbarum, P. corylophilum* and *P. verrucosum,* in a wheat flour-based medium at 25 °C has shown that at least 500 ppm was required for control of growth. Indeed, only clove, thyme, bay and cinnamon completely inhibited growth of all the species studied (Table 24.3). The lag phase prior to growth was also significantly increased in these studies. This is an important indicator of potential for improving shelf-life prior to any visible growth over a range of water activities and temperatures. Similarly, Patkar *et al.* (1993) found that 500 ppm of cinnamon ESO completely inhibited growth of the aflatoxigenic species of *A. flavus* in yeast extract broth and on agar for 7 days, whereas up to 1250 ppm of clove ESO was necessary to exert the same inhibitory activity. Azzouz and Bullerman (1982) also reported a strong anti-mould activity of clove and cinnamon oil against several *Aspergillus* and *Pencillium* species.

Although many other studies have reported the strong effect of thyme and cinnamon ESO on the growth of mould species, results are not always similar. For instance, while 500 ppm of thyme oil completely inhibits growth of *A. ochraceus* on 2% wheat flour agar over 30 days (Arroyo and Magan, unpublished data), Paster *et al.* (1990) reported colonies on PDA (potato dextrose agar) of up to 35 mm diameter in the presence of the same concentration of this ESO.

Özcan (1998) reported growth of *Aspergillus parasiticus* on czapek dox agar in the presence of 1% thyme (wild and black) oil. Although sensitivity to a certain plant essential oil may vary with the species studied, these differences in

Table 24.3 Increase of lag phase (days) of four fungal species growing at 25 °C on 2% wheat flour agar in the presence of 500 ppm (w/w) of different plant essential oils at 0.95 a_w/pH 6 (Arroyo and Magan, unpublished data)

	Increase in lag phase (days)				
	Aspergillus ochraceus	*Cladosporium herbarum*	*Penicillium coryolophilum*	*Penicillium verrucosum* M450	*Penicillium verrucosum* PV3
Lime	0.2	1.5	0.1	0.8	1.0
Nutmeg	1.6	12.6	1.6	0.6	−0.3
Ginger	0.3	0.5	−0.5	−0.2	0.1
Clove	**24.9**	**21.4**	**21.7**	**22.0**	**24.6**
Basil-l	1.9	10.4	1.8	−0.7	1.6
Marjoram	0.2	2.1	0.0	−0.5	0.6
Thyme	**24.9**	**21.4**	**21.7**	**22.0**	**24.6**
Eucalyptus	−1.0	0.2	−0.4	−0.9	0.6
Bay	**24.9**	**21.4**	**21.7**	**22.0**	**24.6**
Sweet fennel	0.9	8.1	1.2	0.9	1.3
Sage	0.8	5.4	−0.6	0.3	1.3
Spearmint	3.3	11.4	7.8	8.9	3.3
Rosemary	−0.3	2.6	−1.6	−0.7	0.6
Lemongrass	4.7	21.4	3.9	4.0	5.3
Basil-m	−0.1	2.5	−0.5	−0.7	−0.2
Peppermint	1.7	7.6	−0.9	0.5	0.0
Pine-syl	−0.7	1.0	−0.6	−0.2	0.0
Grapefruit	−0.5	1.7	−0.5	0.0	1.5
Orange	0.0	6.0	0.5	0.6	0.8
Cinnamon	**24.9**	**21.4**	**21.7**	**22.0**	**24.6**

Note: Bold numerals indicate oils which inhibited moulds for the longest time.

sensitivity may be attributed to the fact that composition of the essential oil can vary with the plant origin and extraction method. In fact while in the current study thyme was a commercial essential oil obtained from Spanish plants, Paster *et al.* (1990) extracted the thyme oil themselves from leaves of Israeli plants.

It is notable that while some work has been done on the potential of using mixtures of anti-oxidants, practically no studies have examined mixtures of anti-oxidants/ESOs, and mixtures of ESOs or deodorised ESOs. However, the evidence from *in vitro* studies suggests that potential does indeed exist for exploiting both anti-oxidants and essential oils for use in bread and other bakery products.

24.3.3 *In situ* control of moulds in bakery products using anti-oxidants and essential oils

The efficacy of anti-oxidants against spoilage moulds in some bakery products has been examined. These studies have shown that the concentrations required for inhibition of growth were generally higher than that shown to be effective *in vitro*. For example, propyl paraben had little effect on growth of four important spoilage moulds, even at 1000 ppm. This suggests that either the anti-oxidants are being bound by ingredients and thus less effective, or that it is difficult to get effective dispersion in the product, providing less direct contact with the contaminant moulds.

Essential oils have been examined in two ways for control of spoilage moulds in bread. The volatiles produced by the essential oils have been used in bread packaging to inhibit spoilage moulds, and attempts have been made to directly incorporate low concentrations with bread ingredients. For the former, Nielsen and Rios (2001) recently examined the efficacy of volatiles in MAP systems for control of rye bread spoilage fungi. Mustard essential oil in the volatile phase at 1–10 $\mu g\,ml^{-1}$ was the most effective against spoilage fungi including *P. commune*, *P. roquefortii*, *Aspergillus flavus* and *Endomyces fibuliger*. Cinnamon, garlic and clove also had high activity in controlling growth *in situ* on slices of bread. Vanilla showed no inhibitory effects, and *A. flavus* was the most resistant of the species tested. Interestingly, the MIC varied with the active component (allylthiocynanate) and at least 3.5 $\mu g\,ml^{-1}$ was required for fungicidal effects on the test fungi. However, the introduction of small sachets or pads into packaging directly could be an effective way of enabling the slow release of volatiles of ESOs to control mould spoilage of bakery products, especially where this can be combined with modified atmosphere systems.

When the ESO is incorporated directly into the substrate, the effectiveness appears to be lower. In fact, up to 1000 ppm of oregano ESO was necessary to completely inhibit growth of *A. flavus* in bakery products for 21 days (Basilico and Basilico, 1999) while only 100 ppm was required to completely inhibit growth of *Aspergillus niger* and several *Penicillium* species over a period of 6 days *in vitro*. Chemically, ESOs consist of a mixture of esters, aldehydes, ketones and terpenes. Although several studies have been carried out on the inhibitory effect of essential oil components (Lachowicz *et al.*, 1998; Saxena and Mathela, 1996; Cosetino *et al.*, 1999; Mahmoud, 1994; Sinha and Gulati, 1990; Bilgrami *et al.*, 1992), the role of these components in the anti-microbial activity of the oil is not clear. Various oil components may act synergistically in the anti-microbial activity while others can stimulate fungal spore germination (French, 1985).

In recent studies over 20 essential oils and 5 anti-oxidants were examined for control of growth of mycotoxigenic species such as *Fusarium culmorum* and *Penicillium verrucosum* on grain for flour and bakery production (Hope and Magan, 2003; Cairns and Magan, 2003). Only three essential oils (bay, clove and cinnamon oil) and two anti-oxidants (propyl paraben and hydroxymethylanisole) were found to be effective in controlling growth in the range 50–200 ppm at 15 and 25 °C over the range 0.995 to 0.955 water activity (a_w). At 500 ppm clove oil and BHA effectively controlled growth and both deoxynivalenol and nivalenol toxin production by *F. culmorum* at 15 and 25 °C. Similarly effective control of *P. verrucosum* and of ochratoxin was achieved at 500 ppm. However, at intermediate concentrations (100 ppm) stimulation of this mycotoxin was sometimes observed.

Direct incorporation of the best ESOs into bread showed that efficacy was less effective than on a wheat flour-based medium, with 50% control of *E. repens* by cinnamon and clove, and only about 25% control of *A. ochraceus* by thyme. Practically no control of *P. verrucosum* and *P. coryolophilum* with any of the best *in vitro* treatments was observed (Arroyo and Magan, unpublished data).

These results suggest that when an essential oil is directly incorporated into the bakery product, less efficacy is observed, probably because specific components of the food product such as proteins or fats can bind essential oil components, inactivating them (McNeil and Schmidt, 1993; Smid and Gorris, 1999). The requirement for higher doses than those required *in vitro* for incorporation into bakery products for effective control means that their typical odour is also noticeable. In some products this flavour may be desirable. For example, in tomato-based bakery products, the addition of basil can be used both as a flavouring and as an anti-fungal agent (Lachowicz *et al.*, 1998). This has particular applications in novel bread products where herbs and spices are often incorporated into the bread. The use of the volatiles produced by essential oils may be a more effective way of using low concentrations of ESOs in conjunction with MAP systems. This needs to be investigated further as an alternative approach to substitute for organic acids.

24.3.4 Biopreservatives

Microorganisms produce several compounds that control the growth of spoilage microorganisms present in their environment. Anti-microbial materials of microbial origin used in food commodities, including sour dough breads, are known as 'biopreservatives'. Nevertheless, when talking about 'biopreservatives' this almost exclusively refers to those anti-microbials produced by lactic acid bacteria (LAB). Lactic acid bacteria produce GRAS anti-bacterial compounds with a broad anti-microbial spectrum, like organic acids or hydrogen peroxide, and compounds with a narrower spectrum called bacteriocins (Smid and Gorris, 1999). Bacteriocins are peptide antibiotics primarily lethal to other strains and species of bacteria (Konings *et al.*, 2000). LABs produce many different types of bacteriocins with possible applications in food preservation, and in bread. Nisin is now the only bacteriocin legally approved for use as a food additive (Gänzle *et al.*, 1999). Recent studies by Lavermicocca *et al.* (2000) showed that an anti-fungal compound obtained from *L. plantarum* 21B isolated from sourdough was effective at controlling germination and growth of a range of spoilage moulds including *Eurotium* spp., *P. coryolophilum*, *P. expansum*, *Endomyces fibuliger*, *A. niger* and *Fusarium graminearum in vitro* at 50 mg ml^{-1}. Sourdough bread produced with this LAB contaminated with *A. niger* was demonstrated to have a better shelf-life than that without this bacterial strain. Thus some metabolites produced by LABs have potential for effective control of spoilage moulds in some types of bakery products.

24.4 Future trends

Since the 1980s the bread industry has been working to reduce the number of additives and so-called synthetic preservatives in a genuine effort to make bread as natural and fresh as possible (Casdagli, 2000). As modern consumers prefer minimally processed high-quality foods with few or no additives, several studies have been focused in the last few years on the use of natural preservatives such as plant essential oils. Novel biotechnology techniques are continuously being developed.

Certain combinations of preservatives, e.g carbon dioxide (in storage and packaging) and ethanol or sorbic acid (Vora *et al.*, 1987; Smith *et al.*, 1988), or vinegar and calcium propionate (0.10% each) (McNaughton *et al.*, 1998), have proved to be useful measures to improve the life of baked goods. Javanainen and co-workers have studied the possibility of producing propionates *in situ* as preservatives through well-balanced and controlled mixed-culture fermentations in order to retard mould growth and thus to extend the shelf life of the product (Javanainen and Linko, 1993; Javanainen and Linko, 1994; Linko *et al.*, 1997).

Overall, the development and use of more natural additives for bread and other bakery products is to a large extent driven by legislation and consumer pressure for the delivery of food with a minimum of chemical additives. However, industry will consider the alternative types of so-called natural preservatives, especially essential oils, only if they are economically viable in the mass production systems used today for the delivery of bread products to the marketplace. The price of essential oils or anti-oxidants must be relatively similar to existing preservatives such as calcium propionate. While no directives or regulations exist at present for the use of essential oils as preservatives, they are used as flavourings in some bakery products. Thus in some bread products they may be appropriate. Another consideration is the concentration at which these newer preservatives may be effective. The limited evidence to date suggests that very high concentrations (>1000 ppm) are required for effective control, and at these concentrations the essential oil odour dominates the product. At the present time, the costs of anti-oxidants are comparable with existing organic acid treatments. However, essential oils are between two and five times more expensive. This also needs to be taken into account. Potential does exist for exploiting natural preservatives but at the present time this may only be in specialised high-value bread products for specific niche markets where they would be economically feasible.

24.5 Sources of further information and advice

DE BOER, E. and NIELSEN, P.V. (1995). Food Preservatives *In*: *Introduction to Food Borne Fungi*, R.A. Samson; E.S. Hoekstra Eds., Ponsen & Looyen, Wageningen, The Netherlands.

GOULD, G.W. (1996a). Methods for preservation and extension of shelf life, *Int J Food Microbiol* **33**, 51–64.

GOULD, G.W. (1996b). *New Methods for Food Preservation*. Blackie Academic & Professional, London.

LEGAN, J.D. (1993). Mould spoilage of bread: the problem and some solutions, *Int Biodet & Biodegr* **32**, 35–53.

MOSSEL, D.A.A., CORRY, J.E.L., STRUIJK, C.B. and BAIRD, R.M. (1995). *Essentials of the Microbiology of Foods. A textbook for Advanced studies.* Academic Press, London.

NYCHAS, G.J.E. (1995). Natural antimicrobials from plants *In*: *New Methods of Food Preservation*, Ed. G.W. Gould, Blackie Academic and Professional, Glasgow, 58–89.

24.6 References

ABELLANA, M., RAMOS, A.J., SANCHIS, V. and NIELSEN, P. (2000). Effect of modified atmosphere packaging and water activity on the growth *of E. amstelodami, E. chevalieri* and *E. herbariorum* on a sponge cake analogue, *J Appl Microbiol*, **88**, 606–616.

ANON (1984). Bread and flour regulations 1984. Statutory Instrument, HMSO, London.

ANON (1992). Household consumption and expenditure on cereal-based foods. *Home Grown Cereals Authority Weekly Digest*, **18**, 2–3.

AZZOUZ, M.A. and BULLERMAN, L.B. (1982). Comparative antimycotic effects of selected herbs, spices, plant components and commercial antifungal agents, *J Food Prot*, **45**, 1298–1301.

BASILICO, M.Z. and BASILICO, J.C. (1999). Inhibitory effects of some spice essential oil on *Aspergillus ochraceus* NRRL 3174 growth and ochratoxin production, *Lett Appl Microbiol*, **29**, 238–241.

BELITZ, H.D. and GROSCH, J.C. (1999) *Food Chemistry*. Springer.

BILGRAMI, K.S., SINHA, K.K. and SINHA, A.K. (1992). Inhibition of aflatoxin production & growth of *Aspergillus flavus* by eugenol & onion & garlic extracts, *Indian J Med Res,* **96**, 171–175.

BINSTOK, G., AMPOS, C., VARELA, O. and GERSCHENSON, L.N. (1998). Sorbate-nitrite reactions in meat products, *Food Res Inter*, **31**(8), 581–585.

CAIRNS, V. and MAGAN, N. (2003). Impact of essential oils on growth and ochratoxin A production by *Penicillium verrucosum* and *Aspergillus ochraceus* on a wheat-based substrate. In *Advances in Pest Control*, Cabi International, Wallingford.

CASDAGLI, T. (2000). Bakery industry 2000+: the challenges ahead, *Food Sci Techn Today*, **14**, 87–90.

COSSENTINO, S., TUBEROSO, C.I.G., PISANO, B., SATTA, M., MASCA, U., ARZED, E. and PALMAS, F. (1999). In vitro antimicrobial activity and chemical composition of sardinian thyme essential oils. *Letters in Applied Microbiology* **29**, 130–135.

DAVISON, P.M. and JUNEJA, V.K. (1990). Antimicrobial agents *In: Food Additives*. Eds. A.L Branen, P.M. Davison and S. Salminen, Marcel Dekker, Inc., New York.

DEENA, M.J. and THOPPIL, J.E. (2000). Antimicrobial activity of the essential oil *of Lantana camara Fitoter, Fitotepapia*, **71**, 453–455.

DWIVEDI, S.K. and DUBEY, N.K. (1993). Potential use of the essential oil of *Trachyspemum ammi* against seed borne fungi of Guar, *Mycopathology*, **121**, 101–104.

EARLE, M.D. and PUTT, G.J. (1984). Microbial spoilage and use of sorbate in bakery products. *Food Technol NZ*, **19**, 25–36.

ETCHEVERRY, M., TORRES, A. RAMIREZ, L.M., CHULZE, S. and MAGAN, N. (2002). In vitro control of growth and fumonisin production by Fusarium verticillioides and F. proliferatum using antioxidants under different water availability and temperature regimes. *Journal of Applied Microbiology* **92**, 624–632.

FARBER, J.M. (1991). Microbiological aspects of modified atmosphere packaging technology: a review, *Journal of Food Protection*, **54**, 58–70.

FRENCH, R.C. (1985). The bio-regulatory action of flavour compounds on fungal spores and other propagules, *Ann Rev. Phytopathol*, **23**, 173–199.

GÄNZLE, M.G., WEBER, S. and HAMMES, W.P. (1999). Effect of ecological factors on the inhibitory spectrum and activity of bacteriocins, *Int J Food Microbiol*, **46**, 207–217.

HOPE, R. and MAGAN, N. (2003). Multi-target environmental approach for control of growth and toxin production by *Fusarium culmorum* using essential oils and antioxidants. *Advances in Pest Control*, Cabi International, Wallingford.

JAVANAINEN, P. and LINKO, Y.Y. (1993). Mixed-culture pre-fermentation of lactic and propionic acid bacteria for improving-wheat bread shelf-life. *Journal of Cereal Science* **18,** 75–88.

JAVAINEN, P. and LINKO, Y.Y. (1994). Mixed culture preferment of lactic and propionic acid bacteria for improving bread shelf-life. *Progress in Biotechnology* **9**, 627–630.

KILLIAN, D. and KRUEGER, J. (1983). Potassium sorbate spray eliminates returns due to mould, *Bak Ind*, **150**, 54–55.

KONINGS, W.N., MURANDI, F., STEIMAN, R. and CREEPY, E.E. (1995) Fungal flora and ochratoxin A production in various food and feeds in Europe. *Systematic and Applied Microbiology* **18**, 455–459.

KUBO, I., XIAO, P. and FUJITA, K. (2001). Antifungal activity of octyl gallati: structural criteria and mode of action. *Bioorganic and Medicinal Chemistry Letters*, **11**, 347–350.

LACHOWICZ, K.J., JONES, G.P., BRIGGS, D.R., BIENVENU, F.E., WAN, J., WILCOCK, A. and

COVENTRY, M.J. (1998). The synergistic preservative effects of the essential oils of sweet basil (*Ocimum basilicum* L.) against acid-tolerant food microflora, *Lett Appl Microbiol*, **26**, 209–214.

LAVERMICOCCA, P., VALERIO, F., EVIDENTE, A., LAZZARONI, S., CORSETTI, A. and GOBBETTI, M. (2000). Purification and characterization of novel antifungal compounds by sourdough *Lactobacillus plantarum* 21B. *Appl Environ Microbiol*, **66**, 4084–4090.

LEGAN, J.D. (1993). Mould spoilage of bread: the problem and some solutions. *Int Biodet & Biodegr*, **32**, 35–53.

LIEWEN, M.B. and MARTH, E.H. (1985). Growth and inhibition of microorganisms in the presence of sorbic acid. *Journal of Food Protection*, **48**.

LINKO, Y., JAVANAINEN, P. and LINKO, S. (1997). Biotechnology of bread baking. *Trends in Food Science and Technology*, **8**, 339–344.

LOPEZ-MALO, A., ALZAMORA, S.M. and ARGAIZ, A. (2002). *Aspergillus flavus* dose-response curves to select natural and synthetic antimicrobials. *Int. J. Food Microbiol*, **73**, 213–218.

LÜCK, E. and JAGER, M. (1997). *Antimicrobial Food Additives. Characteristics. Uses. Effects.* Springer Verlag, Berlin.

MCNAUGHTON, C., TESSENDORF, B.A. and VAN HOLY, A. (1998). Antimicrobial efficacy of preservation combinations in South African brown bread. *Microbios* **93**, 169–178.

MAHMOUD, A.L.E. (1994). Antifungal action and antiaflatoxigenic properties of some essential oil constituents, *Lett Appl Microbiol*, **19**, 111–113.

NIELSEN, P.V. and RIOS, R. (2001). Inhibition of fungal growth on bread by volatile components from spices and herbs, and possible application in active packaging, with special emphasis on mustard essential oil. *Int J Food Microbiol*, **60**, 219–229.

ÖZCAN, M. (1998). Inhibitory effects of spice extracts on the growth of *Aspergillus parasiticus* NRRL2999 strain, *Zeits Lebens-Unters Forsch*, **207**, 253–255.

PASTER, N., JUVEN, B.J. SHAAYA, E., MENASHEROV, M., NITZAN, R. WEISSLOWIWICZ, H. and RAVID, U. (1990). Inhibitory effect of oregano and thyme essential oils on moulds and food borne bacteria. *Letters in Applied Microbiology* **11**, 33–37.

PATKAR, K.L., USHA, C.M., SHETTY, N.S., PASTER, N. and LACEY, .J. (1993). Effect of spice essential oils on growth and aflatoxin B1 production by *Aspergillus flavus. Letters in Applied Microbiology*, **17**, 49–61.

PORTRER, S., PASCOT, B. and BENOUALID, K. (1989). Augmentation de la duree de coservation dun aliment a humidite intermediare, conditionne sous atmosphere modifee ou controlee, *Science desi Aliments*, **9**, 701–712.

RAY, L.L. and BULLERMAN, L.B. (1982). Preventing growth of potentially toxic molds using antifungal agents, *J Food Prot*, **45**(10), 953–963.

REYNOSO, M.M., TORRES, A.M., RAMIREZ, M.L., RODRIGUEZ, M.I., CHULZE, S. and MAGAN, N. (2002). Efficacy of anti-oxidant mixtures on growth, fumonisin production and hydrolytic enzyme production by *Fusarium verticillioides* and *F. proliferatum in vitro* on maize-based media. *Mycological Res*, **106**, 1093–1099.

ROESSLER, P.F. and BALLENGUER, M.C. (1996). Contamination of an unpreserved semisoft baked cookie with a xerophilic *Aspergillus* species, *J Food Prot*, **59**, 1055–1060.

SAXENA, J. and MATHELA, C.S. (1996). Antifungal activity of new compounds from *Nepeta leucophylla* and *Nepeta clarkei*, *Appl Environ Microbiol*, **62**(2), 702–704.

SALMERON, J., JORDANO, R. and POZO, R. (1990). Antimycotic and Antiaflatoxigenic activity of oregano and thyme, *J Food Prot*, **53**(8), 697–700.

SANDERS, S.W. (1991). Using prune juice concentrate in whole wheat bread and other bakery products, *Cereal Food World*, **36**, 280–283.

SAUER, F. (1977). Control of yeast and molds with preservatives, *Food Technol*, **31**(2), 66–67.

SEILER, D.A.L. (1984). Preservation of bakery products. *Institute Food Sci Technol, Proc*, **17**, 31–39.

SHELEF, L.A. (1983). Antimicrobial effects of species, *J Food Saft*, **6**, 29–44.

SINHA, G.K and GULATI, B.C. (1990). Antibacterial and antifungal study of some essential oils and some of their constituents. *Indian Perfumes*, **34**, 126–129.

SMID, E.J. and GORRIS, L.G.M. (1999). Natural antimicrobials for food preservation *In: Handbook of Food Preservation*, M. Shafiur Rahman (ed), Marcel Dekker, Inc., New York.

SMITH, J., KHANIZADEH, S., VAN DE VOORT, F., OORAIKUL, B. and JACKSON, E. (1988). Use of response surface methodology in shelf-life extension studies of a bakery product. *Food Microbiology* **5**, 163–176.

TORRES, A., RAMIREZ, M.L. ARROYO, M., CHULZE, S. and MAGAN, N. (2003) Potential for control of growth and fumonisin production by Fusarium verticillioides and *F. proliferatum* on irradiates maize grain using anti-oxidants. *International Journal of Food Microbiology*. In press.

VORA, H.M. and SIDHU, J.S. (1987). Effect of varying concentrations of ethyl alcohol and carbon dioxide on the shelf-life of bread. *Chemie, Mikrobiologie, Technologie der Lebendmittel*, **11**, 56–59.

WENG, Y.M., CHEN, M.J. and CHEN, W. (1999). Antimicrobial food packaging materials form poly(ethilene-co-methacrylic acid), *Lebebs-Wiss U-Technol.*, **32**, 191–195.

25

Detecting mycotoxin contamination of cereals

C. Waalwijk, Plant Research International, The Netherlands

25.1 Introduction: the problem of mycotoxin contamination

Mycotoxins are secondary metabolites produced by a wide variety of filamentous fungi, including species from the genera *Aspergillus, Fusarium* and *Penicillium*. Mycotoxins have caused major epidemics with thousands of deaths in the Middle Ages due to ergotism. In more recent times, *Fusarium* mycotoxins have been the cause of alimentary toxic aleukia (ATA). Owing to the consumption of overwintered grain, that was contaminated with *Fusarium* spp., an estimated 100 000 people were killed in the USSR between 1942 and 1948 (Joffe, 1978). ATA is also known to have occurred in Russia in 1932 and 1913, and there is little doubt that outbreaks occurred in earlier years as well. Matossian (1981) has argued persuasively that ATA occurred in other countries, including England, in the 14th to 18th centuries. The T-2 toxin, implicated in these cases of ATA, is member of a large group of chemically related compounds named trichothecenes, that also includes deoxynivalenol (DON). DON has been implicated in a human mycotoxicosis, in India, in combination with T-2 toxin and other trichothecenes (Bhat *et al.*, 1989). Other outbreaks of acute human mycotoxicoses, caused by the ingestion of DON and involving large numbers of people, have occurred in rural Japan and China. The Chinese outbreak, in 1984–85, resulted from the ingestion of mouldy maize and wheat. The onset of symptoms occurred within five to thirty minutes and included nausea, vomiting, abdominal pain, diarrhoea, dizziness and headache. Besides these dramatic examples exposure to commodities containing high levels of mycotoxins from the past, chronic exposure to low levels of mycotoxins is of major concern in the cereal food chain (Hussein and Brasel, 2001)

The presence of DON in commodities has a negative impact on their quality. The mycotoxin affects the taste of beer and causes gushing or excess foaming and malting companies will reject barley lots suspected to contain detectable levels of DON. Strains of brewing yeasts show varying levels of sensitivity to DON and other mycotoxins (Boeira *et al.,* 2002). Most malting companies now have a zero tolerance for DON. Due to its relative water solubility highest DON concentrations were found in concentrated steep water and lowest levels were recovered in fibre and gluten fractions (Lauren and Ringrose, 1997)

Breadmaking is also affected by the mycotoxin DON. The flour changes colour and the baking quality is reduced because fusarium-damaged kernels cause sticky, weak dough. Damage to the gluten by a protease from *Fusarium* renders the dough weaker and reduces the ability of the dough to hold gas during fermentation. As a result, the bread does not rise normally, the bread loaf volume is reduced and the breadcrumb becomes coarse.

The baking process does not destroy DON. Naturally contaminated spring wheat specimen were processed using common cleaning, tempering and scouring processes. When the resulting flours were baked into bread or cookies and donuts, no appreciable losses of DON occurred (Scott *et al.*, 1984). Data on DON contamination in wheat, flour, and bakery products from the 1993/94 season in Argentina showed average DON concentrations of 1.8, 1.3, and 0.46 ppm in wheat, flour, and bakery products, respectively (Pacin *et al.*, 1997). A further evaluation of steps in the breadmaking process indicated a decrease in DON of about 21.6% during fermentation of dough and a further decrease of 28.9% during production and baking of the final product (Neira *et al.*, 1997). However, data from other experiments indicated that baking does not reduce DON levels significantly (Creppy, 2002)

25.2 Mycotoxins in the food chain

Fungi develop under certain conditions of humidity and temperature and these conditions are species and isolate specific. Likewise the conditions for mycotoxin production may vary between species and between different individuals of the same species. *Fusarium* spp. are generally considered as field pathogens, that enter the food chain during the growing season (Parry *et al.*, 1995). Fusaria are responsible for pre-harvest contamination of cereals with mycotoxins and improper storage may increase toxin production by existing fungal biomass deteriorating the commodity. *Aspergillus* and *Penicillium* spp. are predominantly food spoilage fungi, that grow and produce mycotoxins under adverse storage conditions.

The role of mycotoxins in fungal metabolism appears to be ambiguous. Field isolates of *F. verticillioides* that do not produce fumonisin are very rare. Yet disruption of genes involved in the biosynthesis of fumonisin does not affect virulence (Desjardins *et al.*, 2002). Trichothecenes, on the other hand, are not produced by all isolates of the Fusarium Head Blight (FHB) complex. Nevertheless

DON and NIV have a reported phytotoxic activity (Eudes *et al.,* 1997, 2000) and disruption of the *tri5* gene, responsible for the first and committing step in the biosynthesis of trichothecenes, significantly reduces the aggressiveness of the *Fusarium* mutant (Proctor *et al.*, 1995; Nicholson *et al.*, 1998).

25.2.1 Which mycotoxins are produced by which fungi?

The primary fungi that produce mycotoxins are members of the genera *Aspergillus, Claviceps, Fusarium* and *Penicillium* (Table 25.1). The same toxin may be formed by a variety of fungal species but not necessarily by all the strains from that species. Likewise, particular isolates of some species may produce several mycotoxins simultaneously. Several of the most common mycotoxins produced by *Fusarium* spp are shown in Fig. 25.1.

Ergot alkaloids

Ergot alkaloids are produced by *Claviceps purpurea*, primarily on rye. Ergotism has historically occurred in areas that used much rye. It is best known for the two frequent extreme forms of the disease: convulsive and gangrenous. Bread made from rye was the staple food two centuries ago in Northern Europe and Russia. Ergot passes readily into a nursing mother's milk. This might explain why, a century ago, Russia had the highest infant mortality rates in Europe (Matossian, 1981).

Aflatoxins

Aflatoxins are produced by several members of the genus *Aspergillus*. Aflatoxins are both acutely and chronically toxic. Extremely high concentrations of aflatoxins caused the death of 100 000 turkeys in an outbreak of 'Turkey-X disease'. Aflatoxins are potent hepato-carginogens and long-term chronic exposure to extremely low levels of aflatoxins in the diet is an important consideration for human health.

Ochratoxins (OTA)

Ochratoxins are synthesized by isolates of *Aspergillus ochraceus, A. carbonarius* and *Penicillium verrucosum* although not all isolates are producers. Ochratoxin A is a potent toxin affecting mainly the kidneys, where it may cause both acute and chronic lesions. Teratogenic and genotoxic effects have also been demonstrated in animal models.

Trichothecenes

Trichothecenes belong to a large family of secondary metabolites, with over 150 members. They are primarily produced by species from the genus *Fusarium*, but also by isolates from the genera *Myrothecium, Stachybotrys and Trichoderma.* Trichothecenes occur in cereal grains such as wheat, barley, maize, oats and rice. Trichothecenes are sesquiterpenoids with a characteristic epoxide group (Fig. 25.1). Type B trichothecenes, including DON, ADON, NIV and ANIV,

Table 25.1 Important mycotoxins, their occurrence, primary producers and their mode of action

Mycotoxin		Commodity	Genus	Species	Disease / mode of action
ergot alkaloids		cereals	*Claviceps*	*purpurea*	St Anthony's fire, vasoconstrictive properties
Aflatoxin	B_1, B_2 G_1, G_2 M_1, M_2	cereals, nuts, spices, figs	*Aspergillus*	*flavus* *parasiticus* *nominus*	carcinogenic
Ochratoxin	A B	cereals, wine, coffee, spices	*Aspergillus* *Penicillium*	*ochraceus* *carbonarius* *verrucosum*	neurotoxic, nephrotoxic, immunotoxic
Type B trichothecenes	DON (Vomitoxin) ADON, NIV, ANIV	cereals	*Fusarium*	*culmorum* *equiseti* *graminearum*	Acute toxicity, immunotoxic ATA (alimentary toxic aleukia)
Type A trichothecenes	T-2, HT-2	small grain cereals	*Fusarium*	*poae* *sporotrichioides*	Acute toxicity, immunotoxic
Zearalenone		cereals	*Fusarium*	*culmorum* *equiseti* *graminearum* *semitectum*	estrogenic effects, infertility, still births, abortion
Moniliformin		cereals	*Fusarium*	*avenaceum* *graminearum* *culmorum e.o.*	cardiotoxic

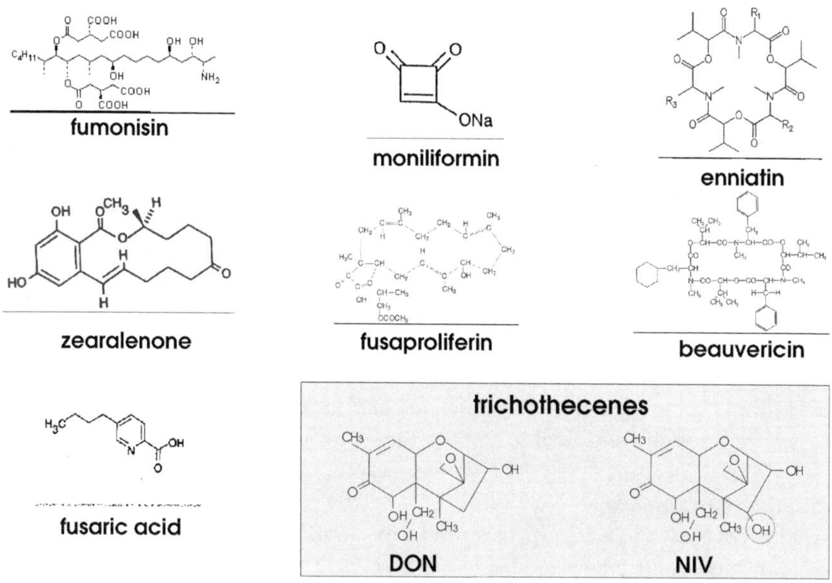

Fig. 25.1 Chemical structure of various secondary metabolites with proven toxic activity produced by *Fusarium* spp.

contain a keto group at the C-8 position of the trichothecene ring, rendering them more polar, whereas the less polar type A trichothecenes, like T-2 and HT-2 have an ester bond at this position. Many naturally occurring toxic derivatives of DON have been reported (reviewed by Mirocha *et al*., 2003) and these mycotoxins frequently co-occur in wheat and barley.

The production of DON is most often associated with *F. graminearum*, but the majority of toxigenic *Fusarium* species encountered in small grain cereals are capable of producing type B trichothecenes (Bottalico and Perrone, 2002). On the other hand only a limited group of species, primarily *F. poae* and *F. sporotrichioides*, produce type A trichothecenes.

Deoxynivalenol (DON)

F. graminearum occurs world-wide and is the most important producer of deoxynivalenol (DON). The outbreaks of emetic (and feed refusal) syndromes in farm animals, due to the presence of DON in their diets, has resulted in the trivial name, vomitoxin, being attributed to this mycotoxin. Humans consuming flour made from scabby wheat or mouldy corn containing DON also have been reported to suffer nausea and headaches which lasted 24 days. DON is probably the most widely distributed *Fusarium* mycotoxin occurring in a variety of cereals, particularly maize and wheat. The immunosuppressive effect, of those concentrations of DON which are naturally occurring, has been reported. There is inadequate evidence in humans and experimental animals, however, for the carcinogenicity of DON. DON is not transferred into milk, meat or eggs.

Nivalenol (NIV)
Deoxynivalenol and nivalenol are produced by the same range of *Fusarium* spp, but isolates never appear to produce DON as well as NIV. NIV appears to be less phytotoxic than DON (Eudes *et al.*, 1997, 2000) but the toxicity for mammals is higher, which led the scientific committee on food of the EU to advise a t-TDI (temporary Total Daily Intake) of 0.7 μg/kg bw/day(EC-SCF, 2000b).

Zearalenone (ZEA)
Zearalenone is an important mycotoxin in temperate and warm regions of the world. It is produced by a large range of *Fusarium* species. The most important effect of zearalenone is on the reproductive system. It has been responsible for outbreaks of hyper-estrogenism in pigs, characterised by genital swelling and infertility. Carcinogenicity of zearalenone has not be demonstrated. Zearalenone is not transmitted from feed to milk to any significant extent.

Moniliformin (MON)
A wide range of *Fusarium* species produces moniliformin, but not all isolates of a given species are MON producers (Schütt *et al.*, 1998). In humans, a possible link between moniliformin ingestion and Keshan disease, a cardiomyopathy endemic to certain areas of China has been suggested (Burmeister *et al.*, 1979). There appears to be a positive correlation between the disease and the ingestion of maize contaminated with moniliformin. Similarities in the ultrastructural changes in the myocardium of rats treated with moniliformin and humans suffering from the disease have been observed (Chen *et al.*,1990).

Electrocardiography was used to examine the acute cardiotoxic effects of moniliformin on 3-week-old broiler chickens and changes in electrocardiogram were monitored for 50 minutes. Moniliformin caused a bradycardia within 15 minutes post-injection. The results indicate that the moniliformin-induced mortality is due primarily to cardiac failure (Nagaraj *et al.*, 1996).

Roasting corn meal spiked with MON at 218 °C for 15 min had a significant effect on the reduction of MON, whereas autoclaving resulted in only 10% reduction. Overall, MON showed heat stability similar to or greater than other *Fusarium* mycotoxins such as deoxynivalenol and fumonisin B_1 (Pineda-Valdes *et al.*, 2003)

25.2.2 When and where are mycotoxins produced

The production of aflatoxin by *A. flavus* and *A. parasiticus* requires a minimum relative humidity of 85% (moisture contents over 17.5% in wheat and 10% in canola) (Pitt, 1995b). The temperature range most favourable for aflatoxin production is 20 to 35 °C, the minimum is 12 °C and the maximum is 40 °C. Production of aflatoxins is reduced by CO_2 concentrations of 60% and inhibited by 80% CO_2 or less than 0.1% O_2. Aflatoxin is not produced by all strains of *A. flavus* nor under all conditions favourable for mould growth.

The production of ochratoxin A by *Aspergillus ochraceus* can occur at 15 to 37 °C and a minimum relative humidity of 83% (Pitt, 1995b), whereas

Penicillium verrucosum produces the mycotoxin at 0 to 31 °C and water activities above 0.8 (Pitt, 1995c).

Fusarium spp. produce mycotoxins in wheat, barley, oats, corn, and rice. No *Fusarium* species grows below a water activity of 0.88. Therefore, the production of *Fusarium* toxins is normally a pre-harvest problem (Bottalico, 1998). However, when high moisture wheat (26%) is stored without drying, *Fusarium* species predominate and may produce mycotoxins (Sinha *et al.* 1991). Several *Fusarium* species can grow at or below 0 °C so cooling wet grain stored in the fall may not prevent the production of *Fusarium* mycotoxins.

25.2.3 What is the economic impact of mycotoxin contamination

Estimates of the direct and secondary economic impacts of FHB infestations during 1998–2000 indicate that FHB is a major problem for US wheat and barley growers. FHB can reduce yield by 30–70% (Bai and Shaner, 1994) but currently the major concern regarding FHB arises from the ability of the majority of species in the complex to produce mycotoxins. The United States Department of Agriculture (USDA) ranks FHB as the worst plant disease to hit the nation since the stem rust epidemics in the 1950s (Windels, 2000). The cumulative direct economic costs for spring wheat, winter wheat, durum wheat and barley amount to \$870 million. Combined direct and secondary economic losses were estimated at \$2.7 billion (Nganje *et al.*, 2002). In developed countries mycotoxins in foods are usually not a public health risk because of their strict surveillance and toxin management procedures (Miller, 1995). For example, from 1980 to 1990, the EU reduced imports of peanut meal by 50% and imports of copra by 75%. The indirect cost of research on aflatoxins in the USA over the last 30 years has been estimated at $\$100 \times 10^6$ (Pitt, 1995a). A study was conducted to estimate the direct costs of aflatoxins in peanuts and corn in Thailand, Indonesia, and the Philippines. Five costs were considered: (*i*) quality deterioration, (*ii*) spoilage, (*iii*) mutagenic and carcinogenic effects on humans, (*iv*) effects on livestock health with emphasis on increased mortality rates and reductions in feed conversion ratios, and (*v*) loss of international markets due to aflatoxin contamination. The study arrived at a total cost of $\$470 \times 10^6$ (Pitt, 1995a). The major losses were the costs incurred in human death or disability (60%) of the total, spoilage (24%), and animal health (16%).

Outbreaks of FHB result in greatly reduced yields and diminished seed quality of wheat and barley. In Manitoba (Canada) it was estimated that between \$50 and \$100 million is lost each year from reduced yields, access to malt and hog feed markets and the added costs of importing mycotoxin free grain. If these imported products are not handled properly they could have the potential to cause significant outbreaks. To meet this threat a management plan was developed in Alberta, Canada, that enforces a zero tolerance policy in cereal seed. Moreover, best management practices must be applied to all feed and straw that has not been tested and declared to be *Fusarium graminearum* free. In 1999 *F. graminearum* was declared a pest under the *Agricultural Pests Act*. This act gives the municipality the authority to enforce control measures for this pest. Appointed pest inspectors have the power to enter land for the purpose of inspection and sample collection. The owner or occupant of the land has the responsibility of preventing the establishment of pests and taking measures to control or destroy all pests on their property.

DON is frequently present at high concentrations (usually >1 ppm, sometimes as high as 20 ppm) in wheat and corn. Although no U.S. government regulation has been made regarding levels of *Fusarium* toxins in human foods, a recommended tolerance level of 1 ppm DON in grains for human use has been set by several countries, including the USA, while higher concentrations are permitted in animal feeds (Chu *et al*, 1997).

25.3 Detecting mycotoxins

The structural relatedness of trichothecenes and their co-occurrence in commodities has led to the development of many different protocols for extraction, clean up and quantification. Despite the fact that DON and NIV are relatively polar, extraction from naturally contaminated samples proved to be more difficult than from samples that were spiked with these mycotoxins (Trenholm *et al.,* 1985). It is therefore recommended that the extraction efficiency be evaluated with naturally contaminated samples (Krska *et al.*, 2001). Depending on the type of detection methods employed, interfering compounds present in the original samples must be removed before quantification. With immunoassays, such as ELISA, cleaning-up is usually not required, but chromatographic methods like **T**hin **L**ayer **C**hromatography (TLC), **H**igh **P**ressure **L**iquid **C**hromatography (HPLC) or **G**as **C**hromatography (GC) require further purification of the samples using liquid-liquid fractionation or clean-up columns (Krska *et al.*, 2001). ELISA is the method of choice for rapid analyses with simple extraction and no clean up. The procedure makes use of specific antibodies that are currently only commercially available for DON and T-2 toxin. The most sensitive ELISA methods have been developed for the tri-acetylated form of DON. When a acetylation step is included in the procedure a level of detection of 0.3–1 ng/g can be obtained. Although ELISAs are fast and simple their specificity can be a drawback when samples are suspected to contain several mycotoxins. When sensitivity and diversity are essential, gas chromatography in combination with mass spectrometry or electron capture (GC/MS or GC/EC) should be used. These methods are accurate, reliable and allow high throughput of samples. Up to eight different mycotoxins can be analysed simultaneously in less than 37 min (Tanaka *et al.*, 2000). Intercomparisons between laboratories (Pieters *et al.*, 2001; Pettersson and Langseth, 2002) using the same set of samples revealed that there is still a need for further improvement in the determination of trichothecenes as well as for the development of certified reference materials (Krska *et al.*, 2001).

25.3.1 Diagnostic methods for mycotoxin producing *Fusarium* spp.

Fusarium head blight (FHB) is caused by a complex of species, but their relative contribution to the disease can vary widely between countries and from year to year. More than a dozen different species have been associated with FHB. Parry *et al.* (1995) reported 17 *Fusarium* species associated with head blight in the U.K. and in China 18 different species were found among 2,450 symptomatic wheat heads (CWSCG, 1984; Wang, 1997). Traditional identification is based on morphology and the lack of clear cut characteristics has confused the taxonomy of the genus *Fusarium* for several decades and reported examples of misidentifications of isolates have added to this confusion. In the 1980s, the morphological species concept (MSC) was complemented with a biological species concept (BSC), that relies on the fact that, individuals that are able to cross, belong to the same species. Although this concept proved a valuable contribution to *Fusarium* taxonomy it has the drawback that many species in the genus are asexual. Most taxonomic controversies could only be resolved with the advent of molecular taxonomy. This technology, now widely applied in fungal taxonomy, is based on the amplification and sequence analysis of specific genomic fragments using polymerase chain reaction (PCR). In the genus *Fusarium* molecular phylogenetics has led to the identification of many new species (e.g. O'Donnell *et al.*, 1998) and several hundreds of new species are expected to be added. Aoki and O'Donnell (1999) were able to discern a group of isolates hitherto known as *F. graminearum* group 1 as a new species. Based on molecular as well as morphological characters this group of isolates could be reclassified as *F. pseudograminearum*. Phylogeography, chemotypes (the profiles of secondary metabolites e.g. the ability to produce mycotoxins), sequence information and fingerprinting (Carter *et al.*, 2000, 2002) from a global set of *F. graminearum* strains from wheat, rice and maize illustrated the genetic diversity within *F. graminearum*. By phylogenetic analyses, seven distinct lineages could be identified (O'Donnell *et al.*, 2000; Ward *et al.*, 2002) and an additional lineage was recently identified among isolates from Brazil. With the exception of the pandemic lineage 7, that is responsible for the current global outbreaks of FHB, these lineages were limited in their geographic distribution. This indicates that they are both temporal and reproductively isolated and may be described as separate species in the future.

25.3.2 Qualitative identification of *Fusarium* species

The predominant fungi involved in FHB are *F. graminearum, F. culmorum, F. avenaceum, F. poae* and *Microdochium nivale* (formerly known as *F. nivale*). The first two species are notorious for their ability to produce DON or NIV along with ZEA (Logrieco *et al.*, 2002). *F. avenaceum* is a MON producer and *F. poae* produces T2 and HT-2 (Logrieco *et al.*, 2002; Bottalico and Perrone, 2002). *M. nivale* is commonly confused with the *Fusarium* species from the FHB complex, but does not produce any known mycotoxin. Using RAPD analyses, Nicholson and co-workers have developed specific PCR conditions that allow the rapid and robust identification of each of the above mentioned species within the FHB complex (Doohan *et al.*, 1998; Nicholson *et al.*, 1996; Parry and Nicholson, 1996; Turner *et al.*, 1999). The disadvantage of performing PCR reactions for each of the species separately was overcome when we combined the ingredients for each of these reactions into a single-tube assay (Waalwijk *et al.*, 2003). This multiplex PCR for the five most commonly found species of the FHB complex was combined with *F. proliferatum* and *M. nivale* var. *nivale*, a variety of *M. nivale* far less aggressive than *M. nivale* var. *majus* usually found on diseased wheat plants (Lees *et al.*, 1995; Diamond and Cooke, 1997). This illustrates that multiplexing can be extended to accommodate additional species.

25.3.3 Quantitative detection of *Fusarium* species

Qualitative identification has the disadvantage that fungi have to be isolated from the diseased plant tissues. This is both time consuming and laborious but a more significant drawback is the fact that a bias will be introduced in favor of the faster growing species e.g. *F. graminearum*. Moreover, the smallest plant part, e.g. a single kernel, may still contain several species from the FHB complex. Quantitative differences between kernels are not easily recognized since heavy infested kernels and hardly colonized kernels will both be overgrown after incubation at appropriate conditions for fungal growth. More accurate reflection of the actual occurrence of different species during the growing season requires a robust and fast quantitative detection. Several researchers have addressed this issue by different approaches. Nicholson *et al.* (1998) used competitive PCR to quantify the amount of *F. graminearum* and *F. culmorum* in winter wheat inoculated with either species. A poor correlation between fungal DNA content and visual symptoms was explained by assuming that symptoms could have been caused by other species of the FHB complex (Nicholson *et al.*, 1998)

25.3.4 Detection of genes involved in the biosynthesis of mycotoxins

Random DNA sequences have been used in many identification and detection protocols, but they all suffer from the fact that the correlation between toxin producing ability and target DNA is based on association. This can be circumvented by using as target sequence (part of) a gene involved in the biosynthesis of the mycotoxin. Trichodiene synthase is the first and committing step in the biosynthesis of all trichothecenes, catalyzing the cyclization of farnesyl pyrophosphate to trichodiene. The *tri5* gene, encoding, trichodiene synthase, has been used by several researchers to detect trichothecene producing fungi. Niessen and Vogel (1998) have used this gene to quantify the fungal biomass in wheat samples and found a good correlation between the intensity of the PCR signal and the amount of DON. The *tri5* gene is common to all trichothecene producers within the FHB complex. The gene is composed of two

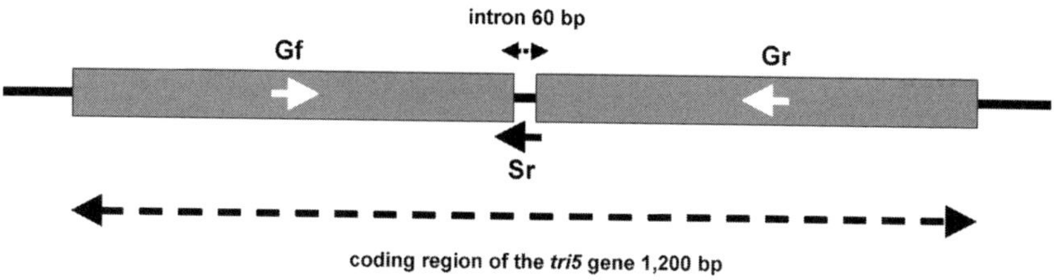

Fig. 25.2 Schematic representation of the *tri5* gene, encoding the first and decisive step in the biosynthetic pathway leading to both type A (T-2 and HT-2 toxin) and type B trichothecenes (DON and NIV and derivatives). Using the generic primers **Gf** and **Gr**, amplicons will be obtained irrespective of the species present in the sample. In RT-PCR reactions, where the same primers are used on RNA the amplicon will be 60 nt shorter. The 60 nt long intron shows sufficient divergence between species to design species specific primers (**Sr** in the picture).

coding regions, exons, that are separated by a single small intron, 60 bp in size (Fig. 25.2). Comparison of the *tri5* sequences from many trichothecene producers revealed that the coding regions contain highly conserved regions, allowing the design of generic primers that will amplify all trichothecene producers. As expected the sequences from the introns show enough divergence to design species-specific primers, enabling both generic as well as specific quantification of trichothecene producers (Schnerr *et al.,* 2001). This real-time PCR procedure showed a high correlation between the amount of total trichothecene-producing *Fusarium* and the amount of DON in field samples (Schnerr *et al.,* 2002). Quantitative PCR using the *tri5* gene was employed by Edwards *et al.* (2001) to study the effects of fungicide applications. A highly significant correlation between DON concentrations and the *tri5* DNA concentration was obtained (Edwards *et al.,* 2001).

Mycotoxins, being secondary metabolites are the end products of complex biosynthetic pathways, involving many enzymatic steps. Hence, the genetic information required to encode these genes encompassing a substantial portion of the genome. The genes involved in the biosynthetic pathways leading to several mycotoxins appear to be clustered in the genome in contrast to several other complex pathways. Interestingly, the majority of genes involved in the biosynthesis of trichothecenes also appears to be clustered (McCormick, 2003; Ward *et al.,* 2002). This facilitated the comparison of these gene-clusters among different *Fusarium* species, e.g. between DON- and NIV- producers within a single species on the one hand (Lee *et al.,* 2001; Ward *et al.,* 2002) and Type A- and Type B-trichothecene producers (Brown *et al.,* 2001), on the other hand. These studies revealed that the gene designated *tri7* is functional in NIV producers but not functional in DON producers. In a subsequent study both groups identified the *tri13* gene as the gene responsible for the chemotype diversity in *F. graminearum* and other trichothecene producers (Lee *et al.,* 2002; Brown *et al.,* 2002).

25.3.5 Expression analyses of the genes involved in the biosynthesis of mycotoxins

The detection of gene(s) involved in the biosynthesis of mycotoxins will give an indication of the capacity to synthesise mycotoxins, but is not a true reflection of biosynthesis. Detection of messenger RNA reflects the expression of target genes and is supposed to be a marker for biosynthesis of the mycotoxin. Reverse transcription of mRNA followed by PCR, e.g. RT-PCR, specifically amplifies mRNA. When primers are designed such that the resulting amplicon spans an intron, products derived from DNA or from mRNA can be discriminated on the basis of the difference in size. By inclusion of a housekeeping gene, with a constitutive expression level, like the β-tubulin gene, the relative expression of the *tri5* gene could be quantified (Doohan *et al.*, 1999). Quantitative RT-PCR was employed by Doohan *et al.* (1999) to detect changes in expression of the *tri5* gene in *in vitro* cultures of *F. culmorum* as well as during infection of wheat.

NASBA is an isothermal amplification method, that exclusively uses single stranded RNA as a template. To improve the applicability of the system for large-scale screenings, NASBA was combined with molecular beacons, enabling a simultaneous real-time amplification and detection in a single tube. This so-called AmpliDet RNA is specifically suited for the detection of gene expression. We have employed the trichodiene synthase gene, *tri5,* to monitor the synthesis of trichothecenes in *Fusarium* spp. with a sensitivity of at least 100 target molecules. Gene specific primers (GSPs) for the *tri5* gene were used to amplify part of the gene from various *Fusarium* spp. (Fig. 25.2) Two sets of primers were designed on the sequence of this amplicon *(i)* part of the T7 promoter region linked to a forward GSP from the 5′ part of the fragment and a reverse GSP located at the 3′-part of the fragment, according to Klerks *et al*., (2000). Using the reverse GSP and AMV reverse transcriptase this material can be transcribed into (+) DNA. This in turn was used to generate so-called *in vitro* RNA that serves as a template for amplification with T7 polymerase. Amplified

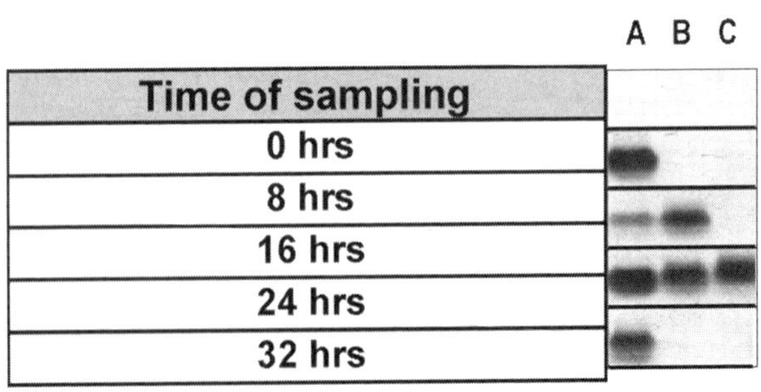

Fig. 25.3 Expression of the *tri5* gene in a young culture of *F. graminearum*. At indicated times samples were harvested from cultures freshly inoculated with 100 000 conidia of *F. graminearum* in potato dextrose broth. After filtration total RNA was extracted from the mycelial mats and NASBA reactions were performed on serial dilutions (A, 10^0; B, 10^{-2} and C, 10^{-4}). Amplicons were separated on agarose gels and detected by hybridization.

RNA was detected on Northern blots or by inclusion of a molecular beacon in the reaction. Using primers derived from the *tri5* gene amplicons were detected in all species tested e.g. *F. culmorum, F. graminearum, F. poae, F. sambucinum, F. sporotrichioides* and *F. venenatum*.

Total RNA was extracted from pure cultures of different *Fusarium* spp. grown in potato dextrose broth at various timepoints after inoculation with 10^5 conidia and subjected to Northern blotting and as early as 8 h after inoculation the first expression of the *tri5* gene was recorded (Fig. 25.3 and Klerks *et al.*, 2000). NASBA reactions on seed samples putatively contaminated with DON performed on the supernatant from kernels incubated at 80 °C for 10 min, showed higher signals in heavily contaminated samples, but this procedure needs further validation.

25.4 The regulatory context

In Canada and the United States maximum tolerated levels of the mycotoxin DON have been determined (FAO survey 1994/1995). In Canada and the USA the maximum tolerated levels of DON in food are respectively 2 mg/kg (in unclean soft wheat) and 1 mg/kg (end product). In Canada the maximum tolerated level (MTL) of DON in feed is determined for cattle and poultry at 5 mg/kg and for swine, young calves and lactating dairy animals at 1 mg/kg. In the USA, MTLs for DON in grains and grain by-products destined for ruminating beef and feedlot cattle older than 4 months and for poultry (not exceeding 50% of their diet) were set at 10 mg/kg, for other animals at 5 mg/kg (not exceeding 40% of their diet, but for swine not exceeding 20% of their diet).

At the European level no regulations have been determined yet for DON and ZEA. However, various countries already set their own regulations. In Austria maximum tolerated levels of DON or ZEA in human food and animal feed have become valid. For humans, maximum tolerated levels of DON in wheat end products are 0.5 mg/kg and for ZEA 0.06 mg/kg. In Italy the Ministry of Health recommended the following maximum levels for ZEA: 0.02 mg/kg in baby food and 0.1 mg/kg in cereals and cereal products.

In the Netherlands, the Ministry of Health together with the Dutch Association of Flour Producers (NVM) agreed that only products containing less than 1 mg/kg could be accepted to be used for trading. This level was reduced to 0.5 mg/kg DON in the end products for human consumption. To achieve this level, only shipments of wheat with a maximum level of DON of 1 mg/kg will become available for commerce. For the 1999 harvest of wheat a level of DON below 0.5 mg/kg could be guaranteed.

The EU has recently recommended action limits of 500 μg/kg of final product and 750 μg/kg in raw materials At the 35th session of the Joint FAO/WHO Expert Committee on Food Additives, JECFA, these recommendations have been adopted (JECFA, 2003). During recent years, the European Commission temporarily assigned a tolerable daily intake (TDI) level for deoxynivalenol

Table 25.2 Temporary tolerable daily intake (tTDI) of *Fusarium* mycotoxins occurring in small grain cereals as recommended by the Scientific Committee on Food of the European Commission (EC-SCF)

	DON	NIV	T-2 and HT-2	ZEA
TDI[a]	1[b]	-	-	-
tTDI	-	0.7	0.06	0.2
NOAEL	100	700	-	40
LOAEL	-	2000	29	-

[a]TDI, tolerable daily intake;
tTDI, temporary total daily intake;
NOAEL, no observed adverse effect level;
LOAEL, lowest observed adverse effect level
[b]Values are expressed as μg/kg bw/day.

(DON) and other mycotoxins (Table 25.2) and there is evidence that the TDI has occasionally been exceeded. Some countries have already established (or are in the process of setting) legislative limits for DON in cereals (JECFA, 2003). In a set of opinions, the EC Scientific Committee on Food (EC-SCF) evaluated the *Fusarium* toxins, e.g. DON (EC-SCF, 1999), zearalenone (EC-SCF, 2000a), nivalenol (EC-SCF, 2000b), and T-2 toxin and HT-2 toxin (EC-SCF, 2001), and performed a group evaluation of T-2 toxin, HT-2 toxin, nivalenol, and deoxynivalenol (EC-SCF, 2002). An FAO Manual on the application of the HACCP system in mycotoxin prevention and control can be accessed: ftp://ftp.fao.org/es/esn/food/mycotoxin_manual.pdf

In Austria maximum tolerated levels of DON are 0.5 mg/kg and for ZEA 0.05 mg/kg feed for pigs kept for meat as well as for breeding. For chickens kept for meat the maximum tolerated level of DON is 1.5 mg/kg feed, whereas for chickens kept for egg production and breeding the level is reduced to 1.0 mg/kg feed. For cows kept for meat production the maximum tolerated level of DON is 1.0 mg/kg feed (Scholten *et al.*, 2002).

25.5 Future trends

25.5.1 Detoxification and decontamination

Prevention of mycotoxin contamination is the primary goal but under adverse conditions formation of mycotoxins is unavoidable. Physical and chemical decontamination methods have been employed with varying success in the past (Visconti, 2002) but biological decontamination methods are of more recent date. Karlovsky (1999) has reviewed degradation of mycotoxins by various micro-organisms. Experiments with microbes from different sources demonstrated that some organisms present in cattle rumen fluid (Hedman and Pettersson, 1997) and soil (Shima *et al.*, 1997) can detoxify DON by deep-oxidation. It may be possible to use one or more of these microbes to reduce toxin levels in feeds or some foods. Lactic acid bacteria and yeasts expressing mycotoxin-degrading enzymes, may also offer a natural way of providing these activities in fermentation processes. Several microbial strains (yeasts, fungi or bacteria) have been proposed for detoxification of deoxynivalenol, T-2 toxin, zearalenone and other trichothecenes, but their practical use has not been shown. (Karlovsky 1999).

25.5.2 Synergism

Evidence is accumulating on the co-occurrence of toxigenic fungi (Waalwijk *et al.*, 2001; Waalwijk, 2002). Moreover, several *Fusarium* species are capable of producing two or more mycotoxins in the same commodity. This finding requires research on the synergism or antagonism between mycotoxins as well as between mycotoxins and other secondary metabolites present.

25.5.3 Endophytes

Apart from toxigenic fungi, plants are colonised by many other microbes including some beneficiary micro-organisms. The DNA fingerprint of the endophytic bacterium *Bacillus mojavensis* has been established and its molecular profile was compared with other isolates and species within the large group of related species. All *B. mojavensis* isolates were antagonistic to fungi, but other species within the group were not. This newly established genus of bacteria forms natural associations with plants, and is both plant friendly and antagonistic to fungi. Members of the genus have potential uses for exploitation to provide plant disease protection and to prevent toxin production.

25.5.4 Plant breeding and transgenic crops

Prevention is still the best way to reduce the mycotoxin problem. Breeding for resistance to *Fusarium* might be accelerated by introduction of new marker technologies, including NBS profiling, that will reveal R-gene homologues in a very efficient manner. DArT technology developed by Jaccoud and co-workers (Jaccoud *et al.*, 2001) might prove a fast method to generate thousands of makers that can be applied in marker assisted breeding (MAS) programs. Transgenic wheat lines are currently developed that over-express endogenous genes involved in defense responses (e.g. Anand *et al.*, 2001). Alternatives to this enhanced intrinsic defense are genes specifically aimed at interfering with the mycotoxin metabolism (Muhitch *et al.*, 2000; Obubara *et al.*, 2002).

25.5.5 Methods of detection of mycotoxins

With the availability of modern equipment additional secondary metabolites will be identified and toxicological data of these compounds will become available. The emergence of new putative mycotoxins demands increased efforts for simultaneous quantification of several mycotoxins. Although much research has already been performed on the development of robust and rapid detection kits, reliable and standardized methods for detection of DON and other trichothecenes is still lacking. Method validation, essential for harmonization between countries requires the availability of certified reference materials

The unexpected preponderance of *F. graminearum* in the latter part of the 1990s in Western Europe (e.g. Waalwijk *et al.,* 2003; Waalwijk, 2002) emphasizes the importance of monitoring for the emergence of hitherto uncommon species. Any of the 18 species identified in China as members of the FHB complex (Wang, 1997) might increase in importance in the near future. Quantitative detection of *Fusarium* spp has been performed by real-time PCR (Schnerr *et al.,* 2001) and competitive PCR (Edwards *et al.,* 2001; Nicholson *et al.,* 1998). Recently, in our lab a quantitative PCR was developed using TaqMan (Waalwijk *et al.,* 2002b), where the samples were spiked with an internal standard to verify amplification. All these methods must be performed for each *Fusarium* spp individually. Multiplexing these reactions will become much more feasible when TaqMan probes can be labeled with reporter molecules based on new fluorescence chemistry.

25.6 References

AOKI, T. and O'DONNELL, K. (1999) Morphological and molecular characterization of *Fusarium pseudograminearum* sp. nov. formerly recognized as Group 1 of *F. graminearum* and PCR primers for its identification. *Mycologia* **91**: 597–609.

ANAND, A., JANAKIRAMAN, V., ZHOU, T., TRICK, H.N., GILL, B.S. and MUTHUKRISHNAN, S. (2001) Transgenic wheat overexpressing PR-proteins shows a delay in Fusarium head blight infection. *2001 National Fusarium Head Blight Forum Proceedings*. 2–6.

BAI, G. and SHANER G. (1994) Scab of wheat: prospects for control. *Pl. Dis.* **78**: 760–766.

BHAT, R.V., SASHIDHAR, R.B., RAMAKRISHNA. Y. and MUNSHI, K.L. (1989) Outbreak of trichothecene mycotoxicosis associated with consumption of mould damaged wheat products in Kashmir Valley, India. *Lancet,* I: 35–37.

BOEIRA, L.S., BRYCE, J.H., STEWART, G.G. and FLANNIGAN, B. (2002) Influence of cultural conditions on sensitivity of brewing yeasts growth to *Fusarium* mycotoxins zearalenone, deoxynivalenol and fumonisin B1. *Int. Biodeter. Biodegr.* **50**: 69–81.

BOTALLICO, A. (1998) Fusarium diseases of cereals: species complex and related mycotoxin profiles, in Europe. *J. Pl. Pathol.* **80**: 85–103.

BOTTALICO, A. and PERRONE, G. (2002) Toxigenic *Fusarium* species and mycotoxins associated with head blight in small-grain cereals in Europe. *Eur. J. Pl. Pathol.* **108**: 611–624.

BROWN, D.W., MCCORMICK, S.P., ALEXANDER, N.J., PROCTOR, R.H. and DESJARDINS, A.E. (2001) A genetic and biochemical approach to study the trichothecene diversity in *Fusarium sporotrichioides* and *Fusarium graminearum*. *Fung. Genet. Biol.* **32**: 121–133.

BROWN, D.W., MCCORMICK, S.P., ALEXANDER, N.J., PROCTOR, R.H. and DESJARDINS, A.E. (2002) Inactivation of a cytochrome P-450 is a determinant of trichothecene diversity in *Fusarium* species. *Fung. Genet. Biol.* **36**: 224–233.

BURMEISTER, H.R., CIEGLER, A. and VESONDER, R.F. (1979) Moniliformin, a metabolite of *Fusarium* moniliforme NRRL 6322: purification and toxicity. *Appl. Env. Microbiol* **37**:11–13.

CARTER, J.P., REZANOOR, H.N., DESJARDINS, A.E. and NICHOLSON, P. (2000) Variation in *Fusarium graminearum* isolates from Nepal associated with their host of origin. *Pl. Pathol.* **49**: 452–460.

CARTER, J.P., REZANOOR, H.N., HOLDEN, D., DESJARDINS, A.E., PLATTNER, R.D. and NICHOLSON, P. (2002) Variation in pathogenicity associated with the genetic diversity of *Fusarium graminearum*. *Eur. J. Pl. Pathol.* **108**: 573–583.

CHEN, L.Y., TIAN, X.L. and YANG, B. (1990) A study on the inhibition of rat myocardium peroxidase and glutathione reductase by moniliformin. *Mycopathologia* **110**: 119–124.

CHU, F.S. (1997) Trichothecene mycotoxins. In: *Encyclopedia of Human Biology*, Second Edition, vol. 8, pp. 511–522. Academic Press.

CREPPY, E.E. (2002) Update of survey, regulation and oxic effects of mycotoxins in Europe. *Tox. Lett.* **127**: 19–28.

CWSCG. (1984) *Fusarium* species, distribution and pathogenicity from scabby heads in China. *J. Shanghai Normal College* **3**: 69–82.

DESJARDINS, A.E., MUNKVOLD, G.P., PLATTNER, R.D. and PROCTOR, R.H. (2002) FUM1 – a gene required for fumonisin biosynthesis but not for maize ear rot and ear infection by *Gibberella moniliformis* in field tests. *Mol Plant Microbe Interact* **15**: 1157–1164.

DIAMOND, H. and COOKE, B.M. (1997) Host specialization in *Microdochium nivale* on cereals. *Cer. Res. Commun.* **25**: 533–538.

DOOHAN, F.M., PARRY, D.W., JENKINSON, P. and NICHOLSON, P. (1998) The use of species-specific PCR-based assays to analyse *Fusarium* ear blight of wheat. *Plant Pathol.* **47**: 197–205.

DOOHAN, F.M., WESTON, G., REZANOOR, H.N., PARRY, D.W. and NICHOLSON, P. (1999) Development and use of a reverse transcription PCR assay to study the expression of *tri5* by *Fusarium* species *in vitro* and *in planta*. *Appl. Environm. Microbiol.* **65**: 3850–3854.

EC-SCF, EUROPEAN COMMISSION, SCIENTIFIC COMMITTEE ON FOOD (1999) Opinion on *Fusarium* toxins, part 1: *Deoxynivalenol (DON)*. Recommendation for a temporary tolerable daily intake (tTDI) for DON.

EC-SCF, EUROPEAN COMMISSION, SCIENTIFIC COMMITTEE ON FOOD (2000a) Opinion on *Fusarium* toxins, part 2: *Zearalenone (ZEA)*. Recommendation for a temporary tolerable daily intake (tTDI) for ZEA.

EC-SCF, EUROPEAN COMMISSION, SCIENTIFIC COMMITTEE ON FOOD, (2000b) Opinion on *Fusarium* toxins, part 4: *Nivalenol (NIV)*. Recommendation for a temporary tolerable daily intake (tTDI) for NIV.

EC-SCF, EUROPEAN COMMISSION, SCIENTIFIC COMMITTEE ON FOOD (2001) Opinion on *Fusarium* toxins, part 5: *HT-2 toxin and T-2 toxin*. Recommendation for a temporary tolerable daily intake.

EC-SCF, EUROPEAN COMMISSION, SCIENTIFIC COMMITTEE ON FOOD (2002) Opinion on *Fusarium* toxins, part 6: Group evaluation of T-2 toxin, HT-2 toxin, nivalenol and deoxynivalenol. Recommendation for a temporary tolerable daily intake.

EDWARDS. S.G., PIRGOZLIEV, S.R., HARE, M.C. and JENKINSON, P. (2001) Quantification of trichothecene-producing *Fusarium* species in harvested grain by competitive PCR to determine efficacies of fungicides against Fusarium Head Blight of winter wheat. *Appl. Environm. Microbiol.* **67**: 1575–1580.

EUDES, F., COMEAU, A., RIOUX, S and COLLIN, J. (2000) Phytotoxicity of eight mycotoxins associated with *Fusarium* head blight in wheat. *Can. J. Pl. Pathol.* **22**: 286–192.

EUDES, F., COLLIN, J., RIOUX, S and COMEAU, A. (1997) The trichothecenes, a major component of wheat scab pathogenesis. *Cer. Res. Commun.* **25**: 495–496.

HEDMAN, R. and PETTERSSON, H. (1997) Transformation of nivalenol by gastrointestinal microbes. *Arch. Anim. Nutr.* **50**: 321–329.

HUSSEIN, H.S. and BRASEL, J.M. (2001) Toxicity, metabolism and impact of mycotoxins on humans and animals. *Toxicol.* **167**: 101–134.

JACCOUD, D., PENG, K., FEINSTEIN, D. and KILIAN, A. (2001) Diversity arrays: a solid state technology for sequence information independent genotyping. *Nucl. Acids Res.* **29**: e25.

JECFA (2003) Codex Alimentarius Commission. 35th session of the Codex Committee on Food Additives and Contaminants. Discussion paper on deoxynivalenol. Tanzania http://www.mykotoxin.de/Dokumente/Codex%20DON%2012.02.pdf

JOFFE, A.Z. (1978) *Fusarium poae* and *F. sporotrichioides* as principal causal agents of alimentary toxic aleukia. In *Mycotoxic Fungi, Mycotoxins, Mycotoxicoses: an Encyclopaedic Handbook.* Vol. 3, eds. T.D. Wyllie and L.G. Morehouse. New York: Marcel Dekker. pp. 21–86.

KARLOVSKY, P. (1999) Biological detoxification of fungal toxins and its use in plant breeding, feed and food production. *Nat. Toxins* **7**: 1–23.

KLERKS, M., SCHOEN C. and WAALWIJK, C. (2000) The detection of trichodiene synthase mRNA in *Fusarium* on seeds using AmpliDet RNA. 5th Eur. *Fusarium* Seminar, Berlin.

KRSKA, R., BAUMGARTNER, S. and JOSEPHS, R. (2001) The state-of-the-art in the analysis of type-A and -B trichothecene mycotoxins in cereals. *Fres. J. Anal.Chem.* **371**: 285–299.

LAUREN, D.R. and RINGROSE, M.A. (1997) Determination of the fate of three *Fusarium* mycotoxins through wet-milling of maize using an improved HPLC analytical technique. *Food Addit. Contam.* **14**, 435–443.

LEE, T., OH, D-W., KIM, H-S., LEE, J., KIM, Y-H, YUN, S-H. and LEE, Y-W. (2001) Identification of deoxynivalenol and nivalenol producing chemotypes of *Gibberella zeae* by using PCR. *Appl. Environm. Microbiol.* **67**: 2966–2972.

LEE, T., HAN, Y.K., KIM, K.H., YUN, S.H. and LEE, Y.W. (2002) *Tri13* and *tri7* determine deoxynivalenol and nivalenol producing chemotypes of *Gibberella zeae*. *Appl. Environ. Microbiol.* **68**: 2148–2154.

LEES, A., NICHOLSON, P., REZANOOR, H.N. and PARRY, D.W. (1995) Analysis of variation within *Microdochium nivale* from wheat: evidence for a distinct sub-group. *Mycol. Res.* **99**: 103–109.

LOGRIECO, A., MULÉ, G., MORETTI, A. and BOTTALICO, A. (2002) Toxigenic *Fusarium* species and mycotoxins associated with maize ear rot in Europe. *Eur. J. Pl. Pathol.* **108**: 597–609.

MATOSSIAN, M.K. (1981) Mold poisoning: an unrecognized English health problem, 1550–1800. *Medical History* **25**: 73–84.

MCCORMICK, S. (2003) The role of DON in pathogenicity. In: *Fusarium head blight of wheat and barley*. Leonard K.J. and Bushnell, W.R. (eds) APS Press, St. Paul. pp. 165–183.

MILLER, J.D. (1995) Mycotoxins in Asia: policies for the future. *ACIAR Postharvest newsletter* (29): 5–15.

MIROCHA, C.J., XIE, W. and FILHO, E.R. (2003) Chemistry and detection of *Fusarium* mycotoxins. In: *Fusarium head blight of wheat and barley*. Leonard K.J. and Bushnell, W.R. (eds) APS Press, St. Paul. pp. 144–164.

MUHITCH, M.J., MCCORMICK, S.P., ALEXANDER, N.J. and HOHN, T.M. (2000) Transgenic expression of the TRI101 or PDR5 gene increases resistance of tobacco to the phytotoxic effects of the trichothecene 4,15-diacetoxyscirpenol. *Pl. Sci.* **157**, 201–207.

NAGARAJ, R.Y., WU, W., WILL, J.A.. and VESONDER, R.F. (1996) Acute cardiotoxicity of moniliformin in broiler chickens as measured by electrocardiography. *Avian Dis* **40**, 223–227.

NEIRA, M.S., PACIN, A.M., MARTÍNEZ, E.J., MOLTÓ, G. and RESNIK, S.L. (1997) The effects of bakery processing on natural deoxynivalenol contamination. *Int. J. Microbiol.* **37**: 2125.

NGANJE, W.E., BANGSUND, D.A., LEISTRITZ, F.L., WILSON, W.W. and TIAPO, N.M. (2002) *National Fusarium Head Blight Forum Proceedings*. 275–281.

NICHOLSON, P., LEES, A.K., MAURIN, N., PARRY, D.W. and REZANOOR, H.N. (1996) Development of a PCR assay to identify and quantify *Microdochium nivale* var. *nivale* and *Microdochium nivale* var. *majus* in wheat. *Physiol. Molec. Plant Pathol.* **48**: 257–271.

NICHOLSON, P., SIMPSON, D.R., WESTON, G., REZANOOR, H.N, LEES, A.K., PARRY, D.W. and JOYCE, D. (1998) Detection and quantification of *Fusarium culmorum* and *Fusarium graminearum* in cereals using PCR assays. *Physiol. Molec. Plant Pathol.* **53**: 17–37.

NIESSEN, L.M. and VOGEL, R.F. (1998) Group specific PCR-detection of potential trichothecene-producing *Fusarium* species in pure cultures and cereal samples. *Syst. Appl. Microbiol.* **21**: 618–631.

O'DONNELL, K., KISTLER, H.C., TACKE, B.K. and CASPER, H.H. (2000) Gene genealogies reveal global phylogeographic structure and reproductive isolation among lineages of *Fusarium graminearum*, the fungus causing wheat scab. *Proc. Natl. Acad. Sci. USA* **97**: 7905–7910.

O'DONNELL, K., CIGELNIK, E. and NIRENBERG, H.I. (1998) Molecular systematics and phylogeography of the *Gibberella fujikuroi* species complex. *Mycologia* **90**: 465–493.

OKUBARA, P.A., BLECHL, A.E., MCCORMICK, S.P., ALEXANDER, N.J., DILL-MACKY, R. and HOHN, T.M. (2002) Engineering deoxynivalenol metabolism in wheat through the expression of a fungal acetyltransferase gene. *Theor. Appl. Genet.* **106**: 74–83.

PACIN, A.M., RESNIK, S.L., NEIRA, M.S., MOLTÓ, G. and MARTíNEZ, E. (1997) Natural occurrence of deoxynivalenol in wheat, wheat flour and bakery products in Argentina. *Food Addit. Contam.* **14**: 327–331.

PARRY, D.W., JENKINSON, P. and MACLEOD, L. (1995) *Fusarium* ear blight (scab) in small grain cereals – a review. *Pl. Pathol.* **44**: 207–238.

PARRY, D.W. and NICHOLSON, P. (1996) Development of a PCR assay to detect *Fusarium poae* in wheat. *Plant Pathol.* **45**: 383–391.

PETTERSSON, H. and LANGSETH, W. (2002) Intercomparison of trichothecene analysis and feasability to produce certified calibrants. Homogeneity and stability studies – final intercomparison. BCR Information, EU report EUR 20285/2 EN 145 pp.

PIETERS, M.N., FREIJER, J., BAARS A.J. and SLOB, W. (2001) Risk assessment of Deoxynivalenol in Food. An assessment of exposure and effects in the Netherlands RIVM Rapport 388802022, 30pp.

PINEDA-VALDES, G., RYU, D., HANNA, M.A. and BULLERMAN, L.B. (2003) Reduction of moniliformin in corn by heat processing *J. of Food Sci.* **68**: 1031–1035.

PITT, J.I. (1995a) Estimating the direct cost of aflatoxins. *Australian mycotoxin newsletter* **6**: 1–12.

PITT, J.I. (1995b) Under what conditions are mycotoxins produced? 2. *Aspergillus* species. *Australian mycotoxin newsletter* **6**: 1–2.

PITT, J.I. (1995c) Under what conditions are mycotoxins produced? 3. *Penicillium* species. *Australian mycotoxin newsletter* **6**: 1–12.

PROCTOR, R.H., HOHN, T.M. and MCCORMICK, S.P. (1995) Reduced virulence of *Gibberella zeae* caused by disruption of a trichothecene toxin biosynthetic gene. *Mol Pl. Microbe Interact.* **8**: 593–601.

SCHNERR, H., NIESSEN, L. and VOGEL, R.F. (2001) Real time detection of the *tri5* gene in *Fusarium* species by lightcycler-PCR using SYBR Green I for continuous fluorescence monitoring. *Int. J. Food Microbiol.* **71**: 53–61.

SCHNERR, H., VOGEL, R.F. and NIESSEN, L. (2002) Correlation between DNA of trichothecene-producing *Fusarium* species and deoxynivalenol concentrations in wheat-samples. *Lett. Appl. Microbiol.* **35**: 121–125.

SCHOLTEN, O.E., RüCKENBAUER, P., VISCONTI, A., VAN OSENBRUGGEN, W.A. and DEN NIJS, A.P.M. (2002) Food safety of cereals: a chain wide approach to reduce *Fusarium* mycotoxins. 94pp.

SCHÜTT, F., NIRENBERG, H.I. and DEML, G. (1998) Moniliformin production in the genus *Fusarium*. *Myxotoxin Research* **14**: 35–40.

SCOTT, P.M., KANHERE, S.R., DEXTER, J.E., BRENNAN, P.W. and TRENHOLM, H.L. (1984) Distribution of the trichothecene mycotoxin deoxynivalenol (vomitoxin) during the milling of naturally contaminated hard red spring wheat and its fate in baked products. *Food Addit. Contam.* **1**: 313–323.

SINHA, R.N., MUIR, W.E., SANDERSON, D.B. and TUMA, D. (1991) Ventilation of bin-stored moist wheat for quality preservation. *Can. Agric. Eng.* **33**: 55–65.

SHIMA, J., TAKASE, S., TAKAHASHI, Y., IWAI, Y., FUJIMOTO, H., YAMAZAKI, M. and OCHI, K. (1997) Novel detoxification of the trichothecene mycotoxin deoxynivalenol by a soil bacterium isolated by enrichment culture. *Appl. Environ. Microbiol.* **63**: 3825–3830.

TANAKA, T., YONEDA, A., INOUE, S., SUGIURA, Y and UENO, Y. (2000) Simultaneous detection of trichothecene mycotoxins and zearalenone in cereals by gas chromatography-mass spectrometry. *J.Chromatogr.* **882**: 23–28.

TRENHOLM, H.L., WARNER, R.M. and PRELUSKY, D.B. (1985) Assessment of extraction procedures in the analysis of naturally contaminated grain products deoxynivalenol (vomitoxin). *J. Assoc. Off. Anal. Chem.* **68**: 645–649.

TURNER, A.S., O'HARA, R.B., REZANOOR, H.N., NUTTALL, M., SMITH, J.N. and NICHOLSON, P. (1999) Visual disease and PCR assessment of stem base disease in winter wheat. *Plant Pathol.* **48**: 742–748.

VISCONTI, A. (2002) Food safety of cereals: a chain wide approach to reduce *Fusarium* mycotoxins In: Scholten, O.E., Rückenbauer, P., Visconti, A., van Osenbruggen, W.A. and den Nijs, A.P.M. (eds). pp. 29–40.

WAALWIJK, C. (2002a) *Fusarium* species on wheat in The Netherlands: inventory and molecular identification. *Journal of Applied Genetics* **43A**: 125–130.

WAALWIJK, C., HESSELINK, T., DE VRIES, PH.M., DE HAAS, B.H., KASTELEIN, P., VERSTAPPEN, E.C.P., VAN DER LEE, T.A.J. and KEMA, G.H.J. (2001) *Fusarium* in Nederland: inventarisatie en identificatie. Plant Research International Report No 54 *(in Dutch)*.

WAALWIJK, C., KASTELEIN, P., HESSELINK, T., DE VRIES, PH.M., LOMBAERS, C., VAN DER HEIDE, R.A. SCHOEN, C.D., VAN DER LEE, T.A.J. KÖHL, J., SCHEPERS, H.T.A.M. and KEMA, G.H.J. (2002b) Inventarisatie van toxigene *Fusarium* spp. en andere ziekten en plagen in Nederlandse wintertarwe in 2001. Plant Research International Report No 57 *(in Dutch)*.

WAALWIJK, C., KASTELEIN, P., DE VRIES, PH.M., KERÉNYI, Z., VAN DER LEE, T.A.J., HESSELINK, T., KÖHL, J. and KEMA, G.H.J. (2003) Major changes in *Fusarium* spp. in wheat in The Netherlands *Eur. J. Pl. Pathol.* (in press).

WANG, Y.Z. (1997) Epidemiology and management of wheat scab in China. In: *Fusarium Head Scab: Global status and future prospects*. Dubin, H.J., Gilchrist, L., Reeves and J., McNab (eds.) CYMMIT, Mexico. P 97–105.

WARD, T.J., BIELAWSKI, J.P., KISTLER, H.C., SULLIVAN, E. and O'DONNELL, K. (2002) Ancestral polymorphism and adaptive evolution in the trichothecene mycotoxin gene cluster of phytopathogenic *Fusarium*. *Proceedings of the National Academy of Sciences USA* **99**: 9278–9283.

WINDELS, C.E. (2000) Economic and social impacts of *Fusarium* head blight: Changing farms and rural communities in the Northern great plains. *Phytopathol.* **90**: 17–21.

26

Improving wheat quality

O. K. Chung, Agricultural Research Service – USDA, USA, S.-H. Park, Kansas State University, USA, M. Tilley and G. L. Lookhart, Agricultural Research Service – USDA, USA

26.1 Introduction

On average for the last three years (1998/99–2000/01), the United States of America (USA) produced 64.2 million metric tons (mmt) of wheat representing about 11% of the world production. Wheat is the most valuable food crop and the major export crop of the US, as 43% (28.8 mmt) enters the export market. The US produces several classes of wheat that have different functional properties and end-uses. The major bread wheat classes, Hard Red Winter (HRW) and Hard Red Spring (HRS) wheats comprise 63–65% of total US wheat production and 62–63% of US wheat exports. There are official US Standards for Wheat, established and maintained by the US Department of Agriculture (USDA).

Wheat quality improvement begins with breeding. Important traits targeted in wheat breeding include both agronomic and end-use qualities. The USDA, Agricultural Research Service (ARS) includes four Regional Wheat Quality Laboratories (RWQLs) that have made paramount contributions to US wheat improvement for all wheat classes. Quality evaluation in the US bread wheat breeding program was once limited to traditional milling and breadbaking tests. It is now rapidly expanding to include a wider range of tests for multiple end-use products. Tremendous growth exists in non-traditional uses, such as Asian products, noodles, frozen dough, par-bake products, tortillas, and pizza crust. To take full advantage of these expanding markets, new quality parameters and quality prediction tests are being developed. Quality evaluation is a valuable approach to retain the competitive edge in world markets while addressing new demands of domestic consumers.

26.2 US wheat classification and grading

The United States Grain Standards Act (USGSA) was implemented in 1916 to bring uniformity to the official US inspection system for marketing and classing of wheat. With approximately 15 revisions since the inception of the USGSA, the most pivotal changes occurred in 1976 with the creation of the Federal Grain Inspection Service (FGIS), now known as the Grain Inspection, Packers and Stockyards Administration (GIPSA), and 1986 with the passage of the Grain Quality Improvement Act (Plaus, 2003).

26.2.1 Wheat classification

Under the authority of the USGSA, GIPSA establishes and maintains the US Standards for grain. The current US Standards categorizes wheat into eight basic classes and nine subclasses based on color, hardness, and growing season.

1. Durum wheat. All varieties of white (amber) durum wheat, divided into three subclasses;
   (i) Hard Amber Durum wheat,
   (ii) Amber Durum wheat, and
   (iii) Durum wheat.
2. Hard Red Spring (HRS) wheat. All varieties of HRS wheat, divided into three subclasses;
   (i) Dark Northern Spring wheat,
   (ii) Northern Spring wheat, and
   (iii) Red Spring wheat.
3. Hard Red Winter (HRW) wheat. All varieties of HRW wheat.
4. Soft Red Winter (SRW) wheat. All varieties of SRW wheat.
5. Hard White (HDWH) wheat. All hard endosperm white wheat.
6. Soft White (SWH) wheat. All soft endosperm white wheat varieties, divided into three subclasses;
   (i) Soft White wheat,
   (ii) White Club wheat, and
   (iii) Western White wheat.
7. Unclassed wheat. Any variety of wheat that is not classifiable under other criteria provided in the wheat standards without subclasses, including any wheat that is other than red or white in color.
8. Mixed wheat. Any mixture of wheat that consists of less than 90% of one class and more than 10% of one other class or a combination of classes that meet the definition of wheat.

26.2.2 Official US Standards for Wheat classes and grading

Under the authority of the USGSA, GIPSA/FGIS inspects all grains exported from the US, and upon request, in the domestic market. The US Standards for Wheat provide for a numerical grading system. For wheat, samples may be graded as US No. 1 through 5 or as US Sample Grade (Table 26.1). In determining what grade to apply to a sample of wheat, inspectors evaluate specific factors which are test weight (TW) per bushel, heat-damage, total damage, foreign material, shrunken and broken kernels, total defects, contrasting classes, and wheat of other classes. All grading factors except for TW (pounds/bushel = lb/bu) are equally applied to all US wheat classes. For US No. 1 grade wheat, the minimum TW is 58.0 lb/bu (74.7 kg/hl) for HRS or White Club wheat and 60.0 lb/bu (77.2 kg/hl) for all other wheat classes/subclasses. The minimum

Table 26.1 US standards for wheat: grades and grade requirement

	US Grade No.				
Grading factors	1	2	3	4	5
Minimum pound/bushel limits of:					
Test weight (lb/bu)[1]					
Hard Red Spring wheat or White Club wheat	58.0	57.0	55.0	53.0	50.0
All other classes/subclasses	60.0	58.0	56.0	54.0	51.0
Maximum percent limits of:					
Defects					
Damaged kernels					
Heat (part of total)	0.2	0.2	0.5	1.0	3.0
Total	2.0	4.0	7.0	10.0	15.0
Foreign material	0.4	0.7	1.3	3.0	5.0
Shrunken and broken kernels	3.0	5.0	8.0	12.0	20.0
Total[2]	3.0	5.0	8.0	12.0	20.0
Wheat of other classes[3]					
Contrasting classes	1.0	2.0	3.0	10.0	10.0
Total[4]	3.0	5.0	10.0	10.0	10.0
Stones	0.1	0.1	0.1	0.1	0.1
Maximum count limits of:					
Other material					
Animal filth	1	1	1	1	1
Castor beans	1	1	1	1	1
Crotalaria seeds	2	2	2	2	2
Glass	0	0	0	0	0
Stone	3	3	3	3	3
Unknown foreign substance	3	3	3	3	3
Total[5]	4	4	4	4	4
Insect-damaged kernels in 100 grams	31	31	31	31	31

US Sample Grade wheat:
(a) Does not meet the requirements for US Nos. 1, 2, 3, 4, or 5; or
(b) Has a musty, sour, or commercially objectionable foreign odor (except smut or garlic odor); or
(c) Is heating or of distinctly low quality.
[1] lb/bu = pounds per bushel and 1lb/bu is equivalent to 1.287 kilogram/hectoliter (kg/hl).
[2] Includes damaged kernels (total), foreign material, and shrunken and broken kernels.
[3] Unclassed wheat of any grade may contain not more than 10.0% of wheat of other classes.
[4] Includes contrasting classes.
[5] Includes any combination of animal filth, castor beans, crotalaria seeds, glass, stones, or unknown foreign substance.

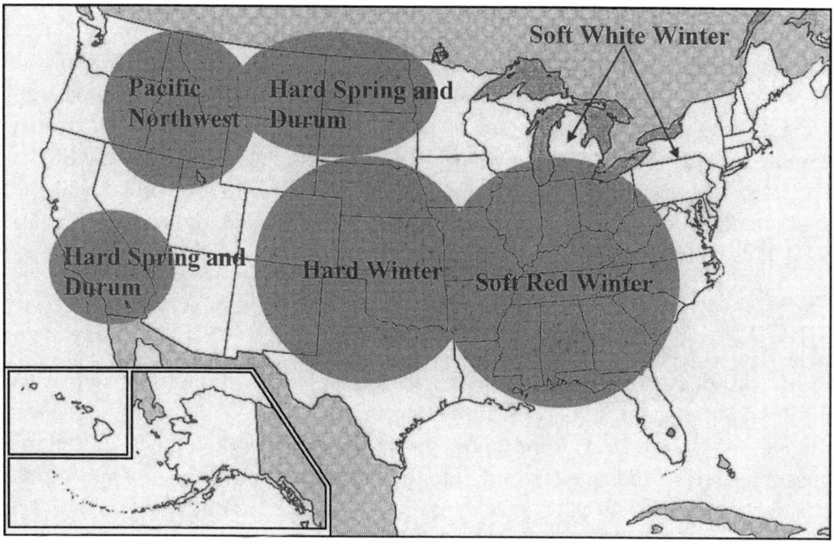

Fig. 26.1 US wheat growing regions. (Adapted from Graybosch, 2003.)

TW are 57.0 lb/bu (73.4 kg/hl) and 55.0 lb/bu (70.8 kg/hl) for HRS or White Club wheat and 58.0 lb/bu (74.7 kg/hl) and 56.0 lb/bu (72.1 kg/hl), respectively, to be graded as US No. 2 and No. 3 grade wheat (Table 26.1).

26.2.3 Usage of wheat by classes

Each wheat class is traditionally grown in a specific region in the US (Fig. 26.1); HRW wheat is grown in the Great Plains, HRS and Durum wheats are grown in the upper Midwest (Northern Plains), SRW wheat in the Ohio Valley and Southeast, SWH wheat in the Pacific Northwest (PNW) and Great Lakes region, and HDWH wheat on the West Coast and recently the Great Plains. In the last decade, the PNW states of Washington, Idaho, and Oregon have annually produced approximately 250–370 million bushels. In 2001, Washington State ranked No. 3 in all wheat production behind Kansas and North Dakota. Their wheat production is very diverse with 81.1% SWH wheat (winter and spring types plus winter type Club), 16.5% HRW, 1.3% HRS and 1.1% SRW wheat (Morris, 2002). Unlike the European terminology of 'hard' and 'soft' wheats, both 'hard' and 'soft' wheats in the US are hexaploid. It is important to note that the European term of 'hard wheat' indicates Durum, tetraploid, wheat and 'soft wheat' defines common wheat including all US hexaploid wheat classes.

The usage of wheat depends on hardness and protein contents (PC). Durum wheat is used for pasta products (macaroni, spaghetti, etc). In general, Durum has the hardest kernel texture with the highest hardness score (HS) measured either by a near-infrared reflectance (NIRR)/near-infrared transmittance (NIRT) or hardness index (HI) value by a Single Kernel Characterization System (SKCS) and has generally high PC (>16%).

Both HRW and HRS wheats are mainly used for breads, hard rolls, and hamburger buns. Soft wheats, including SRW and SWH wheats (White Club and Western White), are traditionally used for biscuits (cookies), cakes, pastries, certain types of Asian noodles and steamed breads, and flat breads. Soft wheats contain low PC (8–11%) and low NIR-HS (<40 with an average 25) and SKCS-HI (<45), whereas hard wheats contain medium PC (13–17%) with HS (50–100 average of 75). In general, HS/HI values and wheat PC are higher for HRS than for HRW wheat class.

26.2.4 Varietal identification

In the US, there are no regulations for varietal release and approximately 1000 different varieties of all classes are grown in the US today. FGIS personnel are trained to class wheat using nine visual kernel morphological descriptive characteristics and/or varietal identification. The nine morphological characteristics that are used are: kernel length, kernel width, slope of the back of the kernel, germ size, germ angle, brush size, cheek shape, crease, and surface texture. Differences in these characteristics generally allow inspectors to properly classify US wheat.

Biochemical methods for identification of wheat varieties are not normally used for official inspection. However, when FGIS inspectors are unable to identify a given variety for classification, that sample is sent to the GMPRC for protein fingerprinting. Separation of the gliadins is accomplished by one of the following methods; polyacrylamide gel electrophoresis at pH 3.1 (Acid- PAGE) (Lookhart *et al.*, 1982; Lookhart 1991), reversed-phase high-performance liquid chromatography (RP-HPLC) (Lookhart *et al.*, 1986), or high-performance capillary electrophoresis (Bean and Lookhart, 2001).

26.3 Breeding and wheat quality

The definition of wheat quality has different meanings to different people, i.e., the 'quality' is in the eye of the consumer. Since wheat quality improvement starts with breeding, we have limited our discussion to US wheat breeding objectives related to major agronomic and end-use quality parameters tested at the HWWQL.

26.3.1 US wheat breeding program

Wheat breeding is a lengthy and costly process requiring 13–15 years with an approximate cost of $750 000 to develop and register a new cultivar. Breeding has three broad stages (Fig. 26.2), including: the first stage (year 1 and 2) of mating; the second stage (year 3 to 7) of inbreeding and selection; and the last (third) stage (year 8 to 12) of evaluation of breeding lines from the germplasm observation nursery, regional and state-wide trials and large-scale quality trial

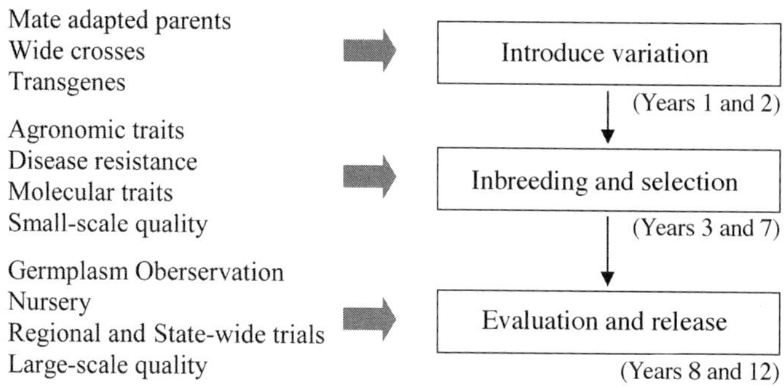

Fig. 26.2 Typical US wheat breeding program. (Adapted from Graybosch, 2003.)

such as Wheat Quality Council (WQC). Final varietal release decisions are made by state or by private firms' release committees (Graybosch, 2003).

Breeders must continue to develop and release new wheat cultivars, as older commercial cultivars tend to lose disease and pest resistance over three to five years. The major objectives of wheat breeding programs are to improve agronomic, disease and pest resistance, and milling and end-use quality traits. Important agronomic traits selected in US bread wheat breeding are high yield, high TW (>60 lb/bu for HRW and >58 lb/bu for HRS), kernel morphology, weight, density and size distribution, and agronomic characteristics such as plant height and maturity. The resistances to be obtained or maintained by breeding include: lodging, weathering, diseases (including stem rust, leaf rust, tan spot, *Fusarium* head blight), and insect resistance, including cereal leaf beetle and wheat stem sawfly (Hareland, 2002). Specific resistance objectives for hard winter wheats (red and white) are towards: barley yellow dwarf and wheat soilborne mosaic viruses for the eastern Great Plains; Russian wheat aphid, wheat streak mosaic virus for the western Great Plains; and green bug, karnal bunt for the southern Great Plains (Graybosch, 2003).

The agronomic appearance of HRW wheat has changed greatly from 'Turkey', the first HRW wheat grown in the Great Plains in 1873, to present day wheats. Modern HRW wheats are much shorter (by 50 cm), mature 2–3 weeks earlier, and have greatly improved straw strength. The number of HRW wheat varieties has increased from five in 1919 to 164 in 1984. The number of primary varieties is about 20, accounting for 90% of wheat acreage in any given year (Chung *et al.*, 2002b). Based on the average data of two regional hard winter wheat performance nurseries (Northern and Southern) for the last three years (2000–2002), Graybosch (2003) reported breeding increased grain yield by 150% over the wheat variety 'Kharkof', that was released in 1931.

The Great Plains, which is traditionally a HRW wheat growing area, is rapidly seeing the introduction of hard white (HDWH) wheat. This is due to economic benefits that are associated with HDWH wheat: primarily higher milling extraction and color advantages over HRW. HDWH wheat can be milled to 1–2% greater extraction with a concomitant increase in PC when milled to the same color. Additional advantages include color properties that are more desirable for Asian products, e.g. bright white color, fewer bran specks, lowered levels of tannins from bran that contribute to color and bitter taste. In addition, specific HDWH wheat varieties contain lower levels of polyphenol oxidase (PPO), an enzyme that is abundant in red wheat, which turns fresh noodles dark upon aging. The planting acreage of HDWH wheat in the Great Plains has steadily increased from 40 000 acres in 1999 to over 350 000 acres in 2002. It is anticipated that HDWH wheat production will increase in the next 5–7 years, even though the number of hard winter wheat breeding lines in the USDA-ARS Regional Germplasm Observation Nursery are 80–85% HRW and 15–20% HDWH wheat (Graybosch, 2003).

26.3.2 Contribution of wheat quality laboratories to varietal improvement

Until the mid 1930s, the end-use quality of all classes of US wheat was evaluated in the USDA laboratory in Washington DC area. Strong concerns about declining wheat quality from both producers and consumers of HRS and HRW wheats led Congress to establish the USDA/ARS/RWQLs. The four RWQLs were established by US congressional mandate during the 1930s–1960s.

The first laboratory was the Soft Wheat Quality Laboratory (SWQL), which was established in 1936 at Wooster, Ohio, followed by the Hard Winter Wheat Quality Laboratory (HWWQL) in 1937 at Manhattan, Kansas. In 1946, the Western Wheat Quality Laboratory (WWQL) was established at Pullman, Washington. Lastly, in 1963 the Hard Red Spring and Durum Wheat Quality Laboratory (HRS & DWQL) was established at Fargo, North Dakota. All four RWQLs have the common missions: work with breeders to improve US wheats by testing end-use quality; develop reliable small-scale tests for evaluating early generation breeding lines; provide research into the contribution of flour biochemical components to observed differences in end-use quality; and develop rapid and objective prediction models for end-use quality.

Although some state universities have wheat quality laboratories (WQLs) for their individual state breeding programs, the USDA/ARS/RWQLs are an essential entity for US wheat quality improvement. At the advanced yield trial stage of each breeding program, a few selected lines with the most promising quality attributes are entered into federally coordinated regional nurseries. Breeding lines submitted by breeders (state, private, and federal) are then grown at multiple locations in their respective growing region. Finally, breeding lines are tested for agronomic and end-use quality attributes affected by different growing locations throughout the growing region. State WQLs study the effects of locations in their own states but not in the growing regions, e.g. the Great Plains for hard winter wheats. All four RWQLs play key roles in WQC activities and also in Overseas Wheat Quality Evaluation Activities, sponsored by the US Wheat Associates and State Wheat Commissions. Thus, the RWQLs make significant contributions to the US wheat quality improvement and to the promotion of US wheat exports. Over 95% of US wheat cultivars been evaluated at one of the RWQLs.

26.4 Quality indices and tests

Each USDA/ARS/RWQL annually evaluates several thousand breeding program samples for grain, milling, flour, dough, and the final product characteristics. The SWQL tests cookies and cake quality for SRW wheat breeding lines, the HWWQL tests bread quality for HRW and HDWH winter wheats, the HRS & DWQL tests bread quality for HRS wheat and pasta quality for Durum wheat lines, and the WWQL conducts diverse tests on various wheat class breeding programs in the PNW, i.e. bread quality for HRW and HRS wheat, cookies and cakes for some soft wheat lines, but the most concerted efforts are in Asian products. US bread wheats are evaluated mainly at the USDA/ARS/HWWQL for winter wheats and HRS & DWQL for spring wheat lines for grain, milling, flour, dough, and bread characteristics.

26.4.1 Grain characteristics

Test weight

TW is the weight per Winchester bushel of cleaned wheat as described in AACC method 55–10 (American Association of Cereal Chemists, AACC, 2000) and is the first quality determinant to be evaluated. TW is affected by kernel density and packing characteristics, which are influenced by kernel characteristics (shape, uniformity of shape and size, moisture content [MC], and surface condition) and foreign materials. TW can be useful to predict flour yield with a wide range of 40 to 64 lb/bu, but not so with a narrow range above 57 lb/bu. Cleaning of wheat is performed on a Carter Dockage Tester, using sieves and riddle specified by the USDA standards. This is the most widely used quality index by US wheat industry, from breeders up to millers, as the price of wheat is influenced by its TW.

Kernel hardness

Hardness is a major factor when classifying wheat. Hardness affects the milling process and, as a result, affects flour characteristics and the final baking or other end-use properties. Slaughter *et al.* (1992) reported significant linear relationships of wheat hardness with PC, loaf volume (LV), and crumb grain, and texture of hard winter and spring wheats. Hardness is determined by various methods. Using NIRT spectroscopy, hardness values can be obtained without destroying the kernels (Norris *et al.*, 1989; Delwiche, 1993). The single kernel (SK) methodology suitable for classification was developed by engineers at the GMPRC in Manhattan, Kansas (Martin *et al.*, 1993; Steele and Martin, 1991). This technology was transferred to commercial production via Perten Instrument Co. The SKCS-4100 measures wheat characteristics of 300 individual kernels and reports the mean and standard deviation values of moisture content (MC), weight (mg), size (diameter, mm), HI and peak force, in addition to wheat classification. The most widely used methods are NIR-HS determination by Norris *et al.* (1989) an AACC Method 39–70A (AACC, 2000) and SKCS-HI by Martin *et al.* (1993) an AACC Method 55–31 (AACC, 2000). Hardness values obtained by NIR are corrected for MC (Windham *et al.*, 1993).

Kernel weight and size

Kernel weight is generally measured as 1000 kernel weight with an electronic seed counter (Halverson and Zeleny, 1988). Kernel size is measured according to average cross-sectional area using the procedure developed by Shuey (1960), who found a significant positive correlation ($r = 0.957$) between actual commercial milling yield and predicted milling yield using kernel size data. Today, kernel weight and size (diameter) are conveniently determined by SKCS.

Wheat protein

It is generally considered a prime factor that determines the quality of wheat flour. Both quantity and quality of proteins affect breadmaking properties such as mixing time (MT) and tolerance, dough handling properties, bake water absorption (WA), oxidation requirement, LV potential, and crumb characteristics of bread (Finney, 1984). Graybosch *et al.* (1993) studied the comparative effect of wheat flour protein, lipids, and pentosan composition in relation to baking and milling quality and found that PC was the primary factor contributing to variation in both dough strength and LV, but loaf textural features appeared to be more highly influenced by protein composition, water-soluble pentosans, and lipids. Wheat PC is determined by a Nitrogen Determinator (Leco Corp, St. Joseph, MI) according to AACC Method 46-30 (AACC, 2000) or by a NIRR or NIRT spectroscopy (AACC Method 39-25). Although wheat PC is still not a required quality parameter of the US wheat grading system, it plays an important role to meet the buyer's specifications.

Wheat moisture

While MC is not a determinant of wheat grade under US Standards, MC has direct economic importance because of its inverse relationship with dry matter. Wheat MC is determined by an oven method (AACC Method 44-15A) or by an NIR method.

Wheat ash

Typically wheat ash content (AC) ranges from about 1.4 to 1.8% with an average of 1.5% (on 14% moisture basis, mb). It is determined by AACC Method 08–01.

26.4.2 Milling characteristics

Milling characteristics are of economical importance to millers. In general, millers' customers desire flours with low AC, white/bright color and higher PC. Since millers have to blend wheats to produce flour quality to meet bakers' specifications, the milling characteristics of individual commercial wheat cultivars are extremely important quality parameters. Therefore, the RWQLs evaluate experimental milling properties of breeding lines.

Flour yield

Of the many milling characteristics, flour yield is generally considered the most important in relation to end-use quality of wheat. Flour yield is determined as percent (%) of straight grade flour. Hard wheat is usually tempered to 15%, milled depending on the sample size, on experimental mills such as Allis-Chalmers, Buhler, Quadrumat Junior or Senior. At the HWWQL, 750 g-1 kg samples of advanced lines are routinely milled on a Quadrumat Senior mill and 100 g of early generation lines are milled on a micro-scale mill. A Miag semi-pilot scale mill is used for larger samples.

Milling score

It is derived from the following equation: milling score = 100 − {TW -60) + (82 − flour yield) + 100× (flour AC − wheat AC/3.9) + 10 × (wheat PC − 1) − flour PC]}. This equation was developed empirically at the HWWQL, Manhattan, KS. Both AC and PC are expressed on 14% mb. At a given nursery, the milling score is compared to that of a standard check wheat and also between the entries. This score may be quite different from the milling industry scores. However, it serves well for a determination of the milling potential index of breeding lines.

26.4.3 Flour characteristics

Flour characteristics are important quality parameters for manufacturers' final products or consumers. The HWWQL evaluates: flour color, MC, PC, AC, sedimentation, sprout damage, PPO activity, and Rapid Visco-analyzer (RVA) pasting properties.

Flour color

Flour color greatly affects the color of the final product. A creamy white color is generally the desirable color of bread flour and that for Asian noodles. It is determined using Agtron according to the AACC Method 14-30 with major modifications. Recently, a Minolta Colorimeter (Model CR-300) method has provided quick results for L*, a*, and b* values.

Proximate analyses

Flour MC and PC are determined by air-oven method (AACC Method 44-15A) and a combustion method (AACC Method 46-30), respectively. Also, flour MC or PC (AACC Method 39-11) can be determined by a NIRR system. The HWWQL determines flour AC by the furnace method (AACC Method 08-01). The goal for hard winter wheat breeding program is to provide flour with at least 12% PC and at most 0.50% AC, both on a 14% mb.

Sprout damage
The 3-min stirring method is applied by using an RVA, according to the AACC Method 22-08. The stirring number is highly correlated to the Falling Number (AACC 56-81B) (Bason *et al.*, 1993).

PPO activity
Darkening and discoloration of noodles are closely related with the PPO activity (Kruger *et al.*, 1994; Baik *et al.*, 1995). Whole kernel PPO is determined by the method of Anderson and Morris (2001) and the modified procedures are used for ground meal and flour PPO activities (Chung *et al.*, 2003b).

RVA pasting properties
Pasting is a result of disruption of starch granule structure, the pasting properties of flour are influenced by other flour components including proteins, pentosans, and lipids as well as cultivar and growing region. Ohm and Chung (1999a) reported that pasting temperature was negatively correlated with LV and an optimum ratio (0.17–0.18) of flour PC (%)/pasting temperature for producing large LV. Pasting characteristics are conducted by AACC Method 76–21, using an RVA.

26.4.4 Dough characteristics

Since dough mixing is the first and most critical stage of breadmaking, it is important to know optimum bake WA and MT requirements and also mixing tolerance of a given flour. Bake WA is determined based on the mixograph WA with adjustments made by the 'feel' of the dough by baking experts. The determination of optimum MT is also indicated by the mixograph. However, the baking expert decides the optimum MT by judging when the dough has a smooth feel and shiny appearance with a minimum mobility or maximum resistance.

A 10 g flour mixograph (AACC Method 54-40A, Shogren and Finney, 1984) has made the greatest contribution to quality improvement of US bread wheat. When the quantity of early generation lines is limited to 50–100 g, wheat is milled on a micro-milling scale and proximate analyses and 10 g flour mixograph testing are conducted. Breeders select for mixing strength based on mixograph parameters, i.e. mixograph MT of 2.0–2.3 min and mixing tolerance as 2.5 (0–6 scale with 4 being satisfactory). Recently, a computerized mixograph is available and its parameters can replace conventional mixograph parameters, using the models developed by Chung *et al.* (2001).

Table 26.2 Formula for pup-straight dough method

Ingredient	Amount
Flour	100 g (14% mb)
Sugar	6.0 g
Salt	1.5 g
Shortening	3.0 g
Malt	0.25 g
Dry yeast	0.8-1.0 g
Water	Optimum
Ascorbic acid	50 ppm

26.4.5 Bread characteristics

The routine breadmaking procedure for breeding lines is a pup-loaf straight dough method (AACC Method 10–10B) using a formulation containing 100 g (14% mb) flour (Table 26.2). Doughs are fermented for 120 min at relative humidity 86% and 30 °C, and loaves are baked at 218 °C (425 °F) for 18 min and weighed immediately after removal from the oven. For special sets of samples, we use a micro-baking procedure using 10 g flour (Shogren and Finney, 1984) or a pound-loaf sponge and dough method using 300 g flour for the large scale testing such as WQC samples. The pound-loaf sponge and dough process utilizes the formulation (Table 26.3) used by the American Institute of Baking (AIB).

26.4.6 External characteristics of bread

LV is one of the most important characteristics of bread. High PC-flour usually produces a large loaf of bread. A large loaf in white-pan bread is generally considered desirable but only when accompanied by good crumb grain and

Table 26.3 Formula for pound sponge and dough method

Ingredients	Baker's %
Sponge	
Flour	70.0
Yeast (instant dry)	0.9
Yeast food (no oxidation)	0.5
Water	42.0
Dough	
Flour	30.0
Granulated Sugar	7.0
Shortening (solid, unemulsified)	3.0
Salt	2.0
Calcium propionate	0.12
Water	Variable (18.0)

texture. LV is usually determined by the rape seed-displacement method and is expressed in cubic centimeters (cm^3) for a given flour. The HWWQL puts an emphasis on LV regression, which is actually slope of regression line of LV on flour PC for pup-loaf using 100 g flour (Finney, 1985). The HWWQL considers this parameter as a LV potential, derived from an equation of (LV – 300)/(flour PC – 3) for samples with flour PC equal to or higher than 12%. For flours with PC lower than 12%, the equation is (LV – 300)/(0.534 × flour PC + 0.018 × square of flour PC), as Finney (1985) suggested to use different equations to estimate LV regression because of curvilinear relationships between PC and LV at the PC level under 12%. Crust color should be deep golden brown for top crust and light golden brown for side and bottom crust, and color should be uniform and free from spots and streaks.

26.4.7 Internal characteristics of bread

Texture is another major quality factor because most American consumers prefer soft, resilient, and short crumb texture in fresh white-pan bread. Texture may be estimated by the sense of touch against the cut surface of the loaf. Crumb grain generally is evaluated visually based on the cell size, cell shape, and cell wall thickness. US white-pan bread crumb commonly contains uniformly small, elongated cells which is considered superior in quality to one containing non-uniform, large, round cells. At the HWWQL, the crumb grain of representative bread slices is graded from 0 to 6, according to the WQC, Hard Winter Wheat Testing Procedure (Wheat Quality Council, 1994). The scores range from poor open grain (0) to outstanding closed grain (6). At the current time, an objective method using a CrumbScan is being developed by AIB. Park *et al.* (2001, 2002) observed a linear correlation between crumb grain score judged by a baking expert and elongation ratio by CrumbScan.

26.4.8 Trends in HRW wheat quality

Major end-use quality attributes of HRW wheats are milling and breadmaking characteristics. In the 1930s, commercial HRW wheats had an undesirably short mixing requirement, and poor mixing tolerance in addition to poor milling yield. Cox *et al.* (1989) demonstrated the historical genetic improvement in end-use quality of HRW wheat cultivars (1919–1988) and that breadmaking parameters (WA, MT, LV, crumb grain score, and overall quality index) have increased 0.1 to 4.6% annually (compared to those of 'Turkey'), despite a decrease in PC of 0.2 percentage points for 20 recent cultivars (1976–1988), as shown in Table 26.4. However, the experimental milling flour yield had not been significantly changed. A recent review of HRW wheat quality (1989–1999 crops) showed increasing trends in kernel weight and milling yield, but decreasing trends in PC for both commercial cultivars and advanced breeding lines submitted to the WQC. Crumb characteristics of bread also showed an increasing trend, resulting from the successful cooperative efforts of hard winter wheat breeders and the

Table 26.4 Historic genetic improvement of HRW wheat[a]

Quality trait	Mean of 20 HRW wheat released	
	1919–1970	1976–1988
Flour protein (%)	11.5	11.3
Bake absorption (%)	55.5	57.4
Mix time (min)	3.0	3.9
Bread volume (cm^3)	929	960
Crumb grain score	7.0	7.6
Quality index	18	23

[a] Adapted from Cox *et al.* (1989).

HWWQL. Since the US bread baking industry has been concerned with decreasing bake WA, increasing the PC or pentosan content, as both increase the WA, may be the urgent task for US bread wheat.

26.5 Predicting the breadmaking quality of wheat

26.5.1 US bread wheat: spring and winter wheat

Irrespective of wheat color, bread wheat cultivars are classified as hard winter or hard spring according to their growth habitats. One of the most obvious differences between winter and spring wheat is the vernalization response of hard winter wheat, which requires over-winter to allow reproductive development. This over-wintering difference creates a strong environmental component making it difficult to compare the genetic factors differentiating end-use properties between hard winter and spring wheats

The annual US wheat production is 64–66 mmt, 41–45% of which is HRW wheat and 19–22% is HRS. Thus, hard winter wheat contributes to 65–70% of US bread wheats. Commercially, spring wheat has been more highly rated for pan-bread characteristics than hard winter wheat.

NIR-HS is a good parameter together with wheat PC for the classification of HRW and HRS wheat (Slaughter *et al.*, 1992). Classification of HRW and HRS has also been investigated by statistical analysis of chromatograms obtained from RP-HPLC (Endo *et al.*, 1990; Lookhart *et al.*, 1993). More recently, Chung *et al.* (2003a) efficiently separated hard winter and spring wheats grown under an over-wintering condition in California with only one canonical variable by including SKCS parameters in addition to NIR-HS and PC (Fig. 26.3).

The quality data of this special set of hard winter and spring wheats grown side by side are summarized in Table 26.5. Even though both winter and spring wheats were grown in the same location, spring wheat showed higher mean values for wheat and flour PC by one percentage point, wheat AC and hardness (both NIR-HS and SKCS-HI). Due to higher PC, wet and dry gluten contents, mixograph and bake WA and LV were higher with spring wheat than with

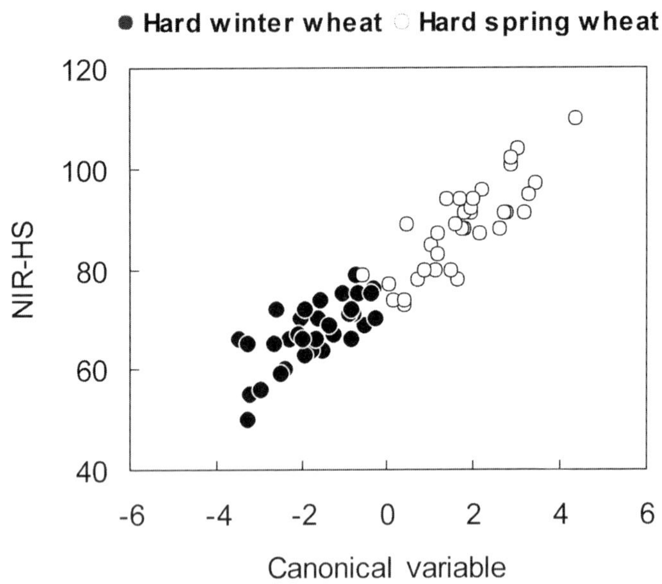

Fig. 26.3 The plot of canonical variable versus near infrared reflectance hardness score (NIR-HS). Canonical variable was calculated using wheat protein and ash contents, N IR-HS, and Single Kernel Characterization System parameters including hardness indices and their standard deviations (SD), peak forces, weight SD, and diameters. (Adapted from Chung *et al.*, 2003a.)

winter wheat. However, there were no significant differences in mean TW, SK weight or size, flour yield, and mixograph and bake MT requirement. The mean gluten index value was significantly higher for winter wheats (Table 26.5). There was no consistent difference in mean LV regression, with the overall mean values of 72.4 for winter and 71.9 cc/% flour PC for spring wheats.

26.5.2 Quality prediction by SKCS

We have evaluated SK wheat quality using a SKCS over the last decade. Some research areas are: (a) the relationship between SK characteristics and end-use quality for soft wheat (Chung *et al.*, 1994) and for hard wheat (Chung *et al.*, 2003a); (b) SK parameters of hard winter wheats in relation to milling and baking quality (Ohm and Chung, 1998; Ohm *et al.*, 1998); (c) SK wheat physical properties and first break grinding (Pasikatan *et al.*, 1998); and (d) prediction of conventional wheat characteristics using SK parameters, using 2890 wheats (1200 and 300 wheats as calibration and validation sets, respectively) to derive the prediction model for flour yield (R^2 =0.635) (Chung *et al.*, 1999). For the special sample set of winter and spring wheat, flour yield (Fig. 26.4) and LV of bread (Fig. 26.5) were well predicted by using the SKCS parameters (Chung *et al.*, 2003a).

Table 26.5 Means and ranges of wheat characteristics of hard winter and spring wheats[a]

Quality	Hard winter wheat			Hard spring wheat		
Parameters	Mean	Minimum	Maximum	Mean	Minimum	Maximum
Wheat protein (%)[b]	13.0	9.0	16.8	14.0 **	10.2	17.6
Wheat ash (%)[b]	1.64	1.48	1.79	1.70**	1.34	1.89
Test weight (lb/bu)	60.0	55.3	64.3	60.0	52.8	64.7
NIR-HS[c]	67.4	50.0	79.0	88.1 **	73.0	110.0
Single kernel parameter						
Hardness index	84.1	70.2	100.8	90.9 **	80.5	102.3
Hardness index SD[c]	16.1 **	11.4	19.1	15.3	11.5	19.3
Peak force[e]	76.7	57.0	111.0	81.8 **	56.0	104.0
Peak force SD[d]	24.6 **	20.0	31.0	24.1	20.0	31.0
Flour yield (%)	70.2	66.1	73.6	70.4	66.7	73.7
Flour protein (%)[b]	12.0	8.0	15.6	13.0**	9.2	16.8
Gluten						
Wet gluten (%)[b]	32.8	20.5	42.9	35.8**	24.5	46.3
Dry gluten (%)[b]	11.2	6.8	15.4	12.3**	8.0	16.4
Gluten index (%)	93.8**	77.3	99.6	89.7	61.7	99.2
Mixograph						
Water absorption (%)[b]	60.6	55.9	65.4	63.5**	58.4	68.0
Mix time (min)	4.8	2.3	10.8	4.7	2.8	10.7
Baking						
Water absorption (%)[b]	60.3	53.5	65.4	63.6**	59.6	68.0
Mix time (min)	6.1	2.6	15.3	6.1	2.3	17.0
Loaf volume (cm^3)[f]	81.7	64.0	103.3	88.4**	70.5	106.5

[a] Adapted from Chung *et al.* (2003a).
**: difference between mean values of winter and spring wheats is significant at $p<0.01$.
[b] On a 14% moisture basis.
[c] NIR-HS:near infrared reflectance hardness score.
[d] SD: standard deviation of single kernel parameters estimated from about 300 kernels.
[e] Maximum load cell force in force-time profile (analog to digital count divided by 100).
[f] Bread baked with 10-g (14% mb) flour.

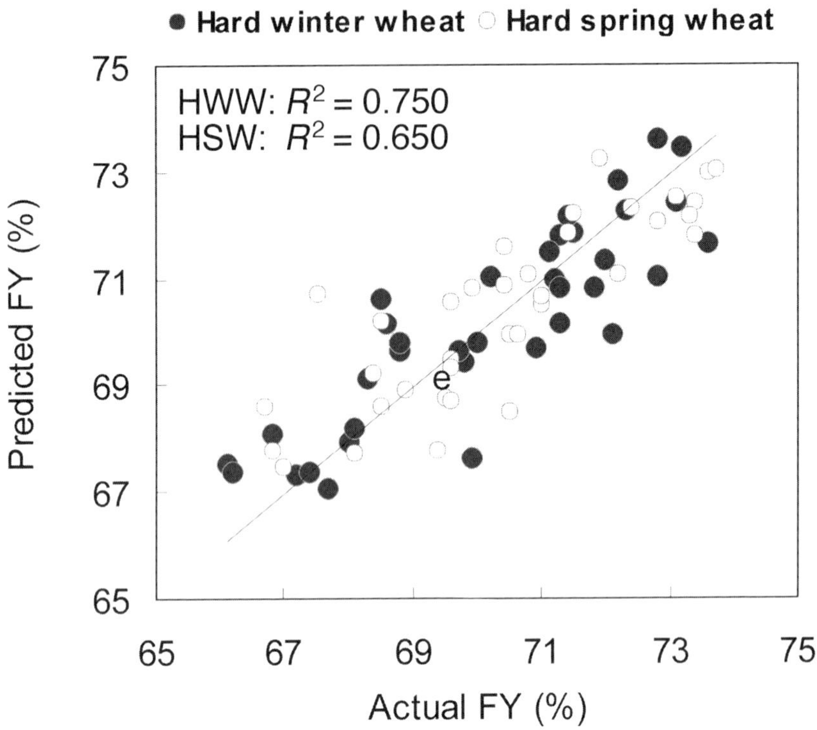

Fig. 26.4 The plot of actual flour yields (FY) and estimated FY by principal component regression, using single kernel parameters of winter and spring wheat, respectively (n = 36 for each class). HWW = hard winter wheat, HSW = hard spring wheat, R^2 = coefficient of determination. (Adapted from Chung *et al.*, 2003a.)

26.5.3 Quality prediction by NIRR/NIRT spectroscopy

The USDA/ARS/GMPRC/HWWQL evaluates quality parameters of hard winter wheat progenies of federal, state, and private nurseries. Our research activities have been carried out with an NIRSystems 6500 scanning spectrometer to construct a spectra database (400–2498 nm) with whole grain, ground meal, milled flour, semi-solid (dough), and liquid samples.

We obtained a spectra database of over 8000 wheats and meals each, and over 5000 flours from numerous nurseries (Seabourn and Chung, 1996). For deriving prediction equations, 833 wheat samples (469 for calibration and 364 for validation) and 831 flour samples (511 for calibration and 320 for validation) were used from the three regional nurseries (1991–1995 crops). NIR technology showed a high potential for rapid estimation of flour quality of early generation breeding lines. For early segregation, the predictive capability with R^2 values >0.60 was suitable (Table 26.6).

Other NIR research activities include: NIRR spectroscopy of flour-water doughs during mixing (Seabourn *et al.*, 1997); protein quality by predicting

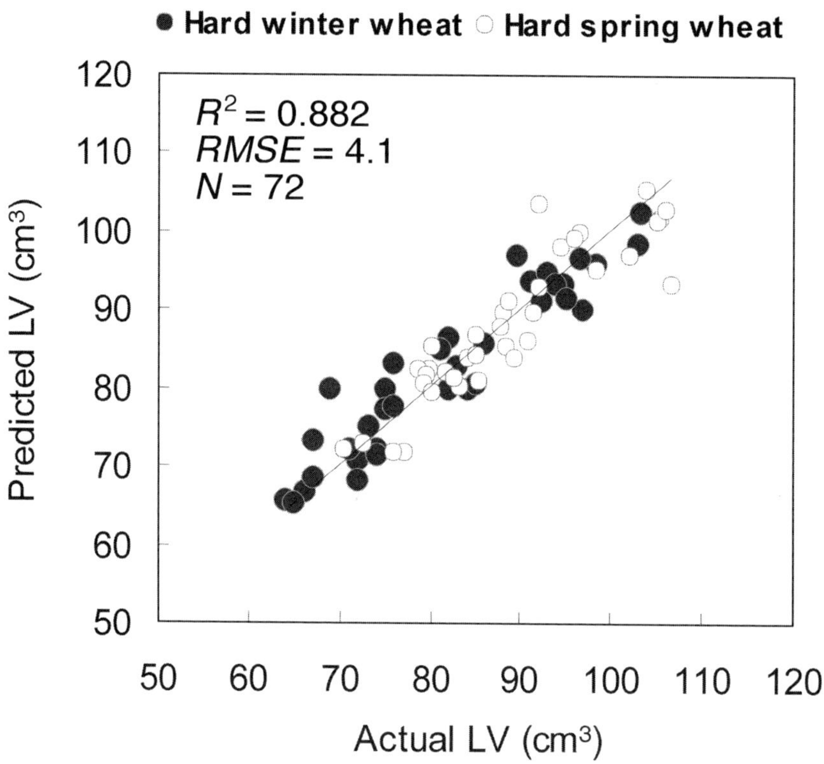

Fig. 26.5 The plot of actual loaf volume (LV) and estimated LV by principal component regression, using wheat quality and single kernel parameters of winter and spring wheats combined (n = 72). R^2 = coefficient of determination. *RMSE* = root mean square of error. (Adapted from Chung *et al.*, 2003a.)

gliadin and soluble/insoluble HMW-glutenin fractions in flours and whole wheat kernels (Seabourn *et al.*, 1999); NIRT estimation of free lipid, glycolipid and digalctosyldiglyceride contents using flour lipid extracts (Ohm and Chung, 2000) with HPLC as a reference method (Ohm and Chung, 1999b); and the collaborative study for PC of HRW wheat by NIR (Delwiche *et al.*, 1998).

26.5.4 Quality prediction by single kernel NIR technology

The SKCS was integrated with a DA-7000 NIR diode-array spectrometer (Perten Instrumentation, Reno, NV) using fiber optics to allow automated collection of VIS and NIR spectra over the range 400–1700 nm (Dowell, 1998). The SK-NIR technology has been used for: wheat color differentiation (Wang *et al.*, 1999); the detection and identification of stored grain insects (Dowell *et al.*, 1999b); prediction of scab, vomitoxin, and ergosterol contents (Dowell *et al.*, 1999a); and differentiation of vitreous and nonvitreous Durum wheat kernels (Dowell, 2000). This technology shows a strong potential for a rapid and objective quality prediction.

Table 26.6 NIRR prediction (R^2) of quality parameters using spectra database of whole wheat and flour samples

Quality parameters	Whole wheat (n=364)	Flour (n=320)
Test weight	0.75	–
SK-Size	0.83	–
Milling flour yield	0.63	–
Wheat PC	0.97	–
Flour PC	0.97	0.99
Mixograph water absorption	0.48	0.73
Bake water absorption	0.46	0.65
Flour color	0.56	0.88
Bread LV	0.62	0.60

(Adapted from Seabourn and Chung, 1996).

26.6 Future trends

In the past 70+ years US wheat breeders have achieved a 50% increase in agronomic yield potential, while improving disease resistance and maintaining quality (Graybosch, 2003). It is expected that HRW wheat (bread wheat) breeding programs will continue to be active in the future.

26.6.1 Expanded uses of bread wheat

Non-conventional breads. Wheat products in the US have changed from traditional white-pan bread to variety breads. Another form of non-conventional bread is 'Fast Bread' that is par-baked to 75–90%. This is flash-frozen product that allows consumer to save time and effort and at the same time to be unable to differentiate from regular bread, while providing a fresh-baked aroma in a baked product.

Non-bread products. Other sweeping changes in US wheat products are from bread to non-bread products, such as tortillas, pizza crust, etc. Sales by the US tortilla industry amounted to $2.8 billion in 1996 and $5.7 billion (projected) in 2002. Flour tortillas are a unique baked product from gluten-structured and chemically-leavened dough. Most tortilla flour is milled from HRW wheat, the same wheat used for bread flour. Specific flours containing 9.5–12.5% PC (ranging from moderate to strong protein quality) yield good dough processability, although the PC of flour alone does not determine its suitability for use in flour tortillas. The role of the HWWQL is being expanded to include testing tortilla quality for hard winter wheat breeding lines. Other non-bread products include pizza crust, pretzels and Asian alkaline noodles.

26.6.2 Expanded quality parameters for varietal improvement

The US wheat export share has drastically decreased from 43.2 mmt in 1987 to 27.2 mmt in 1997. While the US faced a 40% decrease in exports, Australia, had increased their export market over 100% from 9.3 mmt in 1987 to 19.0 mmt in 1997. In order to satisfy our international customers, US breeding programs should strive to meet the needs of export markets, such as Asia and the Middle East. The HWWQL and HRS & DWQL are modifying quality testing for bread wheats, including additional products such as tortillas, pizza crust, variety breads or Asian food-products.

Biochemical components

Gluten proteins play major roles in breadmaking potential of wheat. Gliadins contain mainly single polypeptide chains (monomers). The general roles of gliadins in breadmaking are to: (a) contribute to the extensibility of dough; (b) decrease mixing stability and provide weak and extensible dough; (c) influence bread crumb grain; and (d) control the LV potential of wheat flour. On the other hand, glutenins (inter-molecular disulfide bonded polymers), account for 40–50% of the total flour proteins, with a wide range in molecular weights (MW) from 100 000 to 20 million. Their general roles in breadmaking are to: (a) govern/control mixing requirement and (b) strengthen dough. The relative amount of glutenins as well as the presence of specific glutenin subunits (GS) correlates to breadmaking quality. There have been several reports on the ratios of certain gluten proteins, glutenin fractions such as high- and low MW (HMW/LMW) glutenin or different sized aggregates as an effective breadmaking quality indexes (Southan and MacRitchie, 1999).

Ng and Bushuk (1988) generated prediction equations for baking strength index versus eight different HMW-GS and found HMW-GS composition useful in breeding programs for screening genotypes for breadmaking quality. Dachkevitch and Autran (1989) also suggested that both MW distribution of protein aggregates could improve the accuracy of the prediction and Bean *et al.* (1998) developed a simple, rapid, and reproducible method for quantification of monomeric and polymeric proteins in flour using SE-HPLC. They found a strong correlation between LV and the amount of insoluble (50% 1–propanol) polymeric proteins.

Starch granular size distribution

Wheat starch contains two primary types of granules: large A-granules and small B-granules, with a diameter of 10–35 μm and 1–10 μm, respectively. Bechtel *et al.* (1990, 2000) have suggested a tri-modal distribution of wheat starch granules named type A (>15.9 μm), type B (5.3–15.9 μm), and type C (<5.3 μm), using a quantitative digital image analysis coupled with dark field microscopy. Conflicting results regarding the effect of starch granule size on baking have been reported. Chung and Park (1997) reviewed the literature and categorized four different views on the effects of small B-granules in breadmaking: (a) B-granules are beneficial; (b) B-granules are detrimental; (c) B-granules have little effect; and (d) there is an optimum proportion of B and A-granules for optimum bread LV. Disagreement about the effect of wheat starch granules and their size may be due to the research conducted at several different laboratories using different methods for starch isolation and fractionation in addition to differences in baking tests. Therefore, the starch granular size distribution will be investigated at the HWWQL for hard winter wheat breeding programs using the same methods.

Wheat free lipids (FL)

FL content in flour ranges from 0.8% to 1.0% of flour dry weight. Defatting (removing FL) and reconstituting studies demonstrated the beneficial effects of polar lipids (PoL), especially glycolipids (GL) but no beneficial or detrimental effects of nonpolar lipids (NL) on LV and bread structures (Chung 1989, 1991). Varietal variations in FL composition (GL, NL/PoL, or NL/GL ratios) were significantly correlated with LV (Chung *et al.*, 1982; Békés *et al.*, 1986). Our recent study (Ohm and Chung, 2002) showed that flour FL content and composition (the ratio of mono- or di-galactosyldiglycerides to GL, i.e. MGDG/GL or DGDG/GL) supplemented flour PC to develop prediction equations of WA and LV. Therefore, wheat lipids cannot be the sole quality determinant, but a good supplementary one, especially for screening wheat progenies at early generations (Chung *et al.*, 2002a). There is an urgent need for a rapid, accurate, precise, and environmentally friendly lipid analysis method for developing quality prediction models.

Pentosans and water-soluble enzymes

With regard to processing effects, the native wheat enzymes of primary interest are oxidases, specifically lipoxygenase and peroxidase (Stauffer, 1987). Lipoxygenases are thought to increase mixing tolerance and relaxation times of dough, ultimately leading to increased bread loaf volume (Frazier *et al.*, 1973; Cumbee *et al.*, 1997). Peroxidase, in the presence of hydrogen peroxide, appears to be the most efficient catalyst for the oxidative gelation of wheat flour water-soluble pentosans (Neukom and Markwalder, 1978; Izydorcyk *et al.*, 1990).

Tyrosine crosslinks in flour and dough may play a determinate role in the mixing and baking processes (Tilley *et al.*, 2001). Tyrosine crosslinks can be enzymatically catalyzed *in vitro* using horseradish peroxidase (Michon *et al.*, 1999). Native flour peroxidases, which remain active during the mixing process (Delcros *et al.*, 1998), may cause tyrosine crosslink formation between and among the gluten proteins (Tilley and Tilley, 2002).

Multiple interactions

True differences in quality among different bread wheat varieties depend on the different manner and degree with which wheat flour biochemical components interact with each other, especially in the presence of water and other added ingredients during the dough-mixing stage, and particularly at the elevated temperature during the oven stage. Although the significance of multiple interactions during breadmaking is highly recognized, the proper methodology/techniques of studying these multi-component interactions are not available, especially to be applied for breeding lines. Rapid, accurate and reliable methods of studying multi-component interactions during mixing or baking stages are urgently needed.

26.6.3 Final thoughts

Nearly one half of bread wheat produced in the US is exported to be used to make a variety of products. Therefore, quality evaluation is rapidly expanding to include a wide range of tests for multiple end-use products so that wheat breeding can be focused on specific quality attributes desired by our customers. Although red wheat will remain the predominant class for the next decade, hard white wheat cultivar availability will increase and the increasing acreage of hard white wheat will be very much market-driven by both domestic and export markets. The role of biotechnology will certainly lead to significant advances in wheat improvement in the future. Of course, whether or not a transgenic bread wheat will be released will also be market-driven. Future 'designer' wheat cultivars are being developed for niche markets, such as waxy wheat or wheat with altered gluten PC and composition. One of many issues and concerns is whether the wheat industry will be ready for the upcoming changes to handle the situations, e.g., segregation of wheat, based on color or specialty quality trait, or genetically modified wheat. Both domestic and international customers demand to know about the quality potential of wheat for their own products rather than purchasing them based on US wheat classification and grades. To meet these challenges, a true customer-sensitive system should be a 'Total Quality-Based Wheat Marketing System' for both domestic and international marketing channels. This goal may be obtained from concerted collaborative efforts from all sectors from breeders to customers. Research and technology development are urgently needed to provide accurate prediction models for desired quality attributes demanded by customers. These technologies will be also applied to early generation breeding lines for specific quality trait improvement. The final thought is: Who will provide the major financing for these paramount tasks?

26.7 References

AMERICAN ASSOCIATION OF CEREAL CHEMISTS (2000), *Approved Methods of the AACC, 10th ed*, St. Paul, MN, The Association.

ANDERSON J V and MORRIS C F (2001), 'An improved whole-seed assay for screening wheat germplasm for polyphenol oxidase activity', *Crop Sci*, 41, 1697–1705.

BAIK D K, CZUCHAJOWSKA Z and POMERANZ Y (1995), 'Discoloration of dough for Oriental noodles', *Cereal Chem*, 72, 198–205.

BASON M L, RONALDS J A, WRIGLEY C W and HUBBARD L J (1993), 'Testing for sprout damage in malting barley using the Rapid Visco-Analyser', *Cereal Chem*, 70, 269–272.

BEAN S R and LOOKHART G L (2001), 'Optimizing quantitative reproducibility in high-performance capillary electrophoresis (HPCE) separations of cereal proteins',

Cereal Chem, 78, 530–537.

BEAN S R, LYNE R K, TILLEY K A, CHUNG O K and LOOKHART G L (1998), 'A rapid method for quantification of insoluble polymeric proteins in flour', *Cereal Chem*, 75, 374–379.

BECHTEL D B, ZAYAS I, KALEIKAU L and POMERANZ Y (1990), 'Size-Distribution of Wheat Starch Granules During Endosperm Development', *Cereal Chem*, 67, 59–63.

BECHTEL D B and WILSON J D (2000), 'Variability in a starch isolation method and automated digital image analysis system used for the study of starch size distributions in wheat flour', *Cereal Chem*, 77, 401–405.

BÉKÉS F, ZAWISTOWSKA U, ZILLMAN R R and BUSHUK W (1986), 'Relationship between lipid content and composition and loaf volume of twenty-six common wheats', *Cereal Chem*, 63, 327–331.

CHUNG O K (1989), 'Functional significance of wheat lipids', in Pomeranz Y, *Wheat is Unique*, St. Paul, MN, Am. Assoc. of Cereal Chem, 341–368.

CHUNG O K (1991), 'Cereal lipid', in Lorenz K J and Kulp K, *Handbook of Cereal Science and Technology*, NY, Marcel Dekker, 497–553.

CHUNG O K, FINNEY P L, MARTIN C R, STEELE J L, SEABOURN B W and SMAIL V W (1994), 'Relationship between single kernel characteristics and end use quality. II. Soft wheats', *Cereal Foods World* 39, 604.

CHUNG O K, OHM J B, LOOKHART G L and BRUNS R F (2003a), 'Quality characteristics of hard winter and spring wheat grown under over-wintering condition', *J Cereal Sci*, 37, 91–99.

CHUNG O K, OHM J B and PARK S H (2002a), 'Wheat lipids: A supplementary quality determinant', in Salgo A, *Proceedings of ICC Conference 2002: Novel Raw Materials, Technologies, and Products – New Challenge for the Quality Control*, Budapest, Hungary, Budapest Univ. Technol. and Economics Pub. 167–175.

CHUNG O K, OHM J B, CALEY M S and SEABOURN B W (2001), 'Prediction of baking characteristics of Hard Winter Wheat flours using computer-analyzed mixograph parameters', *Cereal Chem*, 78, 493–497.

CHUNG O K, OHM J B and SEABOURN B W (1999), 'Prediction of conventional wheat characteristics of hard winter wheats using single kernel parameters', in *Abstract Book of 84th AACC Annual Meeting*, 197.

CHUNG O K and PARK S H (1997), 'Functional Properties of Wheat Flour Components and Basic Ingredients in Breadmaking', in ICC-SA symposium: Cereal and grain science technology, in Taylor J R N, *Proceedings of the 14th SAAFoST International Congress & Exhibit, Harness Food Science & Technology for Sustainable Development*, 5–18

CHUNG O K, PARK S H and SEIB P A (2003b), 'Polyphenol oxidase activity in wheat grain kernels, meals, and flours in relation to noodle color', in *Abstract/Program Book of 88th AACC Annual Meeting*.

CHUNG O K, POMERANZ Y and FINNEY K F (1982), 'Relation of polar lipid content to mixing requirement and loaf volume potential of hard red winter wheat flour', *Cereal Chem*, 59, 14–20.

CHUNG O K, TILLEY M, OHM J B, CALEY M S and SEABOURN B W (2002b), 'Hard winter wheats – past, present and future', in Madl R L, *Proceedings of the 5th Annual National Wheat Industry Research Forum*, 20–23.

COX T S, SHOGREN M D, SEARS R G, MARTIN T J and BOLTE L C (1989), 'Genetic improvement in milling and baking quality of hard red winter wheat cultivars, 1919 to 1988', *Crop Sci*, 29, 626–631.

CUMBEE B, HILDEBRAND D F and ADDO K (1997), 'Soybean flour lipoxygenase isozymes. Effects on wheat flour dough rheological and breadmaking properties. *J Food Sci*, *62*,281–294.

DACHKEVITCH T and AUTRAN J C (1989), 'Prediction of baking quality of bread wheats in breeding programs by size-exclusion high-performance liquid chromatography', *Cereal Chem*, 66, 448–456.

DELCROS J -F, RAKOTOZAFY L, BOUSSARD A, DAVIDOU S, PORTE C, POTUS J and NICOLAS J (1998), 'Effect of mixing conditions on the behavior of lipoxygenase, peroxidase and catalase in wheat flour doughs', *Cereal Chem*, *75*, 85–93.

DELWICHE S R (1993), 'Measurement of single-kernel wheat hardness using near-infrared transmittance', *Trans ASAE* 36, 1431–1437.

DELWICHE S R, CHUNG O K and SEABOURN B W (1998), 'Protein content of hard red winter wheat by near-infrared spectroscopy of whole grain: collaborative study', *J AOAC Inter*, 81, 587–603.

DOWELL F E (1998), 'Automated color classification of single wheat kernels using visible and near-infrared reflectance', *Cereal Chem*, 75, 142–144.

DOWELL F E (2000), 'Differentiating vitreous and non-vitreous durum wheat kernels using near-infrared spectroscopy', *Cereal Chem*, 77, 155–158.

DOWELL F E, RAM M S and SEITZ L M (1999a), 'Predicting scab, vomitoxin, and ergosterol in single wheat kernels using near-infrared spectroscopy', *Cereal Chem*, 76, 573–576.

DOWELL F E, THRONE J E, WANG D and BAKER J E (1999b), 'Identifying stored grain insects using near-infrared spectroscopy', *J Econ Entomol*, 92, 165–169.

ENDO D, OKADA K, NAGAO S and D'APPOLONIA B L (1990), 'Quality characteristics of hard red spring and winter wheats. II. Statistical evaluation of reversed phase high-performance liquid chromatography and milling data', *Cereal Chem*, 67, 486–489.

FINNEY K F (1984), 'An optimized, straight-dough, bread-making methods after 44 years', *Cereal Chem*, 61, 20–27.

FINNEY K F (1985), 'Experimental breadmaking studies, functional (breadmaking) properties, and related gluten protein fractions', *Cereal Foods World*, 30, 794–801.

FRAZIER P J, LEIGH DUGMORE F A, DANIELS N W R, RUSSELL EGGITT P W and COPPOCK J B M (1973), 'The effect of lipoxygenase action on the mechanical development of wheat flour doughs', *J Sci Food Agric*, 24, 421–436.

GRAYBOSCH R (2003), 'US hard winter wheat breeding objectives', in Abstract/Program Book of *AACC 2003 Pacific Rim Meeting: Wheat Quality Management and Processing into the 21st Century*, 8.

GRAYBOSCH R A, PETERSON C J, MOORE K J, STEARNS M and GRANT D L (1993), 'Comparative effects of wheat flour protein, lipid, and pentosan composition in relation to baking and milling quality', *Cereal Chem*, 70, 95–101.

HALVERSON J and ZELENY L (1988), 'Criteria of wheat quality', in Pomeranz Y, *Wheat: Chemistry and Technology Vol. I*, St. Paul, MN, Am. Assoc. Cereal Chem., 15–46.

HARELAND G (2002), 'Hard Red Spring and Durum wheat quality-past, present, & future', in Madl R L, *Proceedings of the 5th Annual National Wheat Industry Research Forum*, 24–25.

IZYDORCZYK M S, BILIADERIS C G and BUSHUK W (1990), 'Oxidative gelation studies of water-soluble pentosans from wheat', *J Cereal Sci,* 11, 153–169.

KRUGER J E, HATCHER D W and DEPAUW R (1994), 'A whole seed assay for polyphenol oxidase in Canadian prairie spring wheats and its usefulness as a measure of noodle darkening', *Cereal Chem*, 71, 324–326.

LOOKHART G L (1991), 'Cereal proteins: composition of their major fractions and methods for identification', in Lorenz K J and Kulp K, *Handbook of Cereal Science and Technology*, NY, Marcel Dekker, 441–468.

LOOKHART G L, ALBERS L D and BIETZ J A (1986), 'Comparison of polyacrylamide gel electrophoresis and high-performance liquid chromatography analyses of gliadin polymorphism in the wheat cultivar Newton', *Cereal Chem*, 63, 497–500.

LOOKHART G L, COX T X and CHUNG O K (1993), 'Statistical analysis of gliadin reversed-phase high-performance liquid chromatography patterns of hard red spring and hard red winter cultivars grown in a common environment: Classification indices', *Cereal Chem*, 70, 430–434.

LOOKHART G L, JONES B L, HALL S B and FINNEY K F (1982), 'An improved method for standardizing polyacrylamide gel electrophoresis of wheat gliadin proteins', *Cereal Chem*, 59, 178–181.

MARTIN C R, ROUSSER R and BRABEC D L (1993), 'Development of a single-kernel wheat characterization system', *Trans ASAE*, 36, 1399–1404.

MICHON T, WANG W, FERRASSON E and GUEGUEN J (1999), 'Wheat prolamine crosslinking through dityrosine formation catalyzed by peroxidases: Improvement in the modification of a poorly accessible substrate by "indirect" catalysis', *Biotechnology and Bioengineering*, 63, 449–458.

MORRIS C F (2002), 'The pacific northwest-land of milk and honey (and white wheat), in Madl R L, *Proceedings of the 5th Annual National Wheat Industry Research Forum*, 28–31.

NEUKOM H and MARKWALDER H U (1978), 'Oxidative gelation of wheat flour pentosans: A new way of crosslinking polymers', *Cereal Foods World*, 23, 374–376.

NG P K W and BUSHUK W (1988), 'Statistical relationships between high molecular weight subunits of glutenin and breadmaking quality of Canadian-grown wheat', *Cereal Chem*, 65, 408–413.

NORRIS K H, HRUSCHKA W R, BEAN M M and SLAUGHTER D C (1989), 'A definition of wheat hardness using near infrared reflectance spectroscopy', *Cereal Foods World*, 34, 696–705.

OHM J B and CHUNG O K (1998), 'Single kernel wheat characterization system: its use in estimating end-use properties of hard winter wheats', *Cereal Foods World*, 43, 535.

OHM J B and CHUNG O K (1999a), 'Gluten, pasting, and mixograph parameters of hard winter wheat flours in relation to breadmaking', *Cereal Chem*, 76, 606–613.

OHM J B and CHUNG O K (1999b), 'Estimation of free glycolipids in wheat flour by HPLC', *Cereal Chem*, 76, 873–876.

OHM J B and CHUNG O K (2000), 'NIR transmittance estimation of free lipid content and its glycolipid and digalactosyldiglyceride contents using wheat flour lipid extracts', *Cereal Chem*, 77, 556–559.

OHM J B and CHUNG O K (2002), 'Relationships of free lipids with quality factors in hard winter wheat flours', *Cereal Chem*, 79, 274–278.

OHM J B, CHUNG O K and DEYOE C W (1998), 'Single kernel characteristics of hard winter wheats in relation to milling and baking quality', *Cereal Chem*, 75, 156–161.

PARK S H, CHUNG O K and SEIB P A (2001), 'Effects of flour particle size on loaf volume and internal characteristics of experimental pup-loaf bread', in *Abstract/Program Book of 86th AACC Annual Meeting*, 158.

PARK S H, CHUNG O K and SEIB P A (2002), 'Effects of varying the weight ratio of large and small wheat starch granules on experimental pup-loaf bread. in *Abstract/Program Book of 87th AACC Annual Meeting*, 140.

PASIKATAN M C, HAQUE E, KELLER-MCNULTY S, STEELE J L, FANG Q, SPILLMAN C K and GAO W (1998), 'Single kernel wheat physical properties and first break grinding', ASAE Mid-Central Conf., St. Joseph, MO. Paper No. 98–137.

PLAUS M (2003), 'Evolution of wheat grading factors in the United States', in Abstr/ Program Book of *AACC 2003 Pacific Rim Meeting: Wheat Quality Management and Processing into the 21st Century*, 7.

SEABOURN B W, BEAN, S R, LOOKHART G L and CHUNG O K (1999), 'Prediction of gliadin and soluble/insoluble HMW glutenin fractions in whole kernel wheat by near-infrared reflectance', in *Abstract/Program Book of 84th AACC Annual Meeting*, 177.

SEABOURN B W and CHUNG O K, (1996), 'Rapid estimation of quality parameters of U.S. hard winter wheat breeding lines', *Cereal Foods World*, 41, 592.

SEABOURN B W, CHUNG O K and SEIB P A (1997), 'Near infrared reflectance spectroscopy of flour-water doughs during mixing', *Cereal Foods World*, 42, 612.

SHOGREN M D and FINNEY K F (1984), 'Bread-making test for 10 grams of flour', *Cereal Chem*, 61, 418–423.

SHUEY W C (1960), 'A wheat sizing technique for predicting flour milling yield', *Cereal Sci Today*, 5, 71–72, 75.

SLAUGHTER D C, NORRIS K H and HRUSCHKA W R (1992), 'Quality and classification of hard red wheat', *Cereal Chem*, 69, 428–432.

SOUTHAN M and MACRITCHIE F (1999), 'Review: Molecular weight distribution of wheat proteins', *Cereal Chem*, 76, 827–836.

STAUFFER C E (1987), 'Oxidases', in Kruger J E, Lineback D, and Stauffer C E, *Enzymes and their role in cereal technology*, St. Paul, MN, Am. Assoc. Cereal Chem., 239–263.

STEELE J L and MARTIN C R (1991), 'The effects of kernel weight and size on the performance of a wheat hardness instrument', ASAE. Paper No. 91-3066.

TILLEY K A, BENJAMIN R E, BAGOROGOZA K E, OKOT-KOTBER B M, PRAKASH O and KWEN H (2001), 'Tyrosine Crosslinks: Molecular Basis of Gluten Structure and Function', *J Agric Food Chem*, 49, 2627–2632.

TILLEY M and TILLEY K A (2002), 'Identification of active components from the water-soluble extract of wheat flour that catalyze dityrosine formation', in *Abstract/ Program Book of 87th AACC Annual Meeting*, 127.

WANG D, DOWELL F E and LACEY R E, (1999), 'Single wheat kernel color classification using neural networks', *Trans ASAE*, 42, 233–240.

WHEAT QUALITY COUNCIL (1994), 'Milling and baking test results for hard winter wheats harvested in 1993', Manhattan, KS, Agricultural Experiment Station, Kansas State University.

WINDHAM W R, GAINS C S and LEFFLER R G (1993), 'Effect of wheat moisture content on hardness scores determined by near-infrared reflectance and on hardness score standardization', *Cereal Chem*, 70, 662–666.

27

Preventing bread staling

P. Chinachoti, University of Massachusetts, USA

27.1 Introduction

Quality and shelf-life of bakery products are normally limited by a physico-chemical deterioration called staling, leading to hard and crumbly texture and loss of fresh-bake flavor. If high in water activity (a_w), bakery products are prone to molding. Staling of bread is the major cause for typically very short shelf-life (3–7 days), making bread a processed food with one of the shortest shelf-lives.

Inhibition or delaying staling of bakery products has been of considerable industrial and academic interest. Scientific and technological understanding of the mechanism of staling, however, is far from clear. This is due mainly to the complex physical and chemical phenomena involved. In the past, typical parameters used in studying physico-chemical changes included sensory, starch retrogradation (e.g. loss in starch solubility and re-crystallization), moisture loss and mechanical properties. But these provide only the total or bulk properties or values and not microscopic changes in various molecular or structural domains.

27.2 Economic significance of staling

Understanding the molecular dynamics and functional properties of bread can lead to understanding and solving staling problems. Bread and many other bakery products are processed or finished food products with a short shelf-life. A recent estimate of annual US production of bread has shown that approximately 20 billion pounds of bread are being produced and 3% of this is staled (Zobel and Kulp, 1996). Assuming bread is $2 per pound in market value, annual loss due to bread deterioration can amount to 1.2 billion dollars annually or more (Berkowitz and Oleksyk, 1991).

Loss of revenue also may affect consumers, particularly among those with lower income since bread is one of the most common food staples in many parts of the world. In the military, MRE (Meal, Ready-to-Eat) bread serves as one of the major sources of complex carbohydrates. In the consumer market, regular bread as well as many other bakery products last only 3–5 days. This is a tremendous constraint on the producers in terms of product distribution, shelf turnover rate, and waste. Because of the low profit margin, solutions for making bread with a longer shelf-life are limited by economic feasibility. There are many novel solutions applicable to bread, cakes, cookies, crackers, frozen dough, pizza, microwavable, pre-prepared or pre-baked products (e.g. croissants, pizzas, English muffins, rolls, etc.), and extruded products (e.g. breakfast cereals, puffed snacks, and pet foods).

27.3 The process of bread staling

Consumers are attracted to bread that is freshly baked. It is easily recognized in terms of appealing aroma, crispy crust and moist, soft crumb. The staling process takes place as soon as the bread is taken out of the oven, resulting in a degradation of these desirable attributes. Professional and home bakers know that quality of bread can be influenced by numerous factors from ingredient sources to the conditions in bread dough preparation and baking. There is a strong interdependence of these factors and hence breadmaking is often referred to as an art. The scientific background to bread quality has been described in other chapters in this book and in recent reviews on bread staling (Chinachoti and Vodovotz, 2001). The following discussion focusses mainly on the current issues and future trends on bread staling.

27.4 Factors affecting bread staling

Bread is a heterogeneous and composite system containing miscible and immiscible components with heterogeneous domains (phases). Staling problems can involve various stages in a breadmaking process from formulation to packaging and storage. Complex contributions of all functional components in various stages of dough development, baking, and cooling and storage remain poorly understood at the molecular level. Applying scientific principles to baking faces numerous challenges which require a multidisciplinary approach. The description below shows that solutions to bread staling have evolved over time. The subject still awaits a comprehensive method of investigation that simultaneously and non-invasively monitors not only starch but also gluten, water, and many minor components.

27.4.1 Starch *vs.* gluten

Wheat flour that gives a good bread quality contains an optimum blend of starch (70–80%), proteins (8–18%), lipids (approximately 2%), pentosans (approximately 2%), enzyme and enzyme inhibitors, and other minor components (Pomeranz, 1988). Bread dough development relies in the most part on the capability of the gluten to hydrate and retain gas produced during fermentation and baking. The final structure of bread is fixed after heat gelatinization of starch and disulfide bonding (and other interactions), 'fixing' the gluten in the expanded shape of a bread loaf. The gas retention property affects bread porosity and cell wall viscoelasticity property which in turn influences the texture of bread (Peleg *et al.*, 1989; He and Hoseney, 1991; Kou and Chinachoti, 1991; Nussinovitch *et al.*, 1992). A number of anti-staling additives soften the fresh bread by affecting the surface tension of the dough leading to a finer grain (Rao *et al.*, 1992). The resulting smaller and uniform air cells lead to softer bread and modified firming kinetics.

The major causes of bread staling have been attributed to starch (amylopectin) retrogradation (e.g. Schiraldi and Fessas, 2001) but there are several reports that demonstrate a weak or no correlation (e.g. Sahlstrom and Brathen, 1997; Hallberg and Chinachoti, 2002). Retrogradation is the aggregation of amorphous starch chains in helical forms some of which may develop into crystals (Chinachoti and Vodovotz, 2001). The link between starch retrogradation and staling is based on the common observation that the rate of retrogradation seems to correlate with mechanical firming in conventional bread storage (Conford *et al.*, 1964; McIver *et al.*, 1968; Willhoft, 1971; Kim and D'Appolonia, 1977a). Hence, it is often assumed that starch recrystallization causes bread firming or staling (McIver *et al.*, 1968; Kim and D'Appolonia, 1977a).

However, there have also been contradicting reports (Martin *et al.*, 1991; Martin and Hosensy, 1991). Hardening can be caused by moisture loss and/or redistribution (but not by starch recrystallization). This can be easily demonstrated by rapid drying for fresh bread (Vodovotz *et al.*, 2001). Thus, firming of bread can be significantly influenced by bulk dehydration and local water migration (Hallberg and Chinachoti, 1992, 2002; Baik and Chinachoti, 2001). During storage, water becomes less 'freezable' partly due to dehydration and partly due to redistribution. Some studies have suggested that this involves the transfer of water in the process of starch reorganization from amorphous to crystalline regions (Leung *et al.*, 1983; Wynne-Jones and Blanshard, 1986) whereas others suggested transfer of water among amorphous domains (Kim-Shin *et al.*, 1991). ^{1}H NMR (nuclear magnetic resonance) T_2 distribution analysis has been applied in bread (Ruan *et al.*, 1996; Ruan and Chen, 2001) as an attempt to differentiate water in various components.

Starch granules in bread exhibit a disrupted structure during processing. Starch remnants contain an amylose matrix which can harden, contributing to the overall bread texture (Hug-Iten *et al.*, 1999). Gluten is also changed and aggregated after heating. During storage, the ensemble of domain transformation (local thermal transitions) is linked to local molecular mobility change, which depends highly on the local network, water distribution and spatial partition among components (Leung *et al.*, 1983; Wynne-Jones and Blanshard, 1986; Kim-Shin *et al.*, 1991; Morgan *et al.*, 1992, 1995, 1997; Chinachoti, 1994; Vodovotz, 1996; Baik, 2001).

Water partitioning in bread theoretically speaking should profoundly influence bread polymers and the staling process. How this happens remains an area yet to be further investigated (although there are many conflicting theories in the literature, none has strong supportive evidence). In a shelf-life extension study on bread, local dehydration has been proposed to influence structural rigidity which could change the course of staling (Baik and Chinachoti, 2001). The addition of glycerol, for example, caused more rapid firming despite potential plasticization and reduced amylopectin recrystallization (Baik and Chinachoti, 2001). In this case, it would be important to investigate how water distribution among bread polymers may change with added glycerol and how this is implicated in the staling process.

There has been increasing interest in the role of amorphous starch with respect to glassy–rubbery transition temperature (T_g) and molecular network formation in aging starch polymers (Slade and Levine, 1991; Chinachoti, 1994, 1996a,b; Parker and Ring, 2001; Farhat and Blanshard, 2001; Vodovotz *et al.*, 2001). Thermomechanical transitions of bread have been reported (Hallberg and Chinachoti, 1992; Le Meste *et al.*, 1992, Vodovotz *et al.*, 2001). The great variation in hydration in heterogeneous domains can be a reason for the observed 'range' in T_g (Hallberg and Chinachoti, 1992; Kalichevsky *et al.*, 1992, 1993; Le Meste *et al.*, 1992; Taub *et al.*, 1994; Cherian *et al.*, 1995; Vodovotz, 1996; Vodovotz and Chinachoti, 1996; Baik and Chinachoti, 2000, 2001). Although bread is a composite material, much of the discussion on its glass transition focusses heavily on the starch component. The simplistic view has later been re-evaluated with a clearer description of multi-phase and multi-component situation (Cesaro and Sussich, 2001). Oates (2001) has described a comprehensive model of molecular and microstructural changes of wheat components from dough to final bread products. The most important transformation is that of the gluten matrix from a film-like arrangement into a layer-like structure. Gliadins and glutenins in optimally developed dough arrange in highly fibrous structure and the continuous interconnected gluten matrix is surrounded mostly by starch granules. During baking, starch granules that are entirely enrobed by gluten gelatinize and flexibly fit around air cells. With the gluten sheets providing the basic structure, remnants of relatively intact starch granules reinforce the wall structure with leached amylose. If this leaches inward, the amylose forms a hardened core from the remains of starch granules; if it leaches outward, forms webs of gel layers are formed between granules. Describing the glass–rubbery transition of bread is extremely difficult experimentally due to this complcx multi-phase, composite structure.

Glassy–rubbery states of bread polymers may play a role during staling as a result of annealing and networking of amorphous chains (Slade and Levine, 1991). If this is correct, changing a system T_g and storage temperature could lead to a modified staling rate (Taub *et al.*, 1994). Extended storage of an MRE bread, however, showed no sign of T_g increase (Hallberg and Chinachoti, 1992), suggesting that other changes can may also affect bread shelf-life. Using a simple approach of keeping the amorphous phase plasticized (by addition of more water or glycerol) may work, but it can be complicated by a possible shift in water distribution (favorable hydration of starch at the expense of drying gluten, for instance) or other molecular interaction. A full investigation is needed to study local polymer domain mobility at the molecular level and local hydration of such domains.

Molecular chain rigidity can be vastly different in immiscible, phase-separated systems such as in bread dough. Molecular relaxation times indicate that starch is more rigid than gluten at a given moisture or temperature (Li *et al.*, 1996). The onset temperature for gluten molecular mobility can be many tens of degrees celcius lower than that for starch at a same moisture content (Li *et al.*, 1996). This implies that molecular onset of glass transition temperature (as measured by molecular spectroscopic methods) can be affected by local differences which are generally not detectable by traditional thermal analytical techniques, such as differential scanning calorimetry (DSC).

Some evidence of the role of flour protein in bread staling has been reported by Ponte *et al.* (1962), Kim and D'Appolonia (1977a) and Maleki *et al.* (1980). In this work, different flours produced breads that hardened at different rates. Even though Kim and D'Appolonia (1977a) observed a delay in staling with increasing protein content, they attributed the effect primarily to the dilution of starch by higher protein levels and not to the quality of the protein. In contrast, Maleki *et al.* (1980) suggested that gluten was the major fraction responsible for differences in staling rate. The work of Suhendro *et al.* (1993, 1995) on flour tortilla, a flat bread, has supported the finding by Maleki *et al.* (1980). Flour tortillas produced from higher-quality proteins were less crumbly and more rollable than those produced by lower-quality proteins.

Martin *et al.* (1991) have proposed a model of bread staling that incorporates the role of starch and gluten. They argue that bread firming results from interaction between the continuous protein matrix and discontinuous remnant of starch granules. Poor quality flour has been proposed to have more hydrophilic properties than good quality flour (He and Hoseney, 1991; Martin *et al.*, 1991). Therefore, poor quality gluten would interact more strongly with starch granules in dough; these interactions would also be stronger during and after baking, increasing the tendency of bread firming.

27.4.2 Water

Water is more abundant in the swollen amorphous regions of starch, facilitating local polymer chain mobility (plasticization) and subsequent recrystallization and retrogradation. Water uptake during baking contributes to the moist texture of fresh bread, and its release during storage contributes to firming and crumbliness of aged bread. Migration of water from amorphous to crystalline starch has been suggested (e.g. Leung *et al.*, 1983; Wynne-Jones and Blanshard, 1986; Kim-Shin *et al.*, 1991; Ruan *et al.*, 1996). Since total moisture content and the net moisture loss could not explain mechanical firming during staling without considering other factors (He and Hoseney, 1990; Davidou *et al.*, 1996), it may be necessary to focus also on the water distribution among regions in gluten, amorphous and crystalline starch and its contribution to functionality changes. Water has a pivoted role in controlling starch and gluten rigidity. Its distribution among regions in starch and gluten remains to be investigated for each step of dough development, bread processing and storage. NMR is the technique of choice for investigating the dynamic states of water, starch, and gluten. Recognizing that bread components contain various domains is an important step. NMR can be applied to monitor microscopic distribution of water in various systems (Hills and Quantin, 1993; Hills and Babonneau, 1994; Hills and LeFloc'h, 1994). When applied to starch granules upon swelling and gelatinization, the NMR T_2 results also showed several dynamic states of amylose and amylopectin (e.g. melting transition) and water redistribution from extra-granular to intra-granular and amorphous domains (Tang *et al.*, 2000; Chatakanonda *et al.*, 2003). Similar results in bread have been reported (Ruan and Chen, 2001).

Recrystallized starch contains some water, which may influence both rotational and translational mobility (Leung *et al.*, 1983, Umbach *et al.*, 1992, Ruan *et al.*, 1996). Immobilization of water in aged bread has been attributed to an increase in water of crystallization in the starch crystalline domains (Leung *et al.*, 1983; Wynne-Jones and Blanshard 1986). However, later evidence has also suggested that a decrease in water mobility might be related to changes in the amorphous domains (Kim-Shin *et al.*, 1991).

Water diffusive mobility can be determined in terms of the self-diffusion coefficient of water (D) using pulsed-field gradient NMR. In a study of the storage of glycerol-treated breads, D remained unchanged on storage without crust (with no moisture loss from crumb to crust). When stored with crust (with moisture loss), a decrease in D was observed, linked to a loss in the more mobile water fraction (Baik *et al.*, unpublished results). Baik *et al.* observed that the addition of glycerol in the bread increased the diffusion coefficient (D), indicating the plasticization effect of glycerol, as well as the ability of glycerol to create a more liquid-like domain by drawing water from starch and gluten. Owing to a stronger water interaction with glycerol (as compared with starch and gluten), water is less 'bound' to the macromolecules as a result of added glycerol. On the other hand, whilst recrystallization of the amylopectin during aging may not affect the overall self-diffusion coefficient, it may affect the water distribution within regions (Ruan *et al.*, 1996).

The rigidity increase of starch chains, as observed by C^{13} CP/MAS (cross-polarization magic angle spinning), has been investigated (Baik *et al.*, 2002). According to earlier study (Vodovotz, 1996), the increase in the C^{13} CP/MAS peak intensity indicates a decrease in the mobility, suggesting an increase in rigidity. Such an increase has been observed in aging bread (Baik *et al.*, 2002). Carbon chain mobility increase was also observed upon addition of a plasticizer such as glycerol. But glycerol may promote a local water distribution shift between starch and gluten leading to local stiffness of some amorphous domains and the ability of amorphous starch to organize into helical structures. Since there is competition for water in bread, glycerol affects water dynamics in bread by triggering a shift in the distribution of local viscoelastic properties.

27.4.3 Starch and gluten molecular rigidity

Solid-state ^{13}C NMR investigation of gluten and starch has indicated that gluten and starch experiences different molecular chain mobility (Garbow and Schaefer, 1991). ^{13}C CP/MAS NMR intensity of retrograded starch has been found to increase with storage time, suggesting an increasing chain rigidity following first order kinetics (Morgan *et al.*, 1992, 1995, 1997; Baik *et al.*, 2002). More rigid starch chains are surrounded by mobile (isotropic or anisotropic) water molecules whereas gluten exhibits greater chain mobility at least partly coupled with surrounding water molecules in solids (Li *et al.*, 1998). $T_{1\rho}(^{1}H)$ (relaxation time in the rotating frame) is measured by combining spin-locked CP/MAS. Polarization of carbons occurs from a transfer from nearby protons by static dipolar interactions. $T_{1\rho}(^{1}H)$ is sensitive to motion associated with frequencies in the 10–100 kHz range and thus is appropriate for probing long-range molecular motion (e.g. glassy state of polymers by Schaefer *et al.*, 1977). Additionally, polymer blends can be studied as shown by Li *et al.* (1994), and Dickinson *et al.* (1988), and reviewed by Belton and Colquhoun (1989). To monitor the starch and gluten mobility, $T_{1\rho}(^{1}H)$ as been shown to be sensitive to moisture-heat treatments (Li *et al.*, 1996; Tanner *et al.*, 1991).

Differences in observed starch and gluten dynamics or mobility have raised many interesting questions. If water hydrating gluten affects the gluten rheological behavior (chain mobility) more profoundly than the way water hydrating starch affects the starch chain mobility, how does water migrate between starch and gluten and how this affect the rigidity of each polymer and ultimately of the bread? How do the aggregation of gluten (conformational change) and the retrogradation of the starch influence the water distribution? At a given amount of moisture, does water plasticize gluten more than starch (based on molecular chain mobility)? The solution is to combine the NMR water distribution described above to determine local hydration of starch and gluten domains as well as starch and gluten chain mobility by $T_{1\rho}$ (^{1}H).

27.5 Techniques for preventing bread staling

There are several anti-staling food additives in the market for bakery products. Most have a claim to retard starch retrogradation process. There are also commercial α-amylases (as well as other enzyme systems) and surfactants that are used to soften bread crumbs. α-Amylases and other enzymes depolymerize affecting local viscoelastic properties and the staling process (Min *et al.*, 1998;

Champenois *et al.*, 1999; Jimenez and Martinez-Anaya, 2001). Surfactants (such as sodium stearoyl lactylate) are effective in changing viscoelastic properties of gluten, reducing the grains (gas cells) and increasing porosity, thus affecting the overall firming kinetics (Yamauchi *et al.*, 1994; Armero and Collar, 1998). Although surfactants interact with starch and can potentially retard (although not stop) amylopectin recrystallization, they also interact with gluten through electrostatic forces (Collar *et al.*, 1998, 2002). Surfactants and lipids can reduce inter-granular cohesion by reducing leaching of amylose by forming an amylose–surfactant or amylose–lipid complex. Pentosans are minor components in bread but play a significant role in increasing water absorption and loaf volume (Kim and D'Appolonia, 1977b; Jankiewicz and Michniewicz, 1987; Denli and Ercan, 2001). The effect of pentosans, surfactants, and lipids may involve gluten either by directly interacting with it or influencing starch–gluten interactive forces which is one of the proposed key parameters.

Patented technologies to retard staling that are available involve proprietary formulations, e.g. enzyme cocktails or mixture with selected surfactants in some viscous systems. The challenge is how to introduce a new formulation without creating new problems. For instance, an application of methyl cellulose in the formula has to be done in such a way that the dough absorption and rheology remain relatively unchanged so that there would be no change in mixing requirements. The freshness of shelf-stable MRE bread is preserved by controlling water activity a_w, pH, oxygen content, and initial microbial load (Hallberg *et al.*, 1990; Powers and Berkowitz, 1990). Although the bread is microbiologically safe, physical and chemical changes can still occur during a shelf-life of up to three years.

Because of the specific volume of bakery products, storing bread in refrigerated or frozen conditions does not offer an ideal solution to the staling problem. In addition, a bread stored in a refrigerator stales rapidly after thawing. Freezing can quickly stop staling as long as the freezing rate is reasonably rapid and consumers quickly reheat after thawing (the bread will stale very rapidly otherwise). Developments in packaging can help retard the rate of moisture loss to the air. However, as long as a bread has the crust intact, water migration from crumb to crust remains very significant regardless of packaging properties. In addition, bread that is free of anti-mycotic additives should have a dry crust to prevent surface mold attack. Using impermeable packaging can lead to higher risk of mold spoilage.

Consumers in different geographic regions exhibit different consumption patterns that dictate the regional bread industry. The biggest contrast is the difference among consumers in the USA, Europe, and Asia. Europeans prefer a harder texture with a strong fresh aroma. US customers prefer softer, spongy bread that would pass the 'squeeze' test. Asian customers have been less exposed to a wide variety of wheat breads and they are more accustomed to white and whole wheatbreads with more rubbery crust (not crispy) and very soft crumb. In tropical countries, mold spoilage is of significant threat to bread shelf-life and consumers often store breads in a refrigerator. This accelerates the staling rate of bread once thawed. In this case, use of anti-microbial additives remins a common practice.

27.6 Future trends

From a research point of view, several mechanisms of bread staling and several explanations on modes of actions of anti-staling factors have been suggested. Like many complex problems, there is no single solution or approach to inhibit or retard staling. Implementation of anti-staling additives, processing and packaging modification, is usually accompanied by some modification of the existing formulation, and changes in storage conditions. There are several key important areas requiring further attention in research. Firstly, there is a need to go beyond the over-simplified view of the glass transition concept and to monitor experimentally the changes that occur in each component, including the local microstructure, water distribution and migration, and mechanical properties. Secondly, there is a need to develop relevant methods of analysis of gluten in the form that exists in bread. Work on starch has been ongoing for a century but research on gluten in relation to bread staling has only just started, although it is the most important component in bread dough structure. With amylopectin melting, DSC is the typical method of choice in determining the extent of staling, but this provides a very narrow window of observation. The challenge in the future is how to investigate other components, particularly gluten, in intact bread undergoing staling.

Thirdly, the state of hydration of gluten, starch, pentosans, etc. must be investigated. The modes of action of many ingredients and physical factors remain to be studied in order to understand in more depth the molecular mechanisms of staling. Lastly, the structure of bread with arrangements of starch, gluten, leached amylose, etc., needs to be better described with advanced nanoscopic approaches. Molecular structure and functions should be related more clearly. From an industrial application point of view, there is a real challenge in trying to find a solution to improve bread stability with limited access to information and research tools in a short time frame. With changing market requirements toward fewer additives, and better flavor and texture, current feasible solutions to bread staling can be quite limited.

27.7 Sources of further information and advice

Given the level of complexity of the issue of bread staling, there are many areas readers should investigate further. There are many topics that have not been described in detail, but deserve attention; for instance the structure of bread as a solid foam, mechanical properties, characterization of physical states of polymers, water (freezable and unfreezable), and so on. The literature cited in this paper should serve as an initial reading list. More relevant research can be found in recent books such as Chinachoti and Vodovotz (2001).

27.8 References

ARMERO E and COLLAR C (1998), 'Crumb firming kinetics of wheat breads with anti-staling additives', *J. Cereal Sci.*, **28**, 165–174.

BAIK M-Y (2001), *Effect of water and water migration on starch retrogradation and thermomechanical properties of bread during staling*, Ph.D. Thesis, University of Massachusetts, Amherst, MA.

BAIK M-Y and CHINACHOTI P (2000), 'Moisture redistribution and phase transition during bread staling', *Cereal Chem.,* **77**(4), 484–488.

BAIK M-Y and CHINACHOTI P (2001), 'Effects of glycerol and moisture gradient on thermomechanical properties of white bread', *J. Agric. Food Chem.*, **49**, 4031–4038.

BAIK M-Y, DICKINSON L C and CHINACHOTI P (2002) 'Solid-state ^{13}C CP/MAS NMR studies on aging of starch in white bread', *J. Agric. Food Chem*, **51**, 1242–1248.

BELTON P S and COLQUHOUN I J (1989), 'Nuclear magnetic resonance spectroscopy in food research', *Spectroscopy*, **4**(9), 22–32.

BERKOWITZ D and OLEKSYK L E, (1991). 'Leavened breads with extended shelf-life', *United States Patent* No. 5,059,432.

CESARO A, and SUSSICH F (2001), 'Plasticization: the softening of materials', in Chinachoti P. and Vodovotz Y, *Bread Staling*, CRC Press, New York, pp. 19–59.

CHAMPENOIS Y, DELLA VALLE G, PLANCHOT V, BULEON A and COLONNA P (1999), 'Influence of α-amylases on bread staling and on retrogradation of wheat starch models', *Science des Aliments*, **19**, 471–486.

CHATAKANONDA P, CHINACHOTI P, SRIROTH K, TANG H and HILLS P (2003), 'The influence of time and conditions of harvest on structure–function properties of cassava starch-A proton NMR relaxation study', *Carbohydrate Polymers*, **53**, 233–240.

CHERIAN G, GENNADIOS A, WELLER C and CHINACHOTI P (1995), 'Thermomechanical behavior of wheat gluten films: effect of sucrose, glycerin and sorbitol', *Cereal Chem.*, **72**(1), 1–6.

CHINACHOTI P (1994), 'Probing molecular and structural thermal events in cereal-based products', *Thermochimica Acta*, **246**, 357–369.

CHINACHOTI P (1996a), 'Characterization of thermomechanical properties in starch and cereal products', *Thermal Analysis*, **47**, 195–213.

CHINACHOTI P (1996b). 'NMR dynamic properties of water in relation to thermal characteristics of bread', in D.S. Reid, *The Properties of Water in Foods*: *ISOPOW 6,* Blackie Academic & Professional, New York, pp. 139–159.

CHINACHOTI P. and VODOVOTZ Y. (2001), *Bread Staling*, CRC Press, New York.

COLLAR C, ARMERO E and MARTINEZ J (1998), 'Lipid binding of formula bread doughs. Relationships with dough and bread technological performance', *Z. Lebensm Unters Forsch A*, **207**, 110–121.

COLLAR C, MARTINEZ J C and ROSELL C M (2002), 'Lipid binding of fresh and stored formulated wheat breads. Relationships with dough and bread technological performance', *Food Sci Tech. Int.*, **7**(6), 501–510.

CONFORD S J, AXFORD D W E and ELTON G A H (1964), 'The elastic modulus of bread crumb in linear expression in relation to staling', *Cereal Chem.,* **41**, 216–229.

DAVIDOU S, LE MESTE M, DEBEVER E and BEKAERT D (1996), 'A contribution to the study of staling of white bread; effect of water and hydrocolloid', *Food Hydrocolloids*, **10**(4), 375–383.

DENLI E and ERCAN R. (2001), 'Effect of added pentosans isolated from wheat and rye grain on some properties of bread', *Eur. Food Res. Technol.* **212**, 374–376.

DICKINSON L C, MORGANELLI P, CHU C W, PETROVIC Z, MACKNIGHT W J and CHIEN C W (1988), 'Molecular motions in model network polymers', *Macromolecules*, **21**(2), 338–346.

FARHAT I A and BLANSHARD J M V (2001), 'Modeling the kinetics of starch retrogradation', in Chinachoti P. and Vodovotz Y , *Bread Staling*, CRC Press, New York, pp. 163–172.

GARBOW J R and SCHAEFER J (1991), 'Magic-angle ^{13}C NMR study of wheat flours and doughs', *J. Agric. Food Chem.*, **39**, 877–880.

HALLBERG L M and CHINACHOTI P (1992), 'Dynamic mechanical analysis for glass transitions in long shelf-life bread', *J. Food Sci.*, **57**(5), 1201–1229.

HALLBERG LM and CHINACHOTI P (2002) 'A fresh perspective on staling: the significance of starch recrystallization on the firming of bread', *J. Food Sci.,* 1092–1096.

HALLBERG L M, YANG T and TAUB I A (1990) Optimization of functional ingredients in a low water activity bread using response surface methodology. In *Proceedings of the Third Natick Science Symposium*; Herz, M.L. and Sklarsky, T.A. Eds. US Army Natick Research, Development and Engineering Center. Natick, MA, pp. 277–293.

HE H and HOSENEY R C (1990), 'Changes in bread firmness and moisture during long-term storage', *Cereal Chem.*, **67**(6), 603–605.

HE H and HOSENEY R.C (1991), 'Differences in gas retention, protein solubility, and rheological properties between different baking quality flours'. *Cereal Chem.* **68**(5), 526–530.

HILLS B and BABONNEAU, F. (1994), 'A quantitative study of water proton relaxation in porous-beds of porous particles with varying water content', *Magnetic Resonance Imaging*, **12**, 909.

HILLS B and LEFLOC'H G (1994), 'NMR studies of non-freezing water in cellular plant tissue', *Food Chem.,* **51**, 331–336.

HILLS B and QUANTIN V M (1993), 'Water proton relaxation in dilute and unsaturated suspensions of nonporous particles', *Molecular Physics*, **79**, 77–93.

HUG-ITEN S, HANDSCHIN S, CONDE-PETIT B and ESCHER F (1999), 'Changes in starch microstructure on baking and staling of wheat bread', *Lebensm.-Wiss. U.-Technol*, **32**, 255–260.

JANKIEWICZ M and MICHNIEWICZ J (1987), 'The effect of soluble pantosans isolated from rye grain on staling of bread', *Food Chem.* **25**, 241–246.

JIMENEZ T and MARTINEZ-ANAYA M.A (2001), 'Amylases and hemicellulases in breadmaking. Degradation by-products and potential relationship with functionality', *Food Sci. Tech. Int.,* **7**(1), 5–14.

KALICHEVSKY M T, JAROSZKIEWICZ E M, ABLETT S, BLANSHARD J M V and LILLFORD P J (1992), 'The glass transition of amylopectin measured by DSC, DMTA and NMR', *Carbohydrate Polymers*, **18**(2), 77–88.

KALICHEVSKY M T, BLANSHARD J M V and MARSH R D L (1993), 'Application of mechanical spectroscopy to the study of glassy biopolymers and related systems', in Blanshard J M V and Lillford P J, *The Glassy State in Foods*, University Press, Nottingham, UK.

KIM S K and D'APPOLONIA B L (1977a), 'Bread staling studies. I. Effect of protein content on staling rate and bread crumb pasting properties', *Cereal Chem.,* **54**(2), 207–215.

KIM S K and D'APPOLONIA B L, (1977b), 'Bread staling studies. III. Effect of pentosans on dough, bread, and bread staling rate', *Cereal Chem.* **54**, 225–229.

KIM-SHIN, M S, MARI F, RAO P A, STENGLE T R and CHINACHOTI P, (1991), '^{17}O nuclear magnetic resonance studies of water mobility during bread staling', *J. Agric. Food Chem.*, **39**, 1915–1920.

KOU Y and CHINACHOTI P (1991), 'Structural damage in bread staling as detected by recoverable work and stress-strain analysis', *Cereal Foods World*, **36**, 888–892.

LE MESTE M, HUANG V T, PANAMA J, ANDERSON G and LENTZ R (1992), 'Glass transition of bread', *Cereal Foods World*, **37**, 264–267.

LEUNG H K, MAGNUSON J A and BRUINSMA B L (1983), 'Water binding of wheat flour doughs and breads as studied by deuteron relaxation', *J. Food Sci.*, **48**, 95–99.

LI S, DICKINSON L C and CHIEN J C W (1994), 'Local compositional fluctuations in PPO/HIPS and PPO/SBS blends', *J. Polymer Sci.*, **32**, 607–613.

LI S, DICKINSON C and CHINACHOTI P (1996), 'Proton relaxation of starch and gluten by solid-state nuclear magnetic resonance spectroscopy', *Cereal Chem.*, **73**(6), 736–743.

LI S., DICKINSON L C and CHINACHOTI P (1998), 'Water mobility in waxy corn starch by ^{2}H and ^{1}H NMR', *J. Agric. Food Chem.*, **46**, 62–71.

MALEKI M., HOSENEY R C and MATTERN P J (1980), 'Effects of loaf volume, moisture content, and protein quality on the softness and staling rate of bread', *Cereal Chem.* **57**(2), 138–140.

MARTIN M L and HOSENEY R C (1991), 'A mechanism of bread firming. II. Role of starch hydrolyzing enzymes', *Cereal Chem.*, **68**(5), 503–507.

MARTIN, ML, ZELEZNAK K J and HOSENEY R C (1991), 'A mechanism of bread firming. I. Role of starch swelling', *Cereal Chem.* **68**(5), 498–503.

MCIVER R G, AXFORD D W E, COLWELL K H and ELTON G A H (1968), 'Kinetic study of the retrogradation of gelatinized starch', *J. Sci. Food Agr.*, **19**, 560–563.

MIN B-C, YOON S-H, KIM J-W, LEE Y-W, KIM Y-B and PARK K H (1998), 'Cloning of novel maltooligosaccharide-producing amylases as antistaling agents for bread', *J. Agric. and Food Chem*, **46**, 779–782.

MORGAN K R, FURNEAUX R H and STANLEY R A (1992), Observation by solid-state ^{13}C CP MAS NMR spectroscopy of the transformations of wheat starch associated with the making and staling of bread', *Carbohydr. Res.*, **235**, 15–22.

MORGAN K R, FURNEAUX R H and LARSEN N G (1995), 'Solid-state NMR studies on the structure of starch granules', *Carbohydrate Research*, **276**, 387–399.

MORGAN K R, HUTT L, GERRARD J, EVERY D, ROSS M and GILPIN M (1997), 'Staling in starch breads: The effect of antistaling α-amylase', *Starch*, **49**, 54–59.

NUSSINOVITCH A, STEFFENS M, CHINACHOTI P and PELEG M (1992), 'Effect of strain level and storage time on the recoverable work of compressed breadcrumb', *J. Texture Studies*, 23, 13–24.

OATES, C G (2001), 'Bread microstructure', in Chinachoti, P and Vodovotz, Y, *Bread Staling*, CRC Press, New York, pp. 149–162.

PARKER R and RING S (2001), 'Macromolecular aspects of bread staling', in Chinachoti, P and Vodovotz, Y, *Bread Staling*, CRC Press, New York, pp. 149–162.

PELEG M, ROY I, CAMPANELLA O H and NORMAND M D (1989), 'Mathematical characterization of the compressive stress-strain relationships of spongy baked goods', *J. Food Sci.*, **54**(4), 947–949.

POMERANZ Y (1988), 'Composition and functionality of wheat flour components', in Pomeranz Y, *Wheat Chemistry and Technology*, Vol. 2, American Association of Cereal Chemists, St. Paul, MN, pp. 219–370.

PONTE J G JR., TITCOMB S T and COTTON R H (1962), 'Flour as a factor in bread firming', *Cereal Chem.* **39**, 437–444.

POWERS E M and BERKOWITZ D (1990) 'Efficacy of an oxygen scavenger to modify the atmosphere and prevent mold growth on meal, ready-to-eat pouched bread'. *J. Food Protection.* **53**, 767–771.

RAO P, NUSSINOVITCH A and CHINACHOTI P (1992), 'Effects of surfactants and amylopectin recrystallization and recoverability of bread crumbs during storage', *Cereal Chem.*, **69**, 613–618.

RUAN R and CHEN P L (2001), 'Nuclear magnetic resonance techniques', in Chinachoti P and Vodovotz Y (eds), *Bread Staling*, CRC Press, New York, pp. 113–128.

RUAN R, ALMAER S, HUANG V T, PERKINS P, CHEN P and FULCHER R G (1996), 'Relationship between firming and water mobility in starch-based food systems during storage', *Cereal Chem.*, **73**, 328–332.

SAHLSTROM A and BRATHEN E (1997), 'Effects of enzyme preparations for baking, mixing time and resting time on bread quality and bread staling', *Food Chem.*, **58**(1–2), 75–80.

SCHAEFER J, STEJSKAL E O and BUCHDAHL R (1977), 'Magic-angle carbon-13 NMR analysis of motion in solid glassy polymers', *Macromolecules*, **10**(2), 384–405.

SCHIRALDI A and FESSAS D (2001) 'Mechanism of staling: an overview', in Chinachoti P and Vodovotz Y, *Bread Staling*, CRC Press, New York, pp. 1–17.

SLADE L and LEVINE H (1991) 'Beyond water activity; recent advances based on an alternative approach to the assessment of food quality and safety'. *Crit. Rev. Food Sci. Nutr.*, **30**, 115–360.

SUHENDRO E L, WANISKA R D and ROONEY L W (1993), 'Effects of added proteins in wheat tortillas', *Cereal Chem.* **70**(4), 412–416.

SUHENDRO E L, WANISKA R D, ROONEY L W and GOMEZ M H. (1995), 'Effects of polyols on the processing and qualities of wheat tortillas', *Cereal Chem.* **72**(1): 122–127.

TANG H-R, GODWARD J and HILLS B (2000), 'The distribution of water in native starch granules – a multinuclear NMR study', *Carbohydrate Polymers* **43**, 375–387.

TANNER S F, HILLS B P and PARKER R (1991), 'Interactions of sorbed water with starch studied using proton nuclear magnetic resonance spectroscopy', *J. Chem. Soc. Faraday Trans.*, **87**(16), 2613–2621.

TAUB I A, HALLIDAY J W and KIM Y-K (1994), 'Bread structure and stability: rheological, calorimetric, spectroscopic, and microscopic characterization', *Army Science Conference Proceedings*, Vol. 6, 1391–1398.

UMBACH S L, DAVIS E A, GORDON J and CALLAGHAN P T (1992) 'Water self-diffusion coefficients and dielectric properties determined for starch–gluten water mixtures heated by microwave and by conventional methods', *Cereal Chem.*, **69**, 637–641.

VODOVOTZ Y (1996), *Aging of Starch and Bread as Studied by DSC, DMA, and Confocal Microscopy*, Ph.D. Thesis, University of Massachusetts, Amherst, MA.

VODOVOTZ Y and CHINACHOTI P (1996), 'Thermal transitions in gelatinized wheat starch at different moisture contents by dynamic mechanical analysis', *J. Food Sci.*, **61**, 932–937.

VODOVOTZ Y, BAIK M, VITTADINI E and CHINACHOTI P (2001), 'Instrumental techniques used in bread staling analysis', in Chinachoti P and Vodovotz Y (eds), *Bread Staling*, CRC Press, New York, 93–112.

WILLHOFT E M A (1971), 'Bread staling. II. Theoretical study', *J. Sci. Food Agr.*, **22**, 180–183.

WYNNE-JONES S and BLANSHARD J M V (1986), 'Hydration studies of wheat starch, amylopectin, amylose gels and bread by proton magnetic resonance', *Carbohydrate Polymers*, **6**, 289–306.

YAMAUCHI H, KANESHIGE H, FUJIMURA M, SOSHO Y, HASHIMOTO S, NISHIYAMA T, KODOMA K and KOBAYASHI T (1994), 'A kinetic study on staling of white bread made by straight method', *Nippon Shokuhin Kogyo Gakkaishi*, **41**(1), 1–8.

ZOBEL H F and KULP K (1996), 'The staling mechanism', in Hebeda R E and Zobel H F, *Baked Goods Freshness*, Marcel Dekker, Inc., New York, pp. 1–64.

Index